Studies in Fuzziness and Soft Computing

Volume 420

Series Editor

Janusz Kacprzyk, Systems Research Institute, Polish Academy of Sciences, Warsaw, Poland

The series "Studies in Fuzziness and Soft Computing" contains publications on various topics in the area of soft computing, which include fuzzy sets, rough sets, neural networks, evolutionary computation, probabilistic and evidential reasoning, multi-valued logic, and related fields. The publications within "Studies in Fuzziness and Soft Computing" are primarily monographs and edited volumes. They cover significant recent developments in the field, both of a foundational and applicable character. An important feature of the series is its short publication time and world-wide distribution. This permits a rapid and broad dissemination of research results.

Indexed by SCOPUS, DBLP, WTI Frankfurt eG, zbMATH, SCImago.

All books published in the series are submitted for consideration in Web of Science.

Laxminarayan Sahoo · Tapan Senapati ·
Ronald R. Yager
Editors

Real Life Applications of Multiple Criteria Decision Making Techniques in Fuzzy Domain

Springer

Editors
Laxminarayan Sahoo
Department of Computer and Information Science
Raiganj University
Raiganj, West Bengal, India

Tapan Senapati
School of Mathematics and Statistics
Southwest University
Chongqing, China

Ronald R. Yager
Machine Intelligence Institute
Iona College
New Rochelle, NY, USA

ISSN 1434-9922 ISSN 1860-0808 (electronic)
Studies in Fuzziness and Soft Computing
ISBN 978-981-19-4928-9 ISBN 978-981-19-4929-6 (eBook)
https://doi.org/10.1007/978-981-19-4929-6

This Springer imprint is published by the registered company Springer Nature Singapore Pte Ltd.
The registered company address is: 152 Beach Road, #21-01/04 Gateway East, Singapore 189721, Singapore

To,
My heavenly parents;
my soul mate Anima;
and my daughter Aaratrika

—Laxminarayan Sahoo

To,
My parents;
my better half Swagata;
and my son Ayush

—Tapan Senapati

To,
My lovely wife Rachel to whom I am grateful for all her help

—Ronald R. Yager

Preface

Nowadays, competitiveness in all business sectors is due to the number of companies and the requirements of decision makers and users. Several companies focus on the development and accretion of systems with a high level of reliability and dependability. Also, in the present situation, the compulsion on different manufacturing companies to appraise the most effective ways to control business policies has become a formidable task. Again, with continued economic deregulation, increased safety, rising customer expectations, rising globalization, refined modern technological systems, and a constantly changing mode of business, the company's manager faces a stressful ritual for maintaining the business systems. Moreover, due to uncontrollable factors of the dynamic economic environment, the managers or decision makers of the manufacturing company face a lot of difficulties. Consequently, parameters relating to decision making may arise in terms of impreciseness or lack of proper information, and uncertainty occurs at each and every step leading to suitable decisions. The possibility for decision makers to incorporate and make sense of information is extremely crucial to their decision making at times of emergency. All governments around the globe have been faced with the difficult task of making decisions in the interest of public health and safety during the COVID-19 crisis. In general, officials must deal with a risk whose extent is unclear, and they must do so under time constraints and in the face of significant uncertainties. Various solution methodologies have been proposed over the last decades to deal with the challenges of uncertainty. Recent developments in fuzzy systems and their application to multiple criteria decision making have provided the effectiveness to manage the difficulties and solve real-world problems with imprecise parameters. This edited book, entitled *Real Life Applications of Multiple Criteria Decision Making Techniques in Fuzzy Domain*, includes invited chapters relevant to the subject matter and innovative solution methodologies to grip the research issues in the underlying domains. The contributions of each and every one of them have been highlighted in the following paragraphs.

Chapter "Generalized Dombi Weighted Aggregation Operators for Multi-attribute Decision Making with Hesitant Fuzzy Information" deals with generalized Dombi weighted aggregation operators for multi-attribute decision making with hesitant fuzzy information. Chapter "A Novel Extended Fermatean Fuzzy Framework for Evaluating the Challenges to Sustainable Smart City Development" presents a novel extended Fermatean fuzzy framework for evaluating the challenges to sustainable smart city development. Chapter "Nonlinear Scaled Preferences in Linguistic Multi-criteria Group Decision Making" addresses nonlinear scaled preferences in linguistic multi-criteria group decision making. The Combined Compromise Solution (COCOSO) method and the newly developed Multiple Expert Multiple Criteria Decision Making (MEMCDM) technique have been discussed in Chapter "Fermatean Fuzzy Combined Compromise Solution Multiple Expert Multiple Criteria Decision-Making Approach" in the context of COCOSO and coefficient of variation methods with Fermatean fuzzy information. Chapter "Multi-criteria Decision Analysis and Fuzzy-Decision-Making Trial and Evaluation Laboratory (MCDA and F-DEMATEL) Method for Flipped and Sustainable Mathematics Teaching as a Real-Life Application" explores multi-criteria decision analysis and fuzzy decision-making trial and evaluation laboratory (MCDA and F-DEMATEL) methods for flipped and sustainable mathematics teaching as a real-life application. Accounting for uncertainty and disagreement in multi-criteria decision making using triangular fuzzy numbers and Monte Carlo simulation: A case study about selecting measures for remediation of agricultural land after radioactive contamination has been addressed in Chapter "Accounting for Uncertainty and Disagreement in Multi--criteria Decision Making Using Triangular Fuzzy Numbers and Monte Carlo Simulation: A Case Study About Selecting Measures for Remediation of Agricultural Land After Radioactive Contamination". Chapter "An IT2FBWM Model to Highlight the Significance of Factors Utilized in Determining Pandemic Hospital Site Selection" provides an IT2FBWM model to highlight the significance of factors utilized in determining pandemic hospital site selection. Group decision-making analysis under interval-valued q-rung orthopair fuzzy soft expert sets is addressed in Chapter "Group Decision-Making Analysis Under Interval-Valued q-rung Orthopair Fuzzy Soft Expert Sets". In Chapter "Information Aggregation Approaches for Multi--criteria Applications", the authors present information aggregation approaches for multi-criteria applications. In Chapter "Generalized Interval-Valued Intuitionistic Hesitant Fuzzy Power Bonferroni Means and Their Applications to Multicriteria Decision Making", the authors explain generalized interval-valued intuitionistic hesitant fuzzy power Bonferroni means and their applications to multi-criteria decision making. Modified Senapati and Yager's Fermatean fuzzy distance and its application in students' course placement in the tertiary institution has been studied in Chapter "Modified Senapati and Yager's Fermatean Fuzzy Distance and Its Application in Students' Course Placement in Tertiary Institution". Chapter "A Fuzzy Decision Making Inventory Model for Deteriorating Items Under Discounted Partial Advance–Partial Delayed Payment Strategy" deals with a fuzzy decision-making inventory model for deteriorating items under a discounted partial advance-partial delayed payment strategy. Chapter "The Effect of Solvents on the Oxidation of Acetaldehyde Using TOPSIS Method in a Fuzzy Environment" discusses the effect of solvents on the oxidation of acetaldehyde using the TOPSIS method in a fuzzy environment. In Chapter "Solution of Airport Gate Reassignment Problem Under Uncertain Environment", the solution to the airport gate re-assignment problem under an uncertain environment has been discussed. A bi-level approach for e-bidding auction with regret theory has been discussed in Chapter "Bilevel Approach for E-Bidding Auction with Regret Theory". Chapter "Evaluation of Sustainable Green Building Indicators by Fuzzy Multi-criteria Decision Making" explores the evaluation of sustainable green building indicators by fuzzy multi-criteria decision making. Solving multi-objective linear fractional programming problems utilizing (α, β)-cut in the triangular intuitionistic fuzzy setup has been provided in Chapter "Solving Multi-objective Linear Fractional Programming Problem Utilizing (α, β)-Cut in Triangular Intuitionistic Fuzzy Setup". In Chapter "q-Rung Orthopair Fuzzy 2-Tuple Linguistic Hamy Mean Operators for MAGDM with Modified EDAS Method", the authors present a modified EDAS method to explore q-rung orthopair fuzzy 2-tuple linguistic Hamy mean operators for MAGDM. The spherical fuzzy DEMATEL method is used in Chapter "Determining the Importance Level of Effective Criteria in the Health Information System Selection via Spherical Fuzzy DEMATEL Method" to determine the importance level of effective criteria in the selection of health information systems. Score function-based effective ranking of interval-valued Fermatean fuzzy sets and its applications to multi-criteria decision-making (MCDM) problem have been studied in Chapter "Score Function-Based Effective Ranking of Interval-Valued Fermatean Fuzzy Sets and Its Applications to Multi-criteria Decision Making (MCDM) Problem". Chapter "A Fuzzy EOQ model for Deteriorating Items Under Trade Credit Policy with Unfaithfulness Nature of Customers" presents a fuzzy EOQ model for deteriorating items under a trade credit policy with the unfaithfulness of customers. In Chapter "Comparison of Herd Tracking Systems Using Fuzzy Logic-Based Multi-Criteria Decision Making Methods", the authors provide a comparison of herd tracking systems using fuzzy logic-based multi-criteria decision-making methods. In Chapter "Multi-criteria Decision Making and Its Application to Online Learning Platform Selection During the COVID-19 Pandemic Based on TOPSIS Method", the authors explain multi-criteria decision making and its application to online learning platform selection during the COVID-19 pandemic, based on the TOPSIS method. Chapter "A Comprehensive Study on Neutrosophic Fuzzy Solid Transportation Model and Its Solution Technique" provides a comprehensive study on the neutrosophic fuzzy solid transportation model and its solution technique. Cubic picture hesitant fuzzy linear spaces and their applications in multi-criteria decision making have been reported in Chapter "Cubic Picture Hesitant Fuzzy Linear Spaces and Their Applications in Multi Criteria Decision Making". In Chapter "New Ranking Approach to Solve MCDM Problems with Generalized Intuitionistic Fuzzy Information", the authors provide a new ranking approach to solve MCDM problems with generalized intuitionistic fuzzy information. A note on "Parameter estimation and optimization of multi-objective capacitated stochastic transportation problems for gamma distribution" has been presented in Chapter "A Note on "Parameter Estimation and Optimization of Multi-objective Capacitated Stochastic Transportation Problem for Gamma Distribution"". Assigning Mexican regions to categories of marginality with MR-sort has been described in Chapter "Assigning Mexican Regions to Categories of Marginality with MR-Sort". Clustering faculty members for the betterment of research outcomes: A fuzzy multi-criteria decision-making approach in team formation has been demonstrated in Chapter "Clustering Faculty Members for the Betterment of Research Outcomes: A Fuzzy Multi-criteria Decision-Making Approach in Team Formation". Chapter "Freight Village Efficiency Criteria Evaluation via Fuzzy Multi-criteria Decision-Making Methods" discusses the concept of freight village efficiency criteria evaluation via fuzzy multi-criteria decision-making methods.

This book can be used by researchers, students, engineers, managers of industrial companies, decision-making people, or any other person interested in real-life applications of multiple criteria decision-making problems in fuzzy domains.

The editors are very much grateful to all of the authors from around the world who participated in this editing effort during the difficult COVID-19 pandemic situation. All of the reviewers who have aided in increasing the quality of the chapters by providing comments and ideas under severe pandemic scenarios deserve special recognition. The editors would also like to thank Mr. Aninda Bose, Senior Editor (Applied Sciences, Computational Intelligence, Energy, Water), Springer, New Delhi, for his kind consideration and endless support for this edited book. We hope that the reader will share our exhilaration to present this edited book and will make it beneficial.

Raiganj, India — Laxminarayan Sahoo
Chongqing, China — Tapan Senapati
New Rochelle, USA — Ronald R. Yager

Acknowledgments

The completion of this edited book could not have been possible without the chapter contributors and reviewers whose names may not all be enumerated. Their contributions are highly appreciated and sincerely acknowledged. However, the editors would like to express their deep appreciation and indebtedness, particularly to the following:

Mr. Aninda Bose, Senior Editor (Applied Sciences, Computational Intelligence, Energy, Water), Springer, New Delhi, for his kind consideration and endless support for this edited book.

Sharmila Mary Panner Selvam (Ms.), Project Coordinator, Springer Nature, for her immeasurable assistance for this edited book. Grateful again to Ms. Jayanthi Narayanaswamy, Project Manager at Straive, Chennai, India, for all of her efforts with this edited book.

Springer Nature Singapore Pte Ltd., 152 Beach Road, #21-01/04 Gateway East, Singapore 189721, Singapore, and the publishing teams of Springer Nature for providing the opportunity and professional support for producing the edited book in its present form.

Contents

Editors and Contributors

About the Editors

Laxminarayan Sahoo is currently an Associate Professor of Computer and Information Science, Raiganj University, Raiganj, India. He obtained his M.Sc. from Vidyasagar University, India and his Ph.D. from The University of Burdwan, India. He has received MHRD fellowship from Government of India during his M.Tech. course at ISM, Dhanbad, India and also received Prof. M. N. Gopalan Award for Best Ph.D. thesis in Operations Research from Operational Research Society of India (ORSI). Dr. Sahoo has successfully guided 01 research scholars for Ph.D. degree, five students continuing their Ph.D. degree and has published a good number of articles in international and national journals of repute. He is a reviewer of several international journals and Academic Editor of International Journal *Mathematical Problems in Engineering*, Hindawi Publication. He has also successfully completed one UGC minor research project. His specializations include, Wireless Sensor Network, Distributed Computing, Reliability Optimization, Genetic Algorithms, Particle Swarm Optimization, Graph Theory, Fuzzy Game Theory, Interval Mathematics, Soft Computing, Fuzzy Decision making and Operations Research. Dr. Sahoo is the author of the books *Advanced Operations Research* published by Asian Books, New Delhi, *Advanced Optimization and Operations Research* published by Springer Nature, Singapore.

Tapan Senapati received the B.Sc., M.Sc. and Ph.D. degrees in Mathematics all from the Vidyasagar University, India in 2006, 2008, and 2013 respectively. Presently he works as an Assistant Teacher of Mathematics under Government of West Bengal, India. He has also worked as Postdoctoral fellow at Southwest University, School of Mathematics and Statistics, 400715 Chongqing, China. He has published two books and more than 70 articles in reputed international journals. His research results have been published in *Fuzzy Sets and Systems*, *IEEE Transactions on Fuzzy Systems*, *Applied Soft Computing*, *Engineering Applications of Artificial Intelligence*, *International Journal of Intelligent Systems*, and *International Journal of General*

Systems, among others. He is a reviewer of several international journals and is also an academic editor of *Computational Intelligence and Neuroscience, Discrete Dynamics in Nature and Society, Mathematical Problems in Engineering.* His main scientific interests concentrate on fuzzy sets, fuzzy optimization, soft computing, multi-attribute decision making, aggregation operators.

Ronald R. Yager has worked in the area of machine intelligence for over twenty-five years. He has published over 500 papers and more than thirty books in areas related to artificial intelligence, fuzzy sets, decision-making under uncertainty and the fusion of information. He is among the world's top 1% most highly cited researchers with over 85,000 citations. He was the recipient of the IEEE Computational Intelligence Society's highly prestigious Frank Rosenblatt Award in 2016. He was the recipient of the IEEE Systems, Man and Cybernetics Society 2018 Lotfi Zadeh Pioneer Award. He was also the recipient of the IEEE Computational Intelligence Society Pioneer award in Fuzzy Systems. He received honorary doctorates from the Azerbaijan Technical University, the State University of Information Technologies, Sofia, Bulgaria and the Rostov on the Don University, Russia. Dr. Yager is a fellow of the IEEE and the Fuzzy Systems Association. He was given a lifetime achievement award by the Polish Academy of Sciences for his contributions. He served at the National Science Foundation as program director in the Information Sciences program. He was a NASA/Stanford visiting fellow and a research associate at the University of California, Berkeley. He has been a lecturer at NATO Advanced Study Institutes. He is a Distinguished Adjunct Professor at King Abdulaziz University, Jeddah, Saudi Arabia. He was a distinguished honorary professor at the Aalborg University Denmark. He was distinguished visiting scientist at King Saud University, Riyadh, Saudi Arabia. He received his undergraduate degree from the City College of New York and his Ph.D. from the Polytechnic University of New York. He was editor and chief of the *International Journal of Intelligent Systems.* He serves on the editorial board of numerous technology journals. Currently he is an Emeritus Professor at Iona College and is director of the Machine Intelligence.

Contributors

Abrams Floris Belgian Nuclear Research Centre (SCK CEN), Mol, Belgium; Division of Forest, Nature and Landscape, Faculty of Department of Earth and Environmental Sciences, University of Leuven (KULeuven), Leuven, Belgium

Adak Amal Kumar Department of Mathematics, Ganesh Dutt College, Begusarai, India

Akram Muhammad Department of Mathematics, University of the Punjab, Lahore, Pakistan

Alagöz Nazan Kemaloğlu Digital Technologies in Livestock Sector Joint Application and Research Center, Burdur Mehmet Akif Ersoy University, Burdur, Turkey

Ali Ghous Department of Mathematics, Division of Science and Technology, University of Education, Lahore, Pakistan

Alvarez Pavel Anselmo Department of Management and Economic Sciences, Universidad Autónoma de Occidente, Culiacan, Mexico

Appadoo S. S. Department of Supply Chain Management, Asper School of Business, University of Manitoba, Winnipeg, Canada

Aycin Ejder Department of Business Administration, Faculty of Business and Administrative Sciences, Kocaeli University, Izmit, Turkey

Aytekin Ahmet Department of Business Administration, Faculty of Hopa Economics and Administrative Sciences, Artvin Çoruh University, Hopa, Artvin, Turkey

Bajaj Rakesh Kumar Department of Mathematics, Jaypee University of Information Technology, Solan, Himachal Pradesh, India

Basu Kajla Department of Mathematics, NIT Durgapur, Durgapur, India

Beg Ismat Lahore School of Economics, Lahore, Pakistan

Bhatia Tanveen Kaur School of Mathematics, Thapar Institute of Engineering and Technology (Deemed to Be University), Patiala, Punjab, India

Borkotokey Surajit Department of Mathematics, Dibrugarh University, Dibrugarh, India

Çakır Fatma Sönmez Management Information Systems, Bartın University, Agdacı, Bartın, Turkey

Camps Johan Belgian Nuclear Research Centre (SCK CEN), Mol, Belgium

Cattrysse Dirk Center for Industrial Management, Traffic and Infrastructure, Department of Mechanical Engineering, University of Leuven (KU Leuven), Leuven, Belgium

Das Amrit Department of Mathematics, School of Advanced Sciences, Vellore Institute of Technology, Vellore, India

Demir Gülay Sivas Cumhuriyet University, Sivas, Turkey

Ecer Fatih Sub-Department of Operations Research, Faculty of Economics and Administrative Sciences, Afyon Kocatepe University, Afyonkarahisar, Turkey

Ejegwa Paul Augustine Department of Mathematics, University of Agriculture, Makurdi, Nigeria

Fatima Areej Department of Mathematics, Division of Science and Technology, University of Education, Lahore, Pakistan

García-Zamora Diego University of Jaén, Jaén, Spain

González-Gómez David Departamento de Didáctica de las Ciencias Experimentales y Matemáticas, Universidad de Extremadura, Cáceres, Spain

Guleria Abhishek Department of Physical Sciences and Languages, CSKHPKV, Palampur, Himachal Pradesh, India

Hendrickx Lucas Division of Forest, Nature and Landscape, Faculty of Department of Earth and Environmental Sciences, University of Leuven (KULeuven), Leuven, Belgium

Jana Ranjan Kumar Department of Mathematics and Humanities, Sardar Vallabhbhai National Institute of Technology, Surat, Gujarat, India

Jeong Jin Su Departamento de Didáctica de las Ciencias Experimentales y Matemáticas, Universidad de Extremadura, Cáceres, Spain

Kakati Pankaj Department of Mathematics, Jagannath Barooah College, Jorhat, India

Kar Supriya Bidhan Chandra Institution for Girls', Durgapur, India

Karamaşa Çağlar Faculty of Business, Department of Business Administration, Anadolu University, Eskişehir, Turkey

Kaur Dalbinder Department of Mathematics, Banwarilal Bhalotia College, Asansol, India

Kavyasree P. R. Department of Mathematics, Osmania University, Hyderabad, Telangana, India

Kayapinar Kaya Sema Department of Industrial Engineering, Faculty of Engineering, Munzur University, Tunceli, Turkey

Keleş Nuh Kayseri Customs Directorate, Trade Ministry TR, Kayseri, Turkey

Koca Gözde Department of Business Administration, Faculty of Economics and Administrative Sciences, Bilecik Şeyh Edebali University, Bilecik, Turkey

Kumar Amit School of Mathematics, Thapar Institute of Engineering and Technology (Deemed to Be University), Patiala, Punjab, India

Kumar Ganesh Department of Mathematics, University of Rajasthan, Jaipur, Rajasthan, India

Kureethara Joseph Varghese Christ University, Bangalore, India

Labella Álvaro University of Jaén, Jaén, Spain

Lahiri Dey Jayati Department of Computer and Information Science, Raiganj University, Raiganj, India

Liu Yi School of Mathematics and Information Science, Neijiang Normal University, Neijiang, Sichuan, China

Mahboob Aamir University of Management and Technology, Lahore, Pakistan

Maiti Manas Kumar Department of Mathematics, Mahishadal Raj College, Mahishadal, Purba Medinipur, West Bengal, India

Maitra Subhradeep Department of Computer and Information Science, Raiganj University, Raiganj, India

Martínez Luis University of Jaén, Jaén, Spain

Mesiar Radko Faculty of Civil Engineering, Slovak University of Technology, Bratislava, Slovakia;
Institute of Information Theory and Automation, Czech Academy of Sciences, Prague 8, Czech Republic

Mondal Rituparna Department of Mathematics and Humanities, Sardar Vallabhbhai National Institute of Technology, Surat, Gujarat, India

Nadeem Ahmad School of Mathematics, Minhaj University Lahore, Lahore, Pakistan

Narayanamoorthy Samayan Bharathiar University, Coimbatore, India

Nath Biman Kanti Assam University, Silchar, Assam, India

Naz Sumera Department of Mathematics, Division of Science and Technology, University of Education, Lahore, Pakistan

Nilkamal Lalit Narayan Mithila University, Darbhanga, India

Okoth Basil Oluch Department of Business Administration, Faculty of Economics and Administrative Sciences, Anadolu University, Eskişehir, Turkey

Onyeke Idoko Charles Department of Computer Science, University of Agriculture, Makurdi, Nigeria

Pekkaya Mehmet Business Administration, Zonguldak Bülent Ecevit University, Zonguldak, Turkey

Petry Frederick Naval Research Lab, Center for Geospatial Sciences, Stennis Space Center MS, New Rochelle, NY, USA

Pramanik Prasenjit Department of Applied Mathematics with Oceanology and Computer Programming, Vidyasagar University, Midnapore, Paschim Medinipur, West Bengal, India

Rana Akul Department of Mathematics, Narajole Raj College, Paschim Medinipur, West Bengal, India

Rao Ammilal Department of Chemistry, University of Rajasthan, Jaipur, Rajasthan, India

Rashid Tabasam University of Management and Technology, Lahore, Pakistan

Rodríguez Rosa M. University of Jaén, Jaén, Spain

Rong Yuan School of Management, Shanghai University, Shanghai, China

Saha Abhijit Department of Engineering Mathematics, College of Engineering, Koneru Lakshmaiah Education Foundation, Vaddeswaram, Andhra Pradesh, India

Sahoo Laxminarayan Department of Computer and Information Science, Raiganj University, Raiganj, India

Samanta Aniruddha Department of Basic Science, NSHM Knowledge Campus Durgapur, Durgapur, India

Saraji Mahyar Kamali Kaunas Faculty, Vilnius University, Kaunas, Lithuania

Sen Nabendu Assam University, Silchar, Assam, India

Senapati Tapan Department of Mathematics, Padima Janakalyan Banipith, Kukrakhupi, India;
School of Mathematics and Statistics, Southwest University, Beibei, Chongqing, China

Srivastava Kaushila Nandan Lalit Narayan Mithila University, Darbhanga, India

Srivastava Saurabh Department of Mathematics, Jaypee University of Information Technology, Solan, Himachal Pradesh, India

Streimikiene Dalia Kaunas Faculty, Vilnius University, Kaunas, Lithuania

Surender Reddy B. Department of Mathematics, Osmania University, Hyderabad, Telangana, India

Sweeck Lieve Belgian Nuclear Research Centre (SCK CEN), Mol, Belgium

Tiwary Kalishankar Department of Mathematics, Raiganj University, Raiganj, India

Umme Salma M. Christ University, Bangalore, India

Van Orshoven Jos Division of Forest, Nature and Landscape, Faculty of Department of Earth and Environmental Sciences, University of Leuven (KULeuven), Leuven, Belgium

Yager Ronald R. Machine Intelligence Institute, Iona College, New Rochelle, NY, USA

Yılmaz Merve Kılınç Digital Technologies in Livestock Sector Joint Application and Research Center, Burdur Mehmet Akif Ersoy University, Burdur, Turkey

Yllana-Prieto Félix Departamento de Didáctica de las Ciencias Experimentales y Matemáticas, Universidad de Extremadura, Cáceres, Spain

Yu Liying School of Management, Shanghai University, Shanghai, China

Generalized Dombi Weighted Aggregation Operators for Multi-attribute Decision Making with Hesitant Fuzzy Information

Abhijit Saha, Tapan Senapati, and Radko Mesiar

Abstract Because of two key issues: the presentation of attribute values and the aggregation of available data to form a preference of alternatives, multi-attribute decision making (MADM) is attracting numerous researchers in these days. In contrast to fuzzy set theory, hesitant fuzzy sets (HFSs) are more versatile and comprehensive tools for the representation of attribute values, whereas aggregation operators (AOs) are considered effective tools for the purpose of aggregation in fuzzy set theory. The existence of an aggregation operator (AO) which is capable of providing the necessary generality and versatility in altering perceptions of risk when aggregating attribute data under HFSs has not yet been disclosed. In spite of the fact that weighted AOs were invented earlier in order to attempt to fulfill the above-mentioned scenarios, they are still based on algebraic, Einstein, Hamacher, and Dombi operations. Thus, the primary objective of this study is to construct some comprehensive and adaptable AOs that may be used to address MADM issues with hesitant fuzzy (HF) data. This viewpoint allows us to begin by constructing new operations between HF elements that are based on generalized Dombi (GD) operations. Following that, we construct HF-GD weighted averaging and geometric AOs along with their ordered forms. We will discuss the effectiveness of several of the suggested AOs. Next, we set up a technique for MADM that is based on the HFS operators. Then, we offer an efficient

A. Saha (✉)
Department of Engineering Mathematics, College of Engineering, Koneru Lakshmaiah Education Foundation, Vaddeswaram, Andhra Pradesh 522302, India
e-mail: abhijit84.math@gmail.com

T. Senapati
Department of Mathematics, Padima Janakalyan Banipith, Kukrakhupi 721517, India

School of Mathematics and Statistics, Southwest University, Beibei 400715, Chongqing, China

R. Mesiar
Faculty of Civil Engineering, Slovak University of Technology, Radlinskeho 11, Sk-81005 Bratislava, Slovakia
e-mail: radko.mesiar@stuba.sk

Institute of Information Theory and Automation, Czech Academy of Sciences, Pod Vodárenskou věží 4, 182 08 Prague 8, Czech Republic

L. Sahoo et al. (eds.), *Real Life Applications of Multiple Criteria Decision Making Techniques in Fuzzy Domain*, Studies in Fuzziness and Soft Computing 420,
https://doi.org/10.1007/978-981-19-4929-6_1

instance involving the selection of an e-commerce Web site to illustrate the suggested method's decision phases. We undertake a validity check to ensure the authenticity of our recommended AOs. Additionally, we conduct a sensitivity analysis using a variety of attribute weight sets to ascertain the robustness of our recommended methodology. Furthermore, we add a detailed comparative analysis to demonstrate our model's superiority.

Keywords Hesitant fuzzy set · Generalized Dombi operations · Generalized Dombi weighted aggregation operators · MADM

Abbreviations

HF	Hesitant fuzzy
HFS	HF set
HFE	HF element
FSs	Fuzzy sets
HFE^U	Collection of all HFEs on U
MADM	Multi-attribute decision making
AO	Aggregation operator
DM	Decision-making
GD	Generalized Dombi
HFGDWAA	HF-GD weighted averaging AO
HFGDOWAA	HF-GD ordered weighted averaging AO
HFGDWGA	HF-GD weighted geometric AO
HFGDOWGA	HF-GD ordered weighted geometric AO

1 Introduction

The prevalence of MADM difficulties is well known in a wide range of practical decision-making scenarios. MADM's main goal is to choose the best option(s) from a set of available options based on various attributes, both dependent and impartial variables, in order to achieve the best possible result. Investigate the ways in which considerations, such as inadequate information and awareness, difficulty in information fusion, and so on, will be brought to the forefront, particularly in certain real-world situations, as a result of the increasing ambiguity of financial situation. As a consequence, the decision maker is confronted with a hard challenge of mathematically revealing the attributes' values. As an extension of FSs, Torra's HFS [1, 2] is considered, which allows for the computation of feeling of belonging grades by assuming a finite collection of likely variables rather than a single one. When confronted with unclear or uncertain outcomes, such HFS can be an extremely effective strategy. Rodrguez et al. [3] pioneered the idea of the HF linguistic term set, rather than using a fuzzy model to determine the prevalence of linguistic excerption. Zhang and Wei [4] proposed the aggressive VIKOR approach for solving a succession of MADM difficulties with HF data. Two HF-DM approaches were developed by Zhang and Xu [5, 6] by expanding TOPSIS and TODIM in HF setting. Dong et al. [7] devised an effective method for group DM in HF agreement using linguistic assessments. The HFS and several variants have been widely used in the fields of group DM [8–10], analysis of clusters [11], preference relations based on preferences [12], and other applications [13–16]. Rodrguez et al. [17] specifically cited the HFSs, focusing on the many concepts, techniques, and inclinations associated with the amount of FS. Additionally, Xu [18] dedicated some time to reviewing the evolution of aggregation approaches and algorithms for HFS and highlighting various hot topics in this sector.

Aggregation operators (AOs) are technically equivalent to functions, but when confronted with fairly popular procedures, these are sufficient to combine all of the given unique information into a unique bit of data. Throughout the last twenty years, contextual practitioners and researchers have already focused their efforts on the investigation of AOs due to their significant impact in the immediate proximity of a variety of relevant data processing activities, including such decision making as healthcare diagnosis, information processing, information retrieval, machine learning, and data mining, among others. We should remind ourselves that the study of HF AOs has attracted a great deal of interest because of its significance for data fusion. Xia and Xu [19] contributed several novel AOs and worked on refining their remarkable characteristics. Xia et al. [20] used quasi-arithmetic methods to create a small number of AOs in the presence of a HF surroundings. Wei [21] proposed HF prioritized operators and discussed their application to MADM issues. Wei et al. [22] advanced the HF Choquet ordered geometric and averaging AOs. Dependency among attributes in MADM issues was addressed by Zhu et al. [23] using HF geometric Bonferroni mean operator. Zhang [24] compiled HF data by evaluating a considerable quantity of HF power AOs and then examining the operators' different intriguing characteristics and relationships. Zhang and Wu [25] proposed weighted HFSs and discussed their utilization in the MCDM process. Zhang et al. [26] examined several induced extended HF AOs and related them to MADM problems. Liao and Xu [27] made major contributions to the search for a few HF weighted hybrid AOs and concurrently developed a selection framework to aid with MADM challenges involving HF information. Significantly, Zhou and Xu [28] provide two novel integration tools: The first is an efficient and simpler optimum discrete matching AOs operator for the purpose of examining group DM difficulties. Meanwhile, Wei et al. [29] defined two AOs for the HFLTS, which were used to address MADM difficulties. Yu [30] developed some additional HF AOs based on Einstein's operating rules. Tan et al. [31] created a new class of operators entitled HF Hamacher AOs to deal with MADM difficulties. Qin et al. [32] resolved the issue of interdependency among multiple attributes by introducing the Maclaurin symmetric mean operator under HF setting. The significance of the HF Bonferroni mean operators suggested by Zhu and Xu is that they capture the relationship between two attributes. Qin et al. [34] introduced Frank AOs and applied them to HF MADM issues. Liao and Xu [35] supplied a VIKOR-based method for HF MADM difficulties. Mahmood et al. [36] introduced cubic HFS and applied it in MADM problems. He [37] recently performed the HF Dombi operations in Typhoon disaster evaluation. Xu and Zhou [38] developed HF operations with probabilistic information and applied them to group DM challenges in their research. It is important to mention that the use of a consistently typical HF set (UTHFS), as described by Alcatud and Torra [39], might make a variety of conceptual and applied arguments easier to understand. Wang and Li [40] introduced picture HFS and implemented it in MADM issues. It is noteworthy that Wang et al. [41] created a modified A* algorithm for Arctic Route design that is based on HFS theory as well as other factors. We would want to emphasize that Liao et al. used the HF linguistic preferences value set throughout their technique of evaluating fire rescue operations, which we believe is noteworthy. Yalcin and Pehlivan [43] were capable of developing a technique that connects the fuzzy CODAS approach with the fuzzy envelope of HF linguistic term sets relating to comparative linguistic forms in order to address the employee selection issue. Some recent applications of HFSs can be seen in the works of Wu et al. [44], Alcantud et al. [45], Liu et al. [46], and Liu et al. [47].

The increasing capacity of decision complexity results in the emergence of real-world DM challenges that exploit the applicability of the AOs. Numerous AOs [19, 30, 31, 37] in the HF environment are intended to manage ambiguities during MADM problems, but none of them is formally general and flexible in nature. Thus, as of now, no research can be found related to AOs with the desired generality and flexibility for dealing with the HF DM issues. In this paper, we attempted to investigate a family of weighted AOs based on generalized Dombi operations [48] with HFSs and effectively use them to address MADM problems, inspired and fueled by the preceding fact. The majority of Archimedean operations [49] are generalized in nature because they can be deduced from the majority of t-norms and t-conorms-based operations. On the other hand, Dombi operations (t-norms and t-conorms) [50] are flexible in nature due to the inclusion of a parameter which can be adjusted according to real decision needs. But none of them is generalized as well as flexible. On the contrary, generalized Dombi operations (generalized Dombi t-norms and t-conorms) [48] are not only generalized but flexible too and thus are superior to Archimedean and Dombi operations. As a result, considering generalized Dombi (GD) operations for the development of flexible and generalized AOs to tackle MADM problems in an HF setting is worthwhile.

Therefore, considering the advantages of HFSs and the importance of GD operations, the following objectives are pursued in this article:

(1) Develop a set of novel operational rules and procedures based on GD operations in an HF environment.
(2) Develop a new MADM approach that incorporates HF data in light of the recommended AOs and demonstrate it through a case study.

In these circumstances, based on GD operations, we present in this paper, some operational laws, called GD operational laws for HFEs alongside some HF-GD weighted AOs, such as HFGDWAA, HFGDOWAA, HFGDWGA, and HFGDOWGA. Additionally, we also provide a mechanism for the MADM methodology in the presence of recommended AOs in an HF context. A systematic approach is used to demonstrate the usefulness of the established approach.

To accomplish this, we synthesize the remainder of the work as follows:

Some important and vital notions related to HFSs are given in Sect. 2. In Sect. 3, we develop GD operational laws for HFEs and the associated HF-GD AOs, for example, HFGDWAA, HFGDOWAA, HFGDWGA, and HFGDOWGA. Additionally, all of their key features are presented. Section 4 discusses a novel way of resolving different sorts of MADM concerns involving attribute values represented as HFEs by applying the new AOs. In Sect. 5, an e-commerce Web site selection case study is used to illustrate the established system. Section 6 is devoted entirely to assessing the influence of parameters on prioritized results and performing sensitivity studies on the attribute weights. Additionally, comparative research was conducted to validate the proposed methodology's superiority. Finally, in Sect. 7, we draw some conclusions about the study as a whole and highlight future potential.

2 Preliminaries

Here, we present a comprehensive review related to this study.

Definition 1 *(Torra* [1]*)*
A HFS ξ on a finite set U is offered by

$$\xi = \{< t, \mu_{\xi}^{h}(t) >: t \in U\}$$

where $\mu_{\xi}^{h}(t)$ indicates the collection (finite) of probable belongingness grades of $t \in U$.

Assumed $x \in U$, μ_{ξ}^{h} is referred to as a *HFE* (Xia and Xu [19]). To keep things simple, the HFS ξ is denoted as $\langle \mu_{\xi}^{h} \rangle$. The notation HFE^{U} is utilized to describe the collection of all *HFEs* over U.

Definition 2 *(Xia and Xu* [19]*)*
For $\mu_{\xi_1}^{h}, \mu_{\xi_2}^{h} \in HFE^{U}$ on U, certain fundamental operations can be given as

(1) $\mu_{\xi_1}^{h} \cup \mu_{\xi_2}^{h} = \bigcup_{\alpha_1 \in \mu_{\xi_1}^{h}, \alpha_2 \in \mu_{\xi_2}^{h}} \max\{\alpha_1, \alpha_2\}$

(2) $\mu_{\xi_1}^{h} \cap \mu_{\xi_2}^{h} = \bigcup_{\alpha_1 \in \mu_{\xi_1}^{h}, \alpha_2 \in \mu_{\xi_2}^{h}} \min\{\alpha_1, \alpha_2\}$

(3) $(\mu_{\xi_1}^{h})^{c} = \bigcup_{\alpha_1 \in \mu_{\xi_1}^{h}} \{1 - \alpha_1\}$

(4) $\mu^h_{\xi_1} \oplus \mu^h_{\xi_2} = \underset{\alpha_1 \in \mu^h_{\xi_1}, \alpha_2 \in \mu^h_{\xi_2}}{\cup} \{\alpha_1 + \alpha_2 - \alpha_1\alpha_2\}$

(5) $\mu^h_{\xi_1} \otimes \mu^h_{\xi_2} = \underset{\alpha_1 \in \mu^h_{\xi_1}, \alpha_2 \in \mu^h_{\xi_2}}{\cup} \{\alpha_1\alpha_2\}$

(6) $\lambda\mu^h_{\xi_1} = \underset{\alpha_1 \in \mu^h_{\xi_1}}{\cup} \{1 - (1 - \alpha_1)^\lambda\} \quad (\lambda > 0)$

(7) $\left(\mu^h_{\xi_1}\right)^\lambda = \underset{\alpha_1 \in \mu^h_{\xi_1}}{\cup} \{(\alpha_1)^\lambda\} \quad (\lambda > 0)$

Definition 3 *(Xia and Xu* [19]*)*
The score value of $\mu^h_\xi \in HFE^U$ is given by

$$S(\mu^h_\xi) = \frac{1}{\#\mu^h_\xi} \sum_{\beta \in \mu^h_\xi} \beta$$

($\#\mu^h_\xi$ = number of objects of μ^h_ξ).
A method for comparing HFEs is presented below, depending on their score values:

Definition 4 *(Xia and Xu* [19]*)*
Suppose $\mu^h_{\xi_1}, \mu^h_{\xi_2} \in HFE^U$. Then,

(1) $S(\mu^h_{\xi_1}) > S(\mu^h_{\xi_2}) \Rightarrow \mu^h_{\xi_1} \succ \mu^h_{\xi_2}$
(2) $S(\mu^h_{\xi_1}) < S(\mu^h_{\xi_2}) \Rightarrow \mu^h_{\xi_1} \prec \mu^h_{\xi_2}$
(3) $S(\mu^h_{\xi_1}) = S(\mu^h_{\xi_2}) \Rightarrow \mu^h_{\xi_1} = \mu^h_{\xi_2}$

3 Generalized Dombi Operations and Generalized Dombi Weighted AOs on *HFE*s

3.1 Generalized Dombi (GD) Operations on HFEs

Based on GD operations [48], we introduce some new operations (termed as GD operations) of *HFEs* and demonstrate all desirable characteristics of these operations. Before doing so, let us recall the definition of generalized Dombi operations [48] given below:

Definition 5 *(Dombi* [48]*)*
The generalized Dombi operations $GD^k_\gamma(\alpha_1, \alpha_2)$ are defined by the formula:

$$GD^k_\gamma(\alpha_1, \alpha_2) = \left(1 + \left(\frac{1}{\gamma}\left(\prod_{j=1}^{2} \Delta^k_\gamma(\alpha_j) - 1\right)\right)^{-\frac{1}{k}}\right)^{-1} \quad \text{or}$$

$$GD_{\gamma}^{k}(\alpha_1,\alpha_2)=\left(1+\left(\frac{1}{\gamma}\left(\prod_{j=1}^{2}\nabla_{\gamma}^{k}(\alpha_j)-1\right)\right)^{\frac{1}{k}}\right)^{-1}$$

where $\Delta_{\gamma}^{k}(\alpha_j)=1+\gamma\left(\frac{\alpha_j}{1-\alpha_j}\right)^{k}(\alpha_j\in[0,1),j=1,2)$, $\nabla_{\gamma}^{k}(\alpha_j)=1+\gamma\left(\frac{1-\alpha_j}{\alpha_j}\right)^{k}(\alpha_j\in(0,1],j=1,2)$ with $\gamma>0$ and $k\geq 1$.

Definition 6 Let $\mu_{\xi_1}^{h},\mu_{\xi_2}^{h}\in HFE^{U}$. Then, for $\gamma>0$ and $k\geq 1$, the GD operations on *HFEs* are:

(i) $\mu_{\xi_1}^{h}\tilde{\oplus}\mu_{\xi_2}^{h}=\bigcup_{\alpha_1\in\mu_{\xi_1}^{h},\alpha_2\in\mu_{\xi_2}^{h}}\left\{\left(1+\left(\frac{1}{\gamma}\left(\prod_{j=1}^{2}\Delta_{\gamma}^{k}(\alpha_j)-1\right)\right)^{-\frac{1}{k}}\right)^{-1}\right\}$ where $\Delta_{\gamma}^{k}(\alpha_j)=1+\gamma\left(\frac{\alpha_j}{1-\alpha_j}\right)^{k}(\alpha_j\neq 1)$

(ii) $\mu_{\xi_1}^{h}\tilde{\otimes}\mu_{\xi_2}^{h}=\bigcup_{\alpha_1\in\mu_{\xi_1}^{h},\alpha_2\in\mu_{\xi_2}^{h}}\left\{\left(1+\left(\frac{1}{\gamma}\left(\prod_{j=1}^{2}\nabla_{\gamma}^{k}(\alpha_j)-1\right)\right)^{\frac{1}{k}}\right)^{-1}\right\}$ where $\nabla_{\gamma}^{k}(\alpha_j)=1+\gamma\left(\frac{1-\alpha_j}{\alpha_j}\right)^{k}(\alpha_j\neq 0)$

(iii) $\lambda*\mu_{\xi_1}^{h}=\bigcup_{\alpha_1\in\mu_{\xi_1}^{h}}\left\{\left(1+\left(\frac{1}{\gamma}\left((\Delta_{\gamma}^{k}(\alpha_1))^{\lambda}-1\right)\right)^{-\frac{1}{k}}\right)^{-1}\right\}(\lambda>0,\alpha_j\neq 1))$

(iv) $\lambda\circ\mu_{\xi_1}^{h}=\bigcup_{\alpha_1\in\mu_{\xi_1}^{h}}\left\{\left(1+\left(\frac{1}{\gamma}\left((\nabla_{\gamma}^{k}(\alpha_1))^{\lambda}-1\right)\right)^{\frac{1}{k}}\right)^{-1}\right\}(\lambda>0,\alpha_j\neq 0)$.

Now, we examine several particular cases of the suggested operations.

- **Case-I**: When $\gamma=1$ and $k=1$, the proposed operations are reduced to

  (i) $\mu_{\xi_1}^{h}\tilde{\oplus}\mu_{\xi_2}^{h}=\bigcup_{\alpha_1\in\mu_{\xi_1}^{h},\alpha_2\in\mu_{\xi_2}^{h}}\{\alpha_1+\alpha_2-\alpha_1\alpha_2\}$

  (ii) $\mu_{\xi_1}^{h}\tilde{\otimes}\mu_{\xi_2}^{h}=\bigcup_{\alpha_1\in\mu_{\xi_1}^{h},\alpha_2\in\mu_{\xi_2}^{h}}\{\alpha_1\alpha_2\}$

  (iii) $\lambda*\mu_{\xi_1}^{h}=\bigcup_{\alpha_1\in\mu_{\xi_1}^{h}}\{1-(1-\alpha_1)^{\lambda}\}$

  (iv) $\lambda\circ\mu_{\xi_1}^{h}=\bigcup_{\alpha_1\in\mu_{\xi_1}^{h}}\{\alpha_1^{\lambda}\}$

  These are algebraic operations (Xia and Xu [19]) on *HFEs*.

- **Case-II:** When $k=1$ and $\gamma=2$, the proposed operations are reduced to

  (i) $\mu_{\xi_1}^{h}\tilde{\oplus}\mu_{\xi_2}^{h}=\bigcup_{\alpha_1\in\mu_{\xi_1}^{h},\alpha_2\in\mu_{\xi_2}^{h}}\left\{\frac{\alpha_1+\alpha_2}{1+\alpha_1\alpha_2}\right\}$

(ii) $\mu_{\xi_1}^h \tilde{\otimes} \mu_{\xi_2}^h = \bigcup_{\alpha_1 \in \mu_{\xi_1}^h, \alpha_2 \in \mu_{\xi_2}^h} \left\{ \frac{\alpha_1 \alpha_2}{1+(1-\alpha_1)(1-\alpha_2)} \right\}$

(iii) $\lambda * \mu_{\xi_1}^h = \bigcup_{\alpha_1 \in \mu_{\xi_1}^h} \left\{ \frac{(1+\alpha_1)^\lambda - (1-\alpha_1)^\lambda}{(1+\alpha_1)^\lambda + (1-\alpha_1)^\lambda} \right\}$

(iv) $\lambda \circ \mu_{\xi_1}^h = \bigcup_{\alpha_1 \in \mu_{\xi_1}^h} \left\{ \frac{2(\alpha_1)^\lambda}{(2-\alpha_1)^\lambda + (\alpha_1)^\lambda} \right\}$

These are Einstein operations (Yu [30]) on *HFEs*.

- **Case-III**: When $k = 1$, the proposed operations are reduced to

(i) $\mu_{\xi_1}^h \tilde{\oplus} \mu_{\xi_2}^h = \bigcup_{\alpha_1 \in \mu_{\xi_1}^h, \alpha_2 \in \mu_{\xi_2}^h} \left\{ \frac{\alpha_1+\alpha_2-\alpha_1\alpha_2-(1-\gamma)\alpha_1\alpha_2}{1-(1-\gamma)\alpha_1\alpha_2} \right\}$

(ii) $\mu_{\xi_1}^h \tilde{\otimes} \mu_{\xi_2}^h = \bigcup_{\alpha_1 \in \mu_{\xi_1}^h, \alpha_2 \in \mu_{\xi_2}^h} \left\{ \frac{\alpha_1\alpha_2}{\gamma+(1-\gamma)(\alpha_1+\alpha_2-\alpha_1\alpha_2)} \right\}$

(iii) $\lambda * \mu_{\xi_1}^h = \bigcup_{\alpha_1 \in \mu_{\xi_1}^h} \left\{ \frac{(1+(\gamma-1)\alpha_1)^\lambda - (1-\alpha_1)^\lambda}{(1+(\gamma-1)\alpha_1)^\lambda + (\gamma-1)(1-\alpha_1)^\lambda} \right\}$

(iv) $\lambda \circ \mu_{\xi_1}^h = \bigcup_{\alpha_1 \in \mu_{\xi_1}^h} \left\{ \frac{\gamma(\alpha_1)^\lambda}{(1+(\gamma-1)(1-\alpha_1))^\lambda + (\gamma-1)(\alpha_1)^\lambda} \right\}$

These are Hamacher operations (Tan et al. [31]) on *HFEs*.
Next, we demonstrate the accompanying theorem.

Theorem 1 *Suppose* $\mu_{\xi_1}^h, \mu_{\xi_2}^h \in HFE^U$ *and* $\lambda, \lambda_1, \lambda_2 > 0$. *Then,*

(i) $\mu_{\xi_1}^h \tilde{\oplus} \mu_{\xi_2}^h = \mu_{\xi_2}^h \tilde{\oplus} \mu_{\xi_1}^h$
(ii) $\mu_{\xi_1}^h \tilde{\otimes} \mu_{\xi_2}^h = \mu_{\xi_2}^h \tilde{\otimes} \mu_{\xi_1}^h$
(iii) $\lambda * (\mu_{\xi_1}^h \tilde{\oplus} \mu_{\xi_2}^h) = (\lambda * \mu_{\xi_1}^h) \tilde{\oplus} (\lambda * \mu_{\xi_2}^h)$
(iv) $\lambda \circ (\mu_{\xi_1}^h \tilde{\otimes} \mu_{\xi_2}^h) = (\lambda \circ \mu_{\xi_1}^h) \tilde{\otimes} (\lambda \circ \mu_{\xi_2}^h)$
(v) $(\lambda_1 + \lambda_2) * \mu_{\xi_1}^h = (\lambda_1 * \mu_{\xi_1}^h) \tilde{\oplus} (\lambda_2 * \mu_{\xi_1}^h)$
(vi) $(\lambda_1 + \lambda_2) \circ \mu_{\xi_1}^h = (\lambda_1 \circ \mu_{\xi_1}^h) \tilde{\otimes} (\lambda_2 \circ \mu_{\xi_1}^h)$

Proof (i) and (ii) are straightforward.

(iii) By definition of GD on HFEs, we have

$$\lambda * (\mu_{\xi_1}^h \tilde{\oplus} \mu_{\xi_2}^h)$$
$$= \lambda * \left(\bigcup_{\alpha_1 \in \mu_{\xi_1}^h, \alpha_2 \in \mu_{\xi_2}^h} \left\{ \left(1 + \left(\frac{1}{\gamma} \left(\prod_{j=1}^{2} \Delta_\gamma^k(\alpha_j) - 1 \right) \right)^{-\frac{1}{k}} \right)^{-1} \right\} \right)$$
$$= \bigcup_{\alpha_1 \in \mu_{\xi_1}^h, \alpha_2 \in \mu_{\xi_2}^h} \left\{ \left(1 + \left(\frac{1}{\gamma} \left(\left(\left(1 + \gamma \left(\frac{\left(1 + \left(\frac{1}{\gamma} \left(\prod_{j=1}^2 \Delta_\gamma^k(\alpha_j) - 1 \right) \right)^{-\frac{1}{k}} \right)^{-1}}{1 - \left(1 + \left(\frac{1}{\gamma} \left(\prod_{j=1}^2 \Delta_\gamma^k(\alpha_j) - 1 \right) \right)^{-\frac{1}{k}} \right)^{-1}} \right)^k \right)^\lambda - 1 \right) \right) \right)^{-\frac{1}{k}} \right)^{-1} \right\}$$

$$= \bigcup_{\alpha_1 \in \mu^h_{\xi_1}, \alpha_2 \in \mu^h_{\xi_2}} \left\{ \left(1 + \left(\frac{1}{\gamma} \left(\left(1 + \gamma \times \frac{1}{\gamma} \left(\prod_{j=1}^{2} \Delta^k_\gamma(\alpha_j) - 1 \right) \right)^\lambda - 1 \right) \right)^{-\frac{1}{k}} \right)^{-1} \right\}$$

$$= \bigcup_{\alpha_1 \in \mu^h_{\xi_1}, \alpha_2 \in \mu^h_{\xi_2}} \left\{ \left(1 + \left(\frac{1}{\gamma} \left(\prod_{j=1}^{2} (\Delta^k_\gamma(\alpha_j))^\lambda - 1 \right) \right)^{-\frac{1}{k}} \right)^{-1} \right\}$$

Again, by GD operations (Definition 6), we get

$$(\lambda * \mu^h_{\xi_1}) \tilde{\oplus} (\lambda * \mu^h_{\xi_2})$$
$$= \left(\bigcup_{\alpha_1 \in \mu^h_{\xi_1}} \left\{ \left(1 + \left(\frac{1}{\gamma} ((\Delta^k_\gamma(\alpha_1))^\lambda - 1) \right)^{-\frac{1}{k}} \right)^{-1} \right\} \right) \tilde{\oplus} \left(\bigcup_{\alpha_1 \in \mu^h_{\xi_1}} \left\{ \left(1 + \left(\frac{1}{\gamma} ((\Delta^k_\gamma(\alpha_2))^\lambda - 1) \right)^{-\frac{1}{k}} \right)^{-1} \right\} \right)$$

$$= \bigcup_{\alpha_1 \in \mu^h_{\xi_1}, \alpha_2 \in \mu^h_{\xi_2}} \left\{ \left(1 + \left(\frac{1}{\gamma} \left(\prod_{j=1}^{2} \left(1 + \gamma \left(\frac{\left(1 + \left(\frac{1}{\gamma} ((\Delta^k_\gamma(\alpha_j))^\lambda - 1) \right)^{-\frac{1}{k}} \right)^{-1}}{1 - \left(1 + \left(\frac{1}{\gamma} ((\Delta^k_\gamma(\alpha_j))^\lambda - 1) \right)^{-\frac{1}{k}} \right)^{-1}} \right)^k \right) - 1 \right) \right)^{-\frac{1}{k}} \right)^{-1} \right\}$$

$$= \bigcup_{\alpha_1 \in \mu^h_{\xi_1}, \alpha_2 \in \mu^h_{\xi_2}} \left\{ \left(1 + \left(\frac{1}{\gamma} \left(\prod_{j=1}^{2} \left(1 + \gamma \left(\frac{1}{\gamma} \left((\Delta^k_\gamma(\alpha_j))^\lambda - 1 \right) \right) \right) - 1 \right) \right)^{-\frac{1}{k}} \right)^{-1} \right\}$$

$$= \bigcup_{\alpha_1 \in \mu^h_{\xi_1}, \alpha_2 \in \mu^h_{\xi_2}} \left\{ \left(1 + \left(\frac{1}{\gamma} \left(\prod_{j=1}^{2} (\Delta^k_\gamma(\alpha_j))^\lambda - 1 \right) \right)^{-\frac{1}{k}} \right)^{-1} \right\}$$

Hence, $\lambda * (\mu^h_{\xi_1} \tilde{\oplus} \mu^h_{\xi_2}) = (\lambda * \mu^h_{\xi_1}) \tilde{\oplus} (\lambda * \mu^h_{\xi_2})$.

(iv) Similar to (ii).

(v) According to the GD operational laws (Definition 6), we have

$$(\lambda_1 + \lambda_2) * \mu^h_{\xi_1} = \bigcup_{\alpha_1 \in \mu^h_{\xi_1}} \left\{ \left(1 + \left(\frac{1}{\gamma} ((\Delta^k_\gamma(\alpha_1))^{\lambda_1 + \lambda_2} - 1) \right)^{-\frac{1}{k}} \right)^{-1} \right\}$$

Similarly, using Definition 6, we get

$$(\lambda_1 * \mu^h_{\xi_1}) \tilde{\oplus} (\lambda_2 * \mu^h_{\xi_1})$$
$$= \left(\bigcup_{\alpha_1 \in \mu^h_{\xi_1}} \left\{ \left(1 + \left(\frac{1}{\gamma} ((\Delta^k_\gamma(\alpha_1))^{\lambda_1} - 1) \right)^{-\frac{1}{k}} \right)^{-1} \right\} \right) \tilde{\oplus} \left(\bigcup_{\alpha_1 \in \mu^h_{\xi_1}} \left\{ \left(1 + \left(\frac{1}{\gamma} ((\Delta^k_\gamma(\alpha_1))^{\lambda_2} - 1) \right)^{-\frac{1}{k}} \right)^{-1} \right\} \right)$$

$$= \bigcup_{\alpha_1 \in \mu^h_{\xi_1}} \left\{ \left(1 + \left(\frac{1}{\gamma} \left(\prod_{j=1}^{2} \left(1 + \gamma \left(\frac{\left(1 + \left(\frac{1}{\gamma} ((\Delta^k_\gamma(\alpha_1))^{\lambda_j} - 1) \right)^{-\frac{1}{k}} \right)^{-1}}{1 - \left(1 + \left(\frac{1}{\gamma} ((\Delta^k_\gamma(\alpha_1))^{\lambda_j} - 1) \right)^{-\frac{1}{k}} \right)^{-1}} \right)^k \right) - 1 \right) \right)^{-\frac{1}{k}} \right)^{-1} \right\}$$

$$= \bigcup_{\alpha_1 \in \mu^h_{\xi_1}} \left\{ \left(1 + \left(\frac{1}{\gamma} \left(\prod_{j=1}^{2} \left(1 + \gamma \times \frac{1}{\gamma} ((\Delta^k_\gamma(\alpha_1))^{\lambda_j} - 1) \right) - 1 \right) \right)^{-\frac{1}{k}} \right)^{-1} \right\}$$

$$= \bigcup_{\alpha_1 \in \mu^h_{\xi_1}} \left\{ \left(1 + \left(\frac{1}{\gamma} ((\Delta^k_\gamma(\alpha_1))^{\lambda_1 + \lambda_2} - 1) \right)^{-\frac{1}{k}} \right)^{-1} \right\}$$

Hence, $(\lambda_1 + \lambda_2) * \mu^h_{\xi_1} = (\lambda_1 * \mu^h_{\xi_1}) \tilde{\oplus} (\lambda_2 * \mu^h_{\xi_1})$.
(vi) Similar to (v).

3.2 *Hesitant Fuzzy GD Weighted AOs*

In this part, depending on the GD operational laws (Definition 6), we shall develop HF-GD weighted AOs.

3.2.1 HF-GD Weighted Averaging AOs

This subsection is dedicated to the development of HF-GD weighted averaging AO (*HFGDWAA*) and HF-GD ordered weighted averaging AO (*HFGDOWAA*).

Definition 7 *Let $\mu^h_{\xi_j} \in HFE^U (j = 1(1)n)$. Then, the HFGDWAA operator is therefore considered as a mapping $HFGDWAA : HFE^U \to HFE^U$ in the following manner:*

$$HFGDWAA(\mu^h_{\xi_1}, \mu^h_{\xi_2}, \mu^h_{\xi_3}, \ldots, \mu^h_{\xi_n}) = \tilde{\bigoplus}_{j=1}^{n} (w_j * \mu^h_{\xi_j})$$

$(w_j (> 0)$ = weight of $\mu^h_{\xi_j}$ satisfying $\sum_{j=1}^{n} w_j = 1)$.

Theorem 2 *Consider $\mu^h_{\xi_j} \in HFE^U (j = 1(1)n)$. Then, $HFGDWAA(\mu^h_{\xi_1}, \mu^h_{\xi_2}, \mu^h_{\xi_3}, \ldots, \mu^h_{\xi_n})$ is again a HFE and*

$$HFGDWAA(\mu^h_{\xi_1}, \mu^h_{\xi_2}, \mu^h_{\xi_3}, \ldots, \mu^h_{\xi_n}) = \bigcup_{\alpha_j \in \mu^h_{\xi_j} (j=1(1)n)} \left\{ \left(1 + \left(\frac{1}{\gamma} \left(\prod_{j=1}^{n} (\Delta^k_\gamma(\alpha_j))^{w_j} - 1 \right) \right)^{-\frac{1}{k}} \right)^{-1} \right\} \quad (1)$$

Proof 1st part is obvious. To establish the rest of the portion, let us use the principle of induction on "*n*". By Definition 6, we obtain

$$HFGDWAA(\mu^h_{\xi_1}, \mu^h_{\xi_2}) = (w_1 * \mu^h_{\xi_1}) \tilde{\oplus} (w_2 * \mu^h_{\xi_2})$$

$$= \left(\bigcup_{\alpha_1 \in \mu^h_{\xi_1}} \left\{ \left(1 + \left(\frac{1}{\gamma}((\Delta^k_\gamma(\alpha_1))^{w_1} - 1)\right)^{-\frac{1}{k}}\right)^{-1} \right\} \right) \tilde{\oplus} \left(\bigcup_{\alpha_2 \in \mu^h_{\xi_2}} \left\{ \left(1 + \left(\frac{1}{\gamma}((\Delta^k_\gamma(\alpha_2))^{w_2} - 1)\right)^{-\frac{1}{k}}\right)^{-1} \right\} \right)$$

$$= \bigcup_{\alpha_j \in \mu^h_{\xi_j} (j=1,2)} \left\{ \left(1 + \left(\frac{1}{\gamma}\left(\prod_{j=1}^{2}\left(1 + \gamma\left(\frac{\left(1 + \left(\frac{1}{\gamma}((\Delta^k_\gamma(\alpha_j))^{w_j} - 1)\right)^{-\frac{1}{k}}\right)^{-1}}{1 - \left(1 + \left(\frac{1}{\gamma}((\Delta^k_\gamma(\alpha_j))^{w_j} - 1)\right)^{-\frac{1}{k}}\right)^{-1}}\right)^k\right) - 1\right)\right)^{-\frac{1}{k}}\right)^{-1} \right\}$$

$$= \bigcup_{\alpha_j \in \mu^h_{\xi_j} (j=1,2)} \left\{ \left(1 + \left(\frac{1}{\gamma}\left(\prod_{j=1}^{2}\left(1 + \gamma \times \frac{1}{\gamma}((\Delta^k_\gamma(\alpha_j))^{w_j} - 1)\right) - 1\right)\right)^{-\frac{1}{k}}\right)^{-1} \right\}$$

$$= \bigcup_{\alpha_j \in \mu^h_{\xi_j} (j=1,2)} \left\{ \left(1 + \left(\frac{1}{\gamma}\left(\prod_{j=1}^{2}(\Delta^k_\gamma(\alpha_j))^{w_j} - 1\right)\right)^{-\frac{1}{k}}\right)^{-1} \right\}$$

Therefore, Eq. (1) holds if $n = 2$. Next, suppose Eq. (1) is satisfied for $n = r$, i.e.,

$$HFGDWAA(\mu^h_{\xi_1}, \mu^h_{\xi_2}, \mu^h_{\xi_3}, \ldots, \mu^h_{\xi_r})$$
$$= \bigcup_{\alpha_j \in \mu^h_{\xi_j} (j=1(1)r)} \left\{ \left(1 + \left(\frac{1}{\gamma}\left(\prod_{j=1}^{r}(\Delta^k_\gamma(\alpha_j))^{w_j} - 1\right)\right)^{-\frac{1}{k}}\right)^{-1} \right\}$$

Now, for $n = r + 1$, we have

$$HFGDWAA(\mu^h_{\xi_1}, \mu^h_{\xi_2}, \mu^h_{\xi_3}, \ldots, \mu^h_{\xi_{r+1}})$$
$$= HFGDWAA(\mu^h_{\xi_1}, \mu^h_{\xi_2}, \mu^h_{\xi_3}, \ldots, \mu^h_{\xi_r}) \tilde{\oplus} (\omega_{r+1} * \mu^h_{\xi_{r+1}})$$
$$= \left(\bigcup_{\alpha_j \in \mu^h_{\xi_j} (j=1(1)r)} \left\{ \left(1 + \left(\frac{1}{\gamma}\left(\prod_{j=1}^{r}(\Delta^k_\gamma(\alpha_j))^{w_j} - 1\right)\right)^{-\frac{1}{k}}\right)^{-1} \right\} \right)$$
$$\tilde{\oplus} \left(\bigcup_{\alpha_{r+1} \in \mu^h_{\xi_{r+1}}} \left\{ \left(1 + \left(\frac{1}{\gamma}((\Delta^k_\gamma(\alpha_{r+1}))^{w_{r+1}} - 1)\right)^{-\frac{1}{k}}\right)^{-1} \right\} \right)$$

$$= \bigcup_{\alpha_j \in \mu^h_{\xi_j} (j=1(1)(r+1))} \left\{ \left(1 + \left(\frac{1}{\gamma}\left(1 + \gamma\left(\frac{\left(1 + \left(\frac{1}{\gamma}\left(\prod_{j=1}^{r}(\Delta^k_\gamma(\alpha_j))^{w_j} - 1\right)\right)^{-\frac{1}{k}}\right)^{-1}}{1 - \left(1 + \left(\frac{1}{\gamma}\left(\prod_{j=1}^{r}(\Delta^k_\gamma(\alpha_j))^{w_j} - 1\right)\right)^{-\frac{1}{k}}\right)^{-1}}\right)^k\right)\right.\right.\right.$$
$$\left.\left.\left. \times \left(1 + \gamma\left(\frac{\left(1 + \left(\frac{1}{\gamma}((\Delta^k_\gamma(\alpha_{r+1}))^{w_{r+1}} - 1)\right)^{-\frac{1}{k}}\right)^{-1}}{1 - \left(1 + \left(\frac{1}{\gamma}((\Delta^k_\gamma(\alpha_{r+1}))^{w_{r+1}} - 1)\right)^{-\frac{1}{k}}\right)^{-1}}\right)^k\right) - 1\right)\right)^{-\frac{1}{k}}\right)^{-1} \right\}$$

$$= \bigcup_{\alpha_j \in \mu^h_{\xi_j} (j=1(1)(r+1))} \left\{ \left(1 + \left(\frac{1}{\gamma} \left(\prod_{j=1}^{r} (\Delta^k_\gamma(\alpha_j))^{w_j} \times (\Delta^k_\gamma(\alpha_{r+1}))^{w_{r+1}} - 1 \right) \right)^{-\frac{1}{k}} \right)^{-1} \right\}$$

$$= \bigcup_{\alpha_j \in \mu^h_{\xi_j} (j=1(1)(r+1))} \left\{ \left(1 + \left(\frac{1}{\gamma} \left(\prod_{j=1}^{r+1} (\Delta^k_\gamma(\alpha_j))^{w_j} - 1 \right) \right)^{-\frac{1}{k}} \right)^{-1} \right\}$$

As a result, Eq. (1) is valid for $n = r + 1$. So, Eq. (1) is satisfied according to the doctrine of induction.

Next, we investigate three specific forms of $HFGDWAA$ operator.

Case-I: If $\gamma = 1$ and $k = 1$, then we get the HF weighted averaging AO (*HFWA* [19]).

$$HFWA(\mu^h_{\xi_1}, \mu^h_{\xi_2}, \mu^h_{\xi_3}, \ldots, \mu^h_{\xi_n}) = \bigcup_{\alpha_j \in \mu^h_{\xi_j} (j=1(1)n)} \left\{ 1 - \prod_{j=1}^{n} (1 - \alpha_j)^{w_j} \right\}$$

Case-II: If $\gamma = 2$ and $k = 1$, then we get the HF Einstein weighted averaging AO (*HFEWA* [30]).

$$HFEWA(\mu^h_{\xi_1}, \mu^h_{\xi_2}, \mu^h_{\xi_3}, \ldots, \mu^h_{\xi_n}) = \bigcup_{\alpha_j \in \mu^h_{\xi_j} (j=1(1)n)} \left\{ \frac{\prod_{j=1}^{n} (1 + \alpha_j)^{w_j} - \prod_{j=1}^{n} (1 - \alpha_j)^{w_j}}{\prod_{j=1}^{n} (1 + \alpha_j)^{w_j} + \prod_{j=1}^{n} (1 - \alpha_j)^{w_j}} \right\}$$

Case-III: For $k = 1$, we get the HF Hamachar weighted averaging AO (*HFHWA* [31]).

$$HFHWA(\mu^h_{\xi_1}, \mu^h_{\xi_2}, \mu^h_{\xi_3}, \ldots, \mu^h_{\xi_n}) = \bigcup_{\alpha_j \in \mu^h_{\xi_j} (j=1(1)n)} \left\{ \frac{\prod_{j=1}^{n} (1 + (\gamma - 1)\alpha_j)^{w_j} - \prod_{j=1}^{n} (1 - \alpha_j)^{w_j}}{\prod_{j=1}^{n} (1 + (\gamma - 1)\alpha_j)^{w_j} + (\gamma - 1) \prod_{j=1}^{n} (1 - \alpha_j)^{w_j}} \right\}$$

Theorem 3 (Shift invariance)

Suppose $\mu^h_{\xi_j} \in HFE^U (j = 1(1)n)$ *and* $\mu^h_{\xi_0} \in HFE^U$. *Then*, $HFGDWAA(\mu^h_{\xi_0} \tilde{\oplus} \mu^h_{\xi_1}, \mu^h_{\xi_0} \tilde{\oplus} \mu^h_{\xi_2}, \ldots, \mu^h_{\xi_0} \tilde{\oplus} \mu^h_{\xi_n}) = \mu^h_{\xi_0} \tilde{\oplus} HFGDWAA(\mu^h_{\xi_1}, \mu^h_{\xi_2}, \ldots, \mu^h_{\xi_n})$.

Proof By virtue of generalized operation (Definition 6), we have

$$\mu^h_{\xi_0} \tilde{\oplus} \mu^h_{\xi_j} = \bigcup_{\alpha_0 \in \mu^h_{\xi_0}, \alpha_j \in \mu^h_{\xi_j}} \left\{ \left(1 + \left(\frac{1}{\gamma} (\Delta^k_\gamma(\alpha_0) \times \Delta^k_\gamma(\alpha_j) - 1) \right)^{-\frac{1}{k}} \right)^{-1} \right\}$$

Now, utilizing Theorem 2, we get

$$
\begin{aligned}
& HFGDWAA(\mu^h_{\xi_0}\tilde{\oplus}\mu^h_{\xi_1}, \mu^h_{\xi_0}\tilde{\oplus}\mu^h_{\xi_2}, \ldots, \mu^h_{\xi_0}\tilde{\oplus}\mu^h_{\xi_n}) \\
&= \bigcup_{\alpha_j \in \mu^h_{\xi_j}(j=1(1)n), \alpha_0 \in \mu^h_{\xi_0}} \left\{ \left(1 + \left(\frac{1}{\gamma} \left(\prod_{j=1}^{n} \left(1 + \gamma \left(\frac{\left(1 + \left(\frac{1}{\gamma}\left(\Delta^k_\gamma(\alpha_0) \times \Delta^k_\gamma(\alpha_j) - 1\right)\right)^{-\frac{1}{k}}\right)^{-1}}{1 - \left(1 + \left(\frac{1}{\gamma}\left(\Delta^k_\gamma(\alpha_0) \times \Delta^k_\gamma(\alpha_j) - 1\right)\right)^{-\frac{1}{k}}\right)^{-1}} \right)^{k} \right)^{w_j} - 1 \right) \right)^{-\frac{1}{k}} \right)^{-1} \right\} \\
&= \bigcup_{\alpha_j \in \mu^h_{\xi_j}(j=1(1)n), \alpha_0 \in \mu^h_{\xi_0}} \left\{ \left(1 + \left(\frac{1}{\gamma} \left(\prod_{j=1}^{n} \left(1 + \gamma \times \frac{1}{\gamma}(\Delta^k_\gamma(\alpha_0) \times \Delta^k_\gamma(\alpha_j) - 1) \right)^{w_j} - 1 \right) \right)^{-\frac{1}{k}} \right)^{-1} \right\} \\
&= \bigcup_{\alpha_j \in \mu^h_{\xi_j}(j=1(1)n), \alpha_0 \in \mu^h_{\xi_0}} \left\{ \left(1 + \left(\frac{1}{\gamma} \left(\prod_{j=1}^{n} (\Delta^k_\gamma(\alpha_0))^{w_j} (\Delta^k_\gamma(\alpha_j))^{w_j} - 1 \right) \right)^{-\frac{1}{k}} \right)^{-1} \right\} \\
&= \bigcup_{\alpha_j \in \mu^h_{\xi_j}(j=1(1)n), \alpha_0 \in \mu^h_{\xi_0}} \left\{ \left(1 + \left(\frac{1}{\gamma} \left(\Delta^k_\gamma(\alpha_0) \prod_{j=1}^{n} (\Delta^k_\gamma(\alpha_j))^{w_j} - 1 \right) \right)^{-\frac{1}{k}} \right)^{-1} \right\}
\end{aligned}
$$

Next, using generalized operation from Definition 6, we have from Theorem 2,

$$
\begin{aligned}
& \mu^h_{\xi_0}\tilde{\oplus} HFGDWAA(\mu^h_{\xi_1}, \mu^h_{\xi_2}, \ldots, \mu^h_{\xi_n}) \\
&= \left(\bigcup_{\alpha_0 \in \mu^h_{\xi_0}} \{\alpha_0\} \right) \tilde{\oplus} \left(\bigcup_{\alpha_j \in \mu^h_{\xi_j}(j=1(1)n)} \left\{ \left(1 + \left(\frac{1}{\gamma} \left(\prod_{j=1}^{n} (\Delta^k_\gamma(\alpha_j))^{w_j} - 1 \right) \right)^{-\frac{1}{k}} \right)^{-1} \right\} \right) \\
&= \bigcup_{\alpha_j \in \mu^h_{\xi_j}(j=1(1)n), \alpha_0 \in \mu^h_{\xi_0}} \left\{ \left(1 + \left(\frac{1}{\gamma} \left(\Delta^k_\gamma(\alpha_0) \left(1 + \gamma \left(\frac{\left(1 + \left(\frac{1}{\gamma}\left(\prod_{j=1}^{n} (\Delta^k_\gamma(\alpha_j))^{w_j} - 1\right)\right)^{-\frac{1}{k}}\right)^{-1}}{1 - \left(1 + \left(\frac{1}{\gamma}\left(\prod_{j=1}^{n} (\Delta^k_\gamma(\alpha_j))^{w_j} - 1\right)\right)^{-\frac{1}{k}}\right)^{-1}} \right)^{k} \right) - 1 \right) \right)^{-\frac{1}{k}} \right)^{-1} \right\} \\
&= \bigcup_{\alpha_j \in \mu^h_{\xi_j}(j=1(1)n), \alpha_0 \in \mu^h_{\xi_0}} \left\{ \left(1 + \left(\frac{1}{\gamma} \left(\Delta^k_\gamma(\alpha_0) \prod_{j=1}^{n} (\Delta^k_\gamma(\alpha_j))^{w_j} - 1 \right) \right)^{-\frac{1}{k}} \right)^{-1} \right\}
\end{aligned}
$$

Therefore, $HFGDWAA(\mu^h_{\xi_0}\tilde{\oplus}\mu^h_{\xi_1}, \mu^h_{\xi_0}\tilde{\oplus}\mu^h_{\xi_2}, \ldots, \mu^h_{\xi_0}\tilde{\oplus}\mu^h_{\xi_n}) = \mu^h_{\xi_0}\tilde{\oplus} HFGDWAA(\mu^h_{\xi_1}, \mu^h_{\xi_2}, \ldots, \mu^h_{\xi_n})$.

Theorem 4 (Idempotency)
Assume $\mu^h_{\xi_j} \in HFE^U(j = 1(1)n)$ and $\mu^h_{\xi_0} \in HFE^U$ with $\mu^h_{\xi_0} = \mu^h_{\xi_j}\ \forall j$. Then, we have $HFGDWAA(\mu^h_{\xi_1}, \mu^h_{\xi_2}, \mu^h_{\xi_3}, \ldots, \mu^h_{\xi_n}) = \mu^h_{\xi_0}$.

Proof Follows from Definition 7 and Theorem 2.

Theorem 5 (Boundedness)
Let $\mu^h_{\xi_j} \in HFE^U(j = 1(1)n)$. Then, $\left(\mu^h_\xi\right)^- \prec HFGDWAA(\mu^h_{\xi_1}, \mu^h_{\xi_2}, \mu^h_{\xi_3}, \ldots, \mu^h_{\xi_n}) \prec \left(\mu^h_\xi\right)^+$
where $\left(\mu^h_\xi\right)^+ = \cup_{\alpha_j \in \mu^h_{\xi_j}(j=1(1)n)} \max \alpha_j$ and $\left(\mu^h_\xi\right)^- = \cup_{\alpha_j \in \mu^h_{\xi_j}(j=1(1)n)} \min \alpha_j$.

Proof For $j = 1(1)n$, we have $\min_j(\alpha_j) \leq \alpha_j \leq \max_j(\alpha_j)$ where $\alpha_j \in \mu^h_{\xi_j}$.

$$\prod_{j=1}^{n}\left(\Delta_{\gamma}^{k}(\alpha')\right)^{w_j} \leq \prod_{j=1}^{n}\left(\Delta_{\gamma}^{k}(\alpha_j)\right)^{w_j} \leq \prod_{j=1}^{n}\left(\Delta_{\gamma}^{k}(\alpha'')\right)^{w_j}$$

where $\alpha' \in \left(\mu_{\xi}^{h}\right)^{-}, \alpha'' \in \left(\mu_{\xi}^{h}\right)^{+}$.

This gives,

$$\left(1+\left(\frac{1}{\gamma}\left(\prod_{j=1}^{n}\left(\Delta_{\gamma}^{k}(\alpha')\right)^{w_j}-1\right)\right)^{-\frac{1}{k}}\right)^{-1} \leq \left(1+\left(\frac{1}{\gamma}\left(\prod_{j=1}^{n}\left(\Delta_{\gamma}^{k}(\alpha_j)\right)^{w_j}-1\right)\right)^{-\frac{1}{k}}\right)^{-1}$$

$$\leq \left(1+\left(\frac{1}{\gamma}\left(\prod_{j=1}^{n}\left(\Delta_{\gamma}^{k}(\alpha'')\right)^{w_j}-1\right)\right)^{-\frac{1}{k}}\right)^{-1} \Rightarrow \left(1+\left(\frac{1}{\gamma}\left(\gamma\left(\frac{\alpha'}{1-\alpha'}\right)^{k}\right)\right)^{-\frac{1}{k}}\right)^{-1}$$

$$\leq \left(1+\left(\frac{1}{\gamma}\left(\prod_{j=1}^{n}\left(\Delta_{\gamma}^{k}(\alpha_j)\right)^{w_j}-1\right)\right)^{-\frac{1}{k}}\right)^{-1} \leq \left(1+\left(\frac{1}{\gamma}\left(\gamma\left(\frac{\alpha''}{1-\alpha''}\right)^{k}\right)\right)^{-\frac{1}{k}}\right)^{-1}$$

$$\Rightarrow \left(1+\left(\frac{\alpha'}{1-\alpha'}\right)^{-1}\right)^{-1} \leq \left(1+\left(\frac{1}{\gamma}\left(\prod_{j=1}^{n}\left(\Delta_{\gamma}^{k}(\alpha_j)\right)^{w_j}-1\right)\right)^{-\frac{1}{k}}\right)^{-1}$$

$$\leq \left(1+\left(\frac{\alpha''}{1-\alpha''}\right)^{-1}\right)^{-1} \Rightarrow \alpha' \leq \left(1+\left(\frac{1}{\gamma}\left(\prod_{j=1}^{n}\left(\Delta_{\gamma}^{k}(\alpha_j)\right)^{w_j}-1\right)\right)^{-\frac{1}{k}}\right)^{-1} \leq \alpha''$$

So, by Definition 3, it follows $S\left(\left(\mu_{\xi}^{h}\right)^{-}\right) \leq S(HFGDWAA(\mu_{\xi_1}^{h}, \mu_{\xi_2}^{h}, \mu_{\xi_3}^{h}, \ldots, \mu_{\xi_n}^{h})) \leq S\left(\left(\mu_{\xi}^{h}\right)^{+}\right)$.

Hence, by the ranking rules of *HFE*s, we get, $\left(\mu_{\xi}^{h}\right)^{-} \prec HFGDWAA(\mu_{\xi_1}^{h}, \mu_{\xi_2}^{h}, \mu_{\xi_3}^{h}, \ldots, \mu_{\xi_n}^{h}) \prec \left(\mu_{\xi}^{h}\right)^{+}$.

Theorem 6 (Monotonicity)
Suppose $\mu_{\xi_j}^{h}, \mu_{\xi_j'}^{h} \in HFE^{U}(j = 1(1)n)$ *such that* $\alpha_j \leq \alpha_j'$ where $\alpha_j \in \mu_{\xi_j}^{h}, \alpha_j' \in \mu_{\xi_j'}^{h}$ $(j = 1(1)n)$. *Then we have,* $HFGDWAA(\mu_{\xi_1}^{h}, \mu_{\xi_2}^{h}, \mu_{\xi_3}^{h}, \ldots, \mu_{\xi_n}^{h}) \prec HFGDWAA(\mu_{\xi_1'}^{h}, \mu_{\xi_2'}^{h}, \mu_{\xi_3'}^{h}, \ldots, \mu_{\xi_n'}^{h})$.

Proof Similar to Theorem 5.

Now, utilizing the $HFGDWAA$ operator, we propose the $HFGDOWAA$ operator.

Definition 8 Suppose $\mu_{\xi_j}^{h} \in HFE^{U}(j = 1(1)n)$. Then, the *HFGDOWAA* operator is a function $HFGDOWAA : HFE^{U} \to HFE^{U}$ which is given by

$$HFGDOWAA(\mu_{\xi_1}^{h}, \mu_{\xi_2}^{h}, \mu_{\xi_3}^{h}, \ldots, \mu_{\xi_n}^{h}) = \tilde{\bigoplus}_{j=1}^{n}(w_j * \mu_{\xi_{\sigma(j)}}^{h})$$

where $\{\sigma(j) : j = 1(1)n\}$ is a disposition of $\{j : j = 1(1)n\}$ and $\mu^h_{\xi_{\sigma(j-1)}} \geq \mu^h_{\xi_{\sigma(j)}}$ for all j and $w_j > 0$ is the weight of $\mu^h_{\xi_j}$ with $\sum_{j=1}^n w_j = 1$.

Theorem 7 *The aggregated HFE, i.e., $HFGDOWAA(\mu^h_{\xi_1}, \mu^h_{\xi_2}, \mu^h_{\xi_3}, \ldots, \mu^h_{\xi_n})$ is still a HFE and*

$$HFGDOWAA(\mu^h_{\xi_1}, \mu^h_{\xi_2}, \ldots, \mu^h_{\xi_n})$$
$$= \bigcup_{\alpha_{\sigma(j)} \in \mu^h_{\xi_{\sigma(j)}} (j=1(1)n)} \left\{ \left(1 + \left(\frac{1}{\gamma} \left(\prod_{j=1}^n (\Delta^k_\gamma(\alpha_{\sigma(j)}))^{w_j} - 1 \right) \right)^{-\frac{1}{k}} \right)^{-1} \right\}.$$

Particularly, for $w_j = \frac{1}{n} \forall j = 1\,(1)\,n;$, we obtain the *HFGDWAA* operator.

Next, some specific forms of the *HFGDOWAA* operator are considered below:

Case-I: For $\gamma = 1$ and $k = 1$, we obtain the HF ordered weighted averaging AO (*HFOWA* [19]) as presented below:

$$HFOWA(\mu^h_{\xi_1}, \mu^h_{\xi_2}, \mu^h_{\xi_3}, \ldots, \mu^h_{\xi_n}) = \bigcup_{\alpha_j \in \mu^h_{\xi_j} (j=1(1)n)} \left\{ 1 - \prod_{j=1}^n (1 - \alpha_{\sigma(j)})^{w_j} \right\}$$

Case-II: If $\gamma = 2$ and $k = 1$, then we get the HF Einstein ordered weighted averaging AO (*HFEOWA* [30]) as shown below:

$$HFEOWA(\mu^h_{\xi_1}, \mu^h_{\xi_2}, \mu^h_{\xi_3}, \ldots, \mu^h_{\xi_n})$$
$$= \bigcup_{\alpha_j \in \mu^h_{\xi_j} (j=1(1)n)} \left\{ \frac{\prod_{j=1}^n (1 + \alpha_{\sigma(j)})^{w_j} - \prod_{j=1}^n (1 - \alpha_{\sigma(j)})^{w_j}}{\prod_{j=1}^n (1 + \alpha_{\sigma(j)})^{w_j} - \prod_{j=1}^n (1 - \alpha_{\sigma(j)})^{w_j}} \right\}$$

Case-III: For $k = 1$, the HF Hamachar ordered weighted averaging AO (*HFHOWA* [31]) is obtained as presented below:

$$HFHOWA(\mu^h_{\xi_1}, \mu^h_{\xi_2}, \ldots, \mu^h_{\xi_n})$$
$$= \bigcup_{\alpha_j \in \mu^h_{\xi_j}} \left\{ \frac{\prod_{j=1}^n (1 + (\gamma - 1)\alpha_{\sigma(j)})^{w_j} - \prod_{j=1}^n (1 - \alpha_{\sigma(j)})^{w_j}}{\prod_{j=1}^n (1 + (\gamma - 1)\alpha_{\sigma(j)})^{w_j} + (\gamma - 1) \prod_{j=1}^n (1 - \alpha_{\sigma(j)})^{w_j}} \right\}$$

Theorem 8 (Shift invariance)

Let $\mu^h_{\xi_j} \in HFE^U (j = 1(1)n)$ *and* $\mu^h_{\xi_0} \in HFE^U$. *Then,* $HFGDOWAA(\mu^h_{\xi_0} \tilde{\oplus} \mu^h_{\xi_1}, \mu^h_{\xi_0} \tilde{\oplus} \mu^h_{\xi_2}, \ldots, \mu^h_{\xi_0} \tilde{\oplus} \mu^h_{\xi_n}) = \mu^h_{\xi_0} \tilde{\oplus} HFGDOWAA(\mu^h_{\xi_1}, \mu^h_{\xi_2}, \ldots, \mu^h_{\xi_n})$.

Theorem 9 (Idempotency)

Let $\mu^h_{\xi_j} \in HFE^U(j = 1(1)n)$ *and* $\mu^h_{\xi_0} \in HFE^U$ *such that* $\mu^h_{\xi_0} = \mu^h_{\xi_j}\ \forall j$. *Then, we obtain* $HFGDOWAA(\mu^h_{\xi_1}, \mu^h_{\xi_2}, \ldots, \mu^h_{\xi_n}) = \mu^h_{\xi_0}$.

Theorem 10 (Boundedness)
Let $\mu^h_{\xi_j} \in HFE^U(j = 1(1)n)$. *Then,* $\left(\mu^h_\xi\right)^- \prec HFGDOWAA(\mu^h_{\xi_1}, \mu^h_{\xi_2}, \mu^h_{\xi_3}, \ldots, \mu^h_{\xi_n}) \prec \left(\mu^h_\xi\right)^+$
where $\left(\mu^h_\xi\right)^+ = \bigcup_{\alpha_j \in \mu^h_{\xi_j}(j=1(1)n)} \max \alpha_j$ *and* $\left(\mu^h_\xi\right)^- = \bigcup_{\alpha_j \in \mu^h_{\xi_j}(j=1(1)n)} \min \alpha_j$.

Theorem 11 (Monotonocity)
Assume $\mu^h_{\xi_j}, \mu^h_{\xi'_j} \in HFE^U(j = 1(1)n)$ *such that* $\alpha_j \le \alpha'_j$ where $\alpha_j \in \mu^h_{\xi_j}, \alpha'_j \in \mu^h_{\xi'_j}\ \forall j$. *Then, we have*

$$HFGDOWAA(\mu^h_{\xi_1}, \mu^h_{\xi_2}, \mu^h_{\xi_3}, \ldots, \mu^h_{\xi_n}) \prec HFGDOWAA(\mu^h_{\xi'_1}, \mu^h_{\xi'_2}, \mu^h_{\xi'_3}, \ldots, \mu^h_{\xi'_n}).$$

3.2.2 HF-GD Weighted Geometric AOs

This subsection is dedicated to the development of HF-GD weighted geometric AO (*HFGDWGA*) and HF-GD ordered weighted geometric AO (*HFGDOWGA*).

Definition 9 Let $\mu^h_{\xi_j} \in HFE^U(j = 1(1)n)$. Then, the *HFGDWAA* operator defining a mapping $HFGDWGA : HFE^U \to HFE^U$ is presented as

$$HFGDWGA(\mu^h_{\xi_1}, \mu^h_{\xi_2}, \mu^h_{\xi_3}, \ldots, \mu^h_{\xi_n}) = \tilde{\bigotimes}_{j=1}^{n} (w_j \circ \mu^h_{\xi_j})$$

($w_j(> 0)$ = weight of $\mu^h_{\xi_j}$ satisfying $\sum_{j=1}^n w_j = 1$).

Theorem 12 *Suppose* $\mu^h_{\xi_j} \in HFE^U$. *Then, the aggregated value* $HFGWGA(\mu^h_{\xi_1}, \mu^h_{\xi_2}, \mu^h_{\xi_3}, \ldots, \mu^h_{\xi_n})$ *is again a HFE and*

$$HFGDWGA(\mu^h_{\xi_1}, \mu^h_{\xi_2}, \mu^h_{\xi_3}, \ldots, \mu^h_{\xi_n}) = \bigcup_{\alpha_j \in \mu^h_{\xi_j}(j=1(1)n)} \left\{ \left(1 + \left(\frac{1}{\gamma} \left(\prod_{j=1}^{n} (\nabla^k_\gamma(\alpha_j))^{w_j} - 1 \right) \right)^{\frac{1}{k}} \right)^{-1} \right\}.$$

The following are some specific forms of the $HFGDWAA$ operator:

Case-I: If $\gamma = 1$ and $k = 1$, then we get the HF weighted geometric AO (*HFWG* [19]) as shown below:

$$HFWG(\mu^h_{\xi_1}, \mu^h_{\xi_2}, \mu^h_{\xi_3}, \ldots, \mu^h_{\xi_n}) = \bigcup_{\alpha_j \in \mu^h_{\xi_j}(j=1(1)n)} \left\{ \prod_{j=1}^{n} (\alpha_j)^{w_j} \right\}.$$

Case-II: If $\gamma = 2$ and $k = 1$, then we get the HF Einstein weighted geometric AO (*HFEWG* [30]) as shown below:

$$HFEWG(\mu^h_{\xi_1}, \mu^h_{\xi_2}, \mu^h_{\xi_3}, \ldots, \mu^h_{\xi_n}) = \bigcup_{\alpha_j \in \mu^h_{\xi_j}(j=1(1)n)} \left\{ \frac{2\prod_{j=1}^{n} (\alpha_j)^{w_j}}{\prod_{j=1}^{n} (2-\alpha_j)^{w_j} + \prod_{j=1}^{n} (\alpha_j)^{w_j}} \right\}.$$

Case-III: If $k = 1$, then we get the HF Hamachar weighted geometric AO (*HFHWG* [31]) as shown below:

$$HFHWG(\mu^h_{\xi_1}, \mu^h_{\xi_2}, \mu^h_{\xi_3}, \ldots, \mu^h_{\xi_n}) = \bigcup_{\alpha_j \in \mu^h_{\xi_j}(j=1(1)n)} \left\{ \frac{\gamma \prod_{j=1}^{n} (\alpha_j)^{w_j}}{\prod_{j=1}^{n} (1+(\gamma-1)(1-\alpha_j))^{w_j} + (\gamma-1)\prod_{j=1}^{n} (\alpha_j)^{w_j}} \right\}.$$

Theorem 13 (Shift invariance)
Assume that $\mu^h_{\xi_j} \in HFE^U(j = 1(1)n)$ *and* $\mu^h_{\xi_0} \in HFE^U$. *Then,* $HFGDWGA(\mu^h_{\xi_0}\tilde{\otimes}\mu^h_{\xi_1}, \mu^h_{\xi_0}\tilde{\otimes}\mu^h_{\xi_2}, \ldots, \mu^h_{\xi_0}\tilde{\otimes}\mu^h_{\xi_n}) = \mu^h_{\xi_0}\tilde{\otimes} HFGDWAA(\mu^h_{\xi_1}, \mu^h_{\xi_2}, \ldots, \mu^h_{\xi_n})$.

Theorem 14 (Idempotency)
Suppose $\mu^h_{\xi_j} \in HFE^U(j = 1(1)n)$ *and* $\mu^h_{\xi_0} \in HFE^U$ *satisfying* $\mu^h_{\xi_0} = \mu^h_{\xi_j}\ \forall j$. *Then, we have,* $HFGDWGA(\mu^h_{\xi_1}, \mu^h_{\xi_2}, \mu^h_{\xi_3}, \ldots, \mu^h_{\xi_n}) = \mu^h_{\xi_0}$.

Theorem 15 (Boundedness)
Let $\mu^h_{\xi_j} \in HFE^U(j = 1(1)n)$. *Then,* $(\mu^h_{\xi})^- \prec HFGDWGA(\mu^h_{\xi_1}, \mu^h_{\xi_2}, \mu^h_{\xi_3}, \ldots, \mu^h_{\xi_n}) \prec (\mu^h_{\xi})^+$, *where* $(\mu^h_{\xi})^+ = \cup_{\alpha_j \in \mu^h_{\xi_j}(j=1(1)n)} \max \alpha_j$ *and* $(\mu^h_{\xi})^- = \cup_{\alpha_j \in \mu^h_{\xi_j}(j=1(1)n)} \min \alpha_j$.

Theorem 16 (Monotonicity)
Suppose $\mu^h_{\xi_j}, \mu^h_{\xi'_j} \in HFE^U$ *with* $\alpha_j \le \alpha'_j$ where $\alpha_j \in \mu^h_{\xi_j}, \alpha'_j \in \mu^h_{\xi'_j}\ \forall j$. *Then,* $HFGDWGA(\mu^h_{\xi_1}, \mu^h_{\xi_2}, \mu^h_{\xi_3}, \ldots, \mu^h_{\xi_n}) \prec HFGDWGA(\mu^h_{\xi'_1}, \mu^h_{\xi'_2}, \mu^h_{\xi'_3}, \ldots, \mu^h_{\xi'_n})$.

Following that, depending on $HFGDWAA$ operator, we will design the $HFGDOWGA$.

Definition 10 For $\mu^h_{\xi_j} \in HFE^U(j = 1(1)n)$, the *HFGDOWGA* operator defining a mapping $HFGDOWGA : HFE^U \to HFE^U$ is given by

$$HFGDOWGA(\mu^h_{\xi_1}, \mu^h_{\xi_2}, \mu^h_{\xi_3}, \ldots, \mu^h_{\xi_n}) = \tilde{\bigotimes}_{j=1}^{n} (w_j \circ \mu^h_{\xi_{\sigma(j)}})$$

where $\sigma(j)$ is an arrangement of $\{j : j = 1(1)n\}$ satisfying $\mu^h_{\xi_{\sigma(j-1)}} \geq \mu^h_{\xi_{\sigma(j)}}$ for each j, and $w_j > 0$ is the weight of $\mu^h_{\xi_j}$ with $\sum_{j=1}^{n} w_j = 1$.

Theorem 17 *The aggregated value* $HFGDOWGA(\mu^h_{\xi_1}, \mu^h_{\xi_2}, \mu^h_{\xi_3}, \ldots, \mu^h_{\xi_n})$ *is still a HFE and*

$$HFGDOWGA(\mu^h_{\xi_1}, \mu^h_{\xi_2}, \ldots, \mu^h_{\xi_n})$$
$$= \bigcup_{\alpha_{\sigma(j)} \in \mu^h_{\xi_{\sigma(j)}} (j=1(1)n)} \left\{ \left(1 + \left(\frac{1}{\gamma} \left(\prod_{j=1}^{n} (\nabla^k_\gamma(\alpha_{\sigma(j)}))^{w_j} - 1 \right) \right)^{\frac{1}{k}} \right)^{-1} \right\}.$$

Particularly, if $w_j = \frac{1}{n} \forall j = 1(1)n;$*, we get the HFGDWGA operator.*

The following are some specific forms of the *HFGDOWGA* operator:

Case-I: If $\gamma = 1$ and $k = 1$, then we get the HF ordered weighted geometric AO (*HFOWG* [19]) as shown below:

$$HFOWG(\mu^h_{\xi_1}, \mu^h_{\xi_2}, \mu^h_{\xi_3}, \ldots, \mu^h_{\xi_n}) = \bigcup_{\alpha_{\sigma(j)} \in \mu^h_{\xi_{\sigma(j)}} (j=1(1)n)} \left\{ \prod_{j=1}^{n} (\alpha_{\sigma(j)})^{w_j} \right\}.$$

Case-II: If $\gamma = 2$ and $k = 1$, then we get the HF Einstein ordered weighted geometric AO (*HFEOWG* [30]) as shown below:

$$HFEOWG(\mu^h_{\xi_1}, \mu^h_{\xi_2}, \mu^h_{\xi_3}, \ldots, \mu^h_{\xi_n})$$
$$= \bigcup_{\alpha_{\sigma(j)} \in \mu^h_{\xi_{\sigma(j)}} (j=1(1)n)} \left\{ \frac{2 \prod_{j=1}^{n} (\alpha_{\sigma(j)})^{w_j}}{\prod_{j=1}^{n} (2 - \alpha_{\sigma(j)})^{w_j} + \prod_{j=1}^{n} (\alpha_{\sigma(j)})^{w_j}} \right\}.$$

Case-III: If $k = 1$, then we get the HF Hamachar ordered weighted geometric AO (*HFHOWG* [31]) as shown below:

$$HFHOWG(\mu^h_{\xi_1}, \mu^h_{\xi_2}, \mu^h_{\xi_3}, \ldots, \mu^h_{\xi_n})$$
$$= \bigcup_{\alpha_{\sigma(j)} \in \mu^h_{\xi_{\sigma(j)}} (j=1(1)n)} \left\{ \frac{\gamma \prod_{j=1}^{n} (\alpha_{\sigma(j)})^{w_j}}{\prod_{j=1}^{n} (1 + (\gamma - 1)(1 - \alpha_{\sigma(j)}))^{w_j} + (\gamma - 1) \prod_{j=1}^{n} (\alpha_{\sigma(j)})^{w_j}} \right\}.$$

Theorem 18 (Shift invariance)
Let $\mu^h_{\xi_j} \in HFE^U(j = 1(1)n)$ *and* $\mu^h_{\xi_0} \in HFE^U$. *Then* $HFGDOWGA(\mu^h_{\xi_0} \tilde{\otimes} \mu^h_{\xi_1}, \mu^h_{\xi_0} \tilde{\otimes} \mu^h_{\xi_2}, \ldots, \mu^h_{\xi_0} \tilde{\otimes} \mu^h_{\xi_n}) = \mu^h_{\xi_0} \tilde{\otimes} HFGDOWGA(\mu^h_{\xi_1}, \mu^h_{\xi_2}, \ldots, \mu^h_{\xi_n})$.

Theorem 19 (Idempotency)

For $\mu^h_{\xi_j} \in HFE^U (j = 1(1)n)$ *and* $\mu^h_{\xi_0} \in HFE^U$ *with* $\mu^h_{\xi_0} = \mu^h_{\xi_j}\ \forall j$, *we have* $HFGDOWGA\ (\mu^h_{\xi_1}, \mu^h_{\xi_2}, \ldots, \mu^h_{\xi_n}) = \mu^h_{\xi_0}$.

Theorem 20 (Boundedness)

Suppose $\mu^h_{\xi_j} \in HFE^U (j = 1(1)n)$. *Then,* $\left(\mu^h_{\xi}\right)^- \prec HFGDOWGA(\mu^h_{\xi_1}, \mu^h_{\xi_2}, \mu^h_{\xi_3}, \ldots, \mu^h_{\xi_n}) \prec \left(\mu^h_{\xi}\right)^+$

where $\left(\mu^h_{\xi}\right)^+ = \cup_{\alpha_j \in \mu^h_{\xi_j} (j=1(1)n)} \max \alpha_j$ *and* $\left(\mu^h_{\xi}\right)^- = \cup_{\alpha_j \in \mu^h_{\xi_j} (j=1(1)n)} \min \alpha_j$.

Theorem 21 (Monotonocity)

Suppose $\mu^h_{\xi_j}, \mu^h_{\xi'_j} \in HFE^U (j = 1(1)n)$ *such that* $\alpha_j \leq \alpha'_j$ where $\alpha_j \in \mu^h_{\xi_j}, \alpha'_j \in \mu^h_{\xi'_j}\ \forall j$. *Then, we have*

$$HFGDOWGA(\mu^h_{\xi_1}, \mu^h_{\xi_2}, \mu^h_{\xi_3}, \ldots, \mu^h_{\xi_n}) \prec HFGDOWGA(\mu^h_{\xi'_1}, \mu^h_{\xi'_2}, \mu^h_{\xi'_3}, \ldots, \mu^h_{\xi'_n}).$$

4 Proposed MADM Methodology

Here, we apply the *HFGDWAA* (or *HFGDWGA* or *HFGDOWAA* or *HFGDOWGA*) operator to develop a methodology to deal with *MCDM* issues with HF information.

Let $P = \{P_i : i = 1(1)m\}$ and $Q = \{Q_j : j = 1(1)n\}$ be sets of alternatives and attributes, respectively, and $w = \{w_j : j = 1(1)n\}$ denotes the collection of weights of attributes Q_j $(j = 1\ (1)\ n)$ fulfilling $w_j > 0$ and $\sum_{j=1}^n w_j = 1$. Assume $\tilde{D} = \left[d_{ij}\right]_{m \times n}$ is the assessment matrix, where every d_{ij} is defined using the *HFEs* $\mu^h_{\xi_{ij}}$ $(i = 1\ (1)\ m; j = 1\ (1)\ n)$.

Therefore, the recommended method for resolving MADM concerns with HF information via the HFGDWAA (or HFGDWGA or HFGDOWAA or HFGDOWGA) operator is as follows:

Algorithm

Step-1: Normalize the data used in HF decision making.

In real-world DM scenarios, attributes are frequently classified into two classes, namely profit and cost attributes. The profit attributes have a favorable effect on the MADM outcome, and the cost attributes have a negative effect on the DM problem's ultimate assessment. In this regard, we will bear in mind that it is necessary to standardize and turn cost type characteristics into benefit type attributes in the event that cost type attributes exist. If this circumstance does not happen, we will not address this step further throughout this conversation.

The normalized assessment matrix is $\tilde{D} = \left[d_{ij}\right]_{m \times n}$, in which

$$\tilde{d}_{ij} = \begin{cases} \mu^h_{\xi_{ij}} & \text{if } Q_j \text{ is benefit-type} \\ (\mu^h_{\xi_{ij}})^c & \text{if } Q_j \text{ is cost-type} \end{cases}$$

Step-2: Compute the aggregated HFEs $g_i (i = 1(1)m)$ corresponding to $P_i (i = 1(1)m)$ as

$$g_i = HFGDWAA(\tilde{d}_{i1}, \tilde{d}_{i2}, \ldots, \tilde{d}_{in}) = \bigcup_{\alpha_{i_1} \in \tilde{d}_{i_1}, \alpha_{i_2} \in \tilde{d}_{i_2}, \ldots, \alpha_{i_n} \in \tilde{d}_{i_n}} \left\{ \left(1 + \left(\frac{1}{\gamma} \left(\prod_{j=1}^{n} (\Delta^k_\gamma(\alpha_{ij}))^{w_j} - 1 \right) \right)^{-\frac{1}{k}} \right)^{-1} \right\},$$

or

$$g_i = HFGDWGA(\tilde{d}_{i_1}, \tilde{d}_{i_2}, \ldots, \tilde{d}_{i_n}) = \bigcup_{\alpha_{i_1} \in \tilde{d}_{i_1}, \alpha_{i_2} \in \tilde{d}_{i_2}, \ldots, \alpha_{i_n} \in \tilde{d}_{i_n}} \left\{ \left(1 + \left(\frac{1}{\gamma} \left(\prod_{j=1}^{n} (\nabla^k_\gamma(\alpha_{ij}))^{w_j} - 1 \right) \right)^{\frac{1}{k}} \right)^{-1} \right\},$$

or

$$g_i = HFGDOWAA(\tilde{d}_{i_1}, \tilde{d}_{i_2}, \ldots, \tilde{d}_{i_n}) = \bigcup_{\substack{\alpha_{i\sigma(1)} \in \tilde{d}_{i\,\sigma(1)}, \\ \alpha_{i\sigma(2)} \in \tilde{d}_{i\,\sigma(2)}, \\ \ldots, \alpha_{i\sigma(n)} \in \tilde{d}_{i\,\sigma(n)}}} \left\{ \left(1 + \left(\frac{1}{\gamma} \left(\prod_{j=1}^{n} (\Delta^k_\gamma(\alpha_{i\sigma(j)}))^{w_j} - 1 \right) \right)^{-\frac{1}{k}} \right)^{-1} \right\}$$

or

$$g_i = HFGDOWGA(\tilde{d}_{i_1}, \tilde{d}_{i_2}, \ldots, \tilde{d}_{i_n}) = \bigcup_{\substack{\alpha_{i\sigma(1)} \in \tilde{d}_{i\,\sigma(1)}, \\ \alpha_{i\sigma(2)} \in \tilde{d}_{i\,\sigma(2)}, \\ \ldots, \alpha_{i\sigma(n)} \in \tilde{d}_{i\,\sigma(n)}}} \left\{ \left(1 + \left(\frac{1}{\gamma} \left(\prod_{j=1}^{n} (\nabla^k_\gamma(\alpha_{i\sigma(j)}))^{w_j} - 1 \right) \right)^{\frac{1}{k}} \right)^{-1} \right\}.$$

Step-3: Compute the score values of the aggregated HFEs $g_i (i = 1(1)m)$ corresponding to P_i $(i = 1(1)m)$ utilizing Definition 3.

Step-4: Utilizing Definition 4, obtain the priorities of the alternatives to P_i $(i = 1(1)m)$.

5 Case Study

5.1 Problem Description

As a genuine application of the introduced approach, we bring forth a cabalistic example (adapted from Zhou et al. [51]) that enables us to understand how to evaluate the customer satisfaction of e-commerce Web sites (e-CWs). The motive of the concerned project is driven toward the evaluation of the best suitable e-CW among the several e-CW-based companies that mainly supply options of e-CWs to enterprises. In this example, it has been evaluated the consumer satisfaction of five suitable options of e-CWs P_i $(i = 1(1)\ 5)$. Consider that an investment with e-CWs gets newly identified by an e-commerce enterprise. In view of maximizing the expected profit, it is necessary to devise the consumer satisfaction of the five e-CWs that help to select the optimal one. Below we present four attributes which act significantly during the taking up of a decision by any investment company: Q_1: e-CW platform characteristics; Q_2: e-CW characteristics; Q_3: e-CW pre-sale and after-sale service; Q_4: e-CW trading pay and logistics distribution.

Note that, the decision makers need to assess the five suitable e-CWs P_i $(i = 1(1)\ 5)$ over the four above-mentioned attributes Q_j $(j = 1(1)\ 4)$ for the purpose of avoiding influence of each other, and hence, they form the underlying decision matrix $\tilde{D} = \left[d_{ij}\right]_{5\times 4}$ which is displayed in Table 1, where d_{ij} are in terms of *HFEs*. The attributes' weight vector is $w = (0.3, 0.2, 0.1, 0.4)$.

5.2 Problem Solution

To address the *MADM* issue mentioned earlier, the *HFGDWAA* operator is utilized for assessment of alternatives under HF setting. The developed approach comprises the following steps:

Step-1: Normalization is not required as no cost type criterion has been considered.
Step-2: We determine the aggregated HFEs (with $\gamma = 2$ and $k = 2$) g_i of P_i using the decision data provided in Table 1 and the HFGDWAA operator.

Table 1 HF decision matrix $\tilde{D} = \left[d_{ij}\right]_{5\times 4}$

	Q_1	Q_2	Q_3	Q_4
P_1	<{0.3, 0.4, 0.6}>	<{0.4, 0.7}>	<{0.4, 0.5}>	<{0.5, 0.6, 0.8}>
P_2	<{0.5,0.6, 0.7}>	<{0.4, 0.5}>	<{0.3,0.4}>	<{0.3,0.5,0.6}>
P_3	<{0.5, 0.7}>	<{0.1, 0.2, 0.4}>	<{0.4,0.6}>	<{0.3,0.5}>
P_4	<{0.4, 0.5, 0.6}>	<{0.3, 0.4}>	<{0.2,0.3, 0.4}>	<{0.7,0.8}>
P_5	<{0.5, 0.6}>	<{0.4, 0.6}>	<{0.1,0.2, 0.3}>	<{0.6,0.7}>

To keep this paper concise, an alternative P_1 is offered as a typical example.

$$g_1 = HFGDWAA(d_{11}, d_{12}, d_{13}, d_{14}) = \{0.421630813, 0.432385733, 0.474365613, 0.483553875, 0.596836891, 0.603618339, 0.498763095, 0.507366901, 0.541872438, 0.549593454, 0.647986785, 0.653997397, 0.443753796, 0.453792386, 0.493356227, 0.502083441, 0.610924362, 0.617486622, 0.516573427, 0.524792329, 0.557887436, 0.565317568, 0.660481117, 0.666311153, 0.50798742, 0.516387928, 0.550151454, 0.557720014, 0.654432756, 0.660349977, 0.57036692, 0.577581134, 0.606880745, 0.613505257, 0.699213048, 0.704490029\}$$

Step-3: Find the scores $S(g_i)(i = 1(1)5)$ of the aggregated HFEs g_i corresponding to $P_i(i = 1(1)5)$. Consequently, $S(g_1) = 0.0312$, $S(g_2) = 0.0276$, $S(g_3) = 0.0389$, $S(g_4) = 0.0471$, and $S(g_5) = 0.0328$.

Step-4: The priority order of alternatives is $P_4 \succ P_3 \succ P_5 \succ P_1 \succ P_2$ where "$\succ$" stands for "superior to". Therefore, the most acceptable e-CW is P_4.

6 Discussion

This section has the following four subsections: (1) the impact of parameter variations on the scores of alternatives; (2) a sensitivity analysis of the suggested technique using various sets of criterion weight values; and (3) a comparison between the introduced methods and associated existing strategies.

6.1 The Study of Variation of Parameters k and γ on Score Values

We use the operators HFGDWAA, HFGDWGA, HFGDOWAA, and HFGDOWGA to explore the impact of the parameters 'k' and 'γ' on the score values, and next we synthesize the aggregate score values that reflected in Figs. 1, 2, 3, 4, 5, 6, 7, 8, 9, 10, 11, 12, 13, 14, 15, 16, 17, 18, 19 and 20.

We notice from the Figs. 1, 2, 3, 4, 5, 6, 7, 8, 9, 10, 11, 12, 13, 14, 15, 16, 17, 18, 19 and 20 that the best option is P_4 in all scenarios. Additional examination of these figures provides some important facts:

(1) For a fixed value of γ, we observe the fact of the increasing nature of the score values P_1, P_2, P_3, P_4, P_5 deduced from the *HFGDWAA* operator with the increase of k lying in between 1 and 10. Interestingly, in contrast, for a fixed

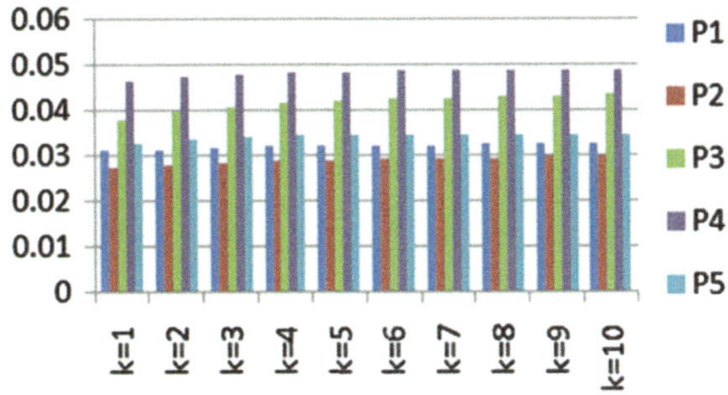

Fig. 1 Score values of alternatives using HFGDWAA operator (taking $\gamma = 1$)

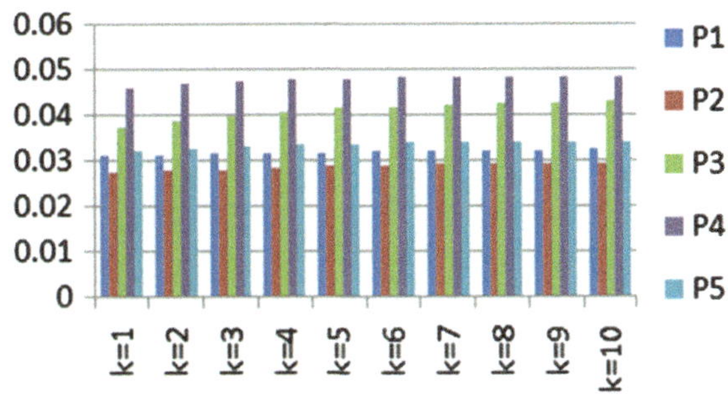

Fig. 2 Score values of alternatives using HFGDWAA operator (taking $\gamma = 2$)

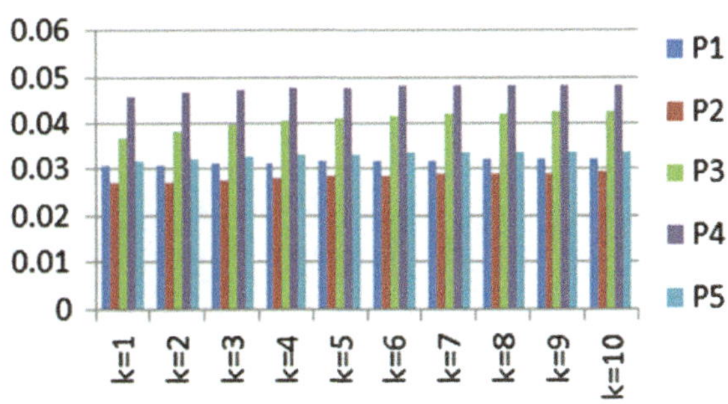

Fig. 3 Score values of alternatives using HFGDWAA operator (taking $\gamma = 3$)

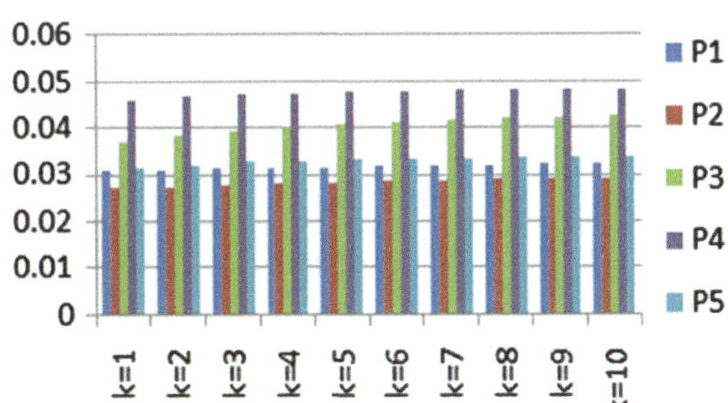

Fig. 4 Score values of alternatives using HFGDWAA operator (taking $\gamma = 4$)

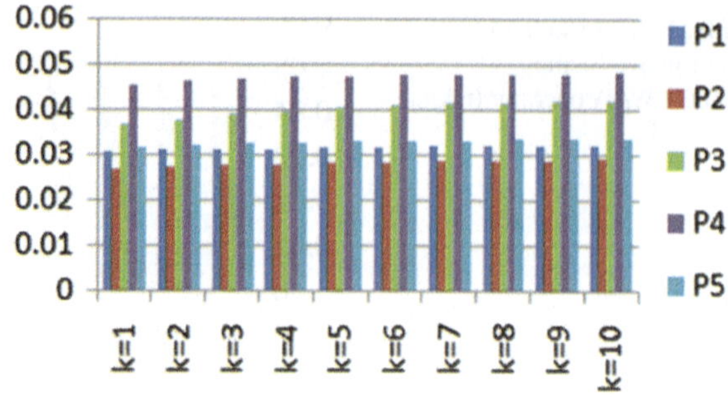

Fig. 5 Score values of alternatives using HFGDWAA operator (taking $\gamma = 5$)

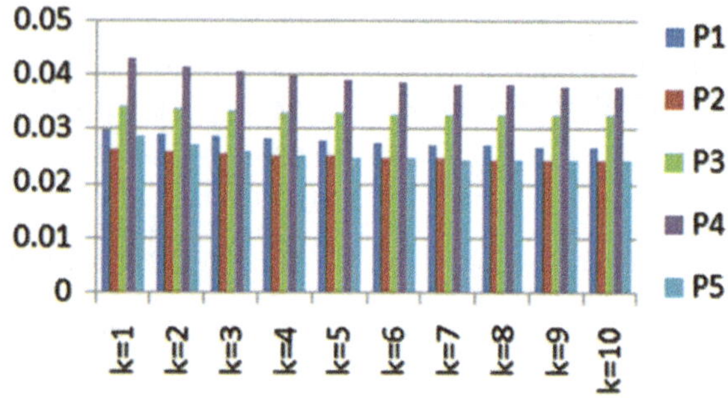

Fig. 6 Score values of alternatives using HFGDWGA operator (taking $\gamma = 1$)

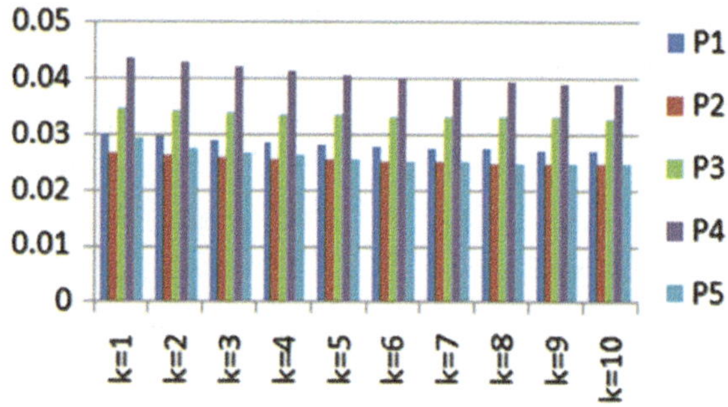

Fig. 7 Score values of alternatives using HFGDWGA operator (taking $\gamma = 2$)

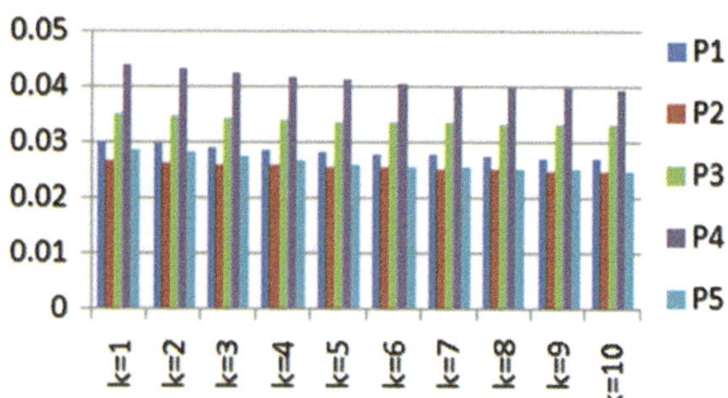

Fig. 8 Score values of alternatives using HFGDWGA operator (taking $\gamma = 3$)

Fig. 9 Score values of alternatives using HFGDWGA operator (taking $\gamma = 4$)

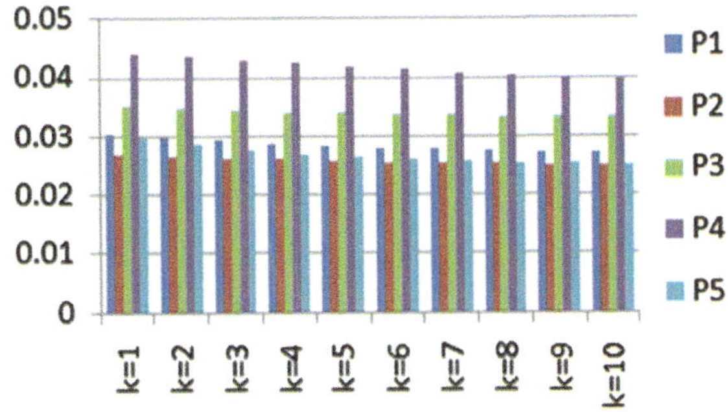

Fig. 10 Score values of alternatives using HFGDWGA operator (taking $\gamma = 5$)

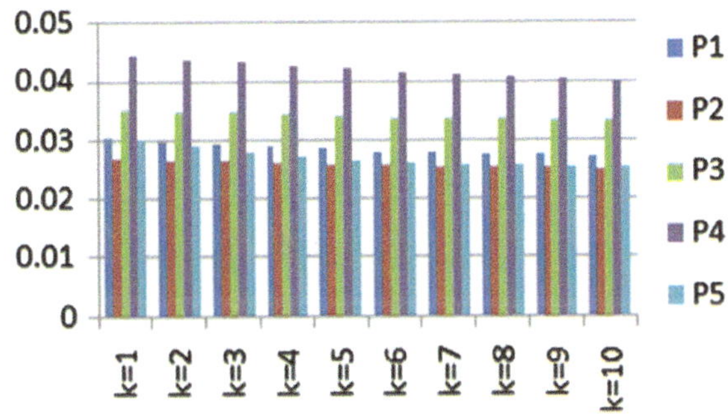

Fig. 11 Score values of alternatives using HFGDOWAA operator (taking $\gamma = 1$)

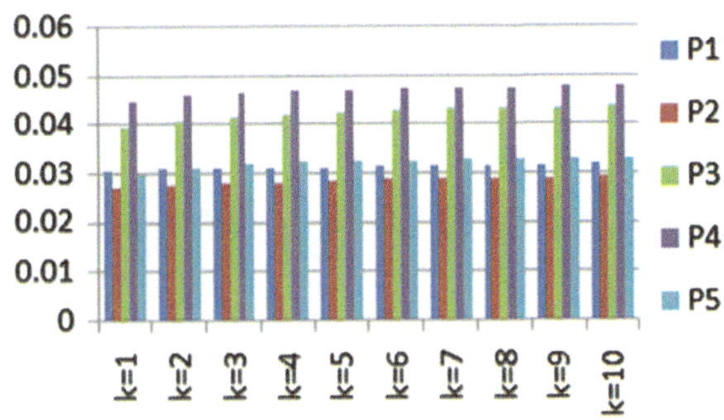

Fig. 12 Score values of alternatives using HFGDOWAA operator (taking $\gamma = 2$)

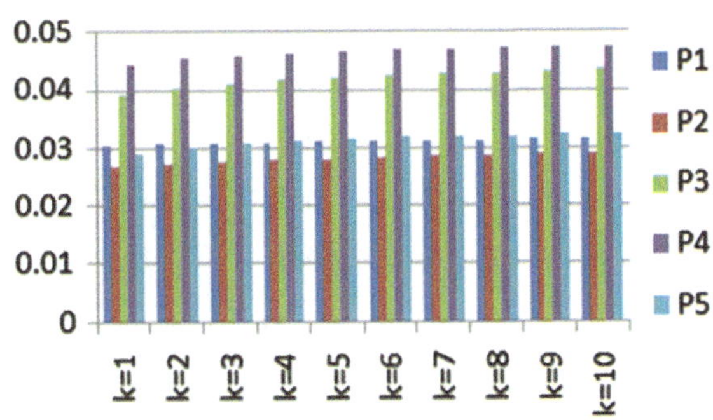

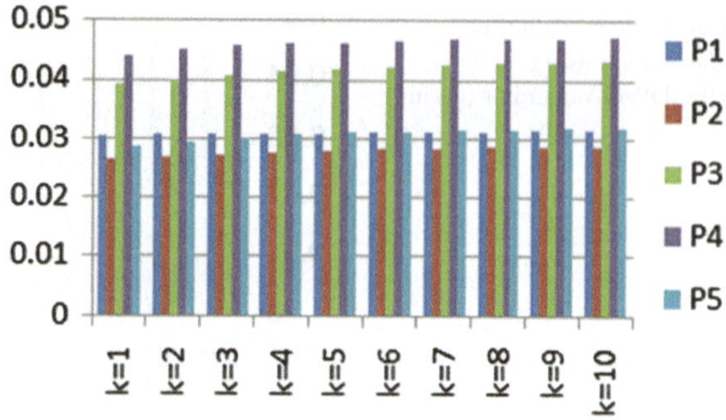

Fig. 13 Score values of alternatives using HFGDOWAA operator (taking $\gamma = 3$)

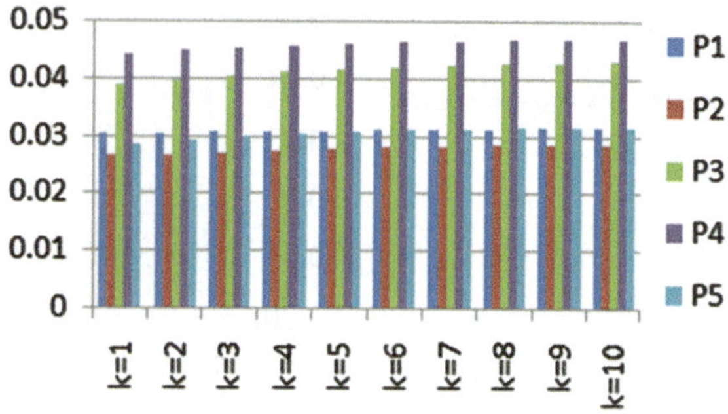

Fig. 14 Score values of alternatives using HFGDOWAA operator (taking $\gamma = 4$)

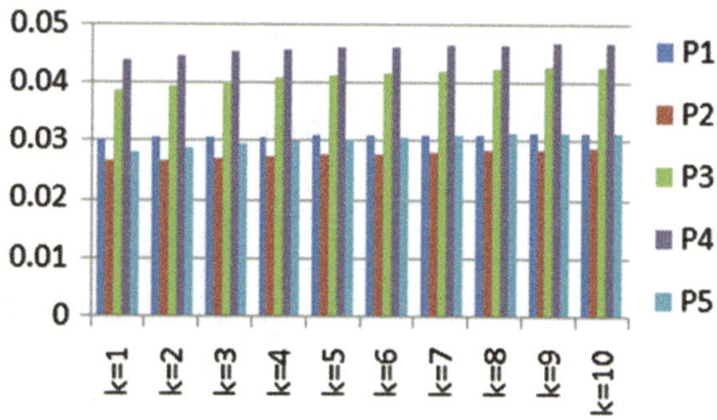

Fig. 15 Score values of alternatives using HFGDOWAA operator (taking $\gamma = 5$)

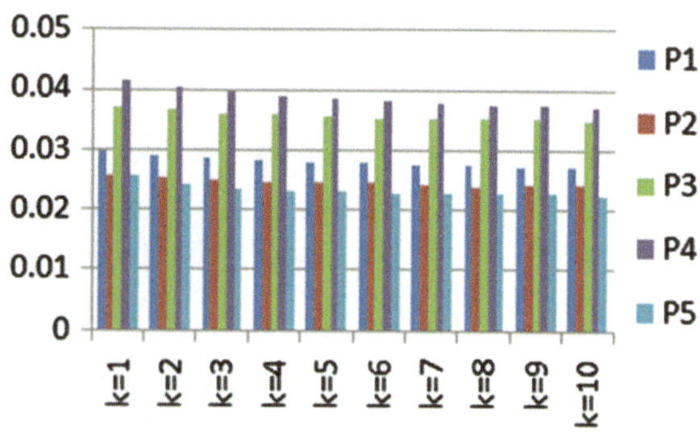

Fig. 16 Score values of alternatives using HFGDOWGA operator (taking $\gamma = 1$)

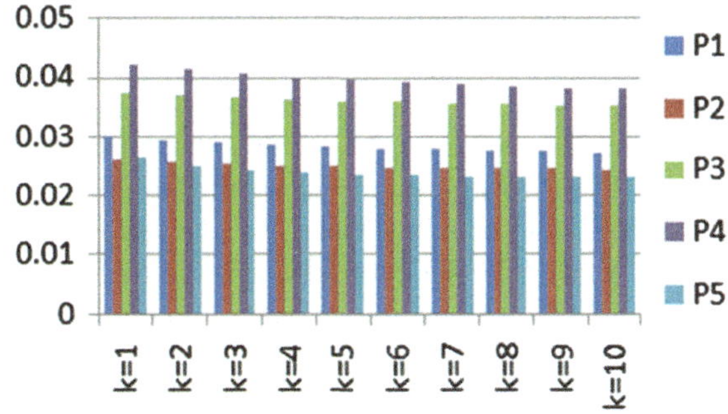

Fig. 17 Score values of alternatives using HFGDOWGA operator (taking $\gamma = 2$)

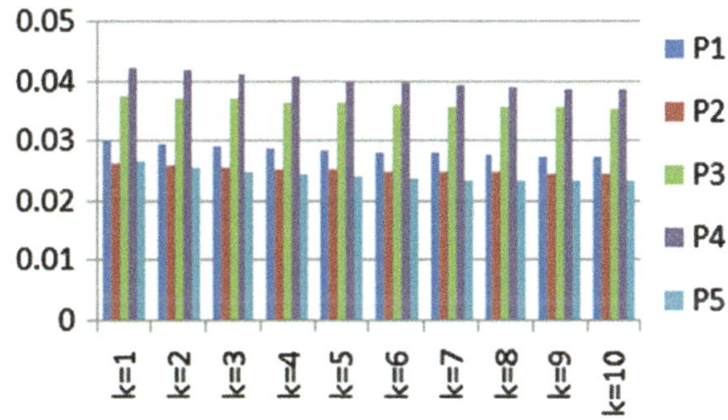

Fig. 18 Score values of alternatives using HFGDOWGA operator (taking $\gamma = 3$)

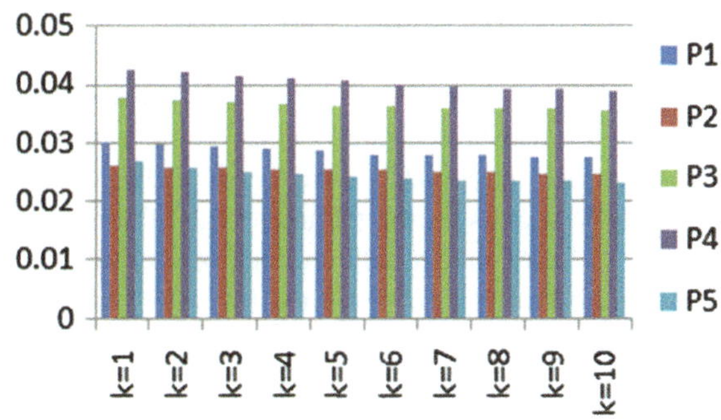

Fig. 19 Score values of alternatives using HFGDOWGA operator (taking $\gamma = 4$)

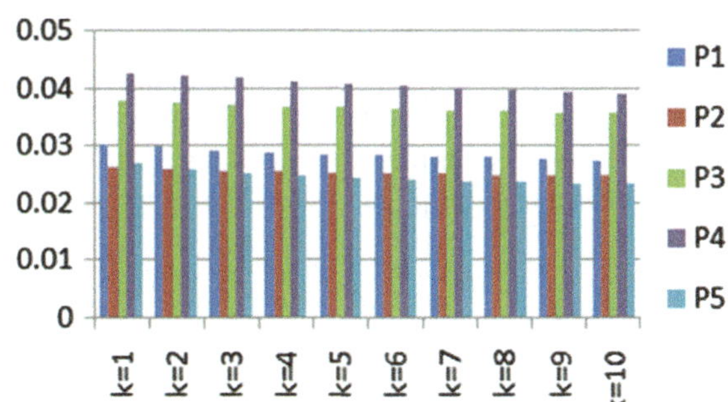

Fig. 20 Score values of alternatives using HFGDOWGA operator (taking $\gamma = 5$)

value of k, the score values P_1, P_2, P_3, P_4, P_5 obtained from the *HFGDWAA* operator are found to decrease with the increase of the value of γ which ranges from 1 to 5. The same characteristic is observed for the operator *HFGDOWAA*.

(2) Similarly, with a fixed value of γ, the HFGDWGA operator's score values for P_1, P_2, P_3, P_4, P_5 progressively drop as the value of k goes from 1 to 10. On the other hand, for a fixed value of k, the score values of P_1, P_2, P_3, P_4, P_5 derived from the *HFGDWGA* operator grow as the value of γ increases from 1 to 5. The HFGDOWGA operator suggests a relatively similar result.

The proposed research on score values shows that the operators HFGDWAA and HFGDOWAA have similar answers, and that the same behavior is observed for the operators HFGDWGA and HFGDOWGA when the parameters k and γ are changed in [1, 10] and [1, 5], respectively, in our suggested MADM method. It should be noted that we treat the parameters k and γ as the evaluation attitudes of the decision maker(s) and that decision makers have the freedom to choose appropriate degrees of parameters k and γ based on their different risk attitudes.

While dealing with the various MADM problems, we follow up on the cases given below:

- **Case-I**: Let us now set the parameter γ. If the DM problem is a risk-preference type, then we must take higher values of the parameter k if we choose any of the operators from *HFGDWGA* and *HFGDOWGA*.
- **Case-II**: Let us fix the parameter γ. If the DM problem is a risk aversion type, then we must take smaller values of the parameter k if we choose any of the operators from *HFGDWGA* and *HFGDOWGA*.
- **Case-III**: Let us fix the parameter k. If the DM problem is a risk-preference type, then we must take higher values of the parameter γ if we choose any of the operators from *HFGDWAA* and *HFGDOWAA*.
- **Case-IV**: Let us fix the parameter k. If the DM problem is a risk aversion type, then we must take smaller values of the parameter γ if we choose any of the operators from *HFGDWAA* and *HFGDOWAA*.

6.2 Sensitivity Assessment (SA)

We use sensitivity analysis (SA) in this part to determine the effect of a suitable attribute on the findings of the presented model. If an attribute has the highest weight value, it is described as the "most significant attribute." Kirkvood [53] and Kahraman [54] proposed using Eq. (2) to show the weight ratio throughout the SA.

$$w_c = (1 - w_s) \times \frac{w_c^0}{W_c^0} = w_c^0 - \alpha_c \times \Delta x \tag{2}$$

where

w_c = Variation in attribute weights in the SA,

w_s = Weight of the most prominent attribute,
w_c^0 = Original values of the attribute weights,
W_c^0 = Sum of actual values of modified attribute weights,
α_c = Weight coefficient of elasticity.

When we associate the variations with the most relevant attribute weight, the relative importance of the other values of the criteria weights is indicated by α_c. The following formula can be used to compute α_c:

$$\alpha_c = \frac{w_c^0}{W_c^0} \tag{3}$$

During the SA, the hypotheses (Kahraman [54]) are just like that:

(1) α_s (Weight coefficient of elasticity of an appropriate attribute) is given;
(2) Throughout the SA process, the ratio of fluctuating weights remains constant.

As shown in Eq. (2), the fluctuation degree applied to a set of weight coefficients is expressed by parameter Δx, which is defined in terms of the respective weighted elastic coefficient. There ought to be a constraint in place in the event that the weight of the most important attribute varies. If this is not performed, related weights can become negative, violating the weight proportionate constraints. The parameter's positive and negative values may suggest an increase or decrease in the degree of importance, respectively. The boundaries for Δx provide the most range in a suitable attribute weight in both the positive and negative directions. We can deduce the limit values of Δx from the following.

$$-w_s^0 \leq \Delta x \leq \min\left\{\frac{w_c^0}{\alpha_c}\right\} \tag{4}$$

Following that, we determine the Δx bounds and then estimate the original weights of attributes using the SA's pre-defined parameters. Equations (5) and (6) are used to derive the values of a set of weight coefficients.

$$w_s = w_s^0 + \alpha_s \times \Delta x \tag{5}$$

$$w_c = w_c^0 - \alpha_c \times \Delta x \tag{6}$$

where

w_s^0 = Original weight of the most influential attribute to SA,
w_c^0 = Original value of variable weights.

It should be remembered that the equation $\sum w_s + \sum w_c = 1$ is regarded as the fundamental condition of the proportion of weight coefficients, and the revised attribute set described above fulfills this formula. Taking the revised values of the attribute weights into account, the rankings of the alternatives are determined.

Q_4 can be identified as one of the most effective attributes in this study simply because it has the greatest value of the weight coefficient $w_4 = 0.4$, which was obtained from the analysis. Following that, the values of weight elasticity (Table 2) are evaluated, and it is discovered that the boundaries for the fluctuation of the weight coefficient (Δx) fall in the range of $-0.4 \leq \Delta x \leq 0.6$. According to the given limits for the variation of weight coefficient values of attributes, various attribute weight sets (SET1, SET2, SET3, and SET4) for SA are calculated. The weight sets for the interval $-0.4 \leq \Delta x \leq 0.6$ are divided into four groups. By employing Eqs. (5) and (6), different perspectives of the weight coefficients are obtained for each set, and these values are displayed in Table 3. As a result, distinct attribute weight sets are used to compute the scores of the alternatives, which are depicted in Fig. 21, and their associated priority orders are provided in Table 4. Evaluating the ranking position of the 1st alternative (P_1) for different sets of attribute weights, we can conclude that P_1 is 4th ranked in 1st and 4th cases and 3rd ranked in the remaining cases. P_2 is 3rd ranked in 1st case and 5th ranked in all remaining cases. P_3 is 1st ranked in 1st case, 3rd ranked in 4th case and 4th ranked in the remaining cases. P_4 is 2nd ranked in 1st case, while in the remaining cases, it is 1st ranked. P_5 is 5th ranked in the 1st case, and in the rest of the cases, it is 2nd ranked. That research revealed that alternative P_4 is the most suitable of the remaining alternatives. Following that, we calculated the SRCC values (r_A) (Ghorabaee et al. [55]) in the results through various weight sets characterized by several attributes that elicited in Table 4. As shown in Table 4, the average of the SRCC (r_A) values is 0.7, indicating a "strong correlation" between alternative ranks (Ghorabaee et al. [55]). As a result of the findings, it is possible to infer that the ranking of the options generated employing our developed approach is trustworthy and dependable.

Table 2 Coefficients of elasticity

Criteria	Q_1	Q_2	Q_3	Q_4
α_c	0.5	0.333333333	0.166666667	1

Table 3 Various attribute weights sets for SA

	SET1	SET2	SET3	SET4
Q_1	0.5	0.04	0.07	0.1
Q_2	0.333333	0.026667	0.046667	0.066667
Q_3	0.166667	0.013333	0.023333	0.033333
Q_4	0	0.92	0.86	0.8

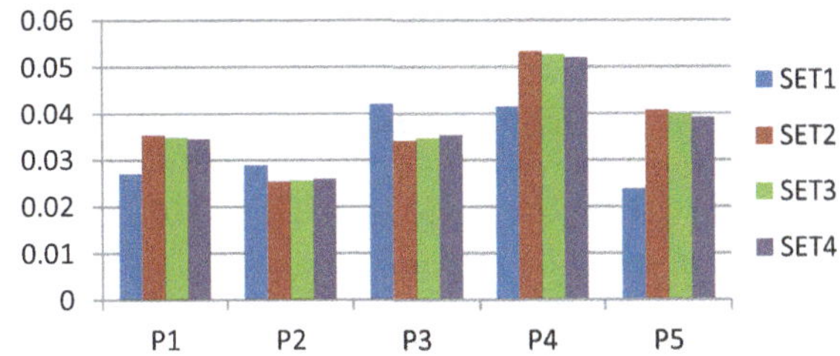

Fig. 21 Alternatives' scores for various weighting schemes for attributes

Table 4 Priority order of alternatives for different attribute weight sets

	SET1	SET2	SET3	SET4	Final ranking
P_1	4	3	3	4	4
P_2	3	5	5	5	5
P_3	1	4	4	3	2
P_4	2	1	1	1	1
P_5	5	2	2	2	3
SRCC values	0.5	0.7	0.7	0.9	–

6.3 Comparative Study

The comparative of the MADM methodology proposed by us mentioned here with all relevant and significant MADM methodologies under HF conditions was carried out using an evaluation of all pertinent current approaches, which included Xia and Xu's method [19], Yu's method [30], Tan et al.'s method [31], and He's method [37]. The results corresponding to these approaches are summarized below.

(i) If we apply the HF weighted arithmetic AO (*HFWA*) proposed by Xia and Xu [19] on the case study, then the final scores of the alternatives are, respectively, 0.0311, 0.0274, 0.0379, 0.0466, and 0.0326. The corresponding preference order is $P_4 \succ P_3 \succ P_5 \succ P_1 \succ P_2$, and so P_4 is the best-suited alternative.

(ii) If we utilize the HF weighted geometric AO (*HFWG*) developed by Xia and Xu [19] on the case study, then the overall scores of the alternatives are, respectively, 0.0300, 0.0264, 0.0344, 0.0430, and 0.0288. The corresponding preference order is $P_4 \succ P_3 \succ P_1 \succ P_5 \succ P_2$, and so P_4 is the best-suited alternative.

(iii) If we use the HF Einstein weighted arithmetic AO (*HFEWA*) defined by Yu [30] on the case study, then the overall scores of the alternatives are, respectively, 0.0310, 0.0272, 0.0374, 0.0462, and 0.0320. The corresponding preference order is $P_4 \succ P_3 \succ P_5 \succ P_1 \succ P_2$, and so P_4 is the best-suited alternative.

(iv) If we apply the HF Einstein weighted geometric AO (*HFEWG*) developed by Yu [30] on the case study, then the overall scores of the alternatives are, respectively, 0.0302, 0.0266, 0.0349, 0.0437, and 0.0294. The corresponding preference order is $P_4 \succ P_3 \succ P_1 \succ P_5 \succ P_2$, and so P_4 is the best-suited alternative.

(v) If we utilize the HF Einstein ordered weighted arithmetic AO (*HFEOWA*) defined by Yu [30] on the case study, then the overall scores of the alternatives are, respectively, 0.0306, 0.0267, 0.0393, 0.0445, and 0.0290. The corresponding preference order is $P_4 \succ P_3 \succ P_1 \succ P_5 \succ P_2$, and so P_4 is the best-suited alternative.

(vi) Applying the HF Einstein ordered weighted geometric AO (*HFEOWG*) developed by Yu [30] on the case study, the overall scores of the alternatives are, respectively, 0.0300, 0.0260, 0.0375, 0.0421, and 0.0263. The corresponding preference order is $P_4 \succ P_3 \succ P_1 \succ P_5 \succ P_2$, and so P_4 is the best-suited alternative.

(vii) Utilizing the HF Hamachar weighted arithmetic AO (*HFHWA*) proposed by Tan et al. [31] on the case study (taking $\gamma = 3$), the overall scores of the alternatives are, respectively, 0.0309, 0.0271, 0.0371, 0.0460, and 0.0317. The corresponding preference order is $P_4 \succ P_3 \succ P_5 \succ P_1 \succ P_2$, and so P_4 is the best-suited alternative.

(viii) Using the HF Hamachar weighted geometric AO (*HFHWG*) defined by Tan et al. [31] on the case study (taking $\gamma = 3$), the overall scores of the alternatives are, respectively, 0.0303, 0.0266, 0.0351, 0.0440, and 0.0270. The corresponding preference order is $P_4 \succ P_3 \succ P_1 \succ P_5 \succ P_2$, and so P_4 is the best best-suited alternative.

(ix) If we apply the HF Hamachar ordered weighted arithmetic AO (*HFHOWA*) proposed by Tan et al. [31] on the case study (taking $\gamma = 3$), then the overall scores of the alternatives are, respectively, 0.0306, 0.0266, 0.0391, 0.0442, and 0.0287. The corresponding preference order is $P_4 \succ P_3 \succ P_1 \succ P_5 \succ P_2$, and so P_4 is the best-suited alternative.

(x) If we utilize the HF Hamachar ordered weighted geometric AO (*HFHOWG*) developed by Tan et al. [31] on the case study (taking $\gamma = 3$), then the overall scores of the alternatives are, respectively, 0.0301, 0.0261, 0.0377, 0.0424, and 0.0266. The corresponding preference order is $P_4 \succ P_3 \succ P_1 \succ P_5 \succ P_2$, and so P_4 is the best-suited alternative.

(xi) If we apply the HF Dombi weighted arithmetic AO (*HFDWA*) introduced by He [37] on the case study, then the overall scores of the alternatives are, respectively, 0.0326, 0.0287, 0.0416, 0.0490, and 0.0370. The corresponding preference order is $P_4 \succ P_3 \succ P_5 \succ P_1 \succ P_2$, and so P_4 is the best-suited alternative.

(xii) Utilizing the HF Dombi weighted geometric AO (*HFDWG*) developed by He [37] on the case study, the overall scores of the alternatives are, respectively, 0.0290, 0.0242, 0.0232, 0.0419, and 0.0243. The corresponding preference order is $P_4 \succ P_1 \succ P_5 \succ P_2 \succ P_3$, and so P_4 is the best-suited alternative.

(xiii) If we apply the HF Dombi ordered weighted arithmetic AO (*HFDOWA*) proposed by He [37] on the case study, then the overall scores of the alternatives are, respectively, 0.0321, 0.0284, 0.0424, 0.0475, and 0.0353. The

Table 5 Characteristic comparison with extant approaches

Approaches	Whether capable of handling MADM issues?	Whether AOs are generalized in nature?	Whether AOs are flexible in nature?	Whether efficient in terms of handling computational complexity?
Xia and Xu [19]	✓	×	×	✓
Yu [30]	✓	×	×	✓
Tan et al. [31]	✓	✓	×	×
He [37]	✓	×	×	×
Proposed	✓	✓	✓	✓

corresponding preference order is $P_4 \succ P_3 \succ P_5 \succ P_1 \succ p_2$, and so P_4 is the best-suited alternative.

(xiv) Applying the HF Dombi ordered weighted geometric AO (*HFDOWG*) introduced by He [37] on the case study, the overall scores of the alternatives are, respectively, 0.0304, 0.0230, 0.0332, 0.0400, and 0.0196. The corresponding preference order is $P_4 \succ P_3 \succ P_1 \succ P_2 \succ P_5$, and so P_4 is the best-suited alternative.

It is abundantly clear from the preceding discussion that, notwithstanding the impression of a slight change in the relative preferences of the alternatives, the optimum, or most acceptable, alternative is identical to that found in prior methods [19, 30, 31, 37].

Along with the comparative study outlined above, we examine some key differences between our created MADM technique and the DM techniques proposed by Xia and Xu [19], Yu [30], Tan et al. [31], and He [37], which are reported in Table 5.

Thus, the proposed AOs can be considered sophisticated tools for dealing with MADM situations since they offer a new flexibility tool for decision experts to select the right values for the parameters "k" and "γ" based on their actual choice demands. Additionally, we may consider our suggested technique to be generalized since numerous existing operators, such as the HF weighted AOs (Xia and Xu [19]), HF Hamacher weighted AOs (Tan et al. [31]), and the HF Einstein weighted AOs (Yu [30]), are particular instances of our created AOs. As a result, the strategy described in this study is one of the most suitable methods established to date for resolving MADM difficulties in HF settings.

7 Conclusion

With the complicated matter of developing confusion and dubiousness around practical DM concerns in mind, the attribute value is appropriately expressed as an HFE during the development of MADM. Conventional fusion methods for aggregating HF data are limited to algebraic operations, Einstein operations, and Hamacher operations, among others, and we detect both a lack of flexibility and sufficient complexity during the optimization procedure. Motivated by this fact, in this paper, using generalized Dombi operations, we have added several new fundamental operations for HFEs. We examine the salient properties of these advanced operations. Additionally, we built many HF AOs depending on such suggested operations, including the HFGDWAA, HFGDOWAA, HFGDWGA, and HFGDOWGA operators. The essential aspects of the created AOs are addressed, including idempotency, boundedness, shift invariance, and monotonicity. Moreover, depending on them, an *MADM* process is illustrated to detect the most desirable alternative in the HF context. An illustrative case study is included here to aid in comprehending the method outlined above, and the results are contrasted against those of a few other traditional methodologies, which clearly illustrate its usefulness and superiority. The comparative research component allows us to state unequivocally that the established methodology is most suitable for usage in MCDM problems in the context of a HF environment. It is pertinent to state here that many existing operators in connection with HF information can be considered particular cases of the advanced AOs. In our future works, we shall extend the developed AOs to handle MADM issues in hesitant intuitionistic fuzzy, hesitant Pythagorean fuzzy, hesitant q-rung orthopair fuzzy, and hesitant single-valued neutrosophic contexts.

References

1. Torra, V.: Hesitant fuzzy sets. Int. J. Intell. Syst. **25**, 529–539 (2010)
2. Torra, V., Narukawa, Y.: On hesitant fuzzy sets and decision. In: The 18th IEEE International Conference on Fuzzy Systems, 20–24 August, Ueju Island, Korea, pp. 1378–1382
3. Rodriguez, R.M., Martinez, L., Herrera, F.: Hesitant fuzzy linguistic term sets for decision making. IEEE Trans. Fuzzy Syst. **20**, 109–119 (2012)
4. Zhang, N., Wei, G.: Extension of VIKOR method for decision making problem based on hesitant fuzzy set. Appl. Math. Modell. **37**, 4938–4947 (2013)
5. Xu, Z., Zhang, X.: Hesitant fuzzy multi-attribute decision making based on TOPSIS with incomplete weight information. Knowl. Based Syst. **52**, 53–64 (2013)
6. Zhang, X., Xu, Z.: The TODIM analysis approach based on novel measured functions under hesitant fuzzy environment. Knowl. Based Syst. **61**, 48–58 (2014)
7. Dong, Y., Chen, X., Herrera, F.: Minimizing adjusted simple terms in the consensus reaching process with hesitant linguistic assessments in group decision making. Inf. Sci. **297**, 95–117 (2015)
8. Chen, N., Xu, Z.: Hesitant fuzzy ELECTRE II approach: a new way to handle multi-criteria decision making problems. Inf. Sci. **292**, 175–197 (2015)
9. Wang, J.Q., Wang, J., Chen, Q.H., Zhang, H.Y., Chen, X.H.: An outranking approach for multi-criteria decision-making with hesitant fuzzy linguistic term sets. Inf. Sci. **280**, 338–351 (2014)
10. Wang, J., Wang, J.Q., Zhang, H.Y., Chen, X.H.: Multi-criteria decision-making based on hesitant fuzzy linguistic term sets: an outranking approach. Knowl. Based Syst. **86**, 224–236 (2015)
11. Zhang, X., Xu, Z.: Hesitant fuzzy agglomerative hierarchical clustering algorithms. Int. J. Syst. Sci. **46**, 562–576 (2015)

12. Zhu, B., Xu, Z., Xu, J.: Deriving a ranking from hesitant fuzzy preference relations under group decision making. IEEE Trans. Cybern. **44**, 1328–1337 (2014)
13. Ye, J.: Correlation coefficient of dual hesitant fuzzy sets and its application to multiple attribute decision making. Appl. Math. Modell. **38**, 659–666 (2014)
14. Liao, H., Xu, Z., Zeng, X.J.: Novel correlation coefficients between hesitant fuzzy sets and their application in decision making. Knowl. Based Syst. **82**, 115–127 (2015)
15. Liu, H., Rodríguez, R.M.: A fuzzy envelope for hesitant fuzzy linguistic term set and its application to multi-criteria decision making. Inf. Sci. **258**, 220–238 (2014)
16. Zhao, N., Xu, Z., Liu, F.: Uncertainty measures for hesitant fuzzy information. Int. J. Intell. Syst. **30**, 818–836 (2015)
17. Rodríguez, R., Martínez, L., Torra, V., Xu, Z., Herrera, F.: Hesitant fuzzy sets: state of the art and future directions. Int. J. Intell. Syst. **29**, 495–524 (2014)
18. Xu, Z.: Hesitant Fuzzy Sets Theory. Springer (2014)
19. Xia, M., Xu, Z.: Hesitant fuzzy information aggregation in decision making. Int. J. Approx. Reason. **52**, 395–407 (2011)
20. Xia, M., Xu, Z., Chen, N.: Some hesitant fuzzy aggregation operators with their application in group decision making. Group Decis. Negotiat. **22**, 259–279 (2013)
21. Wei, G.: Hesitant fuzzy prioritized operators and their application to multiple attribute decision making. Knowl. Based Syst. **31**, 176–182 (2012)
22. Wei, G., Zhao, X., Wang, H., Lin, R.: Hesitant fuzzy Choquet integral aggregation operators and their applications to multiple attribute decision making. Int. Inf. Insti. (Tokyo) Inf. **15**, 441–448 (2012)
23. Zhu, B., Xu, Z., Xia, M.: Hesitant fuzzy geometric Bonferroni means. Inf. Sci. **205**, 72–85 (2012)
24. Zhang, Z.: Hesitant fuzzy power aggregation operators and their application to multiple attribute group decision making. Inf. Sci. **234**, 150–181 (2013)
25. Zhang, Z., Wu, C.: Weighted hesitant fuzzy sets and their application to multi-criteria decision making. Br. J. Math. Comput. Sci. **4**, 1091–1123 (2014)
26. Zhang, Z., Wang, C., Tian, D., Li, K.: Induced generalized hesitant fuzzy operators and their application to multiple attribute group decision making. Comput. Indus. Eng. **67**, 116–138 (2014)
27. Liao, H., Xu, Z.: Extended hesitant fuzzy hybrid weighted aggregation operators and their application in decision making. Soft Comput. 1–14 (2015)
28. Zhou, W., Xu, Z.: Optimal discrete fitting aggregation approach with hesitant fuzzy information. Knowl. Based Syst. **78**, 22–33 (2015)
29. Wei, C., Zhao, N., Tang, X.: Operators and comparisons of hesitant fuzzy linguistic term sets. IEEE Trans. Fuzzy Syst. **22**, 575–585 (2014)
30. Yu, D.: Some hesitant fuzzy information aggregation operators based on Einstein operational laws. Int. J. Intell. Syst. **29**, 320–340 (2014)
31. Tan, C., Yi, W., Chen, X.: Hesitant fuzzy Hamacher aggregation operators for multi-criteria decision making. Appl. Soft Comput. **26**, 325–349 (2015)
32. Qin, J., Liu, X., Pedrycz, W.: Hesitant fuzzy maclaurin symmetric mean operators and its application to multiple-attribute decision making. Int. J. Fuzzy Syst. **17**, 509–520 (2015)
33. Zhu, B., Xu, Z.: Hesitant fuzzy Bonferroni means for multi-criteria decision making. J. Opera. Res. Socie. **64**, 1831–1840 (2013)
34. Qin, J., Liu, X., Pedrycz, W.: Frank aggregation operators and their applications to hesitant fuzzy multiple attribute decision making. Appl. Soft Comput. **41**, 428–452 (2016)
35. Liao, H.C., Xu, Z.: A VIKOR based method for hesitant fuzzy multi-criteria decision making. Fuzzy Optim. Decis. Mak. **12**, 372–392 (2016)
36. Mahmood, T., Mehmood, F., Khan, Q.: Cubic hesitant fuzzy sets and their applications to multi-criteria decision making. Int. J. Algeb. Stat. **5**, 19–51 (2016)
37. He, X.: Typhoon disaster assessment based on Dombi hesitant fuzzy information aggregation operators. Nat. Haza. **90**, 1153–1175 (2018)

38. Xu, Z., Zhou, W.: Consensus building with a group of decision makers under the hesitant probabilistic fuzzy environment. Fuzzy Optim. Decis. Mak. **16**, 481–503 (2017)
39. Alcantud, J.C.R., Torra, V.: Decomposition theorems and extension principles for hesitant fuzzy sets. Inform. Fusn. **41**, 48–56 (2018)
40. Wang, R., Li, Y.: Picture hesitant fuzzy set and its application to multi criteria decision making. Symmet **10** (2018). https://doi.org/10.3390/sym10070295
41. Wang, Y., Zhang, R., Quian, L.: An improved A* algorithm based on hesitant fuzzy set theory for multi criteria acrtic route planning. Symmet **10** (2018). https://doi.org/10.3390/sym10120765
42. Liao, H., Si, G., Xu, Z., Fujita, H.: Hesitant fuzzy linguistic preference utility set and its application in selection of fire rescue plans. Int. J. Envi. Res. Publ. Health **15**(4), 664 (2018). https://doi.org/10.3390/ijerph15040664
43. Yalcin, N., Pehlivan, N.Y.: Application of the fuzzy CODAS method based on fuzzy envelopes for hesitant fuzzy linguistic term sets: a case study on a personnel selection problem. Symmet **11** (2019). https://doi.org/10.3390/sym11040493
44. Wu, J., Liu, X.D., Wang, Z.W., Zhang, S.T.: Dynamic emergency decision making method with probabilistic hesitant fuzzy information based on GM(1,1) and TOPSIS. IEEE Access **7**, 7054–7066 (2018)
45. Alcantud, J.C.R., Garcia, G.S., Peng, X., Zhan, J.: Dual extended hesitant fuzzy sets. Symmet **11** (2019). https://doi.org/10.3390/sym11050714
46. Liu, X.D., Wang, Z.W., Zhang, S.T., Liu, J.S.: A novel approach to fuzzy cognitive map based on hesitant fuzzy sets for modeling risk impact on electric power system. Int. J. Comput. Intell. Syst. **12**, 842–854 (2019)
47. Liu, X., Wang, Z., Zhang, S., Liu, J.: Probabilistic hesitant fuzzy multiple attribute decision making based on regret theory for the evaluation of venture capital projects. Econ. Res. **33**, 672–697 (2020)
48. Dombi, J.: The generalized Dombi operator family and the multiplicative utility function. Stud. Comput. Intell. (2009). https://doi.org/10.1007/978-3-642-00448-3-6
49. Klir, G., Yuan, B.: Fuzzy sets and fuzzy logic: theory and applications. Prentice Hall, Upper Saddle River, NJ (1995)
50. Dombi, J.: A general class of fuzzy operators, the De Morgan class of fuzzy operators and fuzziness measures induced by fuzzy operators. Fuzzy Sets Syst. **8**, 149–163 (1982)
51. Zhou, L., Zhao, X., Wei, G.: Hesitant fuzzy Hamachar aggregation operators and their applications to multiple-attribute decision making. J. Int. Fuzzy Syst. **26**, 2689–2699 (2014)
52. Wang, X., Triantaphyllou, E.: Ranking irregularities when evaluating alternatives by using some ELECTRE methods. Int. J. Manag. Sci. **36**, 45–63 (2008)
53. Kirkwood, C.W.: Strategic Decision Making: Multi-Objective Decision Analysis with Spreadsheets. Duxbury Press, Belmont (1997)
54. Kahraman, Y.R.: Robust sensitivity analysis for multi-attribute deterministic hierarchical value models (No. AFIT/GOR/ENS/02-10). Air Force Inst of Tech Wright-PATTERSONAFB OH (2002)
55. Ghorabaee, M.K., Zavadskas, E.K., Amiri, M., Esmaeili, A.: Multi-criteria evaluation of green suppliers using an extended WASPAS method with interval type-2 fuzzy sets. J. Clean. Prod. **137**, 213–229 (2016)

A Novel Extended Fermatean Fuzzy Framework for Evaluating the Challenges to Sustainable Smart City Development

Mahyar Kamali Saraji and Dalia Streimikiene

Abstract Challenges to sustainable smart city development (SSCD) motivate the present research to propose a novel Fermatean fuzzy multi-criteria framework for evaluating four Iranian cities' performance concerning the identified challenges. To this end, firstly, challenges to SSCD were identified through literature review, and then, an integrated MEREC-TOPSIS model under Fermatean fuzzy environment was proposed to determine the challenges' weight and then rank Iranian cities. Results indicated that "social plausibility" is the most significant challenge to SSCD out of twelve identified challenges, followed by "lack of infrastructure." On top of that, comparative studies and a sensitivity study were conducted to evaluate the efficiency and sensitivity of the proposed framework.

Keywords Digitalization · Sustainable development · Multi-criteria decision analysis (MCDA) · Internet of things (IoT) · Industry 4.0

1 Introduction

A smart city is one in which information and communication technologies (ICT) are employed to increase operational efficiency and enable the system to communicate information autonomously [1]. In other words, ICTs are employed as an enabler to enhance urban services, support city development, and improve stakeholder access to information [2]. Smart cities are swiftly emerging worldwide through city-wide investments in ICT to support the new industries' development, drive technological innovation, foster a robust economy, improve the citizens' life quietly, and maintain a sustainable environment [3]. However, cities have faced many challenges in

M. Kamali Saraji · D. Streimikiene (✉)
Kaunas Faculty, Vilnius University, Muitines 8, 44280 Kaunas, Lithuania
e-mail: dalia.streimikiene@knf.vu.lt

M. Kamali Saraji
e-mail: Mahyar.Kamali@knf.vu.lt

L. Sahoo et al. (eds.), *Real Life Applications of Multiple Criteria Decision Making Techniques in Fuzzy Domain*, Studies in Fuzziness and Soft Computing 420,
https://doi.org/10.1007/978-981-19-4929-6_2

becoming smart, motivating scholars to figure out the challenges and possible solutions to the identified challenges. For instance, Wang et al. [4] studied the challenges of smart cities to elders or people with disabilities to propose an inclusive smart city framework. They concluded that an inclusive smart city should address these people's needs leading governments to develop feasible and appropriate strategies. However, they did not consider other challenges to smart cities related to environmental sustainability, while this study proposes a more holistic framework of challenges to sustainable smart cities development. Also, Goodman et al. [5] studied the public engagement's challenges to smart cities in Canada using semi-structured interviews. Their case studies were the Quebec, Guelph, and Waterloo, and the results indicated that civic participation is still considered a top-down tool though serious attempts were made to engage people. Nevertheless, they only studied one of the challenges to the smart city development while there are several challenges which are studied in the present study; also, there is no comprehensive framework of challenges in Goodman et al. [5], while this study proposed an applicable framework of challenges for possible evaluations.

Sustainable smart city development is an emerging topic, motivating researchers to study the challenges to SSCD. For instance, Rana et al. [6] researched the barriers to sustainable smart cities development in India by reviewing the literature, and they evaluated the identified impediments using a fuzzy analytic hierarchy process (AHP). The findings showed that "governance" is the major impediment to establishing sustainable smart cities, followed by economic and technological impediments. Although the present study is closely related to the study conducted by Rana et al. [6], the present research methodology is more novel in terms of fuzzy environment and integration of MCDM approaches to deal with uncertainties and subjectivity in decision-making. Also, Rana et al. [6] did not apply their proposed framework to evaluate cities concerning their progress toward SSCD. On top of that, Khan et al. [7] conducted survey research in Pakistan to identify SSCD challenges. They interviewed seventeen experts from the executive, operational, and managerial levels of public and private companies. They proposed a conceptual framework of challenges to SSCD, including challenges related to environmental sustainability; however, they did not apply the proposed framework, while this research aims to develop and employ a framework of challenges to evaluate cities in Iran.

On the other hand, researchers have broadly applied fuzzy theory to deal with inaccuracy and uncertainty in dealing with multi-criteria problems [8]. Atanassov [9] proposed intuitionistic fuzzy sets (IFSs) to overcome the conventional fuzzy limitations. The sum of belongingness degree (BD) and non-belongingness degree (NBD) in IFSs must be less than or equal to one [10]. Later [11] developed Pythagorean fuzzy sets (PFSs) to cope with the IFSs limitations and inaccuracy, vagueness, and uncertainty in real-life situations [12]. As a result, several scholars have employed the PFSs in different areas, such as medical waste management [13], medical diagnosis [14], pharmacological therapy selection [15], green supplier selection [16], sustainable energy development [17], and pattern recognition problems [18]. Senapati and Yager [19] have recently developed Fermatean fuzzy sets (FFSs) to deal with the limitations of PFSs. The cube sum of BD and NBD in FFSs is less than or equal to one, making FFSs more accurate in dealing with real-life multi-criteria decision analysis (MCDA) issues. Several researchers have recently shown interest in employing FFSs in different areas. For example, Kamali et al. [20] proposed a Fermatean fuzzy CRITIC-COPRAS to evaluate the challenges to Industry 4.0 adoption toward a sustainable digital transformation.

Shahzadi and Akram [21] developed a novel ranking method under Fermatean fuzzy environment to elect the most efficient antivirus mask in COVID time. Rani and Mishra [22] develop FF-Einstein aggregation operators to select charging stations for electric cars using MULTUMOORA. Mishra and Rani [23] proposed FF- WASPAS based on a novel score function to cope with healthcare waste management issues. Akram et al. [24] developed an evaluation framework under Fermatean fuzzy environment *t* select the most efficient sanitizer to deal with COVID-19. Aydemir and Gunduz [25] developed Dombi operators under Fermatean fuzzy and applied them to a real-life problem using FF-TOPSIS. Garg et al. [26] proposed FF-aggregation operators and applied them to select a facility in COVID time. Liu et al. [27] applied FF-TODIM and FF-TOPSIS based on a new distance measure to deal with real-life problems. Also, Senapati and Yager [28] developed some FF-aggregation operators and applied weighted product models (WPM) to cope with multi-criteria issues. Moreover, Senapati and Yager [29] proposed weighted averaging geometric operators under Fermatean fuzzy and applied them to deal with multi-criteria issues.

According to mentioned research gaps, there are several challenges to SSCD which are needed to be addressed in a comprehensive framework; thus, the present study aims to, firstly, identify the challenges to sustainable smart city development through a rigorous literature review and then determine the importance of the identified challenges using FF-MEREC to figure out which one is the most significant challenges to SSCD. Consequently, FF-TOPSIS ranks four Iranian cities concerning the weighted challenges to determine the best city. In the end, comparative studies and sensitivity analysis are conducted to evaluate the performance and reliability of the proposed method in dealing with real-life multi-criteria problems. As a result, the primary contributions of this study are presented below:

1. To figure out the challenges to SSCD through literature review. There is a lack of study in which a comprehensive framework of challenges to SSCD was presented, motivating the present research to present a holistic framework by reviewing smart city development literature.
2. To extend a novel MEREC-TOPSIS method under Fermatean fuzzy. There is a lack of study in which an evaluating model was presented, especially a novel multi-criteria model under an extended fuzzy environment, which increases the evaluation and avoids uncertainties in decision-making.
3. To assess Iranian cities' progress in coping with identified challenges toward SSCD. To the authors' best knowledge, there is no study in which cities were evaluated concerning their progress in dealing with challenges to SSCD using a multi-criteria evaluation framework under Fermatean fuzzy environment.

4. To do comparative and sensitivity analysis to assess the applicability and reliability of the FF-MEREC method. Although proposing a novel evaluating framework in smart city development could significantly contribute to the literature, the applicability and trustability of the proposed method are crucial aspects of proposing a new framework, motivating the present research to do comparative and sensitivity analysis.

2 Challenges to SSCD

As mentioned, there are several challenges to SSCD stemming from the economy, technology, society, and environment. The identified challenges are presented in the following in detail.

2.1 Lack of Operational Framework (C1)

The smart city is still an emerging concept that motivates scholars to propose conceptual and theoretical frameworks to reach a consensus on theory [30]. However, a lack of an operational framework could be a serious challenge to SSCD as it must have a proper organizational structure and clear management; thus, a practical framework should be distinctly proposed and outlined [7].

2.2 Lack of Sustainable Strategies (C2)

Khan et al. [7] concluded that more sustainable cooperation with environmental protection stakeholders is needed to deal with the challenges to environmental sustainability in smart cities. It should be noted that sustainable strategies are long-term plans leading smart cities to intersect environmental and business interests so that it is possible to deal with social and environmental problems simultaneously. Also, Ahvenniemi et al. [31] mentioned that a non-sustainable city could not be considered a smart city; thus, smart city development should be integrated with sustainability evaluation.

2.3 Lack of Governance Configuration (C3)

Supportive governance configuration could boost smart cities in counties, promoting learning, and innovation. The mentioned system could be established through an experimental attitude in policy application or based on governmental support [32].

More specifically, a practical governance configuration addresses exchanging knowledge, learning, networking through conferences, workshops, etc., in which skills and experiences could be combined, from a wide range of various actors, including manufacturers, including technology providers, developers, network designers, system integrators, cybersecurity specialists, and data analysts [33, 34]. Thus, the mentioned lack could be a serious challenge to SSCD [6].

2.4 *Lack of Inclusivity (C4)*

Another key difficulty emerging smart city development in developing nations must face is the lack of inclusion of the original people, notably the disadvantaged and marginalized groups whose lands are frequently acquired for smart city initiatives [35, 36]. Some of the recorded situations that contradict the idea of inclusion in smart city development include land acquisition by force, relocation of original inhabitants by force, and geographic segregation [37, 38]. A smart city is not only installing a collection of information systems. The primary goal of such a smart city would be to give each individual a more intelligent experience for an event. The cognitive solution will become a critical component in assessing smart cities [39].

2.5 *Lack of Transformation Strategy (C5)*

Khan et al. [7] concluded an urgent need to develop strategies or process for a strategic pathway to smart city development. It may provide a venue for stakeholders to describe their activities. These strategies serve as a road map for the growth of smart cities. Cities cannot be transformed overnight into smart cities; a comprehensive strategy is required [40]. Collective integration of technical, governance, institutional, and transitional components is required for city transformation. Smart cities could provide infrastructure and information-based services and economic growth potential for enterprises [41].

2.6 *Lack of Awareness (C6)*

Effective public awareness and marketing campaigns should be a critical component of any smart city development strategy. The lack of awareness can be solved by advertising the advantages on city radio or print and electronic media awareness campaigns [36]. It is advised that inhabitants be educated about the benefits of smart cities in public spaces. Additionally, service providers must accurately identify their intended user groups and grasp these consumers' particular traits, demographic backgrounds, and interests [7, 42]. On top of that, citizens' knowledge of smart cities is critical for obtaining the most advantages from sustainable smart cities; hence, residents should be provided with a rudimentary understanding to reap the maximum gain [6].

2.7 Lack of Public Engagement (C7)

Another significant driver of smart city development in underdeveloped nations is active public engagement and participation [36, 43]. To guarantee that the main goals of smart city development are realized, active citizen engagement should adopt the concept of citizen empowerment, digital inclusion, collaborative governance, behavioral change, and a benefits-sharing method to ensure that people can enjoy the fruits of smart city development fairly [44]. Building a strong sense of knowledge and ownership among individuals, supporting interactions that involve people, and encouraging active engagement in giving feedback at every level of the policy process are all examples of citizen empowerment [6].

2.8 Lack of Standards (C8)

Standards have made significant contributions in forming and developing smart cities since different aspects could be monitored through clear standards [45]. For instance, standards could boost security issues in smart transporting; other factors, such as resource management and business practices, could be addressed through standards [32]. On top of that, standards could be employed to evaluate the performance of smart cities; however, no clear standards have been proposed so far; thus, many international organizations have been working on proposing clear and comprehensive standards [34, 45].

2.9 Social Plausibility (C9)

Another significant aspect of smart city development in developing nations is establishing a robust regulatory framework that supports investment confidence and public trust [36]. To achieve this, the government must take proactive efforts to establish clear sets of rules and policies that will guide the creation of smart cities. Furthermore, having full and open regulatory agencies would allow the government to trust people and investors. Moreover, Khan et al. [7] mentioned that individuals should be asked to participate in an educational course where they may learn about smart cities to address the issue of social acceptability. It also highlighted the importance of bringing together individuals who have a shared interest to feel supported [7]. On top of that, because creativity is acknowledged as a critical factor in developing a smart city, people, education, learning, and knowledge all play a significant role in developing a smart city. A smart city's wide definition involves cultivating an environment conducive to the growth of a creative class [46].

2.10 *Lack of Infrastructure (C10)*

Infrastructure and technology readiness contribute to the physical grounds of all smart city efforts and serve as critical drivers for developing smart cities in underdeveloped nations [32, 36]. The accessibility of privacy-preserving, wireless and secure, infrastructure and service-oriented information systems that communicate information about public services with citizens would increase managerial and operational efficiency [43]. The Internet of things (IoT) is critical to the architecture of smart cities because it enables the collection and processing of data from scattered sensors and smart devices [47]. The vulnerability of IoT-based applications is intrinsically linked to the network paradigm, in which physical objects, including sensor-based equipment, collect data on critical network interactions and make a connection via wireless distribution connections [48]. Citizens would conduct various transactions and inquiries online using these information systems and service platforms [49]. Additionally, by establishing an integrated e-government portal, the government may leverage these tools to improve resource management and public services [36].

2.11 *Lack of Human Resources (C11)*

Human capital is a critical factor in establishing smart cities in developing nations. It is critical for smart cities to be properly trained and technically capable of exploring the smart city atmosphere, including understanding the potential of various smart city programs [6]. Additionally, boosting human resource technical capacity in terms of the motivation, knowledge, and skills required to deliver IT-enabled services will help authorities accelerate the implementation of different smart city programs [36]. Individuals' technology illiteracy can impede technology acceptance, scaling up adoption, and recognizing that beneficial network effects require a considerable user base comprising a substantial proportion of technology-literate people [49].

2.12 *Financial Support (C12)*

Mishra [50] noted that government funding capacity is perhaps one of the most critical enablers of the smart city. Developing counties are more likely to be fiscally limited, necessitating the use of a diverse array of traditional and novel finance vehicles to boost capital expenditures in the development of smart cities [51]. Vu and Hartley [33] mentioned that budgetary support is one of the most significant drivers to smart city development. Additionally, governments worldwide are constrained by financial limits and old infrastructure, and it is argued that public finance constraints impede the development of smart city projects [52, 53]. While the financial worth of a smart city is not immediately apparent, profit is the primary objective of capital investment; smart city building is a large-scale undertaking with a long-term payoff. As a result, to get financial investment in this sector, we must establish the project's worth through scientific study. Increase investor trust in return on investment by using innovative business models [54].

3 Research Method

The current work presents a unique extended MCDA approach for evaluating the performance of four Iranian cities regarding the barriers to SSCD. To this goal, the issues were first identified through a review of the available literature. Following that, discovered difficulties were rated using the FF-MEREC approach with the help of three decision experts. Then, decision experts evaluated and ranked cities based on their success against weighted tasks using fermatean fuzzy linguistic variables [55], presented in Table 1. The preceding sections discuss the Fermatean fuzzy sets and the suggested method's phases.

3.1 Preliminaries

Definition 1 [19] Assume Δ an incremental limited universe of discourse; thus, a fermatean fuzzy set could be presented using Eq. (1).

$$T = \{t_i, (b_T(t_i), n_T(t_i))|t_i \in \Delta\} \tag{1}$$

Table 1 Terms used in linguistics

FFNs	Terms used in linguistics
(0.9, 0.1)	"Extremely high (EH)"
(0.8, 0.1)	"Very high (VH)"
(0.7, 0.2)	"High (H)"
(0.6, 0.3)	"Medium–high (MH)"
(0.5, 0.4)	"Medium (M)"
(0.4, 0.5)	"Medium–low (ML)"
(0.25, 0.6)	"Low (L)"
(0.1, 0.75)	"Very low (VL)"
(0.1, 0.9)	"Extremely low (EL)"

wherein $b_T : \Delta \rightarrow [0.1]$ indicates the BD of a member $t_i \in \Delta$ in FFS. Also, $b_T : \Delta \rightarrow [0.1]$ represents the NBD of a member $t_i \in \Delta$ in FFS. It should be noted that $0 \leq (b_T(t_i))^3 + (n_T(t_i))^3 \leq 1$ must be true for each $t_i \in \Delta$.

Definition 2 [19] Let $\lambda = (b_\lambda, n_\lambda)$; Eq. (2) represents then the indeterminacy degree of FFS.

$$\pi_\lambda = \sqrt[3]{1 - b_\lambda^3 - n_\lambda^3} \tag{2}$$

where b_λ and $n_\lambda \in [0, 1]$, and $0 \leq b_\lambda^3 + n_\lambda^3 \leq 1$.

Definition 3 [19] assume $\lambda = (b_\lambda, n_\lambda)$ a FFS; then, Eqs. (3) and (4) show the score function and accuracy function of λ.

$$\wp_s(\lambda) = b_\lambda^3 - n_\lambda^3 | - 1 \leq \wp_s(\lambda) \leq 1 \tag{3}$$

$$\Im_a(\lambda) = b_\lambda^3 + n_\lambda^3 | 0 \leq \Im_a(\lambda) \leq 1 \tag{4}$$

Definition 4 [19] assume $\lambda = (b_\lambda, n_\lambda)$, $\lambda_1 = (b_{\lambda_1}, n_{\lambda_1})$, and $\lambda_2 = (b_{\lambda_2}, n_{\lambda_2})$ are three FFSs. Equation (5) to eleven represent some proposed operators for FFSs.

$$\lambda^c = (n_\lambda, b_\lambda) \tag{5}$$

$$\lambda_1 \cup \lambda_2 = \left(\min\{b_{\lambda_1}, b_{\lambda_2}\}, \max\{n_{\lambda_1}, n_{\lambda_2}\}\right) \tag{6}$$

$$\lambda_1 \cup \lambda_2 = \left(\max\{b_{\lambda_1}, b_{\lambda_2}\}, \min\{n_{\lambda_1}, n_{\lambda_2}\}\right) \tag{7}$$

$$\lambda_1 \oplus \lambda_2 = \left(\sqrt[3]{b_{\lambda_1}^3 + b_{\lambda_2}^3 - b_{\lambda_1}^3 b_{\lambda_2}^3}, n_{\lambda_1} n_{\lambda_2}\right) \tag{8}$$

$$\lambda_1 \otimes \lambda_2 = \left(b_{\lambda_1} b_{\lambda_2}, \sqrt[3]{n_{\lambda_1}^3 + n_{\lambda_2}^3 - n_{\lambda_1}^3 n_{\lambda_2}^3}\right) \tag{9}$$

$$l\lambda = \left(\sqrt[3]{1 - \left(1 - b_\lambda^3\right)^l}, (n_\lambda)^l\right), \ l > 0 \tag{10}$$

$$\lambda^l = \left((b_\lambda)^l, \sqrt[3]{1 - \left(1 - n_\lambda^3\right)^l}\right), l > 0 \tag{11}$$

3.2 FF-MEREC-Topsis

As mentioned, the present study applied the FF-MEREC method under the Fermatean fuzzy set to calculate the objective weight of challenges. Keshavarz-Ghorabaee, et al. [56] proposed the conventional MEREC and compared the results with other methods, including CRITIC, entropy, and standard deviation used to calculate the objective weight criteria various studies. Also, this study applied FF-TOPSIS to assess the Iranian cities' performance regarding the identified challenges. Senapati and Yager [19] applied FF-TOPSIS for the first time to deal with a multi-criteria decision-making problem. The proposed method is expressed in the following.

Step 1. Decision matrix construction.

Assume that, $A = \{A_1, A_2, \ldots, A_m\}$ be a set of cities, $C = \{C_1, C_2, \ldots, C_n\}$ a set of challenges, and $E = \left\{E_1, E_2, \ldots, E_p\right\}$ a panel of decision experts (DEs), who support each city O_i with regard to a challenge C_j based on Fermatean fuzzy linguistic variables. The decision matrix (DM) is expressed by $DM = \left(g_{ij}^k\right)$, for$i = i, \ldots, m; j = 1, \ldots, n$, while g_{ij}^k shows the support to the city (i) with regard to the challenge (*j*) by *k*th DE.

Step 2. DEs' significance degree calculation

Suppose the significant degrees of the DEs are presented using FFNs; then, suppose $\omega_k = (b_k, n_k)$ is the significance rating of *k*th DE presented by an expert using FFN. The formula for calculating *k*th DE's weight is presented using Eq. (12).

$$\omega_k = \frac{\left(b_k^3 + \pi_k^3 \times \left(\frac{b_k^3}{b_k^3 + n_k^3}\right)\right)}{\sum_{k=1}^{p}\left(b_k^3 + \pi_k^3 \times \left(\frac{b_k^3}{b_k^3 + n_k^3}\right)\right)}, K = 1, \ldots, p; \omega_k \geq 0, \sum_{1}^{p} \omega_k = 1. \quad (12)$$

Step 3. FF-decision matrix aggregation

Equation (13) should be applied to aggregate the individual decision-making matrices. The Fermatean fuzzy weighted averaging (FFWA) operator is shown in Eq. (13). Let $\mathbb{Z} = \left(z_{ij}\right)_{m \times n}$ be the aggregated FF-decision matrix, where

$$z_{ij} = \left(\sqrt[3]{1 - \prod_{k=1}^{r}\left(1 - \left(b_{ij}^k\right)^3\right)^{\omega_k}}, \prod_{k=1}^{r}\left(n_{ij}^k\right)^{\omega_k}\right) \quad (13)$$

Step 4. Objective weight by MEREC

Step 4.1. Score matrix construction

The first step of MEREC is constructing score matrix $\Xi = \left(k_{ij}\right)_{m \times n}$ by Eq. (14).

$$k_{ij} = \frac{1}{2}\left[\left(b_j^3 - n_j^3 - \ln\left(1 + \pi_j^3\right)\right) + 1\right], for i = 1, \ldots, m \quad (14)$$

Step 4.2. Score matrix normalization

Equation (15) is used to normalize the score matrix.

Equation (15) is used to construct the normalized score matrix $\tilde{\Xi} = (n_{ij})_{m\times n}$.

$$n_{ij} = \begin{cases} \frac{\min_k k_j}{k_{ij}}, & j \in N_b \\ \frac{k_{ij}}{\max_k k_j}, & j \in N_n \end{cases} \quad (15)$$

Step 4.3. Calculating the overall performance of alternatives (OP).

In this stage, a logarithmic metric with equal criteria weights is used to determine the overall performance of the alternatives. Equation (16) is used to calculate the OP.

$$\mathrm{Op}_i = \ln\left(1 + \left(\frac{1}{m}\sum_j \left|\ln\left(n_{ij}^x\right)\right|\right)\right) \quad (16)$$

Step 4.4. Calculating the overall performance of alternatives by removing each criterion.

Similar to the preceding phase, this step makes use of the logarithmic measure. The distinction between this phase and the previous one is that the performance of the alternatives is calculated independently for each criterion. As a result, we have m distinct sets of performances that correspond to m distinct criteria. Assume $OP_{ij}^{'}$, the overall performance of ith alternative concerning the removal of jth criterion. Equation (17) is applied for the calculating $\mathrm{OP}_{ij}^{'}$:

$$\mathrm{OP}_{ij}' = \ln\left(1 + \left(\frac{1}{m}\sum_{k,k\neq j} \left|\ln\left(n_{ij}^x\right)\right|\right)\right) \quad (17)$$

Step 4.5. Calculating the summation of absolute deviations.

The jth criterion's removal impact is computed in this step using the values acquired in Steps 4.3 and 4.5. Allow E_j to express the consequence of eliminating the jth condition. Equation (18) can be used to determine the values of E_j:

$$E_j = \sum_i \left|\mathrm{OP}_{ij}^{'} - \mathrm{Op}_i\right| \quad (18)$$

Step 4.6. Calculate the final weights for each criterion

The objective weight of each criterion is estimated in this phase using E_j. Throughout the remainder of this section, w_j denotes the weight of the jth criterion. Equation (19) is used to calculate w_j:

$$W_j = \frac{E_j}{\sum_k E_k} \quad (19)$$

Step 5. Estimate the Fermatean fuzzy positive and negative ideal solutions.

Equations (20) and (21) are used to calculate the positive and negative ideal solutions.

$$S^+ = \begin{cases} \max_i \text{score}\big(C_j(A_i)\big),\ j \in N_b \\ \min_i \text{score}\big(C_j(A_i)\big),\ j \in N_n \end{cases} \tag{20}$$

$$S^- = \begin{cases} \min_i \text{score}\big(C_j(A_i)\big),\ j \in N_b \\ \max_i \text{score}\big(C_j(A_i)\big),\ j \in N_n \end{cases} \tag{21}$$

Step 6. The relative closeness computing

The relative closeness is computed using Eq. (22) for ideal solutions. The alternative with the highest relative closeness is the best option.

$$R(K_i) = \frac{Y_i^-}{Y_i^- + Y_i^+} \text{ for } i = 1, \ldots, m$$

where

$$Y_i^- = \text{dis}\big(S^-, \lambda_{ij}\big) = \varpi_j \sqrt{\frac{1}{2}\left[\left((b_{ij})^3 - (b_j^-)^3\right)^2 + \left((n_{ij})^3 - (n_j^-)^3\right)^2 + \left((\pi_{ij})^3 - (\pi_j^-)^3\right)^2\right]} \tag{22}$$

$$Y_i^+ = \text{dis}\big(S^+, \lambda_{ij}\big) = \varpi_j \sqrt{\frac{1}{2}\left[\left((b_{ij})^3 - (b_j^+)^3\right)^2 + \left((n_{ij})^3 - (n_j^+)^3\right)^2 + \left((\pi_{ij})^3 - (\pi_j^+)^3\right)^2\right]}$$

4 Results

The decision-making matrix should be constructed in the first step. It should be noted that the importance of the DEs is $\{(0.8, 0.45, 0.67); (0.75, 0.55, 0.74); (0.75, 0.55, 0.74)\}$, and C9 and C12 are considered as benefits in the present research (Table 2).

After computing the significance of the DEs, individual DM should be aggregated, shown in Table 3.

Table 2 Decision-making matrix

Challenges	Experts	CITY 1	CITY 2	CITY 3	CITY 4
Lack of operational framework	"E1"	"H"	"MH"	"ML"	"M"
	"E2"	"H"	"H"	"ML"	"MH"
	"E3"	"MH"	"VH"	"M"	"M"
Lack of sustainable strategies	"E1"	"ML"	"ML"	"H"	"ML"
	"E2"	"M"	"H"	"MH"	"ML"
	"E3"	"ML"	"MH"	"H"	"M"
Lack of governance configuration	"E1"	"ML"	"VL"	"ML"	"H"
	"E2"	"ML"	"VL"	"VL"	"MH"
	"E3"	"ML"	"EL"	"L"	"H"
Lack of inclusivity	"E1"	"EH"	"MH"	"L"	"ML"
	"E2"	"H"	"MH"	"L"	"ML"
	"E3"	"H"	"H"	"VL"	"ML"
Lack of transformation strategy	"E1"	"M"	"M"	"ML"	"ML"
	"E2"	"MH"	"MH"	"ML"	"L"
	"E3"	"MH"	"M"	"M"	"ML"
Lack of awareness	"E1"	"VL"	"ML"	"ML"	"MH"
	"E2"	"VL"	"M"	"ML"	"MH"
	"E3"	"L"	"MH"	"ML"	"MH"
Lack of public engagement	"E1"	"M"	"H"	"L"	"H"
	"E2"	"ML"	"MH"	"L"	"H"
	"E3"	"ML"	"M"	"L"	"H"
Lack of standards	"E1"	"ML"	"MH"	"VL"	"M"
	"E2"	"ML"	"MH"	"L"	"M"
	"E3"	"ML"	"M"	"VL"	"M"
Social plausibility	"E1"	"VL"	"VL"	"ML"	"EH"
	"E2"	"L"	"EL"	"ML"	"H"
	"E3"	"L"	"VL"	"ML"	"H"
Lack of infrastructure	"E1"	"EH"	"L"	"ML"	"VL"
	"E2"	"MH"	"L"	"ML"	"M"
	"E3"	"VH"	"L"	"ML"	"L"
Lack of human resource	"E1"	"VL"	"MH"	"MH"	"ML"
	"E2"	"VL"	"H"	"MH"	"VL"
	"E3"	"L"	"VH"	"EH"	"L"
Financial support	"E1"	"MH"	"ML"	"L"	"VL"
	"E2"	"MH"	"ML"	"M"	"L"
	"E3"	"MH"	"M"	"ML"	"VL"

Table 3 Aggregated decision-making matrix

Challenges	CITY 1	CITY 2	CITY 3	CITY 4
Lack of operational framework	{0.67 0.23 0.88}	{0.71 0.18 0.86}	{0.44 0.47 0.93}	{0.54 0.36 0.93}
Lack of sustainable strategies	{0.44; 0.47 0.93}	{0.60 0.31 0.91}	{0.67 0.23 0.88}	{0.44 0.47 0.93}
Lack of governance configuration	{0.40 0.50 0.93}	{0.10 0.80 0.79}	{0.30 0.61 0.91}	{0.67 0.23 0.88}
Lack of inclusivity	{0.80 0.16 0.78}	{0.64 0.26 0.90}	{0.22 0.65 0.90}	{0.40 0.50 0.93}
Lack of transformation strategy	{0.57 0.33 0.92}	{0.54 0.36 0.93}	{0.44 0.47 0.93}	{0.36 0.53 0.93}
Lack of awareness	{0.18 0.70 0.87}	{0.51 0.39 0.93}	{0.40 0.50 0.93}	{0.60 0.30 0.91}
Lack of public engagement	{0.44 0.46 0.93}	{0.62 0.29 0.90}	{0.25 0.60 0.92}	{0.70 0.20 0.87}
Lack of standards	{0.40 0.50 0.93}	{0.57 0.33 0.92}	{0.18 0.70 0.87}	{0.50 0.40 0.93}
Social plausibility	{0.44 0.46 0.93}	{0.10 0.80 0.79}	{0.40 0.50 0.93}	{0.80 0.16 0.78}
Lack of infrastructure	{0.81 0.14 0.77}	{0.25 0.60 0.92}	{0.40 0.50 0.93}	{0.36 0.57 0.92}
Lack of human resource	{0.18 0.70 0.87}	{0.71 0.18 0.86}	{0.76 0.21 0.82}	{0.30 0.61 0.91}
Financial support	{0.60 0.30 0.91}	{0.44 0.47 0.93}	{0.41 0.50 0.93}	{0.18 0.70 0.87}

Next, the aggregated matrix should be normalized. Afterward, Op_i and OP'_{ij} are calculated, and the results are shown in Table 4.

In the next stage, E_j and W_j are calculated, and the final results are presented in Table 5.

After calculating the weight of the challenges, FF-TOPSIS is applied to rank cities concerning the weighted challenges. S^+ and S^+ are presented in Table 6.

And the final results of FF-TOPSIS are presented in Table 7.

According to table seven, the preference order of the cities concerning the identified challenges to SSCD is $A_3 > A_4 > A_2 > A_4$; thus, A_3 is the best smart city concerning its performance in dealing with the identified challenges.

Table 4 Overall performances

	C1	C2	C3	C4	C5	C6	C7	C8	C9	C10	C11	C12	Op_i
City 1	0.51	0.48	0.48	0.52	0.52	0.45	0.48	0.49	0.48	0.52	0.42	0.45	0.52
City 2	0.41	0.39	0.28	0.38	0.40	0.39	0.39	0.41	0.41	0.31	0.40	0.36	0.41
City 3	0.59	0.63	0.58	0.55	0.61	0.60	0.57	0.57	0.57	0.57	0.63	0.59	0.63
City 4	0.51	0.50	0.53	0.47	0.50	0.53	0.53	0.52	0.40	0.46	0.46	0.53	0.53

Table 5 MEREC results

Challenges	E_j	W_j	Rank
Lack of operational framework	0.07	0.041	11
Lack of sustainable strategies	0.08	0.049	10
Lack of governance configuration	0.22	0.130	3
Lack of inclusivity	0.17	0.099	5
Lack of transformation strategy	0.05	0.032	12
Lack of awareness	0.11	0.066	8
Lack of public engagement	0.11	0.066	7
Lack of standards	0.09	0.056	9
Social plausibility	0.22	0.135	1
Lack of infrastructure	0.22	0.131	2
Lack of human resource	0.17	0.104	4
Financial support	0.15	0.092	6

Table 6 Ideal solutions

	S^+	S^-
Lack of operational framework	{0.44 0.47 0.93}	{0.71 0.18 0.86}
Lack of sustainable strategies	{0.44 0.47 0.93}	{0.67 0.23 0.88}
Lack of governance configuration	{0.10 0.80 0.79}	{0.67 0.23 0.88}
Lack of inclusivity	{0.22 0.65 0.90}	{0.80 0.16 0.78}
Lack of transformation strategy	{0.36 0.53 0.93}	{0.57 0.33 0.92}
Lack of awareness	{0.18 0.70 0.87}	{0.60 0.30 0.91}
Lack of public engagement	{0.25 0.60 0.92}	{0.70 0.20 0.87}
Lack of standards	{0.18 0.70 0.87}	{0.57 0.33 0.92}
Social plausibility	{0.80 0.16 0.78}	{0.10 0.80 0.79}
Lack of infrastructure	{0.25 0.60 0.92}	{0.81 0.14 0.77}
Lack of human resource	{0.18 0.70 0.87}	{0.76 0.21 0.82}
Financial support	{0.60 0.30 0.91}	{0.18 0.70 0.87}

Table 7 FF-TOPSIS results

Cities	Y_i^+	Y_i^-	$R(K_i)$	Rank
City 1	0.24	0.17	0.424	4
City 2	0.20	0.18	0.475	3
City 3	0.18	0.25	0.589	1
City 4	0.16	0.23	0.583	2

5 Comparative Analysis

The proposed framework results are compared with FF-WASPAS, proposed by Keshavarz-Ghorabaee et al. (2020), to validate the proposed framework. The steps of the FF-WASPAS are presented below.

FF-WASPAS

Steps 1–4. Same as the FF-MEREC-TOPSIS.

Step 5. Normalization

The decision-making matrix should be normalized using Eq. (23).

$$\mathbb{N} = \left(\eta_{ij}\right)_{m\times n} = \begin{cases} \left(b_{ij}, n_{ij}\right)\text{for benefit challenges} \\ \left(n_{ij}, b_{ij}\right)\text{for cost challenges} \end{cases} \tag{23}$$

Step 6. Calculating the WSM and WPM

In this step, the weighted sum model (WSM) and weighted product model (WPM) should be calculated using Eqs. (24) and (25). The WSM and SPM are presented in Table 8.

$$Q^S(A_i) = \underset{j=1}{\overset{q}{\oplus}} w_j.\eta_{ij} \text{ for } i = 1, \ldots, m \tag{24}$$

$$Q^P(A_i) = \underset{j=1}{\overset{q}{\otimes}} \eta_{ij}^{W_j} \text{ for } i = 1, \ldots, m \tag{25}$$

Step 7. Calculating WASPAS

Table 8 WPM and WSM

Cities	$Q^S(A_i)$	$Q^P(A_i)$
City 1	{0.50 0.45 0.92}	{0.35 0.61 0.90}
City 2	{0.510.44 0.92}	{0.32 0.61 0.91}
City 3	{0.52 0.39 0.93}	{0.45 0.50 0.92}
City 4	{0.55 0.42 0.91}	{0.40 0.53 0.92}

Table 9 WASPAS results

Cities	Score($Q^S(A_i)$)	Score($Q^P(A_i)$)	$Q^i(A_i)$	Rank
City 1	0.228	0.131	0.1798	4
City 2	0.237	0.125	0.1806	3
City 3	0.248	0.194	0.2213	1
City 4	0.266	0.167	0.2165	2

In this step, the WASPAS is measured using Eq. (26). The obtained outputs of the WASPAS method are shown in Table 9. It should be noted that $\nu = 0.5$.

$$Q^i(A_i) = \nu Q^P(A_i) + (1 - \nu) Q^s(A_i) \quad (26)$$

Table 9 showed that the preference order of the cities concerning based on WASPAS is $A_3 > A_4 > A_2 > A_4$; thus, A_3 is the best smart city, in line with the FF-TOPSIS results.

6 Sensitivity Analysis

This section defines numerous challenges' weight analyze the model sensitivity to the weight change. In other words, if there are m criteria, m sets of criteria weights should be defined for the sensitivity analysis. Each set has the most important (weighted) and least significant criteria, while others have weights between the most and least important criteria [55]. The preceding section should create twelve criteria weights for the sensitivity analysis. Table 10 indicates the relative importance of each criterion in each of the sets.

After calculation challenges' weight in each set, cities are ranked concerning each set. The results of sensitivity analysis are shown in Fig. 1. According to figure one, the proposed framework is weight dependent, but A_3 is the best city in most scenarios.

7 Discussion

According to the results, social plausibility is the most significant challenge to SSCD based on Iranian experts' opinions. In contrast, Khan et al. [7] concluded that budget constraint is the most significant challenge in Pakistan. Also, it should be noted that Khan et al. [7] surveyed to figure out only challenges to sustainable city development in Pakistan, while the present study not only found the challenges but also proposed an evaluation model to evaluate cities' performance in dealing with the identified challenges, making the proposed model more practical. On top of that, the proposed framework comprised various challenges that stemmed from technology,

Table 10 Weight sets

	Set 1	Set2	Set3	Set4	Set5	Set6	Set7	Set8	Set9	Set10	Set11	Set12
C1	0.041	0.049	0.130	0.099	0.032	0.066	0.066	0.056	0.135	0.131	0.104	0.092
C2	0.049	0.130	0.099	0.032	0.066	0.066	0.056	0.135	0.131	0.104	0.092	0.041
C3	0.130	0.099	0.032	0.066	0.066	0.056	0.135	0.131	0.104	0.092	0.041	0.049
C4	0.099	0.032	0.066	0.066	0.056	0.135	0.131	0.104	0.092	0.041	0.049	0.130
C5	0.032	0.066	0.066	0.056	0.135	0.131	0.104	0.092	0.041	0.049	0.130	0.099
C6	0.066	0.066	0.056	0.135	0.131	0.104	0.092	0.041	0.049	0.130	0.099	0.032
C7	0.066	0.056	0.135	0.131	0.104	0.092	0.041	0.049	0.130	0.099	0.032	0.066
C8	0.056	0.135	0.131	0.104	0.092	0.041	0.049	0.130	0.099	0.032	0.066	0.066
C9	0.135	0.131	0.104	0.092	0.041	0.049	0.130	0.099	0.032	0.066	0.066	0.056
C10	0.131	0.104	0.092	0.041	0.049	0.130	0.099	0.032	0.066	0.066	0.056	0.135
C11	0.104	0.092	0.041	0.049	0.130	0.099	0.032	0.066	0.066	0.056	0.135	0.131
C12	0.092	0.041	0.049	0.130	0.099	0.032	0.066	0.066	0.056	0.135	0.131	0.104

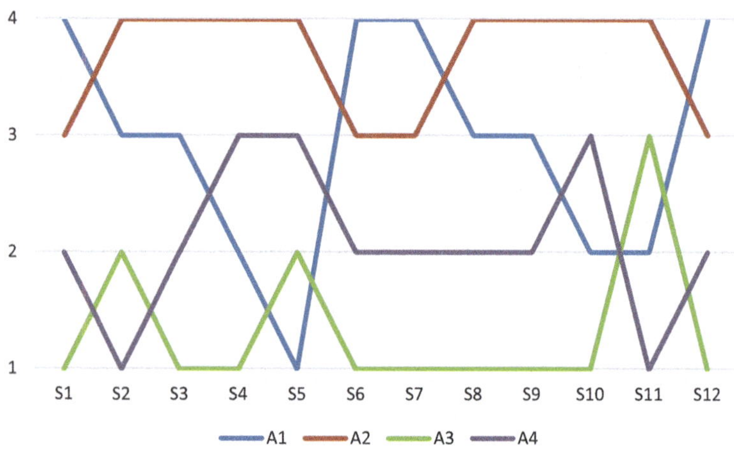

Fig. 1 Cities' rank concerning the challenges' weight change

social, environment, and politic, while previous studies were concerned about only one or two aspects of challenges, such as Goodman et al. [5] and Rana et al. [6].

Moreover, the proposed method is compared with the FF-WASPAS to evaluate the performance of the proposed method and to figure out how results could be changed in terms of changing analyzing methods. Comparative analysis showed that the proposed method's results align with FF-WASPAS, while in the FF-WASPAS, WSM and WPM are calculated, making the FF-WASPAS method more reliable to be compared with the proposed method in the present study. On top of that, the proposed model is partly sensitive to weight changes according to the sensitivity analysis, meaning that if the most significant challenge changed, we would have different results; however, the proposed method's results were the same in most scenarios, showing that how robust the proposed model is, concerning the weight changes.

Furthermore, the current study used a unique extended fermatean fuzzy approach, and the findings suggested that it effectively resolves problems with conflicting criteria. FFNs were applied to cope with ambiguity and uncertainty included in multi-criteria issues, while Rana et al. [6] applied a conventional AHP method. Additionally, the indeterminacy degree was addressed at each stage of the proposed methodology, which increased its accuracy; hence, the proposed method could more adequately deal with complicated multi-criteria situations.

To prevent subjectivity, the objective weights were calculated using the MEREC technique, which is more accurate than other methods, such as CRITIC, which could be considered an advantage over the conventional methods. Thus, the FF-MEREC-TOPSIS may yield more accurate findings than existing methods.

The proposed framework is not limited to MCDA issues using fermatean fuzzy sets; it could apply to problems under conventional, intuitionistic, and Pythagorean fuzzy sets.

8 Conclusions

The present study developed a novel extended framework to assess the challenges to SSCD in Iran. Through a literature review, twelve obstacles were identified. The MEREC-TOPSIS method under FFSs was proposed to assess the challenges and rank four Iranian cites concerning the identified challenges. According to the FF-MEREC-TOPSIS' results, A_3 is the best city, which FF-WASPAS's results also confirmed the proposed method results, indicating the accuracy of the proposed method. On top of that, sensitivity analysis has indicated that the proposed method is sensitive to weight change, while A_3 is the best city in most scenarios, confirming the stability of the proposed method.

On the other hand, according to the results, the most significant challenge to SSCD in Iran is social plausibility, followed by the lack of infrastructure. Therefore, it is recommended that the authorities hold educational events and conferences to boost people's awareness of smart cities, ask communities to join volunteering work related to smart cities, bring people with common interests in smart cities together to share their experiences to increase social acceptability. Also, authorities could overcome the infrastructural issues by developing long-term policies based on developed countries' experiences.

Furthermore, citizens were eager to embrace and employ technology and services as long as they were creative, protected their privacy, and provided a high level of service. Their acceptance is critical: residents have a significant impact on the global smart city model's potential success or failure since they have the opportunity to choose which concepts or technology to accept or reject. As a result, the question arises of whether consumers of smart city services would tolerate a lesser level of privacy if the advantages of integrating such services were plausible. Any security and privacy challenges should be articulated before initiating infrastructure upgrades if smart city technology and services are implemented successfully.

The present research has faced some limitations, including (1) it is assumed that challenges are independent, and there is no inter-connection (2) data gathering was time-consuming due to the unfamiliarity of the experts with MCDA analysis and how to support challenges using the linguistic variables.

Applying the proposed framework in other countries for future research is recommended and asking for more experts' opinions concerning the identified challenges. Also, applying the SWARA method is recommended to calculate the subjective weights and compare the results with the proposed method.

References

1. Zheng, C., Yuan, J., Zhu, L., Zhang, Y., Shao, Q.: From digital to sustainable: a scientometric review of smart city literature between 1990 and 2019. J. Clean. Prod. **258**, 120689 (2020)
2. Laufs, J., Borrion, H., Bradford, B.: Security and the smart city: a systematic review. Sustain. Cities Soc. **55**, 102023 (2020)
3. Law, K.H., Lynch, J.P.: Smart city: technologies and challenges. IT Prof. **21**, 46–51 (2019)
4. Wang, C.H., Steinfeld, E., Maisel, J.L., Kang, B.: Is your smart city inclusive? Evaluating proposals from the US department of transportation's smart city challenge. Sustain. Cities Soc. **74**, 103148 (2021)
5. Goodman, N., Zwick, A., Spicer, Z., Carlsen, N.: Public engagement in smart city development: Lessons from communities in Canada's smart city challenge. The Canadian Geographer/Le Géographe Canadien **64**, 416–432 (2020)
6. Rana, N.P., Luthra, S., Mangla, S.K., Islam, R., Roderick, S., Dwivedi, Y.K.: Barriers to the development of smart cities in Indian context. Inf. Syst. Front. **21**, 503–525 (2019)
7. Khan, H.H., Malik, M.N., Zafar, R., Goni, F.A., Chofreh, A.G., Klemeš, J.J., Alotaibi, Y.: Challenges for sustainable smart city development: a conceptual framework. Sustain. Dev. **28**, 1507–1518 (2020)
8. Saraji, M.K., Mardani, A., Köppen, M., Mishra, A.R., Rani, P.: An extended hesitant fuzzy set using SWARA-MULTIMOORA approach to adapt online education for the control of the pandemic spread of COVID-19 in higher education institutions. Artif. Intell. Rev. 1–26 (2021)
9. Atanassov, K.T.: Intuitionistic fuzzy sets. In: Intuitionistic Fuzzy Sets. Springer pp. 1–137 (1999)
10. Mardani, A., Saraji, M.K., Mishra, A.R., Rani, P.: A novel extended approach under hesitant fuzzy sets to design a framework for assessing the key challenges of digital health interventions adoption during the COVID-19 outbreak. Appl. Soft Comput. **96**, 106613 (2020)
11. Yager, R.R.: Pythagorean membership grades in multicriteria decision making. IEEE Trans. Fuzzy Syst. **22**, 958–965 (2013)
12. Saraji, M.K., Streimikiene, D., Lauzadyte-Tutliene, A.: A novel pythogorean fuzzy-SWARA-CRITIC-COPRAS method for evaluating the barriers to developing business model innovation for sustainability. In: Handbook of Research on Novel Practices and Current Successes in Achieving the Sustainable Development Goals; IGI Global, pp. 1–31 (2021)
13. Liu, P., Rani, P., Mishra, A.R.: A novel Pythagorean fuzzy combined compromise solution framework for the assessment of medical waste treatment technology. J. Clean. Prod. **292**, 126047 (2021)
14. Zhou, Q., Mo, H., Deng, Y.: A new divergence measure of pythagorean fuzzy sets based on belief function and its application in medical diagnosis. Mathematics **8**, 142 (2020)

15. Rani, P., Mishra, A.R., Mardani, A.: An extended Pythagorean fuzzy complex proportional assessment approach with new entropy and score function: application in pharmacological therapy selection for type 2 diabetes. Appl. Soft Comput. **94**, 106441 (2020)
16. Çalık, A.: A novel Pythagorean fuzzy AHP and fuzzy TOPSIS methodology for green supplier selection in the Industry 4.0 era. Soft. Comput. **25**, 2253–2265 (2021)
17. Kamali Saraji, M., Streimikiene, D., Ciegis, R.: A novel Pythagorean fuzzy-SWARA-TOPSIS framework for evaluating the EU progress towards sustainable energy development. Environ. Monit. Assess. **194**, 1–19 (2022)
18. Ejegwa, P.A.: Modified Zhang and Xu's distance measure for Pythagorean fuzzy sets and its application to pattern recognition problems. Neural Comput. Appl. **32**, 10199–10208 (2020)
19. Senapati, T., Yager, R.R.: Fermatean fuzzy sets. J. Ambient. Intell. Humaniz. Comput. **11**, 663–674 (2019)
20. Kamali Saraji, M., Streimikiene, D., Kyriakopoulos, G.L.: Fermatean fuzzy CRITIC-COPRAS method for evaluating the challenges to industry 4.0 adoption for a sustainable digital transformation. Sustainability **13**, 9577 (2021)
21. Shahzadi, G., Akram, M.: Group decision-making for the selection of an antivirus mask under fermatean fuzzy soft information. J. Intell. Fuzzy Syst. 1–16 (2021)
22. Rani, P., Mishra, A.R.: Fermatean fuzzy einstein aggregation operators-based MULTIMOORA method for electric vehicle charging station selection. Expert Syst. Appl. 115267 (2021)
23. Mishra, A.R., Rani, P.: Multi-criteria healthcare waste disposal location selection based on Fermatean fuzzy WASPAS method. Compl. Intell. Syst. 1–16 (2021)
24. Akram, M., Shahzadi, G., Ahmadini, A.A.H.: Decision-making framework for an effective sanitizer to reduce COVID-19 under fermatean fuzzy environment. J. Math. (2020)
25. Aydemir, S.B., Yilmaz Gunduz, S.: Fermatean fuzzy TOPSIS method with dombi aggregation operators and its application in multi-criteria decision making. J. Intell. Fuzzy Syst. **39**, 851–869 (2020)
26. Garg, H., Shahzadi, G., Akram, M.: Decision-making analysis based on Fermatean fuzzy Yager aggregation operators with application in COVID-19 testing facility. Mathe. Prob. Eng. (2020)
27. Liu, D., Liu, Y., Wang, L.: Distance measure for Fermatean fuzzy linguistic term sets based on linguistic scale function: an illustration of the TODIM and TOPSIS methods. Int. J. Intell. Syst. **34**, 2807–2834 (2019)
28. Senapati, T., Yager, R.R.: Some new operations over Fermatean fuzzy numbers and application of Fermatean fuzzy WPM in multiple criteria decision making. Informatica **30**, 391–412 (2019)
29. Senapati, T., Yager, R.R.: Fermatean fuzzy weighted averaging/geometric operators and its application in multi-criteria decision-making methods. Eng. Appl. Artif. Intell. **85**, 112–121 (2019)
30. Taamallah, A., Khemaja, M., Faiz, S.: Towards a framework for participatory strategy design in smart cities. In: Proceedings of the the Proceedings of the Third International Conference on Smart City Applications, pp. 179–192
31. Ahvenniemi, H., Huovila, A., Pinto-Seppä, I., Airaksinen, M.: What are the differences between sustainable and smart cities? Cities **60**, 234–245 (2017)
32. Wu, Y., Zhang, W., Shen, J., Mo, Z., Peng, Y.: Smart city with Chinese characteristics against the background of big data: idea, action and risk. J. Clean. Prod. **173**, 60–66 (2018)
33. Vu, K., Hartley, K.: Promoting smart cities in developing countries: policy insights from Vietnam. Telecommun. Pol. **42**, 845–859 (2018)
34. Chen, Y., Ardila-Gomez, A., Frame, G.: Achieving energy savings by intelligent transportation systems investments in the context of smart cities. Transp. Res. Part D: Transp. Environ. **54**, 381–396 (2017)
35. Hoelscher, K.: The evolution of the smart cities agenda in India. Int. Area Stud. Rev. **19**, 28–44 (2016)
36. Tan, S.Y., Taeihagh, A.: Smart city governance in developing countries: a systematic literature review. Sustainability **12**, 899 (2020)
37. Chang, I.-C.C.: Actually existing sustainabilities. In: Urban China: national initiatives and local contestations. Sustainab. J. Rec. **11**:216–228 (2018)

38. Mboup, G.; Oyelaran-Oyeyinka, B.: Relevance of smart economy in smart cities in Africa. In: Smart Economy in Smart African Cities; Springer, pp. 1–49 (2019)
39. Wenge, R., Zhang, X., Dave, C., Chao, L., Hao, S.: Smart city architecture: a technology guide for implementation and design challenges. China Commun. **11**, 56–69 (2014)
40. Warwick, K.: Beyond industrial policy: emerging issues and new trends (2013)
41. Kumar, H., Singh, M.K., Gupta, M., Madaan, J.: Moving towards smart cities: solutions that lead to the smart city transformation framework. Technol. Forecast. Soc. Chang. **153**, 119281 (2020)
42. Peng, G.C.A., Nunes, M.B., Zheng, L.: Impacts of low citizen awareness and usage in smart city services: the case of London's smart parking system. IseB **15**, 845–876 (2017)
43. Praharaj, S., Han, J.H., Hawken, S.: Innovative civic engagement and digital urban infrastructure: lessons from 100 smart cities mission in India. Proc. Eng. **180**, 1423–1432 (2017)
44. Kumar, A.: Can the smart city allure meet the challenges of Indian urbanization? In: Sustainable Smart Cities in India. Springer, pp. 17–39 (2017)
45. Mohanty, S.P., Choppali, U., Kougianos, E.: Everything you wanted to know about smart cities: the internet of things is the backbone. IEEE Consumer Electron. Magaz. **5**, 60–70 (2016)
46. Nam, T., Pardo, T.A.: Conceptualizing smart city with dimensions of technology, people, and institutions. In: Proceedings of the Proceedings of the 12th Annual International Digital Government Research Conference: Digital Government Innovation in Challenging Times, pp. 282–291 (2011)
47. Ismagilova, E., Hughes, L., Rana, N.P., Dwivedi, Y.K.: Security, privacy and risks within smart cities: Literature review and development of a smart city interaction framework. Inf. Syst. Front. 1–22 (2020)
48. Habibzadeh, H., Soyata, T., Kantarci, B., Boukerche, A., Kaptan, C.: Sensing, communication and security planes: a new challenge for a smart city system design. Comput. Netw. **144**, 163–200 (2018)
49. Peprah, C., Amponsah, O., Oduro, C.: A system view of smart mobility and its implications for Ghanaian cities. Sustain. Cities Soc. **44**, 739–747 (2019)
50. Mishra, A.K.: Henry George and Mohring-Harwitz theorems: lessons for financing smart cities in developing countries. Environ. Urban. ASIA **10**, 13–30 (2019)
51. Vadgama, C.V., Khutwad, A., Damle, M., Patil, S.: Smart funding options for developing smart cities: a proposal for India. Indian J. Sci. Technol. **8**, 1–12 (2015)
52. Lam, P.T., Yang, W.: Factors influencing the consideration of public-private partnerships (PPP) for smart city projects: evidence from Hong Kong. Cities **99**, 102606 (2020)
53. Alim, S., Polak, J.: Public–private partnerships for future urban infrastructure. In: Proceedings of the Institution of Civil Engineers-management, Procurement and Law, vol. 169, pp. 150–158 (2016)
54. Chen, H., Zhao, C., Shen, Z.: Analysis of the problems of current smart city and countermeasures in China. In: Proceedings of the 2018 3rd Technology Innovation Management and Engineering Science International Conference (TIMES-iCON), pp. 1–4 (2018)
55. Keshavarz-Ghorabaee, M., Amiri, M., Hashemi-Tabatabaei, M., Zavadskas, E.K., Kaklauskas, A.: A new decision-making approach based on fermatean fuzzy sets and WASPAS for green construction supplier evaluation. Mathematics **8**, 2202 (2020)
56. Keshavarz-Ghorabaee, M., Amiri, M., Zavadskas, E.K., Turskis, Z., Antucheviciene, J.: Determination of objective weights using a new method based on the removal effects of criteria (MEREC). Symmetry **13**, 525 (2021)

Nonlinear Scaled Preferences in Linguistic Multi-criteria Group Decision Making

Diego García-Zamora, Álvaro Labella, Rosa M. Rodríguez, and Luis Martínez

Abstract Even though the use of data-driven decision models has increased its popularity during recent years, the resolution of some decision making problems still relies on the use of the expertise of specialists in the corresponding area, leading to decision situations characterized by the uncertainty and vagueness of the available information which may also require to model hesitancy between multiple choices. Thus, new approaches have been defined to model decision makers' indecision by means of complex linguistic expressions such as Extended Comparative Linguistic Expressions with Symbolic Translation (ELICIT), based on the 2-tuple linguistic model. Nevertheless, this approach, alike many other linguistic models, uses linear scales to model decision makers' preferences. Recent studies show that humans do not measure the distances between values at different levels of the linear scale in the same way and better decisions are obtained when nonlinear scales are considered. Therefore, this chapter introduces a multi-criteria group decision making model based on fuzzy TOPSIS dealing with ELICIT information which considers the nonlinear scales provided by the recently defined extreme values amplifications. It provides flexibility to express decision makers' preferences and guarantees more reliable results than those obtained with the classical linear scaled preferences.

Keywords Linguistic group decision making · ELICIT information · TOPSIS method · Extreme values amplifications

---

D. García-Zamora · Á. Labella · R. M. Rodríguez · L. Martínez (✉)
University of Jaén, Jaén, Spain
e-mail: martin@ujaen.es

D. García-Zamora
e-mail: dgzamora@ujaen.es

Á. Labella
e-mail: alabella@ujaen.es

R. M. Rodríguez
e-mail: rmrodrig@ujaen.es

L. Sahoo et al. (eds.), *Real Life Applications of Multiple Criteria Decision Making Techniques in Fuzzy Domain*, Studies in Fuzziness and Soft Computing 420,
https://doi.org/10.1007/978-981-19-4929-6_3

1 Introduction

Nowadays, the use of decision processes guided exclusively by data (data-driven) with quantitative modelling is widely extended, and the involvement of human experts, who usually manage qualitative information, has been either neglected or relegated to second place [5, 7]. However, by considering the commitment, cost and relevance of taking into account human stakeholders in economical, social or learning studies, the use of decision approaches guided by experts (expert-driven) is still essential [12, 16] in several fields which demand intelligent, efficient and effective decisions under uncertain conditions. Therefore, in multi-attribute decision making, it is key to consider multi-criteria group decision making (MCGDM) models which allow to make decisions supported by human knowledge.

To model the vagueness of expert knowledge, fuzzy logic and fuzzy linguistic approach were proposed to model the uncertainty by linguistic variables [18] close to the human beings' way of thinking. In this sense, decision makers (DMs) can give their preferences by linguistic terms such as "good", "secure" or "comfortable", arising the linguistic decision making (LDM) approach [9], which allows to accomplish linguistic operations. In the literature, there are different approaches in order to model linguistic information [15]. One of the most well-known linguistic approaches is the 2-tuple linguistic model [8] because it allows to make accurate computations with linguistic information by preserving the interpretability of the obtained results. However, this model allows DMs to elicit his/her knowledge with linguistic information that consists of single linguistic terms that in certain situations in which DMs doubt among several terms may be insufficient. To this regard, other approaches have defined complex linguistic expressions to model DMs' hesitancy, such as the comparative linguistic expressions (CLEs) [15], although the computational processes related to these expressions are limited from the precision and interpretation points of view. To get over these shortcomings, Labella et al. [6] introduced the Extended Comparative Linguistic Expressions with Symbolic Translation (ELICIT) based on the CLEs by incorporating some features of the 2-tuple linguistic model.

These linguistic approaches assume that the DMs who participate in a group decision making (GDM) problem give their preferences by following linear scales. However, recent proposals [3, 10] suggest that when people rate alternatives, we do not measure the distances between values at different levels of the linear scale in the same way. In order to provide more realistic preferences values, closer to the real opinion of the DMs, García-Zamora et al. [4] proposed extreme values amplifications (EVAs) which aim at remapping the original preferences into nonlinear scaled preferences which are more realistic from a psychological point of view.

Keeping in mind the previous issues, this chapter proposes a MCGDM model based on fuzzy TOPSIS [2] in which DMs provide their preferences by using ELICIT information which will be remapped according to a nonlinear scale in order to obtain more realistic results according to human way of thinking. The main features of this contribution are summarized below:

- A MCGDM model is proposed to guide decision situations which require the use of expert knowledge.
- The DMs' preferences will be represented by ELICIT information which allows to model uncertainty and hesitancy by using flexible linguistic expressions.
- The original DMs' preferences will be remapped by nonlinear scales to provide more realistic preferences from a psychological point of view.
- The fuzzy TOPSIS method is combined with nonlinear scaled ELICIT preferences to guarantee more reliable results than the obtained with the classical linear scaled preferences.
- The performance of the proposed method is shown in a real case study.

The remainder sections are summarized below. Section 2 revises some preliminary concepts to facilitate the understanding of the proposal. Section 3 introduces the main proposal which consists of a MCGDM model based on fuzzy TOPSIS with ELICIT information and nonlinear preferences. In Sect. 4, a real-world MCGDM problem is solved to illustrate the working of the model. Section 5 will point out some conclusions.

2 Preliminaries

This section is devoted to revise several basics regarding MCGDM with linguistic information. First, some notation about MCGDM is clarified. Then, 2-tuple linguistic model and ELICIT information, which are key to model complex linguistic preferences, are reviewed. Afterwards, it is analysed the use of nonlinear scales to remap the original preferences obtained from the experts into more realistic values. Finally, the fuzzy TOPSIS method is described.

2.1 Linguistic Decision Making

Human beings face decision situations that consist of evaluating different choices and deciding which is the best one. In GDM, several DMs with different perspectives try to achieve an agreed solution for a decision problem. The alternatives may be evaluated on several criteria, for instance, there are several criteria to consider when we buy a car, speed, security or price. Under these conditions, we talk about MCGDM problems.

Most of MCGDM problems are defined under uncertain environments whose main features are both the lack of information and precision. Modelling such a uncertainty by means of precise numeric assessments may be a really hard task for the DMs, thus they need to use an expression domain able to express uncertainty. Fuzzy modelling and the use of linguistic information have provided successful results arising the LDM. Particularly, the fuzzy linguistic approach has been one of

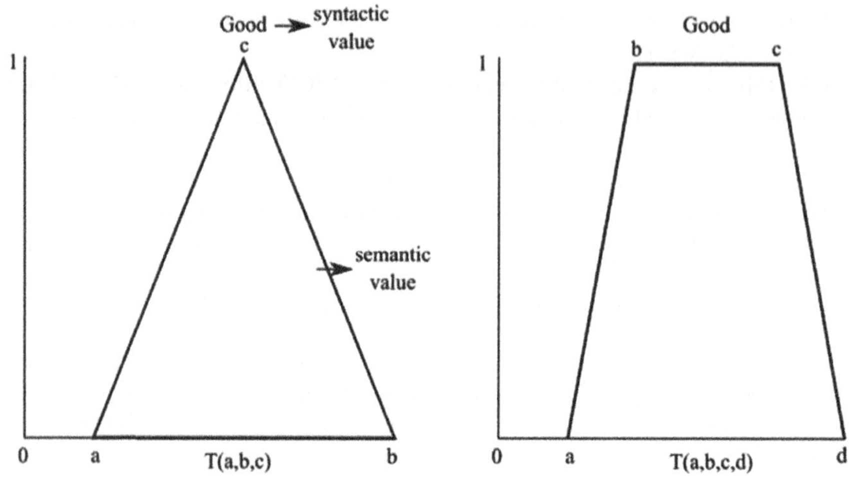

Fig. 1 Linguistic label

the most used methodologies to model uncertainty using means of *linguistic variables* [18] which are characterized by a syntactic value, represented by a common word in natural language, and a semantic value represented by a fuzzy set [19] defined by a membership function with different graphic representations. Triangular and trapezoidal representations are the most common representations (see Fig. 1).

2.2 2-Tuple Linguistic Model

The 2-tuple linguistic model [8] was proposed to improve the performance of classical linguistic computational models. Whereas the classical approaches presented either a lack of accuracy or understanding in their results, the 2-tuple linguistic approach models the linguistic values through a continuous fuzzy representation which allows to overcome such limitations.

Formally, the 2-tuple linguistic values are modelled by a tuple $(s_i, \alpha) \in \overline{S} := S \times [-0.5, 0.5[$ in which s_i represents a single linguistic term that belongs to a predefined linguistic term set $S = \{s_0, s_1, \ldots, s_g\}$ (for a certain even number $g \in \mathbb{N}$) and α refers to the concept symbolic translation, that is, a numerical value that represents the shifting of s_i fuzzy membership function. The symbolic translation concept changed the classical view of a linguistic term set understood as a set of discrete elements defined in a continuous domain which implied approximation processes to derive the results (see Fig. 2). Given a 2-tuple linguistic value $(s_i, \alpha) \in \overline{S}$, the possible values for the symbolic translation α are as follows:

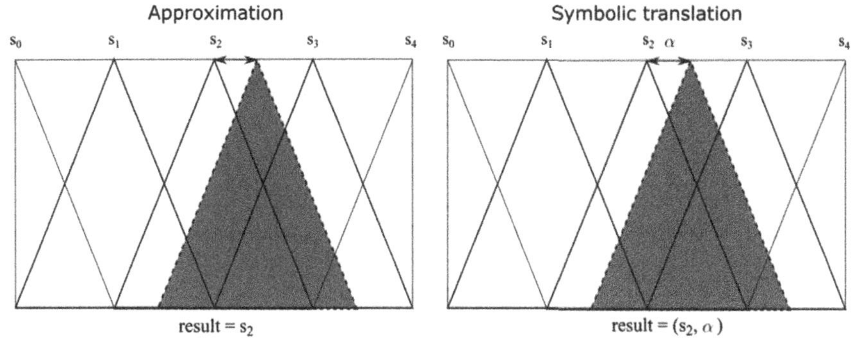

Fig. 2 Symbolic translation

$$\alpha \in \begin{cases} [-0.5, 0.5) & \text{if } s_i \in \{s_1, s_2, \ldots, s_{g-1}\} \\ [0, 0.5) & \text{if } s_i = s_0 \\ [-0.5, 0] & \text{if } s_i = s_g \end{cases}.$$

In addition, a 2-tuple linguistic value can be transformed into an equivalent numerical value $x \in [0, g]$, which makes the computations with 2-tuple linguistic values simple:

Proposition 1 *[8] Let $S = \{s_0, \ldots, s_g\}$ be a linguistic term set. Then, the function $\Delta^{-1} : \overline{S} \rightarrow [0, g]$ defined by*

$$\Delta_S^{-1}(s_i, \alpha) = i + \alpha,\ \forall\, (s_i, \alpha) \in \overline{S}$$

is a bijection whose inverse $\Delta_S : [0, g] \rightarrow \overline{S}$ is given by

$$\Delta_S(x) = (s_{round(x)}, x - round(x))\ \forall\, x \in [0, g],$$

where $round(\cdot)$ is the function that assigns the closest integer number $i \in \{0, \ldots, g\}$.

2.3 Extended Comparative Linguistic Expressions with Symbolic Translation

Labella et al. [6] introduced the ELICIT information to overcome existing limitations related to linguistic preferences' modelling in GDM. These drawbacks concern two main aspects:

- Precision: accuracy in computations is key to obtain reliable results for GDM problems. Some proposals are limited by a discrete interpretation of the linguistic terms set or carry out transformation processes for the input information which lead to loss of information and precision in computations [14, 20].

- Understanding: results should be represented by a format which DMs are able to understand. Some proposals transform the initial linguistic DMs' preferences into other kinds of representations that are more complex from the point of view of comprehension [13, 17].

ELICIT information is based on the CLEs introduced by Rodríguez et al. [15]. These expressions are generated by a context-free grammar which models comparative linguistic structures closer to language human beings such as, *at least bad*, *at most fast* or *between expensive and rather expensive*. However, the CLEs computational processes introduced in the literature are limited by a discrete representation of the expression domain, which inevitably implies approximations to the real results, and consequently, loss of information. To overcome this limitation regarding precision, ELICIT information uses the symbolic translation concept of the 2-tuple linguistic model (see Sect. 2.2). In this way, the context-free grammar used to model CLEs is modified to replace the single linguistic terms that compose these expressions by 2-tuple linguistic values defined in a continuous domain:

Definition 1 [6] Let G_H be a context-free grammar and $S = \{s_0, \dots, s_g\}$ a linguistic term set. The elements of $G_H = (V_N, V_T, I, P)$ are defined as follows:

$$
\begin{aligned}
&V_N = \{\langle\text{continuous primary term}\rangle, \langle\text{composite term}\rangle,\\
&\langle\text{unary relation}\rangle, \langle\text{binary relation}\rangle, \langle\text{conjunction}\rangle\}\\
&V_T = \{\text{at least, at most, between, and}, (s_0, \alpha)^{\gamma},\\
&(s_1, \alpha)^{\gamma}, \dots, (s_g, \alpha)^{\gamma}\}\\
&I \in V_N\\
&P = \{I ::= \langle\text{continuous primary term}\rangle|\\
&\langle\text{composite term}\rangle\\
&\langle\text{composite term}\rangle ::= \langle\text{unary relation}\rangle\\
&\langle\text{continuous primary term}\rangle|\\
&\langle\text{binary relation}\rangle\langle\text{continuous primary term}\rangle\\
&\langle\text{conjunction}\rangle\langle\text{continuous primary term}\rangle\\
&\langle\text{continuous primary term}\rangle ::= (s_0, \alpha)^{\gamma}|\\
&(s_1, \alpha)^{\gamma}|\dots|(s_g, \alpha)^{\gamma}\\
&\langle\text{unary relation}\rangle ::= \text{at least}|\text{at most}\\
&\langle\text{binary relation}\rangle ::= \text{between}\\
&\langle\text{conjunction}\rangle ::= \text{and}\}.
\end{aligned}
$$

Thus, this context-free grammar and the linguistic term set $S = \{$very cheap, cheap, indifferent, expensive, very expensive$\}$ can model linguistic expressions such as, at least $(\text{cheap}, 0.2)^{0.1}$, at most $(\text{expensive}, -0.1)^{0.12}$ or between $(\text{expensive}, 0)^{-0.11}$ and $(\text{very expensive}, 0.32)^{0}$. Notice the γ parameter is used to perform fuzzy computations with ELICIT information. To do so, the ELICIT expressions are transformed into trapezoidal fuzzy numbers by means the fuzzy envelope computation [6]:

Definition 2 [6] Let $H_S = \{\overline{s}_1, \overline{s}_2, \ldots, \overline{s}_m\} \subset \overline{S}$ be a set of ordered 2-tuple linguistic values. The fuzzy envelope of H_S, $env(H_S)$, is defined as the trapezoidal fuzzy number $T(a, b, c, d)$ where:

$$a = \frac{\Delta_S^{-1}(\overline{s}_1) - \frac{1}{g}}{g}, \quad b = \frac{\Delta_S^{-1}(\overline{s}_1)}{g}, \\ c = \frac{\Delta_S^{-1}(\overline{s}_m)}{g}, \quad d = \frac{\Delta_S^{-1}(\overline{s}_m) + \frac{1}{g}}{g}. \tag{1}$$

2.4 Nonlinear Preferences in GDM: Extreme Values Amplifications

Even though classical GDM assumes that DMs give their preferences by using linear scales, several studies [3, 10] reveal that when nonlinear scales are used to remap their original preferences, better decisions are obtained. The reason behind this is a psychological fact: when people make judgements by rating alternatives, we are not completely precise expressing distances between values. For instance, when marking an exam, an A+ test will be required much more quality for getting an S than a C+ test for getting a B–.

In order to be able to model this human beings' behavior in consensus models for GDM, García-Zamora et al. [4] defined extreme values amplifications as those automorphisms on the unit interval which increase or decrease the distances between the most extreme values by preserving the symmetry of the remapped values respect 0.5. Formally:

Definition 3 (*Extreme Values Amplification*) [4] Let $D : [0, 1] \to [0, 1]$ be a function satisfying:

1. D is an automorphism on the interval $[0, 1]$,
2. D is a C^1 function,
3. D satisfies $D(x) = 1 - D(1 - x)\ \forall\, x \in [0, 1]$,
4. $D'(0) > 1$ and $D'(1) > 1$,
5. D is concave in a neighbourhood of 0 and convex in a neighbourhood of 1.

D is called an EVA on the interval $[0, 1]$.

In addition, it was proved that a function satisfying the previous definition also verifies the following properties:

Theorem 1 *[4] Let $D : [0, 1] \to [0, 1]$ be an EVA on $[0, 1]$. Then:*

1. *The function $d_D : [0, 1] \times [0, 1] \to [0, 1]$ given by*

$$d_D(x, y) = |D(x) - D(y)| \; \forall \, x, y \in [0, 1],$$

is a restricted dissimilarity [1] and the function $S_D : [0, 1] \times [0, 1] \to [0, 1]$ defined by

$$S_D(x, y) = 1 - |D(x) - D(y)| \; \forall \, x, y \in [0, 1],$$

is a restricted equivalence function [1].

2. *Three intervals $I_1, I_2, I_3 \subset [0, 1]$ such that $0 \in I_1$, $1 \in I_3$, and $I_1 < I_2 < I_3$ satisfying*

$$\begin{aligned} |D(y) - D(x)| &> |y - x| \; \forall \, x, y \in I_1 \; : \; x \neq y, \\ |D(y) - D(x)| &< |y - x| \; \forall \, x, y \in I_2 \; : \; x \neq y, \\ |D(y) - D(x)| &> |y - x| \; \forall \, x, y \in I_3 \; : \; x \neq y. \end{aligned}$$

can be found.

3. *The graph of D is over the diagonal of the square $[0, 1] \times [0, 1]$ for values close enough to 0 and it is under the same diagonal for those values close enough to 1,*
4. *There are a neighbourhood U_0 containing 0 and a neighbourhood U_1 containing 1 such that for every $x, y \in U_0^\circ$, $x < y$, there exists $h_0 > 0$ such that the inequality $|D(x) - D(x - t)| \geq |D(y) - D(y - t)|$ holds for any $t \in [0, h_0]$ and for every $x, y \in U_1^\circ$, $x < y$, there exists $h_1 > 0$ such that the inequality $|D(x - t) - D(x)| \leq |D(y - t) - D(y)|$ holds for any $t \in [0, h_1]$.*

In other words:

1. EVAs remap the original linear scaled preferences elicited from DMs into non-linear scaled preferences,
2. EVAs amplify the distance between the extreme values, and reduce the distance between the intermediate ones,
3. EVAs have a concrete shape (see Fig. 3).

In this chapter, we will consider the EVA $m : [0, 1] \to [0, 1]$ defined by

$$m(x) = \begin{cases} \frac{1}{2} - \frac{1}{2}(1 - 2x)^2 & 0 \leq x < \frac{1}{2} \\ \frac{1}{2} + \frac{1}{2}(2x - 1)^2 & \frac{1}{2} \leq x \leq 1 \end{cases}, \tag{2}$$

due to its good performance in [4].

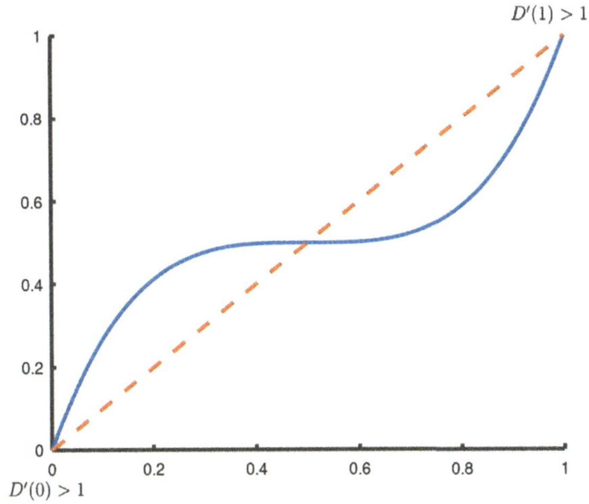

Fig. 3 Sketch of an EVA

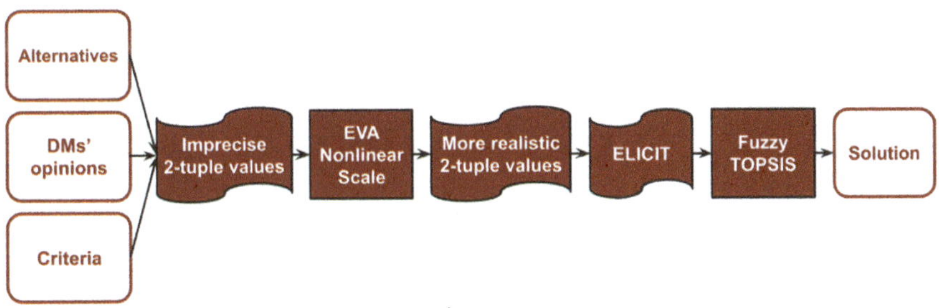

Fig. 4 Scheme of the proposed model

3 Linguistic Nonlinear Scales in Fuzzy TOPSIS

Here, nonlinear scales provided by EVAs are adapted for linguistic preferences. First, the notion of EVA is extended to 2-tuple linguistic values, and subsequently, to ELICIT information. Afterwards, this nonlinear scales are included in fuzzy TOPSIS method to solve a MCGDM problem in order to provide more realistic results from a psychological point of view. Fig. 4 shows the proposal's resolution scheme.

3.1 Nonlinear Preferences for Linguistic Information: 2-Tuple EVAs

This subsection is devoted to adapt the nonlinear scales provided by EVAs, formerly defined for numeric preferences, to linguistic environments (2-tuple and ELICIT information) in order to remap the original values of the DMs' preferences into new values which are more realistic from a psychological point of view [3, 10].

Let us consider a linguistic term set $S = \{s_0, s_1, \ldots, s_g\}$ and the associated set of 2-tuple linguistic values $\overline{S} = S \times [-0.5, 0.5]$. Formally, we aim at defining a nonlinear deformation $D_S : \overline{S} \to \overline{S}$ such that the distances between the extreme 2-tuple values are increased and the distances between intermediate 2-tuple values are reduced.

For the sake of clarity, given two nonempty compact intervals $[a, b], [\alpha, \beta] \subset \mathbb{R}$ we consider the standard affine transformation $h_{a,b}^{\alpha,\beta} : [a, b] \to [\alpha, \beta]$ defined by

$$h_{a,b}^{\alpha,\beta}(x) = \frac{x-a}{b-a}(\beta - \alpha) + \alpha \;\; \forall\, x \in [a, b].$$

In particular, we are interested in the one which transforms the interval [0, 1] into $[0, g]$, $h = h_{0,1}^{0,g} : [0, 1] \to [0, g]$ and in its inverse $h^{-1} = h_{0,g}^{0,1} : [0, g] \to [0, 1]$.

This nomenclature allows to easily define EVAs for 2-tuples values by using the original definition of EVA on [0, 1] [4]:

Definition 4 (*2-Tuple Extreme Value Amplification*) Let $D_S : \overline{S} \to \overline{S}$ be a mapping such that the function $D : [0, 1] \to [0, 1]$ defined by

$$D := h^{-1} \circ \Delta_S^{-1} \circ D_S \circ \Delta_S \circ h, \tag{3}$$

is an EVA on [0, 1]. Then, D_S will be called an EVA on the 2-tuple set $\overline{S}$.

Since both $h : [0, 1] \to [0, g]$ and $\Delta_S : [0, g] \to \overline{S}$ are bijections, the following result is straight-forward:

Proposition 2 *Let* $D : [0, 1] \to [0, 1]$ *be an EVA on* [0, 1]. *Then,* $D_S : \overline{S} \to \overline{S}$ *defined by*

$$\begin{aligned} D_S(s_i, \alpha) &= \Delta_S \circ h \circ D \circ h^{-1} \circ \Delta_S^{-1}(s_i, \alpha) \\ &= \left(s_{round(gD(\frac{i+\alpha}{g}))}, gD\left(\frac{i+\alpha}{g}\right) - round\left(gD\left(\frac{i+\alpha}{g}\right)\right)\right) \;\; \forall\, (s_i, \alpha) \in \overline{S}, \end{aligned}$$

is an EVA on $\overline{S}$.

This extension of the notion of EVA to the 2-tuple environment allows to preserve the properties summarised in Theorem 1. In general terms, the performance of 2-tuple EVAs can be summarised as follows:

- they remap 2-tuple values into 2-tuple values,
- the original preferences are deformed according to a nonlinear scale,
- the distances between 2-tuple values close to the extremes of the linguistic term set, namely s_0 and s_g, are amplified, whereas the distances between 2-tuple values close to the intermediate linguistic term, $s_{\frac{g}{2}}$, are reduced,

- the closer the distance between a point and the extreme values of the linguistic term set, the greater the amplification of distances is.

In addition, it should be highlighted that when using a 2-tuple EVA $D_S : \overline{S} \rightarrow \overline{S}$ there are just three fixed points, i.e. three 2-tuple values which are remapped on themselves, namely $(s_0, 0)$, $(s_{\frac{g}{2}}, 0)$ and $(s_g, 0)$. Furthermore, the deformation provided by 2-tuple EVAs maintains the symmetry towards the median value $(s_{\frac{g}{2}}, 0)$:

$$\Delta^{-1}(D_S(s_i, \alpha)) + \Delta^{-1}(D_S(s_{g-i}, -\alpha)) = g \ \forall\, i = 1, 2, \ldots, g, \ \forall\, \alpha \in [-0.5, 0.5[.$$

3.2 Nonlinear Preferences in ELICIT Information

Even though the 2-tuple linguistic model allows to represent linguistic information and make precise computations with it, this preference structure is not able to model DMs' opinions with some hesitation such as "between good and very good". Therefore, in order to consider richer linguistic expressions, the nonlinear scales provided by EVAs are adapted to ELICIT representation model, based on the 2-tuple linguistic model, which is able to model such hesitation.

In order to consider nonlinear scales in ELICIT information, it suffices to apply 2-tuple EVAs to the 2-tuple linguistic terms that define the ELICIT expression. Formally, given a linguistic term set $S = \{s_0, s_1, \ldots, s_g\}$ any input ELICIT information can be described as the fuzzy envelope of a set of certain 2-tuple linguistic ordered values $\overline{s}_1, \overline{s}_2, \ldots, \overline{s}_m \in \overline{S}$, $m \in \mathbb{N}$. Therefore, to adapt a 2-tuple EVA $D : \overline{S} \rightarrow \overline{S}$ to ELICIT information, we have just to consider the fuzzy envelope associated to $D(\overline{s}_1), D(\overline{s}_2), \ldots, D(\overline{s}_m)$.

3.3 Fuzzy TOPSIS with Linguistic Nonlinear Preferences

This section proposes a MCGDM model based on fuzzy TOPSIS [2] in which DMs' opinions are modelled through the linguistic nonlinear scales given by 2-tuple EVAs and the ELICIT linguistic approach. First, the DMs $E = \{dm_1, dm_2, \ldots, dm_m\}$ $m \in \mathbb{N}$ are asked to give their opinions about the alternatives $X = \{x_1, x_2, \ldots, x_n\}$ and criteria $C = \{c_1, c_2, \ldots, c_r\}$ by using linguistic information. To do so, a linguistic term set $S = \{s_0, s_1, \ldots, s_g\}$ is provided to the DMs whose opinions are elicited by using ELICIT expressions such as "s_1", "at most s_{g-1}" or "at least s_2". Afterwards, the fuzzy envelope for these linguistic expressions will be expressed in terms of linear scaled 2-tuple values, which will be remapped into nonlinear scaled 2-tuple values by using a 2-tuple EVA. Then, the nonlinear scaled preferences will be used to construct the trapezoidal fuzzy numbers corresponding to nonlinear scaled ELICIT information. Finally, the fuzzy TOPSIS method will be applied to obtain the solution for the MCGDM problem [11].

The complete proposal is developed in detail below:

1. Gathering preferences: the DMs $dm_1, dm_2, \ldots, dm_m$ give their linguistic ratings about the criteria importance $c_1, c_2, \ldots, c_r$ and about the alternatives $x_1, x_2, \ldots, x_n$ according to the given criteria by using CLEs.
2. Obtaining ELICIT information: the obtained CLEs are transformed into ELICIT information by assigning $\alpha = \gamma = 0$ for each linguistic term that composes the expression.
3. Remapping using nonlinear scales: the 2-tuple linguistic values associated to the ELICIT information obtained in the previous step are remapped by using an EVA (see Eq. 3). The resulting obtained values define a new nonlinear scaled ELICIT value.
4. Computing fuzzy envelopes: the ELICIT information obtained in the previous step is transformed into trapezoidal fuzzy numbers by using the fuzzy envelope (see Eq. 1). For each alternative x_i and each criteria c_j, the obtained trapezoidal fuzzy numbers are denoted by $\tilde{x}_{ij}^k = (a_{ij}^k, b_{ij}^k, c_{ij}^k, d_{ij}^k)$, and the opinions about the importance for each criteria c_j are denoted by the trapezoidal fuzzy numbers $\tilde{w}_j^k = (w_{j1}^k, w_{j2}^k, w_{j3}^k, w_{j4}^k)$.
5. Aggregating information: the obtained fuzzy numbers are aggregated into a collective opinion.

   - Collective fuzzy decision matrix $\tilde{x}_{ij} = (a_{ij}, b_{ij}, c_{ij}, d_{ij})$

$$\begin{aligned} a_{ij} &= \min_{k=1,2,\ldots,m} \{a_{ij}^k\}, \quad b_{ij} = \frac{1}{m}\sum_{k=1}^{m} b_{ij}^k, \\ c_{ij} &= \frac{1}{m}\sum_{k=1}^{m} c_{ij}^k, \quad d_{ij} = \max_{k=1,2,\ldots,m} \{d_{ij}^k\}. \end{aligned} \tag{4}$$

   - Collective fuzzy weights $\tilde{w}_j = (w_{j1}, w_{j2}, w_{j3}, w_{j4})$.

$$\begin{aligned} w_{j1} &= \min_{k=1,2,\ldots,m} \{w_{j1}^k\}, \quad w_{j2} = \frac{1}{m}\sum_{k=1}^{m} w_{j2}^k, \\ w_{j3} &= \frac{1}{m}\sum_{k=1}^{m} w_{j3}^k, \quad w_{j4} = \max_{k=1,2,\ldots,m} \{w_{j4}^k\}. \end{aligned} \tag{5}$$

6. Computing the normalized fuzzy decision matrix: a normalized fuzzy decision matrix $\tilde{R} = [\tilde{r}_{ij}]$ is built from the aggregated fuzzy information.

$$\begin{aligned} \tilde{r}_{ij} &= \left(\frac{a_{ij}}{d_j^+}, \frac{b_{ij}}{d_j^+}, \frac{c_{ij}}{d_j^+}, \frac{d_{ij}}{d_j^+}\right), \ d_j^+ = \max_i \{d_{ij}\} \text{ (benefit criteria)}, \\ \tilde{r}_{ij} &= \left(\frac{a_j^-}{d_{ij}}, \frac{a_j^-}{c_{ij}}, \frac{a_j^-}{b_{ij}}, \frac{a_j^-}{a_{ij}}\right), \ a_j^- = \min_i \{a_{ij}\} \text{ (cost criteria)}. \end{aligned} \tag{6}$$

7. Computing the weighted normalized decision matrix: the normalized decision ratings are multiplied by the aggregated criteria fuzzy weights.

$$\tilde{T} = [\tilde{t}_{ij}],\ \tilde{t}_{ij} = \tilde{r}_{ij} \times \tilde{w}_j. \tag{7}$$

Note that the multiplication between two trapezoidal fuzzy numbers is defined as $\tilde{A} \times \tilde{B} = (a_1 a_2, b_1 b_2, c_1 c_2, d_1 d_2)$

8. Deriving the fuzzy ideal solution and the fuzzy anti-ideal solution: the ideal (A^+) and anti-ideal (A^-) solutions are computed as follows:

$$A^+ = (\tilde{t}_1^+, \tilde{t}_2^+, \ldots, \tilde{t}_r^+),\ \tilde{t}_j^+ = \max_i \{\tilde{t}_{ij4}\}, \tag{8}$$

$$A^- = (\tilde{t}_1^-, \tilde{t}_2^-, \ldots, \tilde{t}_r^-),\ \tilde{t}_j^- = \min_i \{\tilde{t}_{ij1}\}. \tag{9}$$

9. Deriving distance from ideal and anti-ideal solution: for each alternative x_i the distance from the fuzzy ideal solution (δ_i^+) and the fuzzy anti-ideal solution (δ_i^-) is computed as follows:

$$\begin{aligned} \delta_i^+ &= \sum_{j=1}^{r} d(\tilde{t}_{ij}, \tilde{t}_j^+), \\ \delta_i^- &= \sum_{j=1}^{r} d(\tilde{t}_{ij}, \tilde{t}_j^-). \end{aligned} \tag{10}$$

where $d(\cdot)$ is a distance function between two trapezoidal fuzzy numbers defined as $d(\tilde{A}, \tilde{B}) = \sqrt{\frac{1}{4}[(a_1 - a_2)^2 + (b_1 - b_2)^2 + (c_1 - c_2)^2 + (d_1 - d_2)^2]}$.

10. Computing the closeness coefficient: for each alternative x_i, the closeness coefficient CC_i is derived as follows:

$$\mathrm{CC}_i = \frac{\delta_i^-}{\delta_i^- + \delta_i^+}. \tag{11}$$

11. Ranking the alternatives: the best alternative is the one with the highest CC_i.

The proposal is summarized in the following algorithm:

Algorithm 1: ELICIT fuzzy TOPSIS with nonlinear preferences.

Input : DMs ELICIT preferences
1 Obtain the 2-tuple representation for the ELICIT information;
2 Apply a 2-tuple EVA to the values obtained in the previous step;
3 Use the nonlinear scaled 2-tuple values to reconstruct ELICIT trapezoidal fuzzy numbers;
4 Apply fuzzy TOPSIS to the nonlinear scaled ELICIT information;
Output: Solution to the MCGDM problem.

Table 1 Initial linguistic DMs' preferences for criteria weights (bt stands for between)

	c_1	c_2	c_3	c_4
dm_1	F	F	F	I
dm_2	I	bt LI and F	VI	bt F and I
dm_3	I	I	VI	F
dm_4	VI	I	I	F

4 Case Study

This section is devoted to show the performance of the proposed model in a real-world problem.

Nowadays, the entertaining paradigm is changing for users. New streaming platforms offer lots of advantages before traditional cable TV services such as the possibility to decide what to watch at any time, month-to-month payments, non-permanency subscriptions or add-free contents. For these reasons, more and more people decide to choose one of these streaming platforms instead of traditional TV.

In this context, it is usual for families to face the harder part of this transition process: choosing the best streaming platform among the endless options. Initially, this could be a hard task for novice users of these kinds of services, since there are multiple factors to take into account in order to choose the most suitable alternative such as the cost, exclusive movies and series, variety of contents and frequency of new releases.

Therefore, here it is considered a GDM problem in which a family consisting of 4 members wants to subscribe to an online streaming platform according to the aforementioned criteria.

The $n = 4$ alternatives are x_1 = Netflix, x_2 = Amazon Prime Video, x_3 = HBO and x_4 = Disney+ and they have been rated according to $r = 4$ criteria, namely c_1 = Cost, c_2 = Exclusive movies and series, c_3 = Variety of contents and c_4 = Frequency of new releases. The linguistic expression domain for the rating of the criteria is

$$S_1 = \{\text{Not important (NI), A little important (LI),} \\ \text{Fair (F), Important (I), Very important (VI)}\}$$

and the linguistic terms for rating the alternatives according to the criteria are

$$S_2 = \{\text{Very bad (VB), Bad (B), Fair (F), Good (G), Very good (VG)}\}.$$

The initial DMs' preferences regarding the criteria importance and the rating of the alternatives are given, respectively, in Tables 1 and 2.

Table 2 Initial linguistic DMs' preferences for rating alternatives (bt stands for between)

		c_1	c_2	c_3	c_4
dm_1	x_1	G	G	F	VG
	x_2	VG	G	F	VG
	x_3	F	G	bt F and G	VG
	x_4	F	G	F	VG
dm_2	x_1	G	F	bt F and G	F
	x_2	VG	bt B and F	F	bt B and F
	x_3	bt F and G	bt F and G	F	F
	x_4	bt F and G	At least F	bt F and G	G
dm_3	x_1	G	G	bt F and G	F
	x_2	VG	B	B	F
	x_3	F	F	G	G
	x_4	F	B	F	B
dm_4	x_1	F	G	F	F
	x_2	G	B	B	B
	x_3	VB	B	B	B
	x_4	F	VG	VG	G

Table 3 DMs' preferences for rating alternatives in ELICIT (bt stands for between)

		c_1	c_2	c_3	c_4
dm_1	x_1	$(G, 0^0)$	$(G, 0)^0$	$(F, 0)^0$	$(\text{VG}, 0)^0$
	x_2	$(\text{VG}, 0)^0$	$(G, 0)^0$	$(F, 0)^0$	$(\text{VG}, 0)^0$
	x_3	$(F, 0)^0$	$(G, 0)^0$	bt $(F, 0)^0$ and $(G, 0)^0$	$(\text{VG}, 0)^0$
	x_4	$(F, 0)^0$	$(G, 0)^0$	$(F, 0)^0$	$(\text{VG}, 0)^0$
dm_2	x_1	$(G, 0)^0$	$(F, 0)^0$	bt $(F, 0)^0$ and $(G, 0)^0$	$(F, 0)^0$
	x_2	$(\text{VG}, 0)^0$	bt $(B, 0)^0$ and $(F, 0)^0$	$(F, 0)^0$	bt $(B, 0)^0$ and $(F, 0)^0$
	x_3	bt $(F, 0)^0$ and $(G, 0)^0$	bt $(F, 0)^0$ and $(G, 0)^0$	$(F, 0)^0$	$(F, 0)^0$
	x_4	bt $(F, 0)^0$ and $(G, 0)^0$	At least $(F, 0)^0$	bt $(F, 0)^0$ and $(G, 0)^0$	$(G, 0)^0$
dm_3	x_1	$(G, 0)^0$	$(G, 0)^0$	bt $(F, 0)^0$ and $(G, 0)^0$	$(F, 0)^0$
	x_2	$(\text{VG}, 0)^0$	$(B, 0)^0$	$(B, 0)^0$	$(F, 0)^0$
	x_3	$(F, 0)^0$	$(F, 0)^0$	$(G, 0)^0$	$(G, 0)^0$
	x_4	$(F, 0)^0$	$(B, 0)^0$	$(F, 0)^0$	$(B, 0)^0$
dm_4	x_1	$(F, 0)^0$	$(G, 0)^0$	$(F, 0)^0$	$(F, 0)^0$
	x_2	$(G, 0)^0$	$(B, 0)^0$	$(B, 0)^0$	$(B, 0)^0$
	x_3	$(\text{VB}, 0)^0$	$(B, 0)^0$	$(B, 0)^0$	$(B, 0)^0$
	x_4	$(F, 0)^0$	$(\text{VG}, 0)^0$	$(\text{VG}, 0)^0$	$(G, 0)^0$

Table 4 DMs' preferences for criteria weights in ELICIT (bt stands for between)

	c_1	c_2	c_3	c_4
dm_1	$(F, 0)^0$	$(F, 0)^0$	$(F, 0)^0$	$(I, 0)^0$
dm_2	$(I, 0)^0$	bt $(\text{LI}, 0)^0$ and $(F, 0)^0$	$(VI, 0)^0$	bt $(F, 0)^0$ and $(I; 0)^0$
dm_3	$(I, 0)^0$	$(I, 0)^0$	$(VI, 0)^0$	$(F, 0)^0$
dm_4	$(VI, 0)^0$	$(I, 0)^0$	$(I, 0)^0$	$(F, 0)^0$

Table 5 Nonlinear DMs' preferences for rating alternatives (bt stands for between)

		c_1	c_2	c_3	c_4
dm_1	x_1	$(G, -0.5)^0$	$(G, -0.5)^0$	$(F, 0)^0$	$(\text{VG}, 0)^0$
	x_2	$(\text{VG}, 0)^0$	$(G, -0.5)^0$	$(F, 0)^0$	$(\text{VG}, 0)^0$
	x_3	$(F, 0)^0$	$(G, -0.5)^0$	bt $(F, 0)^0$ and $(G, -0.5)^0$	$(\text{VG}, 0)^0$
	x_4	$(F, 0)^0$	$(G, -0.5)^0$	$(F, 0)^0$	$(\text{VG}, 0)^0$
dm_2	x_1	$(G, -0.5)^0$	$(F, 0)^0$	bt $(F, 0)^0$ and $(G, -0.5)^0$	$(F, 0)^0$
	x_2	$(\text{VG}, 0)^0$	bt $(F, -0.5)^0$ and $(F, 0)^0$	$(F, 0)^0$	bt $(F, -0.5)^0$ and $(F, 0)^0$
	x_3	bt $(F, 0)^0$ and $(G, -0.5)^0$	bt $(F, 0)^0$ and $(G, -0.5)^0$	$(F, 0)^0$	$(F, 0)^0$
	x_4	bt $(F, 0)^0$ and $(G, -0.5)^0$	At least $(F, 0)^0$	bt $(F, 0)^0$ and $(G, -0.5)^0$	$(G, -0.5)^0$
dm_3	x_1	$(G, -0.5)^0$	$(G, -0.5)^0$	bt $(F, 0)^0$ and $(G, -0.5)^0$	$(F, 0)^0$
	x_2	$(\text{VG}, 0)^0$	$(F, -0.5)^0$	$(F, -0.5)^0$	$(F, 0)^0$
	x_3	$(F, 0)^0$	$(F, 0)^0$	$(G, -0.5)^0$	$(G, -0.5)^0$
	x_4	$(F, 0)^0$	$(F, -0.5)^0$	$(F, 0)^0$	$(F, -0.5)^0$
dm_4	x_1	$(F, 0)^0$	$(G, -0.5)^0$	$(F, 0)^0$	$(F, 0)^0$
	x_2	$(G, -0.5)^0$	$(F, -0.5)^0$	$(F, -0.5)^0$	$(F, -0.5)^0$
	x_3	$(\text{VB}, 0)^0$	$(F, -0.5)^0$	$(F, -0.5)^0$	$(F, -0.5)^0$
	x_4	$(F, 0)^0$	$(\text{VG}, 0)^0$	$(\text{VG}, 0)^0$	$(G, -0.5)^0$

Table 6 Nonlinear DMs' preferences for criteria weights (bt stands for between)

	c_1	c_2	c_3	c_4
dm_1	$(F, 0)^0$	$(F, 0)^0$	$(F, 0)^0$	$(I, -0.5)^0$
dm_2	$(I, -0.5)^0$	bt $(F, -0.5)^0$ and $(F, 0)^0$	$(\text{VI}, 0)^0$	bt $(F, 0)^0$ and $(I, -0.5)^0$
dm_3	$(I, -0.5)^0$	$(I, -0.5)^0$	$(\text{VI}, 0)^0$	$(F, 0)^0$
dm_4	$(\text{VI}, 0)^0$	$(I, -0.5)^0$	$(I, -0.5)^0$	$(F, 0)^0$

Firstly, the CLEs are converted into ELICIT expressions (see Tables 3 and 4).

The ELICIT information is still represented by linear preference scales. Then, the preferences are remapped using nonlinear scales by using an EVA. Tables 5 and 6 give the nonlinear preferences by using the EVA $m(x)$ introduced in Eq. 2.

Before starting the fuzzy TOPSIS to obtain the ranking of the alternatives, the nonlinear ELICIT preferences are transformed into fuzzy numbers by using the fuzzy envelope computation (see Eq. 1). The respective fuzzy envelopes are given in Tables 7 and 8.

From this step, the fuzzy TOPSIS starts. First, all the DMs' preferences are aggregated to obtain a collective fuzzy decision matrix and the collective fuzzy criteria weights computed by means of Eqs. 4 and 5, respectively (see Tables 9 and 10).

Afterwards, the fuzzy decision matrix is normalized by using Eq. 6 (see Table 11).

The fuzzy normalized decision matrix is multiplied by the collective fuzzy weights in order to obtain the weighted fuzzy decision matrix (see Eq. 7), which is represented in Table 12.

Therefore, the fuzzy ideal solution (see Eq. 8) is

$$A^{+} = (1, 0.875, 1, 0.875)$$

and the fuzzy anti-ideal solution (see Eq. 9) is

$$A^{-} = (0, 0.015625, 0.03125, 0.03125).$$

Now, the distances from the ideal and anti-ideal solution are computed by using Eq. 10 (see Table 13).

Finally, the closeness coefficient for each alternatives is computed by Eq. 11 (see Table 14).

According to the fuzzy TOPSIS, the best streaming platform is x_4 = Disney+.

Table 7 Fuzzy DMs' preferences

		c_1	c_2	c_3	c_4
dm_1	x_1	$T(0.375, 0.625, 0.625, 0.875)$	$T(0.375, 0.625, 0.625, 0.875)$	$T(0.25, 0.5, 0.5, 0.75)$	$T(0.75, 1, 1, 1)$
	x_2	$T(0.75, 1, 1, 1)$	$T(0.375, 0.625, 0.625, 0.875)$	$T(0.25, 0.5, 0.5, 0.75)$	$T(0.75, 1, 1, 1)$
	x_3	$T(0.25, 0.5, 0.5, 0.75)$	$T(0.375, 0.625, 0.625, 0.875)$	$T(0.25, 0.5, 0.625, 0.875)$	$T(0.75, 1, 1, 1)$
	x_4	$T(0.25, 0.5, 0.5, 0.75)$	$T(0.375, 0.625, 0.625, 0.875)$	$T(0.25, 0.5, 0.5, 0.75)$	$T(0.75, 1, 1, 1)$
dm_2	x_1	$T(0.375, 0.625, 0.625, 0.875)$	$T(0.25, 0.5, 0.5, 0.75)$	$T(0.25, 0.5, 0.625, 0.875)$	$T(0.25, 0.5, 0.5, 0.75)$
	x_2	$T(0.75, 1, 1, 1)$	$T(0.125, 0.375, 0.5, 0.75)$	$T(0.25, 0.5, 0.5, 0.75)$	$T(0.125, 0.375, 0.5, 0.75)$
	x_3	$T(0.25, 0.5, 0.625, 0.875)$	$T(0.25, 0.5, 0.625, 0.875)$	$T(0.25, 0.5, 0.5, 0.75)$	$T(0.25, 0.5, 0.5, 0.75)$
	x_4	$T(0.25, 0.5, 0.625, 0.875)$	$T(0.375, 0.625, 1, 1)$	$T(0.25, 0.5, 0.625, 0.875)$	$T(0.375, 0.625, 0.625, 0.875)$
dm_3	x_1	$T(0.375, 0.625, 0.625, 0.875)$	$T(0.375, 0.625, 0.625, 0.875)$	$T(0.25, 0.5, 0.625, 0.875)$	$T(0.25, 0.5, 0.5, 0.75)$
	x_2	$T(0.75, 1, 1, 1)$	$T(0.125, 0.375, 0.375, 0.625)$	$T(0.125, 0.375, 0.375, 0.625)$	$T(0.25, 0.5, 0.5, 0.75)$
	x_3	$T(0.25, 0.5, 0.5, 0.75)$	$T(0.25, 0.5, 0.5, 0.75)$	$T(0.375, 0.625, 0.625, 0.875)$	$T(0.375, 0.625, 0.625, 0.875)$
	x_4	$T(0.25, 0.5, 0.5, 0.75)$	$T(0.125, 0.375, 0.375, 0.625)$	$T(0.25, 0.5, 0.5, 0.75)$	$T(0.125, 0.375, 0.375, 0.625)$
dm_4	x_1	$T(0.25, 0.5, 0.5, 0.75)$	$T(0.375, 0.625, 0.625, 0.875)$	$T(0.25, 0.5, 0.5, 0.75)$	$T(0.25, 0.5, 0.5, 0.75)$
	x_2	$T(0.375, 0.625, 0.625, 0.875)$	$T(0.125, 0.375, 0.375, 0.625)$	$T(0.125, 0.375, 0.375, 0.625)$	$T(0.125, 0.375, 0.375, 0.625)$
	x_3	$T(0, 0, 0, 0.25)$	$T(0.125, 0.375, 0.375, 0.625)$	$T(0.125, 0.375, 0.375, 0.625)$	$T(0.125, 0.375, 0.375, 0.625)$
	x_4	$T(0.25, 0.5, 0.5, 0.75)$	$T(0.75, 1, 1, 1)$	$T(0.75, 1, 1, 1)$	$T(0.375, 0.625, 0.625, 0.875)$

Table 8 Fuzzy criteria weights

	c_1	c_2	c_3	c_4
dm_1	$T(0.25, 0.5, 0.5, 0.75)$	$T(0.25, 0.5, 0.5, 0.75)$	$T(0.25, 0.5, 0.5, 0.75)$	$T(0.375, 0.625, 0.625, 0.875)$
dm_2	$T(0.375, 0.625, 0.625, 0.875)$	$T(0.125, 0.375, 0.5, 0.75)$	$T(0.75, 1, 1, 1)$	$T(0.25, 0.5, 0.625, 0.875)$
dm_3	$T(0.375, 0.625, 0.625, 0.875)$	$T(0.375, 0.625, 0.625, 0.875)$	$T(0.75, 1, 1, 1)$	$T(0.25, 0.5, 0.5, 0.75)$
dm_4	$T(0.75, 1, 1, 1)$	$T(0.375, 0.625, 0.625, 0.875)$	$T(0.375, 0.625, 0.625, 0.875)$	$T(0.25, 0.5, 0.5, 0.75)$

Table 9 Fuzzy decision matrix

	c_1	c_2	c_3	c_4
x_1	$T(0.25, 0.59375, 0.59375, 0.875)$	$T(0.25, 0.59375, 0.59375, 0.875)$	$T(0.25, 0.5, 0.5625, 0.875)$	$T(0.25, 0.625, 0.625, 1)$
x_2	$T(0.375, 0.90625, 0.90625, 1)$	$T(0.125, 0.4375, 0.46875, 0.875)$	$T(0.125, 0.4375, 0.4375, 0.75)$	$T(0.125, 0.5625, 0.59375, 1)$
x_3	$T(0, 0.375, 0.40625, 0.875)$	$T(0.125, 0.5, 0.53125, 0.875)$	$T(0.125, 0.5, 0.53125, 0.875)$	$T(0.125, 0.625, 0.625, 1)$
x_4	$T(0.25, 0.5, 0.53125, 0.875)$	$T(0.125, 0.65625, 0.75, 1)$	$T(0.25, 0.625, 0.65625, 1)$	$T(0.125, 0.65625, 0.65625, 1)$

Table 10 Fuzzy collective weights

	c_1	c_2	c_3	c_4
Coll. W	$T(0.25, 0.6875, 0.6875, 1)$	$T(0.125, 0.53125, 0.5625, 0.875)$	$T(0.25, 0.78125, 0.78125, 1)$	$T(0.25, 0.53125, 0.5625, 0.875)$

Table 11 Normalized decision matrix

	c_1	c_2	c_3	c_4
x_1	$T(0.25, 0.59375, 0.59375, 0.875)$	$T(0.25, 0.59375, 0.59375, 0.875)$	$T(0.25, 0.5, 0.5625, 0.875)$	$T(0.25, 0.625, 0.625, 1)$
x_2	$T(0.375, 0.90625, 0.90625, 1)$	$T(0.125, 0.4375, 0.46875, 0.875)$	$T(0.125, 0.4375, 0.4375, 0.75)$	$T(0.125, 0.5625, 0.59375, 1)$
x_3	$T(0, 0.375, 0.40625, 0.875)$	$T(0.125, 0.5, 0.53125, 0.875)$	$T(0.125, 0.5, 0.53125, 0.875)$	$T(0.125, 0.625, 0.625, 1)$
x_4	$T(0.25, 0.5, 0.53125, 0.875)$	$T(0.125, 0.65625, 0.75, 1)$	$T(0.25, 0.625.0.65625, 1)$	$T(0.125, 0.65625, 0.65625, 1)$

Table 12 Weighted decision matrix

	c_1	c_2	c_3	c_4
x_1	$T(0.0625, 0.408, 0.408, 0.875)$	$T(0.031225, 0.3150.333, 0.765)$	$T(0.0625, 0.390, 0.439, 0.875)$	$T(0.0625, 0.332, 0.351, 0.875)$
x_2	$T(0.09375, 0.623, 0.623, 1)$	$T(0.015625, 0.232, 0.263, 0.765)$	$T(0.03125, 0.341, 0.341, 0.750)$	$T(0.03125, 0.298, 0.333)$
x_3	$T(0, 0.257, 0.279, 0.875)$	$T(0.015625, 0.265, 0.298, 0.765)$	$T(0.03125, 0.390, 0.415, 0.875)$	$T(0.03125, 0.332, 0.351, 0.875)$
x_4	$T(0.0625, 0.343, 0.365, 0.875)$	$T(0.015625, 0.348, 0.421, 0.875)$	$T(0.0625, 0.488, 0.512, 1)$	$T(0.03125, 0.348, 0.369, 0.875)$

Table 13 Distances to the ideal and anti-ideal solutions

Distances	δ_i^+	δ_i^+
x_1	2.390	1.937
x_2	2.406	1.966
x_3	2.536	1.869
x_4	2.354	2.077

Table 14 Closeness coefficients

Distances	CC_i
x_1	0.447
x_2	0.449
x_3	0.424
x_4	0.468

5 Conclusions

This proposal has introduced the use of nonlinear scaled preferences in linguistic MCGDM by defining an expert-driven fuzzy TOPSIS-based model which is able to consider expert knowledge in the decision process.

To do so, the DMs participating in the decision situation express their opinions by using CLEs, in order to allow them to express their hesitancy about certain ratings, which will be translated to ELICIT information. The corresponding trapezoidal fuzzy numbers of the ELICIT expressions are expressed in terms of 2-tuple values, which are remapped into nonlinear scaled 2-tuple linguistic values by using the novel notion of 2-tuple EVA. From the resulting nonlinear scaled 2-tuple values, new nonlinear scaled ELICIT values are obtained, which will be used in a fuzzy TOPSIS model to obtain more realistic results according to human psychology.

Finally, a real-world problem in which a family wants to decide which streaming platform is most suitable according to their preferences about several criteria has been shown.

References

1. Bustince, H., Barrenechea, E., Pagola, M.: Relationship between restricted dissimilarity functions, restricted equivalence functions and normal en-functions: image thresholding invariant. Pattern Recogn. Lett. **29**(4), 525–536 (2008)
2. Chen, C.: Extensions of the topsis for group decision-making under fuzzy environment. Fuzzy Sets Syst. **114**(1), 1–9 (2000)
3. Delic, A., Ricci, F., Neidhardt, J.: Preference networks and non-linear preferences in group recommendations. In: IEEE/WIC/ACM International Conference on Web Intelligence, WI'19, pp. 403–407. Association for Computing Machinery, New York, NY, USA (2019)

4. García-Zamora, D., Labella, A., Rodríguez, R.M., Martínez, L.: Nonlinear preferences in group decision-making. Extreme values amplifications and extreme values reductions. Int. J. Intell. Syst. **36**(11), 6581–6612 (2021)
5. Kovalchuk, S.V., Krotov, V., Smirnov, P.A., Nasonov, D.A., Yakovlev, A.N.: Distributed data-driven platform for urgent decision making in cardiological ambulance control. Future Gener. Comput. Syst. Int. J. Esci. **79**, 144–154 (2018)
6. Labella, A., Rodríguez, R.M., Martínez, L.: Computing with comparative linguistic expressions and symbolic translation for decision making: ELICIT information. IEEE Trans. Fuzzy Syst. **28**(10), 2510–2522 (2019)
7. Long, Q.Q.: Data-driven decision making for supply chain networks with agent-based computational experiment. Knowl. Based Syst. **141**, 55–66 (2018)
8. Martínez, L., Rodriguez, R.M., Herrera, F.: 2-tuple linguistic model. In: The 2-Tuple Linguistic Model. Springer, Berlin (2015)
9. Martínez, L., Ruan, D., Herrera, F., Herrera-Viedma, E., Wang, P.P.: Linguistic decision making: tools and applications (2009)
10. Masthoff, J.: Group modeling: selecting a sequence of television items to suit a group of viewers. User Model. User-Adap. Inter. **14**, 37–85 (2004)
11. Nădăban, S., Dzitac, S., Dzitac, I.: Fuzzy topsis: a general view. Procedia Comput. Sci. **91**, 823–831 (2016). (Promoting Business Analytics and Quantitative Management of Technology: 4th International Conference on Information Technology and Quantitative Management (ITQM 2016))
12. Palomares, I., Rodriguez, R.M., Martinez, L.: An attitude-driven web consensus support system for heterogeneous group decision making. Expert Syst. Appl. **40**, 139–149 (2013)
13. Pang, Q., Wang, H., Xu, Z.: Probabilistic linguistic term sets in multi-attribute group decision making. Inf. Sci. **369**, 128–143 (2016)
14. Rodriguez, R.M., Martinez, L., Herrera, F.: Hesitant fuzzy linguistic term sets for decision making. IEEE Trans. Fuzzy Syst. **20**(1), 109–119 (2011)
15. Rodríguez, R.M., Martınez, L., Herrera, F.: A group decision making model dealing with comparative linguistic expressions based on hesitant fuzzy linguistic term sets. Inf. Sci. **241**, 28–42 (2013)
16. Sun, Y., He, S., Leu, J.: Syndicating web services: a qos and user-driven approach. Decis. Support Syst. **43** (2007)
17. Tan, C., Jia, Y., Chen, X.: 2-tuple linguistic hesitant fuzzy aggregation operators and its application to multi-attribute decision making. Informatica **28**(2), 329–358 (2017)
18. Zadeh, L.A.: The concept of a linguistic variable and its application to approximate reasoning—I. Inf. Sci. **8**(3), 199–249 (1975)
19. Zadeh, L.A.: Fuzzy sets. In: Zadeh, L.A. (ed.) Fuzzy Sets, Fuzzy Logic, and Fuzzy Systems: Selected Papers, pp. 394–432. World Scientific (1996)
20. Zulueta, Y., Rodríguez, R.M., Bello, R., Martínez, L.: A hesitant heterogeneous approach for environmental impact significance assessment. J. Environ. Inf. **29**(2), 74–87 (2017)

Fermatean Fuzzy Combined Compromise Solution Multiple Expert Multiple Criteria Decision-Making Approach

Yuan Rong, Liying Yu, and Yi Liu

Abstract The Fermatean fuzzy set (FFS) presented is a stronger and usefulness uncertain information representation model for portraying the potential nondeterminacy of data, and the combined compromise solution (COCOSO) method can comprehensively attain a more robust decision outcome through taking into consideration three information fusion strategies. By taking the merits of FFS and COCOSO method, we advance a newly multiple expert multiple criteria decision-making (MEMCDM) technique in light of COCOSO and coefficient of variation methods with Fermatean fuzzy information. Firstly, we review several fundamental conceptions of FFS, including the score function and aggregation operators. Secondly, we put forward an innovative MEMCDM approach through synthesizing the COCOSO method and coefficient of variation method under Fermatean fuzzy environment, where the criterion weight is ascertained through improved Fermatean fuzzy coefficient of variation (COV) method from the visual angle of decision information objectivity. Afterward, a case concerning the assessment of investment enterprises is employed to further confirm the practicability and feasibility of the propounded MEMCDM method, as well as the contrast analysis is conducted to sticking out the flexibility and efficiency of the proffered MEMCDM technique.

Keywords Fermatean fuzzy set · COCOSO method · COV method · MEMCDM

Y. Rong · L. Yu (✉)
School of Management, Shanghai University, Shanghai 200444, China
e-mail: yuliying@shu.edu.cn

Y. Rong
e-mail: rongyuanry@163.com

Y. Liu
School of Mathematics and Information Science, Neijiang Normal University, Neijiang 641100, Sichuan, China
e-mail: liuyiyl@126.com

L. Sahoo et al. (eds.), *Real Life Applications of Multiple Criteria Decision Making Techniques in Fuzzy Domain*, Studies in Fuzziness and Soft Computing 420,
https://doi.org/10.1007/978-981-19-4929-6_4

1 Introduction

The core of decision [1] is the analysis process of prioritizing several alternatives in view of the characteristics of decision issues and scientific and effective methods in a certain time and resources. Traditional deterministic decision-making problems can be effectively solved by combining operations research and decision-making theory. In light of the complicated of decision context and the ambiguity of experts' cognitive thinking, the vague decision on account of fuzzy conceptions has been widely studied in multitude domains [2–7].

The ambiguity of practical problems and conflict psychology of evaluator makes them provide the assessment viewpoint without employing a specific numerical values. Consequently, Zadeh [8] broke through the bondage of "either or" in classical sets, and then proposed a tool to describe uncertainty and named it fuzzy set (FS). However, FS only characterizes the experts' approval of a given goal through membership, ignoring the disapproval and ambiguous opinions. Thus, intuitionistic FS [9] describes vague information more comprehensively by making up the defect which FS fail ponder the non-attachment and hesitation grade [10]. Thereafter, Yager [11, 12] originated the Pythagorean FS through increasing the evaluation freedom degree of evaluator during the course of multiple criterion assessment application [13, 14] Although Pythagorean FS expands the scope of evaluation information provided by evaluators compared with intuitionistic FS, its restrictive conditions still have certain restrictions on the evaluation views provided by evaluators.

Consequently, an innovative uncertainty information expression tool called FFS [15] is put forward to model the fuzziness information via approval and disapproval grades with the condition of cubic of approval, and disapproval grade is less than or equal to 1. Evidently, FFS can provide a larger space for evaluators to express assessment information. In view of the significant power in fuzzy information representation, several operations and information fusion tools of FFS are researched and applied these theories to build up diverse decision models [16–22]. In the aspect of the generations of FFS, Fermatean fuzzy linguistic set and the related operations are advanced to ponder the combination of FFS and linguistic variable [23]. Moreover, Liu et al. [24] further propounded some novel operations with the help of linguistic scale functions and cosine distance measures to construct the innovative TOPSIS and TODIM decision framework. Considering that uncertainty of approval and disapproval grades, the interval-valued Fermatean fuzzy sets theory is built and related operations; ranking and comparison rules are also constructed for developing decision analysis [25]. Besides, several classical decision approaches are propounded under Fermatean fuzzy setting to construct the Fermatean fuzzy decision methodology. Keshavarz-Ghorabaee et al. [26] introduced the Fermatean fuzzy WASPAS method to evaluate the green construction supplier. Further, Mishra and Rani [27] built an extended WASPAS method based upon novel score function and entropy measure to select the optimal healthcare waste disposal location. Gul et al. [28] advanced a risk assessment framework for manufacturing based upon Fermatean fuzzy TOPSIS technique. Aydemir and Gunduz [29] proposed the extended TOPSIS method based on the novel Dombi operators and improved distance under Fermatean fuzzy setting. Afterward, Gul [30] studies some extended Fermatean fuzzy decision method on the basis of SAW, ARAS, and VIKOR method and utilizes these methods to resolve the COVID-19 testing laboratory selection problems. To consider the assessment of group decision experts, Mishra et al. [31] constructed the novel group Fermatean fuzzy decision method by integrating CRITIC and EDAS method based on an improved generalized score function for the selection of sustainable third-party reverse logistics providers. Rani and Mishra [32] suggested the improved MULTI-MOORA method based on the novel Fermatean fuzzy Einstein aggregation operators to select the electric vehicle charging station. Kamali Saraji et al. [33] propounded Fermatean fuzzy COPRAS to evaluate the challenges for Industry 4.0. Also, Deng and Wang [34] studied the evidential reasoning theory and entropy measure of FFS to construct a multi-story decision-making algorithm. However, the extant decision approaches on Fermatean fuzzy theory may cause different sorting outcomes for the same issue owing to different weight distributions of criteria. The outcomes acquired by these approaches may produce the confusion phenomenon during the process of decision analysis. Furthermore, these methods obtain the final rank of alternatives by single aggregation manner which may cause an unstable decision outcome. To conquer the mentioned problem, the COCOSO method which is based on the synthesize utility of alternative from the point of multiple aggregation strategies is extended to Fermatean fuzzy circumstance for strengthening the stability and robustness in dealing with complex decision problems.

Based on the fusion of additive weighted and exponential weighted product models, the COCOSO algorithm was first brought forward by Yazdani et al. [35]. The COCOSO method uses three different aggregation strategies to integrate the evaluation information of schemes under various attributes in the decision-making process and then carries out optimal ranking by calculating the utility value of schemes. Since then, because of its advantages of simple operation and high flexibility, the COCOSO method has been extended to different uncertain environments and used for practical evaluation problem analysis. Yazdani et al. [36] synthesized the COCOSO method with BWM and DEMATEL methods, presented the gray number COCOSO framework, and employed it to choose the construction managers. Peng et al. [37] established the COCOSO method with unknown criteria weight under Pythagorean fuzzy surroundings and established the 5G industrial evaluation framework. Ecer and Pamucar [38] established the COCOSO method with completely unknown weights in triangular fuzzy circumstance based on BWM method and Bonferroni mean. Subsequently, Rani and Mishra [39] built up a group assessment algorithm on account of similarity and COCOSO method, in which the importance grade of expert and criteria is completely unknown. Yazdani et al. [40] proposed the complete consistency model of rough set to determine the subjective weight of criterions and combined with data envelopment analysis method to establish integrated COCOSO method with rough set to select satisfactory logistics center site. However, the application of COCOSO method in FFS has not been explored for the construction of decision models. Therefore, this paper will establish the Fermatean fuzzy COCOSO group decision framework in view of fundamental theories and analyze its advantages in solving complex uncertain decision issues.

The determination of criteria in multiple criteria decision analysis is a vital part to obtaining a reliable and effective decision outcome. Although there exists many weight determination approaches such as MEREC [41], CRITIC [31], and SWARA [42], these methods have different characteristics and effects. The MEREC method considers the removal impact of each criterion on the performance of alternatives to evaluate the criterion weight; the CRITIC method considers the volatility of data and correlation between data. The SWARA method considers the subjective preference and cognition of decision experts. Apart from the mentioned approaches, the coefficient of variation method, which is an objective weighting method, can directly use the information contained in each index to calculate the weight of the index. The basic method of this method is the greater the difference in index values in the evaluation index system, the more difficult it is to achieve, and such indicators are more difficult to reflect the gap between the evaluated units. Considering the objectivity and the correlation of the real decision information, this research is first combined with COV method and COCOSO method in a FFS context to build group decision approach. Based upon the abovementioned analysis, this article put forward an innovative Fermatean fuzzy MEMCDM approach based on improved COCOSO and COV methodology to settle the actual decision issues with unknown importance of criterion and experts. The applicability and validity is tested through empirical and comparative investigation, respectively. Also, we explore the changes of the parameters on the ultimate orders of scheme and test the stability of the propounded Fermatean fuzzy COCOSO decision algorithm. The resultants show that the designed group COCOSO decision algorithm possesses several merits in resolving assessment issues and selection researches.

The organization of this manuscript is displayed as follows. Section 2 looks back several fundamental notions, aggregation operators, and comparison laws of FFS. Section 3 builds a novel Fermatean fuzzy MEMCDM approach through extending the classical COCOSO method. In Sect. 4, an actual case is employed to further validate the practicability of the propounded FF-COCOSO method, and the comparisons are conducted to outstanding the effectiveness, applicability, and efficiency of the decision framework. Several discussions of this research are given in Sect. 5. Conclusion remarks and research prospects are listed in the end.

2 Preliminaries

This section schematic looks back some conceptions of FFS, including the fundamental definition, comparison algorithm the aggregation operators.

Definition 1 [15]. Let U be a fix set. Then, a FFS F on U is stated as below:

$$F = \{< x, \phi_F(u), \kappa_F(u) > |u \in U\}, \tag{1}$$

in which the function $\phi_F(u): U \rightarrow [0, 1]$, $\kappa_F(u): U \rightarrow [0, 1]$ with the following condition $0 \leq (\phi_F(u))^3 + (\kappa_F(u))^3 \leq 1$, for $u \in U$. The function $\phi_F(u), \kappa_F(u)$ severally signify the approval grade and disapproval grade of element u in F. For simplicity, the pair $F = (\phi_F, \kappa_F)$ is called Fermatean fuzzy number (FFN), which holds $\phi_F, \kappa_F \in [0, 1]$ and $0 \leq (\phi_F)^3 + (\kappa_F)^3 \leq 1$. The hesitant grade is depicted as

$$\pi_F(u) = \sqrt[3]{1 - (\phi_F(u))^3 - (\kappa_F(u))^3}. \tag{2}$$

Definition 2 [15]. Let $F = (\phi, \kappa)$ be an FFN. The score and accuracy index of F are stated as

$$S(F) = \phi^3 - \kappa^3, S(F) \in [-1, 1] \tag{3}$$

$$A(F) = \phi^3 + \kappa^3, A(F) \in [0, 1]. \tag{4}$$

Since $S(F) \in [-1, 1]$, an improved score index is defined as:

Definition 3 Assume that $F = (\phi, \kappa)$ be an FFN. Then, the normalized score function is defined as below:

$$S^*(F) = \frac{1}{2}(S(F) + 1), S^*(F) \in [0, 1] \tag{5}$$

Definition 4 Given two FFNs $F_1 = (\phi_1, \kappa_1)$ and $F_2 = (\phi_2, \kappa_2)$. We can compare FFNs F_1 and F_2 according to the following rules:

If $S^*(F_1) > S^*(F_2)$, then $F_1 \succ F_2$,
If $S^*(F_1) = S^*(F_2)$, then
  if $A(F_1) < A(F_2)$, then $F_1 \prec F_2$;
  if $A(F_1) = A(F_2)$, then $F_1 \sim F_2$.

Definition 5 [16, 17]. Suppose that $F_j = (\phi_j, \kappa_j)$, $j = 1(1)n$ are FFNs. Then, the Fermatean fuzzy weighted averaging and geometric operators are depicted by

$$\text{FFWA}_w(F_1, F_2, \ldots, F_n) = \overset{n}{\underset{j=1}{\oplus}} w_j F_j = \left(\sqrt[3]{1 - \prod_{j=1}^{n} \left(1 - (\phi_j)^3\right)^{w_j}}, \prod_{j=1}^{n} \kappa_j^{w_j} \right), \tag{6}$$

$$\text{FFWG}_w(F_1, F_2, \ldots, F_n) = \overset{n}{\underset{j=1}{\otimes}} (F_j)^{w_j} = \left(\prod_{j=1}^{n} \phi_j^{w_j}, \sqrt[3]{1 - \prod_{j=1}^{n} \left(1 - (\kappa_j)^3\right)^{w_j}} \right). \tag{7}$$

where $w_j = (w_1, w_2, ..., w_n)^{\mathrm{T}}$ is the weight vector of $F_j, j = 1(1)n$, with $\sum_{j=1}^{n} w_j = 1$, $w_j \in [0, 1]$.

3 Approach to FF-MEMCDM Based upon COCOSO

In order to settle the complicated practical decision issues, we bring forward the FF-MEMCDM approach on the basis of COCOSO to comprehensively ponder the wisdom of expert group and attain a more scientific and coincident decision outcomes than previous Fermatean fuzzy decision methodologies. The MEMCDM procedure can be portrayed by Fig. 1 [43]. As depicted in Fig. 1, the process main involves three stages, including the information representation expert assessment information fusion, the importance grade determination of evaluators and criterion, and the exploitation by diverse decision techniques. In the presented MEMCDM approach, the FFN is employed to portray uncertain preference opinion of evaluators, and the importance grade of criteria is evaluated through COV approach; the FFWA operator is employed to achieve the information fusion phase; the decision analysis procedure is explored by the generalized COCOSO method FF information. The MEMCDM issues depiction and the detailed decision steps of the proffered COCOSO approach are displayed in subsections, respectively.

3.1 The Decision Issue Statement

For a multiple attribute decision group making problem, it involves the following basic elements which are displayed as below: $\Upsilon = \{\Upsilon_i | i = 1(1)m\}$ denotes the family of decision alternative; $C = \left\{C_j | j = 1(1)n\right\}$ denotes the set of decision criteria. w_j denotes the weight of attribute meeting the condition $w_j \in [0, 1]$, $\sum_{j=1}^{n} w_j = 1$. $\{E_k | k = 1(1)K\}$ denotes the set of decision makers. λ_k denotes the weight of attribute meeting the condition $\lambda_k \in [0, 1]$, $\sum_{k=1}^{K} \lambda_k = 1$. The expert E_k

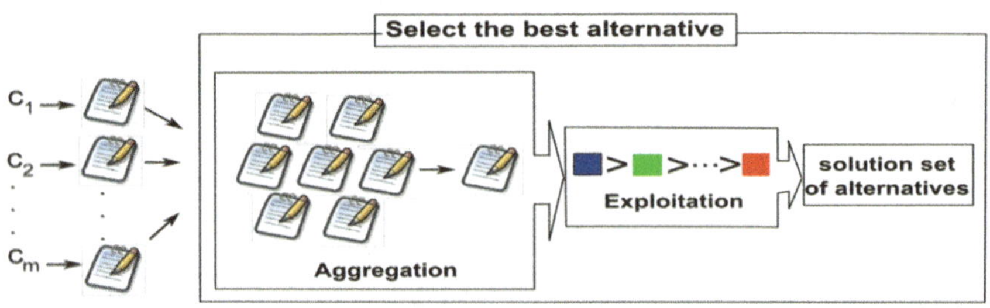

Fig. 1 Procedure of multiple expert multiple criteria decision-making

provides his (her) preference information for alternative Υ_i with respect to attribute C_j by using FFN and denotes as $\overline{\mathbf{A}}^k = \left(\overline{a}_{ij}^k\right)_{m\times n} = \left(\overline{\phi}_{ij}^k, \overline{\kappa}_{ij}^k\right)$. The objective of this problem is to attain the order relation of all alternatives.

$$\overline{\mathbf{A}}^k = \left(\overline{a}_{ij}^k\right)_{m\times n} = \begin{pmatrix} \overline{a}_{11}^k & \overline{a}_{12}^k & \cdots & \overline{a}_{1n}^k \\ \overline{a}_{21}^k & \overline{a}_{22}^k & \cdots & \overline{a}_{2n}^k \\ \vdots & \vdots & \ddots & \vdots \\ \overline{a}_{m1}^k & \overline{a}_{m2}^k & \cdots & \overline{a}_{mn}^k \end{pmatrix}$$

3.2 FF-MEMCDM Based upon COCOSO

In this subsection, we put forward FF-COCOSO to address the MEMCDM problems with unknown weight of criterion and experts. The experts weights are ascertained by subjective weight approach where the expert is given through the FFNs. Considering that the objectivity of criteria weight occupies a crucial location in multiple criteria decision analysis, the weight of criterion is calculated through the COV method based on score function. The final order of alternatives is developed based on the extended FF-COCOSO methodology.

Step 1: Attain the normalized expert decision matrices $\mathbf{A}^k = \left(a_{ij}^k\right)_{m\times n}$ using the following formulation:

$$\mathbf{A}^k = \left(a_{ij}^k\right)_{m\times n} = \begin{pmatrix} a_{11}^k & a_{12}^k & \cdots & a_{1n}^k \\ a_{21}^k & a_{22}^k & \cdots & a_{2n}^k \\ \vdots & \vdots & \ddots & \vdots \\ a_{m1}^k & a_{m2}^k & \cdots & a_{mn}^k \end{pmatrix}$$

where

$$a_{ij}^k = \begin{cases} \overline{a}_{ij}^k = \left(\overline{\phi}_{ij}^k, \overline{\kappa}_{ij}^k\right), & \text{for benefit criterion,} \\ \left(\overline{a}_{ij}^k\right)^c = \left(\overline{\kappa}_{ij}^k, \overline{\phi}_{ij}^k\right), & \text{for cost criterion.} \end{cases} \tag{8}$$

Step 2: Ascertain the weight of evaluators.

Because the fuzziness of experts' cognitive thinking, a mapping relation from linguistic terms to PFNs is provided for experts to give the expert weight. Suppose that $F_k = (\phi_k, \kappa_k)$ is a FFN, then

$$\lambda_k = \frac{(\phi_k)^3 + (\pi_k)^3 \times \left(\frac{(\phi_k)^3}{(\phi_k)^3 + (\kappa_k)^3}\right)}{\sum_{k=1}^{K} \left((\phi_k)^3 + (\pi_k)^3 \times \left(\frac{(\phi_k)^3}{(\phi_k)^3 + (\kappa_k)^3}\right)\right)} \tag{9}$$

In light of the weight information of decision experts, we shall obtain the collective assessment matrix with the aid of FFWA aggregation operator.

Step 3: The aggregated decision matrix $\mathbf{A} = \left(a_{ij}\right)_{m \times n}$ is determined by the FFWA operator as below:

$$a_{ij} = \text{FFWA}_w\left(a_{ij}^1, a_{ij}^2, \ldots, a_{ij}^K\right) = \left(\sqrt[3]{1 - \prod_{j=1}^{n}\left(1 - \left(\phi_{ij}^k\right)^3\right)^{\lambda_k}}, \prod_{j=1}^{n}\left(\kappa_{ij}^k\right)^{\lambda_k}\right) \tag{10}$$

Step 4: Determine the criteria weight information through utilizing COV approach.

Due to the nondeterminacy of decision information and the limitation of evaluators' background, the importance grade of pondered criterion is hard to ascertain forthright in advance. In such situation, we employ the COV method to compute the importance of the predefined criterion objectively owing to its merits and simplicity. Accordingly, the concrete determination steps of the FF-COV method to compute the criteria weight are described as below:

Step 4.1: To compute the mean value of the j th attribute in view of the FFWA operator.

$$\widetilde{a}_j = \frac{1}{m}\sum_{i=1}^{m} a_{ij} = \left(\sqrt[3]{1 - \prod_{i=1}^{m}\left(1 - (\phi_{ij})^3\right)^{\frac{1}{m}}}, \prod_{i=1}^{m} \kappa_{ij}^{\frac{1}{m}}\right) \tag{11}$$

where $\widetilde{a}_j$ is the mean value of all alternatives over criteria C_j.

Step 4.2: To compute the deviation of the criteria C_j.

$$D_j = \sqrt{\frac{1}{m-1}\sum_{i=1}^{m}\left(S^*\left(a_{ij}\right) - S^*\left(\widetilde{a}_j\right)\right)^2}, j = 1(1)n \tag{12}$$

Step 4.3: To compute the coefficient of variation of the jth attribute;

$$Z_j = \frac{D_j}{S^*\left(\widetilde{a}_j\right)}, j = 1(1)n \tag{13}$$

Step 4.4: To normalize the coefficient of variation and obtain the weight of attribute;

$$w_j = \frac{Z_j}{\sum_{j=1}^{n} Z_j} \tag{14}$$

According to the mentioned steps, we can obtain the final criteria weight information objectively.

In what follows, based upon the weight of criteria, we propound FF-COCOSO method on the basis of FFWA operator, FFWG operator, and score function to attain the priority order of alternatives.

Step 5: Compute the weighted sum measure (WSM) and weighted product measure (WPM) through the following formulation:

$$\begin{aligned} \theta_i &= \text{FFWA}_w(a_{i1}, a_{i2}, \ldots, a_{in}) = \mathop{\oplus}_{j=1}^{n} w_j F_{ij} \\ &= \left(\sqrt[3]{1 - \prod_{j=1}^{n} \left(1 - (\phi_{ij})^3\right)^{w_j}}, \prod_{j=1}^{n} \kappa_j^{w_j} \right), \end{aligned} \tag{15}$$

$$\begin{aligned} \rho_i &= \text{FFWG}_w(a_{i1}, a_{i2}, \ldots, a_{in}) = \mathop{\otimes}_{j=1}^{n} \left(F_{ij}\right)^{w_j} \\ &= \left(\prod_{j=1}^{n} \phi_j^{w_j}, \sqrt[3]{1 - \prod_{j=1}^{n} \left(1 - (\kappa_j)^3\right)^{w_j}} \right). \end{aligned} \tag{16}$$

Step 6: Compute the final appraise values of WSM and WPM based on Eq. (5):

$$\begin{aligned} T_i &= S^*(\theta_i), \\ P_i &= S^*(\rho_i), \end{aligned} \tag{17}$$

where $S^*(F) = \frac{1}{2}\left(\phi^3 - \kappa^3 + 1\right)$ for an FFN $F = (\phi, \kappa)$.

Step 7: The relative importance degree of alternatives shall be ascertained with the aid of three appraisal score strategies as below:

$$G_i^1 = \frac{T_i + P_i}{\sum_{i=1}^{m} (T_i + P_i)}, \tag{18}$$

$$G_i^2 = \frac{T_i}{\min_i T_i} + \frac{P_i}{\min_i P_i}, \tag{19}$$

$$G_i^3 = \frac{\varsigma T_i + (1 - \varsigma) P_i}{\varsigma \max_i T_i + (1 - \varsigma) \max_i P_i}, \varsigma \in [0, 1], \tag{20}$$

Step 8: The final assessment index G_i was acquired through Eq. (21):

$$G_i = \left(G_i^1 G_i^2 G_i^3\right)^{\frac{1}{3}} + \frac{1}{3}\left(G_i^1 + G_i^2 + G_i^3\right) \tag{21}$$

Step 9: Ranking the alternatives in view of the following maximum function:

$$\Upsilon_i = \left\{ \Upsilon_i \middle| \max_i G_i \right\}, \tag{22}$$

where Υ_i represents the best alternative attained in light of the value of G_i.

To know about the proposed method better, a concise algorithm involving the steps of the FF-COCOSO method is portrayed as follows.

Algorithm: FF-COCOSO

Input: Alternative set $\Upsilon_i(i = 1(1)m)$, criteria set $C_j(j = 1(1)n)$, expert set $E_k(k = 1(1)K)$
Output: The rank of alternatives
Begin

Step 1: Attaining the normalized expert decision matrices $\mathbf{A}^k = \left(a_{ij}^k\right)_{m\times n}$ by Eq. (8)
Step 2: Ascertaining the weight of decision expert by Eq. (9)
Step 3: Fixing the aggregated decision matrices $\mathbf{A} = \left(a_{ij}\right)_{m\times n}$ by Eq. (10)
Step 4: Ascertaining the criteria weight information by Eqs. (11)–(14)
Step 5: Compute the WSM and WPM by Eqs. (15) and (16)
Step 6: Fixing the final score values of WSM and WPM by Eq. (17)
Step 7: Computing the relative importance degree of alternatives by Eqs. (18)–(20)
Step 8: Computing the final assessment index G_i by Eq. (21)
Step 9: Ranking all alternatives by the value of G_i by Eq. (22)

End

4 Empirical Study

A new energy vehicle company plans to invest some foreign exchange funds in an automobile production enterprise to contract the production of new energy vehicles. After the qualification examination and primary selection of the investment department, the five most potential enterprises $\{\Upsilon_1, \Upsilon_2, \Upsilon_3, \Upsilon_4, \Upsilon_5\}$ are pondered as candidates for the final selection by the expert group of the board of directors of the company. The evaluation group involving three experts assess the investment of five enterprises by using FFN based on their background. After comprehensive analysis and consideration, the following five criterions $\{\mathbb{C}_1, \mathbb{C}_2, \mathbb{C}_3, \mathbb{C}_4, \mathbb{C}_5\}$ are considered in the process of selection and assessment, where $\mathbb{C}_1$: cost required for investment, $\mathbb{C}_2$: financing capacity; $\mathbb{C}_3$: comprehensive management capability, $\mathbb{C}_4$: organization ability, $\mathbb{C}_5$: investment risk. As mentioned, the criteria $\mathbb{C}_1$ and $\mathbb{C}_5$ are cost criteria, and the others are benefit criterions. The weight of experts and attribute is all completely unknown. The evaluators assess the enterprises $\{\Upsilon_1, \Upsilon_2, \Upsilon_3, \Upsilon_4, \Upsilon_5\}$ and assign the

Table 1 Linguistic terms for experts as a refer to assess the alternatives [29]

Linguistic term	Abbreviation	Fermatean fuzzy element
Extremely low	EL	(0.15, 0.95)
Very low	VL	(0.25, 0.85)
Low	L	(0.35, 0.75)
Below middle	BL	(0.45, 0.65)
Middle	M	(0.55, 0.55)
Above middle	AM	(0.65, 0.45)
High	H	(0.75, 0.35)
Very high	VH	(0.85, 0.25)
Extremely high	EH	(0.95, 0.15)

value to the enterprises with the assistance of the linguistic terms shown in Table 1 and give the corresponding evaluation information shown in Table 2.

In light of the aforementioned assessment information evaluated by decision experts, we utilize the proposed FF-COCOSO MEMCDM approach to select the most satisfied investment enterprise for company. The calculation procedure and related outcomes are shown as follows.

Table 2 Assessment information provided by experts

	Alternative	$\mathbb{C}_1$	$\mathbb{C}_2$	$\mathbb{C}_3$	$\mathbb{C}_4$	$\mathbb{C}_5$
E_1	Υ_1	(0.10, 0.90)	(0.70, 0.25)	(0.75, 0.30)	(0.70, 0.30)	(0.20, 0.85)
	Υ_2	(0.35, 0.75)	(0.80, 0.15)	(0.60, 0.40)	(0.90 0.10)	(0.35, 0.65)
	Υ_3	(0.30, 0.70)	(0.70, 0.30)	(0.70, 0.30)	(0.55, 0.40)	(0.10, 0.90)
	Υ_4	(0.20, 0.80)	(0.60, 0.50)	(0.80, 0.15)	(0.70, 0.35)	(0.30, 0.70)
	Υ_5	(0.45, 0.50)	(0.80, 0.10)	(0.60, 0.25)	(0.80, 0.20)	(0.20, 0.80)
E_2	Υ_1	(0.20, 0.85)	(0.80, 0.15)	(0.70, 0.25)	(0.60, 0.20)	(0.30, 0.75)
	Υ_2	(0.30, 0.70)	(0.70, 0.25)	(0.65, 0.30)	(0.80 0.30)	(0.35, 0.60)
	Υ_3	(0.20, 0.60)	(0.50, 0.40)	(0.60, 0.35)	(0.65, 0.20)	(0.20, 0.90)
	Υ_4	(0.10, 0.80)	(0.70, 0.30)	(0.75, 0.10)	(0.60, 0.35)	(0.40, 0.50)
	Υ_5	(0.35, 0.60)	(0.75, 0.15)	(0.65, 0.20)	(0.70, 0.10)	(0.10, 0.90)
E_3	Υ_1	(0.30, 0.75)	(0.70, 0.20)	(0.75, 0.25)	(0.60, 0.20)	(0.20, 0.80)
	Υ_2	(0.10, 0.80)	(0.70, 0.20)	(0.70, 0.35)	(0.70 0.30)	(0.30, 0.75)
	Υ_3	(0.30, 0.60)	(0.55, 0.40)	(0.65, 0.30)	(0.75, 0.20)	(0.20, 0.85)
	Υ_4	(0.30, 0.70)	(0.80, 0.25)	(0.70, 0.20)	(0.60, 0.30)	(0.30, 0.80)
	Υ_5	(0.25, 0.65)	(0.85, 0.10)	(0.60, 0.25)	(0.60, 0.50)	(0.20, 0.90)

4.1 Decision Analysis Procedure

Step 1: Attaining the normalized expert decision matrices by Eq. (8), displayed in Table 3.

Step 2: Determining the weights of decision experts by Eq. (9), displayed in Table 4.

Step 3: Determining the fused assessment matrix $\mathbf{A} = (a_{ij})_{m\times n}$ by Eq. (10), displayed in Table 5.

Step 4: Determining the criteria weight information by Eqs. (11)–(14), displayed in Table 6.

Steps 5 and 6: Compute the weighted sum measure and weighted product measure and their final score values by Eqs. (15) and (16), respectively, displayed in Table 7.

Steps 7 and 8: Computing the relative importance degree of alternatives and the final assessment index G_i, which is displayed in Table 8.

Step 9: Orders of all alternatives are $\Upsilon_1 \succ \Upsilon_5 \succ \Upsilon_2 \succ \Upsilon_3 \succ \Upsilon_4$.

Table 3 Normalized assessment information provided by experts

		$\mathbb{C}_1$	$\mathbb{C}_2$	$\mathbb{C}_3$	$\mathbb{C}_4$	$\mathbb{C}_5$
E_1	Υ_1	(0.90, 0.10)	(0.70, 0.25)	(0.75, 0.30)	(0.70, 0.30)	(0.85, 0.20)
	Υ_2	(0.75, 0.35)	(0.80, 0.15)	(0.60, 0.40)	(0.90 0.10)	(0.65, 0.35)
	Υ_3	(0.70, 0.30)	(0.70, 0.30)	(0.70, 0.30)	(0.55, 0.40)	(0.90, 0.10)
	Υ_4	(0.80, 0.20)	(0.60, 0.50)	(0.80, 0.15)	(0.70, 0.35)	(0.70, 0.30)
	Υ_5	(0.50, 0.45)	(0.80, 0.10)	(0.60, 0.25)	(0.80, 0.20)	(0.80, 0.20)
E_2	Υ_1	(0.85, 0.20)	(0.80, 0.15)	(0.70, 0.25)	(0.60, 0.20)	(0.75, 0.30)
	Υ_2	(0.70, 0.30)	(0.70, 0.25)	(0.65, 0.30)	(0.80 0.30)	(0.60, 0.35)
	Υ_3	(0.60, 0.20)	(0.50, 0.40)	(0.60, 0.35)	(0.65, 0.20)	(0.90, 0.20)
	Υ_4	(0.80, 0.10)	(0.70, 0.30)	(0.75, 0.10)	(0.60, 0.35)	(0.50, 0.40)
	Υ_5	(0.60, 0.35)	(0.75, 0.15)	(0.65, 0.20)	(0.70, 0.10)	(0.90, 0.10)
E_3	Υ_1	(0.75, 0.30)	(0.70, 0.20)	(0.75, 0.25)	(0.60, 0.20)	(0.80, 0.20)
	Υ_2	(0.80, 0.10)	(0.70, 0.20)	(0.70, 0.35)	(0.70 0.30)	(0.75, 0.30)
	Υ_3	(0.60, 0.30)	(0.55, 0.40)	(0.65, 0.30)	(0.75, 0.20)	(0.85, 0.20)
	Υ_4	(0.70, 0.30)	(0.80, 0.25)	(0.70, 0.20)	(0.60, 0.30)	(0.80, 0.30)
	Υ_5	(0.65, 0.25)	(0.85, 0.10)	(0.60, 0.25)	(0.60, 0.50)	(0.90, 0.20)

Table 4 Weight of decision expert

	E_1	E_2	E_3
FFN	(0.85, 0.25)	(0.70, 0.40)	(0.65, 0.45)
λ_k	0.3796	0.3281	0.2923

Table 5 Integrated assessment matrix by FFWA operator

	$\mathbb{C}_1$	$\mathbb{C}_2$	$\mathbb{C}_3$	$\mathbb{C}_4$	$\mathbb{C}_5$
Υ_1	(0.8528, 0.1731)	(0.7393, 0.1981)	(0.7350, 0.2679)	(0.6438, 0.2333)	(0.8083, 0.2285)
Υ_2	(0.7526, 0.2307)	(0.7448, 0.1929)	(0.6497, 0.3500)	(0.8312, 0.1977)	(0.6721, 0.3346)
Υ_3	(0.6438, 0.2626)	(0.6093, 0.3586)	(0.6568, 0.3156)	(0.6576, 0.2602)	(0.8877, 0.1537)
Υ_4	(0.7762, 0.1794)	(0.7087, 0.3453)	(0.7590, 0.1428)	(0.6438, 0.3346)	(0.6968, 0.3297)
Υ_5	(0.5858, 0.3490)	(0.8033, 0.1142)	(0.6177, 0.2324)	(0.7256, 0.2082)	(0.8712, 0.1593)

Table 6 Criteria weight computed by the CVM

	Standard deviation	Coefficient of variation	Weight
C_1	0.0909	0.1296	0.2442
C_2	0.0632	0.0916	0.1726
C_3	0.0484	0.0737	0.1389
C_4	0.0700	0.1035	0.1951
C_5	0.1006	0.1323	0.2493

Table 7 WSM, WPM measures, and their score values

	WSM	Score	WPM	Score
Υ_1	(0.7801, 0.2138)	0.7325	(0.7613, 0.2206)	0.7152
Υ_2	(0.7434, 0.2523)	0.6974	(0.7296, 0.2766)	0.6836
Υ_3	(0.7442, 0.2483)	0.6984	(0.6957, 0.2767)	0.6577
Υ_4	(0.7224, 0.2557)	0.6802	(0.7149, 0.2935)	0.6701
Υ_5	(0.7613, 0.2022)	0.7165	(0.7173, 0.2481)	0.6769

Table 8 Ultimate decision outcomes and order by using FF-COCOSO method

	G_i^1	Order	G_i^2	Order	G_i^3	Order	G_i	Order
Υ_1	0.2090	1	2.1644	1	1.0000	1	1.8920	1
Υ_2	0.1993	3	2.0646	3	0.9539	3	1.8048	3
Υ_3	0.1957	4	2.0269	4	0.9368	4	1.7721	4
Υ_4	0.1949	5	2.0188	5	0.9327	5	1.7647	5
Υ_5	0.2011	2	2.0825	2	0.9624	2	1.8207	2

4.2 Sensitivity Investigation

The fluctuation of parameter will obtain different decision outcomes of all alternatives. This part main discusses the stability of parameters ς on the ultimate decision outcome. Aiming at diverse parameters ς, the corresponding results and orders can be described in Table 9 and Fig. 2. From Table 9, it is obvious that the parameter has a minor impact for the final rank of alternatives. When the parameters $\varsigma \in [0, 0.1]$, the rank of alternatives is $\Upsilon_1 \succ \Upsilon_5 \succ \Upsilon_2 \succ \Upsilon_4 \succ \Upsilon_3$; when the parameters $\varsigma \in [0.3, 1.0]$, the rank of alternatives is $\Upsilon_1 \succ \Upsilon_5 \succ \Upsilon_2 \succ \Upsilon_3 \succ \Upsilon_4$. Decision experts can select a proper parameter to obtain final decision outcomes according to their preference attitude.

Table 9 Decision results and order based on diverse values of parameters

Parameters ς	Assessment index of alternatives					Rank
	Υ_1	Υ_2	Υ_3	Υ_4	Υ_5	
0.0000	1.8920	1.8059	1.7620	1.7671	1.8112	$\Upsilon_1 \succ \Upsilon_5 \succ \Upsilon_2 \succ \Upsilon_4 \succ \Upsilon_3$
0.1000	1.8920	1.8057	1.7640	1.7666	1.8131	$\Upsilon_1 \succ \Upsilon_5 \succ \Upsilon_2 \succ \Upsilon_4 \succ \Upsilon_3$
0.2000	1.8920	1.8055	1.7661	1.7661	1.8150	$\Upsilon_1 \succ \Upsilon_5 \succ \Upsilon_2 \succ \Upsilon_3 \sim \Upsilon_4$
0.3000	1.8920	1.8052	1.7681	1.7657	1.8169	$\Upsilon_1 \succ \Upsilon_5 \succ \Upsilon_2 \succ \Upsilon_3 \succ \Upsilon_4$
0.4000	1.8920	1.8050	1.7701	1.7652	1.8188	$\Upsilon_1 \succ \Upsilon_5 \succ \Upsilon_2 \succ \Upsilon_3 \succ \Upsilon_4$
0.5000	1.8920	1.8048	1.7721	1.7647	1.8207	$\Upsilon_1 \succ \Upsilon_5 \succ \Upsilon_2 \succ \Upsilon_3 \succ \Upsilon_4$
0.6000	1.8920	1.8046	1.7741	1.7642	1.8226	$\Upsilon_1 \succ \Upsilon_5 \succ \Upsilon_2 \succ \Upsilon_3 \succ \Upsilon_4$
0.7000	1.8920	1.8044	1.7761	1.7637	1.8244	$\Upsilon_1 \succ \Upsilon_5 \succ \Upsilon_2 \succ \Upsilon_3 \succ \Upsilon_4$
0.8000	1.8920	1.8041	1.7781	1.7632	1.8263	$\Upsilon_1 \succ \Upsilon_5 \succ \Upsilon_2 \succ \Upsilon_3 \succ \Upsilon_4$
0.9000	1.8920	1.8039	1.7800	1.7627	1.8281	$\Upsilon_1 \succ \Upsilon_5 \succ \Upsilon_2 \succ \Upsilon_3 \succ \Upsilon_4$
1.0000	1.8920	1.8037	1.7820	1.7623	1.8299	$\Upsilon_1 \succ \Upsilon_5 \succ \Upsilon_2 \succ \Upsilon_3 \succ \Upsilon_4$

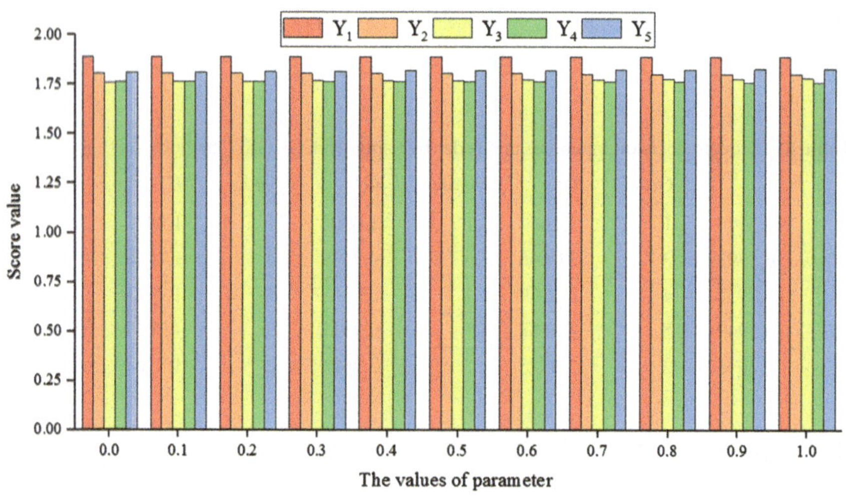

Fig. 2 Change of ranks based on diverse values of parameters

Table 10 Revised comprehensive assessment matrix by FFWA operator

	$\mathbb{C}_1$	$\mathbb{C}_2$	$\mathbb{C}_3$	$\mathbb{C}_4$	$\mathbb{C}_5$
Υ_1	(0.8528, 0.1731)	(0.7393, 0.1981)	(0.7350, 0.2679)	(0.6438, 0.2333)	(0.8083, 0.2285)
Υ_2	(0.7526, 0.2307)	(0.7448, 0.1929)	(0.6497, 0.3500)	(0.8312, 0.1977)	(0.6721, 0.3346)
Υ_3	(0.6438, 0.2626)	(0.6093, 0.3586)	(0.6568, 0.3156)	(0.6576, 0.2602)	(0.8877, 0.1537)
Υ_4	(0.7762, 0.1794)	(0.7087, 0.3453)	(0.7590, 0.1428)	(0.6438, 0.3346)	(0.6968, 0.3297)
Υ_5	(0.3490, 0.5858)	(0.1142, 0.8033)	(0.2324, 0.6177)	(0.2082, 0.7256)	(0.1593, 0.8712)

4.3 Validity Test

The final rank obtained by the FF-COCOSO decision algorithm is $\Upsilon_1 \succ \Upsilon_5 \succ \Upsilon_2 \succ \Upsilon_3 \succ \Upsilon_4$. To confirm the reliability of the FF-COCOSO decision algorithm, the following standards advanced by Wang and Triantaphyllou [46] are applied to test:

Standard 1: When the assessment opinion of non-optimal scheme is replaced by another worse assessment opinion, the best option will not change that proves the decision algorithm is valid.
Standard 2: An effective decision algorithm should obey the transitivity.
Standard 3: When the original decision problem is decomposed into several decision sub-problem, a valid decision approach will ensure that the new order is same with the initial rank.

Next, we conduct the validity analysis on the basis of the aforementioned standards. For the Standard 1, we shift the attribute value of Υ_5 under different criteria into its complement form, which is displayed in Table 10.

Afterward, we recompute this case and obtain the outcome signified as $\{2.5507, 2.4336, 2.3845, 2.3801, 1.4317\}$, the final rank denoted as $\Upsilon_1 \succ \Upsilon_5 \succ \Upsilon_2 \succ \Upsilon_4 \succ \Upsilon_3$, namely the optimal option is not change. Thus, the FF-COCOSO decision algorithm holds the standard 1. With respect to the standard 2 and standard 3, we divide the original into three sub-problems $\{\Upsilon_1, \Upsilon_2, \Upsilon_3\}$, $\{\Upsilon_2, \Upsilon_3, \Upsilon_4\}$ and $\{\Upsilon_3, \Upsilon_4, \Upsilon_5\}$. After computing the sub-problems and aggregated the three ranks, the total rank is same with the initial sorting, which means that the FF-COCOSO decision algorithm holds the standard 2 and standard 3.

4.4 Comparison Analysis

The goal of this part is to verify the significant of the designed FF-COCOSO MEMCDM method; the current subsection will conduct a comparison study and further highlight the superiority and rationally of the designed method. Next, we utilize the FFWA operator [16] and FFWG operator [16] to resolve the mentioned group decision problem. The decision result and the ultimate sorting relation of pondered enterprise are revealed as Table 11. From it, the rank outcomes obtained

Table 11 Comparison outcomes with different Fermatean fuzzy decision methods

Method	Υ_1	Υ_2	Υ_3	Υ_4	Υ_5	Sorting
FFWA operator [16]	0.4650	0.3947	0.3969	0.3603	0.4329	$\Upsilon_1 \succ \Upsilon_5 \succ \Upsilon_3 \succ \Upsilon_2 \succ \Upsilon_4$
FFWG operator [16]	0.4305	0.3672	0.3155	0.3401	0.3538	$\Upsilon_1 \succ \Upsilon_2 \succ \Upsilon_5 \succ \Upsilon_4 \succ \Upsilon_3$
FF-COCOSO	1.8920	1.8048	1.7721	1.7647	1.8207	$\Upsilon_1 \succ \Upsilon_5 \succ \Upsilon_2 \succ \Upsilon_3 \succ \Upsilon_4$

from the FF-COCOSO method are slightly different with FFWA operator and FFWG operator [16], but the best option is always the enterprise Υ_1. Aiming at the reason that produces different ranks, the FFWA operator and FFWG operator only ponder a single integration way to aggregation opinion; the proposed FF-COCOSO method can comprehensively take into account the WSM and WPM measures and utilize three appraisal strategies to obtain the ultimate score of alternative. Thus, the rank acquired through the proffered FF-COCOSO algorithm is more scientific and equitable than FFWA and FFWG operators. Accordingly, the FF-COCOSO method possesses a powerful universal and flexibility than the single aggregation approach.

5 Discussion

This research proffers an innovative group decision method based on COCOSO method and COV technique for dealing with the issues with unknown weight information of experts and criteria under the Fermatean fuzzy environment. We gave the detailed depiction of Fermatean fuzzy group COCOSO method on the basis of Fermatean fuzzy aggregation operators and improved score functions. The empirical of energy vehicle company investment is utilized to verify the feasibility of the propounded method. The effectiveness and applicability of the developed method are confirmed by the validity test and comparisons study, respectively. Based on the comparison analysis, we can find that the decision outcome obtained by the proposed method has a slight difference, but the optimal option is the same with the existing methods. The reason that produces this phenomenon is the proposed FF-COCOSO method acquires the final decision outcomes through aggregating multiple integration strategies and further derives the compromise solution of alternative. While the FFWA operator or FFWG operator only considers the weighted sum method or weighted product method to obtain the final raking value. Accordingly, the proposed FF-COCOSO method in this paper can deduce more robust and effective sorting value of alternatives in comparison with other Fermatean fuzzy decision approaches.

6 Conclusion

Since the FFS has a powerful ability in depicting uncertain and vague information and the COCOSO method possesses a stronger flexibility and robustness to solve the complex decision issues by comprehensively taking into account three different strategies, this essay propounds a FF-COCOSO MEMCDM method to cope with the decision issues with unknown weight information. The criteria weight information through utilizing coefficient of variation method is calculated from the point of objectivity of decision issues. The detailed decision procedure of the FF-COCOSO MEMCDM algorithm is given in view of the score function and FFWA operators. A case concerning the choice of investment companies is applied to validate the practicability and availability of the developed group decision algorithm.

However, there also exists some limitations in the propounded group decision approach. First of all, although we provide linguistic terms for experts as a reference to provide their assessment viewpoint, the number of linguistic terms cannot meet the requirement of experts; secondly, the presented method fails to consider the objective weight of experts in real decision issues and the subjective experience and cognition of experts are also ignored in the procedure of criteria weight determination. Besides, the presented method also fails to ponder the psychological factors of experts in the course of decision analysis especially for the risk decision problems.

In the future, we will continuously explore decision algorithms based on other linguistic information representation models, such as probabilistic linguistic term set, double hierarchy hesitant fuzzy linguistic term set, multi-granularity hesitant fuzzy linguistic term set, and so forth. In addition, the combination of classical decision methods and behavior decision theory will become an important focus to cope with risk decision issues. Furthermore, we will focus on the construction of large-scale intelligent decision algorithm based on the consensus measure [44, 45] feedback mechanism and their applications in emergency management and risk management.

Declarations

Conflict of Interest The authors declare that they have no conflict of interest.

Ethical Statement Articles do not rely on clinical trials.

Human and animal participants: All submitted manuscripts containing research which does not involve human participants and/or animal experimentation.

Acknowledgements The research was funded by the General Program of National Natural Science Foundation of China (No: 12071280).

References

1. Tzeng, G.H., Huang, J.J.: Multiple Attribute Decision Making: Methods and Applications, p. 1981. Springer-Verlag, New York (2011)

2. Liu, Y., Qin, Y., Liu, F., Rong, Y.: GIBWM-MABAC approach for MAGDM under multi-granularity intuitionistic 2-tuple linguistic information model. J. Ambient Intell. Humaniz. Comput., 1–17 (2021). https://doi.org/10.1007/s12652-021-03476-3
3. Liu, Y., Wei, G., Liu, H., Xu, L.: Group decision making for internet public opinion emergency based upon linguistic intuitionistic fuzzy information. Int. J. Mach. Learn. Cybern. 1–16. https://doi.org/10.1007/s13042-020-01262-9
4. Rong, Y., Liu, Y., Pei, Z.: Complex q-rung orthopair fuzzy 2-tuple linguistic Maclaurin symmetric mean operators and its application to emergency program selection. Int. J. Intell. Syst. **35**(11), 1749–1790 (2020)
5. Rong Y, Liu Y, Pei Z.: A novel multiple attribute decision-making approach for evaluation of emergency management schemes under picture fuzzy environment.Int. J. Mach. Learn. Cybern. (2021). https://doi.org/10.1007/s13042-021-01280-1
6. Mardani, A., Saraji, M.K., Mishra, A.R., Rani, P.: A novel extended approach under hesitant fuzzy sets to design a framework for assessing the key challenges of digital health interventions adoption during the COVID-19 outbreak. Appl. Soft Comput. **96**, 106613 (2020)
7. Saraji, M.K., Mardani, A., Köppen, M., Mishra, A.R., Rani, P.:An extended hesitant fuzzy set using SWARA-MULTIMOORA approach to adapt online education for the control of the pandemic spread of COVID-19 in higher education institutions. Artif. Intell. Rev. 1–26 (2021).https://doi.org/10.1007/s10462-021-10029-9
8. Zadeh, L.: Fuzzy sets. Inf. Control **8**(3), 338–353 (1965)
9. Atanssov, K.T.: Intuitionistic fuzzy sets. Fuzzy Sets Syst. **20**(1), 87–96 (1986)
10. Ejegwa, P.A.: Novel correlation coefficient for intuitionistic fuzzy sets and its application to multi-criteria decision-making problems. Int. J. Fuzzy Syst. Appl. **10**(2), 39–58 (2021)
11. Yager, R.R.: Pythagorean membership grades in multicriteria decision making. IEEE Trans. Fuzzy Syst. **22**(4), 958–965 (2013)
12. Yager, R.R.: Pythagorean fuzzy subsets. In: 2013 Joint IFSA World Congress and NAFIPS Annual Meeting (IFSA/NAFIPS), pp. 57–61. IEEE, June 2013
13. Ejegwa, P.A., Jana C., Some new weighted correlation coefficients between Pythagorean fuzzy sets and their applications, In: Garg, H. (eds.) Pythagorean Fuzzy Sets, pp. 39–64, Springer (2021). https://doi.org/10.1007/978-981-16-1989-2-2
14. Ejegwa, P.A., Feng, Y., Zhang, W.: Pattern recognition based on an improved Szmidt and Kacprzyk's correlation coefficient in Pythagorean fuzzy environment, In: Min, H., Sitian, Q., Nian, Z. (eds.) Advances in Neural Networks—17th International Symposium on Neural Networks (ISNN 2020), Lecture Notes in Computer Science (LNCS) 12557, pp. 190–206. Springer (2020). https://doi.org/10.1007/978-3-030-64221-1-17
15. Senapati, T., Yager, R.R.: Fermatean fuzzy sets. J. Ambient Intell. Humaniz. Comput. **11**(2), 663–674 (2020)
16. Senapati, T., Yager, R.R.: Fermatean fuzzy weighted averaging/geometric operators and its application in multi-criteria decision-making methods. Eng. Appl. Artif. Intell. **85**, 112–121 (2019)
17. Senapati, T., Yager, R.R.: Some new operations over Fermatean fuzzy numbers and application of Fermatean fuzzy WPM in multiple criteria decision making. Informatica **30**(2), 391–412 (2019)
18. Garg, H., Shahzadi, G., Akram, M.: Decision-making analysis based on Fermatean fuzzy Yager aggregation operators with application in COVID-19 testing facility. Math. Prob. Eng. **2020** (2020). https://doi.org/10.1155/2020/7279027
19. Shahzadi, G., Muhiuddin, G., Arif Butt, M., Ashraf, A.: Hamacher interactive hybrid weighted averaging operators under Fermatean fuzzy numbers. J. Math. **2021** (2021). https://doi.org/10.1155/2021/5556017
20. Shahzadi, G., Zafar, F., Alghamdi, M.A.: Multiple-attribute decision-making using Fermatean fuzzy Hamacher interactive geometric operators. Math. Prob. Eng. **2021** (2021). https://doi.org/10.1155/2021/5150933
21. Shit, C., Ghorai, G.: Multiple attribute decision-making based on different types of Dombi aggregation operators under Fermatean fuzzy information. Soft. Comput. **25**(22), 13869–13880 (2021)

22. Ejegwa, P.A., Nwankwo, K.N., Ahmad, M., Ghazal, T.M., Khan, M.A.: Composite relation under Fermatean fuzzy context and its application in disease diagnosis. Informatica **32**(10), 87–101 (2021)
23. Liu, D., Liu, Y., Chen, X.: Fermatean fuzzy linguistic set and its application in multicriteria decision making. Int. J. Intell. Syst. **34**(5), 878–894 (2019)
24. Liu, D., Liu, Y., Wang, L.: Distance measure for Fermatean fuzzy linguistic term sets based on linguistic scale function: An illustration of the TODIM and TOPSIS methods. Int. J. Intell. Syst. **34**(11), 2807–2834 (2019)
25. Jeevaraj, S.: Ordering of interval-valued Fermatean fuzzy sets and its applications. Expert Syst. Appl. **185** (2021). https://doi.org/10.1016/j.eswa.2021.115613
26. Keshavarz-Ghorabaee, M., Amiri, M., Hashemi-Tabatabaei, M., Zavadskas, E.K., Kaklauskas, A.: A new decision-making approach based on Fermatean fuzzy sets and WASPAS for green construction supplier evaluation. Mathematics **8**, 2202 (2020). https://doi.org/10.3390/math8122202
27. Mishra, A.R., Rani, P.: Multi-criteria healthcare waste disposal location selection based on Fermatean fuzzy WASPAS method. Complex Intell. Syst. **7**(5), 2469–2484 (2021)
28. Gul, M., Lo, H.-W., Yucesan, M.: Fermatean fuzzy TOPSIS-based approach for occupational risk assessment in manufacturing. Complex Intell. Syst. **7**(5), 2635–2653 (2021)
29. Aydemir, S.B., Gunduz, S.Y.: Fermatean fuzzy TOPSIS method with Dombi aggregation operators and its application in multi-criteria decision making. J. Intell. Fuzzy Syst. **39**(1), 851–869 (2020)
30. Gul, S.: Fermatean fuzzy set extensions of SAW, ARAS, and VIKOR with applications in COVID-19 testing laboratory selection problem. Expert Syst. **38**(8), e12769 (2021). https://doi.org/10.1111/exsy.12769
31. Mishra, A.R., Rani, P., Pandey, K.: Fermatean fuzzy CRITIC-EDAS approach for the selection of sustainable third-party reverse logistics providers using improved generalized score function. J. Ambient Intell. Humaniz. Comput. (2021). https://doi.org/10.1007/s12652-021-02902-w
32. Rani, P., Mishra, A.R.: Fermatean fuzzy Einstein aggregation operators-based MULTIMOORA method for electric vehicle charging station selection. Expert Syst. Appl. **182**, 115267 (2021). https://doi.org/10.1016/j.eswa.2021.115267
33. Kamali Saraji, M., Streimikiene, D., Kyriakopoulos, G.L.: Fermatean fuzzy CRITIC-COPRAS method for evaluating the challenges to industry 4.0 adoption for a sustainable digital transformation. Sustainability **13**(17), 9577 (2021)
34. Deng, Z., Wang, J.: Evidential Fermatean fuzzy multicriteria decision-making based on Fermatean fuzzy entropy. Int. J. Intell. Syst. **36**(10), 5866–5886 (2021)
35. Yazdani, M., Zarate, P., Zavadskas, E.K., Turskis, Z.: A combined compromise solution (CoCoSo) method for multi-criteria decision-making problems. Manag. Decis. **57**(9), 2501–2519 (2019)
36. Yazdani, M., Wen, Z., Liao, H., Banaitis, A., Turskis, Z.: A grey combined compromise solution (COCOSO-G) method for supplier selection in construction management. J. Civ. Eng. Manag. **25**(8), 858–874 (2019)
37. Peng, X., Zhang, X., Luo, Z.: Pythagorean fuzzy MCDM method based on CoCoSo and CRITIC with score function for 5G industry evaluation. Artif. Intell. Rev. **53**(5), 3813–3847 (2020)
38. Ecer, F., Pamucar, D.: Sustainable supplier selection: A novel integrated fuzzy best worst method (F-BWM) and fuzzy CoCoSo with Bonferroni (CoCoSo'B) multi-criteria model. J. Cleaner Prod. **266**, 121981 (2020). https://doi.org/10.1016/j.jclepro.2020.121981
39. Rani, P., Ali, J., Krishankumar, R., Mishra, A.R., Cavallaro, F., Ravichandran, K.S.: An integrated single-valued neutrosophic combined compromise solution methodology for renewable energy resource selection problem. Energies **14**(15), 4594 (2021). https://doi.org/10.3390/en14154594
40. Yazdani, M., Chatterjee, P., Pamucar, D., Chakraborty, S.: Development of an integrated decision making model for location selection of logistics centers in the Spanish autonomous communities. Expert Syst. Appl. **148**, 113208 (2020). https://doi.org/10.1016/j.eswa.2020.113208

41. Rani, P., Mishra, A.R., Saha, A., Hezam, I.M., Pamucar, D.: Fermatean fuzzy Heronian mean operators and MEREC-based additive ratio assessment method: An application to food waste treatment technology selection. Int. J. Intell. Syst. 1–36 (2021). https://doi.org/10.1002/int.22787
42. Mishra, A.R., Rani, P., Pandey, K., et al.: Novel multi-criteria intuitionistic fuzzy SWARA-COPRAS approach for sustainability evaluation of the bioenergy production process. Sustainability **12**(10), 4155 (2020)
43. Roubens, M.: Fuzzy sets and decision analysis. Fuzzy Sets Syst. **90**, 199–206 (1997)
44. Liu, B.S., Zhou, Q., Ding, R.X., Palomares, I., Herrera, F.: Large-scale group decision making model based on social network analysis: Trust relationship-based conflict detection and elimination. Eur. J. Oper. Res. **275**(2), 737–754 (2019). https://doi.org/10.1016/j.ejor.2018.11.075
45. Zhang, Y., Chen, X., Gao, L., Dong, Y., Pedryczc, W.: Consensus reaching with trust evolution in social network group decision making. Expert Syst. Appl. (2021). https://doi.org/10.1016/j.eswa.2021.116022
46. Wang, X., Triantaphyllou, E.: Ranking irregularities when evaluating alternatives by using some ELECTRE methods. Omega **36**(1), 45–63 (2008)

Multi-criteria Decision Analysis and Fuzzy-Decision-Making Trial and Evaluation Laboratory (MCDA and F-DEMATEL) Method for Flipped and Sustainable Mathematics Teaching as a Real-Life Application

Jin Su Jeong, David González-Gómez, and Félix Yllana-Prieto

Abstract Multi-criteria decision analysis (MCDA) method as an instrument has been applied in different fields. Deciding and recognizing the appropriate conditions is important due to adjust the present educational changes, and it is also significant in mathematics education. However, this idea is rarely used in academic discipline. This study tried to examine and categorize proper conditions for mathematics teaching as a real-life utilization with a multi-criteria decision analysis and fuzzy-decision-making trial and evaluation laboratory (MCDA and F-DEMATEL) method. Specifically, adjusting to present educational changes, it is about flipped classroom combined with sustainable development and information, communication, and technology (ICT) in mathematics teaching. In this case, the principal criteria and sub-criteria were carefully chosen after the discussion of an association of specialists as decision-makers with the MCDA and F-DEMATEL methods. Later, the final analyses' findings were measured with simple additive weighting (SAW) throughout their weight coefficients' result and susceptibility study with the chosen criteria and sub-criteria. The findings showed that the most significant criteria and sub-criteria for variation to present educational changes had been revealed to be based on the above-cited evaluation methodologies. Thus, this approach could be used to certify the most important flipped and sustainable mathematics teaching for adjusting to present educational changes with resultant learning environments in more long-term education. Hence, it could be related to decision-making defects in private sectors, at government sectors and more educational areas, which continually requires constant actualization of knowledge and distribution.

J. S. Jeong (✉) · D. González-Gómez · F. Yllana-Prieto
Departamento de Didáctica de las Ciencias Experimentales y Matemáticas, Universidad de Extremadura, Cáceres, Spain
e-mail: jin@unex.es

D. González-Gómez
e-mail: dggomez@unex.es

F. Yllana-Prieto
e-mail: feyllanap@unex.es

L. Sahoo et al. (eds.), *Real Life Applications of Multiple Criteria Decision Making Techniques in Fuzzy Domain*, Studies in Fuzziness and Soft Computing 420,
https://doi.org/10.1007/978-981-19-4929-6_5

Keywords MCDA · Mathematics teaching · Fuzzy-logic · STEM · Sustainability · Flipped classroom

1 Introduction

1.1 The MCDA and F-DEMATEL Method

As an important approach, multi-criteria decision analysis (MCDA) method has been applied in different and several fields combined with analytic hierarchy process (AHP) [12, 25, 42]. Here, it stimulates different decision-makings by specific descriptions with the discrepancy of further decision-makers [29, 53]. Therefore, fuzzy-decision-making trial and evaluation laboratory (F-DEMATEL) system is leveraged [7, 22, 27]. Several characteristics and elements of information, communication, and technology (ICT) combined implementations employing the MCDA and F-DEMATEL have been evaluated to demonstrate complete rates and practices that were comparable by ICTs and traditional methods [11, 49, 61]. A new hybrid model demonstrated with fuzzy-MCDA that the F-DMATEL method was utilized for enhancing a major and efficient association for a system [32, 57]. The MCDA and F-DEMATEL method could evaluate several aspects of flipped and ICTs methods connected with programs to specify complete requirements and experiences that related to both ICTs and traditional methods [7, 22, 33]. In the context of mathematics teaching evaluation, with the pronounced features of this working technique, it is necessary to recognize and select the appropriate criteria because of the adaptation of existing educational changes even though its concept is rarely used and implemented into the academic area [27]. This study tried to categorize and examine suitable criteria for mathematics teaching by a hybrid system.

1.2 Flipped Mathematics Classroom and ICTs

Considering the modifications and changes aforesaid, flipped classroom as an instructional methodology in higher education including mathematics courses causes enhancement in students' learning results [21, 62]. In this context, ICTs could be studied as a central idea to assist the flipped classroom due to have relevant and adaptable student-centered teaching [28, 35, 56]. During last sixty years, ICT's application with constant increase had had important effects in science, technology, engineering, and mathematics (STEM) education [19, 23, 47]. The ICTs support the flipped classroom methodology in virtual education platforms, in which multi-aspect networks can occurr, like instructor to instructor, instructor to learner, and learner to learner [17, 51]. However, further investigation requires to be completed on the capability and competence of specific flipped models with ICTs, even if they are currently being employed [20, 37]. Various recent investigations in postsecondary education have defined the complete work and study of mathematics teaching for sustainable development using flipped e-learning methods [2, 4, 37]. According to Eneroth [13] and McVey [45], combining ICTs with sustainable mathematics teaching that has been affected to flipped e-learning methods that can connect the gap of present education is a challenge [10, 18, 67]. Despite of these reasons and the lack of information and knowledge, this study tried to categorize and analyze the criteria and sub-criteria of flipped and sustainable mathematics teaching with the support of ICTs that adjust present educational changes.

1.3 Pedagogical Adaptation of Sustainable Mathematics Teaching

The European Commission reports that the interest in learning of science and mathematics for elementary and secondary school is decreasing [14]. Here, due to the improvement of the information society, it is essential to increase request for these findings on higher-level learning [14, 27, 59]. Therefore, it exists an attention's increasing of sustainable development in education to supply a long-term teaching course stage as the objectives and practices of sustainability of United Nations Educational, Scientific and Cultural Organization (UNESCO), which manages the Decade of Education for Sustainable Development (DESD) in the United Nations (UN) [50, 54, 60]. In these circumstances, students still have obstacles in mathematics education [15, 24, 58], despite of several authors have recommended new learning approaches for mathematics education and its effectiveness increase adjusting present educational changes [6, 24, 44]. In the construction of a sustainable mathematics teaching adjusting present educational changes, it can be realized and recognized that the topic and procedure of mathematics are important in mathematical constructions, general and detailed educational contents, and focused curricular information [3, 28]. Nevertheless, because mathematics education has a vital function in understanding sustainable mathematics education in a university background is still in its early steps, and an infrequent application existence in several colleges, however the sustainable education has achieved numerous purposes in changing humanities by teaching decision-makers [14, 41, 48]. In these difficult conditions, mathematics education across one's life cycle can produce a chance to supply an educational place for flipped classroom and ICTs with the appropriate categories of criteria (see Fig. 1).

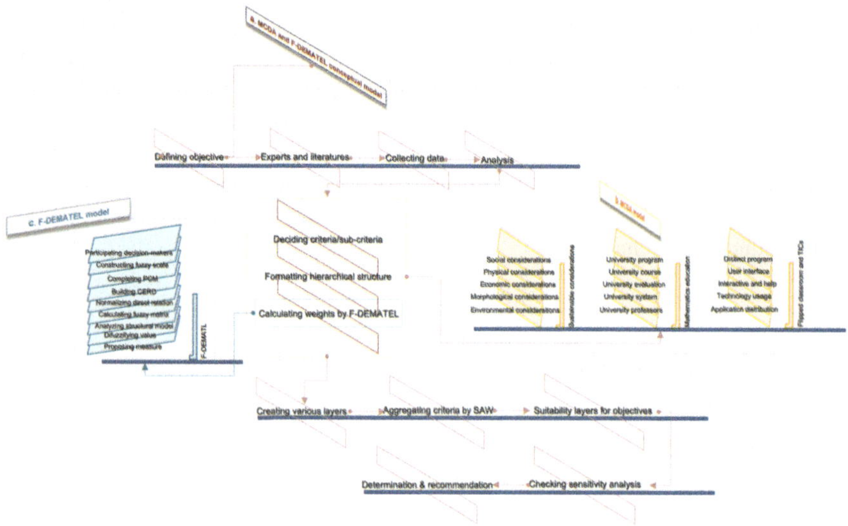

Fig. 1 Multi-criteria decision analysis and fuzzy-decision-making trial and evaluation laboratory (MCDA and F-DEMATEL) theoretical model

2 Methodology

2.1 Study Sample

The study sample was carried out with an initial and compulsory course called "Mathematics and its Didactics". This course belongs to the bachelor's degree in Primary Education in the Universities of Spain. It contains different contents, topics, and ideas for didactics of arithmetic in the context of mathematics education (see Table 1). The course included basic aspects of mathematics education, education skills and technologies practice, mathematics history, and various thoughts and concepts of mathematics. Thus, this course was aiming to instruct the university students about general knowledge of mathematics topics and learning disciplines and study to teach them for primary educations learners.

2.2 Criteria and Sub-criteria Choice Procedure with MCDA and F-DEMATEL

The initial stage of MCDA and F-DEMATEL consists of the appropriate criteria and sub-criteria, which could have had directly influence on the potential criteria evaluation in flipped and sustainable mathematics education for adjusting present educational changes. In Table 2, all the criteria and sub-criteria for this research are

Table 1 Course context for similar case study employed

Chapter title	Chapter description
1. Presentation to mathematics education	Overall contents of mathematics and its education Mathematical instruction of PSTs Didactics of arithmetic in the primary curriculum
2. Natural numbers and enumeration methods	Mathematical science Number notion and coordinative groups Locational enumeration methods Decimal enumeration Information to help in the study of enumeration methods
3. Arithmetic operations	Significance and process of operations Addition and subtraction without return Multiplications without reorganizing Divisions using one and various figures Potentiation and organizing Problem solving Abacuses and multi-base blocks like educational tools
4. Divisibility	Composite and prime numbers Number's divisors Least common multiple and greatest common divisor Divisibility standards
5. Fractions and the decimal numbers	Non-integer sets' numerical expression Fractions like sections of the unit Contrast of calculations Fractional and decimal numbers' operations Educational materials and resources
6. Integer numbers	Positive and negative numbers Integral operations

defined by the authors as a decision-makers with the support of a experts' group. Here, the AHP (an MCDA procedure) was employed to obtain the criteria' and sub-criteria' relative significance loads and to make the assessment approach, as it is completed in certain troubles of decision-making [53, 57].

The decision-makers' defined pair-wise comparison method (PCM) is based on the relevance of every criterion because of specific predetermined rates of scale, 1–9 and 1–1/9 (see Table 3) [66]. The elements $a_{ii} = 1$ and $a_{ij} = 1/a_{ji}$ were analyzed in a nine-point scale in this research. Regarding to the consistency ratio (CR) matrix, every criterion and sub-criterion load with the PCM could be assigned. Specifically, according to Balana et al. [5] and Perpina et al. [52], a CR weight can be organized as to be appropriate only on the condition that the weight is less than 10% in Eq. (1). The decision-makers would change and adjust their weights of conclusions if the condition is not complied.

$$\mathrm{CR} = \frac{\mathrm{CI}}{\mathrm{RI}} \tag{1}$$

Table 2 Criteria and sub-criteria calculated for flipped and sustainable mathematics teaching for adjusting present educational changes

Criteria	Sub-criteria	Sub-criteria framework	Loads
Sustainable consideration	Social considerations	Social considerations include the structure and description in flipped and sustainable mathematics teaching that adjusts to present educational changes that relate to the suggested objectives	0.09
	Physical considerations	Physical considerations include the structure and description in flipped and sustainable mathematics teaching that adjusts to present educational changes that are relate to the suggested objectives	0.28
	Economic considerations	Economic considerations include the structure and description in flipped and sustainable mathematics teaching that adjusts to present educational changes that relate to the suggested objectives	0.05
	Morphological considerations	Morphological considerations include the structure and description in flipped and sustainable mathematics teaching that adjusts to present educational changes that relate to the suggested objectives	0.15
	Environmental considerations	Environmental considerations include the structure and description in flipped and sustainable mathematics teaching that adjusts to present educational changes that relate to the suggested objectives	0.43
Validation	$\lambda_{max} = 5.18$; CI = 0.045; CR = 0.04 < 0.1		
Mathematics education	University program considerations	University program considerations include the structure and description in flipped and sustainable mathematics teaching that adjusts to present educational changes that relate to the suggested objectives	0.12
	University course considerations	University course considerations include the structure and description in flipped and sustainable mathematics teaching that adjusts to present educational changes that relate to the suggested objectives	0.28
	University course evaluations	University course evaluations include the structure and description in flipped and sustainable mathematics teaching that adjusts to present educational changes that relate to the suggested objectives	0.09
	University system updates	University system updates include the structure and description in flipped and sustainable mathematics teaching that adjusts to present educational changes that relate to the suggested objectives	0.06

(continued)

Table 2 (continued)

Criteria	Sub-criteria	Sub-criteria framework	Loads
	University professors	University professors include the structure and description in flipped and sustainable mathematics teaching that adjusts to present educational changes that relate to the suggested objectives	0.45
Validation	$\lambda_{max} = 5.09$; CI = 0.023; CR = 0.02 < 0.1		
Flipped classroom and TICs	Distinct program considerations	Distinct program considerations include the structure and description in flipped and sustainable mathematics teaching that adjusts to present educational changes that relate to the suggested objectives	0.05
	User interface considerations	User interface considerations include the structure and description in flipped and sustainable mathematics teaching that adjusts to present educational changes that relate to the suggested objectives	0.17
	Interactivity and help considerations	Interactivity and help considerations include the structure and description in flipped and sustainable mathematics teaching that adjusts to present educational changes that relate to the suggested objectives	0.07
	Technology usage considerations	Technology usage considerations include the structure and description in flipped and sustainable mathematics teaching that adjusts to present educational changes that relate to the suggested objectives	0.43
	Application distribution considerations	Application distribution considerations include the structure and description in flipped and sustainable mathematics teaching that adjusts to present educational changes that relate to the suggested objectives	0.28
Validation	$\lambda_{max} = 5.19$; CI = 0.048; CR = 0.04 < 0.1		

Moreover, in pair-wise comparison in this equation, the consistency index (CI) can measure the divergences and mean of the CI evaluations that can generate the random index (RI) [43, 53]. Values smaller than 10% of CR are acceptable for its usage, but, if it is superior to 10%, decision-makers think about adjusting their assessment [32, 34].

$$CI = \frac{\lambda_{max} - n}{n - 1} \tag{2}$$

Table 3 Pair-wise comparison of criteria rates and their mathematical ratios [66]

Less relevant numerical value	Category	More relevant numerical value
1/9	More or less immense predilection or significance	9
1/8	More or less very to immensely intense predilection or significance	8
1/7	More or less very intense predilection or significance	7
1/6	More or less intense to very intense predilection or significance	6
1/5	More or less intense predilection or significance	5
1/4	More or less acceptable to intense predilection or significance	4
1/3	More or less acceptable predilection or significance	3
1/2	More or less similar to normal predilection or significance	2
1	Similar predilection or significance	1

2.3 MCDA and F-DEMATEL Method

The F-DEMATEL method based on criteria and sub-criteria chosen is the second stage of MCDA and F-DEMATEL. This is an outranking technique that can use and calculate criteria and sub-criteria coefficients to locate proper choice for flipped and sustainable teachings for modifying present educational changes as a real-world application in the AHP technique environment. In this context, a fuzzy logic that is a strong method that resolves decision-making's uncertainty, which is employed to control and standardize the data of the criteria and sub-criteria in real-life decision-making difficulties (see Fig. 2) [34, 39].

Thus, fuzzy-logic values are important in the process of decision-makings. In Eq. (3), a fuzzified Likert scale was applied for growing criteria of the average matrix [63, 65]. As indicated in Fig. 3, a triangular fuzzy number (TFN) can be presented with the fuzzy-ratings and fuzzy-memberships.

$$P = \left[p_{ij}\right]c_i \times c_i \tag{3}$$

Specifically, a triplet (*l*, *m*, *u*) is a section of a TFN that the component *l* is lower value, the component *m* is medium value, and the component *u* is upper value, which shows a fuzzy set ($x \leq y \leq z$) [1, 65]. The TFN fuzzy-membership opportunity can be defined as Eq. (4).

$$\mu(x) = \begin{cases} 0, & x < l \\ \frac{x-l}{m-l}, & l \leq x \leq m \\ \frac{u-x}{u-m}, & m \leq x \leq u \\ 0, & x \geq u \end{cases} \tag{4}$$

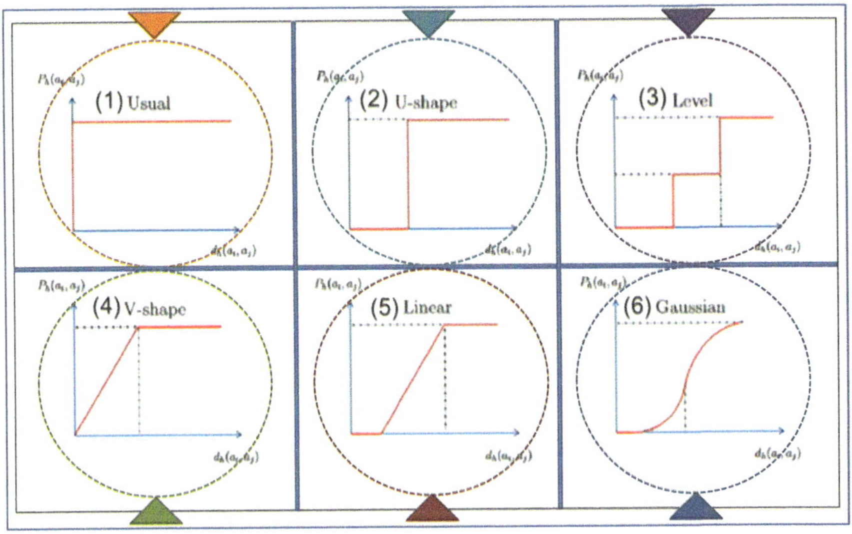

Fig. 2 Several roles that can be applied to increase the criteria and sub-criteria [34]

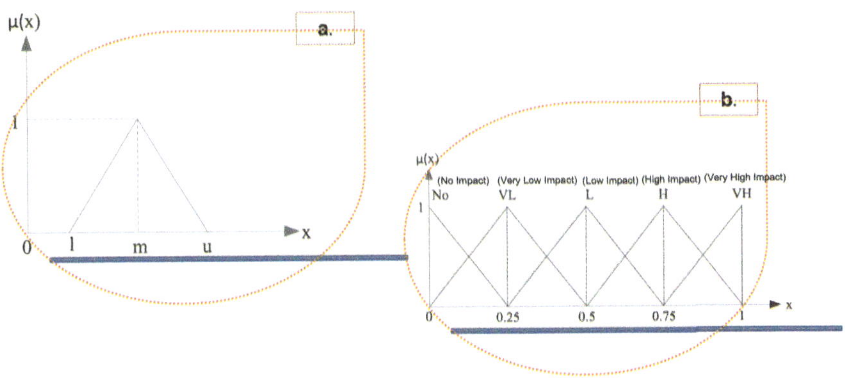

Fig. 3 Fuzzy-logic of TFNs and fuzzy-ratings and fuzzy-memberships function [1, 65]

Hence, the DEMATEL method (the Geneva Battelle Memorial Institute) can assess whole and complex challenges of decision-makings [16]. Lin and Tzeng [39] distinguished that the DEMATEL method is normally an extensive and pragmatic method for assessing the connection between various criteria' cause and effect. The proposed technique based on graph scheme and theory can examine and designate complications through the visualization [1, 38]. In addition, to expose the fuzziness and realize the fuzzy combined DEMATEL, main advantages are determined in terms of the fuzziness circumstance [63]. With this procedure, the associated interconnection can be achieved with the criteria and their robust outcome [22]. To obtain suitable criteria and sub-criteria for flipped and sustainable mathematics teaching for adjusting present educational changes as a real-life application, the fuzzy-logic and the DEMATEL methods are combined. At this point, based on immediate standard interactions, the criteria matrix should be constructed that the decision-makers provided the weighting structure and path and designed the immediate standard associations' matrix in this work. For the groupings of their weights, in Eq. (5), p_{ij}^{e} can show that e implies the decision-maker's predilection and k implies decision-makers' the overall number.

$$p_{ij} = \left(p_{ij}^{(l)}, p_{ij}^{(m)}, p_{ij}^{(u)}\right) = \begin{Bmatrix} p_{ij}^{(l)} = \min\left(p_{ij}^{e}\right) \\ p_{ij}^{(m)} = \sqrt[k]{\prod_{i=1}^{k} p_{ij}^{e}} \\ p_{ij}^{(u)} = \max\left(p_{ij}^{e}\right) \end{Bmatrix} \tag{5}$$

Because of the fuzziness condition, essential advantages are revealing the fuzziness and applying the fuzzy unified DEMATEL, F-DEMATEL [32, 63]. Here, with the TFNs, the cause–effect criteria can create a dialectology version parameterized with the DEMATEL technique seeing these advantages. The F-DEMATEL system can be employed to construct a verdict for organizations facing troubles in a fuzzy-logic background [64]. As indicated in Fig. 1c, the F-DEMATL diagram can contain basic stages suggested [8, 40] and can be employed to achieve the next stages and reach results as shown in Fig. 1b [8]. Equation (6) illustrates a cause-effect relationship diagram (CERD), which is these rates' matrix that is the weight coefficients of uncontrollable criteria.

$$w_i = \sqrt{(D_i + R_i)^2 + (D_i - R_i)^2} \tag{6}$$

Equation (7) shows that the criteria weight coefficients are uniform, and the regular primary effects' matrix is taken.

$$w_i = \frac{w_i}{\sum_{i=1}^{n} w_i} \tag{7}$$

The criteria matrix weight coefficients that criteria and sub-criteria can be defuzzified by process of quotations aforesaid are explained by Eq. (8).

$$A = \left(a^{(l)} + 4a^{(m)} + a^{(u)}\right) \cdot 6^{-1} \tag{8}$$

2.4 SAW and Susceptibility Study for Authentication Procedure

In the third phase of MCDA and F-DEMATEL, it is to apply the SAW technique that can blend the uniform criteria rates to obtain the most appropriate criteria in flipped and sustainable mathematics teaching for adjusting present educational changes. According to Jeong and González-Gómez [26], the SAW as a relatively modest analytical application can be utilized in various MCDA circumstances. Therefore, Eq. (9) can produce the indicator scale of appropriateness (0–1) on multi-criteria troubles that can explain where the most appropriate criteria is for the matter suggested [30].

$$SI_i = \sum_{j=1}^{n} w_i x_{ij} \tag{9}$$

The fourth and final phase of MCDA and F-DEMATEL is to apply a susceptibility study, which is an option to validate the firmness result for the answer data [36, 46]. When the decision-makers give the contributions, a susceptibility study was utilized for flipped and sustainable mathematics teaching that adjusts to present educational changes [31, 55]. The results of the susceptibility study can be shown with five different approaches (see Table 4).

By incorporating various criteria and sub-criteria weights, a susceptibility study could be achieved (see Table 5). This analysis caused five distinct situations. Hence, it was plausible to describe different association groups with the combined criteria by properly choosing a weighting vector. As demonstrated in Eq. (10), a particular

Table 4 Appropriateness catalogue of flipped and sustainable mathematics teaching for adjusting present educational changes

Appropriateness catalogue	Range	Appropriateness description
Na, Not appropriate	0.00–0.20	Appropriateness measure of flipped and sustainable mathematics teaching for adjusting present educational changes does not appear
Sa, Slightly appropriate	0.20–0.40	Appropriateness measure of flipped and sustainable mathematics teaching for adjusting present educational changes is low
Ma, Moderately appropriate	0.40–0.60	Appropriateness measure of flipped and sustainable mathematics teaching for adjusting present educational changes is moderate
A, Appropriate	0.60–0.80	Appropriateness measure of flipped and sustainable mathematics teaching for adjusting present educational changes is high
Ha, Highly appropriate	0.80–1.00	Appropriateness measure of flipped and sustainable mathematics teaching for adjusting present educational changes is very high

Table 5 Susceptibility study blend of three criteria and equivalent weight to criteria along with decisive choice made

Importance to criteria approaches	Criteria weights		
	Sustainable consideration	Mathematics education	Flipped classroom and TICs
IS, Importance to sustainable consideration (I)	0.50	0.25	0.25
IM, Importance to mathematics education (II)	0.25	0.50	0.25
IF, Importance to flipped classroom and TICs (III)	0.25	0.25	0.50
IW, Identical weights to criteria (IV)	0.33	0.33	0.33
ID, Importance to decision-makers' choice (V)	0.14	0.21	0.65

percent change (PC) average can describe the susceptibility assessment; a criterion weight (w_i) can be taken [9, 55]):

$$w_i = w_{i\mp} w_{i0} \times \text{PC} \tag{10}$$

In Eq. (11), the principal criterion weight w_{i0} is different, and the additional criterion weight w_j is changed equitably due to w_i.

$$w_j = (1 - w_i) \times \frac{w_{j0}}{1 - w_{i0}} \tag{11}$$

3 Case Study and Solution

3.1 MCDA and F-DEMATEL Results to Flipped and Sustainable Mathematics Teaching Adjusting Present Educational Changes

Figure 4 shows the diagram example of fifteen sub-criteria created on 0–1 scale with the standardized rates as index appropriateness. Here, it describes the criteria studied with cautions in Table 2, which have significant assets on this evaluation. Specifically, in the weighting procedure, the decision-makers are intricated. Then, the three criteria, sustainable consideration, mathematics education, and flipped classroom and TICs, and fifteen sub-criteria: (1) social considerations, (2) physical considerations, (3) economic considerations, (4) morphological considerations, (5) environmental considerations, (6) university program considerations, (7) university course considerations, (8) university course evaluations, (9) university system updates, (10) university professors, (11) distinct program considerations, (12) user interface considerations, (13) interactivity and help considerations, (14) technology usage considerations, and (15) application distribution considerations. The sustainable consideration criteria are 1–4, the mathematics education criteria are 5–8, and the flipped classroom and TICs criteria are 9–12.

The susceptibility study's results beside the results beyond the use of SAW technique can be seen in Table 5. Figure 5 indicates four distinct (I)–(IV) susceptibility study approaches and the final (V), "ID" approach established in Table 5. The fundamental criteria for flipped and sustainable mathematics teaching that adjusts to present educational changes were recognized in the usual rating scale 0–1, i.e., the higher the rate was, the more significant its criteria and sub-criteria. Regarding to the five distinct situations I–V, it could be nominated as: V was the most significant criteria established on the decision-makers' weightings and rankings, which were refined from expert discussion and exhaustive bibliographical review (0.14 for the sustainable consideration criteria, 0.21 for mathematics education criteria, and 0.65 for flipped classroom and TICs criteria); IV had an equivalent weight for all criteria sets (all 0.33); I was the impact and importance of the sustainable consideration

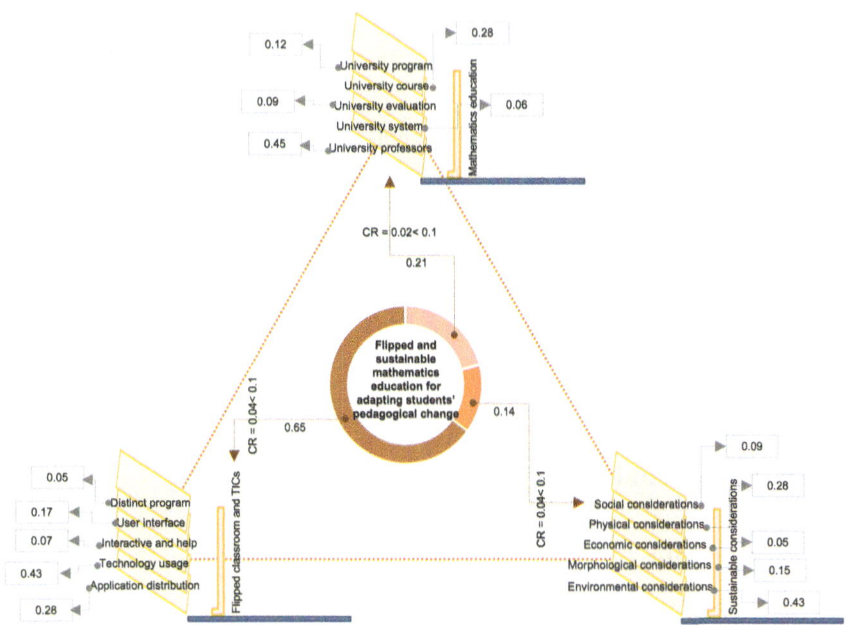

Fig. 4 Findings of a flipped and sustainable mathematic teaching system that adjusts to present educational changes

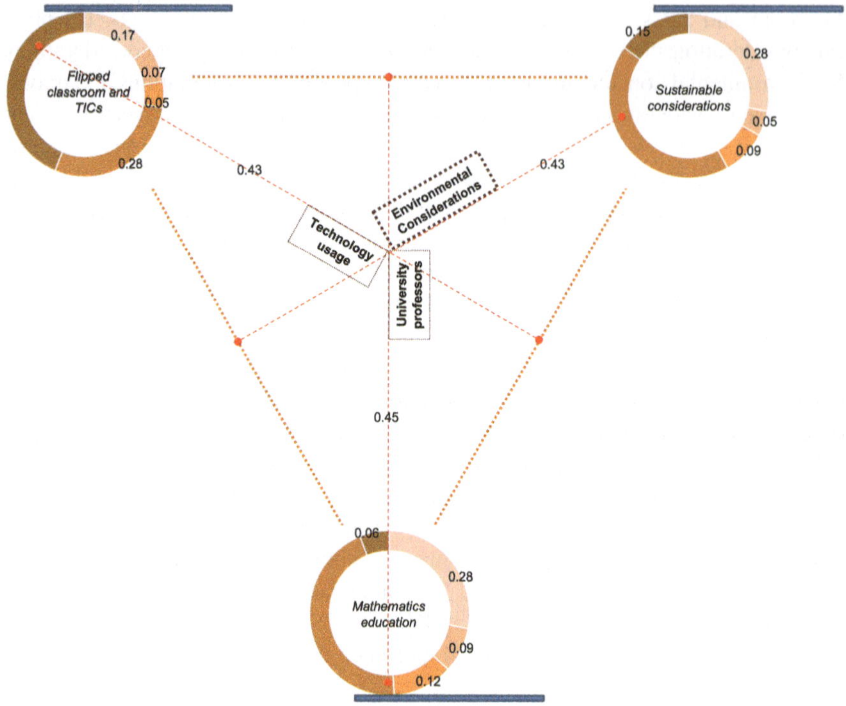

Fig. 5 System configuration of the criteria for flipped and sustainable mathematics teaching for adjusting present educational changes as a real-life application

criteria (0.50 for sustainable consideration criteria, 0.25 for mathematics education criteria, and 0.25 for flipped classroom and TICs criteria); II was the impact and importance of the mathematics education criteria (0.25 for sustainable consideration criteria, 0.50 for mathematics education criteria, and 0.25 for flipped classroom and TICs criteria); III was the impact and importance of flipped classroom and TICs criteria (0.25 for sustainable consideration criteria, 0.25 for mathematics education criteria, and 0.50 for flipped classroom and TICs criteria). All of this is explained in Table 5.

4 Discussion and Conclusion

4.1 Discussion

Because of these incentives and the lack of bibliography and material, this research wanted to categorize and study the criteria and sub-criteria of the schemes of flipped and sustainable mathematics teaching that adjust to present educational changes.

Therefore, the discovered outcomes in mathematics teaching for flipped and sustainable development demonstrated exclusive data about the important criteria and sub-criteria collection of diverse plausible influences that adjusted to present educational changes. This work builds a singular decision-support scheme for flipped and sustainable mathematics teaching and several applications with decision-makers' help. It achieved multi-criteria analyses' functions for decision-making techniques in flipped and sustainable mathematics teaching established on information collected from impartial decision-makers. Hence, this method gives a more unified and general approach that can be supplied by the suggested technique as to decision-making different to techniques that are employing currently. Moreover, a place of decision-making can be covered by the system for improving flipped and sustainable mathematics teaching for adjusting present educational changes based on the ideas and motivation of decision-makers. This approach can improve the design and development of this field through an advanced technique and can be re-applied to different fields.

4.2 Conclusion

The incorporation and organization of MCDA and F-DEMATEL method such as important instrument and strategy can response the ideal and satisfactory criteria and sub-criteria of flipped and sustainable mathematic teaching in the framework of present educational changes as a real-life application. This research describes a combined methodology of MCDA and F-DEMATEL to choose the weight coefficients of criteria and sub-criteria of decision-makers. The SAW technique and a susceptibility study are employed (scale of 0–1) to verify the consistency of this method.

This technique was employed to an obligatory initial mathematics course of the bachelor's degree in Primary Education. Through the MCDA and F-DEMATEL technique, the authors as a decision-maker with the discussion and an exhaustive bibliography review to categorize three criteria—sustainable consideration, mathematics education, and flipped classroom and TICs—and fifteen sub-criteria—(1) social considerations, (2) physical considerations, (3) economic considerations, (4) morphological considerations, (5) environmental considerations, (6) university program considerations, (7) university course considerations, (8) university course evaluations, (9) university system updates, (10) university professors, (11) distinct program considerations, (12) user interface considerations, (13) interactivity and help considerations, (14) technology usage considerations, and (15) application distribution considerations. Before, with SAW and susceptibility study, the most appropriate criteria and sub-criteria were chosen as the flipped classroom and TICs (situation V: 0.65 out of 0–1 in the utilized index suitability scale) and the technology usage considerations (0.43 out of 0–1 in the utilized index appropriateness scale), respectively. Hence, for efficient flipped and sustainable mathematic teaching, the situations were discovered to be viable for adjusting present educational changes in a more long-term education course schedule.

To evaluate the potential for locating best and advantageous criteria and sub-criteria for flipped and sustainable mathematics education for adjusting present educational changes with the MCDA and F-DEMATEL, it essentially differentiates suitable criteria in the scale of 0–1. Thus, this approach can be applied in analyses to prove the most significant flipped and sustainable mathematics teaching criteria that adjust to present educational changes with parallel education environments and accessible, required information. Particularly, it could help lectures and/or curriculum creators for long-term education courses. They could just use these recommendations to their course plans, and/or institution organization could apply some perceptions from final considerations into team organization and curriculum layout.

References

1. Akyuz, E., Celik, E.: A fuzzy DEMATEL method to evaluate critical operational hazards during gas freeing process in crude oil tankers. J. Loss. Prevent. Proc. **38**, 243–253 (2015). https://doi.org/10.1016/j.jlp.2015.10.006
2. Arafat, S., Aljohani, N., Abbasi, R., Hussain, A., Lytras, M.: Connections between e-learning, web science, cognitive computation and social sensing, and their relevance to learning analytics: a preliminary study. Comput. Hum. Behav. **92**, 478–486 (2019). https://doi.org/10.1016/j.chb.2018.02.026
3. Aunio, P., Heiskari, P., Van Luit, J.E., Vuorio, J.: The development of early numeracy skills in kindergarten in low, average and high performance groups. J. Early Child. Res. **13**(1), 3–16 (2015). https://doi.org/10.1177/2F1476718X14538722
4. Azeiteiro, U.M., Bacelar-Nicolau, P., Caetano, F.J., Caeiro, S.: Education for sustainable development through e-learning in higher education: experiences from Portugal. J. Clean. Prod. **106**, 308–319 (2015). https://doi.org/10.1016/j.jclepro.2014.11.056
5. Balana, B.B., Mathijs, E., Muys, B.: Assessing the sustainability of forest management: an application of multi-criteria decision analysis to community forests in northern Ethiopia. J. Environ. Manage. **91**(6), 1294–1304 (2010). https://doi.org/10.1016/j.jenvman.2010.02.005
6. Butterworth, B., Varma, S., Laurillard, D.: Dyscalculia: from brain to education. Science **332**, 1049–1053 (2011)
7. Büyüközkan, G., Çifçi, G.: A novel hybrid MCDM approach based on fuzzy DEMATEL, fuzzy ANP and fuzzy TOPSIS to evaluate green suppliers. Exp. Syst. Appl. **39**, 3000–3011 (2012). https://doi.org/10.1016/j.eswa.2011.08.162
8. Chen-Yi, H., Ke-Ting, C., Gwo-Hshiung, T.: FMCDM with fuzzy DEMATEL approach for customers' choice behavior model. Int. J. Fuzzy Syst. **9**(4), 236–246 (2007)
9. Chen, Y., Yu, J., Khan, S., Spatial sensitivity analysis of multi-criteria weights in GIS based land suitability evaluation. Environ. Model. Softw **25**, 1582–1591 (2010).
10. Çınar, M., Ekici, N., Demir, Ö.: A snapshot of the readiness for e-learning among in-service teachers prior to the pandemic-related transition to e-learning in Turkey. Teach. Teach. Educ. **107**, 103478 (2021). https://doi.org/10.1016/j.tate.2021.103478
11. Delivand, M.K., Cammerino, A.R.B., Garofalo, P., Monteleone, M.: Optimal locations of bioenergy facilities, biomass spatial availability, logistics costs and GHG (greenhouse gas) emissions: a case study on electricity productions in South Italy. J. Clean. Prod. **99**, 129–139 (2015). https://doi.org/10.1016/j.jclepro.2015.03.018

12. Dias, L., Mousseau, V., Figueira, J., Clímaco, J.: An aggregation/disaggregation approach to obtain robust conclusions with ELECTRE TRI. Eur. J. Oper. Res. **138**(2), 332–348 (2002). https://doi.org/10.1016/S0377-2217(01)00250-8
13. Eneroth, C.: E-learning for environment. Improving e-learning as a tool for cleaner production education. Dissertation, Lund University (2000)
14. European Commission: Science Education for Responsible Citizenship (2015). Available online http://goo.gl/Vm3pZ0. Accessed 14 November 2021
15. Fletcher, J.M., Lyon, G.R., Fuchs, L.S., Barnes, M.A.: Learning Disabilities: From Identification to Intervention. Guilford Press, New York (2007)
16. Gabus, A., Fontela, E.: World problems, an invitation to further thought within the framework of DEMATEL. Battelle Geneva Research Centre, Geneva (1972)
17. Garrison, R.: Theoretical challenges for distance education in the 21st century: a shift from structural to transactional issues. IRRODL **1**(1), 1–17 (2000). https://doi.org/10.19173/irrodl.v1i1.2
18. Gaudin, M., Tanguy, G., Plagne, M., Saussac, A., Hansmann, Y., Jaulhac, B., Kelly, M., Ouchchane, L., Lesens, O.: E-learning versus face-to-face training: comparison of two learning methods for Lyme borreliosis. Infect. Dis. Now. (2021). https://doi.org/10.1016/j.idnow.2021.11.001
19. González-Gómez, D., Jeong, J.S.: The flipped learning model in general science: effects on students' learning outcomes and affective dimensions. In: Mintzes, J., Walter, E. (eds.) Active Learning in College Science, pp. 541–549. Springer, Cham (2020)
20. González-Gómez, D., Airado, D., Cañada-Cañada, F., Jeong, J.S.: A comprehensive application to assist in acid-base titration self-learning: an approach for high school and undergraduate students. J. Chem. Educ. **92**(5), 855–863 (2015). https://doi.org/10.1021/ed5005646
21. González-Gómez, D., Jeong, J.S., Airado, D., Cañada-Cañada, F.: Performance and perception in the flipped learning model: an initial approach to evaluate the effectiveness of a new teaching methodology in a general science classroom. J. Sci. Educ. Technol. **25**(3), 450–459 (2016). https://doi.org/10.1007/s10956-016-9605-9
22. Govindan, K., Khodaverdi, R., Vafadarnikjoo, A.: Intuitionistic fuzzy based DEMATEL method for developing green practices and performances in a green supply chain. Exp. Syst. Appl. **42**, 7207–7220 (2015). https://doi.org/10.1016/j.eswa.2015.04.030
23. Hansen, D.E.: Knowledge transfer in on-line learning environments. J. Mark. Educ. **30**(2), 93–105 (2008). https://doi.org/10.1177/2F0273475308317702
24. Hill, H.C., Blunk, M., Charalambous, C., Lewis, J., Phelps, G., Sleep, L., Ball, D.L.: Mathematical knowledge for teaching and the mathematical quality of instruction: an exploratory study. Cogn. Instr. **26**(4), 430–511 (2008). https://doi.org/10.1080/07370000802177235
25. Jeong, J.S.: Design of spatial PGIS-MCDA-based land assessment planning for identifying sustainable land-use adaptation priorities for climate change impacts. Agric. Syst. **167**, 61–71 (2018). https://doi.org/10.1016/j.agsy.2018.09.001
26. Jeong, J.S., González-Gómez, D.: Web-GIS MC-SDSS verification forming the perception of building arrangements in Extremadura landscapes. Investig. Reg.-J. Reg. Res. **41**, 193–213 (2018)
27. Jeong, J.S., González-Gómez, D.: A web-based tool framing a collective method for optimizing the location of a renewable energy facility and its possible application to sustainable STEM education. J. Clean. Prod. **251**, 119747 (2020). https://doi.org/10.1016/j.jclepro.2019.119747
28. Jeong, J.S., González-Gómez, D.: A STEM course analysis during COVID-19: a comparison study in performance and affective domain of PSTs between F2F and F2S flipped classroom. Front. Psychol. **3352** (2021). https://doi.org/10.3389/fpsyg.2021.669855
29. Jeong, J.S., Ramírez-Gómez, A.: A multicriteria GIS-based assessment to optimize biomass facility sites with parallel environment—a case study in Spain. Energies **10**(12), 2095 (2017). https://doi.org/10.3390/en10122095
30. Jeong, J.S., García-Moruno, L., Hernández-Blanco, J.: A site planning approach for rural buildings into a landscape using a spatial multi-criteria decision analysis methodology. Land Use Policy **32**, 108–118 (2013). https://doi.org/10.1016/j.landusepol.2012.09.018

31. Jeong, J.S., García-Moruno, L., Hernández-Blanco, J., Jaraíz-Cabanillas, F.J.: An operational method to supporting siting decisions for sustainable rural second home planning in ecotourism sites. Land Use Policy **41**, 550–560 (2014). https://doi.org/10.1016/j.landusepol.2014.04.012
32. Jeong, J.S., González-Gómez, D., Cañada-Cañada, F.: Prioritizing elements of science education for sustainable development with the MCDA-FDEMATEL method using the flipped e-learning scheme. Sustainability **11**(11), 3079 (2019). https://doi.org/10.3390/su11113079
33. Jeong, J.S., González-Gómez, D., Conde-Núñez, M.C., Gallego-Picó, A.: Examination of students' engagement with R-SPQ-2F of learning approach in flipped sustainable science course. J. Balt. Sci. Educ. **18**(6), 880–891 (2019b). https://doi.org/10.33225/jbse/19.18.880
34. Kablan, M.M.: Decision support for energy conservation promotion: an analytic hierarchy process approach. Energ. Policy **32**, 1151–1158 (2004). https://doi.org/10.1016/S0301-4215(03)00078-8
35. Lee, J., Lee, W.: The relationship of e-learner's self-regulatory efficacy and perception of e-learning environmental quality. Comput. Hum. Behav. **24**(1), 32–47 (2018). https://doi.org/10.1016/j.chb.2006.12.001
36. Leonelli, M., Görgen, C., Smith, J.Q.: Sensitivity analysis in multilinear probabilistic models. Inf. Sci. **411**, 84–97 (2017). https://doi.org/10.1016/j.ins.2017.05.010
37. Lim, C.P., Chai, C.S.: An activity theoretical approach to research of ICT integration in Singapore schools: orienting activities and learner autonomy. Comput. Educ. **43**(3), 215–236 (2004). https://doi.org/10.1016/j.compedu.2003.10.005
38. Lin, R.J.: Using fuzzy DEMATEL to evaluate the green supply chain management practices. J. Clean. Prod. **40**, 32–39 (2013). https://doi.org/10.1016/j.jclepro.2011.06.010
39. Lin, C.L., Tzeng, G.H.: A value-created system of science (technology) park by using DEMETEL. Exp. Syst. Appl. **36**, 9683–9697 (2009). https://doi.org/10.1016/j.eswa.2008.11.040
40. Liou, J.J., Yen, L., Tzeng, G.H.: Building an effective safety management system for airlines. J. Air Transp. Manag. **14**(1), 20–26 (2008). https://doi.org/10.1016/j.jairtraman.2007.10.002
41. Maaß, K., Artigue, M.: Implementation of inquiry-based learning in day-to-day teaching: a synthesis. ZDM Math. Educ. **45**, 779–795 (2013). https://doi.org/10.1007/s11858-013-0528-0
42. Malczewski, J.: GIS and Multicriteria Decision Analysis. Wiley, New York (1999)
43. Malczewski, J., Jackson, M.: Multicriteria spatial allocation of educational resources: an overview. Socio-Econ. Plan. Sci. **34**, 219–235 (2000). https://doi.org/10.1016/S0038-0121(99)00025-7
44. Martínez-Montero, J., Sánchez, C.: Resolución de Problemas Y Método ABN. Wolters Kluwer Educación, The Netherlands (2013)
45. McVey, M.: E-learning and education for sustainability. Int. Rev. Educ. **62**(1), 117–121 (2016). https://doi.org/10.1007/s11159-016-9533-1
46. Meszaros, C., Rapcsak, T.: On sensitivity analysis for a class of decision systems. Decis. Supp. Syst. **16**, 231–240 (1996). https://doi.org/10.1016/0167-9236(95)00012-7
47. Naccarato, E., Karakok, G.: Expectations and implementations of the flipped classroom model in undergraduate mathematics courses. IJEMST **46**(7), 968–978 (2015). https://doi.org/10.1080/0020739X.2015.1071440
48. OECD Library: Global Teaching Insights (2021). https://www.globalteachinginsights.org. Accessed 23 October 2021
49. Parkes, M., Stein, S., Reading, C.: Student preparedness for university e-learning environments. Internet High. Educ. **25**, 1–10 (2014). https://doi.org/10.1016/j.iheduc.2014.10.002
50. Pavlova, M.: Teaching and learning for sustainable development: ESD research in technology education. Int. J. Technol. Des. Educ. **23**, 733–748 (2013). https://doi.org/10.1007/s10798-012-9213-9
51. Pereira, A., Mendes, A.Q., Morgado, L., Amante, L., Bidarra, J.: Universidade Aberta's Pedagogical Model for Distance Education. Universidade Aberta, Lisbon (2008)
52. Perpina, C., Martínez-Llario, J.C., Pérez-Navarro, Á.: Multicriteria assessment in GIS environments for siting biomass plants. Land Use Policy **31**, 326–335 (2013). https://doi.org/10.1016/j.landusepol.2012.07.014

53. Saaty, T.L.: The Analytic Hierarchy Process. McGraw-Hill, New York (1996)
54. Segalàs, J., Ferrer-Balas, D., Svanstrom, M., Lundqvist, U., Mulder, K.F.: What has to be learnt for sustainability? A comparison of bachelor engineering education competences at three European universities. Sustain. Sci. **4**(1), 17–27 (2009). https://doi.org/10.1007/s11625-009-0068-2
55. Selçuk, L.: An avalanche hazard model for Bitlis Province, Turkey, using GIS based multi-criteria decision analysis. Turk. J. Earth Sci. **22**, 523–535 (2013). https://doi.org/10.3906/yer-1201-10
56. Shee, D.Y., Wang, Y.S.: Multi-criteria evaluation of the web-based e-learning system: a methodology based on learner satisfaction and its applications. Comput. Educ. **50**(3), 894–905 (2008). https://doi.org/10.1016/j.compedu.2006.09.005
57. Su, C.H., Tzeng, G.H., Hu, S.K.: Cloud e-learning service strategies for improving elearning innovation performance in a fuzzy environment by using a new hybrid fuzzy multiple attribute decision-making model. Interact. Learn. Environ. **24**(8), 1812–1835 (2016). https://doi.org/10.1080/10494820.2015.1057742
58. Thames, M.H., Ball, D.L.: What mathematical knowledge does teaching require? Knowing mathematics in and for teaching. Teach. Child. Math. **17**(4), 220–229 (2010). https://doi.org/10.5951/TCM.17.4.0220
59. Thomas, I.: Critical thinking, transformative learning, sustainable education, and problem-based learning in universities. J. Transform. Educ. **7**, 245–264 (2009). https://doi.org/10.1177/2F1541344610385753
60. UNESCO: UN Decade of Education for Sustainable Development, 2005–2014: The DESD at a Glance. UNESDOC, New York (2005)
61. Wan, S., Zhong, L., Dong, J.: A new method for group decision making with hesitant fuzzy preference relations based on multiplicative consistency. IEEE Trans. Fuzzy Syst. **28**(7), 1449–1463 (2019). https://doi.org/10.1109/TFUZZ.2019.2914008
62. Wei, X., Cheng, I., Chen, N.S., Yang, X., Liu, Y., Dong, Y., Zhai, X.: Effect of the flipped classroom on the mathematics performance of middle school students. Educ. Tech. Res. Dev. **68**(3), 1461–1484 (2020). https://doi.org/10.1007/s11423-020-09752-x
63. Wu, W.W.: Segmenting critical factors for successful knowledge management implementation using the fuzzy DEMATEL method. Appl. Soft Comput. **12**(1), 527–535 (2012). https://doi.org/10.1016/j.asoc.2011.08.008
64. Wu, W.W., Lee, Y.T.: Developing global managers' competencies using the fuzzy DEMATEL method. Exp. Syst. Appl. **32**(2), 499–507 (2007). https://doi.org/10.1016/j.eswa.2005.12.005
65. Zadeh, L.A.: Fuzzy sets. Inf. Control. **8**, 338–353 (1965)
66. Zeleny, M.: Multiple Criteria Decision Making. McGraw-Hill, New York (1982)
67. Zare, M., Pahl, C., Rahnama, H., Nilashi, M., Mardani, A., Ibrahim, O., Ahmadi, H.: Multi-criteria decision making approach in e-learning: a systematic review and classification. Appl. Soft Comput. **45**, 108–128 (2016). https://doi.org/10.1016/j.asoc.2016.04.020

Accounting for Uncertainty and Disagreement in Multi-criteria Decision Making Using Triangular Fuzzy Numbers and Monte Carlo Simulation: A Case Study About Selecting Measures for Remediation of Agricultural Land After Radioactive Contamination

Floris Abrams, Lucas Hendrickx, Lieve Sweeck, Johan Camps, Dirk Cattrysse, and Jos Van Orshoven

Abstract In many MCDM applications, linguistic evaluation scales are becoming used more often to replace quantitative inputs because words are more natural than numbers to represent the preferences of people (García-Lapresta and González del Pozo in Eur. J. Oper. Res. 279:159–167, 2019 [1]). Most real-world decision making takes place in a complex environment, and therefore, the use of these linguistic scorings can become a tool for dealing with vagueness, imprecision and uncertainty. Both criteria scores for the considered alternatives and criteria weights can be expressed using fuzzy set theory in fuzzy numbers. In addition, when the MCDM is conducted by a group of decision makers, the impact of disagreement among them can be further characterized using these fuzzy numbers. One way of accounting for the uncertainties, related to the fuzziness in the MCDM-inputs, is the substitution of the ordinal

F. Abrams (✉) · L. Sweeck · J. Camps
Belgian Nuclear Research Centre (SCK CEN), Boeretang 200, 2400 Mol, Belgium
e-mail: Floris.abrams@sckcen.be

L. Sweeck
e-mail: Lieve.sweeck@sckcen.be

J. Camps
e-mail: Johan.camps@sckcen.be

F. Abrams · L. Hendrickx · J. Van Orshoven
Division of Forest, Nature and Landscape, Faculty of Department of Earth and Environmental Sciences, University of Leuven (KULeuven), 3001 Leuven, Belgium
e-mail: Lucas.Hendrickx@student.Kuleuven.be

J. Van Orshoven
e-mail: Jos.vanorshoven@kuleuven.be

D. Cattrysse
Center for Industrial Management, Traffic and Infrastructure, Department of Mechanical Engineering, University of Leuven (KU Leuven), 3001 Leuven, Belgium
e-mail: Dirk.cattrysse@kuleuven.be

L. Sahoo et al. (eds.), *Real Life Applications of Multiple Criteria Decision Making Techniques in Fuzzy Domain*, Studies in Fuzziness and Soft Computing 420,
https://doi.org/10.1007/978-981-19-4929-6_6

linguistic scores by triangular fuzzy numbers (TFN) followed by a set of Monte Carlo runs of the MCDM. In this paper, we report on a case study of which the goal is to select the most appropriate of five candidate countermeasures to remediate an agricultural parcel contaminated with the radionuclide Cs-137. The linguistic evaluation scales according to which two of the criteria and the criteria weights are scored were converted in TFN and then sampled using a Monte Carlo procedure to generate 10.000 rankings of the alternative countermeasures. We demonstrate that the ranking resulting from the deterministic approach is increasingly challenged when the uncertainties resulting from (1) the linguistic criteria scores, (2) the linguistic criteria weights and (3) the disagreement between decision makers are additively considered. The outcome is a fuzzy ranking of alternatives that should make decision makers think twice about the apparent superiority or inferiority of alternatives.

Keywords MCDM · Linguistic evaluation scales · Fuzzy number · Group decision-making · Uncertainties · Monte Carlo

1 Introduction

Every decision we ever take requires the balancing of multiple factors, and therefore, a significant part of the problems we address in our daily live can be considered as multi-criteria decision-making (MCDM) problems [2]. More specifically in the field of operation research, decision problems can be defined as "problems, where a group of experts or stakeholders assess alternatives regarding multiple often conflicting criteria for obtaining the best alternative(s) or a ranking of them, are investigated" [3]. Furthermore, when a variety of stakeholders with conflicting beliefs, interests, perceptions and concerns are involved in the decision, it will become difficult to determine a (strict) consensus on the decision characteristics. Hence, these decisions require a (collaborative) group multi-criteria decision-making (group MCDM) approach. The stakeholders included in the decision process can range from a small number of field experts up to representatives of all affected groups. For the purpose of this chapter, all representatives that are included in the process will be referred to as decision makers (DMs). The influence of DMs on the decision process is important in all stages of the MCDM process, and nevertheless, in two stages, the DMs will generally be asked to provide inputs for the MCDM process.

First, the DMs are asked to express their judgments on a predefined set of alternatives. The DMs evaluate the alternatives based on a predefined evaluation scheme, which can be quantitative or qualitative. It is not uncommon that the evaluations or ratings cannot be assessed precisely, because the information may be unquantifiable due to its nature or a lack of precise measurements [4]. To enable the experts to formulate their judgments in a natural way, a limited set of ordinal linguistic labels can be supplied in the form of an evaluation scale (e.g., "low", "medium" and "high"). The number of labels must be small enough to not impose useless precision while it must be rich enough in order to allow a discrimination of the performances in a limited number of grades [4].

Second, to take into account the stakeholders importance weights in the MCDM methodology, the stakeholders are asked to explicitly define their belief structures. A simple but very effective way to retrieve the stakeholders' preferences for a criterion in MCDM is by asking them to directly state the relative importance (weight) of each criteria [5]. However, an important opposition to utilize the direct estimation technique is that stakeholders are not able to quantitatively state their preference [6]. Therefore, in many of these decision-making problems, ordinal qualitative scales formed by linguistic terms are used, and these scales are popular since words are more natural than numbers to represent the preferences of people [7, 8].

These linguistic scorings are an appropriate tool for dealing with the inherent vagueness, imprecision and uncertainty in human thinking and input data [9]. By allowing the DMs to express their opinions in a linguistic form, for both the criterion evaluations and the importance of criteria, the system-expert interaction is simplified. In general, the scales used in these decision-making problems are characterized by ordered response categories with a numerical value assigned to each category, referred to as a Likert scale [10]. From these linguistic evaluation scales, a relation to quantitative values can be determined, using methods from fuzzy set theory. Fuzzy logic and fuzzy set theory was introduced by Zadeh in 1965 and has been extensively used in ambiguity and uncertainty modeling in decision making. Although the recognition and impact of the inherent uncertainty in the MCDM process is considered important and necessary, in practice, it is not implemented on a regular basis [11].

In addition, for group MCDM, an extra source of uncertainty is present when multiple DMs do not find common ground or have diverging opinions even after a participatory and consensus seeking process. The objective of such participatory MCDM is to encourage experts and stakeholders to work together, learn from each other and come to a consensus regarding the criteria to be considered, their scores and relative importance for use in the MCDM [12]. However, achieving consensus between all the decision-making groups might not be possible, especially in complex decision-making problems such as remediation after a nuclear disaster. This is a widely discussed problem described by Nordström [13] as follows: "In a group decision-making context, individual preferences must be brought together in some way in order to obtain a group preference. How this is done is both a practical and a philosophical issue." As stated before, in real situations, humans rarely come to an unanimous agreement, and therefore, it is often needed to synthesize the opinions of the decision makers into one representative view, using a aggregation procedure [14].

This paper proposes a way to account for these uncertainties, related to the fuzziness in the MCDM-inputs, by substitution of the ordinal linguistic scores by triangular fuzzy numbers (TFN), followed by a Monte Carlo simulation of the model. The objective of this process is to allow us to gain better insights in the impact of the uncertainty and investigate the stability of the outcome. Although the application of a sensitivity analysis is not new to the field of nuclear emergencies and emergency preparedness, the stepwise process used in this paper shows a much more detailed impact of each source of uncertainty than currently found in literature. The impact of uncertainties on the decision will be demonstrated based on a case study, concerning the selection of the most appropriate of five candidate countermeasures to remediate a contaminated agricultural parcel.

In Sect. 2, the theoretical background of the MCDM and fuzzy set theory is shown followed by a detailed description of the case study in Sect. 3. Lastly in Sect. 4, the results of the case study are discussed and compared to literature.

2 Preliminaries

In this section, we briefly review the concepts relating to the MCDM framework, linguistic evaluation scales and fuzzy sets, triangular fuzzy numbers, and consensus in group decision making.

2.1 MCDM Framework

In MCDM literature, a variety of terms is used interchangeably to refer to similar concepts. For the MCDM methodology that is adhered to in this chapter, Malczewski [15] uses the term multi-attribute decision making (MADM), and in this chapter, we will further refer to it as MCDM. Although many variants on the framework exist, it can be argued that most approaches involve three main stages: (1) Description of the decision context and structuring of the problem (criteria & alternatives), (2) Analysis of the decision alternatives and (3) The selection of the preferred alternative among the candidate alternatives [16]. The framework as shown in Fig. 1 is used to structure the application of MCDM to the remedial technology selection. In this chapter only the uncertainty related to the introduction of fuzziness for two of the stages: the criteria evaluation and the identification of criteria weights are addressed by the introduction of fuzziness.

2.2 Linguistic Evaluation Scales and Fuzzy Set Theory

A linguistic evaluation scale used in a MCDM application should be a finite and totally ordered discrete scale, with a linguistic labels set $S = \{S_\alpha | \alpha = 0, 1, 2, \ldots, \tau\}$ where S_α represents the value for a linguistic label [17]. The introduction of these linguistic labels should be interpreted as a multi-valued logic where the semantic labels allow values in the entire interval [0,1] [18]. Therefore, there is a need to define the relation between a given label and the numerical scale associated to the label. The label is converted into a fuzzy set, where the fuzzy set represents a membership function to the quantitative scale [0, 1] (Fig. 2). In a finite universe $X = \{x_1, x_2, \ldots, x_n\}$,

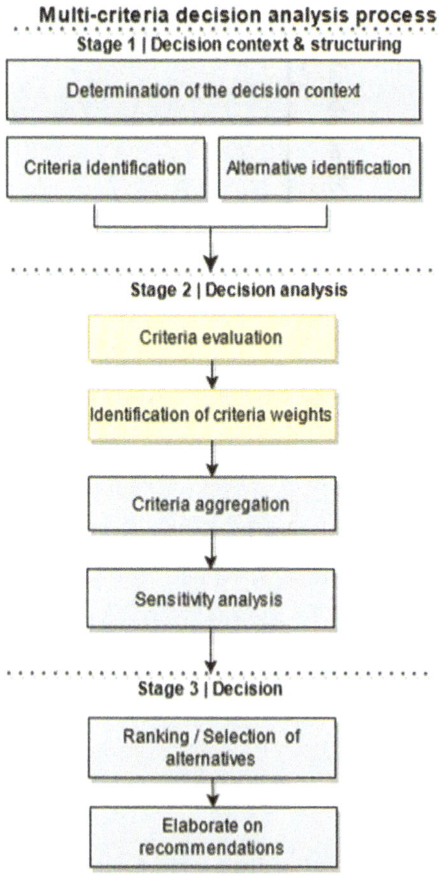

Fig. 1 Generalized MDAM structure adapted from [16]. The stages in yellow represent sources of fuzziness addressed in this paper

a fuzzy set Ã in X is represented as $A = \{(x_i, \mu_A(x_i))|x_i \in X\}$ where μ_A is the membership function of the fuzzy set, which denotes the degree of membership of an element x_i belonging to Ã, with $\mu_A(x_i) \in [0, 1]$ [19].

A commonly used fuzzy set is the triangular fuzzy number, defined by three points [a1, a2, a3], respectively, the lower bound, the mode and the upper bound [20]. The mathematical representation of the membership from the triangular fuzzy number (TFN) is shown in Eq. 1 and Fig. 3. Zardari et al. [5] showed that fuzzy sets are often used to implement vague linguistic scores of experts and stakeholders into the MCDM structure.

$$\mu_A \sim (x) = \begin{cases} 0 & \text{for } x < a_1 \\ \frac{x-a_1}{a_2-a_1} & \text{for } a_1 \le x < a_2 \\ \frac{a_3-x}{a_3-a_2} & \text{for } a_2 \le x \le a_3 \\ 0 & \text{for } x > a_3 \end{cases} \tag{1}$$

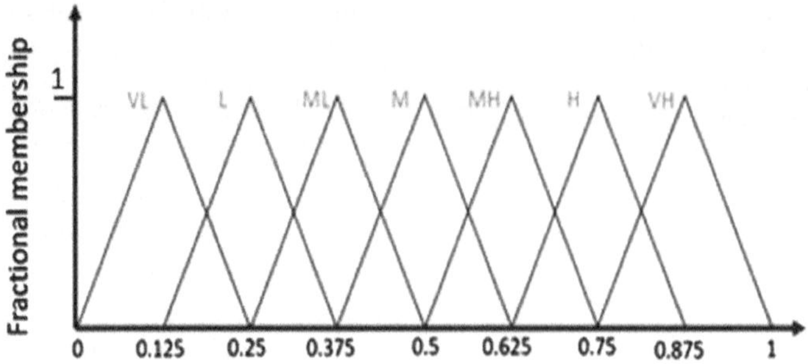

Fig. 2 Triangular fuzzy numbers with the membership functions on the interval 0 to 1. Linguistic values and equivalent integer designations are shown with each membership function. The centroids for the triangular functions evenly divide the interval into seven segments (Very Low = VL, Low = L, Medium Low = ML, Medium = M, Medium High = MH, High = H, Very high = VH)

Fig. 3 Triangular fuzzy number A: [a1, a2, a3]. Adapted from (Han et al. 21)

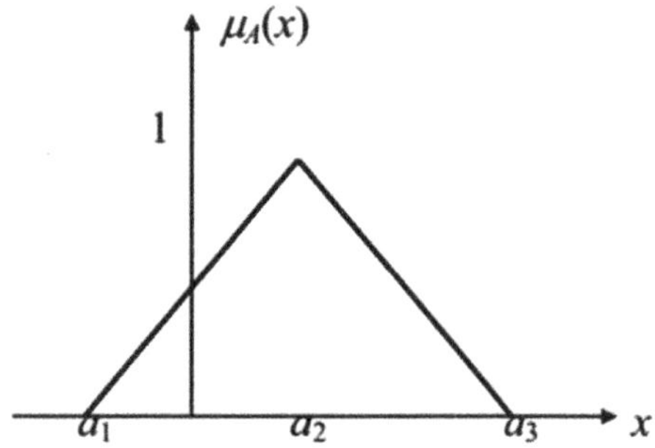

The fuzzy linguistic sets are introduced in two different stages: 1. on the evaluation of the qualitative criteria and 2. for the weight preferences to the criteria. Both stages use a similar linguistic evaluation scale but with different labels as shown in Tables 1 and 2. The values are chosen to evenly divide the interval in equal segments when a 7 point Likert scale is used, the scales are comparable to distributions found in literature [22, 23].

2.3 Group Decision Making

After all stakeholders of the project are determined, it is important to use participatory techniques to align the different stakeholders' perspectives and build consensus on the MCDM parameters. Consensus in the traditional sense is a strict and unanimous agreement; however, in a complex situation, this is unrealistic unattainable goal. Therefore, a softer and human perception of consensus is used, where people are willing to accept that a consensus has been reached when most or the more

Table 1 Linguistic variables with the matching fuzzy number for the criterion evaluation scores for each alternative

Linguistic variable	Abbr.	a_1	a_2	a_3
Very low	VL	0	0.125	0.25
Low	L	0.125	0.25	0.375
Medium low	ML	0.25	0.375	0.5
Medium	M	0.375	0.5	0.625
Medium high	MH	0.5	0.625	0.75
High	H	0.625	0.75	0.875
Very high	VH	0.75	0.875	1

Table 2 Linguistic variables with the matching fuzzy number for the criteria weight

Linguistic variable (Importance)	Abbr.	a_1	a_2	a_3
Low	LI	0	0.125	0.25
Quite low	QLI	0.125	0.25	0.375
Medium low	MLI	0.25	0.375	0.5
Medium	MI	0.375	0.5	0.625
Medium high	MHI	0.5	0.625	0.75
Quite high	QHI	0.625	0.75	0.875
High	HI	0.75	0.875	1

predominant actors agree on the preferences associated with the most relevant alternatives [14]. A technical solution to finding consensus between stakeholders consists of aggregating the different individual views into one generic view, determined from the majority [24] or average view [25]. For a more detailed explanation on aggregation procedures, we recommend article [14].

In this chapter, we propose an approach, in which the information on the initial disagreement is preserved, and can therefore be used to account for uncertainties caused by disagreement between DMs. For each criterion weight, the different TFNs, one per stakeholder, are converted into one TFN, where the width of the resulting TFN is dependent on the degree of (dis)agreement between the decision makers. The peak of the TFN is the average of the initial TFNs, and the upper and lower boundary values are determined by the lowest lower and highest upper boundary values found in the initial TFNs. The width of these TFN's is therefore altered, resulting in a change in its membership, resulting in a wider or more uncertain scoring. An example of the TFN resulting from two disagreeing DMs can be found in Figs. 3 and 4.

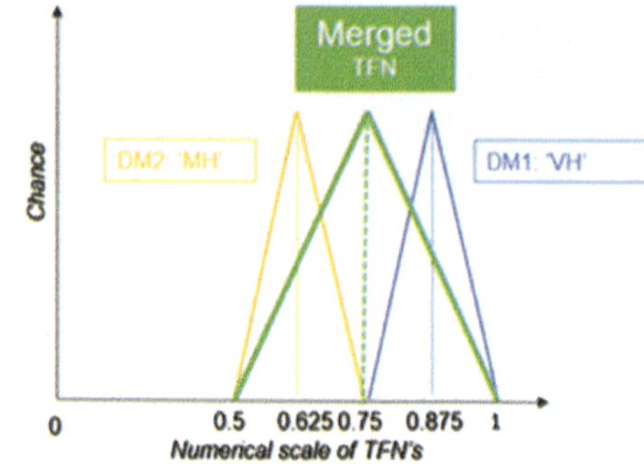

Fig. 4 Example to illustrate the relationship between the opinions of decision makers and the merged TFNs. DM1 stands for the first decision maker and DM2 for the second one

2.4 *Monte Carlo Simulation*

Monte Carlo simulation (MCS) is widely used to propagate (probabilistic) uncertainties from the model inputs to the model outputs [26]. MCS is used to tackle uncertainties that can be described by probability distribution functions (PDFs). A MCS consists of a large number of repeated runs of the model, where each time the input values are randomly sampled from each probabilistic density function (PDF). The sampling of the criteria evaluation values is independent from each other, while the values of the criteria weights should sum to 1. Therefore, after all the weights values are independently sampled, the criteria weights are normalized so that they sum up to 1. The distribution of the output variables can then be examined further using statistical methods [27]. As mentioned earlier, the triangular distributions based on the linguistic scorings are used as the PDFs, defining both the range of values and the likelihood that the weight has a value in specific interval [28]. When this procedure is repeated a large number of times, the MCDM output can be seen as rank acceptability indices that describes the chance of any alternative to be assigned a certain rank [29]. Therefore, the higher the rank acceptability index of an alternative for a specific rank, the more this alternative was ranked at that specific rank during the MCS.

3 Case Study: The Application of Fuzzy MCDM to Support Remediation of Agricultural Land After a Nuclear Accident

3.1 *Materials and Methods*

In line with the generalized MCDM structure presented in Fig. 1, the components of the decision problem are the following:

- Determination of the decision context

Following a large-scale accidental release of radioactivity into the environment, large areas are likely to be contaminated by volatile radionuclides (RN) (e.g., I-131, Cs-137). Due to the persistence and the possible uptake of some long-lived RN (Cs-137) in the environment and food chain, the impact on the food production systems can last for many years [30]. The need for long-term management plans after such a release can be seen in the Fukushima region, where 10 years after the DAIICHI nuclear accident, the remedial activities are still ongoing [31].

The management of the contaminated areas focusses on the restoration of a normal lifestyle for all affected populations. Nevertheless, the recovery strategy to be adopted has to be justified and optimized [32]. The justification principle of a recovery strategy goes far beyond the radiological protection alone, as the implementation of recovery options may also have various economic, environmental, social and psychological impacts [33]. Further, the optimization principle ensures the selection of the best possible strategy and actions under the prevailing circumstances. Taking these principles into account, it becomes clear that for the successful remediation of sites contaminated with variable levels of long-lived RN, priority setting among adequate remediation technologies is of major importance. The selection of the remediation technology is typically conditioned by multiple and often contradicting criteria of an economic, social, technical, environmental and radiological nature. There are inherent trade-offs to be made between these possibly conflicting criteria to ensure a responsible and sustainable remediation process.

As a case study, a hypothetical nuclear emergency at the NPP of Tihange, Belgium, is considered. The accidental release of Cs-137 resulted in the contamination of the surrounding agricultural land. The optimal remedial agricultural measure is sought for a cereal field contaminated with a 453 kBq/m^2 of Cs-137.

- Alternative identification

The possible agricultural countermeasures for the stated problem are listed in Table 3. They are retrieved from the "Generic handbook for assisting in the management of contaminated food productions systems in Europe following a radiological emergency" [33] and expert knowledge.

- Criteria identification

To ensure the selection of the most appropriate remediation measure, a divers set of criteria are to be considered. The combination of the criteria should encompass all dimensions of the decision problem without being redundant. Criteria are either of the benefit or the cost type. Whereas for benefit criteria the higher the score is the better, for cost criteria the opposite is true: lower scores are preferred. For this case study, four criteria are selected, and they are listed in Table 4.

- Criteria evaluation

The score of each of the four criteria for each of the five alternatives are shown in Table 5. The criteria scores of "reduction efficiency" and "direct cost" are expressed

Table 3 Applicable agricultural remedial counter measures for the region around Tihange, applicable on a cereal field

Agricultural countermeasure	Description
Skim and burial ploughing (S&B)	Skimming of the top 5 cm of soil and burry it at 45 cm depth. Lifting the deeper soil layer up without inversion
Deep ploughing (DP)	Inversion of the top 45 cm of soil to decrease Cs-137 root availability due to burial of the most contaminated top layer
Application of potassium fertilizers (PF)	Reduce plant uptake of radio caesium by adding potassium to the soil due to increased competition for plant uptake
Shallow ploughing (SP)	Use of ordinary plough to mix the top 20–30 cm of soil. Resulting in a deeper burial and dilution of the most contaminated top layer
Topsoil removal (TR)	Removal and external waste management of the top 2–5 cm of soil

Table 4 Criteria used in the MCDM process

Criterion	Description	Benefit or cost
Reduction efficiency (RE)	Reduction in radioactivity of agricultural product (compared to doing nothing). [%]	Benefit
Direct cost of remediation technique (DCA)	The total implementation cost. It includes the full remediation cycle, from investigation to monitoring and waste treatment. [€/ha]	Cost
Local impact (LI)	The severity by which the local population is affected due to the implementation of the technique. The impact is based on expert assessment. [Linguistic scale]	Cost
Feasibility (F)	The probability that the remediation strategy is successful. If the successful implementation of a technique is not certain, it highly impacts the preference for such technique. The impact is based on expert assessment and experiences. [Linguistic scale]	Benefit

on a continuous quantitative scale. "Local impact" and "feasibility" are scored qualitatively by experts using a linguistic evaluation score, following the scale in Table 1.

- Criteria Preferences

The criteria weights assigned by three individual decision makers to the four criteria are shown in Table 6. They are expressed using the linguistic scorings from Table 2. Two distinct scenarios are represented to show the impact of reaching a better consensus within the group.

- MCDM method

For the aggregation of the criteria scores, technique of order preference similarity to the ideal solution) was utilized. TOPSIS is a distance-based MCDM method, where

Table 5 Alternative-criteria matrix with the criteria evaluations (L = Low, M = Medium, MH = Medium High, H = High, VH = Very High), with the best alternative score per criterion in bold

Alternatives	Criteria			
	Reduction efficiency	Direct cost of remediation technique	Local impact	Feasibility
S&B	87	95	MH	L
DP	70	54	MH	M
PF	69	66	**L**	H
SP	50	**39**	**L**	**VH**
TR	**93**	2449	VH	L

Table 6 Criteria preferences for decision makers for a scenario with and without consensus: LI = Low importance, QLI = Quite low importance, MI = Medium importance, MHI = Medium High importance, HI = High importance, QHI = Quite High importance

Criterion	Scenario 1 (consensus)			Scenario 2 (less consensus)		
	Decision makers			Decision makers		
	DM 1	DM 2	DM 3	DM 1	DM 2	DM 3
RE	HI	HI	HI	QHI	HI	MHI
DCA	QHI	HI	HI	HI	QHI	MHI
LI	MLI	QLI	LI	LI	QLI	M
F	MI	MI	MHI	QLI	MLI	MHI

the rank of an alternative is based on the distance to the ideal and anti-ideal point [34].

- Uncertainty analysis

To estimate the impact of uncertainty on the ranking of the alternative countermeasures, a Monte Carlo approach was used, sampling the TFN for 10,000 times.

3.2 *Results*

For each of the different scenarios, the output is given as rank acceptability indices, and this index describes the probability of any alternative to be assigned a certain rank. The results are structured so that the uncertainty, related to the imprecision on the input variables, is gradually increased. First, a deterministic run will be performed to set a baseline ranking of the actions. Next, both the uncertainties on the criteria scores and the weights are considered separate and thereafter combined. Then, the impact of a group of stakeholders on the results is considered using two scenarios (low and high consensus).

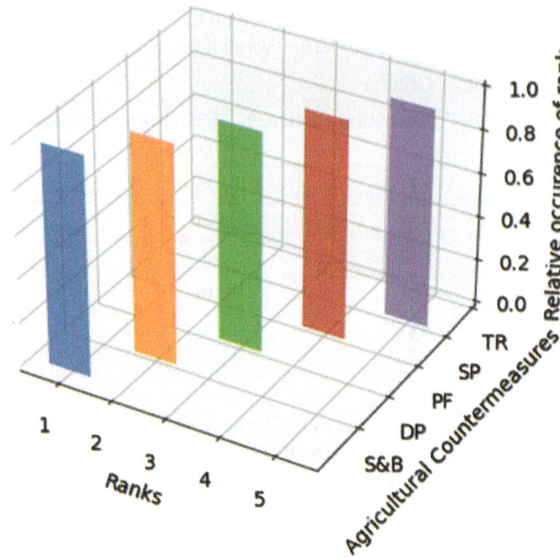

Fig. 5 Ranking of the alternatives based on the deterministic MCDM model

3.2.1 Deterministic Approach

In the deterministic approach, the methodology is applied without considering uncertainty from the fuzzy inputs. Different methods exist for the defuzzification of fuzzy numbers into crisp values, such as the centre of area (COA), mean of maxima (MOM) and maxima-method (MAX). However, for the specific case of a TFN the defuzzification result always equals the peak value: a2 [35]. By using one of the above defuzzification methods, instead of a random sample, the input values will be constant over all the runs.

From Fig. 5, it became clear that skim and burial ploughing (S&B) is ranked highest for this specific decision context (contamination level, parcel size and crop type). The use of topsoil removal is ranked lowest and hence to be considered as the least optimal alternative. This can be explained by its high cost of implementation and the very high local impact. This deterministic approach is used as a baseline ranking, to compare the effect of introducing uncertainty in the fuzzy parameters.

3.2.2 Introduction of Uncertainty on the Criteria Scores (I)

The introduction of uncertainty on the criteria scores, compared to the deterministic case, shows multiple different rankings for S&B, DP and PF. In 2% of the 10.000 runs DP becomes the preferred alternative, and in 3% of the runs PF is the alternative to be preferred. Further, the distinct difference in ranking between PF and DP as shown in the deterministic scenario becomes less straight forward (Fig. 6).

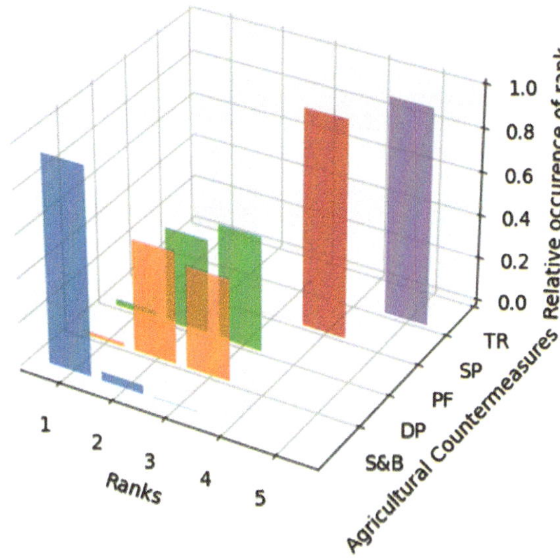

Fig. 6 Rank acceptability index, based on the relative occurrence of the alternatives in that rank, when uncertainty is introduced on the criteria scores

3.2.3 Introduction of Uncertainty on the Criteria Preferences/weights (II)

When the uncertainty on the criteria preferences is introduced, it is observed that, even with uncertain criterion weights, the S&B alternative was ranked first in all cases. Further, in 3% of the runs, the TR alternative outranks alternative SP. This inversion could be a result of a Monte Carlo sample in which the importance for the cost criterion is low and the importance for the reduction efficiency criterion is high (Fig. 7).

3.2.4 Uncertainty on the Criteria Scores and Criteria Preferences Combined (I + II)

Combination of both sources of uncertainty clearly affects discrimination between alternatives DP and PF. The best performing alternative S&B was ranked second in 7% of the runs and even third in 1% of the runs. The alternatives on rank 1, 4 and 5 can be determined with high confidence, while this is not the case for alternatives on rank 2 and 3 anymore (Fig. 8).

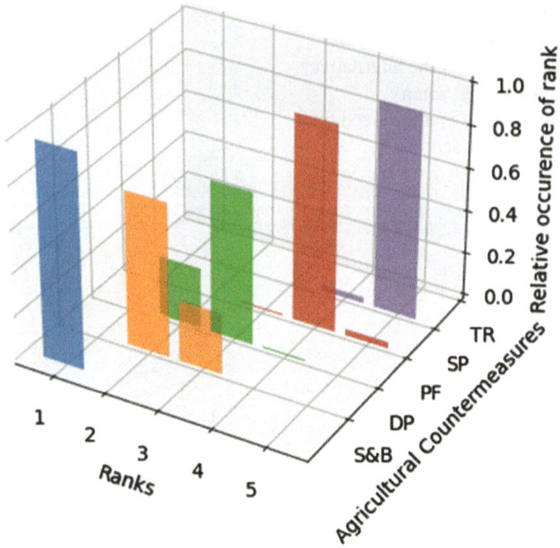

Fig. 7 Rank acceptability index, based on the relative occurrence of the alternatives in that rank, when uncertainty is introduced on the criteria preferences

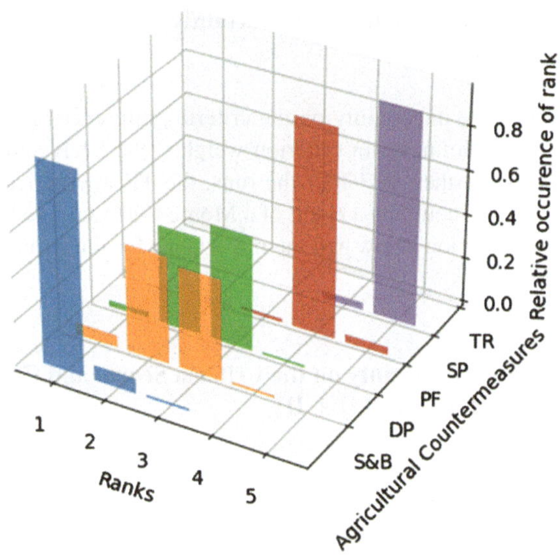

Fig. 8 Rank acceptability index, based on the relative occurrence of the alternatives in that rank, when uncertainty is introduced on both the criteria evaluations and weights

3.2.5 Group Decision Making with Uncertainty (I + II + III)

Two scenarios are considered, differing in the amount of (dis)agreement between DMs. The introduction of disagreement between DMs further increases the number of different ranking options for all alternatives.

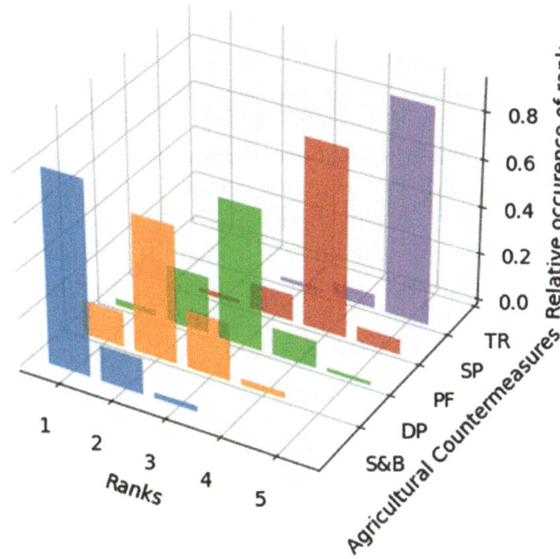

Fig. 9 Rank acceptability index, based on the relative occurrence of the alternatives in that rank, when uncertainty is introduced on both the criteria evaluations and weights but disagreement between the three decision makers is limited

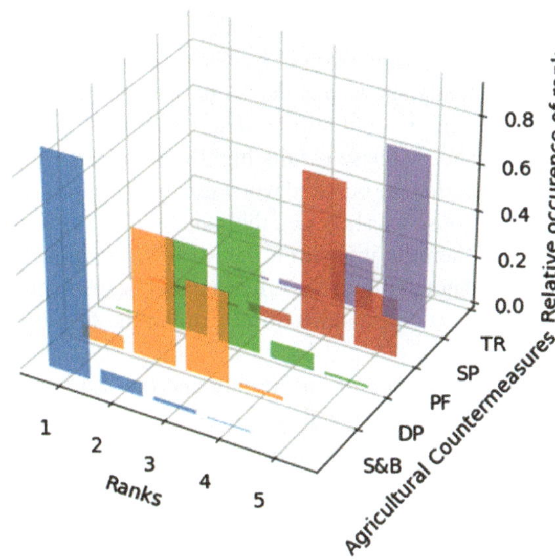

Fig. 10 Rank acceptability index, based on the relative occurrence of the alternatives in that rank, when disagreement between three decision makers is large

All alternatives were distributed over at least four different ranks over the 10.000 runs, and PF was even distributed over all five possible ranks. The impact of limited disagreement between multiple decision makers is rather small. But, when disagreement is larger the distribution of the alternatives over the different ranks also becomes larger. The difference between the ranking of SP and TR becomes smaller so that discriminating between them becomes less clear (Figs. 9 and 10).

4 Discussion

From the graphs and the summary tables (Tables 7 and 8), it became clear that the impact of uncertainty on the decision power cannot be ignored. When using a deterministic method without uncertainty, the user is given a false feeling of confidence in the MCDM. The increased complexity by introducing uncertainty, compared to a deterministic result, should not be seen as a disadvantage since the results are more reasonable and realistic in nature [36]. Other research performing stochastic uncertainty analysis of a MCDM approach to optimize remedial actions has reported similar trends in the rank acceptability indices [37–39]. From introducing uncertainty on the criteria evaluations, it became clear that in some cases the optimal alternative changes. This shows the importance for an accurate criterion evaluation. Nevertheless, it is important to acknowledge the inherent uncertainty of these criteria in the context of large-scale remedial actions. The criteria evaluations should be adapted to each agro-ecological zone to optimally describe and estimate the impacts of each alternative. This can, for example, be done by setting in-situ monitoring schemes can help improve our understanding of the alternatives in the field with time and adapt the decision matrix for consequent decisions. Differences in criteria weights can have similar impacts on the decisions, and this implicates that DM should be trained and prepared to use MCDM, to accurately represent their perspectives through these weights. The introduction of both sources of uncertainties on the criteria scores and weights clearly shows that decisions are less straightforward for rank 2 and 3, while rank 1, 4 and 5 could still be determined with high confidence, as can be seen in Tables 7 and 8. These results highlight the robustness of the MCDM technique to robustly predict the best and worst remedial options, even under uncertain conditions.

When consensus between the stakeholders is reached in a group decision-making problem, then the impact on the rankings is rather small. In comparison, if stakeholders do not agree on the criteria weights, the decision can become less clear. This impacts the models ability to determine the best option between several proposed options.

The introduction of uncertainty and the resulting impact on the rankings should be clearly communicated to the DMs, so that a better informed decision can be made. The proposed technique to use linguistic evaluation scales and corresponding fuzzy numbers to quantify the vagueness and uncertainty of expert assessments could help with a clear communication of the impact of uncertainty. The resistance of stakeholders to give criteria preferences, because of the large influence on the ranking results, could be reduced with the applied method. The methodology is easy to understand, and it can be implemented with a variety of different MCDM aggregation techniques (e.g., PROMETHEE, ELECTRE, WLC and VIKOR). Because of the increase in processing capacity, the use of MCS with even larger number of runs becomes feasible and less time consuming.

Table 7 Summary table of descriptive indicators of the ranking, when uncertainty is additively introduced in both the criteria scores and the criteria weights

	Decision making								
	Fuzzy criteria scores			Fuzzy criteria weights			Fuzzy criteria weights and scores		
	Min rank	Max rank	Median rank	Min rank	Max rank	Median rank	Min rank	Max rank	Median rank
Skim and burial ploughing (S&B)	1	3	1	1	1	1	1	3	1
Deep ploughing (DP)	1	4	2	2	3	2	1	4	3
Potassium fertilization (PF)	1	3	3	2	4	3	1	4	3
Shallow ploughing (SP)	3	4	4	3	5	4	2	5	4
Topsoil removal (TR)	5	5	5	4	5	5	3	5	5

Table 8 Summary table of descriptive indicators of the ranking for a group decision-making process, two scenarios of disagreement between DMs

	Group decision making					
	Consensus (Scenario 1)			No Consensus (Scenario 2)		
	Min rank	Max rank	Median rank	Min rank	Max rank	Median rank
Skim and burial ploughing (S&B)	1	3	1	1	4	1
Deep ploughing (DP)	1	4	2	1	4	2
Potassium fertilization (PF)	1	5	3	1	5	3
Shallow ploughing (SP)	2	5	4	1	5	4
Topsoil removal (TR)	3	5	5	2	5	5

5 Conclusion

The case study clearly showed the impact of uncertainty on the decision-making power, to support the remediation of agricultural land. Although considering the uncertainties in the decision-making process has some disadvantages, such as a less clear ranking of the alternatives, the DMs receive a much more realistic information and advice compared to a deterministic model. It can be concluded that using linguistic evaluation scales and corresponding fuzzy numbers, to incorporate vagueness and uncertainty in the decision process, is helpful. Furthermore, it highlights the need for accurate criteria evaluations and weight settings. The application of fuzzy elements in MCDM could address some issues of DMs to MCDM and help them not to fear uncertainty but embrace its presence in every decision. It can be concluded that the application of fuzzy logic in MCDM to account for the uncertain nature of complex decision problems (e.g., soil pollution remediation projects) is further to be promoted. Future research should include more sources of uncertainty, combined with a careful investigation of possible interactions between them and the effect on the robustness of the MCDM process.

Acknowledgements This work was supported by a PhD grant for Floris Abrams from the Belgian Nuclear Research Centre (SCK CEN).

References

1. García-Lapresta, J.L., González del Pozo, R.: An ordinal multi-criteria decision-making procedure under imprecise linguistic assessments. Eur. J. Oper. Res. 279, 159–167 (2019). https://doi.org/10.1016/j.ejor.2019.05.015
2. Belton, V., Stewart, T.J.: Multiple criteria decision analysis: an integrated approach. (2002)
3. Greco, S., Ehrgott, M., Figueira, J.R.: In: Multiple Analysis Criteria Decision State of the Art Surveys. (2016)
4. Bordogna, G., Fedrizzi, M., Pasi, G.: A linguistic modeling of consensus in group decision making based on OWA operators. IEEE Trans. Syst. Man, Cybern. Part ASystems Humans. **27**, 126–132 (1997). https://doi.org/10.1109/3468.553232
5. Zardari, N.H., Kamal, A., Sharif Monirussaman, S., Zulkifli, Y. B.: In: Weighting Methods and their Effects on Multi-Criteria Decision Making Model Outcomes in Water Resources Management. (2015)
6. Danielson, M., Ekenberg, L.: An improvement to swing techniques for elicitation in MCDM methods. Knowledge-Based Syst. **168**, 70–79 (2019). https://doi.org/10.1016/j.knosys.2019.01.001
7. Zimmer, A.C.: Verbal Vs. numerical processing of subjective probabilities. Adv. Psychol. (1983). https://doi.org/10.1016/S0166-4115(08)62198-6
8. Teigen, K.H.: The language of uncertainty. Acta Psychol. (Amst). (1988). https://doi.org/10.1016/0001-6918(88)90043-1
9. Fasolo, B., Bana e Costa, C.A.: Tailoring value elicitation to decision makers' numeracy and fluency: Expressing value judgments in numbers or words. Omega (United Kingdom) (2014). https://doi.org/10.1016/j.omega.2013.09.006
10. Likert, R.: A technique for the measurement of attitudes. Arch. Psychol. (1932)
11. Scholten, L., Schuwirth, N., Reichert, P., Lienert, J.: Tackling uncertainty in multi-criteria decision analysis—an application to water supply infrastructure planning. Eur. J. Oper. Res. (2015). https://doi.org/10.1016/j.ejor.2014.09.044
12. Oltra, C., Sala, R., Germán, S., López-Asensio, S.: Trust perceptions among residents surrounding nuclear power plants: a descriptive and explanatory study. Prog. Nucl. Energy. (2019). https://doi.org/10.1016/j.pnucene.2018.12.012
13. Nordström, E.M., Eriksson, L.O., Öhman, K.: Integrating multiple criteria decision analysis in participatory forest planning: Experience from a case study in northern Sweden. For. Policy Econ. (2010). https://doi.org/10.1016/j.forpol.2010.07.006
14. Fedrizzi, M., Pasi, G.: Fuzzy logic approaches to consensus modelling in group decision making. In: Da Ruan, Hardeman, F., van der M.K. (eds.) Intelligent Decision and Policy Making Support Systems. Springer, Berlin, Heidelberg, pp. 19–37. (2008)
15. Malczewski, J.: GIS and multicriteria decision analysis (1999)
16. Adem Esmail, B., Geneletti, D.: Multi-criteria decision analysis for nature conservation: a review of 20 years of applications. Methods Ecol. Evol. **9**, 42–53 (2018). https://doi.org/10.1111/2041-210x.12899
17. Martínez, L., Liu, J., Yang, J.B., Herrera, F.: A multigranular hierarchical linguistic model for design evaluation based on safety and cost analysis. Int. J. Intell. Syst. **20**, 1161–1194 (2005). https://doi.org/10.1002/int.20107
18. French, S.: Evaluation and decision models: a critical perspective. J. Oper. Res. Soc. **53**, 809–809 (2002). https://doi.org/10.1057/palgrave.jors.2601380
19. Zadeh, L.A.: Fuzzy sets. Inf. Control. 338−35 (1965). https://doi.org/10.1061/9780784413616.194
20. Kwang, L.: First course on fuzzy theory and applications. In: Advances in Soft Computing Editor-in-chief (2019)
21. Han, E.S., Daniel, G., Richard, B., Mckee, A.: In: First Course on Fuzzy Theory and Applications , Berlin, Heidelberg, New York (2019)

22. Ozsahin, I., Abebe, S.T., Mok, G.S.P.: A multi-criteria decision-making approach for schizophrenia treatment techniques. Arch. Psychiatry Psychother. (2020). https://doi.org/10.12740/APP/111624
23. Moradpour, S., Ebrahimnejad, S., Mehdizadeh, E., Mohamadi, A.: Using hybrid fuzzy PROMETHEE II and fuzzy binary goal programming for risk ranking : a case study of highway construction. Projects **9**, 47–55 (2011)
24. Pasi, G., Yager, R.R.: Modeling the concept of majority opinion in group decision making. In: Information Sciences (2006)
25. Yager, R.R., Filev, D.P.: Induced ordered weighted averaging operators. IEEE Trans. Syst. Man, Cybern. Part B Cybern. (1999). https://doi.org/10.1109/3477.752789
26. Urso, L.: D9.62—Methodology to quantify improvement. Guidance on uncertainty analysis for radioecological models Lead (2019)
27. Qin, X.S., Huang, G.H., Sun, W., Chakma, A.: Optimization of remediation operations at petroleum-contaminated sites through a simulation-based stochastic-MCDA approach. Energy sources, Part A Recover. Util. Environ. Eff. **30**, 1300–1326 (2008). https://doi.org/10.1080/15567030801928623
28. Aral, M.M.: Environmental modeling and health risk analysis (Acts/Risk) (2010)
29. Lahdelma, R., Salminen, P.: Multicriteria decision analysis for choosing the remediation method for a landfill based on mixed ordinal and cardinal information. In: Linkov, I., Ferguson, E., Magar, V. (eds.) Real-Time And Deliberative Decision Making: Application To Emerging Stressors. pp. 379. (2008)
30. Merz, S., Steinhauser, G., Hamada, N.: Anthropogenic radionuclides in Japanese food: environmental and legal implications. Environ. Sci. Technol. (2013). https://doi.org/10.1021/es3037498
31. Evrard, O., Laceby, J.P., Nakao, A.: Effectiveness of landscape decontamination following the Fukushima nuclear accident : a review. 333–350 (2019)
32. ICRP: The 2007 Recommendations of the International Commission on Radiological Protection. ICRP publication, pp. 103. (2007)
33. Nisbet, A.F., Jones, A.: Generic handbook for assisting in the management of contaminated food production systems in Europe following a radiological emergency (2009)
34. Behzadian, M., Khanmohammadi Otaghsara, S., Yazdani, M., Ignatius, J.: A state-of the-art survey of TOPSIS applications (2012)
35. Geldermann, J., Spengler, T., Rentz, O.: Fuzzy outranking for environmental assessment. case study: Iron and steel making industry. Fuzzy Sets Syst. (2000). https://doi.org/10.1016/S0165-0114(99)00021-4
36. Scheffler, A., Roth, T., Ahlf, W.: Sustainable decision making under uncertainty: a case study in dredged material management. Environ. Sci. Eur. (2014). https://doi.org/10.1186/2190-4715-26-7
37. Bonano, E.J., Apostolakis, G.E., Salter, P.F., Ghassemi, A., Jennings, S.: Application of risk assessment and decision analysis to the evaluation, ranking and selection of environmental remediation alternatives. J. Hazard. Mater. 71, 35–57 (2000). https://doi.org/10.1016/S0304-3894(99)00071-0
38. Hokkanen, J., Lahdelma, R., Salminen, P.: Multicriteria decision support in a technology competition for cleaning polluted soil in Helsinki. J. Environ. Manage. **60**, 339–348 (2000). https://doi.org/10.1006/jema.2000.0389
39. Alvarez-Guerra, M., Canis, L., Voulvoulis, N., Viguri, J.R., Linkov, I.: Prioritization of sediment management alternatives using stochastic multicriteria acceptability analysis. Sci. Total Environ. **408**, 4354–4367 (2010). https://doi.org/10.1016/j.scitotenv.2010.07.016

An IT2FBWM Model to Highlight the Significance of Factors Utilized in Determining Pandemic Hospital Site Selection

Ejder Aycin, Sema Kayapinar Kaya, and Fatih Ecer

Abstract The COVID-19 pandemic has spread to the world in a short time since its emergence in Wuhan, China, and mobilized international health authorities with its effects, continues its development seriously. Due to the fast spread of COVID-19, governments have made some decisions to build new temporary hospitals for patients infected with COVID-19. This situation increased the importance of the site selection process for a pandemic hospital during the current critical time. Site selection of a hospital can be considered as multi-criteria decision-making (MCDM) problem since it includes multiple factors. This paper applies an analytical tool for evaluating the criteria of the site selection of a pandemic hospital. For this purpose, the extended best–worst method with trapezoidal interval type-2 fuzzy numbers (IT2FBWM) is employed to determine the importance weights of the pandemic hospital site selection criteria. The paper's findings point out that the IT2FBWM model can help decision-makers determine the importance weights of the criteria used in the pandemic hospital siting decisions.

Keywords BWM · Interval type-2 fuzzy numbers · IT2FBWM · Pandemic hospital siting factors · MCDM

E. Aycin
Department of Business Administration, Faculty of Business and Administrative Sciences, Kocaeli University, Izmit, Turkey
e-mail: ejder.aycin@kocaeli.edu.tr

S. Kayapinar Kaya
Department of Industrial Engineering, Faculty of Engineering, Munzur University, Tunceli, Turkey
e-mail: semakayapinar@munzur.edu.tr

F. Ecer (✉)
Sub-Department of Operations Research, Faculty of Economics and Administrative Sciences, Afyon Kocatepe University, Afyonkarahisar, Turkey
e-mail: fecer@aku.edu.tr

L. Sahoo et al. (eds.), *Real Life Applications of Multiple Criteria Decision Making Techniques in Fuzzy Domain*, Studies in Fuzziness and Soft Computing 420,
https://doi.org/10.1007/978-981-19-4929-6_7

1 Introduction

Traditional fuzzy sets (or type-1 fuzzy sets) are unable to tackle with uncertainty in "prototypical terms", and dealing with meaningless data demonstrates their ambiguity. To different people, the prototypical words denote different things. For example, if the value is 0.5, the element belongs to the type-1 fuzzy set 50% of the time. The value of a membership degree can fluctuate depending on who you ask. Therefore, uncertainty has been indicated as a single membership value [47]. However, the challenges in aggregating expert judgments, as well as noisy data from measurement activation, exposed the ambiguity in the concept of type-1 fuzzy sets. Zadeh [46] presented type-2 fuzzy sets as an expansion of the concept of conventional fuzzy sets. Type-2 fuzzy sets provide for better direct handling of uncertainty than type-2 fuzzy sets [10].

Both of fuzzy sets are mapped to value between 0 and 1. All fuzzy sets are characterized by membership functions. Type-1 fuzzy sets have two-dimensional (2D) membership function, while type-2 fuzzy sets have three-dimensional (3D) membership function. New membership function is called the footprint of uncertainty (FOU) that provides additional degree of freedom to capture more information. On the other hand, type-2 fuzzy sets are difficult to understand and apply due to computational difficulties and excessive mathematical formulas. Interval type-2 fuzzy sets (IT2FSs) developed to reduce the computational complexity of the concept of type-1 fuzzy sets [23]. With the help of a linguistic word, it can reduce the inherent ambiguity or imprecision [9, 18].

Site selection problems are a very interesting research area of researchers, whereas it is an MCDM problem with many qualitative and quantitative criteria conflicting with each other. Due to the economic, social, and ecological reasons, authorities would like to determine where to build their projects. Deciding an optimal site for building a hospital is an example of site selection problems. Various researchers have solved the hospital site selection problem employing MCDM methods or their vague extensions. Senvar et al. [38] proposed a hesitant fuzzy TOPSIS model to decide the best location for hospitals. Hashemkhani Zolfani et al. [16] applied grey CRITIC and CoCoSo framework for deciding the most proper siting for a temporary hospital. Additionally, AHP [36, 37], EDAS [2, 45], TOPSIS [25], interval-valued Pythagorean fuzzy TOPSIS [3], fuzzy ELECTRE [20], fuzzy ANP [39], and hesitant fuzzy EDAS [21] have been preferred for this purpose. The goal of this work is to present an interval type-2 fuzzy best–worst method (IT2FBWM) approach to determine the priority weights of hospital site selection factors. Up to now, IT2FBWM was studied by Komatina et al. [19] in the automotive sector, Wu et al. [44], Qin and Liu [35], and Celik et al. [5] in supply chain management, Celik and Gul [4] in the construction industry, Tang et al. [41] for risk assessment, Gölcük [14] in the project management, and Gong et al. [11] in the energy sector.

This paper applied an integrated trapezoidal IT2FSs method for group decision-making based on BWM, which is named trapezoidal IT2FBWM. Wan et al. [43] and Chen et al. [8] concerned an integration of BWM and IT2FSs for hospital site planning in their papers, which indicates solely few papers have considered the fusion of BWM in IT2F environment. To support and extend the current literature, this work employs this approach to decide the weight of evaluation factors for pandemic hospital site decisions, which is the contribution of this paper.

2 Methodology

2.1 Preliminaries

Some mathematical formulations of IT2FSs are offered in the following:

Definition 1. A *type-2* fuzzy set $\tilde{\tilde{A}}$ is defined by a *type-2* membership function, X can be represented by a type-2 membership function $\mu_{\tilde{\tilde{A}}}(\mathrm{x}, \mathrm{u}) = 1$ where x is the main component in the domain X, given as Eq. (1) [29].

$$\tilde{\tilde{A}} = \left\{ \left((x, u), \mu_{\tilde{\tilde{A}}}(x, u)\right) \middle| \forall x \in X, \forall u \in J_X \subseteq [0, 1], 0 \leq \mu_{\tilde{\tilde{A}}}(x, u) \leq 1 \right\} \quad (1)$$

where $\forall u \in J_X \subseteq [0, 1]$. The type-2 fuzzy set $\tilde{\tilde{A}}$ can be represented as in Eq. 2.

$$\tilde{\tilde{A}} = \int_{x \in X} \int_{u \in J_X} \mu_{\tilde{\tilde{A}}}(x, u)/(x, u) \quad (2)$$

where u is the secondary variable in domain J_X at each $x \in X$. J_X is called the primary membership of x, and the secondary membership grades of $\tilde{\tilde{A}}$ all equal to 1, $J_X \subseteq [0, 1]$ and $\iint$ denote union over all admissible x and u.

Definition 2. An interval type-2 fuzzy set $\tilde{\tilde{A}}$ is a special case of general type-2 fuzzy sets, where all the secondary membership functions of ˜A are equal to 1. If all $\tilde{\tilde{A}}$, then $\tilde{\tilde{A}}$ is called an interval type-2 fuzzy sets that is the special case of a type-2 fuzzy set, represent as Eq. 3

$$\tilde{\tilde{A}} = \int_{x \in X} \int_{u \in J_X} 1/(x, u) \quad \text{where } J_X \subseteq [0, 1] \quad (3)$$

Definition 3. The upper membership function and lower membership function of an interval type-2 fuzzy set are type-1 membership functions, respectively. Figure 1 shows a trapezoidal interval type-2 fuzzy set in this study [29, 28].

$$\begin{aligned} \tilde{\tilde{A}}_i = \left(\tilde{A}_i^U, \tilde{A}_i^L\right) = &\left(\left(a_{i1}^U, a_{i2}^U, a_{i3}^U, a_{i4}^U; H_1\left(\tilde{A}_i^U\right), H_2\left(\tilde{A}_i^U\right)\right),\right. \\ &\left.\left(a_{i1}^L, a_{i2}^L, a_{i3}^L, a_{i4}^L; H_1\left(\tilde{A}_i^L\right), H_2\left(\tilde{A}_i^L\right)\right)\right) \end{aligned} \quad (4)$$

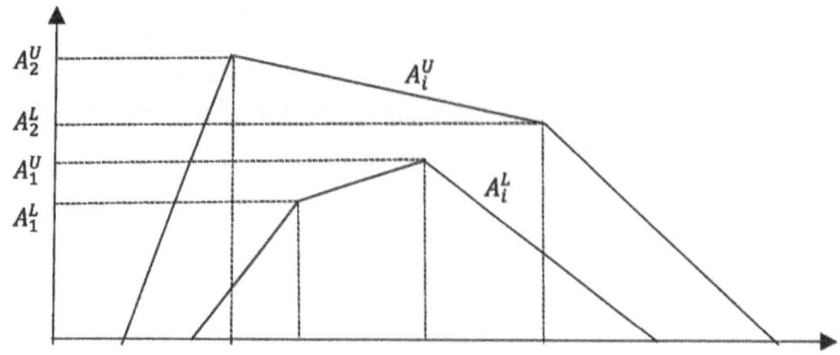

Fig. 1 Two membership functions

where $\widetilde{A}_i^U$ and $\widetilde{A}_i^L$ are type 1 fuzzy sets, $a_{i1}^U, a_{j2}^U, a_{i3}^U, a_{i4}^U, a_{i1}^L, a_{i2}^L, a_{i3}^L, a_{i4}^L$ are the references points of the interval type-2 fuzzy $\widetilde{A}_i^L$ indicates the membership value of the element $a_{i(j+1)}^U$ in the upper trapezoidal membership function $\widetilde{A}_i^U$; $1 \leq j \leq 2$, $H_j(\widetilde{A}_i^L)$ shows the membership value of the element $a_{i(j+1)}^U$ in the upper trapezoidal membership function $\widetilde{A}_i^U$; and $1 \leq j \leq 2, H_j(\widetilde{A}_i^U), H_j(\widetilde{A}_i^L) \in [0, 1], 1 \leq i \leq n[0, 1]$, $1 \leq i \leq n$.

Assume that, two trapezoidal interval type-2 fuzzy numbers as $(\tilde{\tilde{A}}_1, \tilde{\tilde{A}}_2)$, and the mathematical operations of these trapezoidal fuzzy numbers are given as follows [6, 22].

Definition 4. The addition operation between IT2FNs is demonstrated as in Eq. (5)

$$\tilde{\tilde{A}}_1 \oplus \tilde{\tilde{A}}_2 = \left(\widetilde{A}_1^U, \widetilde{A}_1^L\right) \oplus \left(\widetilde{A}_2^U, \widetilde{A}_2^L\right)$$
$$= \begin{pmatrix} a_{11}^U + a_{21}^U, a_{12}^U + a_{22}^U, a_{13}^U + a_{23}^U, a_{14}^U + a_{24}^U; \min\left(H_1\left(\widetilde{A}_1^U\right), H_1\left(\widetilde{A}_2^U\right)\right) \min\left(H_2\left(\widetilde{A}_1^U\right), H_2\left(\widetilde{A}_2^U\right)\right) \\ a_{11}^L + a_{21}^L, a_{12}^L + a_{22}^L, a_{13}^L + a_{23}^L, a_{14}^L + a_{24}^L; \min\left(H_1\left(\widetilde{A}_1^L\right), H_1\left(\widetilde{A}_2^L\right)\right), \min\left(H_2\left(\widetilde{A}_1^L\right), H_2\left(\widetilde{A}_2^L\right)\right) \end{pmatrix} \quad (5)$$

Definition 5. The subtraction operation between IT2FNs is demonstrated as in Eq. (6)

$$\tilde{\tilde{A}}_1 \ominus \tilde{\tilde{A}}_2 = \left(\widetilde{A}_1^U, \widetilde{A}_1^L\right) \ominus \left(\widetilde{A}_2^U, \widetilde{A}_2^L\right)$$
$$= \begin{pmatrix} a_{11}^U - a_{24}^U, a_{12}^U - a_{23}^U, a_{13}^U - a_{22}^U, a_{14}^U - a_{21}^U; \min\left(H_1\left(\widetilde{A}_1^U\right), H_1\left(\widetilde{A}_2^U\right)\right) \min\left(H_2\left(\widetilde{A}_1^U\right), H_2\left(\widetilde{A}_2^U\right)\right) \\ a_{11}^L - a_{24}^L, a_{12}^L - a_{23}^L, a_{13}^L - a_{22}^L, a_{14}^L - a_{21}^L; \min\left(H_1\left(\widetilde{A}_1^L\right), H_1\left(\widetilde{A}_2^L\right)\right), \min\left(H_2\left(\widetilde{A}_1^L\right), H_2\left(\widetilde{A}_2^L\right)\right) \end{pmatrix} \quad (6)$$

Definition 6. The multiplication operation between IT2FNs is defined as in Eq. (7)

$$\tilde{\tilde{A}}_1 \otimes \tilde{\tilde{A}}_2 = \left(\widetilde{A}_1^U, \widetilde{A}_1^L\right) \otimes \left(\widetilde{A}_2^U, \widetilde{A}_2^L\right)$$
$$= \begin{pmatrix} a_{11}^U \times a_{21}^U, a_{12}^U \times a_{22}^U, a_{13}^U \times a_{23}^U, a_{14}^U \times a_{24}^U; \min\left(H_1\left(\widetilde{A}_1^U\right), H_1\left(\widetilde{A}_2^U\right)\right), \min\left(H_2\left(\widetilde{A}_1^U\right), H_2\left(\widetilde{A}_2^U\right)\right) \\ a_{11}^L \times a_{21}^L, a_{12}^L \times a_{22}^L, a_{13}^L \times a_{23}^L, a_{14}^L \times a_{24}^L; \min\left(H_1\left(\widetilde{A}_1^L\right), H_1\left(\widetilde{A}_2^L\right), \min\left(H_2\left(\widetilde{A}_1^L\right), H_2\left(\widetilde{A}_2^L\right)\right)\right) \end{pmatrix} \quad (7)$$

Definition 7. The diversion operation between IT2FNs is demonstrated as in Eq. (8) [6]

$$\frac{\widetilde{\widetilde{A}}_1}{\widetilde{\widetilde{A}}_2} = \frac{(\widetilde{A}_1^U, \widetilde{A}_1^L)}{(\widetilde{A}_2^U, \widetilde{A}_2^L)} = \left(\frac{a_{11}^U}{a_{24}^U}, \frac{a_{12}^U}{a_{23}^U}, \frac{a_{13}^U}{a_{22}^U}, \frac{a_{14}^U}{a_{21}^U}, \min\left(H_1\left(\widetilde{A}_1^U\right), H_1\left(\widetilde{A}_2^U\right)\right)\right)$$
$$\left(\frac{a_{11}^L}{a_{24}^L}, \frac{a_{12}^L}{a_{23}^L}, \frac{a_{13}^L}{a_{22}^L}, \frac{a_{14}^L}{a_{21}^L}, \min\left(H_1\left(\widetilde{A}_1^L\right), H_1\left(\widetilde{A}_2^L\right)\right)\right) \quad (8)$$

Definition 8. [17] offered a method for trapezoidal interval type-2 fuzzy sets were provided by modifying the centre of area $\tilde{\tilde{\beta}}_j$ value for defuzzifying and ranking IT2FSs indicates in Eq. (9):

$$D\left(\tilde{\tilde{\beta}}\right) = \frac{\frac{\left(a_{j4}^U - a_{j1}^U\right) + \left(H_1\left(\tilde{\tilde{A}}\right)_j^U * a_{j2}^U - a_{j1}^U\right) + \left(H_2\left(\tilde{\tilde{A}}\right)_j^U * a_{j3}^U - a_{j1}^U\right)}{4} + a_{j1}^U + \left[\frac{\left[\left(a_{j4}^L - a_{j1}^L\right) + \left(H_1\left(\tilde{\tilde{A}}\right)_j^L * a_{j2}^L - a_{j1}^L\right) + \left(H_2\left(\tilde{\tilde{A}}\right)_j^L * a_{j3}^L - a_{j1}^L\right)\right.}{4} + a_{j1}^L\right]}{2} \quad (9)$$

2.2 *Interval Type-2 Fuzzy Best–Worst Method (IT2FBWM)*

Best–worst method (BWM) has been incorporated with a different type of fuzzy sets such as interval type-2 fuzzy number [35, 44], intuitionistic fuzzy sets [31, 42], triangular fuzzy numbers [9, 13, 15], Z-numbers [1], hesitant fuzzy numbers [24, 30], rough-fuzzy approach [7], Pythagorean hesitant fuzzy sets [26], Bayesian BWM [12], and D numbers [34]. In this study, IT2FBWM, the IT2F pairwise comparisons are applied based on the linguistic terms given in Table 1. Then, the linguistic evaluations are transformed into trapezoidal interval type-2 fuzzy number (TIT2FN).

BWM is more efficient and adaptable than AHP, ANP methods because BWM has the following two advantages [27, 33, 40, 48]:

- Having a smaller number of pairwise comparisons. AHP requires [n (n − 1)/2] comparisons, whereas BWM needs totally [2n − 3] comparisons. If the number of attributes (alternatives and criteria) is more than three, in comparison with AHP, the number of pairwise comparison in BWM can diminish dramatically when

Table 1 Linguistic terms for factor weight [17]

Abb.	Linguistic variables	Crisp number	Trapezoidal interval type-2 fuzzy number
EE	Equally important	1	((1;1;1;1;1;1), (1;1;1;1;0.8;0.8))
SS	Slightly strong	3	((1;2;4;5;1;1), (1.2;2.2;3.8;4.8;0.8;0.8))
FS	Fairly strong	5	((3;4;6;7;1;1), (3.2;4.2;5.8;6.8;0.8;0.8))
VS	Very strong	7	((5;6;8;9;1;1), (5.2;6.2;7.8;8.8;0.8;0.8))
AS	Absolutely strong	9	((7;8;9;9;1;1), (7.2;8.2;8.8;9;0.8;0.8))

attributes (n) is large. Compared with the ANP method, BWM can reduce $(n-2)(n-3)/2$ comparisons when n is greater than two.

- More precise object weights. BWM's item weights are more realistic than AHP method.
- The degree of consistency establishes whether a comparison of criteria is consistent, but the level of consistency in BWM is determined by the level of consistency as BWM output have always been consistent.

The fuzzy pairwise comparisons are implemented based on the interval type-2 linguistic terms supplied in Table 1 in this section, and the fuzzy comparisons are applied based on the interval type-2 linguistic terms given in Table 1. The linguistic evaluation is then converted into fuzzy numbers with a trapezoidal interval type-2 (IT2FNs).

IT2FBWM steps are explained as follows:

Step 1. *Build the decision criteria system.* The decision criteria system consists of a set of decision criteria. n decision criteria set is presented as follows: $\{I_1, I_2, \ldots\ldots\ldots, I_n\}$. The interval type-2 fuzzy comparison matrix ($\tilde{\tilde{M}}$) can be obtained as follows,

$$\tilde{\tilde{M}} = \begin{array}{c} \\ I_1 \\ I_2 \\ \vdots \\ I_n \end{array} \begin{array}{c} I_1 \; I_2 \; \cdots \; I_n \\ \begin{bmatrix} \tilde{\tilde{\eta}}_{11} & \tilde{\tilde{\eta}}_{12} & \cdots & \tilde{\tilde{\eta}}_{1n} \\ \tilde{\tilde{\eta}}_{21} & \tilde{\tilde{\eta}}_{22} & \cdots & \tilde{\tilde{\eta}}_{2n} \\ \vdots & \vdots & \ddots & \vdots \\ \tilde{\tilde{\eta}}_{n1} & \tilde{\tilde{\eta}}_{n2} & \cdots & \tilde{\tilde{\eta}}_{nn} \end{bmatrix} \end{array}$$

where $\tilde{\eta}_{ij}$ indicates the relative interval type-2 fuzzy preference of criterion i over criterion j, which is a trapezoidal interval type-2 fuzzy number; $\tilde{\eta}_{ij} = ((1; 1; 1; 1; 1; 1), (1; 1; 1; 1; 0.8; 0.8))$ when i criterion *is* equal to j *criterion (EE).*

Step 2. *Determine the best and the worst criterion.* Experts select the best and worst criterion in this stage, depending on the created criteria established in Step 1. For each expert's tea, the best and the worst criterion are denoted as Δ_B and Δ_W.

Step 3. *Implement the interval type-2 fuzzy reference comparisons for the best criterion and worst criterion.* In this step, the fuzzy preferences of the best criterion over all the other criteria are decided by experts. Then, the fuzzy comparisons in the linguistic format are converted to trapezoidal interval type-2 fuzzy numbers. The interval type-2 fuzzy best to others vector (($\tilde{\tilde{\Delta}}_{BtO}$) can be obtained as follows:

$$\tilde{\tilde{\Delta}}_{BtO} = \left[\tilde{\tilde{\eta}}_{B1}, \tilde{\tilde{\eta}}_{B2}, \ldots\ldots\ldots\ldots, \tilde{\tilde{\eta}}_{Bn}\right]^T \tag{10}$$

where $\tilde{\tilde{\eta}}_{Bj}$ indicates the interval type-2 fuzzy (IT2F) preference of the best criterion Δ_B over the criterion j and $j = 1, 2, \ldots, n$.

$\tilde{\tilde{\eta}}_{Bj}$ indicates the IT2F preference of the best criterion B over the j criterion. If the best criterion is equal to j criterion, then $\tilde{\tilde{\eta}}_{BB} = ((1, 1, 1, 1; 1; 1), (1, 1, 1, 1; 0.8; 0.8))$.

The fuzzy preferences of all the criteria over the worst criterion are determined. The fuzzy others to worst vector ($\tilde{\tilde{\Delta}}_{OtW}$) can be obtained as:

$$\tilde{\tilde{\Delta}}_{OtW} = \left[\tilde{\tilde{\eta}}_{1W}, \tilde{\tilde{\eta}}_{2W}, \ldots\ldots\ldots\ldots, \tilde{\tilde{\eta}}_{nW}\right]^T \quad (11)$$

where $\tilde{\tilde{\eta}}_{jW}$ denotes the IT2F preference of the criterion j over the worst criterion W. When the worst criterion compared to itself, then $\tilde{\tilde{\eta}}_{WW} = ((1; 1; 1; 1; 1; 1), (1; 1; 1; 1; 0.8; 0.8))$.

Step 4. *Obtain the optimal interval type-2 fuzzy weighting vector* $\tilde{\tilde{W}} = \left[\tilde{\tilde{w}}_1^*, \tilde{\tilde{w}}_2^*, \ldots\ldots.., \tilde{\tilde{w}}_n^*\right]^T$. The optimal IT2F weight for each criterion is found for each fuzzy pair $\left(\frac{\tilde{\tilde{w}}_B}{\tilde{\tilde{w}}_j}\right)$ and $\left(\frac{\tilde{\tilde{w}}_j}{\tilde{\tilde{w}}_W}\right)$. It should have $\left(\frac{\tilde{\tilde{w}}_B}{\tilde{\tilde{w}}_j} = \tilde{\tilde{\eta}}_{Bj}\right)$ and $\left(\frac{\tilde{\tilde{w}}_j}{\tilde{\tilde{w}}_W} = \tilde{\tilde{\eta}}_{jW}\right)$. To achieve the highest consistency, which is obtained that the maximum absolute differences between $\left|\frac{\tilde{\tilde{w}}_B}{\tilde{\tilde{w}}_j} - \tilde{\tilde{\eta}}_{Bj}\right|$ and $\left|\frac{\tilde{\tilde{w}}_j}{\tilde{\tilde{w}}_W} - \tilde{\tilde{\eta}}_{jW}\right|$ for all j should minimized to satisfy these conditions for each j.

$$\begin{aligned}\tilde{\tilde{w}}_B =& \left(w_{B1}^U, w_{B2}^U, w_{B3}^U, w_{B4}^U; H_1(\tilde{w}_{B1}^U), H_2(\tilde{w}_{B2}^U)\right),\\ & \left(w_{B1}^L, w_{B2}^L, w_{B3}^L, w_{B4}^L; H_1(\tilde{w}_{B1}^L), H_2(\tilde{w}_{B2}^L)\right)\end{aligned} \quad (12)$$

$$\begin{aligned}\tilde{\tilde{w}}_W =& \left(w_{W1}^U, w_{W2}^U, w_{W3}^U, w_{W4}^U; H_1(\tilde{w}_{W1}^U), H_2(\tilde{w}_{W2}^U)\right),\\ & \left(w_{W1}^L, w_{W2}^L, w_{W3}^L, w_{W4}^L; H_1(\tilde{w}_{W1}^L), H_2(\tilde{w}_{W2}^L)\right)\end{aligned} \quad (13)$$

$$\begin{aligned}\tilde{\tilde{w}}_j =& (w_{j1}^U, w_{j2}^U, w_{j3}^U, w_{j4}^U; H_1(\tilde{w}_{j1}^U), H_2(\tilde{w}_{j2}^U)),\\ & (w_{j1}^L, w_{j2}^L, w_{j3}^L, w_{j4}^L; H_1(\tilde{w}_{j1}^L), H_2(\tilde{w}_{j2}^L))\end{aligned} \quad (14)$$

$$\begin{aligned}\tilde{\tilde{\eta}}_{jW} =& \left(w_{jW1}^U, w_{jW2}^U, w_{jW3}^U, w_{jW4}^U; H_1(\tilde{w}_{jW1}^U), H_2(\tilde{w}_{jW2}^U)\right),\\ & \left(w_{jW1}^L, w_{jW2}^L, w_{jW3}^L, w_{jW4}^L; H_1(\tilde{w}_{jW1}^L), H_2(\tilde{w}_{jW2}^L)\right)\end{aligned} \quad (15)$$

$$\tilde{\tilde{\eta}}_{BW} = \left(w_{Bj1}^U, w_{Bj2}^U, w_{Bj3}^U, w_{Bj4}^U; H_1(\tilde{w}_{Bj1}^U), H_2(\tilde{w}_{Bj2}^U)\right),$$

$$\left(w^L_{Bj1}, w^L_{Bj2}, w^L_{Bj3}, w^L_{Bj4}; H_1(\tilde{w}^L_{Bj1}), H_2(\tilde{w}^L_{Bj2})\right) \tag{16}$$

Consequently, the constrained optimization problem is constructed for obtaining the optimal fuzzy weights $\left[\tilde{\tilde{w}}^*_1, \tilde{\tilde{w}}^*_2, \ldots\ldots.., \tilde{\tilde{w}}^*_n\right]^T$ as follows:

$$\min \max(j)\left\{\left|\frac{\tilde{\tilde{w}}_B}{\tilde{\tilde{w}}_j} - \eta_{Bj}\right|, \left|\frac{\tilde{\tilde{w}}_j}{\tilde{\tilde{w}}_W} - \tilde{\tilde{\eta}}_{jW}\right|\right\} \tag{17}$$

$$s.t. \begin{cases} \sum_{j=1}^{n} D\left(\tilde{\tilde{\beta}}_j\right) = 1 \\ w^U_{j1} \le w^L_{j1} \\ w^L_{j4} \le w^U_{j4} \\ w^U_{j1} \le w^U_{j2} \le w^U_{j3} \le w^U_{j4} \\ w^L_{j1} \le w^L_{j2} \le w^L_{j3} \le w^L_{j4} \\ w^U_{j1} \ge 0, j = \{1, 2, \ldots\ldots n\} \end{cases} \tag{18}$$

In order to minimize the absolute value as $\varepsilon = (\varepsilon; \varepsilon; \varepsilon; \varepsilon; 1; 1)(\varepsilon; \varepsilon; \varepsilon; \varepsilon; 0.8; 0.8)$, the model is converted to nonlinear programming form.

$$\text{Min}\, \varepsilon$$

$$s.t. \begin{cases} \left|\tilde{\tilde{w}}^U_{B1} - \tilde{\tilde{w}}^U_{Bj1}\tilde{\tilde{w}}^U_{j4}\right| \le \varepsilon; \left|\tilde{\tilde{w}}^U_{B2} - \tilde{\tilde{w}}^U_{Bj2}\tilde{\tilde{w}}^U_{j3}\right| \le \varepsilon; j = \{1, 2, \ldots\ldots\ldots..n\} \\ \left|\tilde{\tilde{w}}^U_{B3} - \tilde{\tilde{w}}^U_{Bj3}\tilde{\tilde{w}}^U_{j2}\right| \le \varepsilon; \left|\tilde{\tilde{w}}^U_{B4} - \tilde{\tilde{w}}^U_{Bj4}\tilde{\tilde{w}}^U_{j1}\right| \le \varepsilon; j = \{1, 2, \ldots\ldots\ldots..n\} \\ \left|\tilde{\tilde{w}}^L_{B1} - \tilde{\tilde{w}}^L_{Bj1}\tilde{\tilde{w}}^L_{j4}\right| \le \varepsilon; \left|\tilde{\tilde{w}}^L_{B2} - \tilde{\tilde{w}}^L_{Bj2}\tilde{\tilde{w}}^L_{j3}\right| \le \varepsilon; j = \{1, 2, \ldots\ldots\ldots..n\} \\ \left|\tilde{\tilde{w}}^L_{B3} - \tilde{\tilde{w}}^L_{Bj3}\tilde{\tilde{w}}^L_{j2}\right| \le \varepsilon; \left|\tilde{\tilde{w}}^L_{B4} - \tilde{\tilde{w}}^L_{Bj4}\tilde{\tilde{w}}^L_{j1}\right| \le \varepsilon; j = \{1, 2, \ldots\ldots\ldots..n\} \\ \left|\tilde{\tilde{w}}^U_{j1} - \tilde{\tilde{w}}^U_{jW1}\tilde{\tilde{w}}^U_{W4}\right| \le \varepsilon; \left|\tilde{\tilde{w}}^U_{j2} - \tilde{\tilde{w}}^U_{jW2}\tilde{\tilde{w}}^U_{W3}\right| \le \varepsilon; j = \{1, 2, \ldots\ldots\ldots..n\} \\ \left|\tilde{\tilde{w}}^U_{j3} - \tilde{\tilde{w}}^U_{jW3}\tilde{\tilde{w}}^U_{W2}\right| \le \varepsilon; \left|\tilde{\tilde{w}}^U_{j4} - \tilde{\tilde{w}}^U_{jW4}\tilde{\tilde{w}}^U_{W1}\right| \le \varepsilon; j = \{1, 2, \ldots\ldots\ldots..n\} \\ \left|\tilde{\tilde{w}}^L_{j1} - \tilde{\tilde{w}}^L_{jW1}\tilde{\tilde{w}}^L_{W4}\right| \le \varepsilon; \left|\tilde{\tilde{w}}^L_{j2} - \tilde{\tilde{w}}^L_{jW2}\tilde{\tilde{w}}^L_{W3}\right| \le \varepsilon; j = \{1, 2, \ldots\ldots\ldots..n\} \\ \left|\tilde{\tilde{w}}^L_{j3} - \tilde{\tilde{w}}^L_{jW3}\tilde{\tilde{w}}^L_{W2}\right| \le \varepsilon; \left|\tilde{\tilde{w}}^L_{j4} - \tilde{\tilde{w}}^L_{jW4}\tilde{\tilde{w}}^L_{W1}\right| \le \varepsilon; j = \{1, 2, \ldots\ldots\ldots..n\} \\ \sum_{j=1}^{n} D\left(\tilde{\tilde{\beta}}_j\right) = 1\ j = \{1, 2, \ldots\ldots\ldots\} \\ w^U_{j1} \le w^L_{j1}, w^L_{j4} \le w^U_{j4}\ j = \{1, 2, \ldots\ldots.n\} \\ w^U_{j1} \le w^U_{j2} \le w^U_{j3} \le w^U_{j4}\ j = \{1, 2, \ldots\ldots.n\} \\ w^L_{j1} \le w^L_{j2} \le w^L_{j3} \le w^L_{j4}\ j = \{1, 2, \ldots\ldots n\} \\ w^U_{j1} \ge 0,\ j = \{1, 2, \ldots\ldots\ldots n\} \end{cases} \tag{19}$$

Step 5. Consistency ratio (CR) is an indicator to measure the degree of consistency of pairwise comparisons. When the equation is satisfied $\tilde{\tilde{\sigma}}_{Bj} \, x \, \tilde{\tilde{\sigma}}_{jW} = \tilde{\tilde{\sigma}}_{BW}$, a fuzzy comparison is fully consistent, whereas when the equality is not achieved $\tilde{\tilde{\sigma}}_{Bj} \, x \, \tilde{\tilde{\sigma}}_{jW} \neq \tilde{\tilde{\sigma}}_{BW}$, it occurs inconsistency. If pairwise comparisons are equal ($\tilde{\tilde{\sigma}}_{Bj} = \tilde{\tilde{\sigma}}_{jW} = \tilde{\tilde{\sigma}}_{BW}$), the inconsistency reaches its highest point, which is represented in ε value. To create the maximum inequality, ($\tilde{\tilde{\sigma}}_{Bj} = \tilde{\tilde{\sigma}}_{jW} = \tilde{\tilde{\sigma}}_{BW}$). Maximum inconsistency value must be subtracted from $\tilde{\tilde{\sigma}}_{Bj}$ and $\tilde{\tilde{\sigma}}_{jW}$ added to $\tilde{\tilde{\sigma}}_{BW}$ given in Eq. (20).

$$\left(\tilde{\tilde{\sigma}}_{Bj} - \varepsilon\right)\left(\tilde{\tilde{\sigma}}_{jW} - \varepsilon\right) = \left(\tilde{\tilde{\sigma}}_{BW} + \varepsilon\right) \tag{20}$$

This equation can be represented as follows:

$$\varepsilon^2 - \left(1 + 2\tilde{\tilde{\sigma}}_{BW}\right)\varepsilon + \left(\tilde{\tilde{\sigma}}_{BW}^2 - \tilde{\tilde{\sigma}}_{BW}\right) = 0 \tag{21}$$

where $\sigma_{BW} = \left(\sigma_{BW1}^U, \sigma_{BW2}^U, \sigma_{BW3}^U, \sigma_{BW4}^U; H_1(A_{BW1}^U), H_2(A_{BW2}^U)\right)$ $\left(\sigma_{BW1}^L, \sigma_{BW2}^L, \sigma_{BW3}^L, \sigma_{BW4}^L; H_1(A_{BW1}^L), H_2(A_{BW2}^L)\right)$ and $\varepsilon = \left(\varepsilon_1^U, \varepsilon_2^U, \varepsilon_3^U, \varepsilon_4^U, H_1(A_1^U), H_2(A_2^U)\right)\left(\varepsilon_1^L, \varepsilon_2^L, \varepsilon_3^L, \varepsilon_4^L, H_1(A_1^L), H_2(A_2^L)\right)$. Accordingly, Eq. (21) can be solved for different values to find the maximum value of ε. Obviously, the greatest value of ε is closely related to σ_{BW4}^U value. In the light of this information, Eq. (21) is transformed as follows [32]:

$$\varepsilon^2 - \left(1 + 2\tilde{\tilde{\sigma}}_{BW}^U\right)\varepsilon + \left(\left(\tilde{\tilde{\sigma}}_{BW}^U\right)^2 - \tilde{\tilde{\sigma}}_{BW}^U\right) = 0 \tag{22}$$

The consistency ratio is calculated using the consistency index that derived from Eq. (22), which is given in Table 2.

When ε = 0, the deviation in Eq. (19) is the minimum value. As a result, the interval type-2 fuzzy pairwise is totally consistent. To check the degree of consistency and the reliability, Eq. (23) is utilized:

$$\mathrm{CR} = \frac{\varepsilon}{\mathrm{CI}} \tag{23}$$

The closer the consistency ratio is to zero, the more consistent the assessments are, and the closer it is to 1, the less consistent [44].

Table 2 Consistency index (CI) for IT2FBWM

Linguistic term	EE	SS	FS	VS	AS
CI	3	8.701	11.274	13.772	13.772

3 IT2FBWM for Determining Criteria Importance of Pandemic Hospital Site Selection

We use the suggested IT2FBWM to deal with hospital site selection problem. For our case study, it was mainly examined three site selection criteria, including social and policy, environmental and economic, and their nine sub-criteria considering the judgements of four experts (EX) (urban planner, medical doctor, transport engineer, and finance manager), represented in Fig. 2.

According to the application of IT2FBWM method, evaluations of experts in linguistic terms are used to obtain the importance weights of hospital siting criteria. The best criteria, the worst criteria, the best to other comparison matrix, and the other to worst comparison matrix of each expert are given in Table 3 by using linguistic terms, respectively.

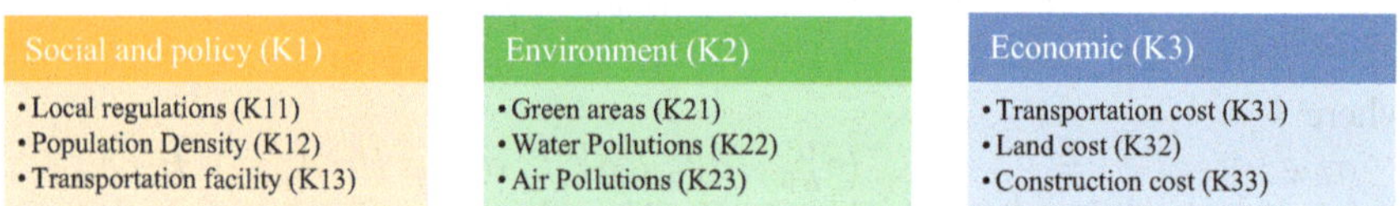

Fig. 2 Hospital site selection criteria and their related sub-criteria

Table 3 Experts' preferences

Expert panel	Best factor	Best to others (BtO)			Worst factor	Others to worst (OtW)		
EX1	*K*1	EE	FS	SS	*K*2	FS	EE	SS
	*K*13	VS	SS	EE	*K*11	EE	SS	FS
	*K*23	FS	SS	EE	*K*21	EE	SS	FS
	*K*32	FS	EE	SS	*K*31	EE	FS	SS
EX2	*K*1	EE	SS	EE	*K*2	SS	EE	SS
	*K*12	FS	EE	SS	*K*11	EE	VS	SS
	*K*22	FS	EE	SS	*K*21	EE	FS	SS
	*K*33	VS	SS	EE	*K*31	EE	SS	VS
EX3	*K*3	SS	FS	EE	*K*2	FS	EE	SS
	*K*12	SS	EE	EE	*K*11	EE	SS	SS
	*K*23	SS	FS	EE	*K*22	SS	EE	FS
	*K*33	SS	EE	EE	*K*31	EE	SS	SS
EX4	*K*1	EE	FS	SS	*K*2	SS	EE	FS
	*K*13	SS	FS	EE	*K*12	SS	EE	FS
	*K*22	SS	EE	EE	*K*21	EE	SS	SS
	*K*32	SS	EE	FS	*K*33	SS	FS	EE

To take evaluations of Expert 1 as an example, the interval type-2 fuzzy preferences of the best factor over all main factors and the other main factors to the worst can be obtained with respect to Table 3 as follows:

$$\tilde{\tilde{\eta}}_{BtO} = \{(1; 1; 1; 1; 1; 1), (1; 1; 1; 1; 0.8; 0.8); (3; 4; 6; 7; 1; 1), (3.2; 4.2; 5.8; 6.8; 0.8; 0.8); (1; 2; 4; 5; 1; 1), (1.2; 2.2; 3.8; 4.8; 0.8; 0.8)\}$$

$$\tilde{\tilde{\eta}}_{OtW} = \{(3; 4; 6; 7; 1; 1), (3.2; 4.2; 5.8; 6.8; 0.8; 0.8); (1; 1; 1; 1; 1; 1), (1; 1; 1; 1; 0.8; 0.8); (1; 2; 4; 5; 1; 1), (1.2; 2.2; 3.8; 4.8; 0.8; 0.8)\}$$

Then, for obtaining the optimal interval type-2 fuzzy weights of Expert 1, the nonlinearly constrained model is solved in LINGO software version 19.0. The following are the remaining comprehensive computing processes of an absolute value mathematical model of the key factors for Expert 1:

$$\text{Min}\,\varepsilon$$

$$\text{s.t.}\begin{cases}
\left|w_{11}^U - 3 * w_{24}^U\right| \le \varepsilon; & \left|w_{12}^U - 4 * w_{23}^U\right| \le \varepsilon; \quad \left|w_{13}^U - 6 * w_{22}^U\right| \le \varepsilon; \quad \left|w_{14}^U - 7 * w_{21}^U\right| \le \varepsilon;\\
\left|w_{11}^L - 3.2 * w_{24}^L\right| \le \varepsilon; & \left|w_{11}^L - 4.2 * w_{23}^L\right| \le \varepsilon; \quad \left|w_{13}^L - 5.8 * w_{22}^L\right| \le \varepsilon; \quad \left|w_{14}^L - 6.8 * w_{21}^L\right| \le \varepsilon;\\
\left|w_{11}^U - 1 * w_{34}^U\right| \le \varepsilon; & \left|w_{12}^U - 2 * w_{33}^U\right| \le \varepsilon; \quad \left|w_{13}^U - 4 * w_{32}^U\right| \le \varepsilon; \quad \left|w_{14}^U - 5 * w_{31}^L\right| \le \varepsilon;\\
\left|w_{11}^L - 1.2 * w_{34}^L\right| \le \varepsilon; & \left|w_{12}^L - 2.2 * w_{33}^L\right| \le \varepsilon; \quad \left|w_{13}^L - 3.8 * w_{32}^L\right| \le \varepsilon; \quad \left|w_{14}^L - 4.8 * w_{31}^L\right| \le \varepsilon;\\
\left|w_{21}^U - 3 * w_{14}^U\right| \le \varepsilon; & \left|w_{22}^U - 4 * w_{13}^U\right| \le \varepsilon; \quad \left|w_{23}^U - 6 * w_{12}^U\right| \le \varepsilon; \quad \left|w_{24}^U - 7 * w_{11}^U\right| \le \varepsilon;\\
\left|w_{21}^L - 3.2 * w_{14}^L\right| \le \varepsilon; & \left|w_{22}^L - 4.2 * w_{13}^L\right| \le \varepsilon; \quad \left|w_{23}^L - 5.8 * w_{12}^L\right| \le \varepsilon; \quad \left|w_{24}^L - 6.8 * w_{11}^L\right| \le \varepsilon;\\
\left|w_{21}^U - 1 * w_{24}^U\right| \le \varepsilon; & \left|w_{22}^U - 1 * w_{23}^U\right| \le \varepsilon; \quad \left|w_{23}^U - 1 * w_{22}^U\right| \le \varepsilon; \quad \left|w_{24}^U - 1 * w_{21}^U\right| \le \varepsilon;\\
\left|w_{21}^L - 1 * w_{24}^L\right| \le \varepsilon; & \left|w_{22}^L - 1 * w_{23}^L\right| \le \varepsilon; \quad \left|w_{23}^L - 1 * w_{22}^L\right| \le \varepsilon; \quad \left|w_{24}^L - 1 * w_{21}^L\right| \le \varepsilon;\\
\left|w_{21}^U - 1 * w_{34}^U\right| \le \varepsilon; & \left|w_{22}^U - 2 * w_{33}^U\right| \le \varepsilon; \quad \left|w_{23}^U - 4 * w_{32}^U\right| \le \varepsilon; \quad \left|w_{24}^U - 5 * w_{31}^U\right| \le \varepsilon;\\
\left|w_{21}^L - 1 * w_{34}^L\right| \le \varepsilon; & \left|w_{22}^L - 2.2 * w_{33}^L\right| \le \varepsilon; \quad \left|w_{23}^L - 3.8 * w_{32}^L\right| \le \varepsilon; \quad \left|w_{24}^L - 4.8 * w_{31}^L\right| \le \varepsilon;\\
\sum_{j=1}^{3} D(\beta_j) = 1 & \\
w_{j1}^U \le w_{j1}^L, \quad w_{j4}^L \le w_{j4}^U & \\
w_{j1}^U \le w_{j2}^U \le w_{j3}^U \le w_{j4}^U & \\
w_{j1}^L \le w_{j2}^L, \quad w_{j3}^L \le w_{j4}^L & \\
w_{j1}^U \le w_{j1}^U, \quad w_{j1}^U \le w_{j1}^U & \\
w_{j1}^U \ge 0, \quad j = \{1, 2, 3\} &
\end{cases}$$

According to result, interval type-2 fuzzy weight score of three key factors as following:

$$\tilde{\tilde{w}}_{K1} = [(0.17, 0.18, 0.26, 0.34; 1, 1)(0.34, 0.34, 0.34, 0.34; 0.8, 0.8)]$$

$$\tilde{\tilde{w}}_{K2} = [(0.17, 0.18, 0.26, 0.34; 1, 1)(0.17, 0.20, 0.28, 0.34; 0.8, 0.8)]$$

$$\tilde{\tilde{w}}_{K3} = [(0.24, 0.28, 0.52, 1.02; 1, 1)(0.25, 0.31, 0.54, 0.99; 0.8, 0.8)]$$

The weights of the key factors and sub-factors obtained for each expert opinion are given in Table 4. The interval type-2 fuzzy global weight of each sub-factor was computed with the help of Eq. 7. The consistency ratio for all Expert-1 is calculated by using Eq. (22). The optimal ε is 0.85, so CR is equal to 0.075. The consistency

Table 4 Optimal IT2F weights for expert 1

Key factor	Local weight	Sub-factor	Local weight	Global weight
*K*1	(0.17,0.18,0.26,0.34;1,1) (0.34,0.34,0.34,0.34;0.8,0.8)	*K*11	(0.17,0.18,0.22,0.26;1,1) (0.17,0.18,0.21,0.25;0.8,0.8)	(0.03,0.03,0.06,0.09;1,1) (0.06,0.06,0.07,0.09;0.8,0.8)
		*K*12	(0.27,0.32,0.32,1.27;1,1) (0.28,0.35,0.58,1.06;0.8,0.8)	(0.05,0.06,0.08,0.43;1,1) (0.05,0.07,0.16,0.36;0.8,0.8)
		*K*13	(0.19,0.22,0.32,0.42;1,1) (0.2,0.23,0.3,0.4;0.8,0.8)	(0.05,0.06,0.17,0.43;1,1) (0.05,0.07,0.16,0.4;0.8,0.8)
*K*2	(0.17,0.18,0.26,0.34;1,1) (0.17,0.2,0.28,0.34;0.8,0.8)	*K*21	(0.19,0.21,0.29,0.38;1,1) (0.19,0.21,0.28,0.36;0.8,0.8)	(0.03,0.04,0.08,0.13;1,1) (0.06,0.07,0.10,0.12;0.8,0.8)
		*K*22	(0.27,0.29,0.29,1.15;1,1) (0.28,0.33,0.53,0.96;0.8,0.8)	(0.05,0.05,0.08,0.39;1,1) (0.05,0.07,0.15,0.33;0.8,0.8)
		*K*23	(0.19,0.21,0.29,0.38;1,1) (0.19,0.21,0.28,0.36;0.8,0.8)	(0.05,0.06,0.15,0.39;1,1) (0.05,0.07,0.15,0.36;0.8,0.8)
*K*3	(0.24,0.28,0.52,1.02;1,1) (0.25,0.31,0.54,0.99;0.8,0.8)	*K*31	(0.19,0.21,0.29,0.38;1,1) (0.19,0.21,0.28,0.36;0.8,0.8)	(0.03,0.04,0.08,0.13;1,1) (0.06,0.07,0.10,0.12;0.8,0.8)
		*K*32	(0.27,0.29,0.29,1.15;1,1) (0.28,0.33,0.53,0.96;0.8,0.8)	(0.05,0.05,0.08,0.39;1,1) (0.05,0.07,0.15,0.33;0.8,0.8)
		*K*33	(0.19,0.21,0.29,0.38;1,1) (0.19,0.21,0.28,0.36;0.8,0.8)	(0.05,0.06,0.15,0.39;1,1) (0.05,0.07,0.15,0.36;0.8,0.8)

ratios of other experts emerged as 0.051, 0.082, and 0.082, respectively, and it is confirmed that the evaluations are more reliable using IT2FBWM.

In the following step, to find the global IT2F weights for sub-factors, the local weights of each key factors are multiplied by each related sub-factor interval type-2 fuzzy local weights, with help of Eq. (7). Results of all global interval type-2 fuzzy weights and average interval type-2 fuzzy optimal fuzzy weights (AIT2FW) for four experts are given in Tables 5 and 6, respectively.

The interval type-2 fuzzy weights are defuzzified with the help of Eq. (9), accordingly the ranking values of the sub-factors are illustrated in Fig. 3. Land cost "$K32$", which is among the economic factors, is the most important indicator to select the optimal hospital site. Other important factors are transportation cost, transportation facility, and construction cost, respectively. Water pollution and population density are the least evaluated characteristics for hospital site selection, according to the findings. Additionally, economic, social and policy, and environment were ranked from the most important to least important.

4 Conclusion

The ongoing COVID-19 outbreak has shown the importance of hospitals both quantitatively and qualitatively. Parallel to the increasing number of patients, the health systems of many countries are experiencing significant difficulties, and the need for hospitals is increasing. Considering the cost and construction time of an ordinary hospital, constructing a new hospital in cases such as an epidemic is far from meeting the urgent need. As such, the construction of temporary hospitals is one of the first practices in the world. In this study, it is tried to determine the importance of the factors considered in the selection of a temporary hospital site. With the help of the IT2FBWM method, it is aimed to make a flexible and correct decision. According to the findings, it can be clearly emphasized that economic factors play the most important role in temporary hospital site planning. Future studies can be carried out by extending different MCDM methods to different environments such as hesitant, Pythagorean, Fermatean, and spherical.

Table 5 IT2F local and global weights for expert-1

Key factor	Local weight	Sub-factor	Local weight	Global weight	Crisp weight
*K*1	(0.17,0.18,0.26,0.34;1,1) (0.34,0.34,0.34,0.34;0.8,0.8)	*K*11	(0.17,0.18,0.22,0.26;1,1) (0.17,0.18,0.21,0.25;0.8,0.8)	(0.03,0.03,0.06,0.09;1,1) (0.06,0.06,0.07,0.09;0.8,0.8)	0.06
		*K*12	(0.27,0.32,0.32,1.27;1,1) (0.28,0.35,0.58,1.06;0.8,0.8)	(0.05,0.06,0.08,0.43;1,1) (0.05,0.07,0.16,0.36;0.8,0.8)	0.15
		*K*13	(0.19,0.22,0.32,0.42;1,1) (0.2,0.23,0.3,0.4;0.8,0.8)	(0.05,0.06,0.17,0.43;1,1) (0.05,0.07,0.16,0.4;0.8,0.8)	0.17
*K*2	(0.17,0.18,0.26,0.34;1,1) (0.17,0.2,0.28,0.34;0.8,0.8)	*K*21	(0.19,0.21,0.29,0.38;1,1) (0.19,0.21,0.28,0.36;0.8,0.8)	(0.03,0.04,0.08,0.13;1,1) (0.06,0.07,0.1,0.12;0.8,0.8)	0.07
		*K*22	(0.27,0.29,0.29,1.15;1,1) (0.28,0.33,0.53,0.96;0.8,0.8)	(0.05,0.05,0.08,0.39;1,1) (0.05,0.07,0.15,0.33;0.8,0.8)	0.14
		*K*23	(0.19,0.21,0.29,0.38;1,1) (0.19,0.21,0.28,0.36;0.8,0.8)	(0.05,0.06,0.15,0.39;1,1) (0.05,0.07,0.15,0.36;0.8,0.8)	0.15
*K*3	(0.24,0.28,0.52,1.02;1,1) (0.25,0.31,0.54,0.99;0.8,0.8)	*K*31	(0.19,0.21,0.29,0.38;1,1) (0.19,0.21,0.28,0.36;0.8,0.8)	(0.03,0.04,0.08,0.13;1,1) (0.06,0.07,0.1,0.12;0.8,0.8)	0.07
		*K*32	(0.27,0.29,0.29,1.15;1,1) (0.28,0.33,0.53,0.96;0.8,0.8)	(0.05,0.05,0.08,0.39;1,1) (0.05,0.07,0.15,0.33;0.8,0.8)	0.14
		*K*33	(0.19,0.21,0.29,0.38;1,1) (0.19,0.21,0.28,0.36;0.8,0.8)	(0.05,0.06,0.15,0.39;1,1) (0.05,0.07,0.15,0.36;0 8,0.8)	0.15

Table 6 Average interval type-2 fuzzy weights

	EX 1	EX 2	EX 3	EX 4	AIT2FW
*K*11	(0.03,0.03,0.06,0.09;1,1) (0.06,0.06,0.07,0.09;0.8,0.8)	(0.04,0.04,0.11,0.28;1,1) (0.04,0.05,0.09,0.22;0.8,0.8)	(0.04,0.04,0.1,0.26;1,1) (0.04,0.04,0.09,0.2;0.8,0.8)	(0.07,0.09,0.31,1.17;1,1) (0.07,0.09,0.26,0.86;0.8,0.8)	(0.04,0.05,0.14,0.45;1,1) (0.05,0.06,0.13,0.34;0.8,0.8)
*K*12	(0.05,0.06,0.08,0.43;1,1) (0.05,0.07,0.16,0.36;0.8,0.8)	(0.03,0.04,0.07,0.17;1,1) (0.03,0.04,0.07,0.14;0.8,0.8)	(0.04,0.04,0.1,0.26;1,1) (0.04,0.04,0.09,0.2;0.8,0.8)	(0.03,0.04,0.08,0.14;1,1) (0.03,0.04,0.07,0.12;0.8,0.8)	(0.04,0.04,0.08,0.25;1,1) (0.04,0.05,0.10,0.205;0.8,0.8)
*K*13	(0.05,0.06,0.17,0.43;1,1) (0.05,0.07,0.16,0.4;0.8,0.8)	(0.05,0.07,0.21,0.84;1,1) (0.06,0.07,0.18,0.59;0.8,0.8)	(0.05,0.06,0.2,0.77;1,1) (0.06,0.07,0.16,0.54;0.8,0.8)	(0.05,0.05,0.08,0.14;1,1) (0.05,0.05,0.08,0.12;0.8,0.8)	(0.05,0.06,0.16,0.54;1,1) (0.05,0.06,0.14,0.41;0.8,0.8)
*K*21	(0.03,0.04,0.08,0.13;1,1) (0.06,0.07,0.1,0.12;0.8,0.8)	(0.04,0.04,0.1,0.26;1,1) (0.04,0.04,0.09,0.2;0.8,0.8)	(0.05,0.06,0.15,0.39;1,1) (0.05,0.06,0.14,0.32;0.8,0.8)	(0.05,0.06,0.2,0.77;1,1) (0.06,0.07,0.16,0.54;0.8,0.8)	(0.04,0.05,0.13,0.39;1,1) (0.05,0.06,0.12,0.30;0.8,0.8)
*K*22	(0.05,0.05,0.08,0.39;1,1) (0.05,0.07,0.15,0.33;0.8,0.8)	(0.04,0.04,0.1,0.26;1,1) (0.04,0.04,0.09,0.21;0.8,0.8)	(0.03,0.04,0.08,0.14;1,1) (0.03,0.04,0.07,0.12;0.8,0.8)	(0.04,0.04,0.1,0.26;1,1) (0.04,0.04,0.09,0.2;0.8,0.8)	(0.04,0.042,0.09,0.26;1,1) (0.04,0.05,0.1,0.21;0.8,0.8)
*K*23	(0.05,0.06,0.15,0.39;1,1) (0.05,0.07,0.15,0.36;0.8,0.8)	(0.05,0.06,0.2,0.77;1,1) (0.06,0.07,0.16,0.54;0.8,0.8)	(0.06,0.07,0.16,0.41;1,1) (0.06,0.07,0.14,0.33;0.8,0.8)	(0.04,0.04,0.1,0.26;1,1) (0.04,0.04,0.09,0.2;0.8,0.8)	(0.05,0.06,0.15,0.46;1,1) (0.05,0.06,0.13,0.36;0.8,0.8)
*K*31	(0.03,0.04,0.08,0.13;1,1) (0.06,0.07,0.1,0.12;0.8,0.8)	(0.04,0.04,0.09,0.2;1,1) (0.04,0.04,0.08,0.16;0.8,0.8)	(0.04,0.04,0.1,0.26;1,1) (0.04,0.04,0.09,0.2;0.8,0.8)	(0.25,0.29,0.55,1.05;1,1) (0.25,0.3,0.5,0.92;0.8,0.8)	(0.09,0.10,0.20,0.41;1,1) (0.10,0.11,0.19,0.35;0.8,0.8)
*K*32	(0.05,0.05,0.08,0.39;1,1) (0.05,0.07,0.15,0.33;0.8,0.8)	(0.05,0.05,0.09,1.01;1,1) (0.07,0.08,0.21,0.71;0.8,0.8)	(0.04,0.04,0.1,0.26;1,1) (0.04,0.04,0.09,0.2;0.8,0.8)	(0.24,0.24,0.28,0.37;1,1) (0.24,0.24,0.28,0.35;0.8,0.8)	(0.09,0.09,0.14,0.51;1,1) (0.1,0.18,0.18,0.40;0.8,0.8)
*K*33	(0.05,0.06,0.15,0.39;1,1) (0.05,0.07,0.15,0.36;0.8,0.8)	(0.04,0.04,0.09,0.2;1,1) (0.04,0.04,0.08,0.16;0.8,0.8)	(0.05,0.06,0.2,0.77;1,1) (0.06,0.07,0.16,0.54;0.8,0.8)	(0.18,0.19,0.28,0.37;1,1) (0.18,0.2,0.26,0.26;0.8,0.8)	(0.08,0.09,0.18,0.43;1,1) (0.08,0.10,0.16,0.33;0.8,0.8)

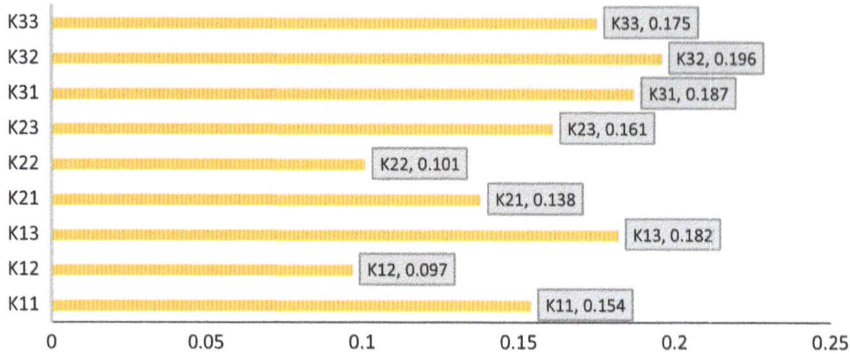

Fig. 3 Optimal crisp value of factor weights

References

1. Aboutorab, H., Saberi, M., Asadabadi, M.R., Hussain, O., Chang, E.: ZBWM: the Z-number extension of best worst method and its application for supplier development. Expert Syst. Appl. **107**, 115–125 (2018)
2. Adalı, E.A., Tuş, A.: Hospital site selection with distance-based multi-criteria decision-making methods. Int. J. Healthc. Manage. **14**(2), 534–544 (2021)
3. Boyacı, A.Ç., Şişman, A.: Pandemic hospital site selection: a GIS-based MCDM approach employing Pythagorean fuzzy sets. Environ. Sci. Pollut. Res. 1–13 (2021)
4. Celik, E., Gul, M.: Hazard identification, risk assessment and control for dam construction safety using an integrated BWM and MARCOS approach under interval type-2 fuzzy sets environment. Autom. Constr. **127**, 103699 (2021)
5. Celik, E., Yucesan, M., Gul, M.: Green supplier selection for textile industry: a case study using BWM-TODIM integration under interval type-2 fuzzy sets. Environ. Sci. Pollut. Res. 1–25 (2021)
6. Chen, S.-M., Lee, L.-W.: Fuzzy multiple attributes group decision-making based on the interval type-2 TOPSIS method. Expert Syst. Appl. **37**, 2790–2798 (2010)
7. Chen, Z., Ming, X.: A rough–fuzzy approach integrating best–worst method and data envelopment analysis to multi-criteria selection of smart product service module. Appl. Soft Comput. **94**, 106479 (2020)
8. Chen, Z.H., Wan, S.P., Dong, J.Y.: An efficiency-based interval type-2 fuzzy multi-criteria group decision making for makeshift hospital selection. Appl. Soft Comput. **115**, 108243 (2021)
9. Ecer, F., Pamucar, D.: Sustainable supplier selection: a novel integrated fuzzy best worst method (F-BWM) and fuzzy CoCoSo with Bonferroni (CoCoSo'B) multi-criteria model. J. Clean. Prod. **266**, 121981 (2020)
10. Ecer, F.: Multi-criteria decision making for green supplier selection using interval type-2 fuzzy AHP: a case study of a home appliance manufacturer. Oper. Res. 1–35 (2020)
11. Gong, X., Yang, M., Du, P.: Renewable energy accommodation potential evaluation of distribution network: a hybrid decision-making framework under interval type-2 fuzzy environment. J. Clean. Prod. **286**, 124918 (2021)
12. Gul, M., Yucesan, M.: Performance evaluation of Turkish Universities by an integrated Bayesian BWM-TOPSIS model. Soc.-Econ. Plan. Sci. **80**, 101173 (2021)
13. Guo, S., Zhao, H.: Fuzzy best-worst multi-criteria decision-making method and its applications. Knowl.-Based Syst. **121**, 23–31 (2017)
14. Gölcük, İ: An interval type-2 fuzzy reasoning model for digital transformation project risk assessment. Expert Syst. Appl. **159**, 113579 (2020)

15. Hafezalkotob, A., Hafezalkotob, A.: A novel approach for combination of individual and group decisions based on fuzzy best-worst method. Appl. Soft Comput. **59**, 316–325 (2017)
16. Hashemkhani Zolfani, S., Yazdani, M., Ebadi Torkayesh, A., Derakhti, A.: Application of a gray-based decision support framework for location selection of a temporary hospital during COVID-19 pandemic. Symmetry **12**(6), 886 (2020)
17. Kahraman, C., Öztay, B., Uçal Sar, İ, Turano, E.: Fuzzy analytic hierarchy process with interval type-2 fuzzy sets. Knowl.-Based Syst. **59**, 48–57 (2014)
18. Kaya, S.K., Aycin, E.: An integrated interval type 2 fuzzy AHP and COPRAS-G methodologies for supplier selection in the era of Industry 4.0. Neural Comput. Appl. 1–21 (2021)
19. Komatina, N., Tadić, D., Aleksić, A., Banduka, N.: The integrated PFMEA approach with interval type-2 fuzzy sets and FBWM: a case study in the automotive industry. Proc. Instit. Mech. Eng., Part D: J. Automobile Eng. 09544070211034799
20. Kumar, P., Singh, R.K., Sinha, P.: Optimal site selection for a hospital using a fuzzy extended ELECTRE approach. J. Manage. Anal. **3**(2), 115–135 (2016)
21. Kutlu Gündo, F., Kahraman, C., Civan, H.N.: A novel hesitant fuzzy EDAS method and its application to hospital selection. J. Intell. Fuzzy Syst. **35**(6), 6353–6365 (2018)
22. Lee, L.-W., Chen, S.-M.: Fuzzy multiple attributes group decision-making based on the extension of TOPSIS method and interval type-2 fuzzy sets. In: 2008 International Conference on Machine Learning and Cybernetics, pp. 3260–3265. IEEE (2008)
23. Liang, Q., Mendel, J.M.: Interval type-2 fuzzy logic systems: theory and design. IEEE Trans. Fuzzy Syst. **8**, 535–550 (2000)
24. Liao, H., Mi, X., Yu, Q., Luo, L.: Hospital performance evaluation by a hesitant fuzzy linguistic best worst method with inconsistency repairing. J. Clean. Prod. **232**, 657–671 (2019)
25. Lin, C.T., Tsai, M.C.: Location choice for direct foreign investment in new hospitals in China by using ANP and TOPSIS. Qual. Quant. **44**(2), 375–390 (2010)
26. Liu, A., Ji, X., Lu, H., Liu, H.: The selection of 3PRLs on self-service mobile recycling machine: interval-valued pythagorean hesitant fuzzy best-worst multi-criteria group decision-making. J. Clean. Prod. **230**, 734–750 (2019)
27. Luo, C., Ju, Y., Santibanez Gonzalez, E.D.R., Dong, P., Wang, A.: The waste-to-energy incineration plant site selection based on hesitant fuzzy linguistic Best-Worst method ANP and double parameters TOPSIS approach: a case study in China. Energy **211**, 118564 (2020). https://doi.org/10.1016/j.energy.2020.118564
28. Mendel, J.M., John, R.I., Liu, F.: Interval type-2 fuzzy logic systems made simple. IEEE Trans. Fuzzy Syst. **14**, 808–821 (2006)
29. Mendel, J., Wu, D.: Perceptual Computing: Aiding People in Making Subjective Judgments. John Wiley & Sons (2010)
30. Mi, X., Liao, H.: An integrated approach to multiple criteria decision making based on the average solution and normalized weights of criteria deduced by the hesitant fuzzy best worst method. Comput. Ind. Eng. **133**, 83–94 (2019)
31. Mou, Q., Xu, Z., Liao, H.: A graph based group decision making approach with intuitionistic fuzzy preference relations. Comput. Ind. Eng. **110**, 138–150 (2017)
32. Norouzi, A.: An interval type-2 hesitant fuzzy best-worst method. J. Intell. Fuzzy Syst. **40**, 11625–11652 (2021). https://doi.org/10.3233/JIFS-202801
33. Pamucar, D., Ecer, F., Cirovic, G., Arlasheedi, M.A.: Application of improved best worst method (BWM) in real-world problems. Mathematics **8**(8), 1342 (2020)
34. Pamučar, D., Puška, A., Stević, Ž, Ćirović, G.: A new intelligent MCDM model for HCW management: The integrated BWM–MABAC model based on D numbers. Expert Syst. Appl. **175**, 114862 (2021)
35. Qin, J., Liu, X.: Interval Type-2 fuzzy group decision making by integrating improved best worst method with COPRAS for emergency material supplier selection. In: Type-2 Fuzzy Decision-Making Theories, Methodologies and Applications, pp. 249–271. Springer (2019)
36. Rahimi, F., Goli, A., Rezaee, R.: Hospital location-allocation in Shiraz using geographical information system (GIS). Shiraz E-Med. J. **18**(8) (2017)

37. Şahin, T., Ocak, S., Top, M.: Analytic hierarchy process for hospital site selection. Health Policy Technol. **8**(1), 42–50 (2019)
38. Senvar, O., Otay, I., Bolturk, E.: Hospital site selection via hesitant fuzzy TOPSIS. IFAC-PapersOnLine **49**(12), 1140–1145 (2016)
39. Soltani, A., Marandi, E.Z.: Hospital site selection using two-stage fuzzy multi-criteria decision making process. J. Urban Environ. Eng. **5**(1), 32–43 (2011)
40. Stević, Ž, Pamučar, D., Subotić, M., Antuchevičiene, J., Zavadskas, E.K.: The location selection for roundabout construction using rough BWM-rough WASPAS approach based on a new rough hamy aggregator. Sustainability **10**, 2817 (2018). https://doi.org/10.3390/su10082817
41. Tang, J., Liu, X., Wang, W.: A hybrid risk prioritization method based on generalized TODIM and BWM for fine-kinney under interval type-2 fuzzy environment. Hum. Ecol. Risk Assess. Int. J. **27**(4), 954–979 (2021)
42. Tian, Z.P., Zhang, H.Y., Wang, J.Q., & Wang, T. L.: Green supplier selection using improved TOPSIS and best-worst method under intuitionistic fuzzy environment. Informatica **29**(4), 773–800 (2018)
43. Wan, S.P., Chen, Z.H., Dong, J.Y.: An integrated interval type-2 fuzzy technique for democratic–autocratic multi-criteria decision making. Knowl.-Based Syst. **214**, 106735 (2021)
44. Wu, Q., Zhou, L., Chen, Y., Chen, H.: An integrated approach to green supplier selection based on the interval type-2 fuzzy best-worst and extended VIKOR methods. Inf. Sci. **502**, 394–417 (2019)
45. Yılmaz, M., Atan, T.: Hospital site selection using fuzzy EDAS method: case study application for districts of Istanbul. J. Intell. Fuzzy Syst. (Preprint), 1–12 (2021)
46. Zadeh, L.A.: The concept of a linguistic variable and its application to approximate reasoning-I Information sciences. Inf. Sci. **8**, 199–249 (1975)
47. Zadeh, L.A.: Fuzzy sets. In: Zadeh, L.A. (ed.) Fuzzy Sets, Fuzzy Logic, and Fuzzy Systems: Selected Papers, pp. 394–432. World Scientific (1996)
48. Zolfani, S.H., Ecer, F., Pamučar, D., Raslanas, S.: Neighborhood selection for a newcomer via a novel BWM-based revised MAIRCA integrated model: a case from the Coquimbo-La Serena conurbation, Chile. Int. J. Strateg. Prop. Manag. **24**(2), 102–118 (2020)

Group Decision-Making Analysis Under Interval-Valued q-rung Orthopair Fuzzy Soft Expert Sets

Ghous Ali and Muhammad Akram

Abstract The pioneer paradigm of q-rung orthopair fuzzy sets was investigated by Yager in 2013, which extended the domains of both membership and non-membership values. Soft expert set theory is free from inherent complexity and a nice mathematical tool for handling uncertainties and vagueness. This paper is devoted to present a novel hybrid model, namely interval-valued q-rung orthopair fuzzy soft expert sets (IV^{q}ROFSE sets), which is a natural combination of soft expert sets and interval-valued q-rung orthopair fuzzy sets. Moreover, some necessary properties and operations on IV^{q}ROFSE sets are investigated with illustrative numerical examples. To show the validity and reliability of the initiated decision-making approach, a group decision-making problem for the ranking of modern renewable energy sources is explored. An algorithm is developed to demonstrate the procedure of the initiated theory. At last, a comparison analysis of the designed model with an existing interval-valued q-rung orthopair fuzzy set model is provided.

Keywords Soft expert set · IV^{q}ROFSE set · Renewable energy source · Group decision-making

1 Introduction

Multi-criteria group decision-making (MCGDM) is one of the reliable approaches of decision analysis in several uncertain situations of different domains. In the literature, a number of fuzzy set-based theories have been developed for group decision-making in different scenarios such as fuzzy soft expert sets, intuitionistic fuzzy soft expert

G. Ali
Department of Mathematics, Division of Science and Technology, University of Education, Lahore 54770, Pakistan
e-mail: ghous.ali@ue.edu.pk

M. Akram (✉)
Department of Mathematics, University of the Punjab, New Campus, Lahore 54590, Pakistan
e-mail: m.akram@pucit.edu.pk

L. Sahoo et al. (eds.), *Real Life Applications of Multiple Criteria Decision Making Techniques in Fuzzy Domain*, Studies in Fuzziness and Soft Computing 420,
https://doi.org/10.1007/978-981-19-4929-6_8

sets, m-polar fuzzy soft expert sets, Fermatean fuzzy soft expert sets, etc. The idea of fuzzy sets was originally presented by Zadeh [38] in 1965, which proved itself an efficient tool for dealing with uncertain problems as compared to probability theory and all other uncertainty theories of that time. In a fuzzy set, a membership function assigns a membership value from the closed unit interval to each object of a given nonempty universal set. On the other side, in fuzzy set theory, the non-membership value totally depends on the membership value which is calculated as '1 minus membership value'. With the inspection of many real-world problems, Atanassov [12] observed that membership and non-membership values may vary from the above-mentioned criterion and then he [12] launched the theory of intuitionistic fuzzy sets (IFSs) as a generalization of fuzzy sets by inserting two parts, that is, the membership $\bar{\mu}$ and non-membership $\bar{\lambda}$ functions with the property that $0 \leq \bar{\mu}(\mathfrak{x}) + \bar{\lambda}(\mathfrak{x}) \leq 1$ where $\mathfrak{x}$ is an arbitrary element of given universal set and $\bar{\mu}(\mathfrak{x}),\ \bar{\lambda}(\mathfrak{x}) \in [0, 1]$. The IFS model fails to tackle problems in which membership and non-membership values sum up greater than 1. In addition, Atanassov [13] proposed another type of IFSs called IFSs of the second type to overcome this issue partially. In an IFS of the second kind, the membership and non-membership values are now satisfied the property $0 \leq \left(\bar{\mu}(\mathfrak{x})\right)^2 + \left(\bar{\lambda}(\mathfrak{x})\right)^2 \leq 1$. Later, Yager [35, 36] presented an alternative way to define this concept in the form of Pythagorean fuzzy sets (PFSs). The theories of IFSs and PFSs fail to deal with different datasets, e.g., in a particular problem where a membership value is 0.75 and a non-membership value is 0.67 of an object, then $(0.75)^2 + (0.67)^2 > 1$. Thus, to provide more space to other datasets as compared to aforementioned models, Senapati and Yager [32] generalized the IFSs and PFSs to Fermatean fuzzy sets (FFSs) which satisfy the property $0 \leq \left(\bar{\mu}(\mathfrak{x})\right)^3 + \left(\bar{\lambda}(\mathfrak{x})\right)^3 \leq 1$. To further enlarged the space for different membership and non-membership values where the above-said theories failed, Yager [34] presented the q-rung orthopair fuzzy set (q-ROFS) theory in which $0 \leq \left(\bar{\mu}(\mathfrak{x})\right)^q + \left(\bar{\lambda}(\mathfrak{x})\right)^q \leq 1$ where $q > 1$. The theories of IFSs, PFSs, and FFSs are particular cases of q-ROFSs when $q = 1, 2$ and $q = 3$, respectively. Shaheen et al. [33] briefly discussed the reasons behind the production of q-ROFSs. This powerful theory attracted many researchers, and a lot of decision-making problems have been solved [3, 21].

Molodtsov [29] was the first who mentioned that the fuzzy set theory [38] is inefficient to perform as a parameterized tool and to cover this issue in fuzzy set theory and other existing models, Molodtsov [29] launched the theory of soft sets, which has the ability to analyze the ambiguities regarding multiple parameters more accurately in a decision-making situation. Nowadays, soft set theory is playing an important role as both individually and collectively, that is, as a combination with other uncertainty theories, e.g., fuzzy set theory [38] and rough set theory [30] in several practical areas ranging from arts to sciences [18, 25]. Afterward, Ali et al. [7] added some novel results in the literature of soft set theory. In addition, Maji et al. [27] firstly implemented soft set model to a real-life application for decision-making. A new era of hybrid soft set models was started when Molodtsov [29] claimed that soft set theory can be combined with other uncertainty theories (i.e., fuzzy and rough sets) to build a more generalized structure. For instance, Maji et al. [28] integrated fuzzy sets with soft sets and produced a generalized hybrid structure, namely fuzzy soft sets. The literature on the theory of soft sets and its generalized hybrid approaches is very rich such as intuitionistic fuzzy soft sets [26], multi-fuzzy N-soft sets [17], etc.

Nowadays, several hybrid extensions of interval-valued fuzzy sets are arising as powerful tools in different decision-making scenarios. Gorzalczany [20] initiated the idea of interval-valued fuzzy sets (IVF sets) as a natural extension of the fuzzy set. In an IVF set, every element of the universe has a membership value in the form of a subset of [0, 1]. To date, many IVF set-based hybrid models have been proposed by researchers all over the globe. For instance, Yang et al. [37] presented interval-valued fuzzy soft sets (IVFS sets) by combining IVF sets with soft sets. Later, Jiang et al. [24] developed an interval-valued intuitionistic fuzzy soft set (IVIFS set) model which is an extension of the IVFS set model. Joshi et al. [23] launched a novel hybrid model, namely interval-valued q-rung orthopair fuzzy (IV^{q}ROF) sets. Recently, Ali et al. [4] introduced a new hybrid model called the IV^{q}ROF soft set (IV^{q}ROFSS) model and some of its parameter reduction methods.

Since hybrid-soft set models only deal with datasets that are evaluated by one expert. Therefore, these existing hybrid-soft-set models fail to deal properly in the case of multiple experts, that is, when a company provides a question bank to their customers for feedback purposes. Alkhazaleh and Salleh [8] presented the idea of soft expert sets (SE sets) to tackle this issue. In addition, Alkhazaleh and Salleh [9] improved their work by initiating another group decision-making approach called fuzzy SE sets. Till today, a lot of novel hybrid models have been added to the literature. For instance, Al-Qudah and Hassan [10] presented the theory of bipolar fuzzy SE sets as an extension of fuzzy SE sets and solved an MCGDM problem. Afterward, Bashir and Salleh [14] developed a novel model, namely fuzzy parameterized SE sets. Broumi and Smarandache [15] extended the fuzzy SE set model to an intuitionistic fuzzy SE set model. In addition, Hassan and Alhazaymeh [22] presented another model for MCGDM called the vague SE set model. Ali and Akram [5] proposed a fuzzy N-SE set model for MCGDM by combining the N-soft sets with fuzzy SE sets. Additionally, Ali et al. [6] initiated another novel hybrid structure called bipolar SE sets.

Motivated by these thriving approaches, a novel MCGDM model called IV^{q}ROFSE sets is initiated, which is an extension of IV^{q}ROF soft sets or IV^{q}ROF sets. Furthermore, some basic properties and operations such as subset, complement, union, intersection, the 'AND' operation, and the 'OR' operation are investigated via corresponding numerical examples. An efficient algorithm is developed under the initiated IV^{q}ROFSE set model and applied to a real situation. Finally, a comparative analysis of the proposed approach with some existing tools is provided.

For other useful terms related to this article, the readers are suggested to [1, 2, 11, 16, 19, 31]. This article is structured as follows. Section 2 first reviews some fundamental results and then presents a novel hybrid model for MCGDM called IV^{q}ROFSE sets. Some basic operations and properties are also investigated in this section. Section 3 provides an algorithm based on the presented mathematical tool and solves a daily-life application of the proposed approach. Section 4 discusses a comparative analysis of the initiated approach with an existing approach. Section 5 concludes this work with main findings and some future directions.

2 Interval-Valued q-rung Orthopair Fuzzy Soft Expert Sets

This section first recalls some fundamental notions, including soft sets, IV^{q}ROF sets, and SE-sets, and then presents a new group decision-making model called IV^{q}FSE sets. Additionally, we investigate some of its basic results(properties and operations) in this section.

Definition 1 [29] Let $\mathcal{X}$ be an initial universal set and $\mathcal{T}$ be a universe of parameters related to the objects in the universal set. For any $\mathcal{S} \subseteq \mathcal{T}$, a soft set $(\tau, \mathcal{S})$ is defined as below:

$$(\tau, \mathcal{S}) = \{(t, \tau(t)) \mid t \in \mathcal{S}, \tau(t) \in \mathcal{P}(\mathcal{X})\},$$

where $\tau : \mathcal{S} \to \mathcal{P}(\mathcal{X})$ is a mapping, and $\mathcal{P}(\mathcal{X})$ represents the power set of $\mathcal{X}$.

Definition 2 [23] Let $\mathcal{X}$ be a universe of discourse. An interval-valued q-ROF set $\mathfrak{Q}_{\mathcal{E}}$ over $\mathcal{X}$ is defined as:

$$\mathfrak{Q}_{\mathcal{E}} = \{\langle \mathfrak{x}, \bar{\mu}_{\mathfrak{Q}_{\mathcal{E}}}(\mathfrak{x}), \bar{\lambda}_{\mathfrak{Q}_{\mathcal{E}}}(\mathfrak{x})\rangle \mid \mathfrak{x} \in \mathcal{X}\}, \tag{1}$$

where $\bar{\mu}_{\mathfrak{Q}_{\mathcal{E}}}(\mathfrak{x}) = [\tilde{\mu}^-_{\mathfrak{Q}_{\mathcal{E}}}(\mathfrak{x}), \tilde{\mu}^+_{\mathfrak{Q}_{\mathcal{E}}}(\mathfrak{x})]$ and $\bar{\lambda}_{\mathfrak{Q}_{\mathcal{E}}}(\mathfrak{x}) = [\tilde{\lambda}^-_{\mathfrak{Q}_{\mathcal{E}}}(\mathfrak{x}), \tilde{\lambda}^+_{\mathfrak{Q}_{\mathcal{E}}}(\mathfrak{x})]$ are the intervals of membership and non-membership values, respectively, $\bar{\mu}_{\mathfrak{Q}_{\mathcal{E}}}(\mathfrak{x}),\ \bar{\lambda}_{\mathfrak{Q}_{\mathcal{E}}}(\mathfrak{x}) \subseteq [0, 1]$ and $0 \leq \left(\sup(\bar{\mu}_{\mathfrak{Q}_{\mathcal{E}}}(\mathfrak{x}))\right)^q + \left(\sup(\bar{\lambda}_{\mathfrak{Q}_{\mathcal{E}}}(\mathfrak{x}))\right)^q \leq 1,\ q \geq 1$.

Definition 3 [8] Let $\mathcal{X}$ be an initial universal set, $\mathcal{T}$ be a universe of parameters (attributes), and $\mathcal{E}$ be a set of experts with their opinions in the form of a set $\mathfrak{O}\mathfrak{P} = \{0 = disagree, 1 = agree\}$. Let $\mathcal{Z} = \mathcal{T} \times \mathcal{E} \times \mathfrak{O}\mathfrak{P}$ and $\mathcal{Y} \subseteq \mathcal{Z}$. A pair $(\zeta, \mathcal{Y})$ is called a soft expert set or SE set on $\mathcal{X}$, where ζ is a mapping given by

$$\zeta : \mathcal{Y} \to \mathcal{P}(\mathcal{X}).$$

Now we are ready to develop the main concept of the current study, i.e., IV^{q}ROFSE sets, and investigate some fundamental properties and operations with explanatory examples.

Definition 4 Let $\mathcal{X}$ be an initial universe of discourse, $\mathcal{T}$ a universe of attributes and $\mathcal{E}$ be a set of experts with their opinions in the form of a set $\mathfrak{O}\mathfrak{P} = \{0 = disagree,\ 1 = agree\}$. For an arbitrary $\mathcal{Y} \subset \mathcal{Z}$ where $\mathcal{Z} = \mathcal{T} \times \mathcal{E} \times \mathfrak{O}\mathfrak{P}$, an *interval-valuedq-rung orthopair fuzzy soft expert set* or IV^{q}ROFSE set over $\mathcal{X}$, denoted by $(\xi, \mathcal{Y})$, is defined as below:

Table 1 Tabular form of the IV^{q}ROFSE set $(\xi, \mathcal{Y})$

$(\xi, \mathcal{Y})$	$\mathfrak{x}_1$	$\cdots$	$\mathfrak{x}_n$
$(\mathfrak{t}_1, \mathfrak{e}_1, 1)$	$\left\langle\left[\mu^{-}_{\mathfrak{Q}^{l1}_{(\mathfrak{x}_1)}}, \mu^{+}_{\mathfrak{Q}^{u1}_{\mathfrak{x}_1}}\right], \left[\lambda^{-}_{\mathfrak{Q}^{l1}_{(\mathfrak{x}_1)}}, \lambda^{+}_{\mathfrak{Q}^{u1}_{(\mathfrak{x}_1)}}\right]\right\rangle$	$\cdots$	$\left\langle\left[\mu^{-}_{\mathfrak{Q}^{l1}_{(\mathfrak{x}_n)}}, \mu^{+}_{\mathfrak{Q}^{u1}_{(\mathfrak{x}_n)}}\right], \left[\lambda^{-}_{\mathfrak{Q}^{l1}_{(\mathfrak{x}_n)}}, \lambda^{+}_{\mathfrak{Q}^{u1}_{(\mathfrak{x}_n)}}\right]\right\rangle$
$(\mathfrak{t}_1, \mathfrak{e}_2, 1)$	$\left\langle\left[\mu^{-}_{\mathfrak{Q}^{l2}_{(\mathfrak{x}_1)}}, \mu^{+}_{\mathfrak{Q}^{u2}_{(\mathfrak{x}_1)}}\right], \left[\lambda^{-}_{\mathfrak{Q}^{l2}_{(\mathfrak{x}_1)}}, \lambda^{+}_{\mathfrak{Q}^{u2}_{(\mathfrak{x}_1)}}\right]\right\rangle$	$\cdots$	$\left\langle\left[\mu^{-}_{\mathfrak{Q}^{l2}_{(\mathfrak{x}_n)}}, \mu^{+}_{\mathfrak{Q}^{u2}_{(\mathfrak{x}_n)}}\right], \left[\lambda^{-}_{\mathfrak{Q}^{l2}_{(\mathfrak{x}_n)}}, \lambda^{+}_{\mathfrak{Q}^{u2}_{(\mathfrak{x}_n)}}\right]\right\rangle$
$\vdots$	$\vdots$	$\ddots$	$\vdots$
$(\mathfrak{t}_q, \mathfrak{e}_k, 1)$	$\left\langle\left[\mu^{-}_{\mathfrak{Q}^{qk}_{(\mathfrak{x}_1)}}, \mu^{+}_{\mathfrak{Q}^{qk}_{(\mathfrak{x}_1)}}\right], \left[\lambda^{-}_{\mathfrak{Q}^{qk}_{(\mathfrak{x}_1)}}, \lambda^{+}_{\mathfrak{Q}^{qk}_{(\mathfrak{x}_1)}}\right]\right\rangle$	$\cdots$	$\left\langle\left[\mu^{-}_{\mathfrak{Q}^{qk}_{(\mathfrak{x}_n)}}, \mu^{+}_{\mathfrak{Q}^{qk}_{(\mathfrak{x}_n)}}\right], \left[\lambda^{-}_{\mathfrak{Q}^{qk}_{(\mathfrak{x}_n)}}, \lambda^{+}_{\mathfrak{Q}^{qk}_{(\mathfrak{x}_n)}}\right]\right\rangle$
$(\mathfrak{t}_1, \mathfrak{e}_1, 0)$	$\left\langle\left[\mu^{-}_{\mathfrak{Q}^{l1}_{(\mathfrak{x}_1)}}, \mu^{+}_{\mathfrak{Q}^{u1}_{(\mathfrak{x}_1)}}\right], \left[\lambda^{-}_{\mathfrak{Q}^{l1}_{(\mathfrak{x}_1)}}, \lambda^{+}_{\mathfrak{Q}^{u1}_{(\mathfrak{x}_1)}}\right]\right\rangle$	$\cdots$	$\left\langle\left[\mu^{-}_{\mathfrak{Q}^{l1}_{(\mathfrak{x}_n)}}, \mu^{+}_{\mathfrak{Q}^{u1}_{(\mathfrak{x}_n)}}\right], \left[\lambda^{-}_{\mathfrak{Q}^{l1}_{(\mathfrak{x}_n)}}, \lambda^{+}_{\mathfrak{Q}^{u1}_{(\mathfrak{x}_n)}}\right]\right\rangle$
$(\mathfrak{t}_1, \mathfrak{e}_2, 0)$	$\left\langle\left[\mu^{-}_{\mathfrak{Q}^{l21}_{(\mathfrak{x}_1)}}, \mu^{+}_{\mathfrak{Q}^{u2}_{(\mathfrak{x}_1)}}\right], \left[\lambda^{-}_{\mathfrak{Q}^{l2}_{(\mathfrak{x}_1)}}, \lambda^{+}_{\mathfrak{Q}^{u2}_{(\mathfrak{x}_1)}}\right]\right\rangle$	$\cdots$	$\left\langle\left[\mu^{-}_{\mathfrak{Q}^{l2}_{(\mathfrak{x}_n)}}, \mu^{+}_{\mathfrak{Q}^{u2}_{(\mathfrak{x}_n)}}\right], \left[\lambda^{-}_{\mathfrak{Q}^{l2}_{(\mathfrak{x}_n)}}, \lambda^{+}_{\mathfrak{Q}^{u2}_{(\mathfrak{x}_n)}}\right]\right\rangle$
$\vdots$	$\vdots$	$\ddots$	$\vdots$
$(\mathfrak{t}_q, \mathfrak{e}_k, 0)$	$\left\langle\left[\mu^{-}_{\mathfrak{Q}^{qk}_{(\mathfrak{x}_1)}}, \mu^{+}_{\mathfrak{Q}^{qk}_{(\mathfrak{x}_1)}}\right], \left[\lambda^{-}_{\mathfrak{Q}^{qk}_{(\mathfrak{x}_1)}}, \lambda^{+}_{\mathfrak{Q}^{qk}_{(\mathfrak{x}_1)}}\right]\right\rangle$	$\cdots$	$\left\langle\left[\mu^{-}_{\mathfrak{Q}^{qk}_{(\mathfrak{x}_n)}}, \mu^{+}_{\mathfrak{Q}^{qk}_{(\mathfrak{x}_n)}}\right], \left[\lambda^{-}_{\mathfrak{Q}^{qk}_{(\mathfrak{x}_n)}}, \lambda^{+}_{\mathfrak{Q}^{qk}_{(\mathfrak{x}_n)}}\right]\right\rangle$

$$(\xi, \mathcal{Y}) = \{\langle \mathfrak{z}, \xi(\mathfrak{z})\rangle \mid \mathfrak{z} \in \mathcal{Y}\}, \tag{2}$$

where $\xi(\mathfrak{z})$ is an IV^{q}ROF subset of $\mathcal{X}$, for all $\mathfrak{z} \in \mathcal{Y}$, that is,

$$\xi(\mathfrak{z}) = \left\{\left\langle \mathfrak{x}, \left(\left[\tilde{\mu}^{-}_{\mathfrak{Q}_{\mathcal{E}}}(\mathfrak{x}), \tilde{\mu}^{+}_{\mathfrak{Q}_{\mathcal{E}}}(\mathfrak{x})\right], \left[\tilde{\lambda}^{-}_{\mathfrak{Q}_{\mathcal{E}}}(\mathfrak{x}), \tilde{\lambda}^{+}_{\mathfrak{Q}_{\mathcal{E}}}(\mathfrak{x})\right]\right)\right\rangle : \mathfrak{x} \in \mathcal{X}\right\},$$

where $0 \leq (\tilde{\mu}^{+}_{\mathfrak{Q}_{\mathcal{E}}}(\mathfrak{x}))^q + (\tilde{\lambda}^{+}_{\mathfrak{Q}_{\mathcal{E}}}(\mathfrak{x}))^q \leq 1$.

Let $\mathcal{X} = \{\mathfrak{x}_1, \mathfrak{x}_2, \ldots, \mathfrak{x}_n\}$ be a universe, $\mathcal{T} = \{\mathfrak{t}_1, \mathfrak{t}_2, \ldots, \mathfrak{t}_q\}$ a set of attributes, and $\mathcal{E} = \{\mathfrak{e}_1, \mathfrak{e}_2, \ldots, \mathfrak{e}_k\}$ be a set of experts with their opinions in the form of set $\mathfrak{DP} = \{0 = \text{disagree},\ 1 = \text{agree}\}$. For each $\mathcal{Y} \subset \mathcal{Z}$ where $\mathcal{Z} = \mathcal{T} \times \mathcal{E} \times \mathfrak{DP}$, the IV^{q}ROFSE set $(\xi, \mathcal{Y})$ is represented by the following tabular format (Table 1).

In the following example, we illustrate the main and novel idea of the current study:

Example 1 Suppose that $\mathcal{X} = \{\mathfrak{x}_1, \mathfrak{x}_2, \ldots, \mathfrak{x}_8\}$ is a universe of eight refrigerators, $\mathcal{T} = \{\mathfrak{t}_1 = \text{design}, \mathfrak{t}_2 = \text{technology}, \mathfrak{t}_3 = \text{energy saving}\}$ is a set of attributes, $\mathcal{E} = \{e_1, e_2\}$ is a set of experts together their evaluations in the form of a set $\mathfrak{DP} = \{0 = \text{disagree},\ 1 = \text{agree}\}$ and $\mathcal{Y} \subseteq \mathcal{Z}$ where $\mathcal{Z} = \mathcal{T} \times \mathcal{E} \times \mathfrak{DP}$. Then, all the experts' evaluations about refrigerators regarding important attributes are accumulated in tabular format as an IV^{q}ROFSE set $(\xi, \mathcal{Y})$ where $q = 4$ (see Table 2).

Each cell of the Table 1 explains the experts' opinions about evaluations in the interval-valued form of each object regarding given attributes. For instance, in the first cell, that is, $\langle[0.41, 0.53], [0.64, 0.74]\rangle$, the expert $\mathfrak{e}_1$ agrees that the least belongingness value regarding parameter 'design' is 0.41 and the most belongingness value is 0.53. Similarly, the expert $\mathfrak{e}_1$ agrees that the least non-belongingness value regarding parameter 'design' is 0.64, and most non-belongingness value is 0.74. In other

Table 2 Tabular arrangement of the IV^{4}ROFSE set $(\xi, \mathcal{Y})$

$(\xi, \mathcal{Y})$	$\mathfrak{r}_1$	$\mathfrak{r}_2$	$\mathfrak{r}_3$	$\mathfrak{r}_4$
$(\mathfrak{t}_1, \mathfrak{e}_1, 1)$	$\langle[0.41, 0.53], [0.64, 0.74]\rangle$	$\langle[0.72, 0.84], [0.60, 0.71]\rangle$	$\langle[0.33, 0.50], [0.55, 0.66]\rangle$	$\langle[0.67, 0.75], [0.18, 0.35]\rangle$
$(\mathfrak{t}_1, \mathfrak{e}_2, 1)$	$\langle[0.42, 0.56], [0.24, 0.33]\rangle$	$\langle[0.43, 0.53], [0.11, 0.27]\rangle$	$\langle[0.19, 0.26], [0.58, 0.69]\rangle$	$\langle[0.84, 0.93], [0.55, 0.63]\rangle$
$(\mathfrak{t}_2, \mathfrak{e}_1, 1)$	$\langle[0.44, 0.55], [0.75, 0.80]\rangle$	$\langle[0.63, 0.72], [0.81, 0.90]\rangle$	$\langle[0.15, 0.23], [0.52, 0.67]\rangle$	$\langle[0.47, 0.60], [0.71, 0.77]\rangle$
$(\mathfrak{t}_2, \mathfrak{e}_2, 1)$	$\langle[0.64, 0.72], [0.23, 0.44]\rangle$	$\langle[0.47, 0.53], [0.81, 0.91]\rangle$	$\langle[0.33, 0.55], [0.61, 0.75]\rangle$	$\langle[0.78, 0.83], [0.18, 0.27]\rangle$
$(\mathfrak{t}_3, \mathfrak{e}_1, 1)$	$\langle[0.49, 0.57], [0.41, 0.46]\rangle$	$\langle[0.91, 0.95], [0.62, 0.73]\rangle$	$\langle[0.65, 0.69], [0.35, 0.52]\rangle$	$\langle[0.13, 0.24], [0.71, 0.83]\rangle$
$(\mathfrak{t}_3, \mathfrak{e}_2, 1)$	$\langle[0.61, 0.70], [0.44, 0.53]\rangle$	$\langle[0.81, 0.89], [0.51, 0.72]\rangle$	$\langle[0.62, 0.68], [0.12, 0.32]\rangle$	$\langle[0.88, 0.95], [0.71, 0.79]\rangle$
$(\mathfrak{t}_1, \mathfrak{e}_1, 0)$	$\langle[0.21, 0.33], [0.80, 0.89]\rangle$	$\langle[0.12, 0.24], [0.40, 0.52]\rangle$	$\langle[0.55, 0.70], [0.05, 0.16]\rangle$	$\langle[0.07, 0.15], [0.78, 0.85]\rangle$
$(\mathfrak{t}_1, \mathfrak{e}_2, 0)$	$\langle[0.22, 0.32], [0.54, 0.63]\rangle$	$\langle[0.13, 0.33], [0.81, 0.87]\rangle$	$\langle[0.69, 0.76], [0.18, 0.29]\rangle$	$\langle[0.14, 0.23], [0.95, 0.96]\rangle$
$(\mathfrak{t}_2, \mathfrak{e}_1, 0)$	$\langle[0.14, 0.26], [0.45, 0.60]\rangle$	$\langle[0.23, 0.32], [0.61, 0.75]\rangle$	$\langle[0.85, 0.93], [0.22, 0.37]\rangle$	$\langle[0.17, 0.30], [0.51, 0.67]\rangle$
$(\mathfrak{t}_2, \mathfrak{e}_2, 0)$	$\langle[0.12, 0.22], [0.73, 0.84]\rangle$	$\langle[0.20, 0.33], [0.60, 0.71]\rangle$	$\langle[0.13, 0.25], [0.80, 0.95]\rangle$	$\langle[0.10, 0.30], [0.52, 0.67]\rangle$
$(\mathfrak{t}_3, \mathfrak{e}_1, 0)$	$\langle[0.14, 0.30], [0.90, 0.95]\rangle$	$\langle[0.52, 0.65], [0.22, 0.33]\rangle$	$\langle[0.15, 0.29], [0.35, 0.42]\rangle$	$\langle[0.23, 0.54], [0.85, 0.93]\rangle$
$(\mathfrak{t}_3, \mathfrak{e}_2, 0)$	$\langle[0.81, 0.90], [0.14, 0.23]\rangle$	$\langle[0.11, 0.19], [0.75, 0.82]\rangle$	$\langle[0.12, 0.28], [0.62, 0.72]\rangle$	$\langle[0.18, 0.65], [0.01, 0.09]\rangle$
$(\xi, \mathcal{Y})$	$\mathfrak{r}_5$	$\mathfrak{r}_6$	$\mathfrak{r}_7$	$\mathfrak{r}_8$
$(\mathfrak{t}_1, \mathfrak{e}_1, 1)$	$\langle[0.39, 0.57], [0.75, 0.80]\rangle$	$\langle[0.25, 0.33], [0.83, 0.91]\rangle$	$\langle[0.39, 0.52], [0.81, 0.87]\rangle$	$\langle[0.77, 0.82], [0.37, 0.47]\rangle$
$(\mathfrak{t}_1, \mathfrak{e}_2, 1)$	$\langle[0.42, 0.55], [0.71, 0.86]\rangle$	$\langle[0.05, 0.20], [0.92, 0.98]\rangle$	$\langle[0.35, 0.48], [0.74, 0.83]\rangle$	$\langle[0.83, 0.90], [0.29, 0.35]\rangle$
$(\mathfrak{t}_2, \mathfrak{e}_1, 1)$	$\langle[0.31, 0.51], [0.77, 0.81]\rangle$	$\langle[0.66, 0.73], [0.11, 0.22]\rangle$	$\langle[0.34, 0.56], [0.61, 0.69]\rangle$	$\langle[0.52, 0.67], [0.81, 0.87]\rangle$
$(\mathfrak{t}_2, \mathfrak{e}_2, 1)$	$\langle[0.25, 0.55], [0.60, 0.70]\rangle$	$\langle[0.01, 0.11], [0.80, 0.95]\rangle$	$\langle[0.33, 0.50], [0.71, 0.83]\rangle$	$\langle[0.45, 0.50], [0.23, 0.40]\rangle$
$(\mathfrak{t}_3, \mathfrak{e}_1, 1)$	$\langle[0.42, 0.50], [0.12, 0.21]\rangle$	$\langle[0.77, 0.79], [0.29, 0.41]\rangle$	$\langle[0.46, 0.53], [0.83, 0.92]\rangle$	$\langle[0.55, 0.60], [0.71, 0.83]\rangle$
$(\mathfrak{t}_3, \mathfrak{e}_2, 1)$	$\langle[0.22, 0.33], [0.74, 0.79]\rangle$	$\langle[0.48, 0.60], [0.75, 0.86]\rangle$	$\langle[0.61, 0.71], [0.09, 0.28]\rangle$	$\langle[0.50, 0.60], [0.71, 0.75]\rangle$
$(\mathfrak{t}_1, \mathfrak{e}_1, 0)$	$\langle[0.19, 0.27], [0.60, 0.70]\rangle$	$\langle[0.75, 0.83], [0.13, 0.31]\rangle$	$\langle[0.69, 0.75], [0.31, 0.47]\rangle$	$\langle[0.17, 0.22], [0.77, 0.87]\rangle$
$(\mathfrak{t}_1, \mathfrak{e}_2, 0)$	$\langle[0.63, 0.75], [0.11, 0.26]\rangle$	$\langle[0.93, 0.97], [0.02, 0.15]\rangle$	$\langle[0.75, 0.88], [0.14, 0.13]\rangle$	$\langle[0.13, 0.40], [0.69, 0.85]\rangle$
$(\mathfrak{t}_2, \mathfrak{e}_1, 0)$	$\langle[0.21, 0.30], [0.90, 0.93]\rangle$	$\langle[0.26, 0.33], [0.70, 0.82]\rangle$	$\langle[0.64, 0.77], [0.11, 0.19]\rangle$	$\langle[0.02, 0.12], [0.51, 0.60]\rangle$
$(\mathfrak{t}_2, \mathfrak{e}_2, 0)$	$\langle[0.56, 0.75], [0.10, 0.30]\rangle$	$\langle[0.51, 0.61], [0.20, 0.35]\rangle$	$\langle[0.73, 0.80], [0.01, 0.13]\rangle$	$\langle[0.35, 0.39], [0.83, 0.90]\rangle$
$(\mathfrak{t}_3, \mathfrak{e}_1, 0)$	$\langle[0.12, 0.20], [0.70, 0.81]\rangle$	$\langle[0.17, 0.29], [0.50, 0.65]\rangle$	$\langle[0.86, 0.92], [0.14, 0.25]\rangle$	$\langle[0.10, 0.35], [0.60, 0.70]\rangle$
$(\mathfrak{t}_3, \mathfrak{e}_2, 0)$	$\langle[0.60, 0.78], [0.04, 0.19]\rangle$	$\langle[0.18, 0.30], [0.90, 0.95]\rangle$	$\langle[0.05, 0.11], [0.79, 0.88]\rangle$	$\langle[0.30, 0.45], [0.78, 0.90]\rangle$

words, $[0.41, 0.53]$ and $[0.64, 0.74]$ are the intervals of belongingness and non-belongingness degrees where 0.41 and 0.64 are lower bounds, and 0.53 and 0.74 are upper bounds, respectively.

The construction of above-initiated model is mainly based on the parameterized family of IV^{q}ROF subsets. That's why, we can accumulate these subsets as a class, and this concept is given in the following definition.

Definition 5 Let $\mathcal{X}$ be a universe and $(\xi, \mathcal{Y})$ be an IV^{q}ROFSE set on $\mathcal{X}$. Then a set of all IV^{q}ROF subsets on $\mathcal{X}$ regarding the IV^{q}ROFSE set $(\xi, \mathcal{Y})$ is called an *IV^{q}ROFSE class* of $(\xi, \mathcal{Y})$, denoted by $\mathcal{CL}_{(\xi,\mathcal{Y})}$, and is defined as:

$$\mathcal{CL}_{(\xi,\mathcal{Y})} = \{\xi(\mathfrak{z}) \mid \mathfrak{z} \in \mathcal{Y}\}.$$

This new concept of IV^{q}ROFSE class is demonstrated via an example given below:

Example 2 Consider IV^{4}ROFSE set in Example 1, then

$$CL_{(\xi,\mathcal{Y})} = \{\xi(\mathfrak{t}_1, \mathfrak{e}_1, 1), \xi(\mathfrak{t}_1, \mathfrak{e}_2, 1), \ldots, \xi(\mathfrak{t}_3, \mathfrak{e}_2, 0)\}$$

where

$$\begin{aligned}
\xi(\mathfrak{t}_1, \mathfrak{e}_1, 1) = \big\{&(\mathfrak{x}_1, \langle[0.41, 0.53], [0.64, 0.74]\rangle), (\mathfrak{x}_2, \langle[0.72, 0.84], [0.60, 0.71]\rangle), (\mathfrak{x}_3, \langle[0.33, 0.50], [0.55, 0.66]\rangle),\\
&(\mathfrak{x}_4, \langle[0.67, 0.75], [0.18, 0.35]\rangle), (\mathfrak{x}_5, \langle[0.39, 0.57], [0.75, 0.80]\rangle), (\mathfrak{x}_6, \langle[0.25, 0.33], [0.83, 0.91]\rangle),\\
&(\mathfrak{x}_7, \langle[0.39, 0.52], [0.81, 0.87]\rangle), (\mathfrak{x}_8, \langle[0.77, 0.82], [0.37, 0.47]\rangle)\big\},\\
\xi(\mathfrak{t}_1, \mathfrak{e}_2, 1) = \big\{&(\mathfrak{x}_1, \langle[0.42, 0.56], [0.24, 0.33]\rangle), (\mathfrak{x}_2, \langle[0.43, 0.53], [0.11, 0.27]\rangle, (\mathfrak{x}_3, \langle[0.19, 0.26], [0.58, 0.69]\rangle),\\
&(\mathfrak{x}_4, \langle[0.84, 0.93], [0.55, 0.63]\rangle), (\mathfrak{x}_5, \langle[0.42, 0.55], [0.71, 0.86]\rangle), (\mathfrak{x}_6, \langle[0.05, 0.20], [0.92, 0.98]\rangle),\\
&(\mathfrak{x}_7, \langle[0.35, 0.48], [0.74, 0.83]\rangle), (\mathfrak{x}_8, \langle[0.83, 0.90], [0.29, 0.35]\rangle)\big\},\\
&\vdots\\
\xi(\mathfrak{t}_3, \mathfrak{e}_2, 0) = \big\{&(\mathfrak{x}_1, \langle[0.81, 0.90], [0.14, 0.23]\rangle), (\mathfrak{x}_2, \langle[0.11, 0.19], [0.75, 0.82]\rangle), (\mathfrak{x}_3, \langle[0.12, 0.28], [0.62, 0.72]\rangle),\\
&(\mathfrak{x}_4, \langle[0.18, 0.65], [0.01, 0.09]\rangle, (\mathfrak{x}_5, \langle[0.60, 0.78], [0.04, 0.19]\rangle), (\mathfrak{x}_6, \langle[0.18, 0.30], [0.90, 0.95]\rangle),\\
&(\mathfrak{x}_7, \langle[0.05, 0.11], [0.79, 0.88]\rangle), (\mathfrak{x}_8, \langle[0.30, 0.45], [0.78, 0.90]\rangle)\big\}.
\end{aligned}$$

We now investigate a few useful properties of the initiated model.

Definition 6 For two IV^{q}ROFSE sets $(\xi_1, \mathcal{Y}_1)$ and $(\xi_2, \mathcal{Y}_2)$ over a universal set $\mathcal{X}$, the IV^{q}ROFSE set $(\xi_1, \mathcal{Y}_1)$ is called an *IV^q ROFSE subset* of $(\xi_2, \mathcal{Y}_2)$, if

1. $\mathcal{Y}_1 \subseteq \mathcal{Y}_2$, and
2. For all $\mathfrak{z} \in \mathcal{Y}_1$, $\xi_1(\mathfrak{z})$ is an IV^{q}ROF subset of $\xi_2(\mathfrak{z})$.

It is represented by $(\xi_1, \mathcal{Y}_1)\tilde{\subseteq}(\xi_2, \mathcal{Y}_2)$. Here the IV^{q}ROFSE set $(\xi_2, \mathcal{Y}_2)$ is also referred to as an IV^{q}ROFSE super-set of the IV^{q}ROFSE set $(\xi_1, \mathcal{Y}_1)$.

The following example illustrates this novel idea of subset-hood among IV^{q}ROFSE sets.

Example 3 Consider $\mathcal{X} = \{\mathfrak{x}_1, \mathfrak{x}_2, \ldots, \mathfrak{x}_8\}$, a universe of eight refrigerators as discussed in Example 1, and $\mathcal{Y}_1,\ \mathcal{Y}_2 \subset \mathcal{Z} = \mathcal{T} \times \mathcal{E} \times \mathfrak{OP}$, which are provided below:

$$\mathcal{Y}_1 = \{(\mathfrak{t}_1, \mathfrak{e}_2, 1), (\mathfrak{t}_2, \mathfrak{e}_1, 1), (\mathfrak{t}_1, \mathfrak{e}_2, 0), (\mathfrak{t}_2, \mathfrak{e}_2, 0)\},$$
$$\mathcal{Y}_2 = \{(\mathfrak{t}_1, \mathfrak{e}_2, 1), (\mathfrak{t}_2, \mathfrak{e}_1, 1), (\mathfrak{t}_2, \mathfrak{e}_2, 1), (\mathfrak{t}_1, \mathfrak{e}_1, 0), (\mathfrak{t}_1, \mathfrak{e}_2, 0), (\mathfrak{t}_2, \mathfrak{e}_2, 0)\}.$$

By Definition 4, two IV^{4}ROFSE sets $(\xi_1, \mathcal{Y}_1)$ and $(\xi_2, \mathcal{Y}_2)$ are, respectively, displayed via Tables 3 and 4.

From Tables 3 and 4, one can readily observe that $\mathcal{Y}_1 \subseteq \mathcal{Y}_2$ and $\xi_1(\mathfrak{z})$ is an IV^{4}ROF subset of $\xi_2(\mathfrak{z})$, for each $\mathfrak{z} \in \mathcal{Y}_1$. Thus, $(\xi_1, \mathcal{Y}_1)\tilde{\subseteq}(\xi_2, \mathcal{Y}_2)$.

Now we explain the equality condition between the IV^{q}ROFSE sets.

Definition 7 Two IV^{q}ROFSE sets $(\xi_1, \mathcal{Y}_1)$ and $(\xi_2, \mathcal{Y}_2)$ on a universal set $\mathcal{X}$ are called equal *IV^{q}ROFSE sets* if

1. $(\xi_1, \mathcal{Y}_1)$ is an IV^{q}ROFSE subset of $(\xi_2, \mathcal{Y}_2)$.
2. $(\xi_2, \mathcal{Y}_2)$ is an IV^{q}ROFSE subset of $(\xi_1, \mathcal{Y}_1)$.

It is denoted by $(\xi_1, \mathcal{Y}_1) = (\xi_2, \mathcal{Y}_2)$.

We now present two new concepts, namely agree-IV^{q}ROFSE sets and disagree-IV^{q}ROFSE sets and explain them via numerical examples.

Definition 8 Let $\mathcal{X}$ be a universe, $\mathcal{T}$ a universe of parameters, and $\mathcal{E}$ be a set of experts with their opinions in the form of a set $\mathfrak{OP} = \{0 = disagree,\ 1 = agree\}$. For every IV^{q}ROFSE set $(\xi, \mathcal{Y})$ where $\mathcal{Y} \subset \mathcal{Z} = \mathcal{T} \times \mathcal{E} \times \mathfrak{OP}$, an agree-IV^{q}ROFSE set $(\xi, \mathcal{Y})_1$ is defined as:

$$(\xi, \mathcal{Y})_1 = \{\xi(\mathfrak{z}) : \mathfrak{z} \in \mathcal{T} \times \mathcal{E} \times \{1\} \subset \mathcal{Z}\}.$$

Note that agree-IV^{q}ROFSE set $(\xi, \mathcal{Y})_1$ is an IV^{q}ROFSE subset of $(\xi, \mathcal{Y})$.

Example 4 Consider IV^{4}ROFSE set again as given in Example 1. Then, an agree-IV^{4}ROFSE set $(\xi, \mathcal{Y})_1$ of the IV^{4}ROFSE set $(\xi, \mathcal{Y})$ over the universal set $\mathcal{X}$ is given in Table 5.

Definition 9 Let $\mathcal{X}$ be a universe, $\mathcal{T}$ a universe of parameters, and $\mathcal{E}$ be a set of experts with their opinions in the form of a set $\mathfrak{OP} = \{0 = disagree,\ 1 = agree\}$. For every IV^{q}ROFSE set $(\xi, \mathcal{Y})$ where $\mathcal{Y} \subset \mathcal{Z} = \mathcal{T} \times \mathcal{E} \times \mathfrak{OP}$, a disagree-IV^{q}ROFSE set $(\xi, \mathcal{Y})_0$ is defined as:

$$(\xi, \mathcal{Y})_0 = \{\xi(\mathfrak{z}) : \mathfrak{z} \in \mathcal{T} \times \mathcal{E} \times \{0\} \subset \mathcal{Z}\}.$$

Note that the disagree-IV^{q}ROFSE set $(\xi, \mathcal{Y})_0$ is an IV^{q}ROFSE subset of $(\xi, \mathcal{Y})$.

Table 3 Tabular arrangement of the IV^4ROFSE set $(\xi_1, \mathcal{Y}_1)$

$(\xi_1, \mathcal{Y}_1)$	$\mathfrak{r}_1$	$\mathfrak{r}_2$	$\mathfrak{r}_3$	$\mathfrak{r}_4$
$(\mathfrak{t}_1, \mathfrak{e}_2, 1)$	$\langle[0.38, 0.51], [0.70, 0.76]\rangle$	$\langle[0.55, 0.68], [0.78, 0.87]\rangle$	$\langle[0.11, 0.20], [0.44, 0.60]\rangle$	$\langle[0.40, 0.55], [0.65, 0.73]\rangle$
$(\mathfrak{t}_2, \mathfrak{e}_1, 1)$	$\langle[0.59, 0.65], [0.20, 0.38]\rangle$	$\langle[0.41, 0.50], [0.79, 0.85]\rangle$	$\langle[0.32, 0.51], [0.58, 0.71]\rangle$	$\langle[0.70, 0.80], [0.15, 0.22]\rangle$
$(\mathfrak{t}_1, \mathfrak{e}_2, 0)$	$\langle[0.12, 0.22], [0.43, 0.55]\rangle$	$\langle[0.20, 0.29], [0.57, 0.69]\rangle$	$\langle[0.80, 0.90], [0.21, 0.36]\rangle$	$\langle[0.15, 0.28], [0.50, 0.65]\rangle$
$(\mathfrak{t}_2, \mathfrak{e}_2, 0)$	$\langle[0.56, 0.66], [0.40, 0.48]\rangle$	$\langle[0.75, 0.80], [0.48, 0.62]\rangle$	$\langle[0.58, 0.64], [0.10, 0.28]\rangle$	$\langle[0.83, 0.91], [0.61, 0.72]\rangle$
$(\xi_1, \mathcal{Y}_1)$	$\mathfrak{r}_5$	$\mathfrak{r}_6$	$\mathfrak{r}_7$	$\mathfrak{r}_8$
$(\mathfrak{t}_1, \mathfrak{e}_2, 1)$	$\langle[0.19, 0.30], [0.75, 0.82]\rangle$	$\langle[0.09, 0.19], [0.35, 0.45]\rangle$	$\langle[0.51, 0.68], [0.01, 0.11]\rangle$	$\langle[0.03, 0.10], [0.73, 0.80]\rangle$
$(\mathfrak{t}_2, \mathfrak{e}_1, 1)$	$\langle[0.18, 0.25], [0.48, 0.59]\rangle$	$\langle[0.10, 0.27], [0.75, 0.85]\rangle$	$\langle[0.65, 0.73], [0.11, 0.25]\rangle$	$\langle[0.11, 0.23], [0.92, 0.94]\rangle$
$(\mathfrak{t}_1, \mathfrak{e}_2, 0)$	$\langle[0.46, 0.52], [0.40, 0.45]\rangle$	$\langle[0.88, 0.93], [0.60, 0.70]\rangle$	$\langle[0.82, 0.88], [0.30, 0.49]\rangle$	$\langle[0.70, 0.81], [0.11, 0.22]\rangle$
$(\mathfrak{t}_2, \mathfrak{e}_2, 0)$	$\langle[0.11, 0.22], [0.70, 0.80]\rangle$	$\langle[0.18, 0.30], [0.58, 0.71]\rangle$	$\langle[0.13, 0.23], [0.78, 0.85]\rangle$	$\langle[0.09, 0.30], [0.52, 0.67]\rangle$

Table 4 Tabular arrangement of the IV^{4}ROFSE set $(\xi_2, \mathcal{Y}_2)$

$(\xi_1, \mathcal{Y}_1)$	$\mathfrak{r}_1$	$\mathfrak{r}_2$	$\mathfrak{r}_3$	$\mathfrak{r}_4$
$(\mathfrak{t}_1, \mathfrak{e}_2, 1)$	$\langle[0.44, 0.55], [0.75, 0.80]\rangle$	$\langle[0.63, 0.72], [0.81, 0.90]\rangle$	$\langle[0.15, 0.23], [0.52, 0.67]\rangle$	$\langle[0.47, 0.60], [0.71, 0.77]\rangle$
$(\mathfrak{t}_2, \mathfrak{e}_1, 1)$	$\langle[0.64, 0.72], [0.23, 0.44]\rangle$	$\langle[0.47, 0.53], [0.81, 0.91]\rangle$	$\langle[0.33, 0.55], [0.61, 0.75]\rangle$	$\langle[0.78, 0.83], [0.18, 0.27]\rangle$
$(\mathfrak{t}_2, \mathfrak{e}_2, 1)$	$\langle[0.25, 0.55], [0.60, 0.70]\rangle$	$\langle[0.01, 0.11], [0.80, 0.95]\rangle$	$\langle[0.33, 0.50], [0.71, 0.83]\rangle$	$\langle[0.45, 0.50], [0.23, 0.40]\rangle$
$(\mathfrak{t}_1, \mathfrak{e}_1, 0)$	$\langle[0.42, 0.50], [0.12, 0.21]\rangle$	$\langle[0.77, 0.79], [0.29, 0.41]\rangle$	$\langle[0.46, 0.53], [0.83, 0.92]\rangle$	$\langle[0.55, 0.60], [0.71, 0.83]\rangle$
$(\mathfrak{t}_1, \mathfrak{e}_2, 0)$	$\langle[0.14, 0.26], [0.45, 0.60]\rangle$	$\langle[0.23, 0.32], [0.61, 0.75]\rangle$	$\langle[0.85, 0.93], [0.22, 0.37]\rangle$	$\langle[0.17, 0.30], [0.51, 0.67]\rangle$
$(\mathfrak{t}_2, \mathfrak{e}_2, 0)$	$\langle[0.61, 0.70], [0.44, 0.53]\rangle$	$\langle[0.81, 0.89], [0.51, 0.72]\rangle$	$\langle[0.62, 0.68], [0.12, 0.32]\rangle$	$\langle[0.88, 0.95], [0.71, 0.79]\rangle$
$(\xi_1, \mathcal{Y}_1)$	$\mathfrak{r}_5$	$\mathfrak{r}_6$	$\mathfrak{r}_7$	$\mathfrak{r}_8$
$(\mathfrak{t}_1, \mathfrak{e}_2, 1)$	$\langle[0.21, 0.33], [0.80, 0.89]\rangle$	$\langle[0.12, 0.24], [0.40, 0.52]\rangle$	$\langle[0.55, 0.70], [0.05, 0.16]\rangle$	$\langle[0.07, 0.15], [0.78, 0.85]\rangle$
$(\mathfrak{t}_2, \mathfrak{e}_1, 1)$	$\langle[0.22, 0.32], [0.54, 0.63]\rangle$	$\langle[0.13, 0.33], [0.81, 0.87]\rangle$	$\langle[0.69, 0.76], [0.18, 0.29]\rangle$	$\langle[0.14, 0.23], [0.95, 0.96]\rangle$
$(\mathfrak{t}_2, \mathfrak{e}_2, 1)$	$\langle[0.12, 0.20], [0.70, 0.81]\rangle$	$\langle[0.17, 0.29], [0.50, 0.65]\rangle$	$\langle[0.86, 0.92], [0.14, 0.25]\rangle$	$\langle[0.10, 0.35], [0.60, 0.70]\rangle$
$(\mathfrak{t}_1, \mathfrak{e}_1, 0)$	$\langle[0.60, 0.78], [0.04, 0.19]\rangle$	$\langle[0.18, 0.30], [0.90, 0.95]\rangle$	$\langle[0.05, 0.11], [0.79, 0.88]\rangle$	$\langle[0.30, 0.45], [0.78, 0.90]\rangle$
$(\mathfrak{t}_1, \mathfrak{e}_2, 0)$	$\langle[0.49, 0.57], [0.41, 0.46]\rangle$	$\langle[0.91, 0.95], [0.62, 0.73]\rangle$	$\langle[0.85, 0.90], [0.35, 0.52]\rangle$	$\langle[0.71, 0.83], [0.13, 0.24]\rangle$
$(\mathfrak{t}_2, \mathfrak{e}_2, 0)$	$\langle[0.12, 0.22], [0.73, 0.84]\rangle$	$\langle[0.20, 0.33], [0.60, 0.71]\rangle$	$\langle[0.13, 0.25], [0.80, 0.95]\rangle$	$\langle[0.10, 0.30], [0.52, 0.67]\rangle$

Table 5 Tabular form of the agree-IV^{4}ROFSE set $(\xi, \mathcal{Y})_1$

$(\xi, \mathcal{Y})_1$	$\mathfrak{r}_1$	$\mathfrak{r}_2$	$\mathfrak{r}_3$	$\mathfrak{r}_4$
$(t_1, e_1, 1)$	$\langle[0.41, 0.53], [0.64, 0.74]\rangle$	$\langle[0.72, 0.84], [0.60, 0.71]\rangle$	$\langle[0.33, 0.50], [0.55, 0.66]\rangle$	$\langle[0.67, 0.75], [0.18, 0.35]\rangle$
$(t_1, e_2, 1)$	$\langle[0.42, 0.56], [0.24, 0.33]\rangle$	$\langle[0.43, 0.53], [0.11, 0.27]\rangle$	$\langle[0.19, 0.26], [0.58, 0.69]\rangle$	$\langle[0.84, 0.93], [0.55, 0.63]\rangle$
$(t_2, e_1, 1)$	$\langle[0.44, 0.55], [0.75, 0.80]\rangle$	$\langle[0.63, 0.72], [0.81, 0.90]\rangle$	$\langle[0.15, 0.23], [0.52, 0.67]\rangle$	$\langle[0.47, 0.60], [0.71, 0.77]\rangle$
$(t_2, e_2, 1)$	$\langle[0.64, 0.72], [0.23, 0.44]\rangle$	$\langle[0.47, 0.53], [0.81, 0.91]\rangle$	$\langle[0.33, 0.55], [0.61, 0.75]\rangle$	$\langle[0.78, 0.83], [0.18, 0.27]\rangle$
$(t_3, e_1, 1)$	$\langle[0.49, 0.57], [0.41, 0.46]\rangle$	$\langle[0.91, 0.95], [0.62, 0.73]\rangle$	$\langle[0.65, 0.69], [0.35, 0.52]\rangle$	$\langle[0.13, 0.24], [0.71, 0.83]\rangle$
$(t_3, e_2, 1)$	$\langle[0.61, 0.70], [0.44, 0.53]\rangle$	$\langle[0.81, 0.89], [0.51, 0.72]\rangle$	$\langle[0.62, 0.68], [0.12, 0.32]\rangle$	$\langle[0.88, 0.95], [0.71, 0.79]\rangle$
$(\xi, \mathcal{Y})_1$	$\mathfrak{r}_5$	$\mathfrak{r}_6$	$\mathfrak{r}_7$	$\mathfrak{r}_8$
$(t_1, e_1, 1)$	$\langle[0.39, 0.57], [0.75, 0.80]\rangle$	$\langle[0.25, 0.33], [0.83, 0.91]\rangle$	$\langle[0.39, 0.52], [0.81, 0.87]\rangle$	$\langle[0.77, 0.82], [0.37, 0.47]\rangle$
$(t_1, e_2, 1)$	$\langle[0.42, 0.55], [0.71, 0.86]\rangle$	$\langle[0.05, 0.20], [0.92, 0.98]\rangle$	$\langle[0.35, 0.48], [0.74, 0.83]\rangle$	$\langle[0.83, 0.90], [0.29, 0.35]\rangle$
$(t_2, e_1, 1)$	$\langle[0.31, 0.51], [0.77, 0.81]\rangle$	$\langle[0.66, 0.73], [0.11, 0.22]\rangle$	$\langle[0.34, 0.56], [0.61, 0.69]\rangle$	$\langle[0.52, 0.67], [0.81, 0.87]\rangle$
$(t_2, e_2, 1)$	$\langle[0.25, 0.55], [0.60, 0.70]\rangle$	$\langle[0.01, 0.11], [0.80, 0.95]\rangle$	$\langle[0.33, 0.50], [0.71, 0.83]\rangle$	$\langle[0.45, 0.50], [0.23, 0.40]\rangle$
$(t_3, e_1, 1)$	$\langle[0.42, 0.50], [0.12, 0.21]\rangle$	$\langle[0.77, 0.79], [0.29, 0.41]\rangle$	$\langle[0.46, 0.53], [0.83, 0.92]\rangle$	$\langle[0.55, 0.60], [0.71, 0.83]\rangle$
$(t_3, e_2, 1)$	$\langle[0.22, 0.33], [0.74, 0.79]\rangle$	$\langle[0.48, 0.60], [0.75, 0.86]\rangle$	$\langle[0.61, 0.71], [0.09, 0.28]\rangle$	$\langle[0.50, 0.60], [0.71, 0.75]\rangle$

Example 5 Consider IV^{4}ROFSE set again as given in Example 1. Then, a disagree-IV^{4}ROFSE set $(\xi, \mathcal{Y})_0$ of the IV^{4}ROFSE set $(\xi, \mathcal{Y})$ on the universal set $\mathcal{X}$ is displayed in Table 6.

Like the importance of complement of a set in crisp set theory, we now discuss the idea of complement for IV^{q}ROFSE sets and demonstrate it through a numerical example:

Definition 10 Let $(\xi, \mathcal{Y})$ be an IV^{q}ROFSE set over a universal set $\mathcal{X}$. Then, the complement of IV^{q}ROFSE set $(\xi, \mathcal{Y})$, denoted by, $(\xi, \mathcal{Y})^{\sim} = (\xi^{\sim}, \neg\mathcal{Y})$, is defined as:

$$(\xi, \mathcal{Y})^{\sim} = (\xi^{\sim}, \neg\mathcal{Y}) = \{\langle \neg\mathfrak{z}, \xi^{\sim}(\neg\mathfrak{z}) \mid \neg\mathfrak{z} \in \neg\mathcal{Y}\rangle\} \tag{3}$$

where $\xi^{\sim}(\neg\mathfrak{z})$ is an IV^{q}ROF subset of $\mathcal{X}$, for all $\neg\mathfrak{z} \in \neg\mathcal{Y}$, which is given as:

$$\xi^{\sim}(\neg\mathfrak{z}) = \left\{\left\langle \mathfrak{x}, \left(\left[\tilde{\lambda}^{-}_{\mathfrak{Q}_{\mathcal{E}}}(\mathfrak{x}), \tilde{\lambda}^{+}_{\mathfrak{Q}_{\mathcal{E}}}(\mathfrak{x})\right], \left[\tilde{\mu}^{-}_{\mathfrak{Q}_{\mathcal{E}}}(\mathfrak{x}), \tilde{\mu}^{+}_{\mathfrak{Q}_{\mathcal{E}}}(\mathfrak{x})\right]\right)\right\rangle : \mathfrak{x} \in \mathcal{X}\right\},$$

for all $\neg\mathfrak{z} \in \neg\mathcal{Y}$ and $\mathfrak{x} \in \mathcal{X}$. Here $\neg\mathcal{Y}$ is the 'Not Set' of attributes which simply involves completely opposite attributes as compared to the set of attributes $\mathcal{Y}$.

Example 6 Reconsider IV^{4}ROFSE set as given in Example 1. Then, by Definition 10, its complement is computed in Table 7.

Now we give a fundamental property of the complement for IV^{q}ROFSE sets.

Proposition 1 Let $(\xi, \mathcal{Y})$ be an IV^{q}ROFSE set on a universe $\mathcal{X}$. Then $((\xi, \mathcal{Y})^{\sim})^{\sim} = (\xi, \mathcal{Y})$.

proof It is directly followed through Definition 10.

We now investigate some more basic operations and properties of IV^{q}ROFSE sets, namely the OR operation, the AND operation, union and intersection, and illustrate them via examples.

Definition 11 The union of two IV^{q}ROFSE sets $(\xi_1, \mathcal{Y}_1)$ and $(\xi_2, \mathcal{Y}_2)$ over a universe $\mathcal{X}$, denoted by $(\xi_1, \mathcal{Y}_1)\hat{\cup}(\xi_2, \mathcal{Y}_2)$, is defined as an IV^{q}ROFSE set $(\mathfrak{J}, \mathcal{Y}) = \{(\mathfrak{z}, \mathfrak{J}(\mathfrak{z})) \mid \mathfrak{z} \in \mathcal{Y} = \mathcal{Y}_1 \cup \mathcal{Y}_2\}$ with the following condition:

$$\mathfrak{J}(\mathfrak{z}) = \begin{cases} \xi_1(\mathfrak{z}), & \text{if } \mathfrak{z} \in \mathcal{Y}_1 - \mathcal{Y}_2, \\ \xi_2(\mathfrak{z}), & \text{if } \mathfrak{z} \in \mathcal{Y}_2 - \mathcal{Y}_1, \\ \xi_1(\mathfrak{z}) \underline{\cup} \xi_2(\mathfrak{z}), & \text{if } \mathfrak{z} \in \mathcal{Y}_1 \cap \mathcal{Y}_2, \end{cases}$$

where

$$\begin{aligned} \xi_1(\mathfrak{z}) \underline{\cup} \xi_2(\mathfrak{z}) = & \left\{\left\langle \mathfrak{x}, \left(\left[\max\left(\tilde{\mu}^{1-}_{\mathfrak{Q}_{\mathcal{E}}}(\mathfrak{x}), \tilde{\mu}^{2-}_{\mathfrak{Q}_{\mathcal{E}}}(\mathfrak{x})\right), \max\left(\tilde{\mu}^{1+}_{\mathfrak{Q}_{\mathcal{E}}}(\mathfrak{x}), \tilde{\mu}^{2+}_{\mathfrak{Q}_{\mathcal{E}}}(\mathfrak{x})\right)\right],\right.\right.\right. \\ & \left.\left.\left. \left[\min\left(\tilde{\lambda}^{1-}_{\mathfrak{Q}_{\mathcal{E}}}(\mathfrak{x}), \tilde{\lambda}^{2-}_{\mathfrak{Q}_{\mathcal{E}}}(\mathfrak{x})\right), \min\left(\tilde{\lambda}^{1+}_{\mathfrak{Q}_{\mathcal{E}}}(\mathfrak{x}), \tilde{\lambda}^{2+}_{\mathfrak{Q}_{\mathcal{E}}}(\mathfrak{x})\right)\right]\right)\right\rangle : \mathfrak{x} \in \mathcal{X}\right\}, \end{aligned}$$

Table 6 Tabular form of the disagree-IV^{4}ROFSE set $(\xi, \mathcal{Y})_0$

$(\xi, \mathcal{Y})_0$	$\mathfrak{r}_1$	$\mathfrak{r}_2$	$\mathfrak{r}_3$	$\mathfrak{r}_4$
$(\mathfrak{t}_1, \mathfrak{e}_1, 0)$	$\langle[0.21, 0.33], [0.80, 0.89]\rangle$	$\langle[0.12, 0.24], [0.40, 0.52]\rangle$	$\langle[0.55, 0.70], [0.05, 0.16]\rangle$	$\langle[0.07, 0.15], [0.78, 0.85]\rangle$
$(\mathfrak{t}_1, \mathfrak{e}_2, 0)$	$\langle[0.22, 0.32], [0.54, 0.63]\rangle$	$\langle[0.13, 0.33], [0.81, 0.87]\rangle$	$\langle[0.69, 0.76], [0.18, 0.29]\rangle$	$\langle[0.14, 0.23], [0.95, 0.96]\rangle$
$(\mathfrak{t}_2, \mathfrak{e}_1, 0)$	$\langle[0.14, 0.26], [0.45, 0.60]\rangle$	$\langle[0.23, 0.32], [0.61, 0.75]\rangle$	$\langle[0.85, 0.93], [0.22, 0.37]\rangle$	$\langle[0.17, 0.30], [0.51, 0.67]\rangle$
$(\mathfrak{t}_2, \mathfrak{e}_2, 0)$	$\langle[0.12, 0.22], [0.73, 0.84]\rangle$	$\langle[0.20, 0.33], [0.60, 0.71]\rangle$	$\langle[0.13, 0.25], [0.80, 0.95]\rangle$	$\langle[0.10, 0.30], [0.52, 0.67]\rangle$
$(\mathfrak{t}_3, \mathfrak{e}_1, 0)$	$\langle[0.14, 0.30], [0.90, 0.95]\rangle$	$\langle[0.52, 0.65], [0.22, 0.33]\rangle$	$\langle[0.15, 0.29], [0.35, 0.42]\rangle$	$\langle[0.23, 0.54], [0.85, 0.93]\rangle$
$(\mathfrak{t}_3, \mathfrak{e}_2, 0)$	$\langle[0.81, 0.90], [0.14, 0.23]\rangle$	$\langle[0.11, 0.19], [0.75, 0.82]\rangle$	$\langle[0.12, 0.28], [0.62, 0.72]\rangle$	$\langle[0.18, 0.65], [0.01, 0.09]\rangle$
$(\xi, \mathcal{Y})_0$	$\mathfrak{r}_5$	$\mathfrak{r}_6$	$\mathfrak{r}_7$	$\mathfrak{r}_8$
$(\mathfrak{t}_1, \mathfrak{e}_1, 0)$	$\langle[0.19, 0.27], [0.60, 0.70]\rangle$	$\langle[0.75, 0.83], [0.13, 0.31]\rangle$	$\langle[0.69, 0.75], [0.31, 0.47]\rangle$	$\langle[0.17, 0.22], [0.77, 0.87]\rangle$
$(\mathfrak{t}_1, \mathfrak{e}_2, 0)$	$\langle[0.63, 0.75], [0.11, 0.26]\rangle$	$\langle[0.93, 0.97], [0.02, 0.15]\rangle$	$\langle[0.75, 0.88], [0.14, 0.13]\rangle$	$\langle[0.13, 0.40], [0.69, 0.85]\rangle$
$(\mathfrak{t}_2, \mathfrak{e}_1, 0)$	$\langle[0.21, 0.30], [0.90, 0.93]\rangle$	$\langle[0.26, 0.33], [0.70, 0.82]\rangle$	$\langle[0.64, 0.77], [0.11, 0.19]\rangle$	$\langle[0.02, 0.12], [0.51, 0.60]\rangle$
$(\mathfrak{t}_2, \mathfrak{e}_2, 0)$	$\langle[0.56, 0.75], [0.10, 0.30]\rangle$	$\langle[0.51, 0.61], [0.20, 0.35]\rangle$	$\langle[0.73, 0.80], [0.01, 0.13]\rangle$	$\langle[0.35, 0.39], [0.83, 0.90]\rangle$
$(\mathfrak{t}_3, \mathfrak{e}_1, 0)$	$\langle[0.12, 0.20], [0.70, 0.81]\rangle$	$\langle[0.17, 0.29], [0.50, 0.65]\rangle$	$\langle[0.86, 0.92], [0.14, 0.25]\rangle$	$\langle[0.10, 0.35], [0.60, 0.70]\rangle$
$(\mathfrak{t}_3, \mathfrak{e}_2, 0)$	$\langle[0.60, 0.78], [0.04, 0.19]\rangle$	$\langle[0.18, 0.30], [0.90, 0.95]\rangle$	$\langle[0.05, 0.11], [0.79, 0.88]\rangle$	$\langle[0.30, 0.45], [0.78, 0.90]\rangle$

Table 7 Tabular form of the complement of IV^{4}ROFSE set $(\xi, \mathcal{Y})$

$(\xi^{\sim}, \neg\mathcal{Y})$	$\mathfrak{r}_1$	$\mathfrak{r}_2$	$\mathfrak{r}_3$	$\mathfrak{r}_4$
$(\neg\mathfrak{t}_1, \mathfrak{e}_1, 1)$	⟨[0.64, 0.74], [0.41, 0.53]⟩	⟨[0.60, 0.71], [0.72, 0.84]⟩	⟨[0.55, 0.66], [0.33, 0.50]⟩	⟨[0.18, 0.35], [0.67, 0.75]⟩
$(\neg\mathfrak{t}_1, \mathfrak{e}_2, 1)$	⟨[0.24, 0.33], [0.42, 0.56]⟩	⟨[0.11, 0.27], [0.43, 0.53]⟩	⟨[0.58, 0.69], [0.19, 0.26]⟩	⟨[0.55, 0.63], [0.84, 0.93]⟩
$(\neg\mathfrak{t}_2, \mathfrak{e}_1, 1)$	⟨[0.75, 0.80], [0.44, 0.55]⟩	⟨[0.81, 0.90], [0.63, 0.72]⟩	⟨[0.52, 0.67], [0.15, 0.23]⟩	⟨[0.71, 0.77], [0.47, 0.60]⟩
$(\neg\mathfrak{t}_2, \mathfrak{e}_2, 1)$	⟨[0.23, 0.44], [0.64, 0.72]⟩	⟨[0.81, 0.91], [0.47, 0.53]⟩	⟨[0.61, 0.75], [0.33, 0.55]⟩	⟨[0.18, 0.27], [0.78, 0.83]⟩
$(\neg\mathfrak{t}_3, \mathfrak{e}_1, 1)$	⟨[0.41, 0.46], [0.49, 0.57]⟩	⟨[0.62, 0.73], [0.91, 0.95]⟩	⟨[0.35, 0.52], [0.65, 0.69]⟩	⟨[0.71, 0.83], [0.13, 0.24]⟩
$(\neg\mathfrak{t}_3, \mathfrak{e}_2, 1)$	⟨[0.44, 0.53], [0.61, 0.70]⟩	⟨[0.51, 0.72], [0.81, 0.89]⟩	⟨[0.12, 0.32], [0.62, 0.68]⟩	⟨[0.71, 0.79], [0.88, 0.95]⟩
$(\neg\mathfrak{t}_1, \mathfrak{e}_1, 0)$	⟨[0.80, 0.89], [0.21, 0.33]⟩	⟨[0.40, 0.52], [0.12, 0.24]⟩	⟨[0.05, 0.16], [0.55, 0.70]⟩	⟨[0.78, 0.85], [0.07, 0.15]⟩
$(\neg\mathfrak{t}_1, \mathfrak{e}_2, 0)$	⟨[0.54, 0.63], [0.22, 0.32]⟩	⟨[0.81, 0.87], [0.13, 0.33]⟩	⟨[0.18, 0.29], [0.69, 0.76]⟩	⟨[0.95, 0.96], [0.14, 0.23]⟩
$(\neg\mathfrak{t}_2, \mathfrak{e}_1, 0)$	⟨[0.45, 0.60], [0.14, 0.26]⟩	⟨[0.61, 0.75], [0.23, 0.32]⟩	⟨[0.22, 0.37], [0.85, 0.93]⟩	⟨[0.51, 0.67], [0.17, 0.30]⟩
$(\neg\mathfrak{t}_2, \mathfrak{e}_2, 0)$	⟨[0.73, 0.84], [0.12, 0.22]⟩	⟨[0.60, 0.71], [0.20, 0.33]⟩	⟨[0.80, 0.95], [0.13, 0.25]⟩	⟨[0.52, 0.67], [0.10, 0.30]⟩
$(\neg\mathfrak{t}_3, \mathfrak{e}_1, 0)$	⟨[0.90, 0.95], [0.14, 0.30]⟩	⟨[0.22, 0.33], [0.52, 0.65]⟩	⟨[0.35, 0.42], [0.15, 0.29]⟩	⟨[0.85, 0.93], [0.23, 0.54]⟩
$(\neg\mathfrak{t}_3, \mathfrak{e}_2, 0)$	⟨[0.14, 0.23], [0.81, 0.90]⟩	⟨[0.75, 0.82], [0.11, 0.19]⟩	⟨[0.62, 0.72], [0.12, 0.28]⟩	⟨[0.01, 0.09], [0.18, 0.65]⟩
$(\xi^{\sim}, \neg\mathcal{Y})$	$\mathfrak{r}_5$	$\mathfrak{r}_6$	$\mathfrak{r}_7$	$\mathfrak{r}_8$
$(\neg\mathfrak{t}_1, \mathfrak{e}_1, 1)$	⟨[0.75, 0.80], [0.39, 0.57]⟩	⟨[0.83, 0.91], [0.25, 0.33]⟩	⟨[0.81, 0.87], [0.39, 0.52]⟩	⟨[0.37, 0.47], [0.77, 0.82]⟩
$(\neg\mathfrak{t}_1, \mathfrak{e}_2, 1)$	⟨[0.71, 0.86], [0.42, 0.55]⟩	⟨[0.92, 0.98], [0.05, 0.20]⟩	⟨[0.74, 0.83], [0.35, 0.48]⟩	⟨[0.29, 0.35], [0.83, 0.90]⟩
$(\neg\mathfrak{t}_2, \mathfrak{e}_1, 1)$	⟨[0.77, 0.81], [0.31, 0.51]⟩	⟨[0.11, 0.22], [0.66, 0.73]⟩	⟨[0.61, 0.69], [0.34, 0.56]⟩	⟨[0.81, 0.87], [0.52, 0.67]⟩
$(\neg\mathfrak{t}_2, \mathfrak{e}_2, 1)$	⟨[0.60, 0.70], [0.25, 0.55]⟩	⟨[0.80, 0.95], [0.01, 0.11]⟩	⟨[0.71, 0.83], [0.33, 0.50]⟩	⟨[0.23, 0.40], [0.45, 0.50]⟩
$(\neg\mathfrak{t}_3, \mathfrak{e}_1, 1)$	⟨[0.12, 0.21], [0.42, 0.50]⟩	⟨[0.29, 0.41], [0.77, 0.79]⟩	⟨[0.83, 0.92], [0.46, 0.53]⟩	⟨[0.71, 0.83], [0.55, 0.60]⟩
$(\neg\mathfrak{t}_3, \mathfrak{e}_2, 1)$	⟨[0.74, 0.79], [0.22, 0.33]⟩	⟨[0.75, 0.86], [0.48, 0.60]⟩	⟨[0.09, 0.28], [0.61, 0.71]⟩	⟨[0.71, 0.75], [0.50, 0.60]⟩
$(\neg\mathfrak{t}_1, \mathfrak{e}_1, 0)$	⟨[0.60, 0.70], [0.19, 0.27]⟩	⟨[0.13, 0.31], [0.75, 0.83]⟩	⟨[0.31, 0.47], [0.69, 0.75]⟩	⟨[0.77, 0.87], [0.17, 0.22]⟩
$(\neg\mathfrak{t}_1, \mathfrak{e}_2, 0)$	⟨[0.11, 0.26], [0.63, 0.75]⟩	⟨[0.02, 0.15], [0.93, 0.97]⟩	⟨[0.14, 0.13], [0.75, 0.88]⟩	⟨[0.69, 0.85], [0.13, 0.40]⟩
$(\neg\mathfrak{t}_2, \mathfrak{e}_1, 0)$	⟨[0.90, 0.93], [0.21, 0.30]⟩	⟨[0.70, 0.82], [0.26, 0.33]⟩	⟨[0.11, 0.19], [0.64, 0.77]⟩	⟨[0.51, 0.60], [0.02, 0.12]⟩
$(\neg\mathfrak{t}_2, \mathfrak{e}_2, 0)$	⟨[0.10, 0.30], [0.56, 0.75]⟩	⟨[0.20, 0.35], [0.51, 0.61]⟩	⟨[0.01, 0.13], [0.73, 0.80]⟩	⟨[0.83, 0.90], [0.35, 0.39]⟩
$(\neg\mathfrak{t}_3, \mathfrak{e}_1, 0)$	⟨[0.70, 0.81], [0.12, 0.20]⟩	⟨[0.50, 0.65], [0.17, 0.29]⟩	⟨[0.14, 0.25], [0.86, 0.92]⟩	⟨[0.60, 0.70], [0.10, 0.35]⟩
$(\neg\mathfrak{t}_3, \mathfrak{e}_2, 0)$	⟨[0.04, 0.19], [0.60, 0.78]⟩	⟨[0.90, 0.95], [0.18, 0.30]⟩	⟨[0.79, 0.88], [0.05, 0.11]⟩	⟨[0.78, 0.90], [0.30, 0.45]⟩

Table 8 Tabular form of the IV^{5}ROFSE set $(\xi_1, \mathcal{Y}_1)$

$(\xi_1, \mathcal{Y}_1)$	$\mathfrak{r}_1$	$\mathfrak{r}_2$	$\mathfrak{r}_3$
$(t_1, e_2, 1)$	$\langle[0.41, 0.60], [0.25, 0.40]\rangle$	$\langle[0.61, 0.64], [0.48, 0.59]\rangle$	$\langle[0.30, 0.50], [0.70, 0.80]\rangle$
$(t_2, e_2, 1)$	$\langle[0.32, 0.46], [0.63, 0.79]\rangle$	$\langle[0.77, 0.89], [0.30, 0.45]\rangle$	$\langle[0.17, 0.33], [0.44, 0.67]\rangle$
$(t_1, e_1, 0)$	$\langle[0.43, 0.55], [0.17, 0.34]\rangle$	$\langle[0.31, 0.45], [0.73, 0.88]\rangle$	$\langle[0.12, 0.32], [0.66, 0.77]\rangle$
$(\xi_1, \mathcal{Y}_1)$	$\mathfrak{r}_4$	$\mathfrak{r}_5$	$\mathfrak{r}_6$
$(t_1, e_2, 1)$	$\langle[0.25, 0.35], [0.64, 0.78]\rangle$	$\langle[0.18, 0.23], [0.67, 0.80]\rangle$	$\langle[0.50, 0.65], [0.10, 0.20]\rangle$
$(t_2, e_2, 1)$	$\langle[0.55, 0.70], [0.23, 0.38]\rangle$	$\langle[0.44, 0.52], [0.87, 0.92]\rangle$	$\langle[0.11, 0.27], [0.65, 0.77]\rangle$
$(t_1, e_1, 0)$	$\langle[0.23, 0.44], [0.10, 0.17]\rangle$	$\langle[0.37, 0.49], [0.82, 0.95]\rangle$	$\langle[0.83, 0.95], [0.61, 0.72]\rangle$

Table 9 Tabular form of the IV^{5}ROFSE set $(\xi_2, \mathcal{Y}_2)$

$(\xi_2, \mathcal{Y}_2)$	$\mathfrak{r}_1$	$\mathfrak{r}_2$	$\mathfrak{r}_3$
$(t_1, e_1, 1)$	$\langle[0.42, 0.57], [0.25, 0.38]\rangle$	$\langle[0.67, 0.74], [0.23, 0.31]\rangle$	$\langle[0.30, 0.40], [0.71, 0.82]\rangle$
$(t_1, e_2, 1)$	$\langle[0.33, 0.45], [0.20, 0.27]\rangle$	$\langle[0.41, 0.55], [0.77, 0.87]\rangle$	$\langle[0.40, 0.50], [0.90, 0.96]\rangle$
$(t_2, e_2, 1)$	$\langle[0.11, 0.25], [0.88, 0.95]\rangle$	$\langle[0.65, 0.80], [0.04, 0.14]\rangle$	$\langle[0.12, 0.20], [0.87, 0.97]\rangle$
$(t_1, e_1, 0)$	$\langle[0.32, 0.48], [0.70, 0.80]\rangle$	$\langle[0.89, 0.97], [0.33, 0.43]\rangle$	$\langle[0.02, 0.11], [0.88, 0.99]\rangle$
$(\xi_2, \mathcal{Y}_2)$	$\mathfrak{r}_4$	$\mathfrak{r}_5$	$\mathfrak{r}_6$
$(t_1, e_1, 1)$	$\langle[0.35, 0.45], [0.70, 0.81]\rangle$	$\langle[0.90, 0.95], [0.55, 0.65]\rangle$	$\langle[0.70, 0.87], [0.03, 0.14]\rangle$
$(t_1, e_2, 1)$	$\langle[0.13, 0.27], [0.60, 0.75]\rangle$	$\langle[0.78, 0.91], [0.12, 0.25]\rangle$	$\langle[0.50, 0.60], [0.12, 0.25]\rangle$
$(t_2, e_2, 1)$	$\langle[0.55, 0.66], [0.34, 0.42]\rangle$	$\langle[0.77, 0.86], [0.40, 0.56]\rangle$	$\langle[0.85, 0.98], [0.05, 0.14]\rangle$
$(t_1, e_1, 0)$	$\langle[0.21, 0.35], [0.60, 0.80]\rangle$	$\langle[0.80, 0.89], [0.43, 0.54]\rangle$	$\langle[0.57, 0.69], [0.13, 0.24]\rangle$

for all $\langle \mathfrak{r}, ([\tilde{\mu}^{1-}_{\mathfrak{Q}\mathcal{E}}(\mathfrak{r}), \tilde{\mu}^{1+}_{\mathfrak{Q}\mathcal{E}}(\mathfrak{r})], [\tilde{\lambda}^{1-}_{\mathfrak{Q}\mathcal{E}}(\mathfrak{r}), \tilde{\lambda}^{1+}_{\mathfrak{Q}\mathcal{E}}(\mathfrak{r})])\rangle \in \xi_1(\mathfrak{z})$, and $\langle \mathfrak{r}, ([\tilde{\mu}^{2-}_{\mathfrak{Q}\mathcal{E}}(\mathfrak{r}), \tilde{\mu}^{2+}_{\mathfrak{Q}\mathcal{E}}(\mathfrak{r})], [\tilde{\lambda}^{2-}_{\mathfrak{Q}\mathcal{E}}(\mathfrak{r}), \tilde{\lambda}^{2+}_{\mathfrak{Q}\mathcal{E}}(\mathfrak{r})])\rangle \in \xi_2(\mathfrak{z})$.

Example 7 Let $\mathcal{X} = \{\mathfrak{r}_1, \mathfrak{r}_2, \mathfrak{r}_3, \mathfrak{r}_4, \mathfrak{r}_5, \mathfrak{r}_6\}$ be a universal set of six houses, $\mathcal{T} = \{t_1 =$ Location, $t_2 =$ Material, $t_3 =$ Design$\}$ be a set of parameters. Let $\mathcal{E} = \{e_1, e_2\}$ be a set of experts with their opinions in the form of a set $\mathfrak{O}\mathfrak{P} = \{1 = agree, 0 = disagree\}$. Let $\mathcal{Y}_1, \mathcal{Y}_2 \subseteq \mathcal{Z} = \mathcal{T} \times \mathcal{E} \times \mathfrak{O}\mathfrak{P}$, which are provided as below:

$$\mathcal{Y}_1 = \{(t_1, e_2, 1), (t_2, e_2, 1), (t_1, e_1, 0)\},$$
$$\mathcal{Y}_2 = \{(t_1, e_1, 1), (t_1, e_2, 1), (t_2, e_2, 1), (t_1, e_1, 0)\}.$$

Now two IV^{5}ROFSE sets $(\xi_1, \mathcal{Y}_1)$ and $(\xi_2, \mathcal{Y}_2)$ are, respectively, given in Tables 8 and 9.

The union of IV^{5}ROFSE sets $(\xi_1, \mathcal{Y}_1)$ and $(\xi_2, \mathcal{Y}_2)$ is determined in Table 10.

Table 10 Tabular form of the union of IV^{5}ROFSE sets $(\xi_1, \mathcal{Y}_1)$ and $(\xi_2, \mathcal{Y}_2)$, which is provided in Tables 8 and 9

$(\mathfrak{J}, \mathcal{Y})$	$\mathfrak{x}_1$	$\mathfrak{x}_2$	$\mathfrak{x}_3$
$(\mathfrak{t}_1, \mathfrak{e}_1, 1)$	$\langle[0.42, 0.57], [0.25, 0.38]\rangle$	$\langle[0.67, 0.74], [0.23, 0.31]\rangle$	$\langle[0.30, 0.40], [0.71, 0.82]\rangle$
$(\mathfrak{t}_1, \mathfrak{e}_2, 1)$	$\langle[0.41, 0.60], [0.20, 0.27]\rangle$	$\langle[0.61, 0.64], [0.48, 0.59]\rangle$	$\langle[0.40, 0.50], [0.70, 0.80]\rangle$
$(\mathfrak{t}_2, \mathfrak{e}_2, 1)$	$\langle[0.32, 0.46], [0.63, 0.79]\rangle$	$\langle[0.77, 0.89], [0.04, 0.14]\rangle$	$\langle[0.17, 0.33], [0.44, 0.67]\rangle$
$(\mathfrak{t}_1, \mathfrak{e}_1, 0)$	$\langle[0.43, 0.55], [0.17, 0.34]\rangle$	$\langle[0.89, 0.97], [0.33, 0.43]\rangle$	$\langle[0.12, 0.32], [0.66, 0.77]\rangle$
$(\mathfrak{J}, \mathcal{Y})$	$\mathfrak{x}_4$	$\mathfrak{x}_5$	$\mathfrak{x}_6$
$(\mathfrak{t}_1, \mathfrak{e}_1, 1)$	$\langle[0.35, 0.45], [0.70, 0.81]\rangle$	$\langle[0.90, 0.95], [0.55, 0.65]\rangle$	$\langle[0.70, 0.87], [0.03, 0.14]\rangle$
$(\mathfrak{t}_1, \mathfrak{e}_2, 1)$	$\langle[0.25, 0.35], [0.60, 0.75]\rangle$	$\langle[0.78, 0.91], [0.12, 0.25]\rangle$	$\langle[0.50, 0.65], [0.10, 0.20]\rangle$
$(\mathfrak{t}_2, \mathfrak{e}_2, 1)$	$\langle[0.55, 0.70], [0.23, 0.38]\rangle$	$\langle[0.77, 0.86], [0.40, 0.56]\rangle$	$\langle[0.85, 0.98], [0.05, 0.14]\rangle$
$(\mathfrak{t}_1, \mathfrak{e}_1, 0)$	$\langle[0.23, 0.44], [0.10, 0.17]\rangle$	$\langle[0.80, 0.89], [0.43, 0.54]\rangle$	$\langle[0.83, 0.95], [0.13, 0.24]\rangle$

Proposition 2 Let $(\xi_1, \mathcal{Y}_1)$, $(\xi_2, \mathcal{Y}_2)$, and $(\xi_3, \mathcal{Y}_3)$ be three IV^{q}ROFSE sets on the universe $\mathcal{X}$. Then

1. $(\xi_1, \mathcal{Y}_1)\hat{\cup}(\xi_2, \mathcal{Y}_2) = (\xi_2, \mathcal{Y}_2)\hat{\cup}(\xi_1, \mathcal{Y}_1)$,
2. $(\xi_1, \mathcal{Y}_1)\hat{\cup}((\xi_2, \mathcal{Y}_2)\hat{\cup}(\xi_3, \mathcal{Y}_3)) = ((\xi_1, \mathcal{Y}_1)\hat{\cup}(\xi_2, \mathcal{Y}_2))\hat{\cup}(\xi_3, \mathcal{Y}_3)$.

proof 1. Using Definition 11, $(\xi_1, \mathcal{Y}_1)\hat{\cup}(\xi_2, \mathcal{Y}_2) = (\mathfrak{J}_1, \mathfrak{V}_1)$ with $\mathfrak{V}_1 = \mathcal{Y}_1 \cup \mathcal{Y}_2$, and for all $\mathfrak{z} \in \mathfrak{V}_1$,

$$\mathfrak{J}_1(\mathfrak{z}) = \begin{cases} \xi_1(\mathfrak{z}), & \text{if } \mathfrak{z} \in \mathcal{Y}_1 - \mathcal{Y}_2, \\ \xi_2(\mathfrak{z}), & \text{if } \mathfrak{z} \in \mathcal{Y}_2 - \mathcal{Y}_1, \\ \xi_1(\mathfrak{z})\underline{\cup}\xi_2(\mathfrak{z}), & \text{if } \mathfrak{z} \in \mathcal{Y}_1 \cap \mathcal{Y}_2, \end{cases}$$

where

$$\begin{aligned} \xi_1(\mathfrak{z})\underline{\cup}\xi_2(\mathfrak{z}) = \Big\{\Big\langle \mathfrak{x}, \Big(\Big[\max\big(\tilde{\mu}^{1-}_{\mathfrak{Q}_\mathcal{E}}(\mathfrak{x}), \tilde{\mu}^{2-}_{\mathfrak{Q}_\mathcal{E}}(\mathfrak{x})\big), \max\big(\tilde{\mu}^{1+}_{\mathfrak{Q}_\mathcal{E}}(\mathfrak{x}), \tilde{\mu}^{2+}_{\mathfrak{Q}_\mathcal{E}}(\mathfrak{x})\big)\Big], \\ \Big[\min\big(\tilde{\lambda}^{1-}_{\mathfrak{Q}_\mathcal{E}}(\mathfrak{x}), \tilde{\lambda}^{2-}_{\mathfrak{Q}_\mathcal{E}}(\mathfrak{x})\big), \min\big(\tilde{\lambda}^{1+}_{\mathfrak{Q}_\mathcal{E}}(\mathfrak{x}), \tilde{\lambda}^{2+}_{\mathfrak{Q}_\mathcal{E}}(\mathfrak{x})\big)\Big]\Big)\Big\rangle : \mathfrak{x} \in \mathcal{X}\Big\}, \end{aligned}$$

for all $\langle \mathfrak{x}, ([\tilde{\mu}^{1-}_{\mathfrak{Q}_\mathcal{E}}(\mathfrak{x}), \tilde{\mu}^{1+}_{\mathfrak{Q}_\mathcal{E}}(\mathfrak{x})], [\tilde{\lambda}^{1-}_{\mathfrak{Q}_\mathcal{E}}(\mathfrak{x}), \tilde{\lambda}^{1+}_{\mathfrak{Q}_\mathcal{E}}(\mathfrak{x})])\rangle \in \xi_1(\mathfrak{z})$, and $\langle \mathfrak{x}, ([\tilde{\mu}^{2-}_{\mathfrak{Q}_\mathcal{E}}(\mathfrak{x}), \tilde{\mu}^{2+}_{\mathfrak{Q}_\mathcal{E}}(\mathfrak{x})], [\tilde{\lambda}^{2-}_{\mathfrak{Q}_\mathcal{E}}(\mathfrak{x}), \tilde{\lambda}^{2+}_{\mathfrak{Q}_\mathcal{E}}(\mathfrak{x})])\rangle \in \xi_2(\mathfrak{z})$. In a similar way, by Definition 11, $(\xi_2, \mathcal{Y}_2)\hat{\cup}(\xi_1, \mathcal{Y}_1) = (\mathfrak{J}_2, \mathfrak{V}_2)$ with $\mathfrak{V}_2 = \mathcal{Y}_2 \cup \mathcal{Y}_1$ and for all $\mathfrak{z} \in \mathfrak{V}_2$,

$$\mathfrak{J}_2(\mathfrak{z}) = \begin{cases} \xi_2(\mathfrak{z}), & \text{if } \mathfrak{z} \in \mathcal{Y}_2 - \mathcal{Y}_1, \\ \xi_1(\mathfrak{z}), & \text{if } \mathfrak{z} \in \mathcal{Y}_1 - \mathcal{Y}_2, \\ \xi_2(\mathfrak{z})\underline{\cup}\xi_1(x), & \text{if } \mathfrak{z} \in \mathcal{Y}_2 \cap \mathcal{Y}_1, \end{cases}$$

$$= \mathfrak{J}_1(\mathfrak{z})$$

where

$$\xi_2(\mathfrak{z})\underline{\cup}\xi_1(\mathfrak{z}) = \Big\{\Big\langle \mathfrak{x}, \Big(\big[\max\big(\tilde{\mu}^{2-}_{\mathfrak{Q}_\mathcal{E}}(\mathfrak{x}), \tilde{\mu}^{1-}_{\mathfrak{Q}_\mathcal{E}}(\mathfrak{x})\big), \max\big(\tilde{\mu}^{2+}_{\mathfrak{Q}_\mathcal{E}}(\mathfrak{x}), \tilde{\mu}^{1+}_{\mathfrak{Q}_\mathcal{E}}(\mathfrak{x})\big)\big], \Big[\min\big(\tilde{\lambda}^{2-}_{\mathfrak{Q}_\mathcal{E}}(\mathfrak{x}), \tilde{\lambda}^{1-}_{\mathfrak{Q}_\mathcal{E}}(\mathfrak{x})\big), \min\big(\tilde{\lambda}^{2+}_{\mathfrak{Q}_\mathcal{E}}(\mathfrak{x}), \tilde{\lambda}^{1+}_{\mathfrak{Q}_\mathcal{E}}(\mathfrak{x})\big)\Big]\Big)\Big\rangle : \mathfrak{x} \in \mathcal{X}\Big\},$$

for all $\langle \mathfrak{x}, ([\tilde{\mu}^{1-}_{\mathfrak{Q}_\mathcal{E}}(\mathfrak{x}), \tilde{\mu}^{1+}_{\mathfrak{Q}_\mathcal{E}}(\mathfrak{x})], [\tilde{\lambda}^{1-}_{\mathfrak{Q}_\mathcal{E}}(\mathfrak{x}), \tilde{\lambda}^{1+}_{\mathfrak{Q}_\mathcal{E}}(\mathfrak{x})])\rangle \in \xi_1(\mathfrak{z})$, and $\langle \mathfrak{x}, ([\tilde{\mu}^{2-}_{\mathfrak{Q}_\mathcal{E}}(\mathfrak{x}), \tilde{\mu}^{2+}_{\mathfrak{Q}_\mathcal{E}}(\mathfrak{x})], [\tilde{\lambda}^{2-}_{\mathfrak{Q}_\mathcal{E}}(\mathfrak{x}), \tilde{\lambda}^{2+}_{\mathfrak{Q}_\mathcal{E}}(\mathfrak{x})])\rangle \in \xi_2(\mathfrak{z})$. Thus, $(\xi_1, \mathcal{Y}_1)\hat{\cup}(\xi_2, \mathcal{Y}_2) = (\xi_2, \mathcal{Y}_2)\hat{\cup}(\xi_1, \mathcal{Y}_1)$.

2. It is same as part (1).

Definition 12 The intersection of two IV^{q}ROFSE sets $(\xi_1, \mathcal{Y}_1)$ and $(\xi_2, \mathcal{Y}_2)$ over a universal set $\mathcal{X}$, denoted by $(\xi_1, \mathcal{Y}_1)\hat{\cap}(\xi_2, \mathcal{Y}_2)$, is defined as an IV^{q}ROFSE set $(\mathfrak{T}, \mathcal{Y}) = \{(\mathfrak{z}, \mathfrak{J}(\mathfrak{z})) \mid \mathfrak{z} \in \mathcal{Y} = \mathcal{Y}_1 \cup \mathcal{Y}_2\}$ with the following condition:

$$\mathfrak{T}(\mathfrak{z}) = \begin{cases} \xi_1(\mathfrak{z}), & \text{if } \mathfrak{z} \in \mathcal{Y}_1 - \mathcal{Y}_2, \\ \xi_2(\mathfrak{z}), & \text{if } \mathfrak{z} \in \mathcal{Y}_2 - \mathcal{Y}_1, \\ \xi_1(\mathfrak{z})\overline{\cap}\xi_2(\mathfrak{z}), & \text{if } \mathfrak{z} \in \mathcal{Y}_1 \cap \mathcal{Y}_2, \end{cases}$$

where

$$\xi_1(\mathfrak{z})\overline{\cap}\xi_2(\mathfrak{z}) = \Big\{\Big\langle \mathfrak{x}, \Big(\big[\min\big(\tilde{\mu}^{1-}_{\mathfrak{Q}_\mathcal{E}}(\mathfrak{x}), \tilde{\mu}^{2-}_{\mathfrak{Q}_\mathcal{E}}(\mathfrak{x})\big), \min\big(\tilde{\mu}^{1+}_{\mathfrak{Q}_\mathcal{E}}(\mathfrak{x}), \tilde{\mu}^{2+}_{\mathfrak{Q}_\mathcal{E}}(\mathfrak{x})\big)\big], \Big[\max\big(\tilde{\lambda}^{1-}_{\mathfrak{Q}_\mathcal{E}}(\mathfrak{x}), \tilde{\lambda}^{2-}_{\mathfrak{Q}_\mathcal{E}}(\mathfrak{x})\big), \max\big(\tilde{\lambda}^{1+}_{\mathfrak{Q}_\mathcal{E}}(\mathfrak{x}), \tilde{\lambda}^{2+}_{\mathfrak{Q}_\mathcal{E}}(\mathfrak{x})\big)\Big]\Big)\Big\rangle : \mathfrak{x} \in \mathcal{X}\Big\},$$

for all $\langle \mathfrak{x}, ([\tilde{\mu}^{1-}_{\mathfrak{Q}_\mathcal{E}}(\mathfrak{x}), \tilde{\mu}^{1+}_{\mathfrak{Q}_\mathcal{E}}(\mathfrak{x})], [\tilde{\lambda}^{1-}_{\mathfrak{Q}_\mathcal{E}}(\mathfrak{x}), \tilde{\lambda}^{1+}_{\mathfrak{Q}_\mathcal{E}}(\mathfrak{x})])\rangle \in \xi_1(\mathfrak{z})$, and $\langle \mathfrak{x}, ([\tilde{\mu}^{2-}_{\mathfrak{Q}_\mathcal{E}}(\mathfrak{x}), \tilde{\mu}^{2+}_{\mathfrak{Q}_\mathcal{E}}(\mathfrak{x})], [\tilde{\lambda}^{2-}_{\mathfrak{Q}_\mathcal{E}}(\mathfrak{x}), \tilde{\lambda}^{2+}_{\mathfrak{Q}_\mathcal{E}}(\mathfrak{x})])\rangle \in \xi_2(\mathfrak{z})$.

Example 8 Consider two IV^{5}ROFSE sets $(\xi_1, \mathcal{Y}_1)$ and $(\xi_2, \mathcal{Y}_2)$ as given by Tables 8 and 9 in Example 7. Then their intersection $(\mathfrak{N}, \mathcal{Y})$ is computed in Table 11.

Proposition 3 Let $(\xi_1, \mathcal{Y}_1)$, $(\xi_2, \mathcal{Y}_2)$, and $(\xi_3, \mathcal{Y}_3)$ be IV^{q}ROFSE sets on the universe $\mathcal{X}$. Then

1. $(\xi_1, \mathcal{Y}_1)\hat{\cap}(\xi_2, \mathcal{Y}_2) = (\xi_2, \mathcal{Y}_2)\hat{\cap}(\xi_1, \mathcal{Y}_1)$,
2. $(\xi_1, \mathcal{Y}_1)\hat{\cap}((\xi_2, \mathcal{Y}_2)\hat{\cap}(\xi_3, \mathcal{Y}_3)) = ((\xi_1, \mathcal{Y}_1)\hat{\cap}(\xi_2, \mathcal{Y}_2))\hat{\cap}(\xi_3, \mathcal{Y}_3)$.

proof 1. By Definition 11, $(\xi_1, \mathcal{Y}_1)\hat{\cap}(\xi_2, \mathcal{Y}_2) = (\mathfrak{T}_1, \mathfrak{U}_1)$ with $\mathfrak{U}_1 = \mathcal{Y}_1 \cup \mathcal{Y}_2$, and for all $\mathfrak{z} \in \mathfrak{U}_1$,

$$\mathfrak{T}_1(\mathfrak{z}) = \begin{cases} \xi_1(\mathfrak{z}), & \text{if } \mathfrak{z} \in \mathcal{Y}_1 - \mathcal{Y}_2, \\ \xi_2(\mathfrak{z}), & \text{if } \mathfrak{z} \in \mathcal{Y}_2 - \mathcal{Y}_1, \\ \xi_1(\mathfrak{z})\overline{\cap}\xi_2(\mathfrak{z}), & \text{if } \mathfrak{z} \in \mathcal{Y}_1 \cap \mathcal{Y}_2, \end{cases}$$

Table 11 Tabular representation of the intersection of IV^{5}ROFSE sets $(\xi_1, \mathcal{Y}_1)$ and $(\xi_2, \mathcal{Y}_2)$, which is given in Tables 8 and 9

$(\mathfrak{T}, \mathcal{Y})$	$\mathfrak{x}_1$	$\mathfrak{x}_2$
$(\mathfrak{t}_1, \mathfrak{e}_1, 1)$	$\langle[0.21, 0.30], [0.35, 0.45], [0.24, 0.35]\rangle$	$\langle[0.15, 0.28], [0.30, 0.40], [0.18, 0.29]\rangle$
$(\mathfrak{t}_2, \mathfrak{e}_1, 1)$	$\langle[0.22, 0.38], [0.15, 0.25], [0.10, 0.26]\rangle$	$\langle[0.68, 0.75], [0.55, 0.65], [0.25, 0.35]\rangle$
$(\mathfrak{t}_3, \mathfrak{e}_2, 1)$	$\langle[0.12, 0.28], [0.10, 0.20], [0.21, 0.36]\rangle$	$\langle[0.25, 0.30], [0.40, 0.52], [0.20, 0.40]\rangle$
$(\mathfrak{t}_2, \mathfrak{e}_2, 0)$	$\langle[0.23, 0.50], [0.25, 0.47], [0.15, 0.27]\rangle$	$\langle[0.80, 0.90], [0.60, 0.70], [0.22, 0.41]\rangle$
$(\mathfrak{t}_3, \mathfrak{e}_1, 0)$	$\langle[0.12, 0.20], [0.25, 0.35], [0.50, 0.60]\rangle$	$\langle[0.79, 0.90], [0.50, 0.70], [0.32, 0.44]\rangle$
$(\mathfrak{T}, \mathcal{Y})$	$\mathfrak{x}_3$	$\mathfrak{x}_4$
$(\mathfrak{t}_1, \mathfrak{e}_1, 1)$	$\langle[0.15, 0.25], [0.10, 0.30], [0.24, 0.38]\rangle$	$\langle[0.60, 0.70], [0.40, 0.52], [0.15, 0.25]\rangle$
$(\mathfrak{t}_2, \mathfrak{e}_1, 1)$	$\langle[0.33, 0.38], [0.27, 0.40], [0.10, 0.20]\rangle$	$\langle[0.60, 0.80], [0.33, 0.53], [0.72, 0.84]\rangle$
$(\mathfrak{t}_3, \mathfrak{e}_2, 1)$	$\langle[0.25, 0.30], [0.21, 0.36], [0.21, 0.38]\rangle$	$\langle[0.27, 0.39], [0.05, 0.15], [0.22, 0.44]\rangle$
$(\mathfrak{t}_2, \mathfrak{e}_2, 0)$	$\langle[0.52, 0.68], [0.25, 0.40], [0.10, 0.30]\rangle$	$\langle[0.45, 0.55], [0.65, 0.75], [0.42, 0.54]\rangle$
$(\mathfrak{t}_3, \mathfrak{e}_1, 0)$	$\langle[0.41, 0.50], [0.51, 0.60], [0.30, 0.40]\rangle$	$\langle[0.60, 0.70], [0.15, 0.25], [0.63, 0.74]\rangle$

where

$$\begin{aligned}\xi_1(\mathfrak{z})\overline{\cap}\xi_2(\mathfrak{z}) = &\Big\{\Big\langle \mathfrak{x}, \Big(\big[\min\big(\tilde{\mu}^{1-}_{\mathfrak{Q}_\mathcal{E}}(\mathfrak{x}), \tilde{\mu}^{2-}_{\mathfrak{Q}_\mathcal{E}}(\mathfrak{x})\big), \min\big(\tilde{\mu}^{1+}_{\mathfrak{Q}_\mathcal{E}}(\mathfrak{x}), \tilde{\mu}^{2+}_{\mathfrak{Q}_\mathcal{E}}(\mathfrak{x})\big)\big],\\ &\Big[\max\big(\tilde{\lambda}^{1-}_{\mathfrak{Q}_\mathcal{E}}(\mathfrak{x}), \tilde{\lambda}^{2-}_{\mathfrak{Q}_\mathcal{E}}(\mathfrak{x})\big), \max\big(\tilde{\lambda}^{1+}_{\mathfrak{Q}_\mathcal{E}}(\mathfrak{x}), \tilde{\lambda}^{2+}_{\mathfrak{Q}_\mathcal{E}}(\mathfrak{x})\big)\Big]\Big)\Big\rangle : \mathfrak{x} \in \mathcal{X}\Big\},\end{aligned}$$

for all $\langle \mathfrak{x}, ([\tilde{\mu}^{1-}_{\mathfrak{Q}_\mathcal{E}}(\mathfrak{x}), \tilde{\mu}^{1+}_{\mathfrak{Q}_\mathcal{E}}(\mathfrak{x})], [\tilde{\lambda}^{1-}_{\mathfrak{Q}_\mathcal{E}}(\mathfrak{x}), \tilde{\lambda}^{1+}_{\mathfrak{Q}_\mathcal{E}}(\mathfrak{x})])\rangle \in \xi_1(\mathfrak{z})$, and $\langle \mathfrak{x}, ([\tilde{\mu}^{2-}_{\mathfrak{Q}_\mathcal{E}}(\mathfrak{x}), \tilde{\mu}^{2+}_{\mathfrak{Q}_\mathcal{E}}(\mathfrak{x})], [\tilde{\lambda}^{2-}_{\mathfrak{Q}_\mathcal{E}}(\mathfrak{x}), \tilde{\lambda}^{2+}_{\mathfrak{Q}_\mathcal{E}}(\mathfrak{x})])\rangle \in \xi_2(\mathfrak{z})$. In a similar way, by Definition 11, $(\xi_2, \mathcal{Y}_2)\hat{\cap}(\xi_1, \mathcal{Y}_1) = (\mathfrak{T}_2, \mathfrak{U}_2)$ with $\mathfrak{V}_2 = \mathcal{Y}_2 \cup \mathcal{Y}_1$, and for all $\mathfrak{z} \in \mathfrak{U}_2$,

$$\begin{aligned}\mathfrak{T}_2(\mathfrak{z}) &= \begin{cases} \xi_2(\mathfrak{z}), & \text{if } \mathfrak{z} \in \mathcal{Y}_2 - \mathcal{Y}_1, \\ \xi_1(\mathfrak{z}), & \text{if } \mathfrak{z} \in \mathcal{Y}_1 - \mathcal{Y}_2, \\ \xi_2(\mathfrak{z})\overline{\cap}\xi_1(\mathfrak{z}), & \text{if } \mathfrak{z} \in \mathcal{Y}_2 \cap \mathcal{Y}_1, \end{cases}\\ &= \mathfrak{T}_1(\mathfrak{z})\end{aligned}$$

where

$$\begin{aligned}\xi_2(\mathfrak{z})\overline{\cap}\xi_1(\mathfrak{z}) = &\Big\{\Big\langle \mathfrak{x}, \Big(\big[\min\big(\tilde{\mu}^{2-}_{\mathfrak{Q}_\mathcal{E}}(\mathfrak{x}), \tilde{\mu}^{1-}_{\mathfrak{Q}_\mathcal{E}}(\mathfrak{x})\big), \min\big(\tilde{\mu}^{2+}_{\mathfrak{Q}_\mathcal{E}}(\mathfrak{x}), \tilde{\mu}^{1+}_{\mathfrak{Q}_\mathcal{E}}(\mathfrak{x})\big)\big],\\ &\Big[\max\big(\tilde{\lambda}^{2-}_{\mathfrak{Q}_\mathcal{E}}(\mathfrak{x}), \tilde{\lambda}^{1-}_{\mathfrak{Q}_\mathcal{E}}(\mathfrak{x})\big), \max\big(\tilde{\lambda}^{2+}_{\mathfrak{Q}_\mathcal{E}}(\mathfrak{x}), \tilde{\lambda}^{1+}_{\mathfrak{Q}_\mathcal{E}}(\mathfrak{x})\big)\Big]\Big)\Big\rangle : \mathfrak{x} \in \mathcal{X}\Big\},\end{aligned}$$

for all $\langle \mathfrak{x}, ([\tilde{\mu}^{1-}_{\mathfrak{Q}_\mathcal{E}}(\mathfrak{x}), \tilde{\mu}^{1+}_{\mathfrak{Q}_\mathcal{E}}(\mathfrak{x})], [\tilde{\lambda}^{1-}_{\mathfrak{Q}_\mathcal{E}}(\mathfrak{x}), \tilde{\lambda}^{1+}_{\mathfrak{Q}_\mathcal{E}}(\mathfrak{x})])\rangle \in \xi_1(\mathfrak{z})$, and $\langle \mathfrak{x}, ([\tilde{\mu}^{2-}_{\mathfrak{Q}_\mathcal{E}}(\mathfrak{x}), \tilde{\mu}^{2+}_{\mathfrak{Q}_\mathcal{E}}(\mathfrak{x})], [\tilde{\lambda}^{2-}_{\mathfrak{Q}_\mathcal{E}}(\mathfrak{x}), \tilde{\lambda}^{2+}_{\mathfrak{Q}_\mathcal{E}}(\mathfrak{x})])\rangle \in \xi_2(\mathfrak{z})$. Thus, $(\xi_1, \mathcal{Y}_1)\hat{\cap}(\xi_2, \mathcal{Y}_2) = (\xi_2, \mathcal{Y}_2)\hat{\cap}(\xi_1, \mathcal{Y}_1)$.

Table 12 Tabular form of the IV^{4}ROFSE set $(\xi_1, \mathcal{Y}_1)$

$(\xi_1, \mathcal{Y}_1)$	$\mathfrak{r}_1$	$\mathfrak{r}_2$
$(t_2, e_1, 1)$	$\langle[0.15, 0.25], [0.71, 0.85]\rangle$	$\langle[0.53, 0.65], [0.25, 0.38]\rangle$
$(t_3, e_1, 1)$	$\langle[0.73, 0.85], [0.41, 0.55]\rangle$	$\langle[0.44, 0.57], [0.14, 0.26]\rangle$
$(\xi_1, \mathcal{Y}_1)$	$\mathfrak{r}_3$	$\mathfrak{r}_4$
$(t_2, e_1, 1)$	$\langle[0.35, 0.43], [0.64, 0.78]\rangle$	$\langle[0.73, 0.89], [0.29, 0.48]\rangle$
$(t_3, e_1, 1)$	$\langle[0.23, 0.41], [0.74, 0.89]\rangle$	$\langle[0.90, 0.95], [0.33, 0.44]\rangle$

Table 13 Tabular form of the IV^{4}ROFSE set $(\xi_2, \mathcal{Y}_2)$

$(\xi_2, \mathcal{Y}_2)$	$\mathfrak{r}_1$	$\mathfrak{r}_2$
$(t_1, e_2, 1)$	$\langle[0.34, 0.51], [0.84, 0.93]\rangle$	$\langle[0.89, 0.96], [0.28, 0.45]\rangle$
$(t_2, e_2, 1)$	$\langle[0.22, 0.38], [0.71, 0.86]\rangle$	$\langle[0.11, 0.29], [0.87, 0.97]\rangle$
$(t_3, e_2, 1)$	$\langle[0.21, 0.45], [0.74, 0.88]\rangle$	$\langle[0.85, 0.93], [0.14, 0.26]\rangle$
$(t_3, e_1, 0)$	$\langle[0.13, 0.26], [0.81, 0.96]\rangle$	$\langle[0.80, 0.89], [0.42, 0.61]\rangle$
$(\xi_2, \mathcal{Y}_2)$	$\mathfrak{r}_3$	$\mathfrak{r}_4$
$(t_1, e_2, 1)$	$\langle[0.32, 0.50], [0.74, 0.82]\rangle$	$\langle[0.35, 0.58], [0.91, 0.96]\rangle$
$(t_2, e_2, 1)$	$\langle[0.41, 0.57], [0.72, 0.86]\rangle$	$\langle[0.77, 0.88], [0.22, 0.34]\rangle$
$(t_3, e_2, 1)$	$\langle[0.46, 0.55], [0.84, 0.88]\rangle$	$\langle[0.76, 0.82], [0.17, 0.28]\rangle$
$(t_3, e_1, 0)$	$\langle[0.76, 0.86], [0.31, 0.46]\rangle$	$\langle[0.68, 0.84], [0.41, 0.53]\rangle$

2. It is same as part (1).

Definition 13 For two IV^{q}ROFSE sets $(\xi_1, \mathcal{Y}_1)$ and $(\xi_2, \mathcal{Y}_2)$ over a universal set $\mathcal{X}$, the 'AND' operation between them denoted by $(\xi_1, \mathcal{Y}_1) \,\overline{\wedge}\, (\xi_2, \mathcal{Y}_2)$, is defined as:

$$(\xi_1, \mathcal{Y}_1) \,\overline{\wedge}\, (\xi_2, \mathcal{Y}_2) = (\Psi, \mathcal{Y}_1 \times \mathcal{Y}_2),$$

where $\Psi(\mathfrak{z}_1, \mathfrak{z}_2) = \xi_1(\mathfrak{z}_1) \overline{\cap} \xi_2(\mathfrak{z}_2)$, for all $(\mathfrak{z}_1, \mathfrak{z}_2) \in \mathcal{Y}_1 \times \mathcal{Y}_2$.

Example 9 Let $\mathcal{X} = \{\mathfrak{r}_1, \mathfrak{r}_2, \mathfrak{r}_3, \mathfrak{r}_4\}$ be a set of four mobile phones, $\mathcal{T} = \{t_1 =$ Technology, $t_2 =$ Dimension, $t_3 = Material\}$ be a set of attributes. Let $\mathcal{E} = \{e_1, e_2\}$ be a set of experts with their opinions in the form of a set $\mathfrak{O}\mathfrak{P} = \{1 = agree, 0 = disagree\}$. Let $\mathcal{Y}_1, \mathcal{Y}_2 \subseteq \mathcal{Z} = \mathcal{T} \times \mathcal{E} \times \mathfrak{O}\mathfrak{P}$ which are provided by

$$\mathcal{Y}_1 = \{(t_2, e_1, 1), (t_3, e_1, 1)\},$$
$$\mathcal{Y}_2 = \{(t_1, e_2, 1), (t_2, e_2, 1), (t_3, e_2, 1), (t_3, e_1, 0)\}.$$

Now two IV^{4}ROFSE sets $(\xi_1, \mathcal{Y}_1)$ and $(\xi_2, \mathcal{Y}_2)$ are, respectively, given in Tables 12 and 13.

The 'AND' operation among the IV^{4}ROFSE sets $(\xi_1, \mathcal{Y}_1)$ and $(\xi_2, \mathcal{Y}_2)$ is displayed in Table 14.

Table 14 Tabular representation of the 'AND' operation among IV^{4}ROFSE sets $(\xi_1, \mathcal{Y}_1)$ and $(\xi_2, \mathcal{Y}_2)$ as provided in Tables 12 and 13

$(\Psi, \mathcal{Y}_1 \times \mathcal{Y}_2)$	$\mathfrak{r}_1$	$\mathfrak{r}_2$
$((t_2, e_1, 1), (t_1, e_2, 1))$	$\langle[0.15, 0.25], [0.84, 0.93]\rangle$	$\langle[0.53, 0.65], [0.28, 0.45]\rangle$
$((t_2, e_1, 1), (t_2, e_2, 1))$	$\langle[0.15, 0.25], [0.71, 0.86]\rangle$	$\langle[0.11, 0.29], [0.87, 0.97]\rangle$
$((t_2, e_1, 1), (t_3, e_2, 1))$	$\langle[0.15, 0.25], [0.74, 0.88]\rangle$	$\langle[0.53, 0.65], [0.25, 0.38]\rangle$
$((t_2, e_1, 1), (t_3, e_1, 0))$	$\langle[0.13, 0.25], [0.81, 0.96]\rangle$	$\langle[0.53, 0.65], [0.42, 0.61]\rangle$
$((t_3, e_1, 1), (t_1, e_2, 1))$	$\langle[0.34, 0.51], [0.84, 0.93]\rangle$	$\langle[0.44, 0.57], [0.28, 0.45]\rangle$
$((t_3, e_1, 1), (t_2, e_2, 1))$	$\langle[0.22, 0.38], [0.71, 0.86]\rangle$	$\langle[0.11, 0.29], [0.87, 0.97]\rangle$
$((t_3, e_1, 1), (t_3, e_2, 1))$	$\langle[0.21, 0.45], [0.74, 0.88]\rangle$	$\langle[0.44, 0.57], [0.14, 0.26]\rangle$
$((t_3, e_1, 1), (t_3, e_1, 0))$	$\langle[0.13, 0.26], [0.81, 0.96]\rangle$	$\langle[0.44, 0.57], [0.14, 0.26]\rangle$
$(\Psi, \mathcal{Y}_1 \times \mathcal{Y}_2)$	$\mathfrak{r}_3$	$\mathfrak{r}_4$
$((t_2, e_1, 1), (t_1, e_2, 1))$	$\langle[0.32, 0.43], [0.74, 0.82]\rangle$	$\langle[0.35, 0.58], [0.91, 0.96]\rangle$
$((t_2, e_1, 1), (t_2, e_2, 1))$	$\langle[0.35, 0.43], [0.72, 0.86]\rangle$	$\langle[0.73, 0.88], [0.29, 0.48]\rangle$
$((t_2, e_1, 1), (t_3, e_2, 1))$	$\langle[0.35, 0.43], [0.84, 0.88]\rangle$	$\langle[0.73, 0.82], [0.29, 0.48]\rangle$
$((t_2, e_1, 1), (t_3, e_1, 0))$	$\langle[0.35, 0.43], [0.64, 0.78]\rangle$	$\langle[0.68, 0.84], [0.41, 0.53]\rangle$
$((t_3, e_1, 1), (t_1, e_2, 1))$	$\langle[0.23, 0.41], [0.74, 0.89]\rangle$	$\langle[0.35, 0.58], [0.91, 0.96]\rangle$
$((t_3, e_1, 1), (t_2, e_2, 1))$	$\langle[0.23, 0.41], [0.74, 0.89]\rangle$	$\langle[0.77, 0.88], [0.33, 0.44]\rangle$
$((t_3, e_1, 1), (t_3, e_2, 1))$	$\langle[0.23, 0.41], [0.84, 0.89]\rangle$	$\langle[0.76, 0.82], [0.33, 0.44]\rangle$
$((t_3, e_1, 1), (t_3, e_1, 0))$	$\langle[0.23, 0.41], [0.74, 0.89]\rangle$	$\langle[0.68, 0.84], [0.41, 0.53]\rangle$

Definition 14 For two IV^{q}ROFSE sets $(\xi_1, \mathcal{Y}_1)$ and $(\xi_2, \mathcal{Y}_2)$ over a universal set $\mathcal{X}$, the 'OR' operation between them denoted by $(\xi_1, \mathcal{Y}_1)\underline{\vee}(\xi_2, \mathcal{Y}_2)$ is defined as:

$$(\xi_1, \mathcal{Y}_1)\underline{\vee}(\xi_2, \mathcal{Y}_2) = (\Omega, \mathcal{Y}_1 \times \mathcal{Y}_2),$$

where $\Omega(\mathfrak{z}_1, \mathfrak{z}_2) = \xi_1(\mathfrak{z}_1)\underline{\cup}\xi_2(\mathfrak{z}_2)$, for all $(\mathfrak{z}_1, \mathfrak{z}_2) \in \mathcal{Y}_1 \times \mathcal{Y}_2$.

Example 10 Reconsider IV^{4}ROFSE sets $(\xi_1, \mathcal{Y}_1)$ and $(\xi_2, \mathcal{Y}_2)$ as provided, respectively, in Tables 12 and 13. The 'OR' operation among the IV^{4}ROFSE sets $(\xi_1, \mathcal{Y}_1)$ and $(\xi_2, \mathcal{Y}_2)$ is computed in Table 15.

We now verify De Morgan's laws concerning operations 'AND' and 'OR' based on IV^{q}ROFSE sets.

Theorem 1 For two IV^{q}ROFSE sets $(\xi_1, \mathcal{Y}_1)$ and $(\xi_2, \mathcal{Y}_2)$ over a universal set $\mathcal{X}$, we have

1. $\left((\xi_1, \mathcal{Y}_1)\underline{\vee}(\xi_2, \mathcal{Y}_2)\right)^{\sim} = (\xi_1, \mathcal{Y}_1)^{\sim}\overline{\wedge}(\xi_2, \mathcal{Y}_2)^{\sim}$.
2. $\left((\xi_1, \mathcal{Y}_1)\overline{\wedge}(\xi_2, \mathcal{Y}_2)\right)^{\sim} = (\xi_1, \mathcal{Y}_1)^{\sim}\underline{\vee}(\xi_2, \mathcal{Y}_2)^{\sim}$.

proof 1. Using Definition 10, we get

$$\begin{aligned}(\xi_1, \mathcal{Y}_1)^{\sim}\underline{\vee}(\xi_2, \mathcal{Y}_2)^{\sim} &= (\xi_1^{\sim}, \neg\mathcal{Y}_1)\underline{\vee}(\xi_2^{\sim}, \neg\mathcal{Y}_2),\\ &= (\Omega^{\sim}, \neg\mathcal{Y}_1 \times \neg\mathcal{Y}_2), \text{ where } \Omega^{\sim}(\neg\mathfrak{z}_1, \neg\mathfrak{z}_2) = \xi_1^{\sim}(\neg\mathfrak{z}_1)\underline{\cup}\xi_2^{\sim}(\neg\mathfrak{z}_2),\end{aligned}$$

Table 15 Tabular form of the 'OR' operation among IV^{4}ROFSE sets $(\xi_1, \mathcal{Y}_1)$ and $(\xi_2, \mathcal{Y}_2)$ as displayed in Tables 12 and 13

$(\Omega, \mathcal{Y}_1 \times \mathcal{Y}_2)$	$\mathfrak{r}_1$	$\mathfrak{r}_2$
$((\mathfrak{t}_2, \mathfrak{e}_1, 1), (\mathfrak{t}_1, \mathfrak{e}_2, 1))$	$\langle[0.34, 0.51], [0.71, 0.85]\rangle$	$\langle[0.89, 0.96], [0.25, 0.38]\rangle$
$((\mathfrak{t}_2, \mathfrak{e}_1, 1), (\mathfrak{t}_2, \mathfrak{e}_2, 1))$	$\langle[0.22, 0.38], [0.71, 0.85]\rangle$	$\langle[0.53, 0.65], [0.25, 0.38]\rangle$
$((\mathfrak{t}_2, \mathfrak{e}_1, 1), (\mathfrak{t}_3, \mathfrak{e}_2, 1))$	$\langle[0.21, 0.45], [0.71, 0.85]\rangle$	$\langle[0.85, 0.93], [0.14, 0.26]\rangle$
$((\mathfrak{t}_2, \mathfrak{e}_1, 1), (\mathfrak{t}_3, \mathfrak{e}_1, 0))$	$\langle[0.15, 0.26], [0.71, 0.85]\rangle$	$\langle[0.80, 0.89], [0.25, 0.38]\rangle$
$((\mathfrak{t}_3, \mathfrak{e}_1, 1), (\mathfrak{t}_1, \mathfrak{e}_2, 1))$	$\langle[0.73, 0.85], [0.41, 0.55]\rangle$	$\langle[0.89, 0.96], [0.14, 0.26]\rangle$
$((\mathfrak{t}_3, \mathfrak{e}_1, 1), (\mathfrak{t}_2, \mathfrak{e}_2, 1))$	$\langle[0.73, 0.85], [0.41, 0.55]\rangle$	$\langle[0.44, 0.57], [0.14, 0.26]\rangle$
$((\mathfrak{t}_3, \mathfrak{e}_1, 1), (\mathfrak{t}_3, \mathfrak{e}_2, 1))$	$\langle[0.73, 0.85], [0.41, 0.55]\rangle$	$\langle[0.85, 0.93], [0.14, 0.26]\rangle$
$((\mathfrak{t}_3, \mathfrak{e}_1, 1), (\mathfrak{t}_3, \mathfrak{e}_1, 0))$	$\langle[0.73, 0.85], [0.41, 0.55]\rangle$	$\langle[0.80, 0.89], [0.14, 0.26]\rangle$
$(\Omega, \mathcal{Y}_1 \times \mathcal{Y}_2)$	$\mathfrak{r}_3$	$\mathfrak{r}_4$
$((\mathfrak{t}_2, \mathfrak{e}_1, 1), (\mathfrak{t}_1, \mathfrak{e}_2, 1))$	$\langle[0.35, 0.50], [0.64, 0.78]\rangle$	$\langle[0.73, 0.89], [0.29, 0.48]\rangle$
$((\mathfrak{t}_2, \mathfrak{e}_1, 1), (\mathfrak{t}_2, \mathfrak{e}_2, 1))$	$\langle[0.41, 0.57], [0.64, 0.78]\rangle$	$\langle[0.77, 0.89], [0.22, 0.34]\rangle$
$((\mathfrak{t}_2, \mathfrak{e}_1, 1), (\mathfrak{t}_3, \mathfrak{e}_2, 1))$	$\langle[0.46, 0.55], [0.64, 0.78]\rangle$	$\langle[0.76, 0.89], [0.17, 0.28]\rangle$
$((\mathfrak{t}_2, \mathfrak{e}_1, 1), (\mathfrak{t}_3, \mathfrak{e}_1, 0))$	$\langle[0.76, 0.86], [0.31, 0.46]\rangle$	$\langle[0.73, 0.89], [0.29, 0.48]\rangle$
$((\mathfrak{t}_3, \mathfrak{e}_1, 1), (\mathfrak{t}_1, \mathfrak{e}_2, 1))$	$\langle[0.32, 0.50], [0.74, 0.82]\rangle$	$\langle[0.90, 0.95], [0.33, 0.44]\rangle$
$((\mathfrak{t}_3, \mathfrak{e}_1, 1), (\mathfrak{t}_2, \mathfrak{e}_2, 1))$	$\langle[0.41, 0.57], [0.72, 0.86]\rangle$	$\langle[0.90, 0.95], [0.22, 0.34]\rangle$
$((\mathfrak{t}_3, \mathfrak{e}_1, 1), (\mathfrak{t}_3, \mathfrak{e}_2, 1))$	$\langle[0.46, 0.55], [0.74, 0.88]\rangle$	$\langle[0.90, 0.95], [0.17, 0.28]\rangle$
$((\mathfrak{t}_3, \mathfrak{e}_1, 1), (\mathfrak{t}_3, \mathfrak{e}_1, 0))$	$\langle[0.76, 0.86], [0.31, 0.46]\rangle$	$\langle[0.90, 0.95], [0.33, 0.44]\rangle$

for all $(\neg\mathfrak{z}_1, \neg\mathfrak{z}_2) \in \neg\mathcal{Y}_1 \times \neg\mathcal{Y}_2$. It follows by Definition 13 that $(\xi_1, \mathcal{Y}_1)\overline{\wedge}(\xi_2, \mathcal{Y}_2) = (\Psi, \mathcal{Y}_1 \times \mathcal{Y}_2)$, which means $\big((\xi_1, \mathcal{Y}_1)\overline{\wedge}(\xi_2, \mathcal{Y}_2)\big)^{\sim} = (\Psi, \mathcal{Y}_1 \times \mathcal{Y}_2)^{\sim} = (\Psi^{\sim}, \neg(\mathcal{Y}_1 \times \mathcal{Y}_2))$. Thus, for any $(\neg\mathfrak{z}_1, \neg\mathfrak{z}_2) \in \neg\mathcal{Y}_1 \times \neg\mathcal{Y}_2$, we get

$$
\begin{aligned}
\Psi^{\sim}(\neg\mathfrak{z}_1, \neg\mathfrak{z}_2) &= (\Psi(\mathfrak{z}_1, \mathfrak{z}_2))^{\sim}, \\
&= (\xi_1(\mathfrak{z}_1)\overline{\sqcap}\xi_2(\mathfrak{z}_2))^{\sim}, \\
&= \xi_1{}^{\sim}(\neg\mathfrak{z}_1)\underline{\sqcup}\xi_2{}^{\sim}(\neg\mathfrak{z}_2).
\end{aligned}
$$

By all these arguments, we have $\big((\xi_1, \mathcal{Y}_1)\overline{\wedge}(\xi_2, \mathcal{Y}_2)\big)^{\sim} = (\xi_1, \mathcal{Y}_1)^{\sim}\underline{\vee}(\xi_2, \mathcal{Y}_2)^{\sim}$.

2. It is similar to part (1).

Theorem 2 For three IV^{q}ROFSE sets $(\xi_1, \mathcal{Y}_1)$, $(\xi_2, \mathcal{Y}_2)$ and $(\xi_3, \mathcal{Y}_3)$ over a universe $\mathcal{X}$, we have

1. $(\xi_1, \mathcal{Y}_1)\overline{\wedge}\big((\xi_2, \mathcal{Y}_2)\overline{\wedge}(\xi_3, \mathcal{Y}_3)\big) = \big((\xi_1, \mathcal{Y}_1)\overline{\wedge}(\xi_2, \mathcal{Y}_2)\big)\overline{\wedge}(\xi_3, \mathcal{Y}_3)$,
2. $(\xi_1, \mathcal{Y}_1)\underline{\vee}\big((\xi_2, \mathcal{Y}_2)\underline{\vee}(\xi_3, \mathcal{Y}_3)\big) = \big((\xi_1, \mathcal{Y}_1)\underline{\vee}(\xi_2, \mathcal{Y}_2)\big)\underline{\vee}(\xi_3, \mathcal{Y}_3)$,
3. $(\xi_1, \mathcal{Y}_1)\overline{\wedge}\big((\xi_2, \mathcal{Y}_2)\underline{\vee}(\xi_3, \mathcal{Y}_3)\big) = \big((\xi_1, \mathcal{Y}_1)\overline{\wedge}(\xi_2, \mathcal{Y}_2)\big)\underline{\vee}\big((\xi_1, \mathcal{Y}_1)\bar{\wedge}(\xi_3, \mathcal{Y}_3)\big)$,
4. $(\xi_1, \mathcal{Y}_1)\underline{\vee}\big((\xi_2, \mathcal{Y}_2)\overline{\wedge}(\xi_3, \mathcal{Y}_3)\big) = \big((\xi_1, \mathcal{Y}_1)\underline{\vee}(\xi_2, \mathcal{Y}_2)\big)\overline{\wedge}\big((\xi_1, \mathcal{Y}_1)\bar{\vee}(\xi_3, \mathcal{Y}_3)\big)$.

proof 1. For every $\mathfrak{z}_1 \in \mathcal{Y}_1, \mathfrak{z}_2 \in \mathcal{Y}_2$ and $\mathfrak{z}_3 \in \mathcal{Y}_3$, $\xi_1(\mathfrak{z}_1)$, $\xi_2(\mathfrak{z}_2)$ and $\xi_3(\mathfrak{z}_3)$ are IV^{q}ROF subsets of $\mathcal{X}$. Thus, by a result of IV^{q}ROFSs, $\xi_1(\mathfrak{z}_1)\overline{\sqcap}(\xi_2(\mathfrak{z}_2)\overline{\sqcap}\xi_3(\mathfrak{z}_3)) = (\xi_1(\mathfrak{z}_1)\overline{\sqcap}\xi_2(\mathfrak{z}_2))\overline{\sqcap}\xi_3(\mathfrak{z}_3)$, we have, $(\xi_1, \mathcal{Y}_1)\overline{\wedge}\big((\xi_2, \mathcal{Y}_2)\overline{\wedge}(\xi_3, \mathcal{Y}_3)\big) = \big((\xi_1, \mathcal{Y}_1)\overline{\wedge}(\xi_2, \mathcal{Y}_2)\big)\overline{\wedge}(\xi_3, \mathcal{Y}_3)$.

2. Same as part (1).
3. For every $\mathfrak{z}_1 \in \mathcal{Y}_1, \mathfrak{z}_2 \in \mathcal{Y}_2$ and $\mathfrak{z}_3 \in \mathcal{Y}_3$, $\xi_1(\mathfrak{z}_1)$, $\xi_2(\mathfrak{z}_2)$ and $\xi_3(\mathfrak{z}_3)$ are IV^{q}ROF subsets of $\mathcal{X}$. Thus, by a result of IV^{q}ROFSs, $\eta(\mathfrak{z}_1)\overline{\sqcap}(\xi_2(\mathfrak{z}_2)\underline{\sqcup}\eta(\mathfrak{z}_3)) = (\eta(\mathfrak{z}_1)\overline{\sqcap}\xi_2(\mathfrak{z}_2))\underline{\sqcup}(\eta(\mathfrak{z}_1)\overline{\sqcap}\eta(\mathfrak{z}_3))$, which easily follows $(\xi_1, \mathcal{Y}_1)\overline{\wedge}\big((\xi_2, \mathcal{Y}_2)\underline{\vee}(\xi_3, \mathcal{Y}_3)\big) = \big((\xi_1, \mathcal{Y}_1)\overline{\wedge}(\xi_2, \mathcal{Y}_2)\big)\underline{\vee}\big((\xi_1, \mathcal{Y}_1)\bar{\wedge}(\xi_3, \mathcal{Y}_3)\big)$.
4. Similar to part (3).

3 Application to MCGDM

This section first introduces some new concepts and then discusses a practical application of the initiated model with an algorithm.

Definition 15 Let $\mathcal{X} = \{\mathfrak{x}_1, \mathfrak{x}_2, \ldots, \mathfrak{x}_n\}$ be a universal set, $\mathcal{T}$ be a universe of attributes. Let $\mathcal{E}$ be a collection of experts with their opinions in the form of a set $\mathfrak{O}\mathfrak{P} = \{0 = disagree,\ 1 = agree\}$. For any IV^{q}ROFSE set $(\xi, \mathcal{Y})$ where $\mathcal{Y} \subset \mathcal{Z} = \mathcal{T} \times \mathcal{E} \times \mathfrak{O}\mathfrak{P}$, the score values of lower and upper bounds of every interval concerning membership and non-membership degrees, represented by $\mathcal{L}^-_{\mathfrak{z}_j}(\mathfrak{x}_i)$, $\mathcal{L}^+_{\mathfrak{z}_j}(\mathfrak{x}_i)$, $\mathcal{U}^-_{\mathfrak{z}_j}(\mathfrak{x}_i)$ and $\mathcal{U}^+_{\mathfrak{z}_j}(\mathfrak{x}_i)$, are, respectively, given as below:

$$\mathcal{L}^-_{\mathfrak{z}_j}(\mathfrak{x}_i) = \sum_{s=1}^{n} \big(\tilde{\mu}^-_{\mathfrak{Q}_\mathcal{E}}(\mathfrak{x}_i) - \tilde{\mu}^-_{\mathfrak{Q}_\mathcal{E}}(\mathfrak{x}_s)\big), \tag{4}$$

$$\mathcal{L}^+_{\mathfrak{z}_j}(\mathfrak{x}_i) = \sum_{s=1}^{n} \big(\tilde{\mu}^+_{\mathfrak{Q}_\mathcal{E}}(\mathfrak{x}_i) - \tilde{\mu}^+_{\mathfrak{Q}_\mathcal{E}}(\mathfrak{x}_s)\big), \tag{5}$$

$$\mathcal{U}^-_{\mathfrak{z}_j}(\mathfrak{x}_i) = \sum_{s=1}^{n} \big(\tilde{\lambda}^-_{\mathfrak{Q}_\mathcal{E}}(\mathfrak{x}_i) - \tilde{\lambda}^-_{\mathfrak{Q}_\mathcal{E}}(\mathfrak{x}_s)\big), \tag{6}$$

$$\mathcal{U}^+_{\mathfrak{z}_j}(\mathfrak{x}_i) = \sum_{s=1}^{n} \big(\tilde{\lambda}^+_{\mathfrak{Q}_\mathcal{E}}(\mathfrak{x}_i) - \tilde{\lambda}^+_{\mathfrak{Q}_\mathcal{E}}(\mathfrak{x}_s)\big), \tag{7}$$

for each $\mathfrak{z}_j \in \mathcal{Y}$.

Definition 16 Let $\mathcal{X} = \{\mathfrak{x}_1, \mathfrak{x}_2, \ldots, \mathfrak{x}_n\}$ be a universal set, $\mathcal{T}$ be a universe of attributes. Let $\mathcal{E}$ be a collection of experts with their opinions in the form of a set $\mathfrak{O}\mathfrak{P} = \{0 = disagree,\ 1 = agree\}$. For any IV^{q}ROFSE set $(\xi, \mathcal{Y})$ where $\mathcal{Y} \subset \mathcal{Z} = \mathcal{T} \times \mathcal{E} \times \mathfrak{O}\mathfrak{P}$, the accumulated scores of available objects, denoted as $\mathcal{AS}_{\mathfrak{z}_j}(\mathfrak{x}_i)$, are defined by

$$\mathcal{AS}_{\mathfrak{z}_j}(\mathfrak{x}_i) = \mathcal{M}_{\mathfrak{z}_j}(\mathfrak{x}_i) - \mathcal{N}_{\mathfrak{z}_j}(\mathfrak{x}_i) \tag{8}$$

where $\mathcal{M}_{\mathfrak{z}_j}(\mathfrak{x}_i) = \mathcal{L}^-_{\mathfrak{z}_j}(\mathfrak{x}_i) + \mathcal{L}^+_{\mathfrak{z}_j}(\mathfrak{x}_i)$ and $\mathcal{N}_{\mathfrak{z}_j}(\mathfrak{x}_i) = \mathcal{U}^-_{\mathfrak{z}_j}(\mathfrak{x}_i) + \mathcal{U}^+_{\mathfrak{z}_j}(\mathfrak{x}_i)$ represent the score values for membership and non-membership parts. Here $\mathcal{L}^-_{\mathfrak{z}_j}(\mathfrak{x}_i)$, $\mathcal{L}^+_{\mathfrak{z}_j}(\mathfrak{x}_i)$,

$\mathcal{U}^-_{\mathfrak{z}_j}(\mathfrak{x}_i)$ and $\mathcal{U}^+_{\mathfrak{z}_j}(\mathfrak{x}_i)$ serve as the scores of lower and upper bounds of both membership and non-membership degree intervals.

In the following, now we develop a new MCGDM algorithm based on IV^{q}ROFSE sets.

Algorithm 1 Ranking of objects based on IV^{q}ROFSE sets

---

1. **Input:**
   $\mathcal{X}$, a universe containing 'n' alternatives,
   $\mathcal{T}$, a universe of parameters,
   $\mathcal{E}$, a collection of experts,
   $\mathfrak{O}\mathfrak{P}$, a set of experts' judgments,
   $(\xi, \mathcal{Y})$, an IV^{q}ROFSE set with $\mathcal{Y} \subset \mathcal{Z} = \mathcal{T} \times \mathcal{E} \times \mathfrak{O}\mathfrak{P}$.
2. For every lower and upper bound of an available IV^{q}ROFSE set, calculate agree- and disagree-scores of membership and non-membership degree intervals' with the help of $\mathcal{L}^-_{\mathfrak{z}_j}(\mathfrak{x}_i)$, $\mathcal{L}^+_{\mathfrak{z}_j}(\mathfrak{x}_i)$, $\mathcal{U}^-_{\mathfrak{z}_j}(\mathfrak{x}_i)$ and $\mathcal{U}^+_{\mathfrak{z}_j}(\mathfrak{x}_i)$, provided from Eqs. (4) to (5).
3. By Definition 16, compute the accumulated scores $\mathcal{AS}_{\mathfrak{z}_j}(\mathfrak{x}_i) = \mathcal{M}_{\mathfrak{z}_j}(\mathfrak{x}_i) - \mathcal{N}_{\mathfrak{z}_j}(\mathfrak{x}_i)$ for the membership $\mathcal{M}_{\mathfrak{z}_j}(\mathfrak{x}_i)$ and non-membership $\mathcal{N}_{\mathfrak{z}_j}(\mathfrak{x}_i)$ parts of both agree- and disagree-IV^{q}ROFSE sets, respectively, for each $\mathfrak{x}_i \in \mathcal{X}$.
4. Determine the agree-score-sum ($\mathfrak{A}_i = \sum_j \mathfrak{x}_{ji}$) and disagree-score-sum ($\mathfrak{D}_i = \sum_j \mathfrak{x}_{ji}$) in the last row of accumulated score tables of agree- and disagree-IV^{q}ROFSE sets, respectively.
5. Calculate the final scores $\mathcal{S}_i = \mathfrak{A}_i - \mathfrak{D}_i$ for all the alternatives $\mathfrak{x}_i \in \mathcal{X}$, $(i = 1, 2, \ldots, n)$.
   Output: Find the numeric υ for which $\mathcal{S}_\upsilon = \max_{i\in\{1,2,\ldots,n\}} \mathcal{S}_i$. Then, $\mathfrak{x}_\upsilon$ will be selected as decision alternative.

Now we provide a practical application of our initiated approach in above Algorithm 1 under IV^{q}ROFSE sets.

3.1 Analysis of Modern Energy Development Methods

Nowadays due to a rapid increase of daily requirements in energy demands by all human-being over the globe, scientists and researchers believe that the future energy demands can be overcome by employing renewable energy sources or by improving existing energy production techniques. Currently, the role of renewable energy is weak to meet the major energy demands, especially, electrical energy. According to the experimental judgments, scientists and experts believe that in near future both developing and developed countries will certainly depend on fossil fuels. The situation is more convenient in developed countries as compared to developing countries. A number of developing countries have been evidently struggling to reorganize their energy sections. Apparently, it is not an easy task to recognize innovations in this domain because market share, policy, and cost are the major hurdles for the development of renewable energy. Modern energy development methods are important to strengthen a country economically and to all the population. Approach to modern energy is necessary for the requirement of health care, sanitation, clean water, efficient and reliable telecommunication, transport, cooking, food, heating, and lighting services. In emerging developing countries, renewable energy technologies provide modern energy services which give different opportunities for the development of off-grid societies. The existence of many technological options using renewable energy sources has proven their adequacy by the responsive behavior to various sustainability problems. However, in recent years, a significant increase has been reported in the utilization of renewable energy techniques, to meet the energy demands of human being and societies but the share of renewable energy technologies remains low than traditional methods. Consequently, there is a need for more powerful efforts to unlock their efficient contributions to lead the development in a more fruitful direction. The modern energy technologies require a lot of effort in testing and developing as well as in the implementation strategies for these innovative energy development methods. With these technical difficulties, economical and social hurdles are also more difficult to overcome. One solution is not present for all the renewable energy technologies successful implementation, but every project provides new findings.

Assume that an analyst wants to find the ranking of modern energy development methods with respect to their successful implementation in a developing country in the presence of different factors(hurdles) which may resist their implementation. Let $\mathcal{X} = \{\mathfrak{x}_1, \mathfrak{x}_2, \ldots, \mathfrak{x}_8\}$ be a set of eight modern energy development sources where

$\mathfrak{x}_1$ denotes the 'Solar Energy',
$\mathfrak{x}_2$ denotes the 'Wind Energy',
$\mathfrak{x}_3$ denotes the 'Geothermal Energy',
$\mathfrak{x}_4$ denotes the 'Hydro Energy',
$\mathfrak{x}_5$ denotes the 'Tidal Energy',
$\mathfrak{x}_6$ denotes the 'Thermal Energy',
$\mathfrak{x}_7$ denotes the 'Biomass Energy',
$\mathfrak{x}_8$ denotes the 'Wave Energy'.

Let $\mathcal{T} = \{\mathfrak{t}_1 = \text{Social Impact(Criterion} - \text{I)}, \mathfrak{t}_2 = \text{Environmental Impact(Criterion} - \text{II)}, \mathfrak{t}_3 = \text{Economic Impact(Criterion} - \text{III)}, \mathfrak{t}_4 = \text{Political Impact(Criterion} - \text{IV)}\}$ be a set of four attributes(criteria). Let $\mathcal{E} = \{\mathfrak{e}_1, \mathfrak{e}_2\}$ be a set of experts who are invited for evaluation and then they provide their judgments in the form of a set $\mathfrak{O}\mathfrak{P} = \{1 = \text{agree}, 0 = \text{disagree}\}$ along with membership and non-membership values as a subset of the closed unit interval. The information collected from the experts after evaluating the given options under consideration is provided in the form of an IV^{4}ROFSE set, which is displayed in Table 16.

Using Eqs. (4), (5), (6) and (7), the score value of each interval's lower and upper bounds are computed in Tables 17, 18, 19 and 20. Note that the first entry of Table 17 is determined as follows:

Table 16 Tabular form of the IV^{4}ROFSE set $(\xi, \mathcal{Y})$

$(\xi, \mathcal{Y})$	$\mathfrak{x}_1$	$\mathfrak{x}_2$	$\mathfrak{x}_3$	$\mathfrak{x}_4$
$(t_1, e_1, 1)$	⟨[0.50, 0.55], [0.33, 0.45]⟩	⟨[0.65, 0.72], [0.26, 0.35]⟩	⟨[0.20, 0.25], [0.75, 0.82]⟩	⟨[0.60, 0.70], [0.17, 0.32]⟩
$(t_1, e_2, 1)$	⟨[0.22, 0.35], [0.73, 0.86]⟩	⟨[0.66, 0.77], [0.47, 0.59]⟩	⟨[0.43, 0.55], [0.61, 0.71]⟩	⟨[0.50, 0.60], [0.37, 0.42]⟩
$(t_2, e_1, 1)$	⟨[0.64, 0.78], [0.23, 0.33]⟩	⟨[0.33, 0.44], [0.76, 0.81]⟩	⟨[0.55, 0.63], [0.80, 0.87]⟩	⟨[0.10, 0.30], [0.27, 0.42]⟩
$(t_2, e_2, 1)$	⟨[0.35, 0.49], [0.74, 0.84]⟩	⟨[0.67, 0.76], [0.26, 0.48]⟩	⟨[0.30, 0.40], [0.72, 0.79]⟩	⟨[0.68, 0.76], [0.27, 0.36]⟩
$(t_3, e_1, 1)$	⟨[0.34, 0.50], [0.71, 0.80]⟩	⟨[0.24, 0.38], [0.67, 0.78]⟩	⟨[0.19, 0.29], [0.69, 0.79]⟩	⟨[0.55, 0.66], [0.37, 0.45]⟩
$(t_3, e_2, 1)$	⟨[0.40, 0.50], [0.70, 0.80]⟩	⟨[0.16, 0.26], [0.86, 0.91]⟩	⟨[0.60, 0.73], [0.28, 0.34]⟩	⟨[0.80, 0.90], [0.10, 0.22]⟩
$(t_4, e_1, 1)$	⟨[0.67, 0.84], [0.23, 0.31]⟩	⟨[0.27, 0.41], [0.78, 0.86]⟩	⟨[0.63, 0.77], [0.15, 0.27]⟩	⟨[0.54, 0.64], [0.37, 0.44]⟩
$(t_4, e_2, 1)$	⟨[0.78, 0.83], [0.43, 0.55]⟩	⟨[0.25, 0.36], [0.56, 0.67]⟩	⟨[0.40, 0.60], [0.17, 0.29]⟩	⟨[0.68, 0.72], [0.37, 0.43]⟩
$(t_1, e_1, 0)$	⟨[0.70, 0.80], [0.23, 0.32]⟩	⟨[0.17, 0.23], [0.66, 0.77]⟩	⟨[0.80, 0.89], [0.15, 0.29]⟩	⟨[0.30, 0.52], [0.67, 0.78]⟩
$(t_1, e_2, 0)$	⟨[0.41, 0.52], [0.82, 0.94]⟩	⟨[0.12, 0.25], [0.67, 0.79]⟩	⟨[0.83, 0.92], [0.31, 0.43]⟩	⟨[0.10, 0.30], [0.55, 0.62]⟩
$(t_2, e_1, 0)$	⟨[0.33, 0.47], [0.63, 0.74]⟩	⟨[0.50, 0.60], [0.19, 0.28]⟩	⟨[0.72, 0.84], [0.21, 0.35]⟩	⟨[0.80, 0.93], [0.50, 0.60]⟩
$(t_2, e_2, 0)$	⟨[0.81, 0.89], [0.25, 0.40]⟩	⟨[0.38, 0.47], [0.68, 0.79]⟩	⟨[0.50, 0.65], [0.32, 0.43]⟩	⟨[0.18, 0.26], [0.57, 0.66]⟩
$(t_3, e_1, 0)$	⟨[0.67, 0.75], [0.21, 0.33]⟩	⟨[0.50, 0.69], [0.17, 0.28]⟩	⟨[0.79, 0.90], [0.28, 0.39]⟩	⟨[0.25, 0.36], [0.77, 0.85]⟩
$(t_3, e_2, 0)$	⟨[0.69, 0.77], [0.22, 0.33]⟩	⟨[0.54, 0.69], [0.37, 0.47]⟩	⟨[0.21, 0.34], [0.58, 0.67]⟩	⟨[0.30, 0.45], [0.70, 0.82]⟩
$(t_4, e_1, 0)$	⟨[0.37, 0.45], [0.73, 0.91]⟩	⟨[0.56, 0.72], [0.18, 0.28]⟩	⟨[0.23, 0.47], [0.85, 0.95]⟩	⟨[0.14, 0.25], [0.77, 0.87]⟩
$(t_4, e_2, 0)$	⟨[0.29, 0.41], [0.83, 0.90]⟩	⟨[0.73, 0.86], [0.36, 0.48]⟩	⟨[0.70, 0.80], [0.39, 0.49]⟩	⟨[0.28, 0.40], [0.65, 0.78]⟩

$(\xi, \mathcal{Y})$	$\mathfrak{x}_5$	$\mathfrak{x}_6$	$\mathfrak{x}_7$	$\mathfrak{x}_8$
$(t_1, e_1, 1)$	⟨[0.68, 0.77], [0.21, 0.30]⟩	⟨[0.16, 0.21], [0.67, 0.75]⟩	⟨[0.77, 0.90], [0.24, 0.30]⟩	⟨[0.27, 0.42], [0.69, 0.80]⟩
$(t_1, e_2, 1)$	⟨[0.39, 0.50], [0.80, 0.90]⟩	⟨[0.11, 0.23], [0.65, 0.78]⟩	⟨[0.82, 0.91], [0.30, 0.41]⟩	⟨[0.09, 0.29], [0.54, 0.60]⟩
$(t_2, e_1, 1)$	⟨[0.32, 0.45], [0.61, 0.75]⟩	⟨[0.52, 0.61], [0.18, 0.29]⟩	⟨[0.70, 0.85], [0.22, 0.32]⟩	⟨[0.81, 0.91], [0.49, 0.62]⟩
$(t_2, e_2, 1)$	⟨[0.80, 0.90], [0.24, 0.42]⟩	⟨[0.37, 0.46], [0.66, 0.78]⟩	⟨[0.48, 0.66], [0.30, 0.41]⟩	⟨[0.19, 0.24], [0.55, 0.66]⟩
$(t_3, e_1, 1)$	⟨[0.64, 0.73], [0.20, 0.31]⟩	⟨[0.48, 0.67], [0.15, 0.26]⟩	⟨[0.77, 0.88], [0.27, 0.37]⟩	⟨[0.24, 0.34], [0.75, 0.83]⟩
$(t_3, e_2, 1)$	⟨[0.66, 0.77], [0.24, 0.32]⟩	⟨[0.52, 0.68], [0.36, 0.45]⟩	⟨[0.20, 0.35], [0.56, 0.68]⟩	⟨[0.32, 0.44], [0.72, 0.80]⟩
$(t_4, e_1, 1)$	⟨[0.36, 0.44], [0.70, 0.90]⟩	⟨[0.54, 0.71], [0.19, 0.27]⟩	⟨[0.21, 0.45], [0.86, 0.94]⟩	⟨[0.15, 0.24], [0.76, 0.86]⟩
$(t_4, e_2, 1)$	⟨[0.28, 0.40], [0.80, 0.91]⟩	⟨[0.70, 0.85], [0.35, 0.45]⟩	⟨[0.72, 0.82], [0.40, 0.50]⟩	⟨[0.29, 0.39], [0.65, 0.80]⟩
$(t_1, e_1, 0)$	⟨[0.52, 0.59], [0.32, 0.46]⟩	⟨[0.63, 0.70], [0.25, 0.35]⟩	⟨[0.21, 0.28], [0.75, 0.80]⟩	⟨[0.62, 0.73], [0.19, 0.30]⟩
$(t_1, e_2, 0)$	⟨[0.23, 0.37], [0.75, 0.87]⟩	⟨[0.68, 0.79], [0.46, 0.58]⟩	⟨[0.44, 0.56], [0.64, 0.72]⟩	⟨[0.51, 0.63], [0.39, 0.44]⟩
$(t_2, e_1, 0)$	⟨[0.65, 0.79], [0.25, 0.35]⟩	⟨[0.35, 0.45], [0.79, 0.84]⟩	⟨[0.54, 0.64], [0.81, 0.88]⟩	⟨[0.12, 0.31], [0.28, 0.40]⟩
$(t_2, e_2, 0)$	⟨[0.37, 0.50], [0.75, 0.86]⟩	⟨[0.68, 0.75], [0.28, 0.49]⟩	⟨[0.35, 0.45], [0.70, 0.78]⟩	⟨[0.69, 0.79], [0.29, 0.39]⟩
$(t_3, e_1, 0)$	⟨[0.35, 0.52], [0.72, 0.83]⟩	⟨[0.25, 0.40], [0.68, 0.79]⟩	⟨[0.20, 0.30], [0.70, 0.80]⟩	⟨[0.58, 0.68], [0.39, 0.49]⟩
$(t_3, e_2, 0)$	⟨[0.41, 0.49], [0.71, 0.81]⟩	⟨[0.18, 0.28], [0.85, 0.94]⟩	⟨[0.61, 0.71], [0.26, 0.33]⟩	⟨[0.82, 0.93], [0.15, 0.25]⟩
$(t_4, e_1, 0)$	⟨[0.66, 0.86], [0.24, 0.32]⟩	⟨[0.28, 0.40], [0.80, 0.89]⟩	⟨[0.64, 0.75], [0.18, 0.28]⟩	⟨[0.55, 0.65], [0.39, 0.45]⟩
$(t_4, e_2, 0)$	⟨[0.75, 0.85], [0.44, 0.56]⟩	⟨[0.24, 0.34], [0.58, 0.69]⟩	⟨[0.45, 0.62], [0.20, 0.28]⟩	⟨[0.69, 0.75], [0.39, 0.45]⟩

$$
\begin{aligned}
\mathcal{L}^{-}_{3_1=(t_1,e_1,1)}(\mathfrak{x}_1) &= \sum_{s=1}^{8} \left(\tilde{\mu}^{-}_{\mathfrak{Q}_{\mathcal{E}}}(\mathfrak{x}_1) - \tilde{\mu}^{-}_{\mathfrak{Q}_{\mathcal{E}}}(\mathfrak{x}_s)\right), \\
&= \left(\tilde{\mu}^{-}_{\mathfrak{Q}_{\mathcal{E}}}(\mathfrak{x}_1) - \tilde{\mu}^{-}_{\mathfrak{Q}_{\mathcal{E}}}(\mathfrak{x}_1)\right) + \left(\tilde{\mu}^{-}_{\mathfrak{Q}_{\mathcal{E}}}(\mathfrak{x}_1) - \tilde{\mu}^{-}_{\mathfrak{Q}_{\mathcal{E}}}(\mathfrak{x}_2)\right) + \ldots + \left(\tilde{\mu}^{-}_{\mathfrak{Q}_{\mathcal{E}}}(\mathfrak{x}_1) - \tilde{\mu}^{-}_{\mathfrak{Q}_{\mathcal{E}}}(\mathfrak{x}_8)\right), \\
&= (0.50 - 0.50) + (0.50 - 0.65) + (0.50 - 0.20) + \ldots + (0.50 - 0.27), \\
&= 0.17.
\end{aligned}
$$

Using Definition 16, the accumulated score values of both agree-IV^{4}ROFSE set $(\xi, \mathcal{Y})_1$ and disagree-IV^{4}ROFSE set $(\xi, \mathcal{Y})_0$ are, respectively, given by Tables 21 and 22.

The final scores of alternatives $\mathfrak{x}_i$ $(i = 1, 2, \ldots, 8)$ are displayed in Table 23.

Table 17 Scores of each membership interval's lower and upper bounds in agree-IV^{4}ROFSE set $(\xi, \mathcal{Y})_1$

	$\mathfrak{x}_1$		$\mathfrak{x}_2$		$\mathfrak{x}_3$		$\mathfrak{x}_4$	
	$\mathcal{L}^-_{3_j}(\mathfrak{x}_1)$	$\mathcal{L}^+_{3_j}(\mathfrak{x}_1)$	$\mathcal{L}^-_{3_j}(\mathfrak{x}_2)$	$\mathcal{L}^+_{3_j}(\mathfrak{x}_2)$	$\mathcal{L}^-_{3_j}(\mathfrak{x}_3)$	$\mathcal{L}^+_{3_j}(\mathfrak{x}_3)$	$\mathcal{L}^-_{3_j}(\mathfrak{x}_4)$	$\mathcal{L}^+_{3_j}(\mathfrak{x}_4)$
$(\mathfrak{t}_1, \mathfrak{e}_1, 1)$	0.17	−0.12	1.37	1.24	−2.23	−2.52	0.97	1.08
$(\mathfrak{t}_1, \mathfrak{e}_2, 1)$	−1.46	−1.40	2.06	1.96	0.22	0.20	0.78	0.60
$(\mathfrak{t}_2, \mathfrak{e}_1, 1)$	1.15	1.27	−1.33	−1.45	0.43	0.07	−3.17	−2.57
$(\mathfrak{t}_2, \mathfrak{e}_2, 1)$	−1.04	−0.75	1.52	1.41	−1.44	−1.47	1.60	1.41
$(\mathfrak{t}_3, \mathfrak{e}_1, 1)$	−0.73	−0.45	−1.53	−1.41	−1.93	−2.13	0.95	0.83
$(\mathfrak{t}_3, \mathfrak{e}_2, 1)$	−0.46	−0.63	−2.38	−2.55	1.14	1.21	2.74	2.57
$(\mathfrak{t}_4, \mathfrak{e}_1, 1)$	1.99	2.22	−1.21	−1.22	1.67	1.66	0.95	0.62
$(\mathfrak{t}_4, \mathfrak{e}_2, 1)$	2.14	1.67	−2.10	−2.09	−0.90	−0.17	1.34	0.79

	$\mathfrak{x}_5$		$\mathfrak{x}_6$		$\mathfrak{x}_7$		$\mathfrak{x}_8$	
	$\mathcal{L}^-_{3_j}(\mathfrak{x}_5)$	$\mathcal{L}^+_{3_j}(\mathfrak{x}_5)$	$\mathcal{L}^-_{3_j}(\mathfrak{x}_6)$	$\mathcal{L}^+_{3_j}(\mathfrak{x}_6)$	$\mathcal{L}^-_{3_j}(\mathfrak{x}_7)$	$\mathcal{L}^+_{3_j}(\mathfrak{x}_7)$	$\mathcal{L}^-_{3_j}(\mathfrak{x}_8)$	$\mathcal{L}^+_{3_j}(\mathfrak{x}_8)$
$(\mathfrak{t}_1, \mathfrak{e}_1, 1)$	1.61	1.64	−2.50	−2.84	2.33	2.68	−1.67	−1.16
$(\mathfrak{t}_1, \mathfrak{e}_2, 1)$	−0.10	−0.20	−2.34	−2.36	3.34	3.08	−2.50	−1.88
$(\mathfrak{t}_2, \mathfrak{e}_1, 1)$	−1.41	−1.37	0.19	−0.09	1.63	1.83	2.51	2.31
$(\mathfrak{t}_2, \mathfrak{e}_2, 1)$	−2.56	2.53	−0.88	−0.99	−0.00	0.61	−2.32	−2.75
$(\mathfrak{t}_3, \mathfrak{e}_1, 1)$	1.67	1.39	0.39	0.91	2.71	2.59	−1.53	−1.73
$(\mathfrak{t}_3, \mathfrak{e}_2, 1)$	1.62	1.53	0.50	0.81	−2.06	−1.83	−1.10	−1.11
$(\mathfrak{t}_4, \mathfrak{e}_1, 1)$	1.62	−0.98	0.95	1.18	−1.69	−0.90	−2.17	−2.58
$(\mathfrak{t}_4, \mathfrak{e}_2, 1)$	−0.49	−1.77	1.50	1.83	1.66	1.59	−1.78	−1.85

It is clear from Table 23 that the best energy development source is $\mathfrak{x}_4$ because $\mathfrak{S}_4 = 78.15$ is the maximum score value as compared to other energy sources' scores. Hence, $\mathfrak{x}_4$ is the most appropriate energy development source from the given options. From Table 23, one can easily observe that the next two most suitable options are $\mathfrak{x}_7$ and $\mathfrak{x}_6$, respectively.

4 Comparative Analysis

Soft computing models, SESs and IV^{q}ROF soft sets, are playing a vital role in dealing with different uncertainties in daily-life datasets. Our initiated model is a more generalized version of these above-mentioned existing hybrid models. The validity and reliability of our initiated hybrid approach to solve different MCGDM problems based on IV^{q}ROFSE sets is verified by comparing it with an existing model called IV^{q}ROFSSs, which is presented by Ali et al. [4]. The presented application in Sect. 3 is solved via the existing model, IV^{q}ROFSSs, and the calculated results are displayed in Table 24. From Table 24, it is clear that the best object is $\mathfrak{x}_4$, which is different when we apply the model, IV^{q}ROFSSs on the Application 3.1. The

Table 18 Scores of each non-membership interval's lower and upper bounds in agree-IV^{4}ROFSE set $(\xi, \mathcal{Y})_1$

	$\mathfrak{x}_1$		$\mathfrak{x}_2$		$\mathfrak{x}_3$		$\mathfrak{x}_4$	
	$\mathcal{U}^-_{\mathfrak{z}_j}(\mathfrak{x}_1)$	$\mathcal{U}^+_{\mathfrak{z}_j}(\mathfrak{x}_1)$	$\mathcal{U}^-_{\mathfrak{z}_j}(\mathfrak{x}_2)$	$\mathcal{U}^+_{\mathfrak{z}_j}(\mathfrak{x}_2)$	$\mathcal{U}^-_{\mathfrak{z}_j}(\mathfrak{x}_3)$	$\mathcal{U}^+_{\mathfrak{z}_j}(\mathfrak{x}_3)$	$\mathcal{U}^-_{\mathfrak{z}_j}(\mathfrak{x}_4)$	$\mathcal{U}^+_{\mathfrak{z}_j}(\mathfrak{x}_4)$
$(\mathfrak{t}_1, \mathfrak{e}_1, 1)$	−0.68	−0.49	−1.24	−1.29	2.68	2.47	−1.96	−1.53
$(\mathfrak{t}_1, \mathfrak{e}_2, 1)$	1.37	1.61	−0.71	−0.55	0.41	0.41	−1.51	−1.91
$(\mathfrak{t}_2, \mathfrak{e}_1, 1)$	−1.72	−1.77	2.52	2.07	2.84	2.55	−1.40	−1.05
$(\mathfrak{t}_2, \mathfrak{e}_2, 1)$	2.18	1.98	−1.66	−0.90	2.02	1.58	−1.58	−1.86
$(\mathfrak{t}_3, \mathfrak{e}_1, 1)$	1.87	1.81	1.55	1.65	1.71	1.73	−0.85	−0.99
$(\mathfrak{t}_3, \mathfrak{e}_2, 1)$	1.78	1.88	3.06	2.76	−1.58	−1.80	−3.02	−2.76
$(\mathfrak{t}_4, \mathfrak{e}_1, 1)$	−2.20	−2.37	2.20	2.03	−2.84	−2.69	−1.08	−1.33
$(\mathfrak{t}_4, \mathfrak{e}_2, 1)$	−0.29	−0.20	0.75	0.76	−2.37	−2.28	−0.77	−1.16
	$\mathfrak{x}_5$		$\mathfrak{x}_6$		$\mathfrak{x}_7$		$\mathfrak{x}_8$	
	$\mathcal{U}^-_{\mathfrak{z}_j}(\mathfrak{x}_5)$	$\mathcal{U}^+_{\mathfrak{z}_j}(\mathfrak{x}_5)$	$\mathcal{U}^-_{\mathfrak{z}_j}(\mathfrak{x}_6)$	$\mathcal{U}^+_{\mathfrak{z}_j}(\mathfrak{x}_6)$	$\mathcal{U}^-_{\mathfrak{z}_j}(\mathfrak{x}_7)$	$\mathcal{U}^+_{\mathfrak{z}_j}(\mathfrak{x}_7)$	$\mathcal{U}^-_{\mathfrak{z}_j}(\mathfrak{x}_8)$	$\mathcal{U}^+_{\mathfrak{z}_j}(\mathfrak{x}_8)$
$(\mathfrak{t}_1, \mathfrak{e}_1, 1)$	−1.64	−1.69	2.04	1.91	−1.40	−1.69	2.20	2.31
$(\mathfrak{t}_1, \mathfrak{e}_2, 1)$	1.93	1.93	0.73	0.97	−2.07	−1.99	−0.15	−0.47
$(\mathfrak{t}_2, \mathfrak{e}_1, 1)$	1.32	1.59	−2.12	−2.09	−1.80	−1.85	0.36	0.55
$(\mathfrak{t}_2, \mathfrak{e}_2, 1)$	−1.82	−1.38	1.54	1.50	−1.34	−1.46	0.66	0.54
$(\mathfrak{t}_3, \mathfrak{e}_1, 1)$	−2.21	−2.11	−2.61	−2.51	−1.65	−1.63	2.19	2.05
$(\mathfrak{t}_3, \mathfrak{e}_2, 1)$	−1.90	−1.96	−0.94	−0.92	0.66	0.92	1.94	1.88
$(\mathfrak{t}_4, \mathfrak{e}_1, 1)$	1.56	2.35	−2.52	−2.69	2.84	2.67	2.04	2.03
$(\mathfrak{t}_4, \mathfrak{e}_2, 1)$	2.67	2.68	−0.93	−1.00	−0.53	−0.60	1.47	1.80

main cause behind these variations is that IV^{q}ROFSSs [4] consider the opinions of only one expert regarding multiple parameters, and in our developed IV^{q}ROFSE set model, two experts are involved in the decision-making process. The other reason is that IV^{q}ROFSSs are a particular case of our initiated model, IV^{q}ROFSE sets, if we consider only one expert. However, IV^{q}ROFSE set model is more feasible because it has the ability to tackle decision-making problems in which all possible properties of given parameters are evaluated by multiple experts in an interval-valued environment than existing IV^{q}ROFSSs [4] (Table 25).

5 Conclusions and Future Directions

Several models have been proposed in the literature to handle the vague and uncertain knowledge individually and collectively in different formats, including memberships, non-memberships, interval-valued representations, and parameterizations. In this paper, we have designed a new hybrid model called interval-valued q-ROFSE sets or (IV^{q}ROFSE sets), which is purely a mixture of all the above-mentioned modalities. We have investigated basic properties of IV^{q}ROFSE sets, including sub-

Table 19 Scores of each membership interval's lower and upper bounds in disagree-IV^{4}ROFSE set $(\xi, \mathcal{Y})_0$

	$\mathfrak{x}_1$		$\mathfrak{x}_2$		$\mathfrak{x}_3$		$\mathfrak{x}_4$	
	$\mathcal{L}^-_{\mathfrak{z}_j}(\mathfrak{x}_1)$	$\mathcal{L}^+_{\mathfrak{z}_j}(\mathfrak{x}_1)$	$\mathcal{L}^-_{\mathfrak{z}_j}(\mathfrak{x}_2)$	$\mathcal{L}^+_{\mathfrak{z}_j}(\mathfrak{x}_2)$	$\mathcal{L}^-_{\mathfrak{z}_j}(\mathfrak{x}_3)$	$\mathcal{L}^+_{\mathfrak{z}_j}(\mathfrak{x}_3)$	$\mathcal{L}^-_{\mathfrak{z}_j}(\mathfrak{x}_4)$	$\mathcal{L}^+_{\mathfrak{z}_j}(\mathfrak{x}_4)$
$(\mathfrak{t}_1, \mathfrak{e}_1, 0)$	1.65	1.66	−2.59	−2.90	2.45	2.38	−1.55	−0.58
$(\mathfrak{t}_1, \mathfrak{e}_2, 0)$	−0.04	−0.18	−2.36	−2.34	3.32	3.02	−2.52	−1.94
$(\mathfrak{t}_2, \mathfrak{e}_1, 0)$	−1.37	−1.27	−0.01	−0.23	1.75	1.69	2.39	2.41
$(\mathfrak{t}_2, \mathfrak{e}_2, 0)$	2.52	2.36	−0.92	−1.00	0.04	0.44	−2.52	−2.68
$(\mathfrak{t}_3, \mathfrak{e}_1, 0)$	1.77	1.40	0.41	0.92	2.73	2.60	−1.59	−1.72
$(\mathfrak{t}_3, \mathfrak{e}_2, 0)$	1.76	1.50	0.56	0.86	−2.08	−1.94	−1.36	−1.06
$(\mathfrak{t}_4, \mathfrak{e}_1, 0)$	−0.47	−0.95	1.05	1.21	−1.59	−0.79	−2.31	−2.55
$(\mathfrak{t}_4, \mathfrak{e}_2, 0)$	−1.81	−1.75	1.71	1.85	1.47	1.37	−1.89	−1.83
	$\mathfrak{x}_5$		$\mathfrak{x}_6$		$\mathfrak{x}_7$		$\mathfrak{x}_8$	
	$\mathcal{L}^-_{\mathfrak{z}_j}(\mathfrak{x}_5)$	$\mathcal{L}^+_{\mathfrak{z}_j}(\mathfrak{x}_5)$	$\mathcal{L}^-_{\mathfrak{z}_j}(\mathfrak{x}_6)$	$\mathcal{L}^+_{\mathfrak{z}_j}(\mathfrak{x}_6)$	$\mathcal{L}^-_{\mathfrak{z}_j}(\mathfrak{x}_7)$	$\mathcal{L}^+_{\mathfrak{z}_j}(\mathfrak{x}_7)$	$\mathcal{L}^-_{\mathfrak{z}_j}(\mathfrak{x}_8)$	$\mathcal{L}^+_{\mathfrak{z}_j}(\mathfrak{x}_8)$
$(\mathfrak{t}_1, \mathfrak{e}_1, 0)$	0.21	−0.02	1.09	0.86	−2.27	−2.5	1.01	1.10
$(\mathfrak{t}_1, \mathfrak{e}_2, 0)$	−1.48	−1.38	2.12	1.98	0.20	0.14	0.76	0.70
$(\mathfrak{t}_2, \mathfrak{e}_1, 0)$	1.19	1.29	−1.21	−1.43	0.31	0.09	−3.05	−2.55
$(\mathfrak{t}_2, \mathfrak{e}_2, 0)$	−1.00	−0.76	1.48	1.24	−1.16	−1.16	1.56	1.56
$(\mathfrak{t}_3, \mathfrak{e}_1, 0)$	−0.79	−0.44	−1.59	−1.40	−1.99	−2.20	1.05	0.84
$(\mathfrak{t}_3, \mathfrak{e}_2, 0)$	−0.48	−0.74	−2.32	−2.42	1.12	1.02	2.80	2.78
$(\mathfrak{t}_4, \mathfrak{e}_1, 0)$	1.85	2.33	−1.19	−1.35	1.69	1.45	0.97	0.65
$(\mathfrak{t}_4, \mathfrak{e}_2, 0)$	1.87	1.77	−2.21	−2.31	−0.53	−0.07	1.39	0.97

set relation, complement, intersection, union, AND and OR operations plus the agree- and disagree-IV^{q}ROFSE sets and illustrated them via numerical examples. Moreover, we have presented a new group decision-making methodology based on the proposed IV^{q}ROFSE model and provided a daily-life application, that is, analysis of modern energy development methods. In the end, we have given a comparison of the initiated model with an existing model to support its validity and feasibility. In future research works, we are planning (a) to develop an interval-valued q-rung orthopair fuzzy N-soft expert set model, (b) to present an interval-valued q-rung orthopair fuzzy bipolar soft expert set model, and (c) to propose an interval-valued q-rung orthopair fuzzy bipolar N-soft expert set model.

Table 20 Scores of each non-membership interval's lower and upper bounds in disagree-IV^{4}ROFSE set $(\xi, \mathcal{Y})_0$

	$\mathfrak{x}_1$		$\mathfrak{x}_2$		$\mathfrak{x}_3$		$\mathfrak{x}_4$	
	$\mathcal{U}^-_{\mathfrak{z}_j}(\mathfrak{x}_1)$	$\mathcal{U}^+_{\mathfrak{z}_j}(\mathfrak{x}_1)$	$\mathcal{U}^-_{\mathfrak{z}_j}(\mathfrak{x}_2)$	$\mathcal{U}^+_{\mathfrak{z}_j}(\mathfrak{x}_2)$	$\mathcal{U}^-_{\mathfrak{z}_j}(\mathfrak{x}_3)$	$\mathcal{U}^+_{\mathfrak{z}_j}(\mathfrak{x}_3)$	$\mathcal{U}^-_{\mathfrak{z}_j}(\mathfrak{x}_4)$	$\mathcal{U}^+_{\mathfrak{z}_j}(\mathfrak{x}_4)$
$(\mathfrak{t}_1, \mathfrak{e}_1, 0)$	−1.38	−1.51	2.06	2.09	−2.02	−1.75	2.14	2.17
$(\mathfrak{t}_1, \mathfrak{e}_2, 0)$	1.97	2.13	0.77	0.93	−2.11	−1.95	−0.19	−0.43
$(\mathfrak{t}_2, \mathfrak{e}_1, 0)$	1.38	1.48	−2.14	−2.20	−1.98	−1.64	0.34	0.36
$(\mathfrak{t}_2, \mathfrak{e}_2, 0)$	−1.84	−1.60	1.60	1.52	−1.28	−1.36	0.72	0.48
$(\mathfrak{t}_3, \mathfrak{e}_1, 0)$	−2.24	−2.12	−2.56	−2.52	−1.68	−1.64	2.24	2.04
$(\mathfrak{t}_3, \mathfrak{e}_2, 0)$	−2.08	−1.98	−0.88	−0.86	0.80	−0.74	1.76	1.94
$(\mathfrak{t}_4, \mathfrak{e}_1, 0)$	1.70	−2.33	−2.70	−2.71	2.66	2.65	2.02	2.01
$(\mathfrak{t}_4, \mathfrak{e}_2, 0)$	2.83	2.57	−0.93	−0.79	−0.69	−0.71	1.39	1.61
	$\mathfrak{x}_5$		$\mathfrak{x}_6$		$\mathfrak{x}_7$		$\mathfrak{x}_8$	
	$\mathcal{U}^-_{\mathfrak{z}_j}(\mathfrak{x}_5)$	$\mathcal{U}^+_{\mathfrak{z}_j}(\mathfrak{x}_5)$	$\mathcal{U}^-_{\mathfrak{z}_j}(\mathfrak{x}_6)$	$\mathcal{U}^+_{\mathfrak{z}_j}(\mathfrak{x}_6)$	$\mathcal{U}^-_{\mathfrak{z}_j}(\mathfrak{x}_7)$	$\mathcal{U}^+_{\mathfrak{z}_j}(\mathfrak{x}_7)$	$\mathcal{U}^-_{\mathfrak{z}_j}(\mathfrak{x}_8)$	$\mathcal{U}^+_{\mathfrak{z}_j}(\mathfrak{x}_8)$
$(\mathfrak{t}_1, \mathfrak{e}_1, 0)$	−0.66	−0.39	−1.22	−1.27	2.78	2.33	−1.70	−1.67
$(\mathfrak{t}_1, \mathfrak{e}_2, 0)$	1.41	1.57	−0.91	−0.75	0.53	0.37	−1.47	−1.87
$(\mathfrak{t}_2, \mathfrak{e}_1, 0)$	−1.66	−1.64	2.66	2.28	2.82	2.60	−1.42	−1.24
$(\mathfrak{t}_2, \mathfrak{e}_2, 0)$	2.16	2.08	−1.60	−0.88	1.76	1.44	−1.52	−1.68
$(\mathfrak{t}_3, \mathfrak{e}_1, 0)$	1.84	1.88	1.52	1.56	1.68	1.64	−0.80	−0.84
$(\mathfrak{t}_3, \mathfrak{e}_2, 0)$	1.84	1.86	−2.96	2.90	−1.76	−1.98	−2.64	−2.62
$(\mathfrak{t}_4, \mathfrak{e}_1, 0)$	−2.22	−2.39	2.26	2.17	−2.70	−2.71	−1.02	−1.35
$(\mathfrak{t}_4, \mathfrak{e}_2, 0)$	−0.53	−0.15	0.83	0.89	−2.21	−2.39	−0.69	−1.03

Table 21 Accumulated score values of alternatives in agree-IV^{4}ROFSE set $(\xi, \mathcal{Y})_1$

	$\mathfrak{x}_1$	$\mathfrak{x}_2$	$\mathfrak{x}_3$	$\mathfrak{x}_4$	$\mathfrak{x}_5$	$\mathfrak{x}_6$	$\mathfrak{x}_7$	$\mathfrak{x}_8$
$(\mathfrak{t}_1, \mathfrak{e}_1, 1)$	1.22	5.14	−9.90	5.54	6.58	−9.34	8.10	−7.34
$(\mathfrak{t}_1, \mathfrak{e}_2, 1)$	−5.84	5.28	−0.40	4.80	−4.16	−6.40	10.48	−3.76
$(\mathfrak{t}_2, \mathfrak{e}_1, 1)$	5.91	−7.37	−4.89	−3.29	−5.69	4.31	7.11	3.91
$(\mathfrak{t}_2, \mathfrak{e}_2, 1)$	−5.95	5.49	−6.51	6.45	8.29	−4.91	3.41	−6.27
$(\mathfrak{t}_3, \mathfrak{e}_1, 1)$	−4.86	−6.14	−7.50	3.62	7.38	6.42	8.58	−7.50
$(\mathfrak{t}_3, \mathfrak{e}_2, 1)$	−4.75	−10.75	5.73	11.09	7.01	3.17	−5.47	−6.03
$(\mathfrak{t}_4, \mathfrak{e}_1, 1)$	8.78	−6.66	8.86	3.98	−5.38	7.34	−8.10	−8.82
$(\mathfrak{t}_4, \mathfrak{e}_2, 1)$	4.30	−5.70	3.58	4.06	−8.98	5.26	4.38	−6.90
Agree-score-sum($\mathfrak{A}_j$)	−1.19	−20.71	−11.03	36.25	5.05	5.85	28.49	−42.71

Table 22 Accumulated score values of alternatives in disagree-IV^{4}ROFSE set $(\xi, \mathcal{Y})_0$

	$\mathfrak{x}_1$	$\mathfrak{x}_2$	$\mathfrak{x}_3$	$\mathfrak{x}_4$	$\mathfrak{x}_5$	$\mathfrak{x}_6$	$\mathfrak{x}_7$	$\mathfrak{x}_8$
$(\mathfrak{t}_1, \mathfrak{e}_1, 0)$	6.20	−9.64	8.60	−6.44	1.24	4.44	−9.88	5.48
$(\mathfrak{t}_1, \mathfrak{e}_2, 0)$	−6.32	−6.40	10.40	−3.84	−5.84	5.76	−0.56	4.80
$(\mathfrak{t}_2, \mathfrak{e}_1, 0)$	−5.50	4.10	7.06	4.10	5.78	−7.58	−5.02	−2.94
$(\mathfrak{t}_2, \mathfrak{e}_2, 0)$	8.32	−5.04	3.12	−6.40	−6.00	5.20	−5.52	6.32
$(\mathfrak{t}_3, \mathfrak{e}_1, 0)$	7.53	6.41	8.65	−7.59	−4.95	−6.07	−7.51	3.53
$(\mathfrak{t}_3, \mathfrak{e}_2, 0)$	7.32	3.16	−5.56	−6.12	−4.92	−10.60	5.88	10.84
$(\mathfrak{t}_4, \mathfrak{e}_1, 0)$	−5.45	7.67	−7.69	−8.89	8.79	−6.97	8.55	3.99
$(\mathfrak{t}_4, \mathfrak{e}_2, 0)$	−8.96	5.28	4.24	−6.72	4.32	−6.24	4.00	4.08
Disagree-score-sum($\mathfrak{D}_j$)	3.14	5.54	28.82	−41.9	−1.58	−22.06	−10.06	36.1

Table 23 Final score values of objects

	Agree-score-sum($\mathfrak{A}_j$)	Disagree-score-sum($\mathfrak{D}_j$)	Final scores($\mathfrak{S}_j$)=$\mathfrak{A}_j - \mathfrak{D}_j$
$\mathfrak{x}_1$	−1.19	3.14	−4.33
$\mathfrak{x}_2$	−20.71	5.54	−26.25
$\mathfrak{x}_3$	−11.03	28.82	−39.85
$\mathfrak{x}_4$	36.25	−41.9	78.15
$\mathfrak{x}_5$	5.05	−1.58	6.40
$\mathfrak{x}_6$	5.85	−22.06	27.91
$\mathfrak{x}_7$	28.49	−10.06	38.55
$\mathfrak{x}_8$	−42.71	36.1	−78.81

Table 24 Comparison between proposed model with existing IV^{q}ROFSS model on Application 3.1

Objects	IV^{q}ROFSSs[4]	Proposed IV^{q}ROFSE sets
$\mathfrak{x}_1$	8.07	−4.33
$\mathfrak{x}_2$	−6.49	−26.25
$\mathfrak{x}_3$	3.19	−39.85
$\mathfrak{x}_4$	28.37	78.15
$\mathfrak{x}_5$	−7.97	6.40
$\mathfrak{x}_6$	24.91	27.91
$\mathfrak{x}_7$	29.55	38.55
$\mathfrak{x}_8$	−9.69	−78.81

Table 25 Comparison between rankings of developed IV^{q}ROFSE set model and existing IV^{q}ROFSSs

Models	Rankings
IV^{q}ROFSSs [4]	$\mathfrak{r}_7 > \mathfrak{r}_4 > \mathfrak{r}_6 > \mathfrak{r}_1 > \mathfrak{r}_3 > \mathfrak{r}_2 > \mathfrak{r}_5 > \mathfrak{r}_8$
Proposed IV^{q}ROFSE sets	$\mathfrak{r}_4 > \mathfrak{r}_7 > \mathfrak{r}_6 > \mathfrak{r}_5 > \mathfrak{r}_1 > \mathfrak{r}_2 > \mathfrak{r}_3 > \mathfrak{r}_8$

Conflict of interest: The authors declare that they have no conflict of interest regarding the publication of this article.
Ethical approval: This article does not contain any studies with human participants or animals performed by any of the authors.

References

1. Akram, M., Ali, G., Butt, M.A., Alcantud, J.C.R.: Novel MCGDM analysis under m-polar fuzzy soft expert sets. Neural Comput. Appl. **33**, 12051–12071 (2021)
2. Akram, M., Ali, G., Alcantud, J.C.R.: Group decision-making with Fermatean fuzzy soft expert knowledge. Art. Intell. Rev. (2022). https://doi.org/10.1007/s10462-021-10119-8
3. Akram, M.: Shumaiza: Multi-criteria decision making based on q-rung orthopair fuzzy promethee approach. Iran. J. Fuzzy Syst. **18**(5), 107–127 (2021)
4. Ali, G., Afzal, M., Asif, M., Shazad, A.: Attribute reduction approaches under interval-valued q-rung orthopair fuzzy soft framework. Appl. Intell. (2021). https://doi.org/10.1007/s10489-021-02853-x
5. Ali, G., Akram, M.: Decision-making method based on fuzzy N-soft expert sets. Arab. J. Sci. Eng. **45**, 10381–10400 (2020)
6. Ali, G., Akram, M., Shahzadi, S., Abidin, M.Z.U.: Group decision-making framework with bipolar soft expert sets. J. Mult.-Valued Log. Soft Comput. **37**(3–4), 211–246 (2021)
7. Ali, M.I., Feng, F., Liu, X.Y., Min, W.K., Shabir, M.: On some new operations in soft set theory. Comput. Math. Appl. **57**(9), 1547–1553 (2009)
8. Alkhazaleh, S., Salleh, A.R.: Soft expert sets. Adv. Decis. Sci. **2011**, 12 pages, Article ID: 757868 (2011). https://doi.org/10.1155/2011/757868
9. Alkhazaleh, S., Salleh, A.R.: Fuzzy soft expert set and its application. Appl. Math. **5**(09), 1349–1368 (2014)
10. Al-Qudah, Y., Hassan, N.: Bipolar fuzzy soft expert set and its application in decision making. Int. J. Appl. Decis. Sci. **10**(2), 175–191 (2017)
11. Arockiarani, I., ArokiaLancy, A.A.: Multi criteria decision making problem with soft expert set. Int. J. Comput. Appl. **78**(15), 1–4 (2013)
12. Atanassov, K.T.: Intuitionistic fuzzy sets. Fuzzy Sets Syst. **20**, 87–96 (1986)
13. Atanassov, K.T.: Intuitionistic Fuzzy Sets-Theory and Applications. Springer, New York, NY, USA (1999)
14. Bashir, M., Salleh, A.R.: Fuzzy parameterized soft expert set. Abstr. Appl. Anal. **2012**, Article ID: 258361 (2012). https://doi.org/10.1155/2012/258361
15. Broumi, S., Smarandache, F.: Intuitionistic fuzzy soft expert sets and its application in decision making. J. New Theor. **1**, 89–105 (2015)
16. Deschrijver, G., Kerre, E.F.: On the relationship between some extensions of fuzzy set theory. Fuzzy Sets Syst. **133**(2), 227–235 (2003)

17. Fatimah, F., Alcantud, J.C.R.: The multi-fuzzy N-soft set and its applications to decision-making. Neural Comput. Appl. **33**, 11437–11446 (2021)
18. Feng, F., Li, Y., Fotea, V.L.: Application of level soft sets in decision-making based on interval-valued fuzzy soft sets. Comput. Math. Appl. **60**, 1756–1767 (2010)
19. Garg, H.: New exponential operation laws and operators for interval-valued q-rung orthopair fuzzy sets in group decision making process. Neural Comput. Appl. **33**, 13937–13963 (2021)
20. Gorzalczany, M.B.: A method of inference in approximate reasoning based on interval-valued fuzzy sets. Fuzzy Sets Syst. **21**(1), 1–17 (1987)
21. Hamid, M.T., Riaz, M., Afzal, D.: Novel MCGDM with q-rung orthopair fuzzy soft sets and TOPSIS approach under q-Rung orthopair fuzzy soft topology. J. Intell. Fuzzy Syst. **39**(3), 3853–3871 (2020)
22. Hassan, N., Alhazaymeh, K.: Vague soft expert set theory. In AIP Conf. Proc. **1522**(1), 953–958 (2013)
23. Joshi, B.P., Singh, A., Bhatt, P.K., Vaisla, K.S.: Interval valued q-rung orthopair fuzzy sets and their properties. J. Intell. Fuzzy Syst. **35**(5), 5225–5230 (2018)
24. Jiang, Y., Tang, Y., Chen, Q., Liu, H., Tang, J.: Interval-valued intuitionistic fuzzy soft sets and their properties. Comput. Math. Appl. **60**(3), 906–918 (2010)
25. Jiang, Y., Tang, Y., Chen, Q.: An adjustable approach to intuitionistic fuzzy soft sets based decision-making. Appl. Math. Model. **35**, 824–836 (2011)
26. Maji, P.K., Biswas, R., Roy, A.R.: Intuitionistic fuzzy soft sets. J. Fuzzy Math. **9**(3), 677–692 (2001)
27. Maji, P.K., Roy, A.R., Biswas, R.: An application of soft sets in a decision-making problem. Comput. Math. Appl. **44**(8), 1077–1083 (2002)
28. Maji, P.K., Biswas, R., Roy, A.R.: Fuzzy soft sets. J. Fuzzy Math. **9**(3), 589–602 (2001)
29. Molodtsov, D.: Soft set theory: first results. Comput. Math. Appl. **37**(4–5), 19–31 (1999)
30. Pawlak, Z.: Rough sets. Int. J. Comput. Inf. Sci. **11**(5), 145–172 (1982)
31. Qayyum, A., Abdullah, S., Aslam, M.: Cubic soft expert sets and their application in decision making. J. Intell. Fuzzy Syst. **31**(3), 1585–1596 (2016)
32. Senapati, T., Yager, R.R.: Fermatean fuzzy sets. J. Ambient Intell. Humaniz. Comput. **11**(2), 663–674 (2020)
33. Shaheen, T., Ali, M.I., Toor, H.: Why do we need q-rung orthopair fuzzy sets? Some evidence established via mass assignment. Int. J. Intell. Syst. **36**(10), 5493–5505 (2021)
34. Yager, R.R.: Generalized orthopair fuzzy sets. IEEE Trans. Fuzzy Syst. **25**(5), 1222–1230 (2016)
35. Yager, R.R., Abbasov, A.M.: Pythagorean membership grades, complex numbers, and decision making. Int J. Intell. Syst. **28**(5), 436–452 (2013)
36. Yager, R.R.: Pythagorean fuzzy subsets. In: 2013 Joint IFSA World Congress and NAFIPS Annual Meeting (IFSA/NAFIPS), IEEE, pp. 57–61 (2013)
37. Yang, X.B., Lin, T.Y., Yang, J.Y., Li, Y., Yu, D.: Combination of interval-valued fuzzy set and soft set. Comput. Math. Appl. **58**(3), 521–527 (2009)
38. Zadeh, L.A.: Fuzzy sets. Inf. Cont. **8**, 338–353 (1965)

Information Aggregation Approaches for Multi-criteria Applications

Frederick Petry and Ronald R. Yager

Abstract The effective use of multi-criteria techniques in specific applications will depend on relevant available data. Often such data arises from several various sources, and it is then typically desirable to aggregate some of this data before use as this could reduce computational complexity. Additionally such data frequently involves uncertainty, and this must be considered in the aggregation. Aggregation approaches ranging from possibilistic conditioning to probability and possibility transformations can be used to produce homogenous representations (probability or possibility) for aggregations. Here we review several approaches to uncertainty aggregation that apply to multi-criteria systems. Additionally, we consider approaches to evaluate the uncertainty of the aggregation to provide guidance of the use of the aggregated data for multi-criteria applications.

Keywords Multi-criteria · Aggregation · Possibility · Probability · Information measures

1 Introduction

A multi-criteria decision function allows one to aggregate an alternative's satisfaction to the individual criteria to obtain the alternative's overall satisfaction to a decision problem. The use of multi-criteria decision functions must be considered in the situation where the individual criteria satisfactions rely on uncertain data [1, 2]. The ordering of the criteria satisfactions causes difficulties in such situations in which there exists a probabilistic/possibilistic type of uncertainty in the knowledge of the criteria satisfactions [3]. For example, multi-criteria issues are of major concern for

F. Petry (✉)
Naval Research Lab, Center for Geospatial Sciences, Stennis Space Center MS, New Rochelle, NY, USA
e-mail: fred.petry@nrlssc.navy.mil

R. R. Yager
Machine Intelligence Institute, Iona College, New Rochelle, NY, USA
e-mail: yager@panix.com

L. Sahoo et al. (eds.), *Real Life Applications of Multiple Criteria Decision Making Techniques in Fuzzy Domain*, Studies in Fuzziness and Soft Computing 420,
https://doi.org/10.1007/978-981-19-4929-6_9

spatial applications such as land use planning and resource allocation [4]. Typically, the spatial data required for such applications can entail several forms of uncertainty which must be considered [5].

The effective use of multi-criteria techniques for specific applications will typically depend on relevant available data. Often such data arises from several various sources and it is then often desirable of aggregate some of these before their use as this could reduce computational complexity. Additionally, such data frequently involves uncertainty and this must be considered in the aggregation. Aggregation approaches ranging from possibilistic conditioning to probability, and possibility transformations used to produce homogenous representations (probability or possibility) can be utilized for aggregations. In this paper, we review several approaches to uncertainty aggregation that apply to multi-criteria systems. Additionally, we consider approaches to evaluate the uncertainty of the aggregation to provide guidance of the use of the aggregated data for multi-criteria applications.

Uncertainty from many sources and of many forms appears in the everyday activities and decisions of humans. We examine approaches that can be used to combine these uncertainties into forms that can become useful for multi-criteria decision-making. Effective decision-making should be able to make use of all the available, and relevant information about such combined uncertainty. Here we consider an overview for approaches for combining separately possibilistic uncertainty, probabilistic uncertainty and situations where both forms of uncertainty appear [6].

2 Aggregation Approaches

2.1 Background

Consider a space X that has both aleatory and epistemic sources of uncertainty [7] Let there be a probability distribution $P : X \to [0, 1]$ such that $p_i \in [0, 1], \sum_{i=1}^{n} p_i = 1$ that models the aleatory uncertainty. Then the epistemic uncertainty can be modeled by a possibility distribution [8] such that $\Pi : X \to [0, 1]$ $[0, 1]$, where $\pi(x_i)$ gives the possibility that x_i is the value, where $i = 1, 2, \ldots, n$. A usual requirement here is the normality condition, $\text{Max}_x[\pi(x)] = 1$, that is at least one element in X must be a fully possible. Abbreviating our notation so that $p_i = p(x_i)$, etc., and $\pi_i = \pi(x_i)$, etc., we have $P = \{p_1, p_2, \ldots, p_n\}$ and $\Pi = \{\pi_1, \pi_2, \ldots, \pi_n\}$.

We consider bounding cases of probability and possibility in evaluation of the aggregations based on various measures such as found in information theory and possibility specificity [6, 9]. For probability, the upper (most specific or certain case) denoted PU is

$$PU = (1, 0, 0, \ldots, 0)$$

Then the lower bounding case is the least specific or certain case, PL

$$PL = (\frac{1}{n}, \frac{1}{n}, \ldots, \frac{1}{n})$$

For possibility distributions, we consider as well the analogous cases ΠU and ΠL

$$\Pi U = (1, 0, 0, \ldots, 0)$$

$$\Pi L = (1, 1, \ldots, 1, 1, 1)$$

2.2 Conditioned Probability

In possibilistic conditioning, a function f dependent on both P and Π is used to find a new conditioned probability distribution $\tilde{P}$ such that

$$\tilde{P} = f(P, \Pi)$$

where $\tilde{P} = \{\tilde{p}_1, \tilde{p}_2, \ldots, \tilde{p}_n\}$ with

$$\tilde{p}_i = \frac{p_i \pi_i}{K}; \; K = \sum_{i=1}^{n} p_i \pi_i$$

Now we consider the conditioned probability in the case of the bounding possibility distributions.

Case 1: ΠU A possibility distribution with exactly one possibility value equal to 1 and all other values equal 0 represents a completely certain distribution. Consider the relationship between such a distribution and the conditioned probability.

If have we the upper possibility, then its conditioned probability isPU. To show this, let Π be completely certain if $\exists k$ such that $\pi_k = 1$, and $\pi_i = 0 \forall i \neq k$. To obtain the conditioned probability, we first calculate K

$$K = p_k \pi_k + \sum_{i \neq k}^{n} p_i \pi_i = p_k * 1 + \sum_{i \neq k}^{n} p_i * 0 = p_k$$

So now we find the conditioned probabilities

$$\tilde{p}_k = \frac{p_k \pi_k}{K} = \frac{p_k * 1}{p_k} = 1$$

$$\tilde{p}_i = \frac{p_i \pi_i}{K} = \frac{p_i * 0}{p_k} = 0,\ i \neq k$$

Thus, the conditioned probability distribution $\tilde{P}$ is

$$\tilde{P} = \{0, \ldots, \tilde{p}_k = 1, \ldots 0\}$$

which is a completely certain probability distribution.

Case 2: Π L

If there is no distinction that is made on the values by the possibility distribution, we say this implies complete uncertainty. If a possibility distribution ΠU is completely uncertain, then its conditioned probability $\tilde{P}$ is identical to the original probability$P$. To show this, let Π be completely uncertain if $\forall i,\ \pi_i = 1$. We first calculate K

$$K = \sum_{i=1}^{n} p_i \pi_i = \sum_{i=1}^{n} p_i * 1 = 1$$

since $\sum_{i=1}^{n} p_i = 1$ for any probability distribution.

So now we find the conditioned probabilities

$$\tilde{p}_i = \frac{p_i \pi_i}{K} = \frac{p_i * 1}{1} = p_i$$

Thus the conditioned probability distribution $\tilde{P}$ is

$$\tilde{P} = \{p_1, p_2, \ldots, p_n\} = P,$$

which is the original probability distribution.

The interpretation of this result is that the possibility distribution shows no preference for any specific value and so the default is that the information to be used in a decision should be that represented by the original probability distribution.

2.3 *Transformation Approaches*

A variety of probability/possibility transformations have been proposed [10–12] and here, without loss of generality, we use well-accepted ones developed [13]. For the transformation, we must assume the possibility distribution Π is ordered such that.

$$\pi_1 \geq \pi_2 \geq \ldots \geq \pi_n \text{ and } \pi_1 = 1.$$

Then an associated probability distribution $P' = (p'_1, p'_2, \ldots, p'_n)$ (denoting $\pi_{n+1} = 0$) can be obtained by using the transformation Γ:

$$P' = \Gamma(\Pi) \text{ where}$$

$$p'_i = \sum_{k=i}^{n} \frac{(\pi_k - \pi_{k+1})}{k}, \text{ for } i = 1 \text{ to } n.$$

The probability transformation,$\Gamma(P)$, also assumes an ordered distribution P where $p_1 \geq p_2 \geq \cdots \geq p_n$. Then we can obtain the possibility distribution Π'

$$\Pi' = \Gamma(P)$$

where $\pi'_1 = 1$ and

$$\begin{aligned} \pi'_i &= \sum_{k=i}^{n} p_k \text{ if } p_{i-1} > p_i. \\ &= \pi'_{i-1} \text{ otherwise.} \end{aligned}$$

3 Aggregation Measures

Here we examine several information measures for probability, a specificity measure for possibility and consistency measure of probability and possibility. We will specify the ranges of these measures by examining the upper and lower bounding cases for probability and possibility.

3.1 Shannon Entropy

Shannon entropy has been the most broadly applied measure of randomness or information content [14, 15]. However, the concept of information is broad so that multiple approaches to the quantification of information are desirable [16] and we will also focus on the Gini measure [17].

For a probability distribution$P = \{p_1, p_2, \ldots, p_n\}$,$H(P) = -\sum_{k=1}^{n} p_k \ln(p_k)$. For the most specific case

$$H(PU) = -(1 \ln(1) + \sum_{k=2}^{n} 0 \ln(0)) = 1 * 0 + \sum_{k=2}^{n} 0 = 0$$

That is when the probability distribution represents complete certainty, there is no uncertainty, i.e. maximum information.

Next for the case of most uncertainty represented by the equi-probable distribution, PL, we obtain

$$H(PL) = -\sum_{k=1}^{n} \frac{1}{n} \ln(\frac{1}{n}) = -\sum_{k=1}^{n} \frac{1}{n}(\ln(1) - \ln(n)) = \sum_{k=1}^{n} \frac{1}{n} \ln(n) = \ln(n).$$

That is, when all probabilities are equi-probable, this is the most unpredictable, uncertain situation and so represents the minimum information. So the range of Shannon's entropy is:

$$0 \leq H(P) \leq \ln(n)$$

3.2 Gini Index

The Gini index, G(P), also known as the Gini coefficient, is a measure of statistical dispersion developed by Gini [17, 18], and is defined as

$$G(P) = 1 - \sum_{k=1}^{n} p_k^2$$

Often $G(P)$ is used versus $H(P)$ since it does not involve a logarithm, making analytic solutions tractable.

The relationship of Shannon entropy and the Gini index has been described [19] and support for this approach comes from an independent consideration of alternative measures of entropy [20], in which the same form for an entropy measure as the Gini measure is developed.

Like Shannon entropy, we evaluate the bounding cases discussed

$$G(PU) = 1 - (p_1^2 + \sum_{k=2}^{n} p_k^2) = 1 - (1^2 + 0) = 0$$

As for the Shannon entropy, this corresponds to no uncertainty.

Then for the equi-probable distributions,

$$G(PL) = 1 - \sum_{k=1}^{n} (\frac{1}{n})^2 = 1 - n(\frac{1}{n})^2 = 1 - \frac{1}{n} = \frac{n-1}{n}$$

and in general, the range for $G(P)$ is

$$0 \leq G(P) \leq \frac{n-1}{n} < 1$$

3.3 Renyi Entropy

Renyi introduced a parameterized family of entropies as a generalization of Shannon entropy [21, 22]. The intention was to have the most general approach that preserved the additivity property and satisfied the probability axioms of Kolmogorov. Renyi entropy is

$$RY_\alpha(P) = \frac{1}{1-\alpha} * \ln \sum_{i=1}^{n} p_i^\alpha$$

Cases of the parameter α:
$\alpha = 0$: $RY(P) = \ln|P|$—Hartley entropy [23]
$\lim \alpha \to 1$: $RY_1(P) = -\sum_{i=1}^{n} p_i \ln(p_i)$—Shannon entropy
$\alpha = 2$: $RY_2(P) = -\sum_{i=1}^{n} \ln(p_i^2)$—Collision or quadratic entropy
$\alpha \to \infty$: $RY_\infty(P) = \underset{i=1}{\overset{n}{\text{Min}}}(\ln(p_i)) = -\underset{i=1}{\overset{n}{\text{Max}}}(\ln(p_i))$

This last case is the smallest entropy in the Renyi family and so is the strongest way to obtain an information content measure. It is never larger than the Shannon entropy. Thus, the possible ranges of α capture the following:

High α	high probability events
Low α	weight possible events more equally
$\alpha = 0, 1$	Hartley or Shannon, respectively

The Hartley entropy is not of great interest here as we have already considered the Shannon entropy. Now we consider the values of the RY_2 measure, for our bounding probabilities. For PU

$$RY_2(PU) = -\ln(p_1 + \sum_{i=2}^{n} p_i^2) = -\ln(1 + \sum_{i=2}^{n} 0^2) = -\ln(1) = 0$$

and for

$$RY_2(PL) = -\ln(\sum_{i=1}^{n} p_i^2) = -\ln(\sum_{i=1}^{n} \frac{1}{n^2}) = -\ln(n * \frac{1}{n^2}) = \ln(n)$$

These are the same as the results for the Shannon entropy.

3.4 Specificity Measure

For possibility theory, the concept of specificity has a measure as for the probability measures above. One approach uses a linear specificity measure that can be viewed as an intuitive measure for a possibility distribution's specificity [24]. Let

$$\pi_m = \operatorname*{Max}_{k=1}^{n} \pi_k,$$

then specificity is formulated as the max value,π_m, minus the average of the rest of the possibility distribution [25],

$$SF(\Pi) = \pi_m - \sum_{k=1,m}^{n} \frac{(\pi_k)}{n}$$

3.5 Consistency Measure

There have been a number of approaches to consistency measures of probability and possibility distributions that have been proposed [26, 27]. In particular [8], Zadeh proposed

$$CZ(\Pi, P) = \sum_{i=1}^{n} \pi_i * p_i$$

This measure does not represent an inherent relationship but rather represents the intuition that a lowering of an event's possibility tends to lower its probability, but not the converse.

Another consistency measure that appears in the literature, $CDP(\Pi, P)$, is due to Dubois and Prade [13, 28]. Here for every subset A of the space X,

$$CDP(\Pi, P) = 1 \quad \text{if}\, \Pi(A) \geq P(A)$$

and is 0 otherwise. This definition is based on the idea that possibility is the weaker representation for a situation than probability. So the stronger support for a situation cannot have a greater value for the situation than the weaker support.

For our purposes, we focus here on CZ as it provides a range of values to evaluate the idea of consistency as it relates to aggregation of probability and possibility. We can note that the maximum value that $CZ(\Pi, P)$ can have in general is 1 as $\sum_{i=1}^{n} p_i = 1$ and π_i is at most 1. Thus, the range of CZ is the interval [0, 1] where 0 can be considered as complete inconsistency and 1 complete consistency.

3.6 Ranges of Measures

In order to apply the results of using the above measures for aggregation, we summarize the range of values for these measures. This can be assessed by using the bounding cases of probability, PU and PL and possibility distributions ΠU and ΠL.

Using these, we have the ranges of measures [29] where 0 is the value for the completely certain cases:

Shannon measure:	$0 \leq H(P) \leq \ln(n)$,
Gini measure:	$0 \leq G(P) \leq \frac{n-1}{n} < 1$,
Specificity measure:	$0 \leq SF(\Pi) \leq 1$,
Zadeh consistency measure:	$0 \leq CZ(\Pi, P) \leq 1$.

3.7 Evaluations of Aggregated Data

There are several important classes of aggregation functions [30–32] and the conjunctive and disjunctive functions corresponding to "and" and "or", respectively, are typical. For conjunctive operators, t-norms and for disjunctive operators, t-conorms are frequently used, of which max and min are representative functions. Fuzzy integrals have also been considered for aggregation [33].

An important combination class is mean operators, with specific examples such as arithmetic average, median, etc. For this study, we take the bounding cases of min and max, and the average as the representative functions to sample across the aggregation functions spectrum.

Now we will aggregate transformed distributions and evaluate the results with the applicable measures. We specifically examine the bounding cases of the upper and lower probability and possibility distributions.

Case 1. Combinations: (a)$(PU, \Pi U)$, (b)$(PL, \Pi L)$

(a) Using $\Gamma(\Pi U)$ for all aggregation operators:

$$\tilde{P} = (1, 0, \ldots, 0); \; G(\tilde{P}) = 1$$

Next using $\Gamma(\Pi U)$ for all aggregation operators:

$$\tilde{\Pi} = (1, 0, \ldots, 0); \; SF(\tilde{\Pi}) = 1$$

(b) Using $\Gamma(\Pi L)$ for all aggregation operators:

$$\tilde{P} = \left(\frac{1}{n}, \frac{1}{n}, \ldots, \frac{1}{n}\right); \; G(\tilde{P}) = \frac{1}{n}$$

Next using $\Gamma(\Pi L)$ for all aggregation operators:

$$\tilde{\Pi} = (1, 1, \ldots, 1);\ SF(\tilde{\Pi}) = 0$$

This is intuitive since when bounding cases are both certain then regardless of the transformation or aggregation operator, we maintain complete certainty. Likewise, for both uncertain, we maintain complete uncertainty. We next consider the case 2 of mixed distributions.

Case 2. Combinations: $(PU, \Pi L)$, $(PL, \Pi U)$

Using $\Gamma(\Pi L)$ and $\Gamma(\Pi U)$

(a) $\tilde{P} = \text{Min}(PU, \Gamma(\Pi L)) = \text{Min}(PL, \Gamma(\Pi U)) = (1, 0, \ldots, 0);\ G(\tilde{P}) = 1$

(b) $\tilde{P} = \text{Max}(PU, \Gamma(\Pi L)) = \text{Max}(PL, \Gamma(\Pi U)) = \left(\frac{n}{2n-1}, \frac{n}{2n-1}, \ldots, \frac{n}{2n-1}\right);\ G(\tilde{P}) = \frac{(n^2+n-1)}{(2n-1)^2}$

(c) $\tilde{P} = \text{Avg}(PU, \Gamma(\Pi L)) = \text{Avg}(PL, \Gamma(\Pi U)) = \left(\frac{n+1}{2n}, \frac{1}{2n}, \ldots, \frac{1}{2n}\right);\ G(\tilde{P}) = \frac{(n+3)}{4n}$

We observe that the values of $G(P)$ for max and avg are both greater than the value for the min operator.

Finally, we consider the probability transforms using $\Gamma(PL)$ and$\Gamma(PU)$.

(d) $\tilde{\Pi} = \text{Min}(\Gamma(PU), \Pi L) = \text{Min}(\Gamma(PL), \Pi U) = (1, 0, \ldots, 0);\ SF(\tilde{\Pi}) = 1$

(e) $\tilde{\Pi} = \text{Max}(\Gamma(PU), \Pi L) = \text{Max}(\Gamma(PL), \Pi U) = (1, 1, \ldots, 1);\ SF(\tilde{\Pi}) = 0$

(f) $\tilde{\Pi} = \text{Avg}(\Gamma(PU), \Pi L) = \text{Avg}(\Gamma(PL), \Pi U) = \left(1, \frac{1}{2}, \ldots, \frac{1}{2}\right);\ SF(\tilde{\Pi}) = \frac{1}{2}$

It is clear that increases in information and specificity can be obtained based on selection of the aggregation function. In general, it appeared that the min operator, representing a conservative stance, was more appropriate. Other choices could be based on operator properties such as fairness and information monotonicity [6].

4 Conclusions

We have addressed the problem of aggregation of heterogeneous uncertain information for multi-criteria decision-making. Many decision-making applications of interest have multiple sources of information with different forms of uncertainty which makes it critical for such applications to be able to effectively utilize all such information. The information evaluations enable the effective use of heterogeneous certain information by establishing the measures of information or specificity to assess the value of the aggregations.

There were two major aspects of the approach that had to be considered, where transformation should be used (probability to possibility and vice-versa) and the selection of the specific aggregation function for aggregating the transformation results. Applying the transformations to these cases and using the aggregation functions provided sets of results that could be evaluated using information theory and specificity measures. These measures allow for comparisons that can guide the selection of which approaches are most suitable for specific applications of multi-criteria approaches.

The choice of using a probability or possibility transformation may depend on whether an application requires a resultant probability versus a possibility distribution. In this we focused on min-to-max range to evaluate what happens across the 'aggregation spectrum'. It is clear that increases in information and specificity can be obtained based on selection of the aggregation function. In general, it appeared that the min operator, representing a conservative stance, was more appropriate, especially for the low consistency cases.

Acknowledgements Petry would like to thank the Naval Research Laboratory's Base Program for sponsoring this research.

References

1. Kahraman, C.: Fuzzy Multi-criteria Decision Making: Theory and Applications with Recent Developments. Springer, New York (2008)
2. Yager, R., Alajlan, N.: Multi-criteria formulations with uncertain satisfactions. Eng. Appl. Artif. Intell. **69**, 104–111 (2018)
3. Yager, R.: Modeling multi-criteria objective functions using fuzzy measures. Inf. Fusion **29**, 105–111 (2016)
4. Thill, J.: Spatial Multicriteria Decision Making and Analysis—A Geographic Information Sciences Approach. Routledge press, London (1999)
5. Jiang, H., Eastman, J.: Applications of fuzzy measures in multi-criteria evaluation in GIS. Int. J. Geogr. Inf. Sci. **14**, 173–184 (2000)
6. Elmore, P., Anderson, D., Petry, F.: Evaluation of heterogeneous uncertain information fusion. Ambient Intell. Humanized Comput. (2019)
7. Parsons, S.: Qualitative Methods for Reasoning under Uncertainty. MIT Press, Cambridge, MA (2001)
8. Zadeh, L.: Fuzzy sets as a basis for a theory of possibility. Fuzzy Sets Syst. **1**, 3–28 (1978)
9. Elmore, P., Petry, F., Yager, R.: (2014) Comparative measures of aggregated uncertainty representations. J. Ambient Intell. Humanized Comput. **5**, 809–819
10. Jumarie, G.: Possibility-probability transformation: a new result via information theory of deterministic functions. Kybernetes **23**, 56–59 (1994)
11. Klir, G., Parviz, B.: Probability-possibility transformations: a comparison. Int. J. Gen. Syst. **21**, 291–310 (1992)
12. Oussalah, M.: On the probability/possibility transformations: a comparative analysis. Int. J. Gen. Syst. **29**, 671–718 (2000)
13. Dubois, D., Prade, H.: Unfair coins and necessity measures: towards a possibilistic interpretations of histograms. Fuzzy Sets Syst **10**, 15–27 (1983)
14. Shannon, C.: A mathematical theory of communication. Bell Syst. Tech. J. **27**, 379–423 (1948)
15. Reza, F.: An Introduction to Information Theory. McGraw Hill, New York (1961)
16. Klir, G.: Uncertainty and Information. Wiley, Hoboken NJ (2006)
17. Gini, C.: (1912) Variabilita e mutabilita (Variability and Mutability), Tipografia di Paolo Cuppini, Bologna, Italy, 156p
18. Aristondo, O., García-Lapresta, J.L., Lasso de la Vega, C., Pereira, R.: The Gini index, the dual decomposition of aggregation functions and the consistent measurement of inequality. Int. J. Intell. Syst. **27**, 132–152 (2012)

19. Eliazar, I., Sokolov, I.: Maximization of statistical heterogeneity: from Shannon's entropy to Gini's index. Phys. A **389**, 3023–3038 (2010)
20. Yager, R.: Measures of entropy and fuzziness related to aggregation operators. Inf. Sci. **82**, 147–166 (1995)
21. Renyi, A.: On measures of information and entropy. In: Proceedings of the 4th Berkeley Symposium on Mathematics, Statistics and Probability, pp. 547–561 (1961)
22. Renyi, A.: Probability Theory. North-Holland, Amsterdam (1970)
23. Hartley, R.: Transmission of information. Bell Syst. Tech. J. **7**, 535–563 (1928)
24. Pedrycz, W., Gomide, F.: An Introduction to Fuzzy Sets: Analysis and Design. MIT Press, Boston MA (1996)
25. Yager, R.: On the specificity of a possibility distribution. Fuzzy Sets Syst. **50**, 279–292 (1992)
26. Delgado, M., Moral. S.: On the concept of possibility-probability consistency. Fuzzy Sets Syst. **21**, 311–318 (1987)
27. Gupta, C.: A note on the transformation of possibilistic information into probabilistic information for investment decisions. Fuzzy Sets Syst. **56**, 175–182 (1993)
28. Dubois, D., Prade, H.: On several representations of an uncertain body of evidence. In: Gupta, M., Sanchez, E. (eds.) Fuzzy Information and Decision Processes, pp. 167–182. North Holland, Amsterdam (1982)
29. Petry, F., Elmore, P., Yager, R.: Combining uncertain information of differing modalities. Inf. Sci. **322**, 237–256 (2015)
30. Beliakov, G., Pradera, A., Calvo, T.: Aggregation Functions: A Guide for Practitioners. Springer, Heidelberg DG (2007)
31. Yager, R.: On mean type aggregation. IEEE Trans. Syst. Man Cybern. **26**, 209–221 (1996)
32. Yager, R., Petry, F.: An intelligent quality based approach to fusing multi-source probabilistic information. Inf. Fusion **31**, 127–136 (2016)
33. Anderson, D., Elmore, P., Petry, F., Havens, T.: Fuzzy Choquet integration of homogenous possibility and probability distributions. Inf. Sci. **363**, 24–39 (2016)

Generalized Interval-Valued Intuitionistic Hesitant Fuzzy Power Bonferroni Means and Their Applications to Multicriteria Decision Making

Pankaj Kakati and Surajit Borkotokey

Abstract In general, the information associated with real-life problems is ambiguous and imprecise, interval-valued intuitionistic hesitant fuzzy sets are an excellent medium to represent the ambiguity of information. The assessment obtained by the Bonferroni mean (BM) can manifest the correlation among the arguments. In this study, the power average operator is fused with the Bonferroni mean under the interval-valued intuitionistic hesitant fuzzy environment to develop some novel power Bonferroni mean operators such as the generalized interval-valued intuitionistic hesitant fuzzy power Bonferroni mean (GIVIHFPBM) and the generalized interval-valued intuitionistic hesitant fuzzy power geometric Bonferroni mean (GIVIHFPGBM). The most prominent characteristic of these proposed operators is that they can not only reflect the correlation among the attributes but also have the unique ability to permit the aggregated values to assist and strengthen each other. Due to this exceptional feature, the GIVIHFPBM and the GIVIHFPGBM operators can reduce the abnormality that may occur in the preference order of the overall aggregated arguments. Further, we investigate some of the special cases and discussed some of the desired properties of the proposed operators. Moreover, an MCDM method is developed based on the proposed operators. Finally, the proposed MCDM method is demonstrated with a numerical example.

1 Introduction

The complexity of solving the real-life MCDM problems is due to the presence of uncertainty or imprecision of information. A fuzzy set [1] which deals with this kind of imprecision by attributing a degree in the unit interval, where the membership grade represents the degree of acceptance. The intuitionistic fuzzy set (IFS) [2,

P. Kakati (✉)
Department of Mathematics, Jagannath Barooah College, Jorhat 785001, India
e-mail: pankaj07kakati@gmail.com

S. Borkotokey
Department of Mathematics, Dibrugarh University, Dibrugarh 786004, India

L. Sahoo et al. (eds.), *Real Life Applications of Multiple Criteria Decision Making Techniques in Fuzzy Domain*, Studies in Fuzziness and Soft Computing 420,
https://doi.org/10.1007/978-981-19-4929-6_10

3] proposed by Atanassov, is described by the non-membership grade besides the membership grade of an object. The non-membership grade of an element in an IFS indicates the index of non-acceptance of criteria in an MCDM problem. The interval-valued intuitionistic fuzzy set (IVIFS) [4] presented by Atanassov and Gargov is a further extension of the IFS for which the membership grades are interval values within the unit interval. In some real-life problems, it's not always possible to depict the membership grades of an object into an exact value. The hesitant fuzzy sets (HFSs) [5, 6] presented by Narukawa and Torra can overcome such difficulties. Chen et al. [7] improved the HFS by extending the membership and non-membership grades of an element to be a collection of various interval numbers and introduced the interval-valued hesitant fuzzy set (IVHFS). Zhang [8] proposed the idea of the interval-valued intuitionistic hesitant fuzzy set (IVIHFS) and defined various aggregation operators for aggregating IVIHFS information. The IVIHFS permits the membership grades of an object to a set of various attainable IVIFNs. Joshi and Kumar [9] proposed the interval-valued intuitionistic hesitant fuzzy Choquet integral (IVIHFCI) and developed a multicriteria group decision making (MCGDM) TOPSIS method. Joshi and Kumar [10] introduced an interval-valued intuitionistic hesitant fuzzy entropy measure and applied it to the MCGDM problem. Narayanamoorthy et al. [11] proposed a new interval-valued intuitionistic hesitant fuzzy entropy for solving IVIHF-VIKOR problems. Bharati [12] introduced an optimization method to handle the multi-objective optimization problem (MOOP) with interval-valued intuitionistic hesitant fuzzy information.

Aggregation operators [13–18] are very powerful tools for decision-making. Aggregation functions usually map several inputs from the interval [0, 1] to a single output in the same interval. Due to this general nature of their definitions, these operators have a wide range of extensions right from ordinary real inputs to those from FSs, IFSs, HFSs, IVHFSs, and many more. Augustine [19], Ejegwa and Onyeke [20], and Ejegwa [21, 22] developed several correlation coefficients for IFSs and applied them to MCDM problems. Xu et al. [23, 24] proposed several aggregation operators with interval-valued intuitionistic fuzzy information, namely IIFWA, IIFWG, IIFHA, and IIFHG, etc. Most of the existing aggregation operators ignore the interrelationship among the criteria that often exists in many real-life problems. The power average (PA) [25] and the power ordered weighted average (POWA) [25] operators proposed by Yager, have the unique characteristic to assist and strengthen each of the accumulated arguments. Also, the weight vector of PA and POWA operators depends on the input values. Further, Xu and Yager [26] introduced some geometric power average operators such as PG and POWG. Zhou and Chen [27] proposed the generalized power average (GPA) operator. Xu [28] and Zhang [29] proposed several PG operators for aggregating Atanassov's intuitionistic fuzzy information. Further, He et al. [30] introduced some generalized PA operators with the interval-valued Atanassov's intuitionistic fuzzy information. Zhang [31] proposed several PA operators for aggregating hesitant fuzzy information. Fahmi et al. [32] fused the PA operator with Einstein operations and introduced some power Einstein aggregation operators under the cubic uncertain linguistic information.

The Bonferroni mean (BM) [33–43] has the potential to reflect the correlation among the input values. Yager [34] presented an interpretation of the BM operator by incorporating a product of each input value with the average of the other inputs. Xu [28] proposed some BM operators under Atanassov's intuitionistic fuzzy environment. Beliakov [36] provided an organized study of a class of composed aggregation operators to generalize the BM. Xia et al. [37] extended the BM operator to propose the geometric Bonferroni mean (GBM). Xia [35] introduced some generalized weighted Bonferroni geometric mean operators based on intuitionistic fuzzy information. He et al. [39] combined the GBM and the PG operators to introduce some geometric power Bonferroni mean operators under the intuitionistic fuzzy environment. Zhu et al. [40, 41] introduced several BM operators for aggregating hesitant fuzzy [5, 6] information. Kakati [42] introduced a new prioritized weighted geometric Bonferroni mean based on Einstein operation under interval neutrosophic fuzzy (INS) [44] environment. Further, He [43] introduced the hesitant fuzzy power Bonferroni mean (HFPBM).

1.1 Motivation

During the assessment of the preferences based on the attributes of an MCDM problem, sometimes due to inadequate skills or limited information, the decision-maker fails to express the assessment to an exact numerical value. Zhang [8] suggested that it is better to select among various possible IVIFNs instead of the exact value, that is, the interval-valued intuitionistic hesitant fuzzy set (IVIHFS). Again, the non-homogeneous nature of the aggregated arguments may result in an unexpected final decision outcome. So power aggregation operators are very useful to deal with such situations. Also, there are situations where input arguments are interrelated, to deal with such situations, the Bonferroni mean (BM) [33] is one of the best tools. Some of the primary causes of motivation to combine the PA with the BM to introduce some new generalized power Bonferroni mean operators under the interval-valued intuitionistic hesitant fuzzy environment are as follows:

(1) Zhang [8] suggested that it is better to select among various possible IVIFNs, i.e., the IVIHFS rather than single values. So it is meaningful to develop a new PBM [39, 43, 45] operator under the interval-valued intuitionistic hesitant fuzzy environment.
(2) Again when aggregated values corresponding to each alternative are fused, due to heterogeneous behaviour of arguments, it often leads to counterfeit impression on the final decision result. In such situations, the only best operator is the power operator (PA) [25, 26], which holds the ability to reduce these counterfeit effects.
(3) The Bonferroni mean (BM) [33] is a suitable tool to tackle the interaction among the criteria of MCDM problems. The BM may be preferred ahead of the fuzzy measure [46] based Choquet integral [47] because the computational complexity in solving fuzzy measure is more when the set of criteria is large.

(4) The PBM operator [39, 43, 45] is an extension of the PA and BM operators. These operators can reflect the correlation among the arguments, moreover, allow the collective argument to assist and strengthen each other. But no such PBM operators are defined under the interval-valued intuitionistic hesitant fuzzy.

Thus motivated by the prominent features of the BM and the PA operators to handle the correlation among the criteria and the ability to permit the aggregated values to assist and strengthen each other, we find it appropriate as well as necessary to combine the BM and the PA operators to introduce some new generalized power Bonferroni mean operators under the interval-valued intuitionistic hesitant fuzzy environment. As already mentioned in the above discussion the interval-valued intuitionistic hesitant fuzzy set being a generalization concept of the FS, IFS, HFS, and IVHFS, etc., and is more suitable to express the ambiguity and imprecision of real-life decision-making problems. Hence it is meaning to extend the PBM operators for decision-making under the interval-valued intuitionistic hesitant fuzzy environment which can tackle the correlation among the criteria and can reduce the abnormality in the overall ranking result due to the heterogeneous nature of the arguments.

The remaining of the study proceeds as follows. Section 2, concisely summarizes some fundamental notions, Sect. 3, proposes the GIVIHFPBM and the GIVIHFPGBM operators. Further, a few particular cases and important features of these proposed operators are explained. Section 4, provides an MCDM method with the proposed operators. Section 5, demonstrates the projected MCDM method by an explanatory example. Moreover, a comparative analysis of the projected MCDM method with an existing method is presented. Section 6, provides the conclusions. Further, some of the advantages and future scopes are presented.

2 Fundamental Concepts

This section provides an overview of some of the basic concepts and notions which are needed for the development of the study.

Definition 1 [5, 6] The hesitant fuzzy set (HFS) F on a set V can be expressed as:

$$F = \{\langle v, \xi_F(v)\rangle \mid v \in V\},$$

where, $\xi_F(v) \subseteq [0, 1]$, which represents the hesitancy membership degree of an element $v \in V$ to the set F.

Definition 2 [8] The interval-valued intuitionistic hesitant fuzzy set (IVIHFS) $\tilde{F}$ on a set V is represented by:

$$\tilde{F} = \{\langle v, \tilde{\xi}_{\tilde{F}}(v)\rangle \mid v \in V\},$$

where, $\tilde{\xi}_{\tilde{F}}(v)$ indicates the membership and non-membership grades of an element $v \in V$, which are portrayed by IVIFNs.

Zhang [8] named $\tilde{\xi} = \tilde{\xi}_{\tilde{F}}(x)$ as interval-valued hesitant fuzzy element (IVIHFE) [8]. Let $\xi \in \tilde{\xi}$, then ξ is an IVIFN, denoted by $\xi = (L_\xi, U_\xi) = ([L_\xi^-, L_\xi^+], [U_\xi^-, U_\xi^+])$.

For any three IVIHFEs $\tilde{\xi}, \tilde{\xi}_1, \tilde{\xi}_2$ and a scalar $\lambda > 0$, we have the following operations:

(i) $\tilde{\xi}_1 \oplus \tilde{\xi}_2 = \{[L_{\xi_1}^- + L_{\xi_2}^- - L_{\xi_1}^- L_{\xi_2}^-, L_{\xi_1}^+ + L_{\xi_2}^+ - L_{\xi_1}^+ L_{\xi_2}^+], [U_{\xi_1}^- U_{\xi_2}^-, U_{\xi_1}^+ U_{\xi_2}^+]\}$
(ii) $\tilde{\xi}_1 \otimes \tilde{\xi}_2 = \{[L_{\xi_1}^- L_{\xi_2}^-, L_{\xi_1}^+ L_{\xi_2}^+], [U_{\xi_1}^- + U_{\xi_2}^- - U_{\xi_1}^- U_{\xi_2}^-, U_{\xi_1}^+ + U_{\xi_2}^+ - U_{\xi_1}^+ U_{\xi_2}^+]\}$
(iii) $\lambda\tilde{\xi} = \{([1-(1-L_\xi^-)^\lambda, 1-(1-L_\xi^+)^\lambda], [(U_\xi^-)^\lambda, (U_\xi^+)^\lambda])\}$
(iv) $\tilde{\xi}^\lambda = \{([(L_\xi^-)^\lambda, (L_\xi^+)^\lambda], [1-(1-U_\xi^-)^\lambda, 1-(1-U_\xi^+)^\lambda])\}$.

Definition 3 [8] Let $\tilde{\xi} = \{([L_\xi^-, L_\xi^+], [U_\xi^-, U_\xi^+]) \mid \xi \in \tilde{\xi}\}$ be an IVIHFE, then the score function $S(\tilde{\xi})$, and the accuracy function of $A(\tilde{\xi})$ are defined as follows,

$$S(\tilde{\xi}) = \frac{\sum_{\xi \in \tilde{\xi}} S(\xi)}{\#\tilde{\xi}}, \tag{1}$$

$$A(\tilde{\xi}) = \frac{\sum_{\xi \in \tilde{\xi}} A(\xi)}{\#\tilde{\xi}} \tag{2}$$

where, $\#\tilde{\xi}$ indicates the order of $\tilde{\xi}$, and $S(\xi)$ and $A(\xi)$ represent the score and the accuracy function of the IVIFEs which are given as [23]:

$$S(\xi) = \frac{L_\xi^- - U_\xi^- + L_\xi^+ - U_\xi^+}{2}, \tag{3}$$

and

$$A(\xi) = \frac{L_\xi^- + U_\xi^- + L_\xi^+ + U_\xi^+}{2}. \tag{4}$$

Consider two IVIHFEs $\tilde{\xi}$ and $\tilde{\xi}'$, then the following rules are hold:

(R1) if $S(\tilde{\xi}) > S(\tilde{\xi}')$, then $\tilde{\xi} \succ \tilde{\xi}'$
(R2) if $S(\tilde{\xi}) < S(\tilde{\xi}')$, then $\tilde{\xi} \prec \tilde{\xi}'$
(R3) if $S(\tilde{\xi}) = S(\tilde{\xi}')$, then the following hold,

(T1) if $A(\tilde{\xi}) > A(\tilde{\xi}')$, then $\tilde{\xi} \succ \tilde{\xi}'$
(T2) if $A(\tilde{\xi}) < A(\tilde{\xi}')$, then $\tilde{\xi} \prec \tilde{\xi}'$
(T3) if $A(\tilde{\xi}) = A(\tilde{\xi}')$, then $\tilde{\xi} = \tilde{\xi}'$.

Definition 4 [25] The power average (PA) operator is expressed as:

$$\mathrm{PA}(\xi_1, \xi_2, \ldots, \xi_l) = \frac{\sum_{i=1}^{l}(1 + T(\xi_i))\xi_i}{\sum_{i=1}^{l}(1 + T(\xi_i))},$$

where,

$$T(\xi_i) = \sum_{\substack{j=1 \\ j \neq i}}^{l} Supp(\xi_i, \xi_j)$$

and $Supp(\xi_i, \xi_j)$ denotes the support function of ξ_i from ξ_j, with the following conditions [25, 48].

(i) $Supp(\xi_i, \xi_j) \in [0, 1]$,
(ii) $Supp(\xi_i, \xi_j) = Supp(\xi_j, \xi_i)$,
(iii) If $|\xi_i - \xi_j| < |\xi_s - \xi_t|$, then $Supp(\xi_i, \xi_j) \geq Supp(\xi_s, \xi_t)$.

Definition 5 [26, 49] The power geometric (PG) operator is expressed as:

$$\mathrm{PG}(\xi_1, \xi_2, \ldots, \xi_l) = \prod_{i=1}^{l} \xi_i^{\frac{1+T(\xi_i)}{\sum_{i=1}^{l}(1+T(\xi_i))}},$$

where,

$$T(\xi_i) = \sum_{\substack{j=1 \\ j \neq i}}^{l} Supp(\xi_i, \xi_j).$$

Definition 6 [33] Let ξ_i $(i = 1, 2, \ldots, l)$ be a set of arguments and $\rho, \sigma > 0$, then the Bonferroni mean (BM) is expressed as:

$$\mathrm{BM}^{\rho,\sigma}(\xi_1, \xi_2, \ldots, \xi_l) = \left(\frac{1}{l(l-1)} \sum_{\substack{j=1 \\ j \neq i}}^{l} \xi_i^{\rho} \xi_j^{\sigma} \right)^{\frac{1}{\rho+\sigma}} \tag{5}$$

Definition 7 [37] Let ξ_i $(i = 1, 2, \ldots, l)$ be a set of arguments $\rho, \sigma > 0$, then the geometric Bonferroni means (GBM) is expressed as:

$$\mathrm{GBM}^{\rho,\sigma}(\xi_1, \xi_2, \ldots, \xi_l) = \frac{1}{\rho + \sigma} \prod_{\substack{j=1 \\ j \neq i}}^{l} (\rho \xi_i + \sigma \xi_j)^{\frac{1}{l(l-1)}} \tag{6}$$

Definition 8 [43] Let ξ_i $(i = 1, 2, \ldots, l)$ be a set of hesitant fuzzy elements (HFEs) and $\rho, \sigma > 0$, then the hesitant fuzzy power Bonferroni mean (HFPBM) is given by,

$$\mathrm{HFPBM}^{\rho,\sigma}(\xi_1, \xi_2, \ldots, \xi_l) = \left(\frac{1}{l(l-1)} \left(\bigoplus_{\substack{i=1 \\ i \neq j}}^{l} \left(\left(\frac{l(1+T(\xi_i))}{\sum_{t=1}^{l}(1+T(\xi_t))} \xi_i \right)^{\rho} \otimes \left(\frac{l(1+T(\xi_j))}{\sum_{t=1}^{l}(1+T(\xi_t))} \xi_j \right)^{\sigma} \right) \right) \right)^{\frac{1}{\rho+\sigma}} \tag{7}$$

where $T(\xi_i) = \sum_{\substack{j=1 \\ j \neq i}}^{l} Supp(\xi_i, \xi_j)$, and $Supp(\xi_i, \xi_j)$ satisfies the three properties mentioned in Definition 4.

3 GIVIHFPBM and GIVIHFPGBM Operators

Definition 9 Let $\tilde{\xi}_i$ $(i = 1, 2, \ldots, l)$ be a collection of IVIHFEs and scalars $\rho, \sigma, \tau > 0$, then the generalized interval-valued intuitionistic hesitant fuzzy power Bonferroni mean (GIVIHFPBM) is given by,

$$\mathrm{GIVIHFPBM}^{\rho,\sigma,\tau}(\tilde{\xi}_1, \tilde{\xi}_2, \ldots, \tilde{\xi}_l)$$
$$= \left(\frac{1}{l(l-1)(l-2)} \left(\bigoplus_{\substack{i,j,k=1 \\ i \neq j \neq k}}^{l} \left(\left(\frac{l(1+T(\tilde{\xi}_i))}{\sum_{t=1}^{l}(1+T(\tilde{\xi}_t))} \tilde{\xi}_i \right)^{\rho} \otimes \left(\frac{l(1+T(\tilde{\xi}_j))}{\sum_{t=1}^{l}(1+T(\tilde{\xi}_t))} \tilde{\xi}_j \right)^{\sigma} \otimes \left(\frac{l(1+T(\tilde{\xi}_k))}{\sum_{t=1}^{l}(1+T(\tilde{\xi}_t))} \tilde{\xi}_k \right)^{\tau} \right) \right) \right)^{\frac{1}{\rho+\sigma+\tau}} \tag{8}$$

where,

$$T(\tilde{\xi}_i) = \sum_{\substack{j=1 \\ j \neq i}}^{l} Supp(\tilde{\xi}_i, \tilde{\xi}_j). \tag{9}$$

The support function $Supp(\tilde{\xi}_i, \tilde{\xi}_j)$ fulfills the conditions in Definition 4.

Theorem 1 Let $\tilde{\xi}_i = \{([L_{\xi_i}^-, L_{\xi_i}^+], [U_{\xi_i}^-, U_{\xi_i}^+]) \mid \xi_i \in \tilde{\xi}_i\}$ $(i = 1, 2, \ldots, l)$ be a collection of (IVIHFEs) and scalars $\rho, \sigma, \tau > 0$, then the collective value determined by utilizing the GIVIHFPBM is again an IVIHFEs. Furthermore,

$$\mathrm{GIVIHFPBM}^{\rho,\sigma,\tau}(\tilde{\xi}_1, \tilde{\xi}_2, \ldots, \tilde{\xi}_l) =$$
$$\bigcup_{\substack{\xi_i \in \tilde{\xi}_i \\ i=1,2,\ldots,l}} \left\{ \left[\left(1 - \prod_{\substack{i,j,k=1 \\ i \neq j \neq k}}^{l} \left(1 - \left(1 - \left(1 - L_{\xi_i}^- \right)^{\lambda_i} \right)^{\rho} \left(1 - \left(1 - L_{\xi_j}^- \right)^{\lambda_j} \right)^{\sigma} \left(1 - \left(1 - L_{\xi_k}^- \right)^{\lambda_k} \right)^{\tau} \right)^{\frac{1}{l(l-1)(l-2)}} \right)^{\frac{1}{\rho+\sigma+\tau}}, \right. \right. \tag{10}$$

$$\left(1-\prod_{\substack{i,j,k=1\\ i\neq j\neq k}}^{l}\left(1-\left(1-\left(1-L_{\xi_i}^{+}\right)^{\lambda_i}\right)^{\rho}\left(1-\left(1-L_{\xi_j}^{+}\right)^{\lambda_j}\right)^{\sigma}\left(1-\left(1-L_{\xi_k}^{+}\right)^{\lambda_k}\right)^{\tau}\right)^{\frac{1}{l(l-1)(l-2)}}\right)^{\frac{1}{\rho+\sigma+\tau}}\Bigg], \tag{11}$$

$$\Bigg[1-\left(1-\prod_{\substack{i,j,k=1\\ i\neq j\neq k}}^{l}\left(1-\left(1-\left(U_{\xi_i}^{-}\right)^{\lambda_i}\right)^{\rho}\left(1-\left(U_{\xi_j}^{-}\right)^{\lambda_j}\right)^{\sigma}\left(1-\left(U_{\xi_k}^{-}\right)^{\lambda_k}\right)^{\tau}\right)^{\frac{1}{l(l-1)(l-2)}}\right)^{\frac{1}{\rho+\sigma+\tau}}, \tag{12}$$

$$1-\left(1-\prod_{\substack{i,j,k=1\\ i\neq j\neq k}}^{l}\left(1-\left(1-\left(U_{\xi_i}^{+}\right)^{\lambda_i}\right)^{\rho}\left(1-\left(U_{\xi_j}^{+}\right)^{\lambda_j}\right)^{\sigma}\left(1-\left(U_{\xi_k}^{+}\right)^{\lambda_k}\right)^{\tau}\right)^{\frac{1}{l(l-1)(l-2)}}\right)^{\frac{1}{\rho+\sigma+\tau}}\Bigg]\Bigg\} \tag{13}$$

where, $\lambda_i = \frac{l(1+T(\tilde{\xi}_i))}{\sum_{t=1}^{l}(1+T(\tilde{\xi}_t))}$, $(i = 1, 2, \ldots, l)$

We have the following special cases for GIVIHFPBM for some particular based on the scalars ρ, σ and τ.

Remark 1 If $\tau \to 0$, then

$$\lim_{\tau\to 0} \text{GIVIHFPBM}^{\rho,\sigma,\tau}(\tilde{\xi}_1, \tilde{\xi}_2, \ldots, \tilde{\xi}_l) = \left(\frac{1}{l(l-1)}\left(\bigoplus_{\substack{i=1\\ i\neq j}}^{l}\left(\left(\frac{l(1+T(\tilde{\xi}_i))}{\sum_{t=1}^{l}(1+T(\tilde{\xi}_t))}\tilde{\xi}_i\right)^{\rho}\otimes\left(\frac{l(1+T(\tilde{\xi}_j))}{\sum_{t=1}^{l}(1+T(\tilde{\xi}_t))}\tilde{\xi}_j\right)^{\sigma}\right)\right)\right)^{\frac{1}{\rho+\sigma}} \tag{14}$$

Further, for the collection $\tilde{\xi}_i = \{([L_{\xi_i}^-, L_{\xi_i}^+], [U_{\xi_i}^-, U_{\xi_i}^+]) \mid \xi_i \in \tilde{\xi}_i\}$ $(i = 1, 2, \ldots, l)$ of IVIHFEs, if we consider $L_{\xi_i}^- = L_{\xi_i}^+ = \gamma_{\xi_i}$ and $U_{\xi_i}^- = U_{\xi_i}^+ = 0, \forall\, i = 1, 2, \ldots, l$. Then the GIVIHFPBM can be restricted to the HFPBM [43] operator.

Remark 2 If $\sigma \to 0$ and $\tau \to 0$, then

$$\lim_{\substack{\sigma\to 0\\ \tau\to 0}} \text{GIVIHFPBM}^{\rho,\sigma,\tau}(\tilde{\xi}_1, \tilde{\xi}_2, \ldots, \tilde{\xi}_l) = \left(\frac{1}{l}\bigoplus_{i=1}^{l}\left(\frac{l(1+T(\tilde{\xi}_i))}{\sum_{t=1}^{l}(1+T(\tilde{\xi}_t))}\tilde{\xi}_i\right)^{\rho}\right)^{\frac{1}{\rho}} \tag{15}$$

Moreover, if $\rho = 1$ and $\frac{(1+T(\tilde{\xi}_i))}{\sum_{t=1}^{l}(1+T(\tilde{\xi}_t))} = \frac{1}{l}$, $(\forall i = 1, 2, \ldots, l)$, then

$$\lim_{\substack{\sigma\to 0\\ \tau\to 0}} \text{GIVIHFPBM}^{\rho,\sigma,\tau}(\tilde{\xi}_1, \tilde{\xi}_2, \ldots, \tilde{\xi}_l) = \frac{1}{l}\left(\bigoplus_{i=1}^{l}\tilde{\xi}_i\right)$$

Thus the IVIHFA [8] operator is a particular instance of the GIVIHFPBM.

Definition 10 Let $\tilde{\xi}_i$ $(i = 1, 2, \ldots, l)$ be a collection of IVIHFEs and scalars $\rho, \sigma, \tau > 0$, then the generalized interval-valued intuitionistic hesitant fuzzy power geometric Bonferroni mean (GIVIHFPGBM) is given by,

$$\text{GIVIHFPGBM}^{\rho,\sigma,\tau}(\tilde{\xi}_1, \tilde{\xi}_2, \ldots, \tilde{\xi}_l)$$
$$= \frac{1}{\rho+\sigma+\tau}\left(\bigotimes_{i=1}^{l}\left(\rho\tilde{\xi}_i^{\frac{l(1+T(\tilde{\xi}_i))}{\sum_{t=1}^{l}(1+T(\tilde{\xi}_t))}} \oplus \bigotimes_{j\neq i}^{l}\left(\sigma\tilde{\xi}_j^{\frac{m(1+T(\tilde{\xi}_j))}{\sum_{t=1}^{l}(1+T(\tilde{\xi}_t))}} \oplus \bigotimes_{k\neq j\neq i}^{l}\left(\tau\tilde{\xi}_k^{\frac{l(1+T(\tilde{\xi}_k))}{\sum_{t=1}^{l}(1+T(\tilde{\xi}_t))}}\right)^{\frac{1}{l-2}}\right)^{\frac{1}{l-1}}\right)^{\frac{1}{l}}\right), \tag{16}$$

where, $T(\tilde{\xi}_i) = \sum_{j\neq i_{j=1}}^{l} Supp(\tilde{\xi}_i, \tilde{\xi}_j)$, and $Supp(\tilde{\xi}_i, \tilde{\xi}_j)$ complies with the conditions in Definition 4.

Theorem 2 Let $\tilde{\xi}_i = \{([L^-_{\xi_i}, L^+_{\xi_i}], [U^-_{\xi_i}, U^+_{\xi_i}]) \mid \xi_i \in \tilde{\xi}_i\}$ $(i = 1, 2, \ldots, l)$ be a collection of (IVIHFEs) and scalars $\rho, \sigma, \tau > 0$, then the collective value determined by utilizing the GIVIHFPGBM is again an IVIHFEs. Furthermore,

$$\text{GIVIHFPGBM}^{\rho,\sigma,\tau}(\tilde{\xi}_1, \tilde{\xi}_2, \ldots, \tilde{\xi}_l) =$$
$$\left\{\left[1-\left(1-\prod_{i=1}^{l}\left(1-\left(1-(L^-_{\xi_i})^{\lambda_i}\right)^{\rho}\left(1-\prod_{j\neq i}^{l}\left(1-\left(1-(L^-_{\xi_j})^{\lambda_j}\right)^{\sigma}\left(1-\prod_{k\neq j\neq i}^{l}\left(1-\left(1-(L^-_{\xi_k})^{\lambda_k}\right)^{\tau}\right)^{\frac{1}{l-2}}\right)\right)^{\frac{1}{l-1}}\right)\right)^{\frac{1}{l}}\right)^{\frac{1}{\rho+\sigma+\tau}},\right.\right.$$
$$\left.1-\left(1-\prod_{i=1}^{l}\left(1-\left(1-(L^+_{\xi_i})^{\lambda_i}\right)^{\rho}\left(1-\prod_{j\neq i}^{l}\left(1-\left(1-(L^+_{\xi_j})^{\lambda_j}\right)^{\sigma}\left(1-\prod_{k\neq j\neq i}^{l}\left(1-\left(1-(L^+_{\xi_k})^{\lambda_k}\right)^{\tau}\right)^{\frac{1}{l-2}}\right)\right)^{\frac{1}{l-1}}\right)\right)^{\frac{1}{l}}\right)^{\frac{1}{\rho+\sigma+\tau}}\right],$$
$$\left[\left(1-\prod_{i=1}^{l}\left(1-\left(1-(1-U^-_{\xi_i})^{\lambda_i}\right)^{\rho}\left(1-\prod_{j\neq i}^{l}\left(1-\left(1-(1-U^-_{\xi_j})^{\lambda_j}\right)^{\sigma}\left(1-\prod_{k\neq j\neq i}^{l}\left(1-\left(1-(1-U^-_{\xi_k})^{\lambda_k}\right)^{\tau}\right)^{\frac{1}{l-2}}\right)\right)^{\frac{1}{l-1}}\right)\right)^{\frac{1}{l}}\right)^{\frac{1}{\rho+\sigma+\tau}},\right.$$
$$\left.\left.\left(1-\prod_{i=1}^{l}\left(1-\left(1-(1-U^+_{\xi_i})^{\lambda_i}\right)^{\rho}\left(1-\prod_{j\neq i}^{l}\left(1-\left(1-(1-U^+_{\xi_j})^{\lambda_j}\right)^{\sigma}\left(1-\prod_{k\neq j\neq i}^{l}\left(1-\left(1-(1-U^+_{\xi_k})^{\lambda_k}\right)^{\tau}\right)^{\frac{1}{l-2}}\right)\right)^{\frac{1}{l-1}}\right)\right)^{\frac{1}{l}}\right)^{\frac{1}{\rho+\sigma+\tau}}\right]\right\}$$

where, $\lambda_i = \frac{l(1+T(\tilde{\xi}_i))}{\sum_{t=1}^{l}(1+T(\tilde{\xi}_t))}$, $(i = 1, 2, \ldots, l)$

Remark 3 If $\sigma \to 0$ and $\tau \to 0$ then, $\lim_{\substack{\sigma\to 0\\ \tau\to 0}} \text{GIVIHFPGBM}^{\rho,\sigma,\tau}(\tilde{\xi}_1, \tilde{\xi}_2, \ldots, \tilde{\xi}_l) = \frac{1}{\rho}\left(\bigotimes_{i=1}^{l}\rho\tilde{\xi}_i^{\frac{l(1+T(\tilde{\xi}_i))}{\sum_{t=1}^{l}(1+T(\tilde{\xi}_t))}}\right)$.

Moreover, if $\rho = 1$ and $\frac{(1+T(\tilde{\xi}_i))}{\sum_{t=1}^{l}(1+T(\tilde{\xi}_t))} = \frac{1}{l}$, $(\forall i = 1, 2, \ldots, l)$, then

$$\lim_{\substack{\sigma\to 0\\ \tau\to 0}} \text{GIVIHFPGBM}^{1,\sigma,\tau}(\tilde{\xi}_1, \tilde{\xi}_2, \ldots, \tilde{\xi}_l) = \bigotimes_{i=1}^{l}(\tilde{\xi}_i)^{\frac{1}{l}} = \text{IVIHFG}(\tilde{\xi}_1, \tilde{\xi}_2, \ldots, \tilde{\xi}_l)$$

Thus the IVIHFG [8] operator is a specific instance of the proposed GIVIHFPGBM operator.

3.1 Properties of GIVIHFPBM and GIVIHFPGBM

Theorem 3 (Idempotency) Let $\tilde{\xi}_i = \{([L^-_{\xi_i}, L^+_{\xi_i}], [U^-_{\xi_i}, U^+_{\xi_i}]) \mid \xi_i \in \tilde{\xi}_i\}$ $(i = 1, 2, \ldots, l)$ be a set of (IVIHFEs) and scalars $\rho, \sigma, \tau > 0$. If all $\tilde{\xi}_i$ are equal. i.e., $\tilde{\xi}_i = \tilde{\xi} = \{([L^-_{\xi}, L^+_{\xi}], [U^-_{\xi}, U^+_{\xi}]) \mid \xi \in \tilde{\xi}\}$, then

$$\text{GIVIHFPBM}^{\rho,\sigma,\tau}(\tilde{\xi},\tilde{\xi},\ldots,\tilde{\xi})=\tilde{\xi};\quad \text{GIVIHFPGBM}^{\rho,\sigma,\tau}(\tilde{\xi},\tilde{\xi},\ldots,\tilde{\xi})=\tilde{\xi}\quad.$$

Proof Since, $\tilde{\xi}_i=\tilde{\xi}$ for all $i=1,2,\ldots,l$, So $L_{\xi_i}=L_{\xi}$, $U_{\xi_i}=U_{\xi}$, $(i=1,2,\ldots,l)$. Also, $Supp(\tilde{\xi}_i,\tilde{\xi}_j)=Supp(\tilde{\xi},\tilde{\xi})=K$ (Constant) for all $i,j=1,2,\ldots,l$.
Therefore $\lambda_i=\frac{l(1+T(\tilde{\xi}_i))}{\sum_{t=1}^{l}(1+T(\tilde{\xi}_t))}$, $(i=1,2,\ldots,l)$.
Then, from Eq. (10)

$$\begin{aligned}
&\text{GIVIHFPBM}^{\rho,\sigma,\tau}(\tilde{\xi}_1,\tilde{\xi}_2,\ldots,\tilde{\xi}_l)\\
&=\text{GIVIHFPBM}^{\rho,\sigma,\tau}(\tilde{\xi},\tilde{\xi},\ldots,\tilde{\xi})\\
&=\bigcup_{\xi\in\tilde{\xi}}\Bigg\{\Bigg[\Bigg(1-\prod_{\substack{i,j,k=1\\ i\neq j\neq k}}^{l}\Big(1-\big(1-(1-L_{\xi}^{-})\big)^{\rho}\big(1-(1-L_{\xi}^{-})\big)^{\sigma}\big(1-(1-L_{\xi}^{-})\big)^{\tau}\Big)^{\frac{1}{l(l-1)(l-2)}}\Bigg)^{\frac{1}{\rho+\sigma+\tau}},\\
&\qquad\Bigg(1-\prod_{\substack{i,j,k=1\\ i\neq j\neq k}}^{l}\Big(1-\big(1-(1-L_{\xi}^{+})\big)^{\rho}\big(1-(1-L_{\xi}^{+})\big)^{\sigma}\big(1-(1-L_{\xi}^{+})\big)^{\tau}\Big)^{\frac{1}{l(l-1)(l-2)}}\Bigg)^{\frac{1}{\rho+\sigma+\tau}}\Bigg],\\
&\qquad\Bigg[1-\Bigg(1-\prod_{\substack{i,j,k=1\\ i\neq j\neq k}}^{l}\Big(1-(1-U_{\xi}^{-})^{\rho}(1-U_{\xi}^{-})^{\sigma}(1-U_{\xi}^{-})^{\tau}\Big)^{\frac{1}{l(l-1)(l-2)}}\Bigg)^{\frac{1}{\rho+\sigma+\tau}},\\
&\qquad 1-\Bigg(1-\prod_{\substack{i,j,k=1\\ i\neq j\neq k}}^{l}\Big(1-(1-U_{\xi}^{+})^{\rho}(1-U_{\xi}^{+})^{\sigma}(1-U_{\xi}^{+})^{\tau}\Big)^{\frac{1}{m(m-1)(m-2)}}\Bigg)^{\frac{1}{\rho+\sigma+\tau}}\Bigg]\Bigg\}\\
&=\bigcup_{\xi\in\tilde{\xi}}\Bigg\{\Bigg[\Bigg(1-\Big(1-\big(1-(1-L_{\xi}^{-})\big)^{\rho+\sigma+\tau}\Big)^{\sum_{i=1}^{l}\frac{1}{l}\sum_{j\neq i}^{l}\frac{1}{l-1}\sum_{k\neq j\neq i}^{l}\frac{1}{l-2}}\Bigg)^{\frac{1}{p+q+r}},\\
&\qquad\Bigg(1-\Big(1-\big(1-(1-L_{\xi}^{+})\big)^{\rho+\sigma+\tau}\Big)^{\sum_{i=1}^{l}\frac{1}{l}\sum_{j\neq i}^{l}\frac{1}{l-1}\sum_{k\neq j\neq i}^{l}\frac{1}{l-2}}\Bigg)^{\frac{1}{\rho+\sigma+\tau}}\Bigg],\\
&\qquad\Bigg[1-\Bigg(1-\Big(1-(1-U_{\xi}^{-})^{\rho+\sigma+\tau}\Big)^{\sum_{i=1}^{l}\frac{1}{l}\sum_{j\neq i}^{l}\frac{1}{l-1}\sum_{k\neq j\neq i}^{l}\frac{1}{l-2}}\Bigg)^{\frac{1}{\rho+\sigma+\tau}},\\
&\qquad 1-\Bigg(1-\Big(1-(1-U_{\xi}^{+})^{\rho+\sigma+\tau}\Big)^{\sum_{i=1}^{l}\frac{1}{l}\sum_{j\neq i}^{l}\frac{1}{l-1}\sum_{k\neq j\neq i}^{m}\frac{1}{l-2}}\Bigg)^{\frac{1}{\rho+\sigma+\tau}}\Bigg]\Bigg\}\\
&=\bigcup_{\xi\in\tilde{\xi}}\Bigg\{\Bigg[\Big(\big(1-(1-L_{\xi}^{-})\big)^{\rho+\sigma+\tau}\Big)^{\frac{1}{p+q+r}},\ \Big(\big(1-(1-L_{\xi}^{+})\big)^{\rho+\sigma+\tau}\Big)^{\frac{1}{\rho+\sigma+\tau}}\Bigg],\\
&\qquad\Bigg[1-\Big((1-U_{\xi}^{-})^{\rho+\sigma+\tau}\Big)^{\frac{1}{\rho+\sigma+\tau}},\ 1-\Big((1-U_{\xi}^{+})^{\rho+\sigma+\tau}\Big)^{\frac{1}{\rho+\sigma+\tau}}\Bigg]\Bigg\}\\
&=\bigcup_{\xi\in\tilde{\xi}}\Big\{[L_{\xi}^{-},L_{\xi}^{+}],[U_{\xi}^{-},U_{\xi}^{+}]\Big\}\\
&=\Big\{[L_{\xi}^{-},L_{\xi}^{+}],[U_{\xi}^{-},U_{\xi}^{+}]\mid\xi\in\tilde{\xi}\Big\}
\end{aligned}\tag{17}$$

Therefore, $\text{GIVIHFPBM}^{\rho,\sigma,\tau}(\tilde{\xi},\tilde{\xi},\ldots,\tilde{\xi})=\tilde{\xi}$.

Similarly, $\text{GIVIHFPGBM}^{\rho,\sigma,\tau}(\tilde{\xi},\tilde{\xi},\ldots,\tilde{\xi})=\tilde{\xi}$.

Theorem 4 (Commutativity) Let $\tilde{\xi}_i=\{[L_{\xi_i}^{-},L_{\xi_i}^{+}],[U_{\xi_i}^{-},U_{\xi_i}^{+}]\mid\xi_i\in\tilde{\xi}_i\}$ $(i=1,2,\ldots,l)$, and $\tilde{\xi}_i^{'}=\{[L_{\xi_i^{'}}^{-},L_{\xi_i^{'}}^{+}],[U_{\xi_i^{'}}^{-},U_{\xi_i^{'}}^{+}]\mid\xi_i^{'}\in\tilde{\xi}_i^{'}\}$ $(i=1,2,\ldots,l)$ be two collections of IVIHFEs and scalars $\rho,\sigma,\tau>0$. If $\tilde{\xi}_i^{'}$ $(i=1,2,\ldots,l)$ be any permutation of

$\tilde{\xi}_i\ (i = 1, 2, \ldots, l)$, then
$\text{GIVIHFPBM}^{\rho,\sigma,\tau}(\tilde{\xi}_1, \tilde{\xi}_2, \ldots, \tilde{\xi}_l) = \text{GIVIHFPBM}^{\rho,\sigma,\tau}(\tilde{\xi}_1', \tilde{\xi}_2', \ldots, \tilde{\xi}_l')$;
$\text{GIVIHFPGBM}^{\rho,\sigma,\tau}(\tilde{\xi}_1, \tilde{\xi}_2, \ldots, \tilde{\xi}_l) = \text{GIVIHFPGBM}^{\rho,\sigma,\tau}(\tilde{\xi}_1', \tilde{\xi}_2', \ldots, \tilde{\xi}_l')$.

Theorem 5 (Boundedness) Let $\tilde{\xi}_i = \{[L_{\xi_i}^-, L_{\xi_i}^+], [U_{\xi_i}^-, U_{\xi_i}^+] \mid \xi_i \in \tilde{\xi}_i\}\ (i = 1, 2, \ldots, l)$, be a set of IVIHFEs and scalars $\rho, \sigma, \tau > 0$, then

$$\tilde{\xi}^- \preceq \text{GIVIHFPBM}^{\rho,\sigma,\tau}(\tilde{\xi}_1, \tilde{\xi}_2, \ldots, \tilde{\xi}_l) \preceq \tilde{\xi}^+,$$
$$\tilde{\xi}_0^- \preceq \text{GIVIHFPGBM}^{\rho,\sigma,\tau}(\tilde{\xi}_1, \tilde{\xi}_2, \ldots, \tilde{\xi}_l) \preceq \tilde{\xi}_0^+,$$

where,

$$\tilde{\xi}^- = \left\{ \left(\begin{array}{l} \left[\min\limits_i \left(1 - \left(1 - L_{\xi_i}^-\right)^{\frac{l(1+T(\tilde{\xi}_i))}{\sum_{t=1}^{l}(1+T(\tilde{\xi}_t))}} \right), \min\limits_i \left(1 - \left(1 - L_{\xi_i}^-\right)^{\frac{l(1+T(\tilde{\xi}_i))}{\sum_{t=1}^{l}(1+T(\tilde{\xi}_t))}} \right) \right], \\ \left[\max\limits_i \left(\left(U_{\xi_i}^-\right)^{\frac{l(1+T(\tilde{\xi}_i))}{\sum_{t=1}^{l}(1+T(\tilde{\xi}_t))}} \right), \max\limits_i \left(\left(U_{\xi_i}^+\right)^{\frac{l(1+T(\tilde{\xi}_i))}{\sum_{t=1}^{l}(1+T(\tilde{\xi}_t))}} \right) \right] \end{array} \right) \middle| \begin{array}{l} \xi_i \in \tilde{\xi}_i \\ i=1,2,\ldots,l \end{array} \right\}, \tag{18}$$

$$\tilde{\xi}^+ = \left\{ \left(\begin{array}{l} \left[\max\limits_i \left(1 - \left(1 - L_{\xi_i}^-\right)^{\frac{l(1+T(\tilde{\xi}_i))}{\sum_{t=1}^{l}(1+T(\tilde{\xi}_t))}} \right), \max\limits_i \left(1 - \left(1 - L_{\xi_i}^-\right)^{\frac{l(1+T(\tilde{\xi}_i))}{\sum_{t=1}^{l}(1+T(\tilde{\xi}_t))}} \right) \right], \\ \left[\min\limits_i \left(\left(U_{\xi_i}^-\right)^{\frac{l(1+T(\tilde{\xi}_i))}{\sum_{t=1}^{l}(1+T(\tilde{\xi}_t))}} \right), \min\limits_i \left(\left(U_{\xi_i}^+\right)^{\frac{m(1+T(\tilde{\xi}_i))}{\sum_{t=1}^{l}(1+T(\tilde{\xi}_t))}} \right) \right] \end{array} \right) \middle| \begin{array}{l} \xi_i \in \tilde{\xi}_i \\ i=1,2,\ldots,l \end{array} \right\} \tag{19}$$

$$\tilde{\xi}_0^- = \left\{ \left(\begin{array}{l} \left[\min\limits_i \left(\left(L_{\xi_i}^-\right)^{\frac{l(1+T(\tilde{\xi}_i))}{\sum_{t=1}^{l}(1+T(\tilde{\xi}_t))}} \right), \min\limits_i \left(\left(L_{\xi_i}^+\right)^{\frac{l(1+T(\tilde{\xi}_i))}{\sum_{t=1}^{l}(1+T(\tilde{\xi}_t))}} \right) \right], \\ \left[\max\limits_i \left(1 - \left(1 - U_{\xi_i}^-\right)^{\frac{l(1+T(\tilde{\xi}_i))}{\sum_{t=1}^{l}(1+T(\tilde{\xi}_t))}} \right), \max\limits_i \left(1 - \left(1 - U_{\xi_i}^+\right)^{\frac{l(1+T(\tilde{\xi}_i))}{\sum_{t=1}^{l}(1+T(\tilde{\xi}_t))}} \right) \right] \end{array} \right) \middle| \begin{array}{l} \xi_i \in \tilde{\xi}_i \\ i=1,2,\ldots,l \end{array} \right\}, \tag{20}$$

$$\tilde{\xi}_0^+ = \left\{ \left(\left(\begin{array}{l} \left[\max_i \left(\left(L^-_{\xi_i} \right)^{\frac{l(1+T(\tilde{\xi}_i))}{\sum_{t=1}^{l}(1+T(\tilde{\xi}_t))}} \right), \max_i \left(\left(L^+_{\xi_i} \right)^{\frac{l(1+T(\tilde{\xi}_i))}{\sum_{t=1}^{l}(1+T(\tilde{\xi}_t))}} \right) \right], \\ \left[\min_i \left(1 - \left(1 - U^-_{\xi_i} \right)^{\frac{l(1+T(\tilde{\xi}_i))}{\sum_{t=1}^{l}(1+T(\tilde{\xi}_t))}} \right), \min_i \left(1 - \left(1 - U^+_{\xi_i} \right)^{\frac{l(1+T(\tilde{\xi}_i))}{\sum_{t=1}^{l}(1+T(\tilde{\xi}_t))}} \right) \right] \end{array} \right) \right) \middle| \begin{array}{l} \xi_i \in \tilde{\xi}_i \\ i=1,2,\ldots,l \end{array} \right\} \tag{21}$$

Proof Let

$$\left. \begin{array}{ll} \gamma^-_{\xi_i} = 1 - (1 - L^-_{\xi_i})^{\lambda_i} & , \gamma^+_{\xi_i} = 1 - (1 - L^+_{\xi_i})^{\lambda_i} \\ \delta^-_{\xi_i} = (U^-_{\xi_i})^{\lambda_i} & , \delta^+_{\xi_i} = (U^+_{\xi_i})^{\lambda_i} \end{array} \right\} \tag{22}$$

where, $\lambda_i = \frac{l(1+T(\tilde{\xi}_i))}{\sum_{t=1}^{l}(1+T(\tilde{\xi}_t))}$, for all $i = 1, 2, \ldots, l$.
Now using the Equation set (22) and Eq. (10) we have,

$$\begin{aligned} &\text{GIVIHFPBM}^{\rho,\sigma,\tau}(\tilde{\xi}_1, \tilde{\xi}_2, \ldots, \tilde{\xi}_l) \\ &= \bigcup_{\substack{\xi_i \in \tilde{\xi}_i \\ i=1,2,\ldots,l}} \left\{ \left[\left[\left(1 - \prod_{\substack{i,j,k=1 \\ i \neq j \neq k}}^{l} \left(1 - \left(\gamma^-_{\xi_i}\right)^{\rho} \left(\gamma^-_{\xi_j}\right)^{\sigma} \left(\gamma^-_{\xi_k}\right)^{\tau} \right)^{\frac{1}{l(l-1)(l-2)}} \right)^{\frac{1}{\rho+\sigma+\tau}}, \right. \right. \right. \\ &\qquad \left. \left(1 - \prod_{\substack{i,j,k=1 \\ i \neq j \neq k}}^{l} \left(1 - \left(\gamma^+_{\xi_i}\right)^{\rho} \left(\gamma^+_{\xi_j}\right)^{\sigma} \left(\gamma^+_{\xi_k}\right)^{\tau} \right)^{\frac{1}{l(l-1)(l-2)}} \right)^{\frac{1}{\rho+\sigma+\tau}} \right], \\ &\qquad \left[1 - \left(1 - \prod_{\substack{i,j,k=1 \\ i \neq j \neq k}}^{m} \left(1 - \left(1 - \delta^-_{\xi_i}\right)^{\rho} \left(1 - \delta^-_{\xi_j}\right)^{\sigma} \left(1 - \delta^-_{\xi_k}\right)^{\tau} \right)^{\frac{1}{l(l-1)(l-2)}} \right)^{\frac{1}{\rho+\sigma+\tau}}, \right. \\ &\qquad \left. \left. \left. 1 - \left(1 - \prod_{\substack{i,j,k=1 \\ i \neq j \neq k}}^{l} \left(1 - \left(1 - \delta^+_{\xi_i}\right)^{\rho} \left(1 - \delta^+_{\xi_j}\right)^{\sigma} \left(1 - \delta^+_{\xi_k}\right)^{\tau} \right)^{\frac{1}{l(l-1)(l-2)}} \right)^{\frac{1}{\rho+\sigma+\tau}} \right] \right] \right\}. \end{aligned} \tag{23}$$

Let

$$\left. \begin{array}{l} \gamma^-_{\min} = \min_i(\gamma^-_{\xi_i}) \ , \ \gamma^+_{\min} = \min_i(\gamma^+_{\xi_i}) \ , \ \gamma^-_{\max} = \max_i(\gamma^-_{\xi_i}) \ , \ \gamma^+_{\max} = \max_i(\gamma^+_{\xi_i}) \ , \\ \delta^-_{\min} = \min_i(\delta^-_{\xi_i}) \ \ , \ \delta^+_{\min} = \min_i(\delta^+_{\xi_i}) \ \ , \ \delta^-_{\max} = \max_i(\delta^-_{\xi_i}) \ \ , \ \delta^+_{\max} = \max_i(\delta^+_{\xi_i}) \ \ , \end{array} \right\}. \tag{24}$$

Using equation set (24) from Eqs. (18), and (19) we have,

$$\tilde{\xi}^- = \left\{ \left([\gamma^-_{\min}, \gamma^+_{\min}], [\delta^-_{\max}, \delta^+_{\max}] \right) \right\} \tag{25}$$

$$\tilde{\xi}^{+} = \left\{\left([\gamma_{\max}^{-}, \gamma_{\max}^{+}], [\delta_{\min}^{-}, \delta_{\min}^{+}]\right)\right\} \tag{26}$$

Now,

$$\begin{aligned}\prod_{\substack{i,j,k=1\\ i\neq j\neq k}}^{l}\left(1-\left(\gamma_{\xi_i}^{-}\right)^{\rho}\left(\gamma_{\xi_j}^{-}\right)^{\sigma}\left(\gamma_{\xi_k}^{-}\right)^{\tau}\right)^{\frac{1}{l(l-1)(l-2)}} &\leq \prod_{\substack{i,j,k=1\\ i\neq j\neq k}}^{l}\left(1-\left(\gamma_{\min}^{-}\right)^{\rho+\sigma+\tau}\right)^{\frac{1}{l(l-1)(l-2)}}\\ &= \left(1-\left(\gamma_{\min}^{-}\right)^{\rho+\sigma+\tau}\right)^{\sum_{i=1}^{l}\frac{1}{l}\sum_{j\neq i}^{l}\frac{1}{l-1}\sum_{k\neq j\neq i}^{l}\frac{1}{l-2}}\\ &= 1-\left(\gamma_{\min}^{-}\right)^{\rho+\sigma+\tau}\end{aligned}$$

Thus,

$$\left(\gamma_{\min}^{-}\right)^{\rho+\sigma+\tau} \leq 1-\prod_{\substack{i,j,k=1\\ i\neq j\neq k}}^{m}\left(1-\left(\gamma_{\xi_i}^{-}\right)^{\rho}\left(\gamma_{\xi_j}^{-}\right)^{\sigma}\left(\gamma_{\xi_k}^{-}\right)^{\tau}\right)^{\frac{1}{l(l-1)(l-2)}}$$

So,

$$\gamma_{\min}^{-} \leq \left(1-\prod_{\substack{i,j,k=1\\ i\neq j\neq k}}^{l}\left(1-\left(\gamma_{\xi_i}^{-}\right)^{\rho}\left(\gamma_{\xi_j}^{-}\right)^{\sigma}\left(\gamma_{\xi_k}^{-}\right)^{\tau}\right)^{\frac{1}{l(l-1)(l-2)}}\right)^{\frac{1}{\rho+\sigma+\tau}} \tag{27}$$

Again,

$$\begin{aligned}\prod_{\substack{i,j,k=1\\ i\neq j\neq k}}^{l}\left(1-\left(\gamma_{\xi_i}^{-}\right)^{\rho}\left(\gamma_{\xi_j}^{-}\right)^{\sigma}\left(\gamma_{\xi_k}^{-}\right)^{\tau}\right)^{\frac{1}{l(l-1)(l-2)}} &\geq \prod_{\substack{i,j,k=1\\ i\neq j\neq k}}^{l}\left(1-\left(\gamma_{\max}^{-}\right)^{\rho+\sigma+\tau}\right)^{\frac{1}{l(l-1)(l-2)}}\\ &= 1-\left(\gamma_{\max}^{-}\right)^{\rho+\sigma+\tau}\end{aligned}$$

So,

$$\left(1-\prod_{\substack{i,j,k=1\\ i\neq j\neq k}}^{l}\left(1-\left(\gamma_{\xi_i}^{-}\right)^{\rho}\left(\gamma_{\xi_j}^{-}\right)^{\sigma}\left(\gamma_{\xi_k}^{-}\right)^{\tau}\right)^{\frac{1}{l(l-1)(l-2)}}\right)^{\frac{1}{\rho+\sigma+\tau}} \leq \gamma_{\max}^{-} \tag{28}$$

Therefore, from Inequalities (27) and (28) we have

$$\gamma_{\min}^{-} \leq \left(1-\prod_{\substack{i,j,k=1\\ i\neq j\neq k}}^{l}\left(1-\left(\gamma_{\xi_i}^{-}\right)^{\rho}\left(\gamma_{\xi_j}^{-}\right)^{\sigma}\left(\gamma_{\xi_k}^{-}\right)^{\tau}\right)^{\frac{1}{l(l-1)(l-2)}}\right)^{\frac{1}{\rho+\sigma+\tau}} \leq \gamma_{\max}^{-} \tag{29}$$

Similarly, we have

$$\gamma_{\min}^{+} \leq \left(1-\prod_{\substack{i,j,k=1\\ i\neq j\neq k}}^{l}\left(1-\left(\gamma_{\xi_i}^{+}\right)^{\rho}\left(\gamma_{\xi_j}^{+}\right)^{\sigma}\left(\gamma_{\xi_k}^{+}\right)^{\tau}\right)^{\frac{1}{l(l-1)(l-2)}}\right)^{\frac{1}{\rho+\sigma+\tau}} \leq \gamma_{\max}^{+} \tag{30}$$

Now,

$$\prod_{\substack{i,j,k=1\\ i\neq j\neq k}}^{l}\left(1-\left(1-\delta_{\xi_i}^{-}\right)^{\rho}\left(1-\delta_{\xi_j}^{-}\right)^{\sigma}\left(1-\delta_{\xi_k}^{-}\right)^{\tau}\right)^{\frac{1}{l(l-1)(l-2)}} \geq \prod_{\substack{i,j,k=1\\ i\neq j\neq k}}^{l}\left(1-\left(1-\delta_{\min}^{-}\right)^{\rho+\sigma+\tau}\right)^{\frac{1}{l(l-1)(l-2)}}$$

$$= 1-\left(1-\delta_{\min}^{-}\right)^{\rho+\sigma+\tau}$$

Thus,

$$\left(1-\prod_{\substack{i,j,k=1\\ i\neq j\neq k}}^{l}\left(1-\left(1-\delta_{\xi_i}^{-}\right)^{\rho}\left(1-\delta_{\xi_j}^{-}\right)^{\sigma}\left(1-\delta_{\xi_k}^{-}\right)^{\tau}\right)^{\frac{1}{l(l-1)(l-2)}}\right)^{\frac{1}{\rho+\sigma+\tau}} \leq 1-\delta_{\min}^{-}$$

So,

$$\delta_{\min}^{-} \leq 1-\left(1-\prod_{\substack{i,j,k=1\\ i\neq j\neq k}}^{l}\left(1-\left(1-\delta_{\xi_i}^{-}\right)^{\rho}\left(1-\delta_{\xi_j}^{-}\right)^{\sigma}\left(1-\delta_{\xi_k}^{-}\right)^{\tau}\right)^{\frac{1}{l(l-1)(l-2)}}\right)^{\frac{1}{\rho+\sigma+\tau}} \tag{31}$$

Again,

$$\prod_{\substack{i,j,k=1\\ i\neq j\neq k}}^{l}\left(1-\left(1-\delta_{\xi_i}^{-}\right)^{\rho}\left(1-\delta_{\xi_j}^{-}\right)^{\sigma}\left(1-\delta_{\xi_k}^{-}\right)^{\tau}\right)^{\frac{1}{l(l-1)(l-2)}} \leq \prod_{\substack{i,j,k=1\\ i\neq j\neq k}}^{l}\left(1-\left(1-\delta_{\max}^{-}\right)^{\rho+\sigma+\tau}\right)^{\frac{1}{l(l-1)(l-2)}}$$

$$= 1-\left(1-\delta_{\max}^{-}\right)^{\rho+\sigma+\tau}$$

Thus,

$$1-\delta_{\max}^{-}\leq\left(1-\prod_{\substack{i,j,k=1\\ i\neq j\neq k}}^{l}\left(1-\left(1-\delta_{\xi_i}^{-}\right)^{\rho}\left(1-\delta_{\xi_j}^{-}\right)^{\sigma}\left(1-\delta_{\xi_k}^{-}\right)^{\tau}\right)^{\frac{1}{m(m-1)(m-2)}}\right)^{\frac{1}{\rho+\sigma+\tau}}$$

So,

$$1-\left(1-\prod_{\substack{i,j,k=1\\ i\neq j\neq k}}^{l}\left(1-\left(1-\delta_{\xi_i}^{-}\right)^{\rho}\left(1-\delta_{\xi_j}^{-}\right)^{\sigma}\left(1-\delta_{\xi_k}^{-}\right)^{\tau}\right)^{\frac{1}{l(l-1)(l-2)}}\right)^{\frac{1}{\rho+\sigma+\tau}}\leq\delta_{\max}^{-} \tag{32}$$

Therefore from Inequality (31), we have

$$\delta_{\min}^{-}\leq 1-\left(1-\prod_{\substack{i,j,k=1\\ i\neq j\neq k}}^{l}\left(1-\left(1-\delta_{\xi_i}^{-}\right)^{\rho}\left(1-\delta_{\xi_j}^{-}\right)^{\sigma}\left(1-\delta_{\xi_k}^{-}\right)^{\tau}\right)^{\frac{1}{l(l-1)(l-2)}}\right)^{\frac{1}{\rho+\sigma+\tau}}\leq\delta_{\max}^{-} \tag{33}$$

Similarly, we have

$$\delta_{\min}^{+}\leq 1-\left(1-\prod_{\substack{i,j,k=1\\ i\neq j\neq k}}^{l}\left(1-\left(1-\delta_{\xi_i}^{+}\right)^{\rho}\left(1-\delta_{\xi_j}^{+}\right)^{\sigma}\left(1-\delta_{\xi_k}^{+}\right)^{\tau}\right)^{\frac{1}{l(l-1)(l-2)}}\right)^{\frac{1}{\rho+\sigma+\tau}}\leq\delta_{\max}^{+} \tag{34}$$

Let $\text{GIVIHFPBM}^{\rho,\sigma,\tau}(\tilde{\xi}_1,\tilde{\xi}_2,\ldots,\tilde{\xi}_l)=\tilde{\xi}$.

Using the score function of IVIHFEs mentioned in Definition 3, we have

$$S\left(\text{GIVIHFPBM}^{\rho,\sigma,\tau}(\tilde{\xi}_1,\tilde{\xi}_2,\ldots,\tilde{\xi}_l)\right)$$
$$=\frac{1}{2}\left[\sum_{\substack{\xi_i\in\tilde{\xi}_i\\ i=1,2,\ldots,l}}\left(\begin{array}{l}\left(1-\prod_{\substack{i,j,k=1\\ i\neq j\neq k}}^{l}\left(1-\left(\gamma_{\xi_i}^{-}\right)^{\rho}\left(\gamma_{\xi_j}^{-}\right)^{\sigma}\left(\gamma_{\xi_k}^{-}\right)^{\tau}\right)^{\frac{1}{l(l-1)(l-2)}}\right)^{\frac{1}{\rho+\sigma+\tau}}\\ -\left(1-\left(1-\prod_{\substack{i,j,k=1\\ i\neq j\neq k}}^{l}\left(1-\left(1-\delta_{\xi_i}^{-}\right)^{\rho}\left(1-\delta_{\xi_j}^{-}\right)^{\sigma}\left(1-\delta_{\xi_k}^{-}\right)^{\tau}\right)^{\frac{1}{l(l-1)(l-2)}}\right)^{\frac{1}{\rho+\sigma+\tau}}\right)\\ +\left(1-\prod_{\substack{i,j,k=1\\ i\neq j\neq k}}^{l}\left(1-\left(\gamma_{\xi_i}^{+}\right)^{\rho}\left(\gamma_{\xi_j}^{+}\right)^{\sigma}\left(\gamma_{\xi_k}^{+}\right)^{\tau}\right)^{\frac{1}{l(l-1)(l-2)}}\right)^{\frac{1}{\rho+\sigma+\tau}}\\ -\left(1-\left(1-\prod_{\substack{i,j,k=1\\ i\neq j\neq k}}^{l}\left(1-\left(1-\delta_{\xi_i}^{+}\right)^{\rho}\left(1-\delta_{\xi_j}^{+}\right)^{\sigma}\left(1-\delta_{\xi_k}^{+}\right)^{\tau}\right)^{\frac{1}{l(l-1)(l-2)}}\right)^{\frac{1}{\rho+\sigma+\tau}}\right)\end{array}\right)\right]\times(\#\tilde{\xi})^{-1} \tag{35}$$

Also,

$$S\left(\tilde{\xi}^{-}\right)=\frac{1}{2}\left(\gamma_{\min}^{-}-\delta_{\max}^{-}+\gamma_{\min}^{+}-\delta_{\max}^{+}\right) \tag{36}$$

$$S\left(\tilde{\xi}^{+}\right)=\frac{1}{2}\left(\gamma_{\max}^{-}-\delta_{\min}^{-}+\gamma_{\max}^{+}-\delta_{\min}^{+}\right) \tag{37}$$

Now using Inequalities (29),(30),(33) and (34), we have,

$$S\left(\tilde{\xi}^{-}\right)\leq S\left(\text{GIVIHFPBM}^{\rho,\sigma,\tau}(\tilde{\xi}_1,\tilde{\xi}_2,\ldots,\tilde{\xi}_l)\right)\leq S\left(\tilde{\xi}^{+}\right)$$

Therefore,

$$\tilde{\xi}^{-}\preceq \text{GIVIHFPBM}^{\rho,\sigma,\tau}(\tilde{\xi}_1,\tilde{\xi}_2,\ldots,\tilde{\xi}_l)\preceq\tilde{\xi}^{+}$$

Similarly,

$$\tilde{\xi}_0^{-}\preceq \text{GIVIHFPGBM}^{\rho,\sigma,\tau}(\tilde{\xi}_1,\tilde{\xi}_2,\ldots,\tilde{\xi}_l)\preceq\tilde{\xi}_0^{+}$$

4 An Approach to Multicriteria Decision Making with GIVIHFPBM and GIVIHFPGBM

This section proposes a decision-making model with the projected GIVIHFPBM and GIVIHFPGBM operators in the line of HFPBM [43].

Let $V = \{v_1, v_2, \ldots, v_l\}$ be a collection of preferences and $\mathcal{E} = \{\mathcal{E}_1, \mathcal{E}_2, \ldots, \mathcal{E}_p\}$ be a group of criteria. Consider that the assessment results of the preferences $v_i \in V$ based on the criteria $\mathcal{E}_j \in \mathcal{E}$ $(j = 1, 2, \ldots, p)$ is represented by an IVIHFE. Let $\tilde{M}^{(s)} = \left(\tilde{\xi}_{ij}^{(s)}\right)_{l\times p}$ be an IVIHF decision matrix [8], so that each $\tilde{\xi}_{ij}^{(s)}$ $(s = 1, 2, \ldots, q)$ is an IVIHFE, and is given by

$$\tilde{\xi}_{ij}^{(s)} = \left\{\left([L^-_{\xi_{ij}^{(s)}}, L^+_{\xi_{ij}^{(s)}}], [U^-_{\xi_{ij}^{(s)}}, U^+_{\xi_{ij}^{(s)}}]\right) | \xi_{ij}^{(s)} \in \tilde{\xi}_{ij}^{(s)}\right\}.$$

During assessment of the alternatives, DMs often need to tackle two types of criteria, namely, benefit type [8] and cost type [8] criteria. The benefit type criteria prefers larger assessment values, while cost type criteria prefers smaller assessment values of the alternatives. However, the cost type criteria can be converted to the benefit type criteria. Therefore the IVIHF decision matrix [8] $\tilde{M}^{(s)} = \left(\tilde{\xi}_{ij}^{(s)}\right)_{l\times p}$ is converted into the normalized IVIHF decision matrix [8] $\tilde{N} = \left(\tilde{\rho}_{ij}^{(s)}\right)_{l\times p}$, using the Xu and Hu's [50] method, for which

$$\tilde{\rho}_{ij}^{(s)} = \begin{cases} \tilde{\xi}_{ij}^{(s)}, & \text{for benefit criteria } T_j \\ \left(\tilde{\xi}_{ij}^{(s)}\right)^c, & \text{for cost criteria } T_j \end{cases}, \tag{38}$$

where, $\left(\tilde{\xi}_{ij}^{(s)}\right)^c$ $(i = 1, 2, \ldots, l; j = 1, 2, \ldots, p; s = 1, 2, \ldots, q)$ is the complement of $\tilde{\xi}_{ij}^{(s)}$ such that,

$$\left(\tilde{\xi}_{ij}^{(s)}\right)^c = \left\{\left([U^-_{\xi_{ij}^{(s)}}, U^+_{\xi_{ij}^{(s)}}], [L^-_{\xi_{ij}^{(s)}}, L^+_{\xi_{ij}^{(s)}}]\right) | \xi_{ij}^{(s)} \in \tilde{\xi}_{ij}^{(s)}\right\} \tag{39}$$

In the following, we develop an MCDM method by utilizing the proposed GIVIHFPBM (or GIVIHFPGBM) operator with IVIHF [8] information. The algorithm comprises of the subsequent steps:

Step 1. Derive the normalized IVIHF decision matrix [8] $\tilde{N} = \left(\tilde{\rho}_{ij}^{(s)}\right)_{l\times p}$, using Equation set (38), where, $\tilde{\rho}_{ij}^{(s)} = \left\{\left([L^-_{\xi_{ij}^{(s)}}, L^+_{\xi_{ij}^{(s)}}], [U^-_{\xi_{ij}^{(s)}}, U^+_{\xi_{ij}^{(s)}}]\right) | \xi_{ij}^{(s)} \in \tilde{\xi}_{ij}^{(s)}\right\}$ $(i = 1, 2, \ldots, l)$ $(j = 1, 2, \ldots, p; s = 1, 2, \ldots, q)$.

Step 2. Calculate $Supp\left(\tilde{\rho}_{ij}^{(s_1)}, \tilde{\rho}_{ij}^{(s_2)}\right)$ using,

$$Supp\left(\tilde{\rho}_{ij}^{(s_1)}, \tilde{\rho}_{ij}^{(s_2)}\right) = 1 - d\left(\tilde{\rho}_{ij}^{(s_1)}, \tilde{\rho}_{ij}^{(s_2)}\right) \tag{40}$$

and $d\left(\tilde{\rho}_{ij}^{(s_1)}, \tilde{\rho}_{ij}^{(s_2)}\right)$ denotes the normalized Hamming distance [9] between two IVIHFEs.

Step 3. Evaluate the companion weights of the IVIHFE $\tilde{\rho}_{ij}^{(s_d)}$ ($s_d = 1, 2, \ldots, q; d = 1, 2, 3$) using

$$\lambda_{ij}^{(s_d)} = \frac{l\left(1 + T\left(\tilde{\rho}_{ij}^{(s_d)}\right)\right)}{\sum_{s_{d^*}=1}^{l}\left(1 + T\left(\tilde{\rho}_{ij}^{(s_{d^*})}\right)\right)}, \tag{41}$$

($s_d, s_{d^*} = 1, 2, \ldots, q; d, d^* = 1, 2, 3; d \neq d^*$)
where,

$$T\left(\tilde{\rho}_{ij}^{(s_d)}\right) = \sum_{d \neq d' \, s_{d'}=1}^{l} Supp\left(\tilde{\rho}_{ij}^{(s_d)}, \tilde{\rho}_{ij}^{(s_{d'})}\right) \tag{42}$$

and $d, d' = 1, 2, 3; d \neq d'$, and $s_d, s_{d'} = 1, 2, \ldots, q$. Use Eqs. (41) and (42) to calculate the values of $\lambda_{ij}^{s_1}, \lambda_{ij}^{s_2}, \lambda_{ij}^{s_3}; T\left(\tilde{\rho}_{ij}^{s_1}\right), T\left(\tilde{\rho}_{ij}^{s_2}\right)$, and $T\left(\tilde{\rho}_{ij}^{s_3}\right)$ ($s_1, s_2, s_3 = 1, 2, \ldots, l; s_1 \neq s_2 \neq s_3$)

Step 4. Utilize GIVIHFPBM or GIVIHFPGBM operators expressed in Eqs. (10) and (17), in order to aggregate each of the IVIHF decision matrix [8] $\tilde{N}^{(s)} = \left(\tilde{\rho}_{ij}^{(s)}\right)_{l \times p}$ ($s = 1, 2, \ldots, q$) into the aggregated IVIHF decision matrix [8] $\tilde{N} = \left(\tilde{\rho}_{ij}\right)_{l \times p}$.

Step 5. Determine the supports of $\tilde{\rho}_{ij}$ from $\tilde{\rho}_{it}$, ($i = 1, 2, \ldots, l; j, t = 1, 2, \ldots, p$), using

$$Supp\left(\tilde{\rho}_{ij}, \tilde{\rho}_{it}\right) = 1 - d\left(\tilde{\rho}_{ij}, \tilde{\rho}_{it}\right), \tag{43}$$

where, $d\left(\tilde{\rho}_{ij}, \tilde{\rho}_{it}\right)$ ($i = 1, 2, \ldots, l; j, t = 1, 2, \ldots, p; j \neq t$) denotes the Hamming distance [9] between two IVIHFEs.

Step 6. Evaluate the companion weights of the IVIHFEs $\tilde{\rho}_{ij}$ ($i = 1, 2, \ldots, l; j = 1, 2, \ldots, p$), using

$$\lambda_{ij} = \frac{q\left(1 + T\left(\tilde{\rho}_{ij}\right)\right)}{\sum_{s=1}^{q}\left(1 + T\left(\tilde{\rho}_{is}\right)\right)}, \quad (j, s = 1, 2, \ldots, q; j \neq s) \tag{44}$$

$$T\left(\tilde{\rho}_{ij}\right) = \sum_{t \neq j_{t=1}}^{q} Supp\left(\tilde{\rho}_{ij}, \tilde{\rho}_{it}\right) \tag{45}$$

Step 7. Utilize GIVIHFPBM or GIVIHFPGBM to accumulate the IVIHFEs $\tilde{\rho}_{ij}$ $(i = 1, 2, \ldots, l; j = 1, 2, \ldots, p)$ to the collective IVIHFEs $\tilde{\rho}_i$ $(i = 1, 2, \ldots, l)$, similar to Step 4

Step 8. Obtain the score [8] and accuracy [8] values of the aggregated IVIHFEs $\tilde{\rho}_i$ $(i = 1, 2, \ldots, l)$, by using Definition 3.

Step 9. Rank the aggregated IVIHFEs $\tilde{\rho}_i$ $(i = 1, 2, \ldots, l)$ by decreasing order based on the score and the accuracy values [8].

Step 10. Obtain the best one(s) among the alternatives.

5 Illustrative Example

Assume that a manufacturing company intends to select a suitable production partner for an international joint venture. Suppose that there are four available alternatives $\mathcal{B}_i$ $(i = 1, 2, 3, 4)$. A committee of three experts d_s $(s = 1, 2, 3)$ acts as decision makers (DMs) to provide their preference on the alternatives $\mathcal{B}_i$ $(i = 1, 2, 3, 4)$, based on the following criteria, (i) $\mathcal{E}_1$: Investment Skills (ii) $\mathcal{E}_2$: Technical Skills (iii) $\mathcal{E}_3$: Delegation Skills Assume that the four alternatives $\mathcal{B}_i$ $(i = 1, 2, 3, 4)$, which are to be evaluated by three decision makers d_s $(s = 1, 2, 3)$ under the criteria $\mathcal{E}_j$ $(j = 1, 2, 3)$ are represented by the three interval-valued intuitionistic hesitant fuzzy decision matrix [8] $\tilde{M}^{(s)} = \left(\tilde{\xi}_{ij}^{(s)}\right)_{4\times 3}$ $(s = 1, 2, 3)$ as shown in Tables 1, 2 and 3.

Step 1. Clearly, $\mathcal{E}_2$ and $\mathcal{E}_3$ are the benefit type criteria [51], $\mathcal{E}_1$ the cost type criteria [51]. We use Equation set (38) to find the normalized interval-valued intuitionistic hesitant fuzzy decision matrices $\tilde{N}^{(s)}$, $(s = 1, 2, 3)$ as shown from Tables 4, 5 and 6.

Step 2. Using Eq. (40) the supports $Supp\left(\tilde{\rho}_{ij}^{(s_1)}, \tilde{\rho}_{ij}^{(s_2)}\right)$, are calculate as:

Table 1 The IVIHF decision matrix $\tilde{M}^{(1)}$ due to d_1

	$\mathcal{E}_1$	$\mathcal{E}_2$	$\mathcal{E}_3$
$\mathcal{B}_1$	{([0.4, 0.5], [0.3, 0.4])}	{([0.1, 0.2], [0.2, 0.3])}	{([0.4, 0.5], [0.3, 0.4]), ([0.2, 0.3], [0.5, 0.6])}
$\mathcal{B}_2$	{([0.1, 0.2], [0.2, 0.3]), ([0.2, 0.3], [0.3, 0.4])}	{([0.5, 0.6], [0.3, 0.4])}	{([0.5, 0.6], [0.2, 0.3]), ([0.6, 0.7], [0.4, 0.5])}
$\mathcal{B}_3$	{([0.6, 0.7], [0.5, 0.6]), ([0.5, 0.6], [0.4, 0.5])}	{([0.6, 0.7], [0.2, 0.3]), ([0.7, 0.8], [0.1, 0.2])}	{([0.6, 0.7], [0.5, 0.6])}
$\mathcal{B}_4$	{([0.2, 0.4], [0.3, 0.6])}	{([0.4, 0.6], [0.3, 0.5])}	{([0.5, 0.7], [0.5, 0.6]), ([0.4, 0.5], [0.2, 0.3])}

Table 2 The IVIHF decision matrix $\tilde{M}^{(2)}$ due to d_2

	$\mathcal{E}_1$	$\mathcal{E}_2$	$\mathcal{E}_3$
$\mathcal{B}_1$	{([0.1, 0.2], [0.2, 0.3])}	{([0.3, 0.4], [0.4, 0.5]), ([0.2, 0.3], [0.5, 0.6])	{([0.2, 0.3], [0.3, 0.4])}
$\mathcal{B}_2$	{([0.5, 0.6], [0.1, 0.2]), ([0.6, 0.7], [0.2, 0.3])}	{([0.4, 0.5], [0.1, 0.2]), ([0.5, 0.6], [0.4, 0.5])}	{([0.3, 0.5], [0.1, 0.2])}
$\mathcal{B}_3$	{([0.4, 0.7], [0.3, 0.4]), ([0.3, 0.6], [0.1, 0.3])}	{([0.6, 0.7], [0.4, 0.5])}	{([0.3, 0.6], [0.3, 0.4]), ([0.6, 0.7], [0.1, 0.2])}
$\mathcal{B}_4$	{([0.6, 0.7], [0.7, 0.8])}	{([0.6, 0.7], [0.5, 0.8])}	{([0.3, 0.4], [0.2, 0.5])}

Table 3 The IVIHF decision matrix $\tilde{M}^{(3)}$ due to d_3

	$\mathcal{E}_1$	$\mathcal{E}_2$	$\mathcal{E}_3$
$\mathcal{B}_1$	{([0.7, 0.8], [0.4, 0.5]), ([0.5, 0.7], [0.3, 0.4])}	{([0.4, 0.6], [0.2, 0.5])}	{([0.6, 0.7], [0.7, 0.8]), ([0.6, 0.8], [0.3, 0.4])}
$\mathcal{B}_2$	{([0.5, 0.8], [0.3, 0.7])}	{([0.4, 0.6], [0.5, 0.6]), ([0.6, 0.7], [0.7, 0.8])}	{([0.3, 0.4], [0.1, 0.2]), ([0.5, 0.6], [0.7, 0.8])}
$\mathcal{B}_3$	{([0.7, 0.8], [0.1, 0.3]), ([0.5, 0.8], [0.1, 0.2])}	{([0.3, 0.6], [0.3, 0.4])}	{([0.6, 0.9], [0.8, 0.9]), ([0.7, 0.8], [0.2, 0.4])}
$\mathcal{B}_4$	{([0.5, 0.7], [0.7, 0.8]), ([0.3, 0.5], [0.4, 0.7])}	{([0.7, 0.9], [0.4, 0.5])}	{([0.1, 0.2], [0.1, 0.7])}

Table 4 Normalized IVIHF decision matrix $\tilde{N}^{(1)}$

	$\mathcal{E}_1$	$\mathcal{E}_2$	$\mathcal{E}_3$
$\mathcal{B}_1$	{([0.3, 0.4], [0.4, 0.5])}	{([0.1, 0.2], [0.2, 0.3])}	{([0.4, 0.5], [0.3, 0.4]), ([0.2, 0.3], [0.5, 0.6])}
$\mathcal{B}_2$	{([0.2, 0.3], [0.1, 0.2]), ([0.3, 0.4], [0.2, 0.3])}	{([0.5, 0.6], [0.3, 0.4])}	{([0.5, 0.6], [0.2, 0.3]), ([0.6, 0.7], [0.4, 0.5])}
$\mathcal{B}_3$	{([0.5, 0.6], [0.6, 0.7]), ([0.4, 0.5], [0.5, 0.6])}	{([0.6, 0.7], [0.2, 0.3]), ([0.7, 0.8], [0.1, 0.2])}	{([0.6, 0.7], [0.5, 0.6])}
$\mathcal{B}_4$	{([0.3, 0.6], [0.2, 0.4])}	{([0.4, 0.6], [0.3, 0.5])}	{([0.5, 0.7], [0.5, 0.6]), ([0.4, 0.5], [0.2, 0.3])}

Table 5 Normalized IVIHF decision matrix $\tilde{N}^{(2)}$

	$\mathcal{E}_1$	$\mathcal{E}_2$	$\mathcal{E}_3$
$\mathcal{B}_1$	{([0.2, 0.3], [0.1, 0.2])}	{([0.3, 0.4], [0.4, 0.5]), ([0.2, 0.3], [0.5, 0.6])}	{([0.2, 0.3], [0.3, 0.4])}
$\mathcal{B}_2$	{([0.1, 0.2], [0.5, 0.6]), ([0.2, 0.3], [0.6, 0.7])}	{([0.4, 0.5], [0.1, 0.2]), ([0.5, 0.6], [0.4, 0.5])}	{([0.3, 0.5], [0.1, 0.2])}
$\mathcal{B}_3$	{([0.3, 0.4], [0.4, 0.7]), ([0.1, 0.3], [0.3, 0.6])}	{([0.6, 0.7], [0.4, 0.5])}	{([0.3, 0.6], [0.3, 0.4]), ([0.6, 0.7], [0.1, 0.2])}
$\mathcal{B}_4$	{([0.7, 0.8], [0.6, 0.7])}	{([0.6, 0.7], [0.5, 0.8])}	{([0.3, 0.4], [0.2, 0.5])}

Table 6 Normalized IVIHF decision matrix $\tilde{N}^{(3)}$

	$\mathcal{E}_1$	$\mathcal{E}_2$	$\mathcal{E}_3$
$\mathcal{B}_1$	{([0.4, 0.5], [0.7, 0.8]), ([0.3, 0.4], [0.5, 0.7])}	{([0.4, 0.6], [0.2, 0.5])}	{([0.6, 0.7], [0.7, 0.8]), ([0.6, 0.8], [0.3, 0.4])}
$\mathcal{B}_2$	{([0.3, 0.7], [0.5, 0.8])}	{([0.4, 0.6], [0.5, 0.6]), ([0.6, 0.7], [0.7, 0.8])}	{([0.3, 0.4], [0.1, 0.2]), ([0.5, 0.6], [0.7, 0.8])}
$\mathcal{B}_3$	{([0.1, 0.3], [0.7, 0.8]), ([0.1, 0.2], [0.5, 0.8])}	{([0.3, 0.6], [0.3, 0.4])}	{([0.6, 0.9], [0.8, 0.9]), ([0.7, 0.8], [0.2, 0.4])}
$\mathcal{B}_4$	{([0.7, 0.8], [0.5, 0.7]), ([0.4, 0.7], [0.3, 0.5])}	{([0.7, 0.9], [0.4, 0.5])}	{([0.1, 0.2], [0.1, 0.7])}

$$Supp\left(\tilde{\rho}_{ij}^{(1)}, \tilde{\rho}_{ij}^{(2)}\right) = \begin{bmatrix} 0.9000 & 0.9000 & 0.9500 \\ 0.7500 & 0.9500 & 0.9000 \\ 0.8250 & 0.9250 & 0.9000 \\ 0.8375 & 0.9000 & 0.9187 \end{bmatrix}, \; Supp\left(\tilde{\rho}_{ij}^{(1)}, \tilde{\rho}_{ij}^{(3)}\right) = \begin{bmatrix} 0.9312 & 0.8875 & 0.7375 \\ 0.8375 & 0.9062 & 0.7750 \\ 0.7750 & 0.9000 & 0.9062 \\ 0.9000 & 0.9125 & 0.8437 \end{bmatrix},$$

$$Supp\left(\tilde{\rho}_{ij}^{(2)}, \tilde{\rho}_{ij}^{(3)}\right) = \begin{bmatrix} 0.8312 & 0.9125 & 0.8437 \\ 0.9000 & 0.7750 & 0.9000 \\ 0.8500 & 0.9250 & 0.7250 \\ 0.9375 & 0.9125 & 0.9000 \end{bmatrix}.$$

Also, $Supp\left(\tilde{\rho}_{ij}^{(s_1)}, \tilde{\rho}_{ij}^{(s_2)}\right) = Supp\left(\tilde{\rho}_{ij}^{(s_2)}, \tilde{\rho}_{ij}^{(s_1)}\right)$, $(s_1, s_2 = 1, 2, 3;\, s_1 \neq s_2)$.

Step 3. The weights associated with the IVIHFE $\tilde{\rho}_{ij}^{(s_d)}$ $(s_d = 1, 2, 3;\, d = 1, 2, 3)$, are calculated using the Eqs. (41) and (42) as follows:

$$\lambda_{ij}^{(1)} = \begin{bmatrix} 1.0203 & 0.9955 & 1.0000 \\ 0.9734 & 1.0371 & 0.9847 \\ 0.9873 & 0.9971 & 1.0442 \\ 0.9835 & 0.9985 & 0.9955 \end{bmatrix}, \; \lambda_{ij}^{(2)} = \begin{bmatrix} 0.9842 & 1.0045 & 1.0395 \\ 0.9969 & 0.9894 & 1.0307 \\ 1.0158 & 1.0059 & 0.9768 \\ 0.9970 & 0.9985 & 1.0158 \end{bmatrix},$$

$$\lambda_{ij}^{(3)} = \begin{bmatrix} 0.9955 & 0.1000 & 0.9605 \\ 1.0298 & 0.9735 & 0.9847 \\ 0.9968 & 0.9971 & 0.9791 \\ 1.0195 & 1.0030 & 0.9887 \end{bmatrix}.$$

Step 4. For sake of simplicity, we assume that $\rho = \sigma = \tau = 1$ in Eq. (10), then the collective interval-valued intuitionistic hesitant fuzzy decision matrix $\tilde{N} = \left(\tilde{\rho}_{ij}\right)_{4\times 3}$ obtained by aggregating all the decision matrices $\tilde{N}^{(s)} = \left(\tilde{\rho}_{ij}^{(s)}\right)_{4\times 3}$, $(s = 1, 2, 3)$ is shown in Table 7.

Step 5. Using Eqs. (43)–(45), the companion weights of the IVIHFEs $\tilde{\rho}_{ij}$ $(i = 1, 2, 3, 4;\, j = 1, 2, 3)$ are determined as:

Table 7 The overall IVIHF decision matrix $\tilde{N} = (\tilde{\rho}_{ij})_{4\times 3}$

	$\mathcal{E}_1$	$\mathcal{E}_2$	$\mathcal{E}_3$
$\mathcal{B}_1$	{([0.1592, 0.2172], [0.6922, 0.7599]), ([0.1445, 0.2013], [0.6282, 0.7234])}	{([0.0629, 0.1054], [0.7693, 0.8420]), ([0.0550, 0.0957], [0.7833, 0.8534])}	{([0.2020, 0.2646], [0.7047, 0.7697]), ([0.2020, 0.2778], [0.5928, 0.6589]), ([0.1598, 0.2218], [0.7377, 0.7995]), ([0.1598, 0.2326], [0.6425, 0.7055])}
$\mathcal{B}_2$	{([0.1000, 0.1918], [0.6521, 0.7769]), ([0.1146, 0.2116], [0.6674, 0.7871]), ([0.1261, 0.2202], [0.6795, 0.7978]), ([0.1446, 0.2432], [0.6933, 0.8069])}	{([0.2396, 0.3190], [0.6066, 0.6735]), ([0.2589, 0.3411], [0.6631, 0.7243]), ([0.2763, 0.3378], [0.6758, 0.7451]), ([0.2991, 0.3617], [0.7110, 0.7834])}	{([0.1971, 0.2762], [0.4582, 0.5453]), ([0.2349, 0.3195], [0.6602, 0.7313]), ([0.2098, 0.2918], [0.5295, 0.6045]), ([0.2503, 0.3382], [0.6943, 0.7612])}
$\mathcal{B}_3$	{([0.1361, 0.2314], [0.7689, 0.8554]), ([0.1361, 0.2015], [0.7240, 0.8554]), ([0.0943, 0.2098], [0.7564, 0.8408]), ([0.0943, 0.1828], [0.7087, 0.8408]), ([0.1262, 0.2173], [0.7503, 0.8407]), ([0.1262, 0.1893], [0.7012, 0.8407]), ([0.0875, 0.1971], [0.7367, 0.8245]), ([0.0875, 0.1718], [0.6844, 0.8245])}	{([0.2662, 0.3834], [0.5958, 0.6622]), ([0.2810, 0.4041], [0.5760, 0.6449])}	{([0.2656, 0.4241], [0.7710, 0.8408]), ([0.2806, 0.4039], [0.6224, 0.7048]), ([0.3411, 0.4529], [0.7498, 0.8243]), ([0.3616, 0.4304], [0.5821, 0.6719])}
$\mathcal{B}_4$	{([0.2962, 0.4258], [0.6936, 0.7902]), ([0.2439, 0.4038], [0.6540, 0.7501])}	{([0.3113, 0.4236], [0.6623, 0.7957])}	{([0.1360, 0.2122], [0.5846, 0.7829]), ([0.1262, 0.1892], [0.4889, 0.7365])}

$$(\lambda_{ij})_{4\times 3} = \begin{bmatrix} 0.9808 & 1.0448 & 0.9744 \\ 0.9744 & 1.0182 & 1.0075 \\ 0.9589 & 1.0441 & 0.9971 \\ 1.0000 & 1.0205 & 0.9796 \end{bmatrix} \tag{46}$$

Step 6. Using Eq. (17), the IVIHFEs $\tilde{\rho}_{ij}$ are aggregated into the collective IVIHFEs $\tilde{\rho}_i$ $(i = 1, 2, 3, 4)$. Large components of each $\tilde{\rho}_i$ $(i = 1, 2, 3, 4)$'s are not displayed here.

Step 7. Using Definition 3, the score values of the collective IVIHFEs $\tilde{\rho}_i$ are determined as follows, $S(\rho_1) = 0.3897,\ S(\rho_2) = 0.3491,\ S(\rho_3) = 0.3493,\ S(\rho_4) = 0.3419$. Clearly, $S(\rho_1) > S(\rho_3) > S(\rho_2) > S(\rho_4)$. Although $S(\rho_2) \approx S(\rho_3) \approxeq 0.35$, using Definition 3, the accuracy values of ρ_2 and ρ_3 are as follows $A(\rho_2) = 0.4801,\ A(\rho_3) = 0.4914$. So, $A(\rho_3) > A(\rho_2)$. Therefore the ranking order of the collective IVIHFEs ρ_i $(i = 1, 2, 3, 4)$ is $\rho_1 \succ \rho_3 \succ \rho_2 \succ \rho_4$

Step 8. Since $\rho_1 \succ \rho_3 \succ \rho_2 \succ \rho_4$, so the alternatives $\mathcal{B}_i$ $(i = 1, 2, 3, 4)$ can be re-ordered as $\mathcal{B}_1 \succ \mathcal{B}_3 \succ \mathcal{B}_2 \succ \mathcal{B}_4$. Hence, $\mathcal{B}_1$ is the optimum choice.

5.1 A Comparative Study

This section provides a comparative analysis of the projected MCDM method with Zhang [8] method. To establish the authenticity of the projected model, we investigate the same example used by Zhang [8]. Using Eq. (10), with $\rho = \sigma = \tau = 1$ the individual normalized IVIHF decision matrices are aggregated into the collective IVIHF decision matrix $\tilde{N}$, as shown in Table 8.
Using Eqs.(43)–(45), the companion weights of the IVIHFEs $\tilde{\rho}_{ij}$ are calculated as:

$$(\lambda_{ij})_{3\times 3} = \begin{bmatrix} 1.0573 & 1.0547 & 0.8880 \\ 0.9770 & 1.0308 & 0.9922 \\ 0.9773 & 0.9705 & 1.0523 \end{bmatrix}$$

The aggregated IVIHFEs $\tilde{\rho}_i$ $(i = 1, 2, 3)$ are obtained on using Eq. 10 has the following score values [8], $S(\tilde{\rho}_1) = -0.7099,\ S(\tilde{\rho}_2) = -0.6812,\ S(\tilde{\rho}_3) = -0.6367$. Clearly, $S(\tilde{\rho}_3) > S(\tilde{\rho}_2) > S(\tilde{\rho}_1)$. So, $\tilde{\rho}_3 \succ \tilde{\rho}_2 \succ \tilde{\rho}_1$. Therefore, the alternatives can be re-ordered as: $\mathcal{B}_3 \succ \mathcal{B}_2 \succ \mathcal{B}_1$. Hence, $\mathcal{B}_3$ is the optimum choice. Again, the best alternative obtained by Zhang [8] is $\mathcal{A}_3$. Thus our proposed method has the same ranking results as the method by Zhang [8], which verifies the validity of our proposed method. Moreover, our proposed method handles the interrelationship among the criteria and incorporates the significance of the alternatives, which were ignored in the method proposed by Zhang [8].

Further, we have provided an extended comparative study of the proposed MCDM method in this study with several other existing methods by solving the above example

Table 8 The collective IVIHF decision matrix $\tilde{N}$

	$\mathcal{E}_1$	$\mathcal{E}_2$	$\mathcal{E}_3$
$\mathcal{B}_1$	{([0.1259, 0.1610], [0.7110, 0.8390]), ([0.1259, 0.2036], [0.7110, 0.7964]), ([0.1259, 0.1610], [0.6895, 0.7964])}	{([0.1391, 0.2102], [0.7071, 0.7548]), ([0.1103, 0.1449], [0.7872, 0.8471]), ([0.1502, 0.2220], [0.6920, 0.7548]), ([0.1190, 0.1529], [0.7767, 0.8471])}	{([0.4241, 0.5110], [0.4573, 0.4890]), ([0.4241, 0.4798], [0.5202, 0.5202]), ([0.4524, 0.5427], [0.4573, 0.4573])}
$\mathcal{B}_2$	{([0.1602, 0.2329], [0.6909, 0.7671]), ([0.1902, 0.2515], [0.6909, 0.7485]), ([0.1363, 0.2018], [0.7042, 0.7982]), ([0.1617, 0.2177], [0.7042, 0.7823])}	{([0.1752, 0.2877], [0.6808, 0.6808]), ([0.2007, 0.3239], [0.6043, 0.6212]), ([0.1888, 0.3239], [0.6043, 0.6212])}	{([0.2736, 0.3896], [0.5894, 0.6104]), ([0.1871, 0.2771], [0.6634, 0.6793]), ([0.2736, 0.3571], [0.5894, 0.6429]), ([0.2917, 0.4106], [0.5694, 0.5894]), ([0.1990, 0.2904], [0.6486, 0.6634]), ([0.2917, 0.3755], [0.5694, 0.6245])}
$\mathcal{B}_3$	{([0.1699, 0.1802], [0.1699, 0.1802]), ([0.1699, 0.2286], [0.6089, 0.7071]), ([0.1699, 0.1900], [0.7106, 0.8005]), ([0.1699, 0.2412], [0.5900, 0.7071])}	{([0.2922, 0.4443], [0.5557, 0.5557]), ([0.3244, 0.4114], [0.5557, 0.5886]), ([0.2922, 0.4114], [0.5557, 0.5886]), ([0.3087, 0.4675], [0.5325, 0.5325]), ([0.3433, 0.4315], [0.5325, 0.5685]), ([0.3087, 0.4315], [0.5325, 0.5685])}	{([0.1669, 0.2895], [0.6572, 0.6913]), ([0.2111, 0.2512], [0.6913, 0.7327]), ([0.2426, 0.3210], [0.6572, 0.6572])}

Table 9 A Comparative Study of different MCDM methods

MCDM method	Aggregation operator	Ranking order of alternatives
Zhang [8]	IVIHFWA [8]	$x_3 \succ x_2 \succ x_1$
Joshi and Kumar [9]	IVIHFCI [9]	$x_3 \succ x_1 \succ x_2$
Joshi and Kumar [10]	IVIHF-Entropy [10]	$x_3 \succ x_2 \succ x_1$
Narayanamoorthy et al. [11]	IVIHF-TOPSIS [11]	$x_3 \succ x_2 \succ x_1$
Narayanamoorthy et al. [11]	IVIHF-VIKOR [11]	$x_3 \succ x_1 \succ x_2$
Proposed method	GIVIHFPBM	$x_3 \succ x_2 \succ x_1$
Proposed method	GIVIHFPGBM	$x_3 \succ x_1 \succ x_2$

applying the existing methods [8–11], thus obtaining the following ranking results as shown in Table 9.

Thus the ranking results due to Zhang [8], Joshi and Kumar [9, 10], and Narayanamoorthy et al. [11] have the same best alternative x_3 as of the proposed method. However, the existing methods [8, 10, 11] beside the method by Joshi and Kumar [9] don't take into consideration interrelationship among the criteria whereas handling of the interrelationship among the criteria is a prominent characteristic of the proposed method in this study. However, the method by Joshi and Kumar [9] involves the computational difficulty of solving fuzzy measures for a large set of attributes. The proposed method based on GIVIHFPBM and GIVIHFPGBM operators can capture the interrelationship among the arguments, moreover can assist and strengthen aggregated to provide more accurate results as compared to the existing methods [8–11].

6 Conclusions

This study introduces two new PBM operators under the interval-valued intuitionistic hesitant fuzzy environment, namely the GIVIHFPBM and the GIVIHFPGBM operators. Further, some of their special cases and properties are analyzed. An MCDM model using the proposed operators is formulated. To demonstrate the proposed MCDM model a suitable example is presented. Moreover, a comparative investigation of the proposed MCDM approach is provided.

6.1 Advantages of the Proposed MCDM Method

Some of the main advantages of the proposed method in this study can be briefly summarized as follows:

- The purpose of the proposed MCDM method is to obtain the best choice from the ranking order of the alternatives when the evaluation information is expressed in terms of the interval-valued intuitionistic hesitant fuzzy sets (IVIHFS). The IVIHFS is an effective extension of the FS, IFS, HFS, and IVHFS, etc., and is more appropriate to express the uncertainty and imprecision of real-life decision-making problems. Since the proposed method is based on the interval-valued intuitionistic hesitant fuzzy information, hence the proposed method can become a useful MCDM approach under the interval-valued intuitionistic hesitant fuzzy environment.
- The proposed MCDM method in this study is based on some novel power Bonferroni means, namely the GIVIHFPBM and the GIVIHFPGBM operators. The GIVIHFPBM and the GIVIHFPGBM operators are effective tools to reflect interrelationship among the criteria. Moreover, the proposed operators GIVIHFPBM and the GIVIHFPGBM operators have less computational complexity as compared to the interval-valued intuitionistic hesitant fuzzy Choquet integral (IVIHFCI) [9] operator. Hence the GIVIHFPBM and the GIVIHFPGBM operators are more suitable than the IVIHFCI operator.
- The proposed operators GIVIHFPBM and the GIVIHFPGBM operators are obtained by the fusion of the power average (PA) [25, 34] operator and the Bonferroni mean (BM) [33]. The main characteristic of the GIVIHFPBM and the GIVIHFPGBM operators is that they not only deal with the interaction among the criteria but also can assist and strengthen the aggregated argument during the process of obtaining the best alternative. Moreover, during the decision-making process, due to the heterogeneous nature of the aggregated arguments which are being fused, may often have some abnormal impact on the final ranking result. The GIVIHFPBM and the GIVIHFPGBM operators proposed in this study can nullify such abnormal ranking behaviour of alternatives. These unique characteristics of can make the proposed GIVIHFPBM and the GIVIHFPGBM operators furthermore appropriate than some of the existing operators such as the IVIHFWA [8], IVIHFCI [9], and IVIHF-Entropy [10], etc.

6.2 Future Scopes

Real-life decision-making problems are rather complex due to the heterogeneous behaviour of arguments being fused and interaction among the criteria. So it becomes a very challenging task to obtain the genuine outcome of a real-life MCDM problem. The proposed MCDM method can be very helpful for decision-makers (DMs) in obtaining a genuine and best outcome. Thus the proposed MCDM method can play an important role in solving real-life decision-making problems. Besides MCDM problems, it can be further extended in the following application fields:

- The proposed method may be extended to study Big data analytic.

- The proposed method may be implemented to solve Engineering equipment selection problems.
- The proposed method may be utilized to solve Medical diagnosis problems.
- The proposed method may be extended solve to signal processing and image processing problems.

References

1. Lotfi, A.Z.: Fuzzy sets. Inf. Control **8**(3), 338–353 (1965)
2. Atanassov, K.T.: Intuitionistic fuzzy sets. Fuzzy Sets Syst. **20**(1), 87–96 (1986)
3. Atanassov, K.T.: More on intuitionistic fuzzy sets. Fuzzy Sets Syst. **33**(1), 37–45 (1989)
4. Atanassov, K., Gargov, G.: Interval valued intuitionistic fuzzy sets. Fuzzy Sets Syst. **31**(3), 343–349 (1989)
5. Torra, V., Narukawa, Y.: On hesitant fuzzy sets and decision. In: IEEE International Conference on Fuzzy Systems, FUZZ-IEEE 2009, pp 1378–1382 (2009)
6. Torra, V.: Hesitant fuzzy sets. Int. J. Intelli. Syst. **25**(6), 529–539 (2010)
7. Chen, N., Xu, Z., Xia, M.: Interval-valued hesitant preference relations and their applications to group decision making. Knowl.-Based Syst. **37**, 528–540 (2013)
8. Zhang, Z.: Interval-valued intuitionistic hesitant fuzzy aggregation operators and their application in group decision-making. J. Appl. Math. 1-33 (2013)
9. Joshi, D., Kumar, S.: Interval-valued intuitionistic hesitant fuzzy Choquet integral based TOPSIS method for multi-criteria group decision making. Eur. J. Oper. Res. **248**(1), 183–191 (2016)
10. Joshi, D.K., Kumar, S.: Entropy of interval-valued intuitionistic hesitant fuzzy set and its application to group decision making problems. Granular Comput. **3**(4), 367–381 (2018)
11. Narayanamoorthy, S., Geetha, S., Rakkiyappan, R., Joo, Y.H.: Interval-valued intuitionistic hesitant fuzzy entropy based VIKOR method for industrial robots selection. Expert Syst. Appl. **121**, 28–37 (2019)
12. Bharati, S.K.: An interval-valued intuitionistic hesitant fuzzy methodology and application. New Generation Comput. **39**(2), 377–407 (2021)
13. Beliakov, G., Pradera, A., Calvo, T.: Aggregation Functions: A Guide for Practitioners, vol. 221. Springer, Heidelberg (2007)
14. Choquet, G.: Theory of capacities. In Annales de l'institut Fourier **5**, 131–295 (1954)
15. Even, Y., Lehrer, E.: Decomposition-integral: unifying Choquet and the concave integrals. Econ. Theor. **56**(1), 33–58 (2014)
16. Grabisch, M., Marichal, J.L., Mesiar, R., Pap, E.: Aggregation Functions (127). Cambridge University Press. ISBN: 9781139644150 (2009). https://doi.org/10.1017/CBO9781139644150
17. Mesiar, R., Borkotokey, S., Jin, L., Kalina, M.: Aggregation under uncertainty. IEEE Trans. Fuzzy Syst. **26**(4), 2475–2478 (2017)
18. Mesiar, R., Borkotokey, S., Jin, L., Kalina, M.: Aggregation functions and capacities. Fuzzy Sets Syst. **346**, 138–146 (2018)
19. Augustine, E.P.: Novel correlation coefficient for intuitionistic fuzzy sets and its application to multi-criteria decision-making problems. Int. J. Fuzzy Syst. Appl. **10**(2), 39–58 (2021)
20. Ejegwa, P.A., Onyeke, I.C.: Intuitionistic fuzzy statistical correlation algorithm with applications to multicriteria-based decision-making processes. Int. J. Intelli. Syst. **36**(3), 1386–1407 (2021)
21. Ejegwa, P.A.: An improved correlation coefficient between intuitionistic fuzzy sets and its applications to real-life decision-making problems. Note Intuitionistic Fuzzy Sets **26**(2), 1–14 (2020)
22. Ejegwa, P.A.: Modified and generalized correlation coefficient between intuitionistic fuzzy sets with applications. Notes Intuitionistic Fuzzy Sets **26**(1), 8–22 (2020)

23. ZeShui, X.: Methods for aggregating interval-valued intuitionistic fuzzy information and their application to decision making. Control Decis. **22**(2), 215–219 (2007)
24. Xu, Z., Chen, J.: On geometric aggregation over interval-valued intuitionistic fuzzy information. In: Fourth International Conference on Fuzzy Systems and Knowledge Discovery, FSKD 2007. IEEE, Vol. 2, pp. 466–471 (2007)
25. Yager, R.R.: The power average operator. IEEE Trans. Syst. Man Cybern.-Part A: Syst. Humans **31**(6), 724–731 (2001)
26. Xu, Z., Yager, R.R.: Power-geometric operators and their use in group decision making. IEEE Trans. Fuzzy Syst. **18**(1), 94–105 (2010)
27. Zhou, L., Chen, H., Liu, J.: Generalized power aggregation operators and their applications in group decision making. Comput. Ind. Eng. **62**(4), 989–999 (2012)
28. Xu, Z.: Approaches to multiple attribute group decision making based on intuitionistic fuzzy power aggregation operators. Knowl.-Based Syst. **24**(6), 749–760 (2011)
29. Zhang, Z.: Generalized Atanassov's intuitionistic fuzzy power geometric operators and their application to multiple attribute group decision making. Inf. Fusion **14**(4), 460–486 (2013)
30. He, Y.D., Chen, H., Zhou, L., Liu, J., Tao, Z.: Generalized interval-valued Atanassov's intuitionistic fuzzy power operators and their application to group decision making. Int. J. Fuzzy Syst. **15**(4), 401–411 (2013)
31. Zhang, Z.: Hesitant fuzzy power aggregation operators and their application to multiple attribute group decision making. Inf. Sci. **234**, 150–181 (2013)
32. Fahmi, A., Abdullah, S., & Amin, F.: Cubic uncertain linguistic powered Einstein aggregation operators and their application to multi-attribute group decision making. In: Mathematical Sciences, pp. 1–24 (2019)
33. Bonferroni, C.: Sulle medie multiple di potenze. Bollettino dell'Unione Matematica Italiana **5**(3–4), 267–270 (1950)
34. Yager, R.R.: On generalized Bonferroni mean operators for multi-criteria aggregation. Int. J. Approx. Reasoning **50**(8), 1279–1286 (2009)
35. Xia, M., Xu, Z., Zhu, B.: Generalized intuitionistic fuzzy Bonferroni means. Int. J. Intelli. Syst. **27**(1), 23–47 (2012)
36. Beliakov, G., James, S., Mordelova, J., Rueckschlossova, T., Yager, R.R.: Generalized Bonferroni mean operators in multi-criteria aggregation. Fuzzy Sets Syst. **161**(17), 2227–2242 (2010)
37. Xia, M., Xu, Z., Zhu, B.: Geometric Bonferroni means with their application in multi-criteria decision making. Knowl.-Based Syst. **40**, 88–100 (2013)
38. Li, D., Zeng, W., Li, J.: Geometric Bonferroni mean operators. Int. J. Intelli. Syst. **31**(12), 1181–1197 (2016)
39. He, Y., He, Z., Jin, C., Chen, H.: Intuitionistic fuzzy power geometric Bonferroni means and their application to multiple attribute group decision making. Int. J. Uncertainty Fuzziness Knowl.-Based Syst. **23**(2), 285–315 (2015)
40. Zhu, B., Xu, Z., Xia, M.: Hesitant fuzzy geometric Bonferroni means. Inf. Sci. **205**, 72–85 (2012)
41. Zhu, B., Xu, Z.S.: Hesitant fuzzy Bonferroni means for multi-criteria decision making. J. Oper. Res. Soc. **64**(12), 1831–1840 (2013)
42. Kakati, P.: Interval Neutrosophic Einstein prioritized normalized weighted geometric Bonferroni mean operator and its application to multicriteria decision making. Neural Process Lett. **53**, 3395–3425 (2019). https://doi.org/10.1007/s11063-021-10553-3
43. He, Y., He, Z., Wang, G., Chen, H.: Hesitant fuzzy power Bonferroni means and their application to multiple attribute decision making. IEEE Trans. Fuzzy Syst. **23**(5), 1655–1668 (2015)
44. Wang, H., Smarandache, F., Sunderraman, R., Zhang, Y.Q.: Interval neutrosophic sets and logic: theory and applications in computing: theory and applications in computing. In: Infinite Study, vol. 5 (2005)
45. Liu, P., Li, H.: Interval-valued intuitionistic fuzzy power Bonferroni aggregation operators and their application to group decision making. Cogn. Comput. **9**(4), 494–512 (2017)

46. Sugeno, M.: Theory of Fuzzy Integrals and Its Applications. Doct, Thesis, Tokyo Institute of Technology (1974)
47. Choquet, G.: Theory of capacities. Annales De L'institut Fourier **5**, 131–295 (1954)
48. Liang, D., Darko, A.P., Zeng, J.: Interval-valued pythagorean fuzzy power average-based MULTIMOORA method for multi-criteria decision-making. J. Experimental Theor. Artif. Intelli. **32**(5), 845–874 (2020)
49. Mahmood, T., Ali, W., Ali, Z., Chinram, R.: Power aggregation operators and similarity measures based on improved intuitionistic hesitant fuzzy sets and their applications to multiple attribute decision making. Comput. Modeling Eng. Sci. **126**(3), 1165–1187 (2021)
50. Xu, Z., Hu, H.: Projection models for intuitionistic fuzzy multiple attribute decision making. Int. J. Inf. Technol. Decis. Making **9**(2), 267–280 (2010)
51. Gou, X., Xu, Z., Liao, H.: Multiple criteria decision making based on Bonferroni means with hesitant fuzzy linguistic information. Soft Comput. **21**(21), 6515–6529 (2017)

Modified Senapati and Yager's Fermatean Fuzzy Distance and Its Application in Students' Course Placement in Tertiary Institution

Idoko Charles Onyeke and Paul Augustine Ejegwa

Abstract Distance measure under Fermatean fuzzy domain is a soft computing technique for resolving problems of decision-making where decisions are made based on the closest proximity. Senapati and Yager initiated the study of Fermatean fuzzy distance measure and proposed a Fermatean fuzzy distance technique capturing the orthodox parameters of Fermatean fuzzy sets to guarantee reasonable output. Nonetheless, the technique lacks accuracy and yields results that contradict the axiomatic description of distance measure. Because of these limitations, this chapter modifies Senapati and Yager's Fermatean fuzzy distance to ameliorate the setbacks and improve reliability. Some properties of the modified Senapati and Yager's Fermatean fuzzy distance are discussed, and the distance is numerically illustrated to ascertain its merit. To demonstrate the application of the modified Senapati and Yager's Fermatean fuzzy distance, we deploy the tool to address students' course placement in tertiary institution. A comparative analysis of the Senapati and Yager's Fermatean fuzzy distance and its modified counterpart is presented to justify the present work.

Keywords Course placement · Fermatean fuzzy distance · Fermatean fuzzy sets · Intuitionistic fuzzy set · Pythagorean fuzzy set

1 Introduction

The determination of suitable courses for tertiary institution's students is a day to day experience engrossed with uncertainties. The notion of fuzzy sets [36] brought some leverages to the problem of uncertainties embedded in real-life problems. Fuzzy set is characterized with membership degree (MD) mapped to a closed interval containing

I. C. Onyeke
Department of Computer Science, University of Agriculture, P.M.B. 2373, Makurdi, Nigeria

P. A. Ejegwa (✉)
Department of Mathematics, University of Agriculture, P.M.B. 2373, Makurdi, Nigeria
e-mail: ejegwa.augustine@uam.edu.ng

L. Sahoo et al. (eds.), *Real Life Applications of Multiple Criteria Decision Making Techniques in Fuzzy Domain*, Studies in Fuzziness and Soft Computing 420,
https://doi.org/10.1007/978-981-19-4929-6_11

0 and 1. In a fuzzy set, the notion of non-membership degree (NMD) is depending on MD, and so, the usefulness of NMD is neglected. On the contrary, many real-life experiences capture MD and NMD with some degree of hesitation. As such, fuzzy set has been generalized to various classes to represent real-life experiences accurately.

One of such generalizations is intuitionistic fuzzy set (IFS) proposed in [1]. IFS is characterized with MD and NMD, both mapped to a closed interval [0, 1] with the possibility of the existence of the grade of indeterminacy. With the aid of some information measures, IFSs have been deployed to solve myriad of problems ranges from pattern recognition, image segmentation, career determination, medical diagnosis, and other decision-making cases [7, 8, 14, 18, 32]. The notion of intuitionitic fuzzy distance was earlier studied in [5], which is the extension of fuzzy distance [9]. Intuitionitic fuzzy distances/similarities have been deployed to solve several problems [4, 6, 22, 31, 33].

The idea of IFSs is restricted in such a way that it cannot be useful if the aggregate of MD and NMD is beyond one. For instance, whenever a decision-maker intends to make a decision in which MD is 0.5 and NMD is 0.6, the idea of IFS cannot represent such a scenario. As a necessity, IFS was generalized by extending its scope to accommodate the situation when the aggregate of MD and NMD is beyond one such that the sum of the aggregate of each MD and NMD is at most one. This expression is captured as intuitionistic fuzzy set of type-2 [2] or Pythagorean fuzzy set (PFS) [34]. PFS is a special IFS with additional properties, and thus has more capacity to control vagueness more appropriately compare to IFS. The concept of PFSs have been deployed to solve some practical problems [13, 16, 20, 21, 24, 35, 37, 38]. Pythagorean fuzzy similarities and dissimilarities have been explored and applied in myriad decision-making problems [10–12, 15, 17, 23, 39].

Akin to the limitation of IFS, PFS could not model problem when MD is $\frac{\sqrt[3]{6}}{2}$ and NMD is $\frac{1}{2}$. PFS is constrained to model cases (i) when the sum of MD and NMD is at most one, (ii) when the sum of MD and NMD is greater than one, and (iii) when the sum of square of both MD and NMD is at most one only. In an attempt to resolve the setback of PFS, the space of PFS was inflated, and thus, a concept called intuitionistic fuzzy set of type-3 is proposed [3]. This same construct is also known as Fermatean fuzzy sets (FFSs) [28]. In a FFS, the sum of square of both MD and NMD can be greater than one, but the aggregate of their cubes is at most one. Certain operators on FFSs and their properties were discussed in [30]. Also, in [28], some basic operations on FFSs, score/accuracy functions for the ranking of FFSs, and the Euclidean Fermatean fuzzy distance with application were discussed. The idea of Fermatean fuzzy linguistic term sets, certain basic operational laws and Fermatean fuzzy linguistic weighted operator, and a distance measure for Fermatean fuzzy linguistic sets with application were explored in [25]. Continuous Fermatean fuzzy information and their properties were discussed in [35]. Some new weighted aggregated operators on FFSs and their corresponding necessary properties were discussed, and their application in solving decision-making problems was addressed in [26, 27]. The notion of Fermatean fuzzy composite relation was introduced in

[19] and deployed to solve the problem of medical diagnosis. The application of interval-valued FFSs in capital budgeting techniques has been discussed [29].

Distance between FFSs was first studied in [28] by modifying the Euclidean distance measure under intuitionistic fuzzy context. However, this measure lacks accuracy and often time violates the axiomatic description of distance measure. To this end, we are motivated to introduce a modified Senapati and Yager's distance measure [28] which overrides the limitations of the Senapati and Yager's distance measure and enhances the application of FFSs. In a nutshell, the modified technique seeks to strengthened the Fermatean fuzzy Euclidean distance in [28]. The objectives of this chapter include:

- re-examination of Senapati and Yager's distance measure between FFSs,
- modification of Senapati and Yager's distance measure between FFSs to fulfill the axiomatic description of distance function and enhance accuracy,
- application of Senapati and Yager's distance measure and its modified variant in solving students' course allocation problems, and
- comparative analysis to showcase the superiority of the modified variant over Senapati and Yager's distance measure between FFSs.

The rest of the chapter is organized as follows; Sect. 2 revisits the concept of FFSs and Senapati and Yager's distance measure between FFSs; Sect. 3 presents the modified Senapati and Yager's Frematean fuzzy distance and elucidates some of its properties with numerical illustrations; Sect. 4 discusses the application of Senapati and Yager's Fermatean fuzzy distance and its modified variant in solving the problem of course allocations in tertiary institution, and Sect. 5 summarizes the chapter and provides areas of future researches.

2 Preliminaries

This section presents some basic concepts of FFS and elucidates the Senapati and Yager's distance measure between FFSs.

2.1 Fermatean Fuzzy Sets

As established, FFS is the generalization of both IFS and PFS, respectively. Take X to be a non-empty set throughout the chapter.

Definition 1 [1] An IFS denoted by C in X is of the form

$$C = \{\langle x, \mu_C(x), \nu_C(x)\rangle \mid x \in X\}, \tag{1}$$

where $\mu_C : X \to [0, 1]$ and $\nu_C : X \to [0, 1]$ represent MD and NMD of $x \in X$ to C with the property $0 \leq \mu_C(x) + \nu_C(x) \leq 1$. Since $\mu_C \neq 1 - \nu_C$ and $\nu_C \neq 1 - \mu_C$, then the index of C is defined by

$$\pi_C(x) \in [0, 1] = 1 - \mu_C(x) - \nu_C(x). \tag{2}$$

The parameter $\pi_C(x)$ describes the degree of non-determinacy of $x \in X$. Certainly, $\mu_C(x) + \nu_C(x) + \pi_C(x) = 1$.

Definition 2 [34] A PFS denoted by $\tilde{C}$ in X is defined by

$$\tilde{C} = \{\langle x, \mu_{\tilde{C}}(x), \nu_{\tilde{C}}(x)\rangle \mid x \in X\}, \tag{3}$$

where $\mu_{\tilde{C}} : X \to [0, 1]$ and $\nu_{\tilde{C}} : X \to [0, 1]$ depict MD and NMD of $x \in X$ to $\tilde{C}$ with the property $0 \leq \mu^2_{\tilde{C}}(x) + \nu^2_{\tilde{C}}(x) \leq 1$. The index of PFS $\tilde{C}$ is defined by

$$\pi_{\tilde{C}}(x) \in [0, 1] = \sqrt{1 - \left(\mu^2_{\tilde{C}}(x) + \nu^2_{\tilde{C}}(x)\right)}. \tag{4}$$

Clearly, $\mu^2_{\tilde{C}}(x) + \nu^2_{\tilde{C}}(x) + \pi^2_{\tilde{C}}(x) = 1$.

Definition 3 [28] A FFS denoted by $\hat{C}$ in X is characterized by

$$\hat{C} = \{\langle x, \mu_{\hat{C}}(x), \nu_{\hat{C}}(x)\rangle | x \in X\}, \tag{5}$$

where $\mu_{\hat{C}}$ and $\nu_{\hat{C}} : X \to [0, 1]$ describe MD and NMD of $x \in X$ for $0 \leq \mu^3_{\hat{C}}(x) + \nu^3_{\hat{C}}(x) \leq 1$. For a FFS $\hat{C}$ in X, $\pi_{\hat{C}}(x) \in [0, 1] = \sqrt{3}1 - \left(\mu^3_{\hat{C}}(x) + \nu^3_{\hat{C}}(x)\right)$ represents the hesitation margin or FFS index of $\hat{C}$. Certainly, $\mu^3_{\hat{C}}(x) + \nu^3_{\hat{C}}(x) + \pi^3_{\hat{C}}(x) = 1$.

Definition 4 [27] Suppose $\hat{C}$ and $\hat{D}$ are FFSs in X, then

(i) $\overline{\hat{C}} = \{\langle x, \nu_{\hat{C}}(x), \mu_{\hat{C}}(x)\rangle | x \in X\}$.
(ii) $\hat{C} \cup \hat{D} = \{\langle x, \max\left(\mu_{\hat{C}}(x), \mu_{\hat{D}}(x)\right), \min\left(\nu_{\hat{C}}(x), \nu_{\hat{D}}(x)\right)\rangle | x \in X\}$.
(iii) $\hat{C} \cap \hat{D} = \{\langle x, \min\left(\mu_{\hat{C}}(x), \mu_{\hat{D}}(x)\right), \max\left(\nu_{\hat{C}}(x), \nu_{\hat{D}}(x)\right)\rangle | x \in X\}$.

In addition, $\hat{C} = \hat{D}$ iff $\mu_{\hat{C}}(x) = \mu_{\hat{D}}(x)$, $\nu_{\hat{C}}(x) = \nu_{\hat{D}}(x)$ for all $x \in X$, and $\hat{C} \subseteq \hat{D}$ iff $\mu_{\hat{C}}(x) \leq \mu_{\hat{C}}(x)$, $\nu_{\hat{C}}(x) \geq \nu_{\hat{D}}(x)$ for all $x \in X$. Also, $\hat{C} \subset \hat{D}$ iff $\hat{C} \subseteq \hat{D}$ and $\hat{C} \neq \hat{D}$.

Definition 5 [19] FFV is characterized by $\langle c, d\rangle$ for $c^3 + d^3 \leq 1$, where $c, d \in [0, 1]$. FFV assesses the FFS for which the constituents (c and d) are taken as MD and NMD, respectively.

2.2 Senapati and Yager's Distance Between Fermatean Fuzzy Sets

We start by reiterating the axiomatic description of distance function between FFSs.

Definition 6 Suppose $\hat{C}$ *and* $\hat{D}$ *are FFSs in* $X = \{x_1, x_2, \ldots, x_n\}$ *for* $n < \infty$. *Then the distance depicted by* $\Omega(\hat{C}, \hat{D})$ *between* $\hat{C}$ *and* $\hat{D}$ *is a measuring function*

$$\Omega: FFS \times FFS \rightarrow [0, 1]$$

which satisfies

(i) $\Omega(\hat{C}, \hat{D}) \in [0, 1]$,
(ii) $\Omega(\hat{C}, \hat{D}) = \Omega(\hat{D}, \hat{C})$,
(iii) $\Omega(\hat{C}, \hat{D}) = 0$ *if and only if* $\hat{C} = \hat{D}$,
(iv) $\Omega(\hat{C}, \hat{E}) \leq \Omega(\hat{C}, \hat{D}) + \Omega(\hat{D}, \hat{E})$, *where* $\hat{E}$ *is a FFS in* X.

If $\Omega(\hat{C}, \hat{D})$ approaches 0, then $\hat{C}$ and $\hat{D}$ are more related. Similarly, if $\Omega(\hat{C}, \hat{D})$ approaches 1, then $\hat{C}$ and $\hat{D}$ are not related. In other words, as $\Omega(\hat{C}, \hat{D})$ decreases, the more the FFSs are closer. Also, as $\Omega(\hat{C}, \hat{D})$ increases, the less related are the FFSs.

The Senapati and Yager's distance based on the principle of Euclidean distance between FFSs $\hat{C}$ and $\hat{D}$ in $X = \{x_1, x_2, \ldots, x_n\}$ is given in Eq. (6).

$$\Omega(\hat{C}, \hat{D}) = \Big(\frac{1}{2}\sum_{i=1}^{n}\big([\mu^3_{\hat{C}}(x_i) - \mu^3_{\hat{D}}(x_i)]^2 + [\nu^3_{\hat{C}}(x_i) - \nu^3_{\hat{D}}(x_i)]^2 + [\pi^3_{\hat{C}}(x_i) - \pi^3_{\hat{D}}(x_i)]^2\big)\Big)^{0.5} \quad (6)$$

Numerical Illustrations. Some examples are considered to illustrate the computational process of the distance measure technique in Eq. (6).

Example 1 Suppose the FFSs $\hat{C}$ and $\hat{D}$ in $X = \{x_1, x_2, x_3\}$ are defined by

$$\hat{C} = \{\langle x_1, 0.5, 0.2\rangle, \langle x_2, 0.4, 0.3\rangle, \langle x_3, 0.3, 0.5\rangle\},$$

$$\hat{D} = \{\langle x_1, 0.44, 0.3\rangle, \langle x_3, 0.35, 0.4\rangle\},$$

where

$$\pi_{\hat{C}}(x_1) = 0.9535, \pi_{\hat{C}}(x_2) = 0.9687, \pi_{\hat{C}}(x_3) = 0.9465,$$

$$\pi_{\hat{D}}(x_1) = 0.9611, \pi_{\hat{C}}(x_3) = 0.9630.$$

Hence, the distance between $\hat{C}$ and $\hat{D}$ is:

$$\begin{aligned}\Omega(\hat{C},\hat{D}) &= \Big(\frac{1}{2}((0.5^3-0.44^3)^2+(0.2^3-0.3^3)^2+(0.9535^3-0.9611^3)^2\\ &= (0.4^3-0^3)^2+(0.3^3-1^3)^2+(0.9687^3-0^3)^2\\ &= (0.3^3-0.35^3)^2+(0.5^3-0.4^3)^2+(0.9465^3-0.963^3)^2)\Big)^{0.5}\\ &= 0.6479.\end{aligned}$$

Similarly, $\Omega(\hat{C},\hat{D}) = \Omega(\hat{D},\hat{C}) = 0.6479$.

Example 2 Taken $\hat{C}$, $\hat{D}$ and $\hat{E}$ as FFSs in $X = \{x_1, x_2, x_3, x_4\}$ defined by

$$\hat{C} = \{\langle x_1, 0.95, 0.5\rangle, \langle x_2, 0.8, 0.65\rangle, \langle x_3, 0.7, 0.5\rangle, \langle x_4, 0.1, 0.9\rangle\},$$

$$\hat{D} = \{\langle x_1, 0.92, 0.55\rangle, \langle x_2, 0.8, 0.62\rangle, \langle x_3, 0.5, 0.4\rangle, \langle x_4, 0.8, 0.4\rangle\},$$

$$\hat{E} = \{\langle x_1, 0.6, 0.8\rangle, \langle x_2, 0.8, 0.6\rangle, \langle x_3, 0.9, 0.6\rangle, \langle x_4, 0.85, 0.7\rangle\},$$

where

$$\pi_{\hat{C}}(x_1) = 0.2602, \pi_{\hat{C}}(x_2) = 0.5976, \pi_{\hat{C}}(x_3) = 0.8103, \pi_{\hat{C}}(x_4) = 0.6463,$$

$$\pi_{\hat{D}}(x_1) = 0.3801, \pi_{\hat{D}}(x_2) = 0.6297, \pi_{\hat{D}}(x_3) = 0.9326, \pi_{\hat{D}}(x_4) = 0.7513,$$

$$\pi_{\hat{E}}(x_1) = \pi_{\hat{E}}(x_2) = 0.6479, \pi_{\hat{E}}(x_3) = 0.3803, \pi_{\hat{E}}(x_4) = 0.35.$$

The distances among the FFSs $\hat{C}$, $\hat{D}$ and $\hat{E}$ are as follow:

$$\begin{aligned}\Omega(\hat{C},\hat{D}) &= \Big(\frac{1}{2}((0.95^3-0.92^3)^2+(0.5^3-0.55^3)^2+(0.2602^3-0.3801^3)^2\\ &= (0.8^3-0.8^3)^2+(0.65^3-0.62^3)^2+(0.5976^3-0.6297^3)^2\\ &= (0.7^3-0.5^3)^2+(0.5^3-0.4^3)^2+(0.8103^3-0.9326^3)^2\\ &= (0.1^3-0.8^3)^2+(0.9^3-0.4^3)^2+(0.6463^3-0.7513^3)^2)\Big)^{0.5}\\ &= 0.9862.\end{aligned}$$

$$\begin{aligned}\Omega(\hat{C},\hat{E}) &= \Big(\frac{1}{2}((0.95^3-0.6^3)^2+(0.5^3-0.8^3)^2+(0.2602^3-0.6479^3)^2\\ &= (0.8^3-0.8^3)^2+(0.65^3-0.6^3)^2+(0.5976^3-0.6479^3)^2\\ &= (0.7^3-0.9^3)^2+(0.5^3-0.6^3)^2+(0.8103^3-0.3803^3)^2\\ &= (0.1^3-0.85^3)^2+(0.9^3-0.7^3)^2+(0.6463^3-0.35^3)^2)\Big)^{0.5}\\ &= 0.8928.\end{aligned}$$

$$\Omega(\hat{D}, \hat{E}) = \Big(\frac{1}{2}\big((0.92^3 - 0.6^3)^2 + (0.55^3 - 0.8^3)^2 + (0.3801^3 - 0.6479^3)^2$$
$$= (0.8^3 - 0.8^3)^2 + (0.62^3 - 0.6^3)^2 + (0.6297^3 - 0.6479^3)^2$$
$$= (0.5^3 - 0.9^3)^2 + (0.4^3 - 0.6^3)^2 + (0.9326^3 - 0.3803^3)^2$$
$$= (0.8^3 - 0.85^3)^2 + (0.4^3 - 0.7^3)^2 + (0.7513^3 - 0.35^3)^2\big)\Big)^{0.5}$$
$$= 0.9158.$$

From the outputs, the relations among $\hat{C}$, $\hat{D}$ and $\hat{E}$ are weak; however $\hat{C}$ and $\hat{E}$ are more related because $\Omega(\hat{C}, \hat{E})$ is the least.

3 Modified Senapati and Yager's Distance Between Fermatean Fuzzy Sets

The distance between FFSs in [28] extended the distances between IFSs [5, 31] by considering the properties of FFS. Notwithstanding, the distance technique [28] does not considered (i) the number of the conventional parameters of FFS, viz., (i.e., 3 represents MDs, NMDs, and hesitation margins) and (ii) the cardinality of the elements of X, which is n. By putting these setbacks into consideration, a new modified Fermatean fuzzy distance technique is proposed in Eq. (7):

$$\tilde{\Omega}(\hat{C}, \hat{D}) = \Big(\frac{1}{3n}\sum_{i=1}^{n}\big([\mu^3_{\hat{C}}(x_i) - \mu^3_{\hat{D}}(x_i)]^2 + [\nu^3_{\hat{C}}(x_i) - \nu^3_{\hat{D}}(x_i)]^2 + [\pi^3_{\hat{C}}(x_i) - \pi^3_{\hat{D}}(x_i)]^2\big)\Big)^{0.5} \quad (7)$$

between FFSs $\hat{C}$ and $\hat{D}$ in $X = \{x_i\}$ for $i = 1, \ldots, n$, where $n < \infty$. This novel technique takes into account the number of the considered parameters (i.e., MD, NMD, and hesitation margin) and the cardinality (i.e., n) of the ground set X to foster efficiency and accuracy.

3.1 *Some Properties of Modified Senapati and Yager's Fermatean Fuzzy Distance*

Here, some properties of the modified Senapati and Yager's Fermatean fuzzy distance are considered to validate the new technique.

Theorem 1 *For FFSs $\hat{C}$ and $\hat{D}$ in X, the function $\tilde{\Omega}(\hat{C}, \hat{D})$ satisfies $\tilde{\Omega}(\hat{C}, \hat{D}) = 0$ iff $\hat{C} = \hat{D}$.*

Proof Suppose that $\tilde{\Omega}(\hat{C}, \hat{D}) = 0$, then we have

$$\Big(\frac{1}{3n}\sum_{i=1}^{n}([\mu^3_{\hat{C}}(x_i) - \mu^3_{\hat{D}}(x_i)]^2 + [\nu^3_{\hat{C}}(x_i) - \nu^3_{\hat{D}}(x_i)]^2 + [\pi^3_{\hat{C}}(x_i) - \pi^3_{\hat{D}}(x_i)]^2)\Big)^{0.5} = 0$$

which implies $[\mu^3_{\hat{C}}(x_i) - \mu^3_{\hat{D}}(x_i)]^2 = 0$, $[\nu^3_{\hat{C}}(x_i) - \nu^3_{\hat{D}}(x_i)]^2 = 0$ and $[\pi^3_{\hat{C}}(x_i) - \pi^3_{\hat{D}}(x_i)]^2 = 0, \Longrightarrow \mu^3_{\hat{C}}(x_i) = \mu^3_{\hat{D}}(x_i)$, $\nu^3_{\hat{C}}(x_i) = \nu^3_{\hat{D}}(x_i)$ and $\pi^3_{\hat{C}}(x_i) = \pi^3_{\hat{D}}(x_i)$. By taking the cube root of each, we deduce that $\hat{C} = \hat{D}$.

Conversely, if $\hat{C} = \hat{D}$, then clearly $\tilde{\Omega}(\hat{C}, \hat{D}) = 0$.

Theorem 2 *For FFSs $\hat{C}$ and $\hat{D}$ in X, we have*

(i) $\tilde{\Omega}(\hat{C}, \hat{D}) = \tilde{\Omega}(\hat{D}, \hat{C})$,
(ii) $\tilde{\Omega}(\hat{C}, \hat{D}) = \tilde{\Omega}(\overline{\hat{D}}, \overline{\hat{C}})$.

Proof Now, we get

$$\begin{aligned}\tilde{\Omega}(\hat{C}, \hat{D}) &= \Big(\frac{1}{3n}\sum_{i=1}^{n}([\mu^3_{\hat{C}}(x_i) - \mu^3_{\hat{D}}(x_i)]^2 + [\nu^3_{\hat{C}}(x_i) \\ &\quad -\nu^3_{\hat{D}}(x_i)]^2 + [\pi^3_{\hat{C}}(x_i) - \pi^3_{\hat{D}}(x_i)]^2)\Big)^{0.5} \\ &= \Big(\frac{1}{3n}\sum_{i=1}^{n}([\mu^3_{\hat{D}}(x_i) - \mu^3_{\hat{C}}(x_i)]^2 + [\nu^3_{\hat{D}}(x_i) \\ &\quad -\nu^3_{\hat{C}}(x_i)]^2 + [\pi^3_{\hat{D}}(x_i) - \pi^3_{\hat{C}}(x_i)]^2)\Big)^{0.5} \\ &= \tilde{\Omega}(\hat{C}, \hat{D}),\end{aligned}$$

which proves (i).

The proof of (ii) is straightforward.

Theorem 3 *The distance function $\tilde{\Omega}(\hat{C}, \hat{D})$ for FFSs $\hat{C}$ and $\hat{D}$ in X satisfies $\tilde{\Omega}(\hat{C}, \hat{D}) \in [0, 1]$.*

Proof To proof $\tilde{\Omega}(\hat{C}, \hat{D}) \in [0, 1]$, we show that $0 \leq \tilde{\Omega}(\hat{C}, \hat{D}) \leq 1$. For any two FFSs $\hat{C}$ and $\hat{D}$, we must have

$$[\mu^3_{\hat{C}}(x_i) - \mu^3_{\hat{D}}(x_i)]^2 \geq 0, [\nu^3_{\hat{C}}(x_i) - \nu^3_{\hat{D}}(x_i)]^2 \geq 0, \text{ and } [\pi^3_{\hat{C}}(x_i) - \pi^3_{\hat{D}}(x_i)]^2 \geq 0.$$

Hence $\tilde{\Omega}(\hat{C}, \hat{D}) \geq 0$.

We rewrite Eq. (7) as

$$\tilde{\Omega}(\hat{C},\hat{D})=\frac{\sqrt{\sum_{i=1}^{n}\left([\mu_{\hat{C}}^{3}(x_i)-\mu_{\hat{D}}^{3}(x_i)]^2+[\nu_{\hat{C}}^{3}(x_i)-\nu_{\hat{D}}^{3}(x_i)]^2+[\pi_{\hat{C}}^{3}(x_i)-\pi_{\hat{D}}^{3}(x_i)]^2\right)}}{\sqrt{3n}}.$$

Assume

$$\sum_{i=1}^{n}\left([\mu_{\hat{C}}^{3}(x_i)-\mu_{\hat{D}}^{3}(x_i)]^2\right)=\alpha,\ \sum_{i=1}^{n}\left([\nu_{\hat{C}}^{3}(x_i)-\nu_{\hat{D}}^{3}(x_i)]^2\right)=\beta$$

and

$$\sum_{i=1}^{n}\left([\pi_{\hat{C}}^{3}(x_i)-\pi_{\hat{D}}^{3}(x_i)]^2\right)=\gamma,$$

then we have

$$\begin{aligned}\tilde{\Omega}(\hat{C},\hat{D})&=\frac{\sqrt{\sum_{i=1}^{n}\left([\mu_{\hat{C}}^{3}(x_i)-\mu_{\hat{D}}^{3}(x_i)]^2+[\nu_{\hat{C}}^{3}(x_i)-\nu_{\hat{D}}^{3}(x_i)]^2+[\pi_{\hat{C}}^{3}(x_i)-\pi_{\hat{D}}^{3}(x_i)]^2\right)}}{\sqrt{3n}}\\&\leq\frac{\sqrt{\sum_{i=1}^{n}\left([\mu_{\hat{C}}^{3}(x_i)-\mu_{\hat{D}}^{3}(x_i)]^2\right)+\sum_{i=1}^{n}\left([\nu_{\hat{C}}^{3}(x_i)-\nu_{\hat{D}}^{3}(x_i)]^2\right)+\sum_{i=1}^{n}\left([\pi_{\hat{C}}^{3}(x_i)-\pi_{\hat{D}}^{3}(x_i)]^2\right)}}{\sqrt{3n}}\\&=\frac{\sqrt{\alpha+\beta+\gamma}}{\sqrt{3n}}.\end{aligned}$$

But

$$\tilde{\Omega}^2(\hat{C},\hat{D})\leq\frac{\alpha+\beta+\gamma}{3n}.$$

Then

$$\begin{aligned}\tilde{\Omega}^2(\hat{C},\hat{D})-1&\leq\frac{\alpha+\beta+\gamma}{3n}-1\\&=\frac{\alpha+\beta+\gamma-3n}{3n}\\&=\frac{-(3n-\alpha-\beta-\gamma)}{3n}\\&\leq 0.\end{aligned}$$

Thus $\tilde{\Omega}^2(\hat{C},\hat{D})\leq 1$, and so $\tilde{\Omega}(\hat{C},\hat{D})\leq 1$. Hence $0\leq\tilde{\Omega}(\hat{C},\hat{D})\leq 1$.

Theorem 4 *Suppose $\hat{C}$, $\hat{D}$ and $\hat{E}$ are FFSs in X with $\hat{C}\subseteq\hat{D}\subseteq\hat{E}$, then*

(i) $\tilde{\Omega}(\hat{C},\hat{E})\geq\tilde{\Omega}(\hat{C},\hat{D})$,
(ii) $\tilde{\Omega}(\hat{C},\hat{E})\geq\tilde{\Omega}(\hat{D},\hat{E})$,
(iii) $\tilde{\Omega}(\hat{C},\hat{E})\geq\max\left(\tilde{\Omega}(\hat{C},\hat{D}),\tilde{\Omega}(\hat{D},\hat{E})\right)$.

Proof For the fact that $\hat{C}\subseteq\hat{D}\subseteq\hat{E}$, we have

$$[\mu_{\hat{C}}(x_i)^3 - \mu_{\hat{E}}(x_i)^3]^2 \geq [\mu_{\hat{C}}(x_i)^3 - \mu_{\hat{D}}(x_i)^3]^2,$$

$$[\nu_{\hat{C}}(x_i)^3 - \nu_{\hat{E}}(x_i)^3]^2 \geq [\nu_{\hat{C}}(x_i)^3 - \nu_{\hat{D}}(x_i)^3]^2,$$

$$[\pi_{\hat{C}}(x_i)^3 - \pi_{\hat{E}}(x_i)^3]^2 \geq [\pi_{\hat{C}}(x_i)^3 - \pi_{\hat{D}}(x_i)^3]^2.$$

Hence

$$\begin{aligned}[\mu_{\hat{C}}(x_i)^3 - \mu_{\hat{E}}(x_i)^3]^2 &+ [\nu_{\hat{C}}(x_i)^3 - \nu_{\hat{E}}(x_i)^3]^2 + [\pi_{\hat{C}}(x_i)^3 - \pi_{\hat{E}}(x_i)^3]^2 \\ &\geq [\mu_{\hat{C}}(x_i)^3 - \mu_{\hat{D}}(x_i)^3]^2 + [\nu_{\hat{C}}(x_i)^3 - \nu_{\hat{D}}(x_i)^3]^2 \\ &+ [\pi_{\hat{C}}(x_i)^3 - \pi_{\hat{D}}(x_i)^3]^2,\end{aligned}$$

and so the proof of (i) holds.

Similarly,

$$[\mu_{\hat{C}}(x_i)^3 - \mu_{\hat{E}}(x_i)^3]^2 \geq [\mu_{\hat{D}}(x_i)^3 - \mu_{\hat{E}}(x_i)^3]^2,$$

$$[\nu_{\hat{C}}(x_i)^3 - \nu_{\hat{E}}(x_i)^3]^2 \geq [\nu_{\hat{D}}(x_i)^3 - \nu_{\hat{E}}(x_i)^3]^2,$$

$$[\pi_{\hat{C}}(x_i)^3 - \pi_{\hat{E}}(x_i)^3]^2 \geq [\pi_{\hat{D}}(x_i)^3 - \pi_{\hat{E}}(x_i)^3]^2.$$

Hence

$$\begin{aligned}[\mu_{\hat{C}}(x_i)^3 - \mu_{\hat{E}}(x_i)^3]^2 &+ [\nu_{\hat{C}}(x_i)^3 - \nu_{\hat{E}}(x_i)^3]^2 + [\pi_{\hat{C}}(x_i)^3 - \pi_{\hat{E}}(x_i)^3]^2 \\ &\geq [\mu_{\hat{D}}(x_i)^3 - \mu_{\hat{E}}(x_i)^3]^2 + [\nu_{\hat{D}}(x_i)^3 - \nu_{\hat{E}}(x_i)^3]^2 \\ &+ [\pi_{\hat{D}}(x_i)^3 - \pi_{\hat{E}}(x_i)^3]^2,\end{aligned}$$

which proves (ii).

The proof of (iii) follows from (i) and (ii).

3.2 Numerical Verifications

Now, we exemplify the modified Senapati and Yager's distance between Fermatean fuzzy sets to ascertain its compliance with the description of distance measure in literature. By using Example 1, we get the distance between $\hat{C}$ and $\hat{D}$ as

$$\begin{aligned}\Omega(\hat{C}, \hat{D}) &= \Big(\frac{1}{9}\big((0.5^3 - 0.44^3)^2 + (0.2^3 - 0.3^3)^2 + (0.9535^3 - 0.9611^3)^2 \\ &= (0.4^3 - 0^3)^2 + (0.3^3 - 1^3)^2 + (0.9687^3 - 0^3)^2 \\ &= (0.3^3 - 0.35^3)^2 + (0.5^3 - 0.4^3)^2 + (0.9465^3 - 0.963^3)^2\big)\Big)^{0.5} \\ &= 0.3054.\end{aligned}$$

Similarly, $\Omega(\hat{C}, \hat{D}) = \Omega(\hat{D}, \hat{C}) = 0.3054$.

Using Example 2, we have the following distances between $\hat{C}$, $\hat{D}$ and $\hat{E}$:

$$\begin{aligned}\Omega(\hat{C}, \hat{D}) &= \Big(\frac{1}{12}((0.95^3 - 0.92^3)^2 + (0.5^3 - 0.55^3)^2 + (0.2602^3 - 0.3801^3)^2 \\ &= (0.8^3 - 0.8^3)^2 + (0.65^3 - 0.62^3)^2 + (0.5976^3 - 0.6297^3)^2 \\ &= (0.7^3 - 0.5^3)^2 + (0.5^3 - 0.4^3)^2 + (0.8103^3 - 0.9326^3)^2 \\ &= (0.1^3 - 0.8^3)^2 + (0.9^3 - 0.4^3)^2 + (0.6463^3 - 0.7513^3)^2)\Big)^{0.5} \\ &= 0.4026.\end{aligned}$$

$$\begin{aligned}\Omega(\hat{C}, \hat{E}) &= \Big(\frac{1}{12}((0.95^3 - 0.6^3)^2 + (0.5^3 - 0.8^3)^2 + (0.2602^3 - 0.6479^3)^2 \\ &= (0.8^3 - 0.8^3)^2 + (0.65^3 - 0.6^3)^2 + (0.5976^3 - 0.6479^3)^2 \\ &= (0.7^3 - 0.9^3)^2 + (0.5^3 - 0.6^3)^2 + (0.8103^3 - 0.3803^3)^2 \\ &= (0.1^3 - 0.85^3)^2 + (0.9^3 - 0.7^3)^2 + (0.6463^3 - 0.35^3)^2)\Big)^{0.5} \\ &= 0.3645.\end{aligned}$$

$$\begin{aligned}\Omega(\hat{D}, \hat{E}) &= \Big(\frac{1}{12}((0.92^3 - 0.6^3)^2 + (0.55^3 - 0.8^3)^2 + (0.3801^3 - 0.6479^3)^2 \\ &= (0.8^3 - 0.8^3)^2 + (0.62^3 - 0.6^3)^2 + (0.6297^3 - 0.6479^3)^2 \\ &= (0.5^3 - 0.9^3)^2 + (0.4^3 - 0.6^3)^2 + (0.9326^3 - 0.3803^3)^2 \\ &= (0.8^3 - 0.85^3)^2 + (0.4^3 - 0.7^3)^2 + (0.7513^3 - 0.35^3)^2)\Big)^{0.5} \\ &= 0.3739.\end{aligned}$$

These results validate the reliability of the new distance technique since the conditions in Definition 6 are completely satisfied. Again, the outputs from the new distance technique show that better relationship exists among $\hat{C}$, $\hat{D}$ and $\hat{E}$, thereby $\hat{C}$ and $\hat{E}$ relate more against the stance when the technique in [28] was deployed.

The results from the Senapati and Yager's distance technique and the new technique are represented by Figs. 1 and 2.

SYDT is for Senapati and Yager's distance technique and MSYDT is for modified Senapati and Yager's distance technique, respectively. From Examples 1 and 2, and Figs. 1 and 2, it clear that MSYDT yields better results than SYDT.

4 Course Allocations Based on Fermatean Fuzzy Distance Technique

The process of allocating study-course to university applicants is always tedious due to uncertainties. This is where the idea of Fermatean fuzzy model is necessary since FFSs is equipped with the facilities to curb uncertainties. The exercise is carried out through the approach of machine learning by classifying the applicants into their various suitable courses. Assume there is an alternative sample represented in FFV $\hat{F}$ to be classified with any of FFVs $\hat{G}_i$, where $i = 1, \ldots, n$. The value of Eq. (8)

$$\tilde{\Omega}(\hat{G}_i, \hat{F}) = \min\left(\tilde{\Omega}(\hat{G}_1, \hat{F}), \ldots, \tilde{\Omega}(\hat{G}_n, \hat{F})\right), \quad (8)$$

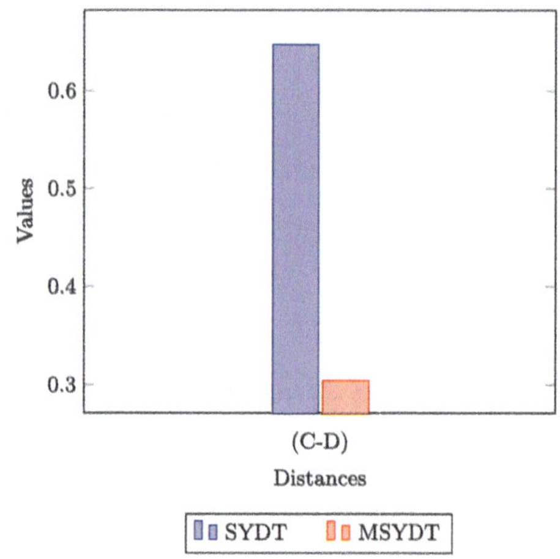

Fig. 1 Representation of results of Example 1

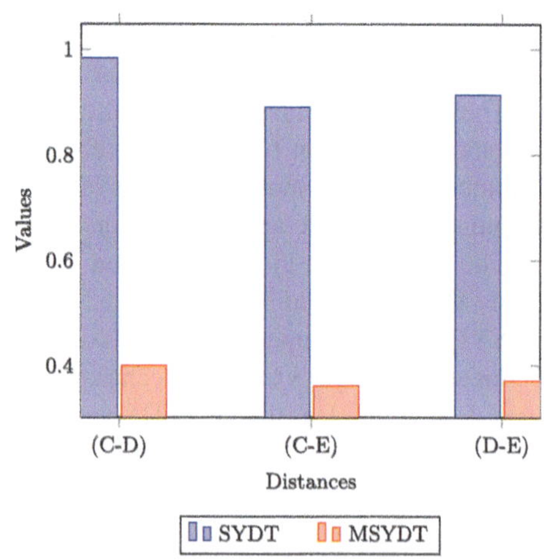

Fig. 2 Representation of results of Example 2

indicates the suitable classification for $\hat{F}$.

4.1 Experimental Example

Here, a hypothetical scenario of students vying for admission to study some set of courses are presented to buttress the Fermatean fuzzy course allocations based on distance technique. Assuming ten students $\hat{S} = \{\hat{S}_1, \ldots, \hat{S}_{10}\}$ are vying for admission to study some courses denoted by $\hat{C} = \{\hat{C}_1, \ldots, \hat{C}_6\}$ in a certain university, where $\hat{C}_1$ = medicine, $\hat{C}_2$ = pharmacy, $\hat{C}_3$ = nursing, $\hat{C}_4$ = anatomy, $\hat{C}_5$ = physiology, and $\hat{C}_6$ = midwifery, respectively. The expected subject-criteria require to study C_i, $i = 1, \ldots, 6$ are represented by a set $\rho = \{\rho_1, \rho_2, \rho_3, \rho_4, \rho_5, \rho_6\}$, where ρ_1 = English language, ρ_2 = Maths, ρ_3 = Biology, ρ_4 = Physics, ρ_5 = Chemistry, ρ_6 = Health science, respectively. After the qualification examination on ρ, the students performance is represented in Table 1. The university benchmarks for course allocations to read each of the available courses are represented in Table 2.

The MD (μ) represents the mark scored by the student, and NMD (ν) represents the mark allocated to the questions the student failed, respectively.

After computing the Fermatean fuzzy index for the FFVs in Tables 1 and 2, we apply both the Senapati and Yager's distance technique and the new distance technique to determine which course is suitable to be study by each of the students. The results for the computations are in Tables 3 and 4, respectively.

By juxtaposing Tables 3 and 4, one can confidently says that the new technique yields more appropriate results far reliable than the results from Senapati and Yager's method [28]. The method in [28] does not satisfied the conditions in Definition 6. Although it was stated in [28] that Eq. (6) satisfies $0 \leq \Omega(\hat{C}, \hat{D}) \leq \sqrt{2}$, but the assertion is not true from Table 3. On the other hand, the new technique satisfies the conditions in Definition 6 satisfactorily. Because of the unreliable results of the method in [28], we process the course allocations based on the results in Table 4.

Using Table 4 and Eq. (8), we deduce the following possibilities: $\hat{S}_1$, $\hat{S}_2$, $\hat{S}_4$, and $\hat{S}_9$ are suitable to study medicine, pharmacy, nursing, and anatomy in that order; $\hat{S}_3$ and $\hat{S}_{10}$ are suitable to study medicine, pharmacy, nursing, anatomy, and physiology in that order; $\hat{S}_7$ is suitable to study medicine and pharmacy in that order; $\hat{S}_5$, $\hat{S}_6$, and $\hat{S}_8$ are suitable to study medicine, pharmacy, nursing, anatomy, physiology, and midwifery in that order. Applicants $\hat{S}_5$, $\hat{S}_6$, and $\hat{S}_8$ are the most suitable to study any of the enlisted course of their choice because of their closeness to all the courses due to their academic performances.

5 Conclusions

Fermatean fuzzy set is an important soft computing tool. In this chapter, we have modified a Fermatean fuzzy distance technique proposed in [28] by taking into account

Table 1 Students performance

$\hat{S}$	ρ_1	ρ_2	ρ_3	ρ_4	ρ_5	ρ_6
	Subjects					
$\mu_{\hat{S}_1}$	0.6000	0.5000	0.6000	0.5000	0.5000	0.6000
$\nu_{\hat{S}_1}$	0.3000	0.4000	0.3000	0.3000	0.5000	0.2000
$\mu_{\hat{S}_2}$	0.5000	0.6000	0.5000	0.4000	0.7000	0.7000
$\nu_{\hat{S}_2}$	0.3000	0.3000	0.3000	0.5000	0.2000	0.3000
$\mu_{\hat{S}_3}$	0.7000	0.7000	0.7000	0.5000	0.4000	0.6000
$\nu_{\hat{S}_3}$	0.3000	0.2000	0.1000	0.4000	0.5000	0.4000
$\mu_{\hat{S}_4}$	0.6000	0.8000	0.6000	0.6000	0.5000	0.7000
$\nu_{\hat{S}_4}$	0.4000	0.2000	0.3000	0.2000	0.3000	0.3000
$\mu_{\hat{S}_5}$	0.8000	0.7000	0.8000	0.7000	0.6000	0.8000
$\nu_{\hat{S}_5}$	0.1000	0.2000	0.2000	0.1000	0.1000	0.1000
$\mu_{\hat{S}_6}$	0.9000	0.5000	0.8000	0.5000	0.6000	0.8000
$\nu_{\hat{S}_6}$	0.1000	0.4000	0.1000	0.2000	0.4000	0.2000
$\mu_{\hat{S}_7}$	0.5000	0.9000	0.5000	0.8000	0.6000	0.6000
$\nu_{\hat{S}_7}$	0.2000	0.1000	0.4000	0.2000	0.4000	0.3000
$\mu_{\hat{S}_8}$	0.7000	0.6000	0.8000	0.6000	0.6000	0.8000
$\nu_{\hat{S}_8}$	0.3000	0.3000	0.2000	0.4000	0.3000	0.1000
$\mu_{\hat{S}_9}$	0.4000	0.9000	0.8000	0.7000	0.7000	0.8000
$\nu_{\hat{S}_9}$	0.5000	0.1000	0.2000	0.3000	0.2000	0.1000
$\mu_{\hat{S}_{10}}$	0.5000	0.3000	0.9000	0.7000	0.6000	0.8000
$\nu_{\hat{S}_{10}}$	0.3000	0.7000	0.1000	0.2000	0.3000	0.2000

Table 2 College benchmarks

$\hat{C}$	ρ_1	ρ_2	ρ_3	ρ_4	ρ_5	ρ_6
	Subjects					
$\mu_{\hat{C}_1}$	0.8000	0.9000	0.8000	0.7000	0.7000	0.8000
$\nu_{\hat{C}_1}$	0.1000	0.1000	0.2000	0.3000	0.3000	0.1000
$\mu_{\hat{C}_2}$	0.7000	0.8000	0.4000	0.5000	0.5000	0.4000
$\nu_{\hat{C}_2}$	0.2000	0.1000	0.2000	0.4000	03000	0.3000
$\mu_{\hat{C}_3}$	0.9000	0.8000	0.9000	0.8000	0.9000	0.9000
$\nu_{\hat{C}_3}$	0.1000	0.2000	0.1000	0.2000	0.1000	0.1000
$\mu_{\hat{C}_4}$	0.6000	0.5000	0.5000	0.6000	0.6000	0.5000
$\nu_{\hat{C}_4}$	0.3000	0.2000	0.3000	0.2000	0.3000	0.2000
$\mu_{\hat{C}_5}$	0.8000	0.7000	0.7000	0.8000	0.7000	0.6000
$\nu_{\hat{C}_5}$	0.2000	0.2000	0.3000	0.1000	0.3000	0.4000
$\mu_{\hat{C}_6}$	0.8000	0.8000	0.8000	0.9000	0.8000	0.8000
$\nu_{\hat{C}_6}$	0.2000	0.1000	0.2000	0.1000	0.2000	0.1000

Table 3 Results via Senapati and Yager's technique [28]

Admission process	$\hat{C}_1$	$\hat{C}_2$	$\hat{C}_3$	$\hat{C}_4$	$\hat{C}_5$	$\hat{C}_6$
$\hat{S}_1$	0.7640	1.0805	1.2784	1.4548	1.586	1.7166
$\hat{S}_2$	0.7591	1.0883	1.3178	1.4557	1.6027	1.7690
$\hat{S}_3$	0.6452	0.8649	1.0660	1.2674	1.3928	1.5099
$\hat{S}_4$	0.7216	0.9565	1.2331	1.4236	1.5831	1.7519
$\hat{S}_5$	0.3858	0.5352	0.6876	0.8282	0.9132	1.0011
$\hat{S}_6$	0.4661	0.5991	0.6893	0.8612	0.9761	1.0268
$\hat{S}_7$	0.9291	1.2823	1.6735	1.9302	2.1531	2.4020
$\hat{S}_8$	0.4103	0.6474	0.7390	0.8265	0.8686	0.9306
$\hat{S}_9$	0.6303	0.9335	1.2510	1.4315	1.5878	1.7935
$\hat{S}_{10}$	0.6342	1.0072	1.2230	1.3685	1.4770	1.6252

Table 4 Results via modified Senapati and Yager's technique

Admission process	$\hat{C}_1$	$\hat{C}_2$	$\hat{C}_3$	$\hat{C}_4$	$\hat{C}_5$	$\hat{C}_6$
$\hat{S}_1$	0.2547	0.3602	0.4261	0.4849	0.5287	0.5722
$\hat{S}_2$	0.253	0.3628	0.4393	0.4853	0.5342	0.5897
$\hat{S}_3$	0.2151	0.2883	0.3553	0.4225	0.4642	0.5033
$\hat{S}_4$	0.2406	0.3188	0.411	0.4746	0.5277	0.5839
$\hat{S}_5$	0.1286	0.1784	0.2292	0.2761	0.3044	0.3337
$\hat{S}_6$	0.1554	0.1997	0.2297	0.2871	0.3254	0.3423
$\hat{S}_7$	0.3097	0.4275	0.5578	0.6434	0.7177	0.8007
$\hat{S}_8$	0.1368	0.2158	0.2463	0.2755	0.2896	0.3102
$\hat{S}_9$	0.2101	0.3112	0.417	0.4772	0.5293	0.5979
$\hat{S}_{10}$	0.2114	0.3358	0.4077	0.4562	0.4923	0.5417

the number of the considered parameters and the cardinality of the ground set to foster efficiency and accuracy. The novel Fermatean fuzzy distance technique was deployed to determine course allocations in tertiary institution based on the applicants academic performances. While exploring the Fermatean fuzzy distance technique in [28], we discovered that the approach takes no cognizance of the number of the conventional parameters of FFSs and neglected the cardinality of the set upon which the FFSs are derivable, and thereby violating the conditions of distance function. But, the modified Fermatean fuzzy distance technique completely satisfied the conditions of distance function and hence yields reasonable and accurate results compare to the approach in [28]. Some theoretical results on the novel Fermatean fuzzy distance were deduced and proved to authenticate the validity of the technique. This modified Fermatean fuzzy distance could be deployed to solve multi-attributes group decision-making problems in future and could be re-modified to capture more than three parameters for the case of picture fuzzy sets.

References

1. Atanassov, K.T.: Intuitionistic fuzzy sets. Fuzzy Set Syst. **20**, 87–96 (1986)
2. Atanassov, K.T.: Geometrical interpretation of the elements of the intuitionistic fuzzy objects. Preprint IM-MFAIS-1-89, Sofia (1989)
3. Begum, S.S., Srinivasan, R.: Some properties on intuitionistic fuzzy sets of third type. Ann. Fuzzy Math. Inform. **10**(5), 799–804 (2015)
4. Boran, F.E., Akay, D.: A biparametric similarity measure on intuitionistic fuzzy sets with applications to pattern recognition. Inf. Sci. **255**(10), 45–57 (2014)
5. Burillo, P., Bustince, H.: Entropy on intuitionistic fuzzy sets and on interval-valued fuzzy sets. Fuzzy Set Syst. **78**, 305–315 (1996)
6. Chen, S.M., Chang, C.H.: A novel similarity measure between Atanassov's intuitionistic fuzzy sets based on transformation techniques with applications to pattern recognition. Inf. Sci. **291**, 96–114 (2015)
7. Dhar, S., Kundu, M.K.: Interval type-2 fuzzy set and theory of weak continuity constraints for accurate multiclass image segmentation. IEEE Trans. Fuzzy Syst. **28**(9), 2151–2163 (2019)
8. De, S.K., Biswas, R., Roy, A.R.: An application of intuitionistic fuzzy sets in medical diagnosis. Fuzzy Set Syst. **117**(2), 209–213 (2001)
9. Diamond, P., Kloeden, P.: Metric Spaces of Fuzzy Sets Theory and Applications. Word Scientific, Singapore (1994)
10. Ejegwa, P.A.: Distance and similarity measures for Pythagorean fuzzy sets. Granul. Comput. **5**(2), 225–238 (2020)
11. Ejegwa, P.A.: Modified Zhang and Xu's distance measure of Pythagorean fuzzy sets and its application to pattern recognition problems. Neural Comput. Appl. **32**(14), 10199–10208 (2020)
12. Ejegwa, P.A.: New similarity measures for pythagorean fuzzy sets with applications. Int. J. Fuzzy Comput. Modelling **3**(1), 75–94 (2020)
13. Ejegwa, P.A.: Generalized triparametric correlation coefficient for Pythagorean fuzzy sets with application to MCDM problems. Granul. Comput. **6**(3), 557–566 (2021)
14. Ejegwa, P.A.: Novel correlation coefficient for intuitionistic fuzzy sets and its application to multi-criteria decision-making problems. Int. J. Fuzzy Syst. Appl. **10**(2), 39–58 (2021)
15. Ejegwa, P.A., Awolola, J.A.: Novel distance measures for Pythagorean fuzzy sets with applications to pattern recognition problems. Granul. Comput. **6**, 181–189 (2021)
16. Ejegwa, P.A., Awolola, J.A.: Real-life decision making based on a new correlation coefficient in Pythagorean fuzzy environment. Ann. Fuzzy Math. Inform. **21**(1), 51–67 (2021)
17. Ejegwa, P.A., Onyeke, I.C.: A robust weighted distance measure and its applications in decision making via Pythagorean fuzzy information. J. Inst. Elect. Comput. **3**, 87–97 (2021)
18. Ejegwa, P.A., Onyeke, I.C.: A novel intuitionistic fuzzy correlation algorithm and its applications in pattern recognition and student admission process. Int. J. Fuzzy Syst. Appl. **11**(1) (2022). https://doi.org/10.4018/IJFSA.285984
19. Ejegwa, P.A., Nwankwo, K.N., Ahmad, M., Ghazal, T.M., Khan, M.A.: Composite relation under Fermatean fuzzy context and its application in disease diagnosis. Informat. **32**(10), 87–101 (2021)
20. Garg, H.: A new improved score function of an interval-valued Pythagorean fuzzy set based TOPSIS method. Int. J. Uncertainty Quantif. **7**(5), 463–474 (2017)
21. Garg, H.: A linear programming method based on an improved score function for interval-valued Pythagorean fuzzy numbers and its application to decision-making. Int. J. Uncert. Fuzz. Knowl.-Based Syst. 29(1), 67–80 (2018)

22. Hatzimichailidis, A.G., Papakostas, A.G., Kaburlasos, V.G.: A novel distance measure of intuitionistic fuzzy sets and its application to pattern recognition problems. Int. J. Intell. Syst. **27**, 396–409 (2012)
23. Li, D.Q., Zeng, W.Y.: Distance measure of Pythagorean fuzzy sets. Int. J. Intell. Syst. **33**, 348–361 (2018)
24. Liang, D., Xu, Z.: The new extension of TOPSIS method for multiple criteria decision making with hesitant Pythagorean fuzzy fets. Appl. Soft Comput. **60**, 167–179 (2017)
25. Liu, D., Liu, Y., Chen, X.: Fermatean fuzzy linguistic set and its application in multicriteria decision making. Int. J. Intell. Syst. **34**(5), 878–894 (2019)
26. Senapati, T., Yager, R.R.: Fermatean fuzzy weighted averaging/geometric operators and its application in multi-criteria decision-making methods. Eng. Appl. Artif. Intell. **85**, 112–121 (2019)
27. Senapati, T., Yager, R.R.: Some new operations over Fermatean fuzzy numbers and application of Fermatean fuzzy WPM in multiple criteria decision making. Informat. **30**(2), 391–412 (2019)
28. Senapati, T., Yager, R.R.: Fermatean fuzzy sets. J. Amb. Intell. Human. Comput. **11**, 663–674 (2020)
29. Sergi, D., Sari, I.U., Senapati, T.: Extension of capital budgeting techniques using interval-valued Fermatean fuzzy sets. J. Intell. Fuzzy Syst. **42**(1), 365–376 (2022)
30. Silambarasan, I.: New operators for Fermatean fuzzy sets. Ann. Commun. Math. **3**(2), 116–131 (2020)
31. Szmidt, E., Kacprzyk, J.: Distances between intuitionistic fuzzy sets. Fuzzy Set Syst. **114**, 505–518 (2000)
32. Szmidt, E., Kacprzyk, J.: Medical diagnostic reasoning using a similarity measure for intuitionistic fuzzy sets. Note IFS **10**(4), 61–69 (2004)
33. Wang, W., Xin, X.: Distance measure between intuitionistic fuzzy sets. Pattern Recog. Lett. **26**, 2063–2069 (2005)
34. Yager, R.R.: Pythagorean membership grades in multicriteria decision making. Technical Report MII-3301 Machine Intelligence Institute, Iona College, New Rochelle, NY (2013)
35. Yang, Z., Garg, H., Li, X.: Differential calculus of Fermatean fuzzy functions: continuities, derivatives, and differentials. Int. J. Comput. Intell. Syst. **14**(1), 282–294 (2021)
36. Zadeh, L.A.: Fuzzy sets. Inform. Cont. **8**, 338–353 (1965)
37. Zhang, X.: A novel approach based on similarity measure for Pythagorean fuzzy multiple criteria group decision making. Int. J. Intell. Syst. **31**, 593–611 (2016)
38. Zhang, X.L., Xu, Z.S.: Extension of TOPSIS to multiple criteria decision making with Pythagorean fuzzy sets. Int. J. Intell. Syst. **29**(12), 1061–1078 (2014)
39. Zeng, W., Li, D., Yin, Q.: Distance and similarity measures of Pythagorean fuzzy sets and their applications to multiple criteria group decision making. Int. J. Intell. Syst. **33**(11), 2236–2254 (2018)

A Fuzzy Decision Making Inventory Model for Deteriorating Items Under Discounted Partial Advance-Partial Delayed Payment Strategy

Biman Kanti Nath and Nabendu Sen

Abstract This article presents a buyer's EOQ model under discounted partial advance payment and partial delayed payment strategies for a single supplier–single buyer supply chain. The durations of advance payment and delayed payment are set by the supplier. Supplier also offers cash discount to the buyer due to advance payment. The rate of cash discount depends on the amount of advance payment made at the time of placing the order, which seems to be realistic in the present business world. It is often observed in the market that a large pile of goods on the shelf has positive affect on the demand; this model considers demand depending on instantaneous stock level of the commodity. Further, to deal with the uncertainty, several inventory parameters are taken as fuzzy numbers (parabolic-flat fuzzy number). For defuzzification of the total average profit of the system, Graded Mean Integration Representation (GMIR) method is used. The aim of this paper is to determine optimal cycle time and optimal ordering quantity for the buyer under no shortages in order to maximize the profit of the system. The necessary theoretical results have been derived in connection with optimality of the total average profit function. To illustrate the model, numerical examples describing a hypothetical system are considered. The concavity of the profit function is visible from graph. Finally, in order to investigate the effects of changes of different system parameters, post optimality analysis is performed.

Keywords Stock-dependent demand · Partial advance payment · Partial delayed payment · Price discount · Parabolic-fuzzy number · Graded mean integration representation

B. K. Nath · N. Sen (✉)
Assam University, Silchar, Assam 788011, India
e-mail: nsen08@yahoo.com

B. K. Nath
e-mail: bimannath2012@gmail.com

L. Sahoo et al. (eds.), *Real Life Applications of Multiple Criteria Decision Making Techniques in Fuzzy Domain*, Studies in Fuzziness and Soft Computing 420,
https://doi.org/10.1007/978-981-19-4929-6_12

1 Introduction

In today's competitive business world, suppliers put major effort to extend facilities in order to attract and turn prospective customers into actual buyers. To draw the attention of the customers and to stimulate sales, suppliers undertake different marketing strategies from time to time as market behaviour does not always remain stable for longer periods. Trade credit policy is one amongst such facilities which is popularly used in supplier–buyer contract. The buyer may settle the entire amount or a portion of it during the credit period. During the delay period also known as credit period, no interest on the amount due will be charged, but after that fixed interest-free period, supplier charges interest on the remaining dues. This trade credit policy is very beneficial for small buyers as it allows buyers some grace period to settle the account. In this permissible interest-free period, buyers collect the revenues on the sales and get chance to earn interest on the collected revenue. From supplier's perspective, it results in easy product disposal and boasting up customers' demand. For these reasons, trade credit policy has been one of popular marketing strategies often adopted in business to stimulate sales. Goyal [18] was the first to introduce the concept of credit policy by developing an economic order quantity (EOQ) model deeming for permissible delay in payments. Aggarwal and Jaggi [2] studied ordering policy with constant demand with trade credit policy. Jamal et al. [27] also put forward ordering policy for deteriorating items allowing shortages. Teng [63] extended Goyal [18] model considering dissimilarity amongst purchase cost and selling price. Abad and Jaggi [1] considered seller–buyer channel with price-dependent demand and trade credit policy. Huang [22] studied an ordering policy with cash discount facility under delay in payment strategy and constant demand. Ouyang et al. [44] further proposed a model for non-instantaneous deteriorating items under permissible delay in payment strategy. Chang et al. [8], Chen and Kang [11], Sarkar [51], Taleizadeh et al. [60], Tiwari et al. [65] and Jaggi et al. [24] are some worth mentioning works with constant demand under delayed payment strategy. Mashud et al. [41] developed a pricing model under the assumption of trade credit policy and preservation technology investment. Recently, Das et al. [13] proposed a multiple credit period-based trade credit policy with price- and stock-dependent demand for non-instantaneous deteriorating items under preservation technology investment.

Apart from credit policy, another popular tactic prevailing in real business and market scenario is advance payment strategy. In advance payment policy, supplier often asks for full payment or partial payment of the total purchasing cost from the buyer at the time of placement of supply order of the commodities. Advance payment policy is useful for buyers because it brings out various discount facilities (viz., trade discounts, promotional discounts, quantity discounts, seasonal discounts, cash discounts etc.) for them. Moreover, it reduces the risk of cancellation of orders. It is also beneficial from supplier perspective as it does not stop cash flow and it motives the customers to purchase more, thereby increases the sales. Out of the many discount facilities offered due to advance payment, cash discount facility has become very popular in real business world. It is also observed that supplier allows cash discount due to advance payment where the rate of discount is fixed or sometimes it depends upon amount of advance payment. This paper considers the case of cash discount due to advance payment where the discount rate depends upon the amount of prepayment. The effect of prepayment (advance payment) on total cost was studied for the first time by Zhang [69] where the author found that advance payment has positive effects on total cost of the system. Maiti et al. [40] proposed an inventory model to study the effect of advance payment on the inventory system. They developed the model with stochastic lead time and price-dependent demand over finite time horizon. Their study also showed positive effect of advance payment on inventory system. Gupta et al. [19] further extended their model by assuming all inventory cost parameters as interval numbers. Taleizadeh et al. [61] proposed optimal ordering policy addressing multiple partial prepayment scheme with constant demand under three conditions—no shortage, partial backlogging and complete backlogging. This work was further extended by Taleizadeh [62] for deteriorating items. Zhang et al. [71] suggested a two-stage supply chain model under prepayment strategy. Khan et al. [34] discussed a two-warehouse EOQ model for price-dependent demand where multiple prepayment scheme and partial backlogging were allowed. This work had been further generalized by Shaikh et al. [56] allowing all inventory cost parameters as interval numbers. Khan et al. [33] proposed a model with price- and stock-dependent demand for deteriorating items to study the effect of advance payment on total cost of the system. In their model, cash discount facility as a result of advance payment (full and partial) was contemplated. Khan et al. [29] also studied an ordering policy for deteriorating items with price-sensitive demand and order size-dependent discount policy. Khan et al. [32] discussed a two-warehouse system for non-instantaneous deteriorating items under advance payment policy to find optimal ordering decision. Chaudhari et al. [10] developed an inventory model for deteriorating items with quadratic time-dependent demand with multiple advance payment scheme. Khan et al. [31] demonstrated inventory model for perishable items with price- and advertisement-dependent demand under the effect of advance payment policy. Recently, Rahman et al. [47] also proposed an inventory model for perishable goods under advance payment policy. They assumed hybrid price (a convex combination of linear and non-linear price) and stock-dependent demand.

Though the trade credit policy is a popular plan of action for supplier to elevating sales, it brings financial complication for supplier, whereas, although advance payment policy brings various discount facilities, it does not allow extra grace period to buyers for settling the account. Thus, to meet the flaws and to facilitate the sales, the concept of partial advance-partial delayed payment strategy has been popularly used in supplier–buyer contract in recent times. In case of partial advance payment-partial trade credit policy, buyer pays some portion of the entire purchase cost whilst placing the order and pays off the remaining dues or a part of remaining dues at the end of fixed interest-free period (credit period) offered by the supplier to the buyer. This policy has drawn considerable attention of researchers in recent times. In this direction, Zhang et al. [70] developed an EOQ model for constant demand with the incorporation of full advance payment, and partial advance-partial delayed payment conditions. They observed that buyer's optimal replenishment cycle is not affected by the length of advance payment period. Zia and Taleizadeh [72] proposed an EOQ model with constant demand under partial advance-partial delayed payment strategy linked to ordering quantity. Agrawal et al. [3] developed a model with iso-elastic-price-dependent demand under partial advance-partial delayed payment strategy. They allowed price discount due advance payment. Mashud et al. [42] studied a two-warehouse system for price- and advertisement-dependent demand under multiple advance payment and trade credit policy. They assumed that the rate of discount depends on number of instalments. For more details, one may refer to Thangam [64], Li et al. [39], Krommyda et al. [35] and Khan et al. [30].

Market demand is another most important factor in any inventory system. In inventory literature, lots of models have been developed and studied by assuming different types of demand patterns for different types of products to represent real market scenarios. Demand depends on plethora of factors like price, quality, time, credit period, advertisement etc. On hand displayed stock is one of such parameters which also largely affects market demand. It is noticed that a large pile of goods displayed attracts customers to buy more, thereby raising customers' demand. This phenomenon is usually observed in real market. Thus, considering the practical aspect of market, it will be more appropriate to consider demand to be stock-dependent. Wu et al. [67], Sajadieh et al. [50], Pando et al. [45], Annadurai and Uthayakumar [4] and Singh et al. [57, 58] are some worth mentioning works related to stock-dependent demand.

The assumption that inventory parameters namely, holding cost, ordering cost, deterioration rate, demand of the product etc. which control an inventory system will have fixed/crisp values do not represent real market behaviour. Indeed, these parameters should be imprecise, because these parameters might fluctuate due to dynamic market nature. To handle such uncertain/imprecise parameters, researchers use interval, fuzzy, stochastic approaches or a combination of them. In interval approach, imprecise parameters are represented by interval numbers whilst in stochastic approach these imprecise parameters are taken as random variables having known probability distribution. In fuzzy approach, the imprecise parameters are assumed to be fuzzy numbers with suitable membership functions. And in the combined approach, some are taken in one form and other remaining parameters are taken in other form. Kalpakam and Sapan [28] developed a model for perishable items with stochastic lead time. Shah and Shah [54] presented a probabilistic model under delay in payment. Roy et al. [48] developed an inventory model with stock-dependent demand and random planning horizon. Bhunia and Shaikh [6] used Weibull distribution for deterioration in their paper. Duari and Chakrabarti [16] also developed a two-warehouse system with Weibull distribution for deterioration rate and considered stock-dependent demand. Saha and Sen [49] proposed an inventory model where deterioration follows uniform, beta and triangular distributions. Tripathi and Chaudhary [66] presented an EOQ model with Weibull distribution for deterioration under inflation and trade credit policy. Gupta et al. [20] developed an inventory model under interval uncertainty with demand dependent on selling price, displayed stock level and frequency of advertisement. Bhunia and Shaikh [7] proposed a two-warehouse inventory model under inflationary condition and linear time-dependent demand. They assumed interval valued inventory cost and obtained the optimal solution using Particle Swarm Optimization (PSO) technique. A partially integrated production-inventory model was proposed by Bhunia et al. [5] for selling price and marketing cost-dependent demand, and interval valued inventory cost. Shaikh et al. [56] proposed a two-warehouse model with advance payment scheme, price-dependent demand and interval valued cost parameters. Chang et al. [9] proposed an economic reordering point system for fuzzy backordered quantity. Yao et al. [68] developed a fuzzy EOQ inventory model with fuzzy order quantity and fuzzy total demand quantity. Priyan et al. [46] developed an EOQ model with fuzzy parameters. Dutta and Kumar [17] also proposed a fuzzy model with no-shortages. Sen et al. [52] discussed a fuzzy inventory model considering triangular fuzzy numbers. De and Mahata [14] proposed a fuzzy EOQ model under cloudy fuzzy environment. Jain et al. [26] proposed an imperfect and repair model under fuzzy impreciseness time-dependent demand. Supakar and Mahato [59] used fuzzy-stochastic approach to develop their model under advance payment policy with constant demand. They allowed lead time to be a random variable and other inventory parameters like demand, unit purchase cost, holding cost, ordering cost etc. as parabolic-flat-fuzzy numbers. Indrajitsingha et al. [23] discussed a two-warehouse inventory system with fuzzy approach, where holding cost, deterioration rate, shortage cost and lost sale cost are imprecise in nature and represented by triangular fuzzy numbers. Hemalatha and Annadurai [21] proposed an inventory model under advance payment policy. The impreciseness of system parameters (cost components) has been represented by four types of fuzzy numbers—triangular, trapezoidal, pentagonal and hexagonal fuzzy numbers.

In the existing inventory literature, lots of models have been developed in crisp environment considering different demand patterns (like constant demand or demand dependent on time or selling price or credit period or instantaneous stock level or advertisement or combination of them). Some of these papers have considered delay in payment [13, 24, 60] and some have regarded advance payment scheme [32–34] or both advance and delayed payment strategies [3, 42]. But very few have regarded uncertainty in inventory parameters under delayed and/or advance payment [40, 56, 59]. Although the current literature review does not include all works, as far as the survey is concerned, it is visible that there is a gap in terms of developing inventory model for deteriorating items with stock-dependent demand under partial advance-partial delayed payment strategy and under imprecise environment. So, in this paper an attempt has been made to develop an inventory model for deteriorating items with stock-dependent demand in the context of partial advance-partial delayed payment policy, where cash discount for advance payment is taken under consideration. Here, the discount rate depends upon the amount of initial payment made whilst placing the order. Further, the deteriorated items are salvaged to earn some revenue. The demand parameters, ordering cost, holding cost, lead time, deterioration rate, salvage co-efficient are taken to be fuzzy numbers (parabolic-flat fuzzy number) to represent their impreciseness. Hence, the model transforms into a fuzzy model and it is more realistic with regard to uncertainty prevailing in the real market scenario.

For defuzzification of fuzzy numbers, the very effective Graded Mean Integration Representation (GMIR) method is used. The main objective of this model is to obtain optimal replenish policy so that profit of system is maximized. The necessary theoretical results have been derived for optimality of the profit function (see Appendix 2). Moreover, the concavity of the profit function has been shown graphically. A comparison of the present work with some of the related works is presented in Table 1.

The structure of rest of this paper is as follows. Section 2 provides detailed description about the assumptions and notations to be used throughout this paper. In Sect. 3, some basic definitions related to fuzzy numbers are provided. Section 4 contains the mathematical models (case 1 and case 2) under crisp and fuzzy environments. Section 5 contains solution procedure to select optimal case, and hence to find the optimal ordering quantity. Section 6 deals with numerical examples to illustrate the developed models. This is followed by sensitivity analysis performed in Sect. 7. Section 8 offers some managerial implications for ordering decision. Lastly, Sect. 9 exhibits conclusions and further research scopes.

2 Assumptions and Notations

To develop the model, following assumptions and notations are used.

2.1 Assumptions

- The model considers single deteriorating item.
- Demand depends on instantaneous stock level of items.
- Replenishment rate is infinite and lead time is non-zero.
- There is no repair or replacement of the deteriorating items for the period under consideration.
- Occurrence of shortages is not allowed.
- Partial Advance-Partial Delayed payment scheme is adopted by the buyer. Here, the buyer makes payment for a certain portion of total quantity ordered before receiving the goods. The supplier also renders cash discount for early payment to keep the customers in hand. The cash discount rate depends on the amount of prepayment. Moreover, the supplier extends a credit period to settle the remaining due.

2.2 Notations

$I(t)$	inventory level at any time t.

Table 1 Related research works and their contributions

Author	Payment	Demand	Deterioration	Salvage value	Cash discount	Inventory parameter(s)
Gupta et al. [19]	Advance payment	Constant	No	No	Yes	Interval
Maiti et al. [40]	Advance payment	Price-dependent	No	No	No	Stochastic
Zhang et al. [70]	Full advance, and partial advance-partial delayed payment	Constant	No	No	Yes	Crisp
Zia and Taleizadeh [72]	Multiple advance payments	Constant	No	No	No	Crisp
Lashgari et al. [38]	Partial advance	Constant	No	No	No	Crisp
Shabani et al. [53]	Delay in payment	Constant	Yes	No	No	Fuzzy
Diabat et al. [15]	Full advance, and partial advance-partial delayed payment	Time-dependent	Constant	No	No	Crisp
Jaggi et al. [25]	Delay in payment	Constant	Constant	No	No	Crisp
Das and Roy [12]	No	Time-dependent	Constant	Yes	No	Interval
Shaikh et al. [55]	Delay in payment	Price- and advertisement-dependent	Constant	No	No	Fuzzy
Supakar and Mahato [59]	Advance payment	Constant	No	No	Yes	Fuzzy-stochastic

(continued)

Table 1 (continued)

Author	Payment	Demand	Deterioration	Salvage value	Cash discount	Inventory parameter(s)
Khan et al. [34]	Multiple prepayment	Price-dependent	Constant	No	No	Crisp
Kumar et al. [37]	No	Time-dependent	Constant	Yes	No	Crisp
Shaikh et al. [56]	Advance payment	Price-dependent	Constant	No	No	Interval
Agrawal et al. [3]	Partial advance–partial delayed payment	Price-dependent	No	No	Yes	Crisp
Khan et al. [31]	Multiple prepayment	Advertisement-and price-dependent	Expiration date-dependent	No	No	Crisp
Khan et al. [33]	Full advance and partial advance payment	Price- and stock-dependent	Constant	No	Yes	Crisp
Kumar et al. [36]	Delay in payment	Time-dependent	No	No	No	Fuzzy
Mashud et al. [42]	Advance and delayed in payment	Price- and advertisement-dependent	Constant	No	Yes	Crisp
This paper	Partial advance–partial delayed payment	Stock-dependent	constant	Yes	Yes	Fuzzy

$D(I(t))$	demand rate of the form $a + b\,I(t)$, where $a, b > 0$ are constants but fuzzy in nature.
Q	quantity to be ordered.
T	cycle length.
θ	deterioration rate, $0 < \theta < 1$.
$\tilde{\theta}$	fuzzy deterioration rate.
A	ordering cost in \$ per order.
$\widetilde{A}$	fuzzy ordering cost in \$ per order.
x	per unit purchasing cost of the item in \$, $p > x$.
h	holding cost per unit item per unit time.
$\tilde{h}$	fuzzy holding cost per unit item per unit time.
L	lead time.
$\widetilde{L}$	fuzzy lead time.
M	credit period offered by the supplier to the buyer.
m_1	a fraction of Q for which payment is made at the time of order $0 \le m_1 \le 1$.
m_2	a fraction of Q for which payment is made at time $t = M$, $0 \le m_2 \le 1$ and $0 \le m_1 + m_2 \le 1$.
p	per unit selling price of the items, where $x < p$.
η	per unit salvage coefficient.
$\tilde{\eta}$	fuzzy per unit salvage coefficient.
I_e	interest rate earned per unit time by the buyer.
I_B	loan interest rate per unit time to be paid by the buyer to the financer or bank.
I_c	loan interest rate paid per unit time to the supplier by the buyer.
d_1	discount rate in case of full advance payment or maximum discount rate to be rendered by the supplier.
$d_2 = f(m_1)d_1$	discount rate in case of partial advance payment, where $f(m_1)$ is an increasing function of m_1 and $0 < f(m_1) < 1$.
TC_1	total cost of the system when $M < T$.
TC_2	total cost of the system when $M \ge T$.
TPU_1	total profit per unit time when $M < T$.
$\widetilde{\mathrm{TPU}}_1$	fuzzy total profit per unit time when $M < T$.
TPU_2	total profit per unit time when $M \ge T$.
$\widetilde{\mathrm{TPU}}_2$	fuzzy total profit per unit time when $M \ge T$.
$d_F(\widetilde{\mathrm{TPU}}_1)$	defuzzified total profit per unit time when $M < T$.
$d_F(\widetilde{\mathrm{TPU}}_2)$	defuzzified total profit per unit time when $M \ge T$.

*represents optimal case.

2.3 Decision Variable

T independent decision variables.

TPU_i dependent decision variables, $i = 1, 2$.

3 Some Definitions

This section provides some basic definitions related to fuzzy numbers. For definitions in this section, we refer to Supakar and Mahato [59].

3.1 Fuzzy Sets

A fuzzy set is a class that acknowledges possibility of grade membership. Let Y be a space of objects. A fuzzy set $\widetilde{F}$ in Y is a set of pairs $\widetilde{F} = \{(y, \mu_{\widetilde{F}}(y)) : y \in Y\}$, where $\mu_{\widetilde{F}}(y) : Y \to [0, 1]$ is a function called 'membership function or grade of membership of y in F'.

3.2 α-Cut

The α-cut of a fuzzy set $\widetilde{F}$ is defined by $F_\alpha = \{y : \mu_{\widetilde{F}}(y) = \alpha, \alpha \geq 0\}$.

3.3 Convex Fuzzy Set

A fuzzy set $\widetilde{F}$ is said to be convex if for all $y_1, y_2 \in Y$,

$$\mu_{\widetilde{F}}(\lambda y_1 + (1 - \lambda) y_2) \geq \min\{\mu_{\widetilde{F}}(y_1), \mu_{\widetilde{F}}(y_2)\},$$

where $\lambda \in [0, 1]$.

3.4 Normal Fuzzy Set

A normal fuzzy set $\widetilde{F}$ is a fuzzy set with the property that there exists at least one $y \in Y$ such that $\mu_{\widetilde{F}}(y) = 1$.

3.5 Fuzzy Number

A fuzzy set defined on the universal set of real numbers is said to be fuzzy number if it is normal as well as convex. If $\widetilde{F}$ is a fuzzy number then its membership function is denoted by

$$\mu_{\widetilde{F}}(y) = \begin{cases} L(y), & \text{if } y < U \\ 1, & \text{if } U \le y \le V \\ R(y), & \text{if } y > V \end{cases}$$

where $L(x)$ and $R(x)$ are left and right shape functions respectively.

3.6 Parabolic-Flat Fuzzy Number (PfFN)

The parabolic-flat fuzzy number is denoted by $\widetilde{F} = (f_1, f_2, f_3, f_4)$ and is defined by the membership function as;

$$\mu_{\widetilde{F}}(x) = \begin{cases} 1 - \frac{(f_2-x)^2}{(f_2-f_1)^2} & \text{if } f_1 \le x \le f_2 \\ 1 & \text{if } f_2 \le x \le f_3 \\ 1 - \frac{(x-f_3)^2}{(f_4-f_3)^2} & \text{if } f_3 \le x \le f_4 \\ 0 & \text{otherwise} \end{cases}$$

3.7 Arithmetic Operations

Let $\widetilde{F} = (f_1, f_2, f_3, f_4)$ and $\widetilde{G} = (g_1, g_2, g_3, g_4)$ be two parabolic-flat fuzzy numbers. The arithmetic operations between them are given as below;

Addition:

$$\widetilde{F} + \widetilde{G} = (f_1, f_2, f_3, f_4) + (g_1, g_2, g_3, g_4) = (f_1 + g_1, f_2 + g_2, f_3 + g_3, f_4 + g_4)$$

Subtraction:

$$\widetilde{F} - \widetilde{G} = (f_1, f_2, f_3, f_4) - (g_1, g_2, g_3, g_4) = (f_1 - g_4, f_2 - g_3, f_3 - g_2, f_4 - g_1)$$

Multiplication:

$$\widetilde{F}\widetilde{G} = (f_1, f_2, f_3, f_4)(g_1, g_2, g_3, g_4) = (f_1 g_1, f_2 g_2, f_3 g_3, f_4 g_4)$$

Division:

$$\frac{\widetilde{F}}{\widetilde{G}} = \left(\frac{f_1}{g_4}, \frac{f_2}{g_3}, \frac{f_3}{g_2}, \frac{f_4}{g_1}\right)$$

Multiplication by a scalar:

If $\lambda \neq 0$ is any real number then

$$\lambda\widetilde{F} = (f_1, f_2, f_3, f_4) = \begin{cases} (\lambda f_1, \lambda f_2, \lambda f_3, \lambda f_4), & \lambda > 0 \\ (\lambda f_4, \lambda f_3, \lambda f_2, \lambda f_1), & \lambda < 0 \end{cases}.$$

3.8 Graded Mean Integration Representation (GMIR) Method for Defuzzification

For the generalized fuzzy number $\widetilde{F}$ with membership function $\mu_{\widetilde{F}}$, the Graded Mean Integration Value (GMIV) of $\widetilde{F}$ is

$$\begin{aligned} d_F(\widetilde{F}) &= \frac{\int_0^1 \left[y(1-w_F)L^{-1}(y) + w_F R^{-1}(y)\right] \mathrm{d}y}{\int_0^1 y\,\mathrm{d}y} \\ &= 2\int_0^1 \left[y(1-w_F)L^{-1}(y) + w_F R^{-1}(y)\right] \mathrm{d}y \end{aligned}$$

where $w_F \in [0, 1]$ signifies the degree of optimism. $w_F = 1$ refers to optimistic decision makers' point of view, $w_F = 0$ refers to pessimistic point of view, and $w_F = 0.5$ indicates a moderately optimistic point of view.

If $\widetilde{F}$ is a parabolic-flat fuzzy number, the inverses of $L(y)$ and $R(y)$ are given by

$$L^{-1}(y) = f_2 - (f_2 - f_1)\sqrt{1-y}, \quad R^{-1}(y) = f_3 + (f_4 - f_3)\sqrt{1-y}$$

Therefore, the GMIV of $\widetilde{F}$ is

$$\begin{aligned} d_F(\widetilde{F}) &= 2\int_0^1 \left[y(1-w_F)L^{-1}(y) + w_F R^{-1}(y)\right] \mathrm{d}y \\ &= 2\int_0^1 \left[y(1-w_F)\left(f_2 - (f_2 - f_1)\sqrt{1-y}\right)\right. \end{aligned}$$

$$+w_F\left(f_3 + (f_4 - f_3)\sqrt{1-y}\right)\Big]\mathrm{d}y$$
$$= \frac{(1-w_F)}{15}(8f_1 + 7f_2) + \frac{w_F}{15}(7f_3 + 8f_4)$$

Taking $w_F = 0.5$, we get

$$d_F(\widetilde{F}) = \frac{1}{30}(8f_1 + 7f_2 + 7f_3 + 8f_4)$$

4 Problem Description and Model Formulation

Consider a situation where a buyer places an order L time units before beginning of selling season. In order to confirm the order, the buyer makes payment for a portion m_1 of total quantity Q. The supplier offers a cash discount $d_2 = f(m_1)d_1 = m_1 d_1$ for early payment to keep the customers in hand. Therefore, the payment made at the time of order is $m_1 Qx(1 - d_2)$. The supplier also allows a permissible period M to the buyer to settle the account. During this period no interest will be charged but after which a charge on the remaining due at the rate I_c will be applicable. The buyer sells the goods and the revenue collected are kept in an interest-bearing account to earn interest at the rate I_e. To make payment at time of order and/or at the end of credit period M, the buyer takes loan from financer (like banks) at a certain interest rate I_B. The loan amount along with interest charged are paid back to the financer at the end of selling period that is, at $t = T$ form the revenue collected. Now we encounter with following two cases:

4.1 Case 1: Credit Period is Less Than Cycle Length ($M < T$)

In this case the buyer makes a payment at the rate x for the portion m_2 of Q at $t = M$ by taking loan from the financer at the interest rate I_B. No interest is charged by the supplier during the period $[0, M]$. The remaining payment for the portion $(1 - m_1 - m_2)$ of Q is paid at $t = T$ (Fig. 1) along with the interest charged by the supplier on the due amount $(1 - m_1 - m_2)xQ$ for the time period $[M, T]$ at the rate I_c. The governing differential equation along with the conditions is given by:

$$\frac{\mathrm{d}}{\mathrm{d}t}I(t) + \theta I(t) = -\{a + bI(t)\},\quad I(0) = Q,\quad I(T) = 0 \tag{1}$$

The solution to the Eq. (1) and the ordering quantity are given as below:

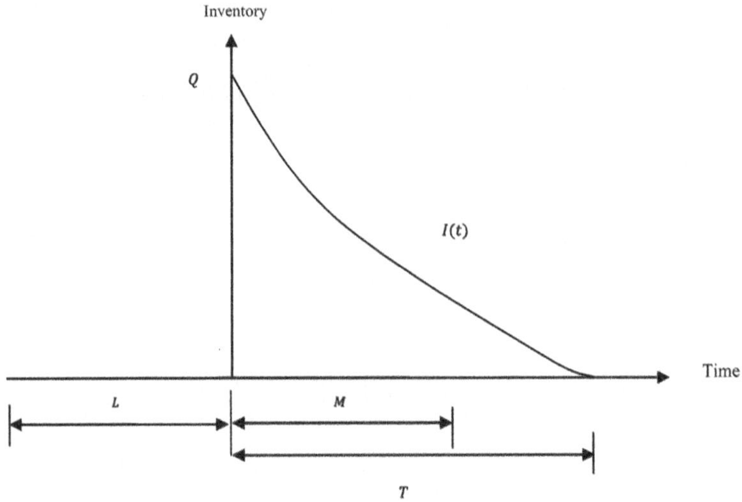

Fig. 1 Inventory level when $M < T$

$$I(t) = \frac{a}{\theta + b}\left[e^{(\theta+b)(T-t)} - 1\right], \quad 0 \le t \le T \tag{2}$$

and

$$Q = \frac{a}{\theta + b}\left[e^{(\theta+b)T} - 1\right] \approx a\left[T + \frac{\theta + b}{2}T^2\right], \text{ neglecting higher powers of } \theta. \tag{3}$$

The different components associated with the profit function are calculated as below:

- Ordering cost

$$\text{OC} = A$$

- Holding cost

$$\text{HC} = h\int_0^T I(t)\text{d}t = h\int_0^T \frac{a}{\theta + b}\left[e^{(\theta+b)(T-t)} - 1\right]\text{d}t$$
$$= \frac{ah}{(\theta + b)^2}\left[e^{(\theta+b)(T-t)} - (\theta + b)T - 1\right] \approx \frac{1}{2}ahT^2$$

- Deterioration cost

$$\mathrm{DC} = x\theta\int_0^T I(t)\mathrm{d}t = x\theta\int_0^T \frac{a}{\theta+b}\left[\mathrm{e}^{(\theta+b)(T-t)} - 1\right]\mathrm{d}t$$
$$= \frac{ax\theta}{(\theta+b)^2}\left[\mathrm{e}^{(\theta+b)(T-t)} - (\theta+b)T - 1\right] \approx \frac{1}{2}ax\theta T^2$$

Salvage value of the deteriorated items

$$\mathrm{SV} = \eta\mathrm{DC} \approx \frac{1}{2}\eta ax\theta T^2$$

- Purchase cost to be paid by the buyer

$$\mathrm{PC} = xm_1Q(1-d_2) + xm_2Q + x(1-m_1-m_2)Q = xQ - xm_1^2d_1Q$$

- Interest paid by the buyer

$$\mathrm{IP} = xm_1Q(1-d_2)I_B(L+T) + xm_2QI_B(T-M)$$
$$+ x(1-m_1-m_2)QI_c(T-M)$$
$$= xm_1Q(1-m_1d_1)I_B(L+T) + xm_2QI_B(T-M)$$
$$+ x(1-m_1-m_2)QI_c(T-M)$$

- Interest earned by the buyer

$$\mathrm{IE} = pI_e\int_0^T tD\,\mathrm{d}t = pI_e\left[\frac{a\theta T^2}{2(\theta+b)} - \frac{ab}{(\theta+b)^3}\left\{(\theta+b)T + 1 - \mathrm{e}^{(\theta+b)T}\right\}\right]$$
$$\approx \frac{1}{2}pI_eaT^2$$

- Sales revenue

$$\mathrm{SR} = p\int_0^T D\,\mathrm{d}t = p\int_0^T (a + bI(t))\,\mathrm{d}t \approx p\left(aT + \frac{1}{2}abT^2\right)$$

The total cost of the system is

$$\mathrm{TC}_1 = \text{Ordering cost + Holding cost + Deterioration cost + Purchase cost + Interest Paid}$$
$$= \mathrm{OC} + \mathrm{HC} + \mathrm{DC} + \mathrm{PC} + \mathrm{IP}$$

The total revenue is

$$\mathrm{TR} = \text{Sales revenue} + \text{Interest earned} + \text{Savage value} = \mathrm{SR} + \mathrm{IE} + \mathrm{SV}$$

Thus, the total average profit that is, the total profit per time unit is

$$\mathrm{TPU}_1(T) = (\mathrm{TR} - \mathrm{TC}_1)/T$$
$$= \frac{1}{2T}\begin{bmatrix} -2A - ahT^2 + aI_e pT^2 + apT(2+bT) - ax\theta T^2 \\ +ax\eta\theta T^2 + ax(d_1 m_1^2 - 1)(2 + bT + \theta T)T \\ -2ax\{I_c(m_1 + m_2 - 1)(M - T) + I_B m_1(1 - d_1 m_1)(L + T) \\ +I_B m_2(T - M)\}\{T + \frac{1}{2}T^2(b+\theta)\} \end{bmatrix} \tag{4}$$

Our aim is to find maximum variable profit per unit time. The necessary and sufficient conditions to maximize $\mathrm{TPU}_1(T)$ are respectively,

$$\frac{\mathrm{dTPU}_1}{\mathrm{d}T} = 0 \tag{5}$$

and

$$\frac{\mathrm{d}^2\mathrm{TPU}_1}{\mathrm{d}T^2} < 0 \tag{6}$$

Now we discuss the fuzzy inventory model with the following fuzzy parameters:

$$\tilde{A} = (A_1, A_2, A_3, A_4), \tilde{a} = (a_1, a_2, a_3, a_4), \tilde{b} = (b_1, b_3, b_3, b_4),$$
$$\tilde{\theta} = (\theta_1, \theta_2, \theta_3, \theta_4), \tilde{h} = (h_1, h_2, h_3, h_4) \text{ and } \tilde{\eta} = (\eta_1, \eta_2, \eta_3, \eta_4).$$

The fuzzy total average profit is

$$\widetilde{\mathrm{TPU}}_1(T) = (\mathrm{TPU}_1^1, \mathrm{TPU}_1^2, \mathrm{TPU}_1^3, \mathrm{TPU}_1^4)$$

where

$$\mathrm{TPU}_1^i = \frac{1}{2T}\begin{bmatrix} -2A_{5-i} - a_{5-i}h_{5-i}T^2 + a_i I_e pT^2 + a_i pT(2 + b_i T) \\ -a_{5-i}x\theta_{5-i}T^2 + a_i x\eta_i\theta T^2 \\ +a_{5-i}x(d_1 m_1^2 - 1)(2 + b_{5-i}T + \theta_{5-i}T)T \\ -2a_{5-i}x\left\{T + \frac{1}{2}T^2(b_{5-i} + \theta_{5-i})\right\}\left\{\begin{array}{l} I_c(m_1 + m_2 - 1)(M - T) \\ +I_B m_1(1 - d_1 m_1)(L_{5-i} + T) \\ +I_B m_2(T - M) \end{array}\right\} \end{bmatrix}$$

$i = 1, 2, 3, 4$

Therefore, using GMIR method, the defuzzified value of total average profit is

$$d_F(\widetilde{\mathrm{TPU}_1})(T) = \frac{1}{30}\left(8\,\mathrm{TPU}_1^1 + 7\,\mathrm{TPU}_1^2 + 7\,\mathrm{TPU}_1^3 + 8\,\mathrm{TPU}_1^4\right)$$

$$
= \frac{1}{60T}\left[\begin{array}{l}
7\left(\begin{array}{l}-2A_2 - a_2h_2T^2 + a_3I_epT^2 + a_3pT(2+b_3T)\\ -a_2xT^2\theta_2 + a_2x(-1+d_1m_1^2)T(2+b_2T+T\theta_2)\\ -2a_2x(I_c(-1+m_1+m_2)(M-T) + I_Bm_1(1-d_1m_1)(L_2+T)\\ +I_Bm_2(-M+T))(T+\frac{1}{2}T^2(b_2+\theta_2)) + a_3x\eta_3T^2\theta_3\end{array}\right)\\
+7\left(\begin{array}{l}-2A_3 - a_3h_3T^2 + a_2I_epT^2 + a_2pT(2+b_2T)\\ +a_2x\eta_2T^2\theta_2 - a_3xT^2\theta_3 + a_3x(-1+d_1m_1^2)T(2+b_3T+T\theta_3)\\ -2a_3x(I_c(-1+m_1+m_2)(M-T) + I_Bm_1(1-d_1m_1)(L_3+T)\\ +I_Bm_2(-M+T))(T+\frac{1}{2}T^2(b_3+\theta_3))\end{array}\right)\\
+8\left(\begin{array}{l}-2A_1 - a_1h_1T^2 + a_4I_epT^2 + a_4pT(2+b_4T)\\ -a_1xT^2\theta_1 + a_1x(-1+d_1m_1^2)T(2+b_1T+T\theta_1)\\ -2a_1x(I_c(-1+m_1+m_2)(M-T) + I_Bm_1(1-d_1m_1)(L_1+T)\\ +I_Bm_2(-M+T))(T+\frac{1}{2}T^2(b_1+\theta_1)) + a_4x\eta_4T^2\theta_4\end{array}\right)\\
+8\left(\begin{array}{l}-2A_4 - a_4h_4T^2 + a_1I_epT^2 + a_1pT(2+b_1T) + a_1x\eta_1T^2\theta_1\\ -a_4xT^2\theta_4 + a_4x(-1+d_1m_1^2)T(2+b_4T+T\theta_4)\\ -2a_4x(I_c(-1+m_1+m_2)(M-T) + I_Bm_1(1-d_1m_1)(L_4+T)\\ +I_Bm_2(-M+T))(T+\frac{1}{2}T^2(b_4+\theta_4))\end{array}\right)
\end{array}\right] \tag{7}
$$

The necessary and sufficient condition to maximize $d_F(\widetilde{\mathrm{TPU}}_1)$ are respectively,

$$\frac{\mathrm{d}(d_F\widetilde{\mathrm{TPU}}_1)}{\mathrm{d}T} = 0 \tag{8}$$

and

$$\frac{\mathrm{d}^2(d_F\widetilde{\mathrm{TPU}}_1)}{\mathrm{d}T^2} < 0 \tag{9}$$

The defuzzified value of ordering quantity is

$$
\begin{aligned}
d_F\widetilde{Q}_1 = \frac{1}{30}\Bigg[& 8a_1\left(T + \frac{\theta_1+b_1}{2}T^2\right) + 7a_2\left(T + \frac{\theta_2+b_2}{2}T^2\right)\\
&+7a_3\left(T + \frac{\theta_3+b_3}{2}T^2\right) + 8a_4\left(T + \frac{\theta_4+b_4}{2}T^2\right)\Bigg]
\end{aligned} \tag{10}
$$

4.2 *Case 2: When the Credit Period is Greater Than or Equal to Cycle Length ($M \geq T$)*

Here, the entire remaining amount $(1-m_1)Q\,x$ is paid at M rather than at T (Fig. 2) to take the advantage of credit period. No loan is required to be taken by the buyer from financer as the buyer pays the due amount at $M(M \geq T)$. Also, no interest will be charged by the supplier on the due amount. The differential equation describing the above situation together with boundary conditions, and the corresponding solution is

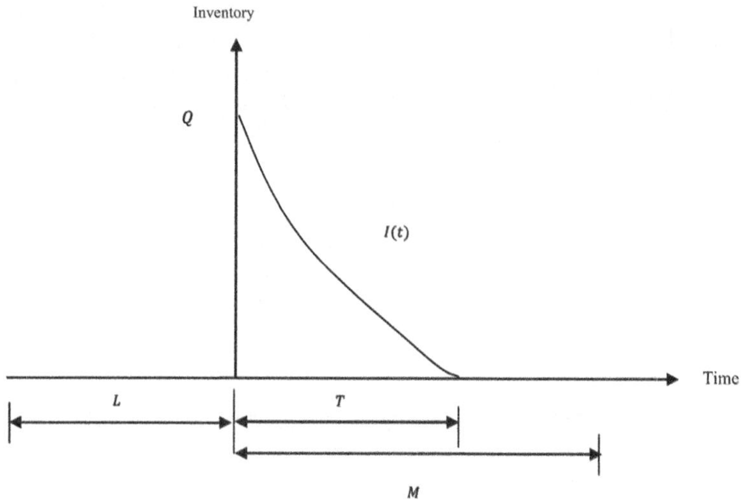

Fig. 2 Inventory level when $M \geq T$

same as Eqs. (1) and (2). The expression for ordering quantity is same as in Eq. (3). Here, the expressions for ordering cost, holding cost, deterioration cost, salvage value and sales revenue are same as obtained in case I. The other relevant costs and total revenue are calculated as below:

- Purchase cost

$$\mathrm{PC} = xm_1Q(1-d_2) + x(1-m_1)Q = xQ(1-m_1^2d_1)$$

- Interest paid by the buyer

$$\begin{aligned}\mathrm{IP} &= xm_1Q(1-d_2)I_B(L+T)\\ &= xm_1Q(1-m_1d_1)I_B(L+T)\end{aligned}$$

- Interest earned by the buyer

$$\begin{aligned}\mathrm{IE} &= pI_e\int_0^T tD\,\mathrm{d}t + (M-T)pI_e\int_0^T D\,\mathrm{d}t\\ &\approx \frac{1}{2}pI_eaT^2 + (M-T)pI_e\left(aT + \frac{1}{2}abT^2\right)\end{aligned}$$

Therefore, the total cost of the system is

$$\begin{aligned}\mathrm{TC}_2 &= \text{Ordering cost} + \text{Holding cost} + \text{Deterioration cost} \\ &\quad + \text{Purchase cost} + \text{Interest Paid} \\ &= \mathrm{OC} + \mathrm{HC} + \mathrm{DC} + \mathrm{PC} + \mathrm{IP}\end{aligned}$$

The total revenue of the system is

$$\mathrm{TR} = \text{Sales revenue} + \text{Interest earned} + \text{Savage value} = \mathrm{SR} + \mathrm{IE} + \mathrm{SV}$$

As in case 1, the total average profit of the system is

$$\mathrm{TPU}_2(T) = (\mathrm{TR} - \mathrm{TC}_2)/T$$
$$= \frac{1}{2T}\left[\begin{array}{l} -2A - ahT^2 + ap(2+bT)T - apI_e(T+bT^2) + apI_e(2M+bMT)T \\ +xT\left\{\begin{array}{l} -2a + d_1m_1^2a(2+bT+\theta T) - (ab+2a\theta)T + a\eta\theta T \\ +I_Bm_1(d_1m_1-1)(aL+aT)(2+bT+\theta T) \end{array}\right\} \end{array}\right] \tag{11}$$

As before, the necessary and sufficient conditions to maximize $\mathrm{TPU}_1(T)$ are respectively,

$$\frac{\mathrm{d}\mathrm{TPU}_2}{\mathrm{d}T} = 0 \tag{12}$$

and

$$\frac{\mathrm{d}^2\mathrm{TPU}_2}{\mathrm{d}T^2} < 0 \tag{13}$$

The fuzzy total average profit is

$$\widetilde{\mathrm{TPU}}_2(T) = (\mathrm{TPU}_2^1, \mathrm{TPU}_2^2, \mathrm{TPU}_2^3, \mathrm{TPU}_2^4)$$

where

$$\mathrm{TPU}_2^i = \frac{1}{2T}\left[\begin{array}{l} -2A_{5-i} - a_{5-i}h_{5-i}T^2 + a_ip(2+b_iT)T \\ -a_{5-i}pI_e(T+b_{5-i}T^2) + a_ipI_e(2M+b_iMT)T \\ +xT\left\{\begin{array}{l} -2a_{5-i} + d_1m_1^2a_i(2+b_iT+\theta_iT) \\ -(a_{5-i}b_{5-i} + 2a_{5-i}\theta_{5-i})T + a_i\eta_i\theta_iT \\ +I_Bm_1(d_1m_1-1)(a_{5-i}L_{5-i} + a_{5-i}T)(2+b_{5-i}T+\theta_{5-i}T) \end{array}\right\} \end{array}\right], i = 1, 2, 3, 4$$

Therefore, using GMIR method, the defuzzified value of total average profit is

$$d_F(\widetilde{\mathrm{TPU}}_2)(T) = \frac{1}{30}\left(8\,\mathrm{TPU}_2^1 + 7\,\mathrm{TPU}_2^2 + 7\,\mathrm{TPU}_2^3 + 8\,\mathrm{TPU}_2^4\right)$$

$$= \frac{1}{60T}\left[\begin{array}{l} 7\left(\begin{array}{l}-2A_2 - a_2h_2T^2 - a_2I_e pT^2(1+b_2T)\\ +a_3pT(2+b_3T)+a_3I_eMpT(2+b_3T)\\ +xT(-2a_2 - a_2T(b_2+2\theta_2)\\ +I_Bm_1(-1+d_1m_1)(a_2L_2+a_1T)(2+b_2T+T\theta_2)\\ +a_3\eta_3T\theta_3+a_3d_1m_1^2(2+b_3T+T\theta_3))\end{array}\right)\\ +7\left(\begin{array}{l}-2A_3 - a_3h_3T^2 + a_2pT(2+b_2T)\\ +a_2I_eMpT(2+b_2T) - a_3I_epT^2(1+b_3T)\\ +xT(-2a_3+a_2\eta_2T\theta_2+a_2d_1m_1^2(2+b_2T+T\theta_2) - a_3T(b_3+2\theta_3)\\ +a_3I_Bm_1(-1+d_1m_1)(L_3+T)(2+b_3T+T\theta_3))\end{array}\right)\\ +8\left(\begin{array}{l}-2A_1 - a_1h_1T^2 - a_1I_epT^2(1+b_1T)\\ +a_4pT(2+b_4T)+a_4I_eMpT(2+b_4T)\\ +xT(-2a_1 - a_1T(b_1+2\theta_1)\\ +a_1I_Bm_1(-1+d_1m_1)(L_1+T)(2+b_1T+T\theta_1)\\ +a_4\eta_4T\theta_4+a_4d_1m_1^2(2+b_4T+T\theta_4))\end{array}\right)\\ +8\left(\begin{array}{l}-2A_4 - a_4h_4T^2 + a_1pT(2+b_1T)+a_1I_eMpT(2+b_1T)\\ -a_4I_epT^2(1+b_4T)\\ +xT(-2a_4+a_1\eta_1T\theta_1+a_1d_1m_1^2(2+b_1T+T\theta_1) - a_4T(b_4+2\theta_4)\\ +a_4I_Bm_1(-1+d_1m_1)(L_4+T)(2+b_4T+T\theta_4))\end{array}\right)\end{array}\right] \tag{14}$$

The necessary and sufficient conditions to maximize $d_F(\widetilde{\text{TPU}_1})$ are respectively,

$$\frac{\mathrm{d}(d_F\widetilde{\text{TPU}_2})}{\mathrm{d}T} = 0 \tag{15}$$

and

$$\frac{\mathrm{d}^2(d_F\widetilde{\text{TPU}_2})}{\mathrm{d}T^2} < 0 \tag{16}$$

As in case 1, the defuzzified value of ordering quantity is

$$d_F\widetilde{Q}_2 = \frac{1}{30}\left[8a_1\left(T+\frac{\theta_1+b_1}{2}T^2\right)+7a_2\left(T+\frac{\theta_2+b_2}{2}T^2\right)\right.$$
$$\left.+7a_3\left(T+\frac{\theta_3+b_3}{2}T^2\right)+8a_4\left(T+\frac{\theta_4+b_4}{2}T^2\right)\right] \tag{17}$$

5 Solution Procedure

The optimal solution has been obtained using the following steps:

Step 1: Set the parameters

Step 2: Find the optimal value T_1^* from Eq. (5). If $M < T_1^*$ then evaluate $\text{TPU}_1(T^*)$ from Eq. (4).

Step 2: Find the optimal value T_2^* from Eq. (12). If $T_2^* \leq M$ then evaluate $\text{TPU}_2(T^*)$ from Eq. (11).

Step 3: If $T_2^* \le M < T_1^*$ then move to step 4 otherwise step 5.
Step 4: Compare $\text{TPU}_1(T^*)$ and $\text{TPU}_2(T^*)$ and find the maximum profit. Move to step 6.
Step 5: If $M < T_1^*$ but $T_2^* > M$ then $\text{TPU}_1(T^*)$ is maximum profit, else if $T_2^* \le M$ but $M \ge T_1^*$ then $\text{TPU}_2(T^*)$ is maximum profit. Move to step 6.
Step 6: Find the corresponding optimal ordered quantity.

The same steps are followed to find the optimal values in fuzzy case.

6 Numerical Illustration

For numerical illustration and validation, numerical examples depicting a hypothetical system are considered. The examples cover the possible cases that may arise in this study.

Example 1: (Case 1) For Crisp model, the following values of the parameters are taken;

$$\begin{aligned}
A &= \$100 \text{ per order}, d_1 = 20\%, a = 4500 \text{ units per year}, \\
h &= \$0.65 \text{ per unit per year}, b = 0.2, \theta = 0.2, \eta = 0.1, \\
m_1 &= 0.2, m_2 = 0.4, x = \$5 \text{ per unit}, p = \$8 \text{ per unit}, \\
I_B &= \$0.09 \text{ per annum}, I_e = \$0.06 \text{ per annum}, \\
I_c &= \$0.1 \text{ per annum}, M = 0.083 \text{ year}, L = 0.25 \text{ year}
\end{aligned}$$

Solving Eq. (5) we get $T_1 = 0.135150$ year and $\frac{\text{d}^2\text{TPU}_1}{\text{d}T^2} = -81857 < 0$. Again solving Eq. (12) we get $T_2 = 0.130208$ year and $\frac{\text{d}^2\text{TPU}_2}{\text{d}T^2} = -91185 < 0$. So, $T_1 = T_1^* = 0.135150$ year and $T_2 = T_2^* = 0.130208$ year. Here, $T_2^* > M$ contradicts case 2. So, only case 1 holds as $M < T_1^*$. That is, case 1 is optimal case and maximum average profit is $\text{TPU}_1(T_1^*) = 12252.6$.

For Fuzzy case, the following parameters are set as fuzzy numbers and other parameters are kept same as in crisp case;

$$\begin{aligned}
\widetilde{A} &= (90, 97, 104, 111), \tilde{a} = (4495, 4505, 4515, 4525), \\
\tilde{b} &= (0.15, 0.19, 0.23, 0.27), \tilde{h} = (0.55, 0.62, 0.69, 0.76), \\
\tilde{\theta} &= (0.15, 0.19, 0.22, 0.30), \tilde{\eta} = (0.05, 0.08, 0.11, 0.14), \\
\widetilde{L} &= (0.20, 0.24, 0.27, 0.29)
\end{aligned}$$

Applying similar argument, optimal solution under fuzzy case is obtained under case 1. The optimal solutions for crisp and fuzzy cases with respect to Example 1 have been presented in Table 2 (Fig. 3).

Table 2 Optimal solutions with respect to Example 1

Model	Optimal values			Optimal case
	T^*	Q^*	Total average profit	
Crisp	0.135150	624.614	12,252.6	Case 1
Fuzzy	0.132071	612.393	12,241.2	Case 1

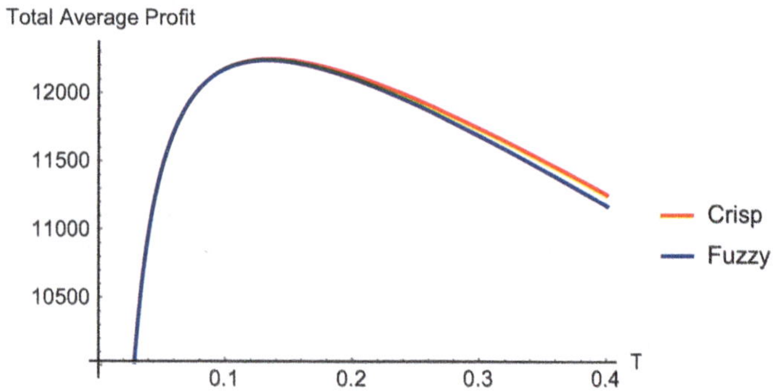

Fig. 3 Plot of total average profit versus time with respect Example 1

Example 2: (Case 2) For Crisp model the following values of the parameters are taken;

$$
\begin{aligned}
A &= \$100 \text{ per order}, d_1 = 20\%, a = 17{,}000 \text{ units per year}, \\
h &= \$0.65 \text{ per unit per year}, b = 0.2, \theta = 0.2, \eta = 0.1, \\
m_1 &= 0.2, m_2 = 0.4, x = \$5 \text{ per unit}, p = \$8 \text{ per unit}, \\
I_B &= \$0.09 \text{ per annum}, I_e = \$0.06 \text{ per annum}, \\
I_c &= \$0.1 \text{ per annum}, M = 0.083 \text{ year}, L = 0.25 \text{ year}
\end{aligned}
$$

Solving Eq. (5) we get $T_1 = 0.0698846$ year and $\frac{\mathrm{d}^2\mathrm{TPU}_1}{\mathrm{d}T^2} = -589,156 < 0$. Again solving Eq. (12) we get $T_2 = 0.0672024$ year and $\frac{\mathrm{d}^2\mathrm{TPU}_2}{\mathrm{d}T^2} = -661,205 < 0$. So, $T_1 = T_1^* = 0.0698846$ year and $T_2 = T_2^* = 0.0672024$ year. Here, $T_1^* < M$ contradicts case 1. So, only case 2 holds as $T_2^* < M$. That is, case 2 is optimal case and maximum average profit is $\mathrm{TPU}_2(T_2^*) = 49,019$.

For Fuzzy case, the following parameters are set as fuzzy numbers and other parameters are kept same as in crisp case;

$$
\begin{aligned}
&\widetilde{A} = (90, 97, 104, 111), \tilde{a} = (16995, 17005, 17015, 17025), \\
&\tilde{b} = (0.15, 0.19, 0.23, 0.27), \tilde{h} = (0.55, 0.62, 0.69, 0.76), \\
&\tilde{\theta} = (0.15, 0.19, 0.22, 0.30), \tilde{\eta} = (0.05, 0.08, 0.11, 0.14), \\
&\widetilde{L} = (0.20, 0.24, 0.27, 0.29)
\end{aligned}
$$

With similar argument, optimal solution for fuzzy case is obtained under case 2. The optimal solutions for crisp and fuzzy cases with respect to Example 2 are presented in Table 3 (Fig. 4).

Example 3: (Case 1 and Case 2) For Crisp model, the following values of the parameters are taken;

$$A = \$100 \text{ per order}, d_1 = 20\%, a = 11{,}500 \text{ units per year},$$
$$h = \$0.65 \text{ per unit per year}, b = 0.2, \theta = 0.2,$$
$$\eta = 0.1, m_1 = 0.2, m_2 = 0.4, x = \$5 \text{ per unit},$$
$$p = \$8 \text{ per unit}, I_B = \$0.09 \text{ per annum},$$
$$I_e = \$0.06 \text{ per annum}, I_c = \$0.1 \text{ per annum},$$
$$M = 0.083 \text{ year}, L = 0.25 \text{ year}$$

Solving Eq. (5) we get $T_1 = 0.0848698$ year and $\frac{d^2\mathrm{TPU}_1}{dT^2} = -329,312 < 0$. Again solving Eq. (12) we get $T_2 = 0.0816482$ year and $\frac{d^2\mathrm{TPU}_2}{dT^2} = -368,945 < 0$. So, $T_1 = T_1^* = 0.0848698$ year and $T_2 = T_2^* = 0.0816482$ year. Here, $T_2^* < M < T_1^*$ so both case 1 and case 2 hold together. Now $\mathrm{TPU}_1(T_1^*) = 32,726 > \mathrm{TPU}_2(T_2^*) = 32,725$. Hence the optimal case is case 1.

For Fuzzy case, the following parameters are set as fuzzy numbers and other parameters are kept same as in crisp case;

Table 3 Optimal solutions with respect to Example 2

Model	Optimal values			Optimal case
	T^*	Q^*	Total average profit	
Crisp	0.0672024	1157.80	49,019.0	Case 2
Fuzzy	0.0658514	1135.83	48,973.9	Case 2

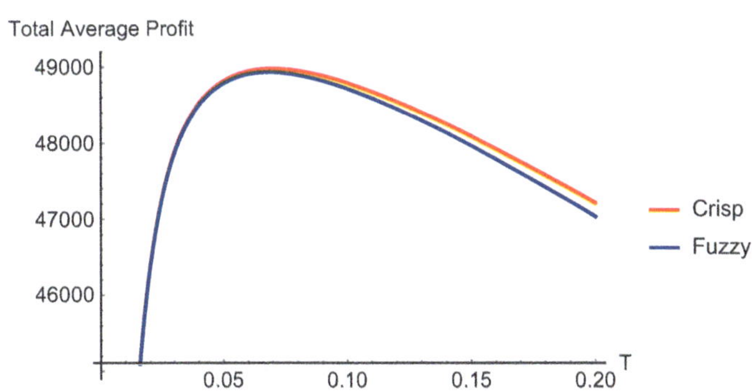

Fig. 4 Plot of total average profit versus time with respect to Example 2

Table 4 Optimal solutions with respect to Example 3

Model	Optimal values			Optimal case
	T^*	Q^*	Total average profit	
Crisp	0.0848698	992.569	32,726	Case 1
Fuzzy	0.0799943	936.414	32,693	Case 2

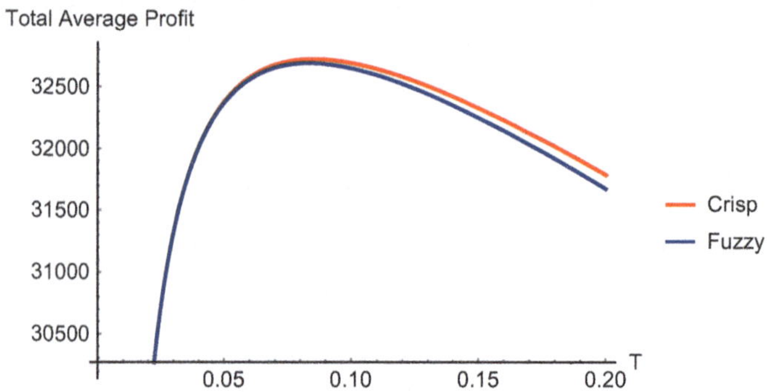

Fig. 5 Plot of total average profit versus time with respect to Example 3

$$\widetilde{A} = (90, 97, 104, 111), \tilde{a} = (11495, 11505, 11515, 11525),$$
$$\tilde{b} = (0.15, 0.19, 0.23, 0.27), \tilde{h} = (0.55, 0.62, 0.69, 0.76),$$
$$\tilde{\theta} = (0.15, 0.19, 0.22, 0.30), \tilde{\eta} = (0.05, 0.08, 0.11, 0.14),$$
$$\widetilde{L} = (0.20, 0.24, 0.27, 0.29)$$

Here, $T_2^* = 0.079943 < M < T_1^* = 0.083101$ and $\mathrm{TPU}_1(T_1^*) = 32,691.3 < \mathrm{TPU}_2(T_2^*) = 32,693$. Hence case 2 is optimal.

The optimal solutions for both crisp and fuzzy cases with respect to Example 3 have been presented in Table 4 (Fig. 5).

7 Sensitivity Analysis

To study the effects of changes of different system parameters on total average cost under fuzzy case, here, in this section sensitivity analyses are carried out with regard to Example 1. For carrying out the analysis, each parameter is changed by − 20, − 10, − 5, 5, 10 and 20% at a time whilst the other parameters are left unchanged.

The following observations can be made from Table 5;

- T_1^* and T_2^* both decrease when the values of $\tilde{a}$, $\tilde{h}$, $\tilde{\theta}$, $\widetilde{L}$ and x increase. T_1^* decreases with increase in I_c whereas T_2^* does not show any change in its value. On the other

Table 5 Sensitivity analysis table for Example 1

Parameter	% Change	% Change in optimal values						Optimal case
		T_1^*	$d_F\widetilde{\text{TPU}}_1$	$d_F\widetilde{Q}_1$	T_2^*	$d_F\widetilde{\text{TPU}}_2$	$d_F\widetilde{Q}_2$	
$\widetilde{A}$	– 20%	– 10.461	1.311963	– 10.7173	– 10.497	1.361591	– 10.7454	Case 1
	– 10%	– 5.08211	0.638009	– 5.21397	– 5.10065	0.661975	– 5.22866	Case 1
	– 5%	– 2.50698	0.315329	– 2.57384	– 2.51659	0.326487	– 2.58148	Case 1
	+ 5%	2.443383	– 0.30716	2.511949	2.453834	– 0.3183	2.520173	Case 1
	+ 10%	4.828463	– 0.60697	4.966909	4.848832	– 0.63006	4.983108	Case 1
	+ 20%	9.438105	– 1.18698	9.720719	9.478011	– 1.23149	9.752165	Case 1
$\tilde{a}$	– 20%	11.66948	– 21.1745	– 10.3791	11.72004	– 21.2187	– 10.3471	Case 1
	– 10%	5.35091	– 10.6052	– 5.04529	5.37286	– 10.6284	– 5.02985	Case 1
	– 5%	2.570587	– 5.30667	– 2.48941	2.580918	– 5.31871	– 2.48173	Case 1
	+ 5%	– 2.38584	5.314838	2.428016	– 2.395	5.326896	2.42043	Case 1
	+ 10%	– 4.60888	10.63621	4.798226	– 4.62526	10.66034	4.78396	Case 1
	+ 20%	– 8.63172	21.29938	9.383027	– 8.66216	21.3485	9.354377	Case 1
$\tilde{b}$	– 20%	– 2.30331	– 0.30144	– 2.6225	– 2.14867	– 0.29212	– 2.45396	Case 1
	– 10%	– 1.17058	– 0.15195	– 1.33395	– 1.0912	– 0.14647	– 1.24739	Case 1
	– 5%	– 0.59059	– 0.07597	– 0.67326	– 0.54992	– 0.07364	– 0.62894	Case 1
	+ 5%	0.600435	0.07679	0.685344	0.55933	0.074462	0.640119	Case 1
	+ 10%	1.21147	0.15358	1.383262	1.127289	0.148106	1.290738	Case 1
	+ 20%	2.466855	0.30961	2.819431	2.291448	0.298666	2.626182	Case 1
$\tilde{h}$	– 20%	2.637975	0.323498	2.711984	2.461678	0.311758	2.527963	Case 1
	– 10%	1.293244	0.160932	1.329049	1.208089	0.15547	1.240443	Case 1
	– 5%	0.640565	0.080058	0.658237	0.598553	0.077735	0.614718	Case 1
	+ 5%	– 0.62845	– 0.07924	– 0.6455	– 0.58836	– 0.07692	– 0.60354	Case 1

(continued)

Table 5 (continued)

Parameter	% Change	% Change in optimal values						Optimal case
		T_1^*	$d_F\widetilde{\text{TPU}}_1$	$d_F\widetilde{Q}_1$	T_2^*	$d_F\widetilde{\text{TPU}}_2$	$d_F\widetilde{Q}_2$	
	+ 10%	– 1.24554	– 0.15848	– 1.27941	– 1.16573	– 0.15302	– 1.19658	Case 1
	+ 20%	– 2.44641	– 0.31451	– 2.51162	– 2.29302	– 0.30439	– 2.35218	Case 1
$\tilde{\theta}$	– 20%	9.090565	1.040748	9.031945	8.352291	1.000736	8.276843	Case 1
	– 10%	4.249987	0.508937	4.220329	3.926291	0.49014	3.889825	Case 1
	– 5%	2.058741	0.251609	2.044112	1.907056	0.243024	1.88886	Case 1
	+ 5%	– 1.93911	– 0.24671	– 1.92458	– 1.80429	– 0.23811	– 1.78624	Case 1
	+ 10%	– 3.76994	– 0.48851	– 3.74074	– 3.51444	– 0.32403	– 3.47951	Case 1
	+ 20%	– 7.14388	– 0.95906	– 7.08614	– 6.68372	– 0.92873	– 6.61541	Case 1
$\tilde{\eta}$	– 20%	– 0.43159	– 0.05392	– 0.44334	– 0.404	– 0.05237	– 0.41421	Case 1
	– 10%	– 0.21655	– 0.02696	– 0.22241	– 0.20239	– 0.02618	– 0.20778	Case 1
	– 5%	– 0.10828	– 0.01307	– 0.1112	– 0.1012	– 0.01309	– 0.10398	Case 1
	+ 5%	0.108275	0.013888	0.111693	0.101982	0.013092	0.104485	Case 1
	+ 10%	0.218065	0.027775	0.224039	0.203963	0.026184	0.209309	Case 1
	+ 20%	0.436886	0.054733	0.448895	0.408711	0.053187	0.419634	Case 1
$\widetilde{L}$	– 20%	0.036344	0.1642	0.037394	0.034517	0.163653	0.035393	Case 1
	– 10%	0.018172	0.081691	0.018779	0.017258	0.081826	0.017781	Case 1
	– 5%	0.009086	0.040846	0.009308	0.008629	0.040913	0.008806	Case 1
	+ 5%	– 0.00909	– 0.04085	– 0.00947	– 0.00863	– 0.04091	0.008975	Case 1
	+ 10%	– 0.01817	– 0.08169	– 0.01894	– 0.01726	– 0.08183	– 0.01744	Case 1
	+ 20%	– 0.0371	– 0.16338	– 0.03788	– 0.03452	– 0.16365	– 0.03539	Case 1
M	– 20%	– 0.05224	– 0.46237	– 0.05356	– 0.02981	– 0.29785	– 0.03065	Case 1
	– 10%	– 0.0265	– 0.11927	– 0.02678	– 0.01491	– 0.14892	– 0.01541	Case 1
	– 5%	– 0.01287	– 0.05963	– 0.01339	– 0.00706	– 0.07446	– 0.00762	Case 1

(continued)

Table 5 (continued)

Parameter	% Change	% Change in optimal values						Optimal case
		T_1^*	$d_F\widetilde{\text{TPU}}_1$	$d_F\widetilde{Q}_1$	T_2^*	$d_F\widetilde{\text{TPU}}_2$	$d_F\widetilde{Q}_2$	
	+ 5%	0.012872	0.059635	0.013227	0.007845	0.074462	0.007959	Case 1
	+ 10%	0.025744	0.120086	0.026617	0.015689	0.148924	0.015749	Case 1
	+ 20%	0.052245	0.239356	0.053397	0.030594	0.297848	0.031329	Case 1
x	– 20%	20.77595	38.62448	21.46204	14.46569	38.40848	14.90326	Case 1
	– 10%	8.932317	19.26118	9.198831	6.507209	19.17683	6.690262	Case 1
	– 5%	4.180327	9.619972	4.299853	3.101025	9.581867	3.185186	Case 1
	+ 5%	– 3.71543	– 9.59955	– 3.8134	– 2.83744	– 9.57041	– 2.91051	Case 1
	+ 10%	– 7.04545	– 19.1826	– 7.22494	– 5.44739	– 19.1305	– 5.5836	Case 1
	+ 20%	– 12.7863	– 38.3027	– 13.0914	– 10.0946	– 38.2225	– 10.3345	Case 1
I_c	– 20%	1.650627	0.075973	1.696133	0	0	0	Case 1
	– 10%	0.814713	0.037578	0.837207	0	0	0	Case 1
	– 5%	0.405085	0.018789	0.415909	0	0	0	Case 1
	+ 5%	– 0.39978	– 0.01797	– 0.41085	0	0	0	Case 1
	+ 10%	– 0.79503	– 0.03676	– 0.81647	0	0	0	Case 1
	+ 20%	– 1.57037	– 0.07271	– 1.61302	0	0	0	Case 1
I_e	– 20%	– 1.80963	– 0.23119	– 1.85845	1.853711	– 0.06382	1.903593	Case 1
	– 10%	– 0.91693	– 0.116	– 0.94204	0.913912	– 0.03273	0.938164	Case 1
	– 5%	– 0.46187	– 0.058	– 0.47437	0.453426	– 0.01637	0.465865	Case 1
	+ 5%	0.46793	0.058818	0.4809	– 0.44715	0.016365	– 0.45892	Case 1
	+ 10%	0.942675	0.117636	0.968822	– 0.88881	0.033549	– 0.91175	Case 1
	+ 20%	1.912608	0.236088	1.965731	– 1.7533	0.067916	– 1.7986	Case 1
p	– 20%	– 7.45584	– 59.959	– 7.64444	– 4.03533	– 59.8913	– 4.13757	Case 1
	– 10%	– 3.94258	– 29.9891	– 4.04593	– 2.07886	– 29.9484	– 2.13238	Case 1

(continued)

Table 5 (continued)

Parameter	% Change	% Change in optimal values						Optimal case
		T_1^*	$d_F\widetilde{\text{TPU}}_1$	$d_F\widetilde{Q}_1$	T_2^*	$d_F\widetilde{\text{TPU}}_2$	$d_F\widetilde{Q}_2$	
	– 5%	– 2.02997	– 14.9969	– 2.08445	– 1.05512	– 14.975	– 1.08295	Case 1
	+ 5%	2.160202	15.00343	2.22047	1.089634	14.97668	1.118346	Case 1
	+ 10%	4.465023	30.01176	4.592476	2.213785	29.95418	2.273947	Case 1
	+ 20%	9.578181	60.04803	9.865397	4.578188	59.91572	4.704876	Case 1
m_1	– 20%	– 0.20368	– 0.34065	– 0.20951	0.564821	– 0.30685	0.579833	Case 1
	– 10%	– 0.10676	– 0.18626	– 0.1099	– 0.276919	– 0.16938	– 0.283989	Case 1
	– 5%	– 0.05452	– 0.0964	– 0.05634	0.137283	– 0.08837	0.140555	Case 1
	+ 5%	0.056788	0.105382	0.058786	– 0.13415	0.096555	– 0.13734	Case 1
	+ 10%	0.116604	0.218116	0.120021	– 0.26515	0.200475	– 0.27214	Case 1
	+ 20%	0.243808	0.466458	0.250819	– 0.51932	0.432861	– 0.53292	Case 1

% Change in optimal value = [(Changed optimal value – Base example optimal value)/Base example optimal value] × 100

side T_1^* and T_2^* both increase when the values of the parameters $\tilde{b}$, $\widetilde{A}$, $\tilde{\eta}$, M and p increase. T_1^* increases with the increase in the value of I_e and m_1 but T_2^* decreases whilst I_e and m_1 increase. It is observed that both T_1^* and T_2^* have high sensitivity with respect to parameter x as both T_1^* and T_2^* decrease rapidly with the increase in x. T_1^* and T_2^* have moderate sensitivity with regard to $\tilde{a}$, $\widetilde{A}$, $\tilde{\theta}$ and $\tilde{p}$ as both T_1^* and T_2^* increase or decrease gradually when the mentioned parameters increase. T_1^* and T_2^* are less sensitive with respect to $\tilde{b}$, $\tilde{h}$, $\tilde{\eta}$, L, M, I_c, I_e and m_1 as the values of T_1^* and T_2^* change slowly or remain static if the values of the mentioned parameters are changed.

- $d_F\widetilde{\text{TPU}}_1$ and $d_F\widetilde{\text{TPU}}_2$ increase as the values of the parameters $\tilde{a}$, $\tilde{b}$, $\tilde{\eta}$, M, I_e and p increase. And their values decrease with the increasing values of $\widetilde{A}$, $\tilde{h}$, $\tilde{\theta}$, $\widetilde{L}$ and c. $d_F\widetilde{\text{TPU}}_1$ decreases as I_c increases whereas $d_F\widetilde{\text{TPU}}_2$ remains static with increase in I_c. Both $d_F\widetilde{\text{TPU}}_1$ and $d_F\widetilde{\text{TPU}}_2$ show high sensitivity towards the parameters $\tilde{a}$, x and p whereas they are less sensitive with respect to rest of the parameters.
- $d_F\widetilde{Q}_1$ and $d_F\widetilde{Q}_2$ increase with the increase in $\tilde{a}$, $\tilde{b}$, $\widetilde{A}$, $\tilde{\eta}$, M and p. On the other hand, both $d_F\widetilde{Q}_1$ and $d_F\widetilde{Q}_2$ decrease with the increase in $\tilde{h}$, $\tilde{\theta}$, $\widetilde{L}$ and x. $d_F\widetilde{Q}_1$ decreases with increase in I_c whereas $d_F\widetilde{Q}_2$ $d_F\widetilde{\text{TPU}}_2$ remains static with increase in I_c. $d_F\widetilde{Q}_1$ increases with the increase in the value of I_e and m_1 but $d_F\widetilde{Q}_2$ decreases with increase in the value of I_e and m_1. $d_F\widetilde{Q}_1$ and $d_F\widetilde{Q}_2$ are highly

sensitive with respect to parameter c, moderately sensitive with respect to $\tilde{a}$, $\widetilde{A}$, $\tilde{\theta}$ and $\tilde{p}$ and less sensitive with regard to $\tilde{b}$, $\tilde{h}$, $\tilde{\eta}$, L, M, I_c, I_e and m_1.

8 Managerial Implications

In accordance with the sensitivity analyses performed and observations made in Sect. 7, the following measures may be taken for increasing the total average profit of the system.

- When per unit purchase cost raises, the profit of the system falls significantly. So, it is advisable for decision maker to negotiate with the supplier to reduce unit purchasing cost so that the buyer procures more items and increases the profit.
- Increase in portion of advance payment results in increasing total profit of the system. Therefore, it is beneficial to pay bigger portion of money in advance to earn more discount and thereby gets a chance to increase the profit of the system.
- Increase in initial market demand increases the total average profit and ordering quantity. Thus, in this case, buyer orders more quantity to earn more profit.
- Decrease in length of advance payment period and increase in credit period offered by supplier have positive effects on total average profit. Thus, selecting a supplier, who offers smaller advance payment period and larger credit period, is more economical from buyer perspective.

9 Conclusion

This work deals with a buyer's EOQ model for deteriorating items under partial advance-partial delayed payment policy. The cash discount facility is rendered by the supplier due to advance payment and discount rate depends on the portion of advance payment. The corresponding crisp model has been developed. Due to uncertain nature, some of system parameters like demand parameters, ordering cost, holding cost, deterioration rate, salvage coefficient and lead time have been considered as fuzzy in nature. The corresponding fuzzy profit function has also been formulated. Parabolic-flat fuzzy numbers are taken to tackle the impreciseness of the said parameters. For defuzzification, Graded Mean Integration Representation (GMIR) method is used. The numerical values of optimal results obtained in both environments that is, crisp and fuzzy environments have been compared. It is noticed that the total average profit in fuzzy case is less as compared to crisp case. Further, sensitivity analysis for fuzzy case is carried out to study the effects of variation of system parameters on the total average profit. It has been observed that initial demand (a) and unit purchase cost (x) play significant role on the total average profit of system. Elevating customers' demand by executing proper marketing strategies and allaying unit purchasing cost by negotiating with the supplier will raise the profit of the system.

The present study does not consider shortages, which is a limitation of this model. Thus, consideration of shortages (completely backlogged or partially backlogged) in this present model will give a scope of extending the model. As the selling price is one of the decisive factors that affects customers' demand, it will be interesting to study the effect of selling price on the profit of the system. Thus, another interesting extension of this model is possible with price-dependent demand. As it is observed in the real market that credit period facility helps in increasing sales and the present work considered the up-stream credit period that is, the credit period offered by the supplier to buyer. So, one may also think of extending this work incorporating down-stream credit period that is, the credit period offered by buyers to their customers. Infect, down-stream credit period may be considered in the demand function.

Appendix 1

This appendix presents the explicit expressions of Eqs. (5), (8), (12) and (15).

The explicit expression for Eq. (5) is

$$\frac{\mathrm{dTPU}_1}{\mathrm{d}T} = 0 \text{ that is,}$$

$$2A + aT^2 \begin{bmatrix} -h - 2xI_c - 2xI_B m_1 + 2xI_c m_1 + 2xd_1 I_B m_1^2 \\ -2xI_B m_2 + 2xI_c m_2 + I_e p \\ +b(p + x(-1 + d_1 m_1^2 - I_c(-1 + m_1 + m_2)(M - 2T)) \\ +xI_B(Lm_1(-1 + d_1 m_1) + Mm_2 - 2m_1 T + 2d_1 m_1^2 T - 2m_2 T)) \\ -2x\theta + x\eta\theta + xI_c M\theta - xI_B Lm_1\theta - xI_c Mm_1\theta \\ +xd_1 m_1^2\theta + xd_1 I_B Lm_1^2\theta + xI_B Mm_2\theta - xI_c Mm_2\theta - 2xI_c T\theta \\ -2xI_B m_1 T\theta + 2xI_c m_1 T\theta + 2xd_1 I_B m_1^2 T\theta \\ -2xI_B m_2 T\theta + 2xI_c m_2 T\theta \end{bmatrix} = 0 \tag{18}$$

The explicit expression for Eq. (8) is

$$\frac{\mathrm{d}\, d_F\widetilde{\mathrm{TPU}}_1}{\mathrm{d}T} = 0 \text{ that is,}$$

$$
\left[\begin{array}{l}
7T\left(\begin{array}{l}
-2a_2h_2T + a_3b_3pT + 2a_3I_epT + a_3p(2+b_3T) \\
-2a_2xT\theta_2 + a_2x(-1+d_1m_1^2)T(b_2+\theta_2) \\
+a_2x(-1+d_1m_1^2)(2+b_2T+T\theta_2) \\
-2a_2x(I_c(-1+m_1+m_2)(M-T)+I_Bm_1(1-d_1m_1)(L_2+T) \\
+I_Bm_2(-M+T))(1+T(b_2+\theta_2)) \\
-2a_2x(-I_c(-1+m1+m2)+I_B(m_1-d_1m_1^2+m_2)) \\
\left(T+\frac{1}{2}T^2(b_2+\theta_2)\right) + 2a_3x\eta_3T\eta_3
\end{array}\right) \\
-7\left(\begin{array}{l}
-2A_2 - a_2h_2T^2 + a_3I_epT^2 + a_3pT(2+b_3T) - a_2xT\theta_2 \\
+a_2x(-1+d_1m_1^2)T(2+b_2T+T\theta_2) \\
-2a_2x(I_c(-1+m_1+m_2)(M-T)+I_Bm_1(1-d_1m_1)(L_2+T) \\
+I_Bm_2(-M+T))\left(T+\frac{1}{2}T^2(b_2+\theta_2)\right) + a3x\eta_3T^2\theta_3
\end{array}\right) \\
+7T\left(\begin{array}{l}
-2a_3h_3T + a_2b_2pT + 2a_2I_epT + a_2p(2+b_2T) + 2a_2x\eta_2T\theta_2 \\
-2a_3xT\theta_3 + a3x(-1+d_1m_1^2)T(b_3+\theta_3) \\
+a_3x(-1+d_1m_1^2)(2+b_3T+T\theta_3) \\
-2a_3x(I_c(-1+m_1+m_2)(M-T)+I_Bm_1(1-d_1m_1)(L_3+T) \\
+I_Bm_2(-M+T))(1+T(b_3+\theta_3)) - 2a_3c(-I_c(-1+m_1+m_2) \\
+I_B(m_1-d_1m_1^2+m_2))\left(T+\frac{1}{2}T^2(b_3+\theta_3)\right)
\end{array}\right) \\
-7\left(\begin{array}{l}
-2A_3 - a_3h_3T^2 + a_2I_epT^2 + a_2pT(2+b_2T) + a_2x\eta_2T^2\theta_2 \\
-a_3xT^2\theta_3 + a_3x(-1+d_1m_1^2)T(2+b_3T+T\theta_3) \\
-2a_3x(I_c(-1+m_1+m_2)(M-T)+I_Bm_1(1-d_1m_1)(L_3+T) \\
+I_Bm_2(-M+T))\left(T+\frac{1}{2}T^2(b_3+\theta_3)\right)
\end{array}\right) \\
+8T\left(\begin{array}{l}
-2a_1h_1T + a_4b_4pT + 2a_4I_epT + a_4p(2+b_4T) - 2a_1xT\theta_1 \\
+a_1x(-1+d_1m_1^2)T(b_1+\theta_1) + a_1x(-1+d_1m_1^2)(2+b_1T+T\theta_1) \\
-2a_1x(I_c(-1+m_1+m_2)(M-T)+I_Bm_1(1-d_1m_1)(L_1+T) \\
+I_Bm_2(-M+T))(1+T(b_1+\theta_1)) \\
-2a_1x(-I_c(-1+m_1+m_2)+I_B(m_1-d_1m_1^2+m_2)) \\
\left(T+\frac{1}{2}T^2(b_1+\theta_1)\right) + 2a_4x\eta_4T\theta_4
\end{array}\right) \\
-8\left(\begin{array}{l}
-2A_1 - a_1h_1T^2 + a_4I_epT^2 + a_4pT(2+b_4T) - a_1xT^2\theta_1 \\
+a_1x(-1+d_1m_1^2)T(2+b_1T+T\theta_1) \\
-2a_1x(I_c(-1+m_1+m_2)(M-T)+I_Bm_1(1-d_1m_1)(L_1+T) \\
+I_Bm_2(-M+T))\left(T+\frac{1}{2}T^2(b_1+\theta_1)\right) + a_4x\eta_4T^2\theta_4
\end{array}\right) \\
+8T\left(\begin{array}{l}
-2a_4h_4T + a_1b_1pT + 2a_1I_epT + a_1p(2+b_1T) + 2a_1x\eta_1T\theta_1 \\
-2a_4xT\theta_4 + a_4x(-1+d_1m_1^2)T(b_4+\theta_4) \\
+a_4x(-1+d_1m_1^2)(2+b_4T+T\theta_4) - 2a_4x(I_c(-1+m_1+m_2) \\
(M-T)+I_Bm_1(1-d_1m_1)(L_4+T) \\
+I_Bm_2(-M+T))(1+T(b_4+\theta_4)) - 2a_4x(-I_c(-1+m_1+m_2) \\
+I_B(m_1-d_1m_1^2+m_2))\left(T+\frac{1}{2}T^2(b_4+\theta_4)\right)
\end{array}\right) \\
-8\left(\begin{array}{l}
-2A_4 - a_4h_4T^2 + a_1I_epT^2 + a_1pT(2+b_1T) + a_1x\eta_1T^2\theta_1 \\
-a_4xT^2\theta_4 + a_4x(-1+d_1m_1^2)T(2+b_4T+T\theta_4) \\
-2a_4x(I_c(-1+m_1+m_2)(M-T)+I_Bm_1(1-d_1m_1)(L_4+T) \\
+I_Bm_2(-M+T))\left(T+\frac{1}{2}T^2(b_4+\theta_4)\right)
\end{array}\right)
\end{array}\right] = 0 \quad (19)
$$

The explicit expression for Eq. (12) is

$$\frac{\mathrm{dTPU}_2(T)}{\mathrm{d}T} = 0 \text{ that is,}$$

$$2A + aT^2 \begin{bmatrix} -h - 2xI_Bm_1 + 2xd_1I_Bm_1^2 - I_ep + b\{p((1 + I_e(M - 2T)) \\ +x(-1 + d_1m_1^2 + I_Bm_1(-1 + d_1m_1)(L + 2T))\} \\ -2x\theta + x\eta\theta - xI_BLm_1\theta + xd_1m_1^2\theta \\ +xd_1I_BLm_1^2\theta - 2xI_Bm_1T\theta + 2xd_1I_Bm_1^2T\theta \end{bmatrix} = 0 \tag{20}$$

The explicit expression for Eq. (15) is

$$\frac{\mathrm{d}d_F\widetilde{\mathrm{TPU}_2}}{\mathrm{d}T} = 0 \text{ that is,}$$

$$\begin{bmatrix} 16A_1 + 14A_2 + 14A_3 + 16A_4 - 8a_1b_1xT^2 - 7a_2b_2xT^2 \\ -7a_3b_3xT^2 - 8a_4b_4xT^2 - 8a_1h_1T^2 - 7a_2h_2T^2 \\ -7a_3h_3T^2 - 8a_4h_4T^2 - 30a_1xI_Bm_1T^2 - 14a_3xI_Bm_1T^2 \\ -16a_4xI_Bm_1T^2 - 8a_1b_1xI_BL_1m_1T^2 \\ -7a_2b_2xI_BL_2m_1T^2 - 7a_3b_3xI_BL_3m_1T^2 - 8a_4b_4xI_BL_4m_1T^2 \\ +8a_1b_1xd_1m_1^2T^2 + 7a_2b_2xd_1m_1^2T^2 \\ +7a_3b_3xd_1m_1^2T^2 + 8a_4b_4xd_1m_1^2T^2 + 30a_1xd_1I_Bm_1^2T^2 \\ +14a_3xd_1I_Bm_1^2T^2 + 16a_4xd_1I_Bm_1^2T^2 \\ +8a_1b_1xd_1I_BL_1m_1^2T^2 + 7a_2b_2xd_1I_BL_2m_1^2T^2 \\ +7a_3b_3xd_1I_BL_3m_1^2T^2 + 8a_4b_4xd_1I_BL_4m_1^2T^2 \\ +8a_1b_1pT^2 + 7a_2b_2pT^2 + 7a_3b_3pT^2 + 8a_4b_4pT^2 \\ -8a_1I_e\,pT^2 - 7a_2I_e\,pT^2 - 7a_3I_epT^2 \\ -8a_4I_epT^2 + 8a_1b_1I_eMpT^2 + 7a_2b_2I_eMpT^2 + 7a_3b_3I_eMpT^2 \\ +8a_4b_4I_eMpT^2 - 16a_1b_1xI_Bm_1T^3 \\ -14a_1b_2xIBm_1T^3 - 14a_3b_3xI_Bm_1T^3 - 16a_4b_4xI_Bm_1T^3 \\ +16a_1b_1xd_1I_Bm_1^2T^3 + 14a_1b_2xd_1I_Bm_1^2T^3 \\ +14a_3b_3xd_1I_Bm_1^2T^3 + 16a_4b_4xd_1I_Bm_1^2T^3 - 16a_1b_1I_epT^3 \\ -14a_2b_2I_epT^3 - 14a_3b_3I_epT^3 \\ -16a_4b_4I_epT^3 - 16a_1xT^2\theta_1 + 8a_1x\eta_1T^2\theta_1 - 8a_1xI_BL_1m_1T^2\theta_1 \\ +8a_1xd_1m_1^2T^2\theta_1 + 8a_1xd_1I_BL_1m_1^2T^2\theta_1 \\ -16a_1xI_Bm_1T^3\theta_1 + 16a_1xd_1I_Bm_1^2T^3\theta_1 \\ -14a_2xT^2\theta_2 + 7a_2x\eta_2T^2\theta_2 - 7a_2xI_BL_2m_1T^2\theta_2 \\ +7a_2xd_1m_1^2T^2\theta_2 + 7a_2xd_1I_BL2m_1^2T^2\theta_2 - 14a_1xI_Bm_1T^3\theta_2 \\ +14a_1xd_1I_Bm_1^2T^3\theta_2 - 14a_3xT^2\theta_3 \\ +7a_3x\eta_3T^2\theta_3 - 7a_3xI_BL_3m_1T^2\theta_3 + 7a_3xd_1m_1^2T^2\theta_3 \\ +7a_3xd_1I_BL_3m_1^2\,T^2\theta_3 - 14a_3xI_Bm_1T^3\theta_3 \\ +14a_3xd_1I_Bm_1^2T^3\theta_3 - 16a_4xT^2\theta_4 + 8a_4x\eta_4T^2\theta_4 \\ -8a_4xI_B\,L_4m_1T^2\theta_4 + 8a_4xd_1m_1^2T^2\theta_4 \\ +8a_4xd_1I_BL_4m_1^2T^2\theta_4 - 16a_4xI_Bm_1T^3\theta_4 + 16a_4xd_1I_Bm_1^2T^3\theta_4 \end{bmatrix} = 0 \tag{21}$$

Appendix 2

This appendix presents some results on optimality of the profit functions.

Lemma 1 *If a function* $G(y) = \frac{K(y)}{y}$, *where* $K(y)$ *is twice differentiable function of* y, *then maximum value of* $G(y)$ *exists at* $y = y^*$ *obtained by solving* $\frac{\mathrm{d}G}{\mathrm{d}y} = 0$ *if* $\frac{1}{y}\frac{\mathrm{d}^2 K}{\mathrm{d}y^2} < 0$ *at* $y = y^*$.

Proof We have $G(y) = \frac{K(y)}{y}$. For maximum or minimum, the necessary condition is $\frac{\mathrm{d}G}{\mathrm{d}y} = 0$.

Now,

$$\frac{\mathrm{d}G}{\mathrm{d}y} = \frac{1}{y^2}\left[y\frac{\mathrm{d}K(y)}{\mathrm{d}y} - K(y)\right], \frac{\mathrm{d}^2G}{\mathrm{d}y^2} = \frac{1}{y^3}\left[y^2\frac{\mathrm{d}^2K(y)}{\mathrm{d}y^2} - 2\left\{y\frac{\mathrm{d}K(y)}{\mathrm{d}y} - K(y)\right\}\right].$$

$$\frac{\mathrm{d}G}{\mathrm{d}y} = 0 \Rightarrow y\frac{\mathrm{d}K(y)}{\mathrm{d}y} - K(y) = 0 \tag{22}$$

Let $y = y^*$ which satisfies Eq. (22). Therefore, at $y = y^*$ we have

$$\frac{\mathrm{d}^2G}{\mathrm{d}y^2} = \frac{1}{y^3}\left[y^2\frac{\mathrm{d}^2K(y)}{\mathrm{d}y^2} - 2\left\{y\frac{\mathrm{d}K(y)}{\mathrm{d}y} - K(y)\right\}\right] = \frac{1}{y}\frac{\mathrm{d}^2K(y)}{\mathrm{d}y^2}.$$

using Eq. (22).

Thus, $G(y)$ is maximum at $y = y^*$ if $\frac{\mathrm{d}^2G}{\mathrm{d}y^2} < 0$ at $y = y^*$ that is, if $\frac{1}{y}\frac{\mathrm{d}^2K(y)}{\mathrm{d}y^2} < 0$ at $y = y^*$. Hence the lemma.

Result 1

$$\mathrm{TPU}_1(T)$$

is maximum if

$$a\begin{bmatrix} -h - 2x\big(-I_c(-1 + m_1 + m_2) + I_B(m1 - d_1 m_1^2 + m_2)\big) \\ +bp + I_e p - x\theta + x\eta\theta \\ -xI_B L m_1(1 - d_1 m_1)(b + \theta) + x(-1 + d_1 m_1^2)(b + \theta) \\ +xI_B M m_2(b + \theta) - xI_c M(-1 + m_1 + m_2)(b + \theta) \end{bmatrix}$$

$$+ aT\begin{bmatrix} -xI_B m_1(1 - d_1 m_1)(b + \theta) - xI_B m_2(b + \theta) \\ +xI_c(-1 + m_1 + m_2)(b + \theta) \\ -2x\big(-I_c(-1 + m_1 + m_2) + I_B(m_1 - d_1 m_1^2 + m_2)\big)(b + \theta) \end{bmatrix} < 0$$

at $T = T^* > 0$ which is obtained by solving the Eq. (18).

Proof Let $\mathrm{TPU}_1(T) = \frac{K(T)}{T}$, where

$$K(T) = \frac{1}{2}\begin{bmatrix} -2A - ahT^2 + aI_e pT^2 + apT(2+bT) - ax\theta T^2 \\ +ax\eta\theta T^2 + ax(d_1 m_1^2 - 1)(2 + bT + \theta T)T \\ -2ax\{I_c(m_1 + m_2 - 1)(M - T) + I_B m_1(1 - d_1 m_1)(L + T) \\ +I_B m_2(T - M)\}\{T + \frac{1}{2}T^2(b + \theta)\} \end{bmatrix}$$

Let $T = T^* > 0$ be obtained by solving the Eq. (18). Since $T > 0$, by Lemma 1, it follows that average profit will be optimal if $\frac{\mathrm{d}^2 K(T)}{\mathrm{d}T^2} < 0$ at $T = T^*$. That is if,

$$a\begin{bmatrix} -h - 2x\big(-I_c(-1 + m_1 + m_2) + I_B(m1 - d_1 m_1^2 + m_2)\big) \\ +bp + I_e p - x\theta + x\eta\theta \\ -xI_B L m_1(1 - d_1 m_1)(b + \theta) + x(-1 + d_1 m_1^2)(b + \theta) \\ +xI_B M m_2(b + \theta) - xI_c M(-1 + m_1 + m_2)(b + \theta) \end{bmatrix}$$

$$+aT\begin{bmatrix} -xI_B m_1(1 - d_1 m_1)(b + \theta) - xI_B m_2(b + \theta) \\ +xI_c(-1 + m_1 + m_2)(b + \theta) \\ -2x\big(-I_c(-1 + m_1 + m_2) + I_B(m_1 - d_1 m_1^2 + m_2)\big)(b + \theta) \end{bmatrix} < 0$$

Result 2 $d_F(\widetilde{\mathrm{TPU}_1})(T)$ is maximum if

$$\begin{bmatrix} -14a_3h_3 - 16a_4h_4 + 14a_3b_3p + 16a_4b_4p + 14a_3I_ep \\ +16a_4I_ep - 14a_3b_3x - 16a_4b_4x - 28a_3I_Cx - 32a_4I_Cx \\ +14a_3b_3I_CMx + 16a_4b_4I_CMx - 2a_3I_Bm_1x - 32a_4I_Bm_1x \\ +28a_3I_Cm_1x + 32a_4I_Cm_1x - 14a_3b_3I_BL_3m_1x \\ -16a_4b_4I_BL_4m_1x - 14a_3b_3I_CMm_1x - 16a_4b_4I_CMm_1x \\ +14a_3b_3d_1m_1^2x + 16a_4b_4d_1m_1^2x + 28a_3d_1m_1^2x \\ +32a_4I_Bd_1m_1^2x + 14a_3b_3I_BL_3d_1m_1^2x + 16a_4b_4I_BL_4d_1m_1^2x \\ -28a_3I_Bm_2x - 32a_4I_Bm_2x + 28a_3I_Cm_2x \\ +32a_4I_Cm_2x + 14a_3b_3I_BMm_2x + 16a_4b_4I_BMm_2x \\ -14a_3b_3I_CMm_2x - 16a_4b_4I_CMm_2x - 42a_3b_3I_CTx \\ -48a_4b_4I_CTx - 42a_3b_3I_Bm_1Tx - 48a_4b_4I_Bm_1Tx \\ +42a_3b_3I_Cm_1Tx + 48a_4b_4I_Cm_1Tx + 42a_3b_3I_Bd_1m_1^2Tx \\ +48a_4b_4I_Bd_1m_1^2Tx - 42a_3b_3I_Bm_2Tx - 48a_4b_4I_Bm_2Tx \\ +42a_3b_3I_Cm_2Tx + 48a_4b_4I_Cm_2Tx - 28a_3\theta_3x \\ +14a_3\eta_3\theta_3x + 14a_3I_CM\theta_3x - 14a_3I_BL_3m_1\theta_3x \\ -14a_3I_CMm_1\theta_3x + 14a_3d_1m_1^2\theta_3x + 14a_3I_BL_3d_1m_1^2\theta_3x \\ +14a_3I_BMm_2\theta_3x - 14a_3I_CMm_2\theta_3x - 42a_3I_CT\theta_3x \\ -42a_3I_Bm_1T\theta_3x + 42a_3I_Cm_1T\theta_3x + 42a_3I_Bd_1m_1^2T\theta_3x \\ -42a_3I_Bm_2T\theta_3x + 42a_3I_Cm_2T\theta_3x - 32a_4\theta_4x + 16a_4\eta_4\theta_4x \\ +16a_4I_CM\theta_4x - 16a_4I_BL_4m_1\theta_4x - 16a_4I_CMm_1\theta_4x \\ +16a_4d_1m_1^2\theta_4x + 16a_4I_BL_4d_1m_1^2\theta_4x + 16a_4I_BMm_2\theta_4x \\ -16a_4I_CMm_2\theta_4x - 48a_4I_CT\theta_4x - 48a_4I_Bm_1T\theta_4x \\ +48a_4I_Cm_1T\theta_4x + 48a_4I_Bd_1m_1^2T\theta_4x - 48a_4I_Bm_2T\theta_4x \\ +48a_4I_Cm_2T\theta_4x + 8a_1(-2h_1 + 2(b_1 + I_e)p \\ +(-2I_C(-1 + m_1 + m_2)(-2 + b_1(M - 3T) + M\theta_1 - 3T\theta_1) \\ +2(-b_1 - 2\theta_1 + \eta_1\theta_1 + d_1m_1^2(b_1 + \theta_1)) \\ +I_B(2m_2(-2 + b_1(M - 3T) + M\theta_1 - 3T\theta_1) \\ -2m_1(2 + b_1(L_1 + 3T) + L_1\theta_1 + 3T\theta_1) + 2d_1m_1^2(2 + b_1(L_1 + 3T) \\ +L_1\theta_1 + 3T\theta_1)))x) + 7a_2(-2h_2 + 2(b_2 + I_e)p \\ +(-2I_C(-1 + m_1 + m_2)(-2 + b_2(M - 3T) + M\theta_2 - 3T\theta_2) \\ +2(-b_2 - 2\theta_2 + \eta_2\theta_2 + d_1m_1^2(b_2 + \theta_2)) \\ +I_B(2m_2(-2 + b_2(M - 3T) + M\theta_2 - 3T\theta_2) \\ -2m_1(2 + b_2(L_2 + 3T) + L_2\theta_2 + 3T\theta_2) \\ +2d_1m_1^2(2 + b_2(L_2 + 3T) + L_2\theta_2 + 3T\theta_2)))x) \end{bmatrix} < 0$$

at $T = T^* > 0$ which is obtained by solving the Eq. (19).

Proof Similar to Result 1.

Result 3 $\text{TPU}_2(T)$ is maximum if

$$aT[6bxI_Bm_1(-1+d_1m_1)-6bI_ep+6xI_Bm_1(-1+d_1m_1)\theta] + a\begin{bmatrix} -2b-x-2h+4xI_Bm_1(-1+d_1m_1) \\ +2bxI_BLm_1(-1+d_1m_1)+2bp-2I_ep \\ +2bI_eMp-4c\theta+2x\eta\theta+2xI_BLm_1(-1+d_1m_1)\theta \\ +2xd_1m_1^2(b+\theta) \end{bmatrix} < 0$$

at $T = T^* > 0$ which is obtained by solving the Eq. (20).

Proof $\text{TPU}_2(T) = \frac{K(T)}{T}$, where

$$K(T) = \frac{1}{2}\begin{bmatrix} -2A-ahT^2-aI_epT^2+apT(2+bT) \\ +aI_ep(M-T)T(2+bT)-axT^2\theta+ax\eta T^2\theta \\ +ax(-1+d_1m_1^2)T(2+bT+T\theta) \\ +axI_Bm_1(-1+d_1m_1)T(L+T)(2+bT+T\theta) \end{bmatrix}$$

Let $T = T^* > 0$ be obtained by solving the Eq. (20). Since $T > 0$, by Lemma 1, it follows that average profit will be optimal if $\frac{\mathrm{d}^2K(T)}{\mathrm{d}T^2} < 0$ at $T = T^*$. That is if,

$$aT[6bxI_Bm_1(-1+d_1m_1)-6bI_ep+6xI_Bm_1(-1+d_1m_1)\theta] + a\begin{bmatrix} -2b-x-2h+4xI_Bm_1(-1+d_1m_1) \\ +2bxI_BLm_1(-1+d_1m_1)+2bp-2I_ep \\ +2bI_eMp-4c\theta+2x\eta\theta+2xI_BLm_1(-1+d_1m_1)\theta \\ +2xd_1m_1^2(b+\theta) \end{bmatrix} < 0$$

Result 4 $d_F(\widetilde{\text{TPU}_2})(T)$ is maximum if

$$\begin{bmatrix} -14a_3h_3-16a_4h_4+14a_3b_3p+16a_4b_4p-14a_3I_ep-16a_4I_ep \\ +14a_3b_3I_eMp+16a_4b_4I_eMp-42a_3b_3I_epT-48a_4b_4I_epT \\ -14a_3b_3x-16a_4b_4x-28a_3I_Bm_1x-32a_4I_Bm_1x-14a_3b_3I_BL_3m_1x \\ -16a_4b_4I_BL_4m_1x+14a_3b_3d_1m_1^2x+16a_4b_4d_1m_1^2x+28a_3I_Bd_1m_1^2x \\ +32a_4I_Bd_1m_1^2x+14a_3b_3I_BL3d_1m_1^2x+16a_4b_4I_BL_4d_1m_1^2x \\ -42a_3b_3I_Bm_1Tx-48a_4b_4I_Bm_1Tx+42a_3b_3I_Bd_1m_1^2Tx \\ +48a_4b_4I_Bd_1m_1^2Tx-28a_3\theta_3x+14a_3\eta_3\theta_3x-14a_3I_BL_3m_1\theta_3x \\ +14a_3d_1m_1^2\theta_3x+14a_3I_BL_3d_1m_1^2\theta_3x-42a_3I_Bm_1T\theta_3x \\ +42a_3I_Bd_1m_1^2T\theta_3x-32a_4\theta_4x+16a_4\eta_4\theta_4x-16a_4I_BL_4m_1\theta_4x \\ +16a_4d_1m_1^2\theta_4x+16a_4I_BL_4d_1m_1^2\theta_4x-48a_4I_Bm_1T\theta_4x \\ +48a_4I_Bd_1m_1^2T\theta_4x+14a_2(-h_2+p(b_2-I_e+b_2I_eM-3b_2I_eT) \\ +(-b_2-2\theta_2+\eta_2\theta_2d_1m_1^2(b_2+\theta_2)+I_BL_2m_1(-1+d_1m_1)(b_2+\theta_2))x) \\ +a_1(-16h_1+16p(b_1-I_e+b_1I_eM-3b_1I_eT)+(16(-b_1-2\theta_1 \\ +\eta_1\theta_1+d_1m_1^2(b_1+\theta_1))+2I_Bm_1(-1+d_1m_1)(30+21b_2T \\ +8b_1(L_1+3T)+8L_1\theta_1+24T\theta_1+21T\theta_2))x) \end{bmatrix} < 0$$

at $T = T^* > 0$ which is obtained by solving the Eq. (21).

Proof Similar to Result 3.

References

1. Abad, P.L., Jaggi, C.K.: A joint approach for setting unit price and the length of the credit period for a seller when end demand is price sensitive. Int. J. Prod. Econ. **83**(2), 115–122 (2003)
2. Aggarwal, P.S., Jaggi, K.C.: Ordering policies of deteriorating items under permissible delay in payment. J. Oper. Res. Soc. **46**(5), 658–662 (1995)
3. Agrawal, S., Gupta, R., Banerjee, S.: EOQ model under discounted partial advance partial trade credit policy with price-dependent demand. In: Shah, N.H., Mittal, M. (eds.) Optimization and Inventory Management, pp. 219–237. Springer, Singapore (2020)
4. Annadurai, K., Uthayakumar, R.: Decaying inventory model with stock-dependent demand and shortages under two-level trade credit. Int. J. Adv. Manuf. Technol. **77**, 525–543 (2015)
5. Bhunia, A.K., Shaikh, A.A., Cárdenas-Barrón, L.E.: A partially integrated production-inventory model with interval valued inventory costs, variable demand and flexible reliability. Appl. Soft Comput. **55**, 491–502 (2017)
6. Bhunia, A.K., Shaikh, A.A.: A deterministic model for deteriorating items with displayed inventory level dependent demand rate incorporating marketing decisions with transportation cost. Int. J. Ind. Eng. Comput. **2**(3), 547–562 (2011)
7. Bhunia, A.K., Shaikh, A.A.: Investigation of two-warehouse inventory problems in interval environment under inflation via particle swarm optimization. Math. Comput. Model. Dyn. Syst. **22**(2), 160–179 (2016)
8. Chang, C.T., Ouyang, L.Y., Teng, J.T., Cheng, M.C.: Optimal ordering policies for deteriorating items using a discounted cash-flow analysis when a trade credit is linked to order quantity. Comput. Ind. Eng. **59**(4), 770–777 (2010)
9. Chang, S.C., Yao, J.S., Lee, H.M.: Economic reorder point for fuzzy backorder quantity. Eur. J. Oper. Res. **109**(1), 183–202 (1998)
10. Chaudhari, U., Shah, N.H., Jani, M.Y.: Inventory modelling of deteriorating item and preservation technology with advance payment scheme under quadratic demand. In: Shah, N.H., Mittal, M. (eds.) Optimization and Inventory Management, pp. 69–79. Springer, Singapore (2020)
11. Chen, L.H., Kang, F.S.: Integrated inventory models considering the two-level trade credit policy and a price-negotiation scheme. Eur. J. Oper. Res. **205**, 47–58 (2010)
12. Das, A.K., Roy, T.K.: An imprecise EOQ model for non-instantaneous deteriorating item with imprecise inventory parameters using interval number. Int. J. Appl. Comput. Math. **4**(2), 79–94 (2018)
13. Das, S.C., Manna, A.K., Rahman, M.S., Shaikh, A.A., Bhunia, A.K.: An inventory model for non-instantaneous deteriorating items with preservation technology and multiple credit periods-based trade credit financing via particle swarm optimization. Soft Comput. **25**(7), 5365–5384 (2021)
14. De, S.K., Mahata, G.C.: Decision of a fuzzy inventory with fuzzy backorder model under cloudy fuzzy demand rate. Int. J. Appl. Comput. Math. **3**(3), 2593–2609 (2017)
15. Diabat, A., Taleizadeh, A.A., Lashgari, M.: A lot sizing model with partial downstream delayed payment, partial upstream advance payment, and partial backordering for deteriorating items. J. Manuf. Syst. **45**, 322–342 (2017)
16. Duari, N.K., Chakrabarti, T.: Optimal ordering policy for deteriorating items with power-form stock dependent demand and shortages under two-warehouse facility. Int. J. Inf. Res. Rev. **3**(12), 3347–3360 (2016)

17. Dutta, D., Kumar, P.: Fuzzy inventory model without shortages using trapezoidal fuzzy number with sensitivity analysis. IOSR J. Math. **4**(3), 32–37 (2012)
18. Goyal, S.K.: Economic order quantity under conditions of permissible delay in payments. J. Oper. Res. Soc. **36**(4), 335–338 (1985)
19. Gupta, R.K., Bhunia, A.K., Goyal, S.K.: An application of genetic algorithm in solving an inventory model with advance payment and interval valued inventory costs. Math. Comput. Model. **49**(5–6), 893–905 (2009)
20. Gupta, R.K., Bhunia, A.K., Goyal, S.K.: An application of genetic algorithm in a marketing oriented inventory model with interval valued inventory costs and three-component demand rate dependent on displayed stock level. Appl. Math. Comput. **192**(2), 466–478 (2007)
21. Hemalatha, S., Annadurai, K.: A fuzzy EOQ inventory model with advance payment and various fuzzy numbers. Mater. Today Proc. (2020). https://doi.org/10.1016/j.matpr.2020.10.769
22. Huang, Y.F.: A note on EOQ model under cash discount and payment delay. Int. J. Inf. Manag. Sci. **16**(3), 97–107 (2005)
23. Indrajitsingha, S.K., Samanta, P., Raju, L.K., Misra, U.: Two-storage inventory model for deteriorating items with price dependent demand and shortages under partial backlogged in fuzzy approach. LogForum **15**(4), 487–499 (2019)
24. Jaggi, C.K., Cárdenas-Barrón, L.E., Tiwari, S., Shafi, A.A.: Two-warehouse inventory model for deteriorating items with imperfect quality under the conditions of permissible delay in payments. Sci. Iranica **24**(1), 390–412 (2017)
25. Jaggi, C.K., Tiwari, S., Goel, S.K.: Credit financing in economic ordering policies for non-instantaneous deteriorating items with price dependent demand and two storage facilities. Ann. Oper. Res. **248**(1–2), 253–280 (2017)
26. Jain, S., Tiwari, S., Cárdenas-Barrón, L.E., Shaikh, A.A., Singh, S.R.: A fuzzy imperfect production and repair inventory model with time dependent demand, production and repair rates under inflationary conditions. RAIRO-Oper. Res. **52**(1), 217–239 (2018)
27. Jamal, A.M.M., Sarker, B.R., Wang, S.: An ordering policy for deteriorating items with allowable shortage and permissible delay in payment. J. Oper. Res. Soc. **48**(8), 826–833 (1997)
28. Kalpakam, S., Sapan, K.P.: Stochastic inventory systems; a perishable inventory model. In: International Conference on Stochastic Process and Their Applications, pp. 246–253. Anna University, India (1996)
29. Khan, M.A.A., Ahmed, S., Babu, M.S., Sultana, N.: Optimal lot-size decision for deteriorating items with price-sensitive demand, linearly time-dependent holding cost under all-units discount environment. Int. J. Syst. Sci. Oper. Log. **9**, 1–14 (2020)
30. Khan, M.A.A., Shaikh, A.A., Cárdenas-Barrón, L.E.: An inventory model under linked-to-order hybrid partial advance payment, partial credit policy, all-units discount and partial backlogging with capacity constraint. Omega **103**, 102418 (2021)
31. Khan, M.A.A., Shaikh, A.A., Konstantaras, I., Bhunia, A.K., Cárdenas-Barrón, L.E.: Inventory models for perishable items with advanced payment, linearly time-dependent holding cost and demand dependent on advertisement and selling price. Int. J. Prod. Econ. **230**, 107804 (2020)
32. Khan, M.A.A., Shaikh, A.A., Panda, G.C., Bhunia, A.K., Konstantaras, I.: Non-instantaneous deterioration effect in ordering decisions for a two-warehouse inventory system under advance payment and backlogging. Ann. Oper. Res. **289**, 243–275 (2020)
33. Khan, M.A.A., Shaikh, A.A., Panda, G.C., Konstantaras, I., Cárdenas-Barrón, L.E.: The effect of advance payment with discount facility on supply decisions of deteriorating products whose demand is both price and stock dependent. Int. Trans. Oper. Res. **27**(3), 1343–1367 (2020)
34. Khan, M.A.A., Shaikh, A.A., Panda, G.C., Konstantaras, I.: Two-warehouse inventory model for deteriorating items with partial backlogging and advance payment scheme. RAIRO-Oper. Res. **53**(5), 1691–1708 (2019)
35. Krommyda, I.P., Skouri, K., Lagodimos, A.G.: A unified EOQ model with financial constraints and market tolerance. Appl. Math. Model. **65**, 89–105 (2019)
36. Kumar, B.A., Paikray, S.K., Mishra, U.: Two-storage fuzzy inventory model with time dependent demand and holding cost under acceptable delay in payment. Math. Model. Anal. **25**(3), 441–460 (2020)

37. Kumar, P.: An inventory planning problem for time-varying linear demand and parabolic holding cost with salvage value. Croat. Oper. Res. Rev. **10**, 187–199 (2019)
38. Lashgari, M., Taleizadeh, A.A., Ahmadi, A.: Partial up-stream advanced payment and partial down-stream delayed payment in a three-level supply chain. Ann. Oper. Res. **238**(1–2), 329–354 (2016)
39. Li, R., Liu, Y., Teng, J.T., Tsao, Y.C.: Optimal pricing, lot-sizing and backordering decisions when a seller demands an advance-cash-credit payment scheme. Eur. J. Oper. Res. **278**(1), 283–295 (2019)
40. Maiti, A.K., Maiti, M.K., Maiti, M.: Inventory model with stochastic lead-time and price dependent demand incorporating advance payment. Appl. Math. Model. **33**(5), 2433–2443 (2009)
41. Mashud, A.H.M., Wee, H.M., Huang, C.V.: Preservation technology investment, trade credit and partial backordering model for a non-instantaneous deteriorating inventory. RAIRO-Oper. Res. **55**, S51–S77 (2021)
42. Mashud, A.H.M., Wee, H.M., Sarkar, B., Li, Y.H.C.: A sustainable inventory system with the advanced payment policy and trade-credit strategy for a two-warehouse inventory system. Kybernetes **50**(5), 1321–1348 (2020)
43. Ouyang, L.Y., Chen, M.S., Chuang, K.W.: Economic order quantity model under cash discount and payment delay. Int. J. Inf. Manag. Sci. **13**(1), 1–10 (2002)
44. Ouyang, L.Y., Wu, K.S., Yang, C.T.: A study on an inventory model for non-instantaneous deteriorating items with permissible delay in payments. Comput. Ind. Eng. **51**, 637–651 (2006)
45. Pando, V., Garca-Lagunaa, J., San-Jos, L.A., Sicilia, J.: Maximizing profits in an inventory model with both demand rate and holding cost per unit time dependent on the stock level. Comput. Ind. Eng. **62**(2), 599–608 (2012)
46. Priyan, S., Palanivel, M., Uthayankumar, R.: Mathematical modeling for EOQ inventory system with advance payment and fuzzy parameters. Int. J. Supply Oper. Manag. **1**(3), 260–278 (2014)
47. Rahman, M.S., Khan, M.A.A., Halim, M.A., Nofal, T.A., Shaikh, A.A., Mahmoud, E.E.: Hybrid price and stock dependent inventory model for perishable goods with advance payment related discount facilities under preservation technology. Alex. Eng. J. **60**(3), 3455–3465 (2021)
48. Roy, A., Maiti, M.K., Kar, S., Maiti, M.: An inventory model for a deteriorating item with displayed stock dependent demand under fuzzy inflation and time discounting over a random planning horizon. Appl. Math. Model. **33**(2), 744–759 (2009)
49. Saha, S., Sen, N.: A study on inventory model with negative exponential demand and probabilistic deterioration under backlogging. Uncertain Supply Chain Manag. **5**(2), 77–88 (2017)
50. Sajadieh, M.S., Thorstenson, A., Jokar, M.R.K.: An integrated vendor–buyer model with stock dependent demand. Transp. Res. E Logist. Transp. Rev. **46**(6), 963–974 (2010)
51. Sarkar, B.: An EOQ model with delay in payments and time varying deterioration rate. Math. Comput. Model. **55**(3–4), 367–377 (2012)
52. Sen, N., Nath, B.K., Saha, S.: A fuzzy inventory model for deteriorating items based on different defuzzification techniques. Am. J. Math. Stat. **6**(3), 128–137 (2016)
53. Shabani, S., Mirzazadeh, A., Sharifi, E.: A two-warehouse inventory model with fuzzy deterioration rate and fuzzy demand rate under conditionally permissible delay in payment. J. Ind. Prod. Eng. **33**(2), 134–142 (2016)
54. Shah, N.H., Shah, Y.K.: A discrete-in-time probabilistic inventory model for deteriorating items under conditions of permissible delay in payment. Int. J. Syst. Sci. **29**, 121–126 (1997)
55. Shaikh, A.A., Bhunia, A.K., Cárdenas-Barrón, L.E., Sahoo, L., Tiwari, S.: A fuzzy inventory model for a deteriorating item with variable demand, permissible delay in payments and partial backlogging with shortage follows inventory (SFI) policy. Int. J. Fuzzy Syst. **20**(5), 1606–1623 (2018)
56. Shaikh, A.A., Das, S.C., Bhunia, A.K., Panda, G.C., Khan, M.A.A.: A two-warehouse EOQ model with interval-valued inventory cost and advance payment for deteriorating item under particle swarm optimization. Soft Comput. **23**(24), 13531–13546 (2019)

57. Singh, S.R., Khurana, D., Tayal, S.: An economic order quantity model for deteriorating products having stock dependent demand with trade credit period and preservation technology. Uncertain Supply Chain Manag. **4**, 29–42 (2016)
58. Singh, S.R., Rastogi, M., Tayal, S.: An inventory model for deteriorating items having seasonal and stock dependent demand with allowable shortages. In: Pant, M., Deep, K., Bansal, J., Nagar, A., Das, K. (eds.) Proceedings of Fifth International Conference on Soft Computing for Problem Solving. Advances in Intelligent Systems and Computing, vol. 437, pp. 501–513. Springer, Singapore (2016)
59. Supakar, P., Mahato, S.K.: Fuzzy-stochastic advance payment inventory model having no shortage and with uniform demand using ABC algorithm. Int. J. Appl. Comput. Math. **4**(4), 1–19 (2018)
60. Taleizadeh, A.A., Lashgari, M., Akram, R., Heydari, J.: Imperfect economic production quantity model with upstream trade credit periods linked to raw material order quantity and downstream trade credit periods. Appl. Math. Model. **40**(19–20), 8777–8793 (2016)
61. Taleizadeh, A.A., Pentico, D.W., Jabalameli, M.S., Aryanezhad, M.: An economic order quantity model with multiple partial prepayments and partial backordering. Math. Comput. Model. **57**(3–4), 311–323 (2013)
62. Taleizadeh, A.A.: An economic order quantity model for deteriorating item in a purchasing system with multiple prepayments. Appl. Math. Model. **38**(23), 5357–5366 (2014)
63. Teng, J.T.: On the economic order quantity under conditions of permissible delay in payment. J. Oper. Res. Soc. **53**(8), 915–918 (2002)
64. Thangam, A.: Optimal price discounting and lot-sizing policies for perishable items in a supply chain under advance payment scheme and two-echelon trade credits. Int. J. Prod. Econ. **139**(2), 459–472 (2012)
65. Tiwari, S., Cárdenas-Barrón, L.E., Khanna, A., Jaggi, C.K.: Impact of trade credit and inflation on retailer's ordering policies for non-instantaneous deteriorating items in a two-warehouse environment. Int. J. Prod. Econ. **176**, 154–169 (2016)
66. Tripathi, R.P., Chaudhary, S.K.: Inflationary induced EOQ model for Weibull distribution deterioration and trade credits. Int. J. Appl. Comput. Math. **3**(4), 3341–3353 (2017)
67. Wu, K.S., Ouyang, L.Y., Yang, C.T.: An optimal replenishment policy for non-instantaneous deteriorating items with stock-dependent demand and partial backlogging. Int. J. Prod. Econ. **101**(2), 369–384 (2006)
68. Yao, J.S., Chang, S.C., Su, J.S.: Fuzzy inventory without backorder for fuzzy order quantity and fuzzy total demand quantity. Comput. Oper. Res. **27**(10), 935–962 (2000)
69. Zhang, A.X.: Optimal advance payment scheme involving fixed per-payment costs. Omega **24**(5), 577–582 (1996)
70. Zhang, Q., Tsao, Y.C., Chen, T.H.: Economic order quantity under advance payment. App. Math. Model. **38**(24), 5910–5921 (2014)
71. Zhang, Q., Zhang, D., Tsao, Y.C., Luo, J.: Optimal ordering policy in a two-stage supply chain with advance payment for stable supply capacity. Int. J. Prod. Econ. **177**, 34–43 (2016)
72. Zia, N.P., Taleizadeh, A.A.: A lot-sizing model with backordering under hybrid linked-to-order multiple advance payments and delayed payment. Transp. Res. E Logist. Transp. Rev. **82**, 19–37 (2015)

The Effect of Solvents on the Oxidation of Acetaldehyde Using TOPSIS Method in a Fuzzy Environment

Ammilal Rao and Ganesh Kumar

Abstract Researchers and practitioners have focused on MCDM methodologies for analyzing, rating, and ranking alternatives in a variety of industries. In this paper, acetaldehyde is oxidized in several solvents, including CF, DCE, DCM, DMSO, ACETONE, and DMF, with BIFC, BPCC, PCC, MCC, and QFC. Due to uncertainty and ambiguity in measuring at different temperatures, it is not possible to classify the situation as crisp. We used a fuzzy approach to deal with the scenario. The TOPSIS technique is used to select the strongest solvent from a large number of available alternatives. With the help of the TOPSIS technique, we discovered that the declining sequence of solvents for rapid reactions is DMSO > DMF > DCE > ACETONE > DCM > CF.

1 Introduction

There have been numerous studies on the oxidation of organic compounds [1–5] that have made use of chromium-containing substances. An extensive range of metal complexes [6–19] has been used to demonstrate that aldehydes (both aliphatic and aromatic) can oxidize in DMSO, as have a number of other investigations. Several research [20–26] examined the effect of the solvent on the oxidation of acetaldehyde by halochromes. The results were mixed. When investigating the influence of the solvent on the model, the fuzzy TOPSIS approach may be utilized to evaluate the results. Using a fuzzy TOPSIS approach for robot selection, a study of the ratings of numerous options in comparison with a large number of subjective criteria was carried out. Using alpha-level sets and a nonlinear programming (NLP) solution methodology, Wang and Eihag performed research [27] on the topic and created a fuzzy TOPSIS method for resolving it. The fuzzy TOPSIS method was examined

A. Rao
Department of Chemistry, University of Rajasthan, Jaipur, Rajasthan 302004, India

G. Kumar (✉)
Department of Mathematics, University of Rajasthan, Jaipur, Rajasthan 302004, India
e-mail: ganeshmandha1988@gmail.com

L. Sahoo et al. (eds.), *Real Life Applications of Multiple Criteria Decision Making Techniques in Fuzzy Domain*, Studies in Fuzziness and Soft Computing 420,
https://doi.org/10.1007/978-981-19-4929-6_13

in [28] using the fuzzy-weighted average (FWA) approach, and the results were published. As previously stated, the interval-valued fuzzy TOPSIS approach was built on the concept of interval-valued fuzzy sets and was used to deal with MCDM challenges caused by unequal weighting of the criteria in the MCDM problem. The use of MCDM approaches in energy supply systems is becoming increasingly prevalent, particularly in renewable energy supply systems that are graded according to the multi-criteria decision support framework. It was decided to incorporate multi-criteria decision techniques into the study in order to solve the conflicting issues that arise during the process of selecting renewable energy sources (Fuzzy TOPSIS). Using intuitionistic fuzzy sets (IFS), intuitionistic fuzzy sets of the second type (IFS2) and neutrosophic fuzzy sets (NFS), it was discovered in [29] that these three-dimensional extensions of regular fuzzy sets were all intuitionistic fuzzy sets. TOPSIS was used to demonstrate in [30] that the concept of generalized three-dimensional spherical fuzzy sets (SFS), as well as some major distinctions from prior fuzzy sets, could be introduced using an example. The TOPSIS approach, which was later developed into the spherical fuzzy environment. Reference [31] contains an additional comparison analysis that includes intuitionistic fuzzy TOPSIS (IF-TOPSIS), as well as other results. Reference [32] applied the TOPSIS method on a real-world problem. The effect of various solvents on acetaldehyde oxidation reactions involving the BIFC, BPCC, PCC, MCC, and QFC is investigated using the fuzzy TOPSIS technique. The solvents investigated include dimethyl sulfoxide (DMSO), chloroform (CF), dichloromethane (DCM), 1,2-dichloroethane (DCE), acetone, and dimethyl formamide (DMF) in Table 1.

To determine the best solvents for crisp cases, there are various papers available in the literature. However, it is impossible to create realistic representations of real-life situations in a clear and concise manner. As a result, we used the fuzzy technique to deal with the real-life scenarios that occurred.

Fuzzy TOPSIS is a famous and frequently used technique for selecting options in an uncertain environment, invented by Hwang and Yoon in 1981. Fuzzy TOPSIS is a strategy for ordering options in a fuzzy environment that uses mechanism composition MCDM.

Table 1 Investigation of various solvents

	BIFC	BPCC	PCC	MCC	QFC
CF	(14.6,15.0,15.4)	(14.2,15.5,16.2)	(5.74,6.76,7.20)	(25.2,27.6,29.3)	(30.2,31.6,31.8)
DCE	(18.5,19.2,19.8)	(18.8,19.5,20.2)	(82.8,85.1,88.2)	(29.2,31.6,32.2)	(40.6,42.7,43.8)
DCM	(17.4,17.8,18.2)	(16.7,17.4,18.3)	(72.3,74.1,76.2)	(33.2,34.7,35.4)	(34.8,36.3,36.8)
DMSO	(58.5,60.7,61.8)	(53.3,56.2,58.6)	(244,262,273)	(94.2,99.5,107)	(121,123,126)
ACETONE	(14.2,15.8,16.4)	(17.6,18.2,18.8)	(74.2,77.6,79.4)	(27.4,28.8,29.5)	(36.7,38.9,39.4)
DMF	(31.8,33.6,34.5)	(27.6,28.2,29.2)	(121,126,130)	(54.5,56.2,57.8)	(62.2,64.6,66.4)

2 The Fuzzy TOPSIS Approach is Comprised of Several Steps

2.1 When Making a Choice, Draw a Decision Matrix

In this study, a total of 5 criterion and 6 choices are taken into consideration, with the outcomes being ranked utilizing the FUZZY TOPSIS ranking system. The kinds of criterion that was utilized, as well as the weight that was assigned to each factor, are mentioned in Table 2.

The fuzzy scale that was employed in the model is depicted in Table 3. The scales are composed in accordance with the information in Table 1 and are assigned values ranging from 0 to 10 in accordance with their solvent effects.

The choices are analyzed in terms of a variety of criteria, and the findings of the decision matrix are presented in Table 4. Remember that if more than one expert participates in the evaluation, then the matrix below reflects the arithmetic mean of all the data.

Table 2 Features of criteria

	Name	Type	Weight
1	BIFC	+	(0.200,0.200,0.200)
2	BPCC	+	(0.200,0.200,0.200)
3	PCC	+	(0.200,0.200,0.200)
4	MCC	+	(0.200,0.200,0.200)
5	QFC	+	(0.200,0.200,0.200)

Table 3 Fuzzy scale

S. No.	Linguistic terms	Lower	Medium	Upper
1	VL	0	0	1
2	L	0	1	3
3	ML	1	3	5
4	M	3	5	7
5	MH	5	7	9
6	H	7	9	10
7	VH	9	10	10

Table 4 Decision matrix

	BIFC	BPCC	PCC	MCC	QFC
CF	(0.000,1.000,3.000)	(0.000,1.000,3.000)	(0.000,0.000,1.000)	(1.000,3.000,5.000)	(3.000,5.000,7.000)
DCE	(0.000,1.000,3.000)	(0.000,1.000,3.000)	(9.000,10.000,10.000)	(1.000,3.000,5.000)	(3.000,5.000,7.000)
DCM	(0.000,1.000,3.000)	(0.000,1.000,3.000)	(9.000,10.000,10.000)	(3.000,5.000,7.000)	(3.000,5.000,7.000)
DMSO	(7.000,9.000,10.000)	(7.000,9.000,10.000)	(9.000,10.000,10.000)	(9.000,10.000,10.000)	(9.000,10.000,10.000)
ACETONE	(0.000,1.000,3.000)	(0.000,1.000,3.000)	(9.000,10.000,10.000)	(0.000,1.000,3.000)	(3.000,5.000,7.000)
DMF	(3.000,5.000,7.000)	(1.000,3.000,5.000)	(9.000,10.000,10.000)	(7.000,9.000,10.000)	(9.000,10.000,10.000)

2.2 Create a Normalized Decision Matrix to Use in Your Decision-Making

The following equation may be used to build a normalized choice matrix based on the ideal solutions that are both positive and negative in nature:

$$\tilde{r}_{ij} = \left(\frac{a_{ij}}{c_j^*}, \frac{b_{ij}}{c_j^*}, \frac{c_{ij}}{c_j^*}\right); c_j^* = \max_i c_{ij}; \text{ Positive ideal solution}$$

$$\tilde{r}_{ij} = \left(\frac{a_j^-}{c_{ij}}, \frac{a_j^-}{b_{ij}}, \frac{a_j^-}{a_{ij}}\right); a_j^- = \min_i a_{ij}; \text{ Negative ideal solution}$$

As indicated in Table 5, the normalized decision matrix is used to make decisions.

2.3 Make a Decision Matrix that is Weighted and Normalized

In order to compute the criteria weights decision model, multiplying the criteria weights in the normalized fuzzy decision problem by the weight of each criterion in the normalized fuzzy decision matrix, according with following formula, while taking into consideration the different scales of each criterion, is necessary.

$$\tilde{v}_{ij} = \tilde{r}_{ij}.\tilde{w}_{ij}$$

where $\tilde{w}_{ij}$ represents weight of criterion c_j.

Table 6 summarizes the weighted normalized decision matrix.

2.4 For the Problem Statement, Compute the Fuzzy Favorable Ideal Solution (FPIS, A*) as Well as the Fuzzy Negative Ideal Alternative (FNIS, A−)

As an explanation, the following is how the FPIS and FNIS of the alternatives are defined:

$$A^* = \{\tilde{v}_1^*, \tilde{v}_2^*, \ldots, \tilde{v}_n^*\} = \left\{\left(\max_j v_{ij} | i \in B\right), \left(\min_j v_{ij} | i \in C\right)\right\}$$

$$A^- = \{\tilde{v}_1^-, \tilde{v}_2^-, \ldots, \tilde{v}_n^-\} = \left\{\left(\min_j v_{ij} | i \in B\right), \left(\max_j v_{ij} | i \in C\right)\right\}$$

Table 7 lists the best solution, both favorable and unfavorable.

Table 5 Normalized decision matrix

	BIFC	BPCC	PCC	MCC	QFC
CF	(0.000,0.100,0.300)	(0.000,0.100,0.300)	(0.000,0.000,0.100)	(0.100,0.300,0.500)	(0.300,0.500,0.700)
DCE	(0.000,0.100,0.300)	(0.000,0.100,0.300)	(0.900,1.000,1.000)	(0.100,0.300,0.500)	(0.300,0.500,0.700)
DCM	(0.000,0.100,0.300)	(0.000,0.100,0.300)	(0.900,1.000,1.000)	(0.300,0.500,0.700)	(0.300,0.500,0.700)
DMSO	(0.700,0.900,1.000)	(0.700,0.900,1.000)	(0.900,1.000,1.000)	(0.900,1.000,1.000)	(0.900,1.000,1.000)
ACETONE	(0.000,0.100,0.300)	(0.000,0.100,0.300)	(0.900,1.000,1.000)	(0.000,0.100,0.300)	(0.300,0.500,0.700)
DMF	(0.300,0.500,0.700)	(0.100,0.300,0.500)	(0.900,1.000,1.000)	(0.700,0.900,1.000)	(0.900,1.000,1.000)

Table 6 Weighted and normalized decision matrix

	BIFC	BPCC	PCC	MCC	QFC
CF	(0.000,0.020,0.060)	(0.000,0.020,0.060)	(0.000,0.000,0.020)	(0.020,0.060,0.100)	(0.060,0.100,0.140)
DCE	(0.000,0.020,0.060)	(0.000,0.020,0.060)	(0.180,0.200,0.200)	(0.020,0.060,0.100)	(0.060,0.100,0.140)
DCM	(0.000,0.020,0.060)	(0.000,0.020,0.060)	(0.180,0.200,0.200)	(0.060,0.100,0.140)	(0.060,0.100,0.140)
DMSO	(0.140,0.180,0.200)	(0.140,0.180,0.200)	(0.180,0.200,0.200)	(0.180,0.200,0.200)	(0.180,0.200,0.200)
ACETONE	(0.000,0.020,0.060)	(0.000,0.020,0.060)	(0.180,0.200,0.200)	(0.000,0.020,0.060)	(0.060,0.100,0.140)
DMF	(0.060,0.100,0.140)	(0.020,0.060,0.100)	(0.180,0.200,0.200)	(0.140,0.180,0.200)	(0.180,0.200,0.200)

Table 7 Optimal solutions, both favorable and unfavorable

	PI	NI
BIFC	(0.140,0.180,0.200)	(0.000,0.020,0.060)
BPCC	(0.140,0.180,0.200)	(0.000,0.020,0.060)
PCC	(0.180,0.200,0.200)	(0.000,0.000,0.020)
MCC	(0.180,0.200,0.200)	(0.000,0.020,0.060)
QFC	(0.180,0.200,0.200)	(0.060,0.100,0.140)

2.5 *Identify the Gap Between Each Choice and the Fuzzy Optimistic Best Solution A*, as Well as the Gap with Each Choice and the Fuzzy Negative Ideal Way A−*

The distance with each choice and FPIS, as well as the distance with each choice and FNIS, is computed as follows using the formulae provided here:

$$S_i^* = \sum_{j=1}^{n} d(\tilde{v}_{ij}, \tilde{v}_j^*) \quad i = 1, 2, \ldots, m$$

$$S_i^- = \sum_{j=1}^{n} d(\tilde{v}_{ij}, \tilde{v}_j^-) \quad i = 1, 2, \ldots, m$$

d is the distance between two TFNs (a_1, b_1, c_1) and (a_2, b_2, c_2), can be computed as follows:

$$d_v\left(\tilde{M}_1, \tilde{M}_2\right) = \sqrt{\frac{1}{3}\left[(a_1 - a_2)^2 + (b_1 - b_2)^2 + (c_1 - c_2)^2\right]}$$

where $d\left(\tilde{v}_{ij}, \tilde{v}_j^*\right)$ and $d\left(\tilde{v}_{ij}, \tilde{v}_j^-\right)$ are crisp.

The distance between PI and NI is provided in Table 8.

Table 8 Distance from PIS and NIS

	Distance from PI	Distance from NI
CF	0.713	0.035
DCE	0.526	0.222
DCM	0.487	0.261
DMSO	0	0.745
ACETONE	0.558	0.187
DMF	0.213	0.539

2.6 Calculation of Closeness Coefficient

The closeness coefficient can be computed using the following formula:

$$CC_i = \frac{S_i^-}{S_i^+ + S_i^-}$$

If you want to be as close to the FPIS as possible, you should choose the option that is the furthest away from the FNIS. Table 9 shows the coefficients of closeness for each option, as well as the order in which the options are ordered, for your convenience.

The closeness coefficient of each possibility is shown in Fig. 1.

Table 9 Closeness coefficient

	C_i	Rank
CF	0.046	6
DCE	0.296	4
DCM	0.349	3
DMSO	1	1
ACETONE	0.251	5
DMF	0.716	2

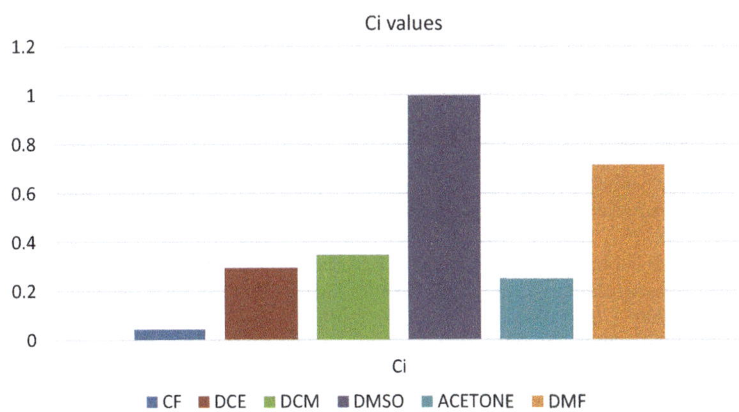

Fig. 1 Closeness coefficient graph

3 Conclusion

The fuzzy TOPSIS approach can be used to compare solvent effects. This method assists to find out the suitable solvent for acetaldehyde oxidation by various oxidants. In this investigation, we found the following preferences of solvents for fast reactions: DMSO > DMF > DCE > ACETONE > DCM > CF. The most significant contribution of this work is the use of a fuzzy technique in a real-world industrial situation, which has not been done or has been done infrequently previously, on this subject.

References

1. Corey, E.J., Suggs, W.J.: Pyridinium chlorochromate. An efficient reagent for oxidation of primary and secondary alcohols to carbonyl compounds. Tetrahedron Lett. 2647 (1975)
2. Guziec, F.S., Luzio, F.A.: Aquachloroiridium (III)-catalyzed oxidation of some unsaturated acids in acetone by acidic quinolinium fluorochromate. Synthesis 691 (1980)
3. Bhattacharjee, M.N., Choudhuri, M.K., Dasgupta, H.S., Roy, N., Khathing, D.T.: Pyridinium fluorochromate; a new and efficient oxidant for organic substrates. Synthesis 588 (1982)
4. Balasubramanian, K., Prathiba, V.: Quinolinium dichromate—A new reagent for oxidation of alcohols. Indian J. Chem. **25B**, 326 (1986)
5. Pandurangan, A., Murugesan, V., Palamichamy, P.: Quinolinium bromochromate: A new, selective and efficient reagent for the oxidation of alcohols in anhydrous acetic acid. J. Indian Chem. Soc. **72**, 479 (1995)
6. Agarwal, S., Choudhary, K., Banerji, K.K.: Kinetics and mechanism of the oxidation of the aromatic aldehydes by pyridinium fluorochromate. J. Org. Chem. **56**, 5111 (1991)
7. Vadera, K., Yajurvedi, D., Purohit, P., Mishra, P., Sharma, P.K.: Structure-rate relationship in the oxidation of substituted benzaldehydes by pyridinium bromochromate: A Kinetic and mechanistic study. Proc. React. Kinet. Mech. **35**, 265 (2010)
8. Gehlot, M., Prasadrao, P., Sharma, V.: Structure-reactivity correlation in the oxidation of substituted benzaldehydes by tetraethylammonium chlorochromate. Asian J. Chem. **23**(3), 1173 (2011)
9. Pohani, S., Sharma, D., Panchariya, P., Sharma, P.K.: Structure-reactivity correlation in the oxidation of substituted benzaldehydes by quinolinium bromochromate. J. Ind. Council Chem. **27**(2), 122 (2010)
10. Purohit, T., Banerji, J., Kotai, L., Banerji, K.K., Sharma, P.K.: Kinetics and mechanism of the oxidation of substituted benzaldehydes with bis-(pyridine) silver permanganate. J. Indian Chem. Soc. **89**(8), 1045 (2012)
11. Barthora, S., Baghmar, D., Gilla, M., Choudhary, A., Sharma, V.: Structure-reactivity correlation in the oxidation of substituted benzaldehydes by benzyltriethylammonium chlorochromate. J. Chem. Biol. Phys. Sc. **1**(1), 07 (2011)
12. Daiya, A., Purohit, P., Kumbhat, R., Kotai, L., Sharma, V.: Structure-rate-reactivity correlation in the oxidation of substituted benzaldehydes by imidazolium fluorochromate. Int. J. Chem. **1**(2), 230 (2012)
13. Chouhan, K., Prasadrao, P., Sharma, P.K.: The kinetics and mechanism of oxidation of aliphatic aldehydes by benzyltriethylammonium chlorochromate. J. Indian Chem. Soc. **83**, 191 (2006)
14. Kumbhat, R., Sharma, V., Banerji, K.K.: Kinetics and mechanism of oxidation of aliphatic aldehydes by quinolinium bromochromate. Oxid. Commun. **30**(1), 97 (2007)
15. Patel, M., Poonam, Jha, K., Baghmar, M., Kothari, A., Shastri, I., Sharma, P.K.: Oxidation of some aliphatic aldehydes by tetrakis(pyridine) silver dichromate. Kinetics and mechanism of the (TPSD). J. Indian Chem. Soc. **89**(8), 1149 (2012)

16. Soni, U., Yajurvedi, D., Vyas, S., Prakash, O., Sharma, P.K.: Correlation analysis of reactivity in the oxidation of substituted benzaldehydes by bis[dipyridinesilver(i)]dichromate. Eur. Chem. Bull. **4**(9), 442 (2015)
17. Sharma, D., Panchariya, P., Purohit, P., Sharma, P.K.: Oxidation of aliphatic aldehydes by imidazolium fluorochromate (IFC): A kinetic and mechanistic study. Oxid Commun. **35**(4), 821 (2012)
18. Panwar, S., Pohani, S., Swami, P., Vyas, S., Sharma, P.K.: Kinetics and mechanism of the oxidation of aliphatic aldehydes quinolinium chlorochromate. Eur. Chem. Bull. **2**(10), 904 (2013)
19. Swami, P., Malani, N., Agarawal, S., Sharma, P.K.: Oxidation of aliphatic aldehydes by tetraethylammonium chlorochromate: A kinetic study. Prog. React. Kinet. Mech. **35**, 309 (2010)
20. Sharma, P.K.: Structure-reactivity correlation in the oxidation of substituted benzaldehydes by 2, 2′-bipyridinium chlorochromate. J. Indian Chem. Soc. **85**, 1281 (2008)
21. Choudhary, A., Malani, N., Agarwal, S., Sharma, M., Sharma, V.: Correlation analysis of reactivity in the oxidation of substituted benzaldehydes by morpholinium chlorochromate. J. Indian Chem. Soc. **86**, 927 (2009)
22. Khurana, M., Sharma, P.K., Banerji, K.K.: Kinetics and mechanism of oxidation of aliphatic aldehydes by quinolinium fluorochromate. React. Kinet. Catal. Lett. **67**, 341 (1999)
23. Kumbhat, V., Sharma, P.K., Banerji, K.K.: Kinetics and mechanism of oxidation of aliphatic aldehydes by 2, 2′-bipyridinium chlorochromate. Indian J. Chem. **39A**, 1169 (2000)
24. Saraswat, S., Sharma, V., Banerji, K.K.: Kinetics and mechanism of oxidation of aliphatic aldehydes by pyridinium chlorochromate. Indian J. Chem. **40A**, 583 (2001)
25. Soni, N., Kumbhani, S., Shastri, I., Sharma, V.: Kinetics and mechanism of the oxidation of aliphatic aldehydes by morpholinium chlorochromate. J. Indian Chem. Soc. **85**, 857 (2008)
26. Asghar, B.H., Mansoor, S.S., Hussain, A.M., Malik, V.S., Aswin, K., Sudhan, S.P.N.: Oxidation of aliphatic aldehydes by benzimidazolium fluorochromate in non aqueous medium—A kinetic and mechanistic study. Arab. J. Chem. **10**, S2115 (2017)
27. Chu, T.C., Lin, Y.C.: A fuzzy TOPSIS method for robot selection. Int. J. Adv. Manuf. Technol. **21**(4), 284–290 (2003)
28. Wang, Y.M., Elhag, T.M.: Fuzzy TOPSIS method based on alpha level sets with an application to bridge risk assessment. Expert Syst. Appl. **31**(2), 309–319 (2006)
29. Ashtiani, B., Haghighirad, F., Makui, A., Montazer, G.A.: Extension of fuzzy TOPSIS method based on interval-valued fuzzy sets. Appl. Soft Comput. **9**(2), 457–461 (2009)
30. Şengül, U., Eren, M., Shiraz, S.E., Gezder, V., Şengül, A.B.: Fuzzy TOPSIS method for ranking renewable energy supply systems in Turkey. Renew. Energy **75**, 617–625 (2015)
31. Gündoğdu, F.K., Kahraman, C.: Spherical fuzzy sets and spherical fuzzy TOPSIS method. J. Intell. Fuzzy Syst. **36**(1), 337–352 (2019)
32. Rao, A., Kumar, G.: Comparative study of solvent effect by TOPSIS method in the oxidation of acetaldehyde. J. Appl. Chem. **8**(4), 1798–1804 (2019)

Solution of Airport Gate Reassignment Problem Under Uncertain Environment

Aniruddha Samanta, Supriya Kar, Dalbinder Kaur, and Kajla Basu

Abstract In day-to-day schedule, airports are highly involved in a tight schedule of operations of air traffic system. However, sometimes, several unfavorable conditions like bad weather, technical and mechanical failures, air traffic control, and human errors create a big problem for airport authorities and causing flight delays. Therefore, considering these unfavorable situations, this present paper suggests a multi-objective optimization model of airport gate reassignment for these delayed flights with three novel objectives, namely time inconsistency, space inconsistency, and economic loss value. Also, to address the impreciseness of information, fuzzy concept has been introduced in the model. Furthermore, the formulated model has been illustrated with a numerical example. The solution is performed by the traditional fuzzy programming technique and genetic algorithm to obtain the results of the optimization problem. Finally, the results indicate the maximum financial benefit of airport authorities with the minimization of other objectives.

Keywords Airport gate reassignment problem · Multi-objective optimization · Fuzzy programming technique · Genetic algorithm

---

A. Samanta (✉)
Department of Basic Science, NSHM Knowledge Campus Durgapur, Durgapur, India
e-mail: animath22@gmail.com

S. Kar
Bidhan Chandra Institution for Girls', Durgapur, India

D. Kaur
Department of Mathematics, Banwarilal Bhalotia College, Asansol, India

K. Basu
Department of Mathematics, NIT Durgapur, Durgapur, India

L. Sahoo et al. (eds.), *Real Life Applications of Multiple Criteria Decision Making Techniques in Fuzzy Domain*, Studies in Fuzziness and Soft Computing 420,
https://doi.org/10.1007/978-981-19-4929-6_14

1 Introduction

Nowadays, the market demand of airplane operation has zoomed up due to the increment of tight schedules, convenient transportation type, and lack of time. Addressing this demand, airport authorities have developed infrastructures like several runways, gates, and multiple terminals. They have also included new airlines, aircrafts as well as increased the frequency of airlines. These resulted in every day's tight schedules of operations of air traffic system. So, for avoiding the operational disturbance, airport gate assignment problem (AGAP) has been introduced to assign feasible flights to the available gates at large and busy hub airports. This assignment minimizes the walking distances, overall spending time of passengers for connecting flights, number of flights not assigned to gates, etc. However, such assigned schedules are disrupted frequently due to some unavoidable circumstances like breakdown of flights or gates, early or delayed flights, some urgent or emergency flights, bad weather, technical and mechanical failures, and air traffic control, human errors. Such disruption causes not only time and economic losses but also inconvenience to airlines and passengers. So, a calculative reassignment of flights to the available gates become necessary when the gate assignments are highly influenced by the disruptions and cannot be re-scheduled normally as per the pre-assigned gate. Therefore, the requirement of airport gate re-assignment problem (AGRAP) has increased for these delayed and transferred flights in different gates. These problems are necessarily significant for proper utilization of airport resources and improvement of airport service quality for passengers.

Realizing the necessity of AGRAP, several models have been developed. Steuart [5] suggested a simple stochastic model for finding the requirements of the gate positions. Furthermore, different airport gate assignment models have been introduced based on different objectives. The objectives are considered depending on different perspectives. In airport owner's viewpoint, the aims are to expand the operation of available gates [5–7], decrease the confliction at the number of gates [13], minimize the flights delays [11], and reduce the number of un-gated flights [4, 7–10]. In airlines owners' perspective, the goals are to increase the customer satisfaction and reduce the passenger walking distance [1, 4, 7, 8, 14].

AGAP is an integer or mixed integer linear models. For solving such models, several exacts or meta-heuristic methods have been used in literatures. Li proposed a binary mixed integer AGAP model and solved this model by CPLEX software [13]. Bihr [1] established a primal–dual simplex algorithm to search the solution of AGAP. Li [12] used branch and bound algorithm for solving AGAP. Zheng et al. [2] constructed a gate assignment problem as a mixed binary quadratic problem and solved this problem using Tabu search technique. Gu and Chung [3] also solved AGAP using genetic algorithm. Ding et al. [4] used mixed algorithm of Tabu search and simulated annealing to solve the AGAP model. Kar et al. also solved generalized assignment problem using fuzzy programming technique and genetic algorithm [15].

The airport authorities need to schedule and reschedule flights in case of any delayed or transferred flights. The main aim should be the minimization of time and space inconsistencies and the minimization of economic loss of airport taking into account the limited resources like gates and time points. Addressing these requirements, the present paper focuses on the solution of AGRAP for the delayed flights with three novel objective functions, namely time inconsistency, space inconsistency, and economic loss value. For the first time, fuzzy theory has been introduced to such problems taking into account the impreciseness of data. The formulated multi-objective model has been represented with the assistance of a numerical illustration. Finally, the problem has been solved utilizing traditional fuzzy programming technique and genetic algorithm.

The remaining portion of the paper is arranged as follows: Sect. 2 discusses the required preliminaries. Formulated model is described in Sect. 3 whose solution techniques are presented in Sect. 4. Section 5 discusses the formulated problem by a numerical example. The obtained result is shown and discussed in Sect. 6. Finally, Sect. 7 provides a conclusion on the overall work.

2 Preliminaries

1. **Fuzzy Set** [15]: A fuzzy set is an extension of an ordinary set whose elements are characterized by their grade of membership. Specifically, a fuzzy set can explain as follows:

$$\tilde{A} = \left\{x : \mu_{\tilde{A}}(x) : x \in X\right\}$$

where $\mu_{\tilde{A}}(x)$ is the membership value or degree of membership of $x \in X$ in the fuzzy set $\tilde{A}$ and the function $\mu_{\tilde{A}} : X \to M$ is called the membership function from X to the membership space $M \in [0, 1]$.
2. **Triangular Fuzzy Number** [15]: A fuzzy number $X = (a, b, c)$ is considered as a triangular fuzzy number if its membership function is given by:

$$(x : a, b, c) = \begin{cases} 0 & \text{if } x < a; \\ \frac{x-a}{b-a} & \text{if } a \le x \le b; \\ \frac{c-x}{c-b} & \text{if } b \le x \le c; \\ 0 & \text{if } c < x. \end{cases}$$

3. **Defuzzification of TFN** [15]: The concept of α-cut method is used to convert triangular fuzzy number into a crisp value. The α-cut value is denoted by $Y(\tilde{C})$ where:

$$Y(\tilde{C}) = \int_{0}^{1} 0.5\left(C_{\alpha}^{L} + C_{\alpha}^{U}\right)\mathrm{d}\alpha$$

4. **Pareto Front**: Pareto front is the set of optimal solutions in the multi-objective optimization problem that are non-dominated from each other.
5. **L_P-norm Strategy for Best Non-dominated Solution**: Generally, multi-objective optimization problems create non-dominated solutions and Pareto optimal front. However, it is tough for decision maker to choose a unique solution. Therefore, for finding the best non-dominated solution, the most appreciated method is L_P norm strategy. Here, the normalized distance of Pareto set and an ideal solution (Utopia point) is minimized using the formula:

$$\text{Minimize}\left(\sum_{i=1}^{m}\left(\frac{f_i(x)-f_i^{\min}}{f_i^{\max}-f_i^{\min}}\right)^P\right)^{\frac{1}{P}} \quad P=1,2,3,\ldots,\infty.$$

$f_i^{\min}$ = minimum value of ith objective in the Pareto optimal set.
$f_i^{\max}$ = maximum value of ith objective in the Pareto optimal set.

Here, all objective functions must be considered in minimization form.

3 Model Formulation

The AGRAP for delayed flights is formulated under fuzzy environment with three objectives functions, namely time inconsistency (Z_1), space inconsistency (Z_2), and economic loss value (Z_3) which are to be minimized. The problem is formulated as

$$\text{Min } Z_1 = \sum_{i\in I}\sum_{j\in E_i}\sum_{k\in D_{ij}} \widetilde{d_{ikj}}x_{ijk} \tag{1}$$

$$\text{Min } Z_2 = \sum_{i\in I}\sum_{j\in E_i}\sum_{k\in D_{ij}} \widetilde{w_{ikj}}x_{ijk} \tag{2}$$

$$\text{Min } Z_3 = \sum_{i\in I}\sum_{j\in E_i}\sum_{k\in D_{ij}} \widetilde{L_{ikj}}x_{ijk} \tag{3}$$

Subject to,

$$\sum_{j\in E_i}\sum_{k\in D_{ij}} x_{ijk} = 1 \,\forall i \in I \tag{4}$$

$$\sum_{i\in F_{jsi}}\sum_{k\in H_{is}} x_{ijk} \le 1 \,\forall j \in G, \forall s \in S \tag{5}$$

$$x_{ijk} = 0 \text{ or } 1\ \forall k \in D_{ij}, \forall j \in E_i, \forall i \in I \tag{6}$$

where

I = Set of considered (transferred) flights.
E_i = Set of gates that ith flight can be assigned.
D_{ij} = Set of time points that ith flight can be assigned to jth gate.
S = Set of available gates.
H_{is} = Set of three points assigned to ith flight for sth time point.
$\widetilde{d_{ikj}}$ = Fuzzy time inconsistency for ith flight at jth gate for kth time point.
$\widetilde{w_{ijk}}$ = Space inconsistency for ith flight at jth gate for kth time point.
$\widetilde{L_{ijk}}$ = Economic loss value for ith flight at jth gate for kth time point.
x_{ijk} = ith transferred flight has been assigned to the jth available gate at kth time point.

4 Solution Methodology

To solve AGRAP traditional fuzzy programming technique and genetic algorithm have been used here.

For the fuzzy programming technique [15], the following steps have been performed:

Lower (L_k) and upper (U_k) bound for all the three objective functions are calculated. L_k is the aspired level; U_k is the highest acceptable level for the kth objective. Then, the fuzzy model has been formed using linear membership function. After that, the fuzzy model is converted into a crisp model using defuzzification process and solved by LINGO software.

4.1 Algorithm

Step 1: Find the solution of AGRAP as a single objective assignment problem utilizing each time just a single objective ignoring other objectives subject to constraints.

Step 2: For each set of solutions, determine the corresponding values of every objective function.

Step 3: Thereafter, a pay-off matrix is constructed with the different values of the objective functions as in Table 1.

From the pay-off matrix, search the best (L_k) and the worst (U_k) values for each objective function.

Table 1 Pay-off matrix

Pay-off matrix			
	$Z_1(x)$	$Z_2(x)$	$Z_3(x)$
$X^{(1)}$	Z_{11}	Z_{12}	Z_{13}
$X^{(2)}$	Z_{21}	Z_{22}	Z_{23}
$X^{(3)}$	Z_{31}	Z_{32}	Z_{33}

Step 4: Determine a linear membership function $\mu(Z_k)$ for the kth objective function as follow:

$$\mu^L(Z_k) = \begin{cases} 1 & \text{if } Z_R \leq L_R \\ 1 - \frac{Z_K - L_K}{U_K - L_K} & \text{if } L_K \leq Z_K \leq U_K \\ 0 & \text{if } Z_K \geq U_K \end{cases}$$

Based on this membership function, the crisp model can be clarified as:

4.2 Model

Maximize λ.

S.t, $Z_R + \lambda(U_R - L_R) \leq U_K, R = 1, 2, 3$ and other constraints.

4.3 MOGA Methodology [15]

The pseudo-code of multi-objective genetic algorithm is given below.

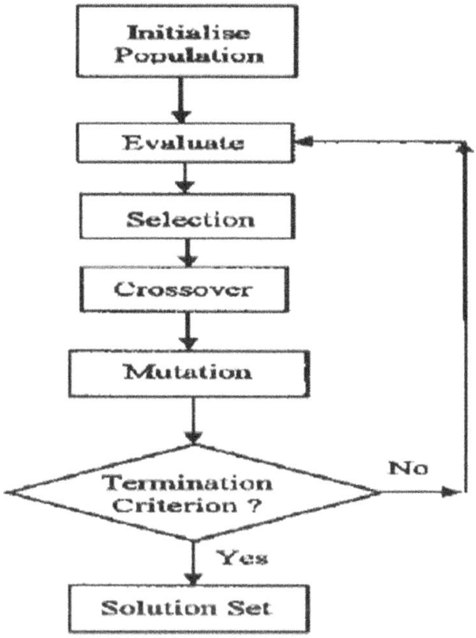

Algorithm: Pseudo code for multi-objective GA

5 Numerical Illustration

5.1 Problem Description

The formulated model is a multi-objective AGRAP under uncertain environment. Here, three objective functions (Time inconsistency, space inconsistency, and economic loss value) have been considered, and all of them are to be minimized.

The numerical data on available gates and time point for transferred flights are shown in Table 2. Tables 3, 4 and 5 contain the fuzzy data on time inconsistency, space inconsistency, and economic loss, respectively.

From Table 2, it is clear that two gates (1st and 2nd) and three time points ($K = 1, 2, 3$) are available for 1st flight. Similarly, for other flights, the availabilities are given in Table 2.

Table 2 Available gates and time point for transferred flights

Transferred number of flight (i)	Available gates (i)	Time points (K)
$i = 1$	$j = 1, 2$	$K = 1, 2, 3$
$i = 2$	$j = 1$	$K = 1, 2$
$i = 3$	$j = 1, 2$	$K = 1, 2$
$i = 4$	$j = 1, 2, 3$	$K = 1$
$i = 5$	$j = 1$	$K = 1, 2, 3$

Table 3 Fuzzy data on time inconsistency

$i = 1$ $[d_{11}\ d_{12}]$ $k = 1, 2, 3$	$i = 2$ $[d_{21}]$ $k = 1, 2$	$i = 3$ $[d_{31}\ d_{32}]$ $k = 1, 2$	$i = 4$ $[d_{41}\ d_{42}\ d_{43}]$ $k = 1$	$i = 5$ $[d_{51}]$ $k = 1, 2, 3$
[(2, 4, 6) (1, 4, 7)] $k = 1$	[(7, 8, 10)] $k = 1$	[(6, 7, 9) (2, 3, 5)] $k = 1$	[(1, 2, 6) (1, 4, 7) (3, 6, 8)] $k = 1$	[(6, 7, 8)] $k = 1$
[(8, 10, 13) (10, 12, 14)] $K = 2$	[(8, 9, 12)] $k = 2$	[(3, 7, 11) (9, 11, 12)] $K = 2$		[(3, 5, 7)] $k = 2$
[(2, 3, 5) (6, 7, 9)] $K = 3$				[(8, 9, 10)] $k = 3$

Table 4 Fuzzy data on space inconsistency

$i = 1$ $[w_{11}\ w_{12}]$ $k = 1, 2, 3$	$i = 2$ $[w_{21}]$ $k = 1, 2$	$i = 3$ $[w_{31}\ w_{32}]$ $k = 1, 2$	$i = 4$ $[w_{41}\ w_{42}\ w_{43}]$ $k = 1$	$i = 5$ $[w_{51}]$ $k = 1, 2, 3$
[(1, 2, 3) (3, 5, 7)] $k = 1$	[(2, 4, 6)] $k = 1$	[(4, 5, 6) (6, 7, 8)] $k = 1$	[(2, 3, 4) (5, 7, 9) (5, 7, 9)] $k = 1$	[(2, 3, 5)] $k = 1$
[(3, 5, 7) (2, 4, 6)] $K = 2$	[(4, 6, 8)] $k = 2$	[(1, 3, 5) (3, 6, 9)] $K = 2$		[(4, 5, 6)] $k = 2$
[(6, 8, 10) (4, 6, 8)] $K = 3$				[(7, 9, 11)] $k = 3$

Table 5 Fuzzy data on economic loss

$i = 1$ $[L_{11}\ L_{12}]$ $k = 1, 2, 3$	$i = 2$ $[L_{21}]$ $k = 1, 2$	$i = 3$ $[L_{31}\ L_{32}]$ $k = 1, 2$	$i = 4$ $[L_{41}\ L_{42}\ L_{43}]$ $k = 1$	$i = 5$ $[L_{51}]$ $k = 1, 2, 3$
[(1, 3, 5) (4, 6, 8)] $k = 1$	[(2, 4, 6)] $k = 1$	[(5, 7, 9) (6, 8, 10)] $k = 1$	[(4, 6, 8) (1, 4, 7) (2, 3, 4)] $k = 1$	[(4, 5, 6)] $k = 1$
[(1, 2, 3) (2, 4, 6)] $K = 2$	[(3, 5, 7)] $k = 2$	[(4, 5, 6) (2, 3, 4)] $K = 2$		[(4, 6, 8)] $k = 2$
[(3, 5, 7) (3, 4, 5)] $K = 3$				[(1, 4, 7)] $k = 3$

6 Results and Discussion

6.1 Result by Fuzzy Programming Technique

Solving the above problem by fuzzy programming technique, we got the solution as $x_{123} = 1$, $x_{211} = 1$, $x_{322} = 1$, $x_{421} = 1$, and $x_{512} = 1$ others are zero. With respect to this solution, the objective function values are as follows:

$$Z_1 = 35.25, Z_2 = 28, Z_3 = 21.$$

where x_{123} means 1st transferred flight has been assigned to the 2nd available gate at 3rd time point. Similarly for other variables also.

6.2 Results for MOGA

Initially, the above problem has been considered as a multi-objective optimization problem and solved using multi-objective genetic algorithm. So, as a result, a Pareto optimal front (Fig. 1) is obtained which contains non-dominated solutions of the objective values.

Although the non-dominated solutions are very useful for decision, but the decision makers are highly passionate to obtain a single optimal. Therefore, after obtaining the non-dominated solutions, L_p norm strategy has been applied to obtain a single optimal solution from Pareto optimal front.

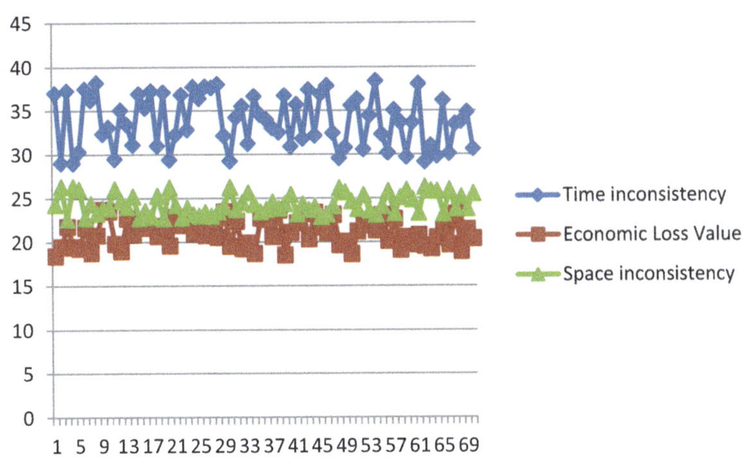

Fig. 1 Pareto front for non-dominated solutions using MOGA

The solution is $x_{123} = 1$, $x_{211} = 1$, $x_{322} = 1$, $x_{421} = 1$ and $x_{511} = 1$ and others are zero. Therefore, the objective function values at this solution are $Z_1 = 37.25$, $Z_2 = 26$, $Z_3 = 20$.

From the above results (Z_1, Z_2, Z_3) obtained by two methods, the aim of the paper to minimize the three mentioned objective functions is fulfilled. In addition, it can be observed that the obtained value of the 3rd objective function, namely economic loss value, is the least. Therefore, the proposed model is appropriate for the airport authorities to accomplish the financial benefits due to airport gate reassignments. Along with this, time inconsistency and space inconsistency are also minimized.

7 Conclusion

Addressing the flight delays caused by bad weather, technical and mechanical failures, air traffic control, and human errors, a multi-objective optimization-based airport gate reassignment model has been formulated here. Three important objectives have been minimized here, namely time inconsistencies, space inconsistencies, and economic loss value. Also, the vagueness of the data has been addressed using fuzzy theory. Finally, the formulated problem has been solved using fuzzy programming technique and genetic algorithm. Furthermore, based on the obtained results, it is established that the proposed model is highly helpful for financial benefit of the airport authority with the minimization of time inconsistency and space inconsistency.

In future, different new objectives based on different perspectives of airport owner and airlines owner can be considered with this formulated model. By increasing the number of flight types and defining the different gate types more restrictively, new constraints and preferences can be included in future research. Improvement of meta-heuristic algorithms is also a significant future purview for this perspective.

References

1. Bihr, R.A.: A conceptual solution to the aircraft gate assignment problem using 0, 1 linear programming. Comput. Ind. Eng. **19**(1–4), 280–284 (1990)
2. Zheng, P., Hu, S., Zhang, C.: Airport gate assignments model and algorithm. In: Proceedings of the 3rd IEEE International Conference on Computer Science and Information Technology (ICCSIT), vol. 2, pp. 457–461 (2010)
3. Gu, Y., Chung, C.A.: Genetic algorithm approach to aircraft gate reassignment problem. J. Transp. Eng. **125**(5), 384–389 (1999)
4. Ding, H., Lim, A., Rodrigues, B., Zhu, Y.: The over constrained airport gate assignment problem. Comput. Oper. Res. **32**(7), 1867–1880 (2005)
5. Steuart, G.N.: Gate position requirements at metropolitan airports. Transp. Sci. **8**(2), 169–189 (1974)
6. Yan, S., Huo, C.M.: Optimization of multiple objective gate assignments. Transp. Res. A Policy Prac. **35**(5), 413–432 (2001)

7. Drexl, A., Nikulin, Y.: Multi criteria airport gate assignment and Pareto simulated annealing. IIE Trans. **40**(4), 385–397 (2008)
8. Ding, H., Lim, A., Rodrigues, B., Zhu, Y.: New heuristics for over-constrained flight to gate assignments. J. Oper. Res. Soc. **55**(7), 760–768 (2004)
9. Dorndorf, U., Jaehn, F., Pesch, E.: Flight gate scheduling with respect to a reference schedule. Ann. Oper. Res. **194**, 177–187 (2012)
10. Seker, M., Noyan, N.: Stochastic optimization models for the airport gate assignment problem. Transp. Res. E Logist. Transp. Rev. **48**(2), 438–459 (2012)
11. Yan, S., Tang, C.H.: A heuristic approach for airport gate assignments for stochastic flight delays. Eur. J. Oper. Res. **180**(2), 547–567 (2007)
12. Li, W.: Optimized assignment of civil airport gate. In: Proceedings of the International Conference on Intelligent System Design and Engineering Application (ISDEA'10), pp. 33–38 (October 2010)
13. Li, C.: Airport gate assignment—A hybrid model and implementation. http://arxiv.org/abs/0903.2528
14. Lam, S., Cao, J., Fan, H.: Development of an intelligent agent for airport gate assignment. J. Airport Transp. **7**(2), 103–114 (2002)
15. Kar, S., Samanta, A., Basu, K.: Solution of fuzzy multi-objective generalised assignment problem. Int. J. Math. Oper. Res. **13**(1), 33–54 (2019)

Bilevel Approach for E-Bidding Auction with Regret Theory

Tabasam Rashid, Aamir Mahboob, and Ismat Beg

Abstract In e-price bid auctions, we construct an auction model using regret decision theory to explain how a bidder decides his bidding price. The fundamental distinction between regret theory and other decision theories under uncertainty is clarified in this article. The suggested regret decision theory is scenario-based rather than the other theories. The process of the suggested theory is split into two stages. The first stage is to consider every possible lowest bidding offered by the auctioneers and select the one action point from the bidding price. After selecting the action point in the second stage, a decision-maker evaluates the best bidding price that produces the best results for him. Our model captures the various preferences for risk in such auction issues, which are considerably different from the typical auction model based on probability distribution in which maximizing expected utility is the aim. The study results indicate that the suggested decision models should give valuable framework for understanding the different behaviors of decision makers and assist the decision maker in finding the optimal solution according to his attitude regarding partly known information.

1 Introduction

The right decision at the right time is the key to success, and the wrong decision will definitely cause failure and loss in life. Everyone needs the right decision irrespective of a common person, a manager, an owner, a bidder or seller, etc. Without making good decisions, tactical, operational, and strategic success is impossible. Decisions also play a vital role in short-term, medium-term, and long-term success. However, due to confusion, limited information and lack of confidence might lead to wrong decisions, so decision-making is a complex process that can be evaluated by two different methodologies. i.e. rational decision-making and behavioral decision-making.

T. Rashid · A. Mahboob
University of Management and Technology, Lahore 54770, Pakistan

I. Beg (✉)
Lahore School of Economics, Lahore, Pakistan
e-mail: ibeg@lahoreschool.edu.pk

L. Sahoo et al. (eds.), *Real Life Applications of Multiple Criteria Decision Making Techniques in Fuzzy Domain*, Studies in Fuzziness and Soft Computing 420,
https://doi.org/10.1007/978-981-19-4929-6_15

Rational decision-making comes from the expected utility [27], but the individual may take the wrong decision due to uncertain situations, spoiling all possible benefits. That's why the different logical correlations of the expected utility have been identified. For example, Arias's paradox and Ellsberg's paradox [1, 12, 14]. Whereas Behavioral decision-making theories have been devised as "option to expected utility" like prospect and regret theories [2, 17, 22, 29].

Interestingly regret theory [2, 22] analyzes the emotions of decision-makers when the result of decision comes; if the outcome is better, the individual will feel happy, otherwise it will have regret. This theory is less complicated because no frame of reference is needed and fewer parameters are required in the calculation process. That's why, in multiple criteria decision-making problems, the regret theory is more acceptable and it has a wide application, like as risk evaluation of new energy resources investing [28], a bestseller publication [29] and selection of air-fighter [42]. Even then, it has some research gaps. In previous research articles, the decision-maker (DM) generally utilized linear utility functions and concave utility functions that cannot reflect the subjective views of DM [9, 20]. When evaluation values approach the target value, DMs commonly become sensitive. As a result, achieving the goal with the target value approaching will demand unnecessary work. Jahan et al. [16] suggested a target-based approach using normalization in this regard; however, their functions cannot explain the psychology of the decision-makers. So, the purpose of this article is to establish behavioral models based on regret theory for solving reverse auctions problems.

Many businesses have discovered cost-saving opportunities and improved their efficiency using online acquisition with the rapid development of internet technology and data innovation. This way of defining, determining, and optimizing buyers and suppliers best grouping in a supply chain that responds to changing customer's demands can be referred to as an online auction.

Modern communication and technology have created multiple online selling and buying opportunities through the internet. It has undoubtedly increased the level of competition and connections between the sellers and the buyers [4]. An online auction or e-auction is an agreement (negotiation) that takes place on an electronic marketplace between the sellers and the buyers. A "reverse auction" is the most dominant kind of e-auction in which there is only one buyer and multiple sellers. In a forward auction or English auction, a bidder gives the highest bid for a commodity. The seller holds the market when a product demands the number of items being sold. A bidder continuously increases the price of the product until a reasonable price is received in the market [11, 35].

In the current situation, online auction has been promoted and encouraged by the multiple advantages to sellers and customers. Suppliers have a wide range of customers and markets, low cost of organizing and managing the auction [38]. A faster purchasing cycle and a wider range of products are also the benefits of e-auction. For all these reasons, an online auction is getting the popularity day by day [35, 38].

Auctions are also categorized as single round auctions and multiple round auctions. Sealed bid auctions are one-shot or single-round auctions and multiple round

[31] auctions can be sealed bid or open bid which have more than one round bid. After every interval, the auctioneer provides the information to the bidders so that bidders can prepare themselves for the next bid. Request for quotation is the fundamental of a reverse auction process. In the context of the product's technical specification, a buyer declares the demands with criteria and attracts the supplier to send his bids for open discussion. To secure the order, the bidder offers a bid against someone other on the internet. These types of bids are one-shot or multi-stage bids (Sealed bid or open bid) [6, 8, 23, 39].

Online auctions are famous nowadays in every field in which new techniques depend upon the quantity and quality of the products to be purchased [18]. In the discount-based auction, the supplier discounts on the purchase price for buying a large number of particular products by reducing many economic benefits, including decreasing the cost of production, manufacturing, the storage operating, shipping, etc. In these types of discount based auctions, the auctioneers bidding is based on the cost function. These functions reflect the discount given by the seller on the quantity purchased. In real-life circumstances, the price for bargaining with availability by a more productive knowledge sharing of customer desires and supplier deals, it is essential to understand multiple contract criteria. These types of auctions are called multi-criteria auctions [5, 36]. The auctioneer selects several bargaining criteria of an item in a multi-criteria reverse auction and determines a ranking rule depending on the preference using the utility function. The auctioneer receives bids from interested bidder as per the bidding system's rules during the online auction process [36, 38].

Generally, online auction depends on reverse auctions where the price is the most important common element and non-price criteria are set before the agreement. The majority of the people observe that the reverse auction is an efficient online tool for negotiating with the distributor to minimize the unit price of the purchased products, keeping in mind the strategies as mentioned above without access the other characteristics like as processing time, standard payment conditions, after-sale support and so on [6, 13, 18]. While price auctions have been mentioned to minimize the buyer's cost [36], possibly when they have differences in quality and the auctioneers demonstrate competitive advantages in production cost. The qualitative characteristics of multi-criteria decision-making problems depend on the assessment of the decision-makers [24, 25]. The choice could be based on insufficient information or personal judgement due to vagueness, and the results will be intolerant decisions [36].

In an e-bid auction, all the auctioneers offer their bids, and who present the lowest bidding price, the bidder purchases that object. If the bidder fails to submit the bidding price, his payoff will be zero. Otherwise, the payoff will be offered according to the situation. Hypothetically, structures of optimal bidding price have been studied by many researchers [7, 19], and online reverse auctions are examined [30]. We suggest a behavioural model for the single bidder in the e-bid auction model in his decision-making process. There are two stages; the first stage is: The bidder evaluates every lowest auctioneer price given by other auctioneers for each possible bidding price. Let us consider the probability of auctioneers and outcomes generated to all possible lowest bidding prices provided by other auctioneers. The bidder selects the one with the greatest probability. We say these types of selection points are known as the regret action points of bidding price. The bidder selects the best bidding price based on the action points to produce the best results in the second stage. In comparison with the other e-auction decision theories, our suggested regret e-auction decision theory is scenario-based. The other e-auction decision theories calculate each bidding price using the weighted average of all feasible bidding price provided by the suppliers.

The remainder of this article is organized as follows: In Sect. 2, we go through concise literature review of regret and possibility theories. For the re-matching scenario, we discuss our experimental procedures and experimental structure in Sect. 3. In Sect. 4, we present the application of our experiment and a comprehensive comparison analysis for the concept of continuous improvement is presented. The conclusion is presented in fifth section.

2 Preliminaries

This section introduces basic regret theory and e-auction theory to the multi-criteria auction for both supplier and bidder. Some people struggle with making decisions in unpredictable situations because they are afraid of making the "wrong decision", meaning that the result of their preferred alternative will be changed [3]. Sugden [37] described this type of situation with the word "regret" which is "the painful sensation of recognizing that 'what is' compares unfavorably with 'what might have been'". In an analysis, verbal representations of feelings in daily speech found that regret is the commonly called negative emanation [34]. Therefore, regret is a very important component that works on the decision-makers behavior.

Definition 1 [14] Let X be a set of sample space and function π is a mapping from X to $[0, 1]$. Then $\pi(x)$ is called the possibility grades, where for all $x \in X$ and $\pi(x) = 1$ is the maximum possibility grades and $\pi(x) = 0$ shows that x does not occur.

The procedure to calculate the degree of possibility is as follows: Suppose that $X = \{x_i\}, \forall x \in X, i = 1, 2, \ldots, n$ be the sample space then the possibility grade of x_i will be calculate as $\frac{x_i}{\max x_i}$, this shows that each event has equally opportunity to occur.

Definition 2 [14] Let X be a set of state of nature and all c_i's belongs to X where b_j is the set of alternatives. The outcomes of the combination of state of nature and alternatives is known as payoff, it is denoted by $u(c_i, b_j)$, where $1 \le i \le n$, $1 \le j \le m$.

The bidder with the lowest bidding price wins the e-auction. It is however, often followed by the disclosure of all the supplier's prices. In this case, the decision-maker is suffering in regret from his selection. The evaluation of the relative position of value, the regret function will be defined as:

Definition 3 [14] Assume that when a decision-maker selects an alternative B, then the condition of existence of c_i will be appeared. The decision-maker may come to regret his decision. The regret value is $v(c_i, b_j) = \max_{j=1}^{m} u(c_i, b_j) - u(c_i, b_j)$, $1 \leq i \leq n$. Then the relative position of regret value is written as $R(c_i, b_j)$ and it is calculated as:

$$R(c_i, b_j) = \frac{v(c_i, b_j)}{\max_{d_j \in B} v(c_i, d_j)} \tag{1}$$

For simplicity $R(c_i, b_j)$ can be written as R. A regret function can be used to express a decision maker's regret level for a relative position of regret value, which is defined as:

Let $r : R \to [0, 1]$ be a regret function with satisfy the condition $r(R_1) > r(R_2)$, if $R_1 > R_2$, where, $\forall\ R_1, R_2 \in R$.

3 E-Bid Auction Models with Regret Theory

This research looks at the specific scenario of an e-bid auction in which bids are presented with several attributes. In this auction, there is one bidder and several suppliers. There is one and only one supplier who wins the bid. The whole demand is assigned to an individual supplier. The e-auction process is shown schematically in Fig. 1. In e-bid auction models with regret, we believe that a decision-maker makes a decision at once based on a specific situation with the help of possibility grades and grades of regret. The decision-makers behavior influences the situation he chooses. A decision-maker may be active and another may be passive. The proposed scheme of the system is as follows. In the first step, a decision-maker selects a specific scenario, which refers to the regret action point of each alternative according to the decision maker's behavior. In the second step, the valid regret action points will be determined based on first step. He calculates the optimal bidding price based on regret levels of reasonable regret action points in the previous step. A decision-maker must evaluate a state of nature that should be weighted when determined a regret action point. Each state of nature is the combination of possibility grades and regret grades. The accepted state of nature is referred to as a regret action point. The selection of the action point is dependent on the attitude of the decision-maker. The four types of action points are discussed which show the behavior of DM. The first type of regret action point consists of higher possibility grades and the higher regret level. The state of nature with lower possibility grades and lower regret level is referred to as the second type of regret action point. The third type of regret action point is the action point which consists of the higher level of possibility and the lower level of regret. The fourth type of regret action point is the point that includes a low level of possibility grades and a higher level of regret. These types of regret action points are as follows:

$$X_1(B_j) = \arg\max_{c_i \in X} \min\left[\pi(c_i), r(R(c_i, B_j))\right], 1 \leq i \leq n, 1 \leq j \leq m \tag{2}$$

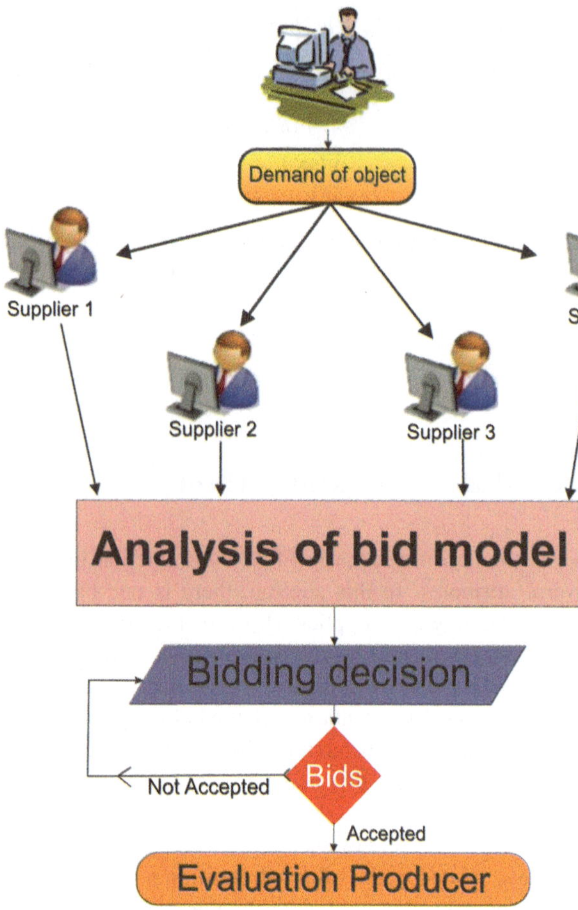

Fig. 1 Framework of e-auction process

$$X_2(B_j) = \arg\min_{c_i \in X} \max\left[\pi(c_i), r(R(c_i, B_j))\right], 1 \le i \le n, 1 \le j \le m \tag{3}$$

$$X_3(B_j) = \arg\min_{c_i \in X} \max\left[1 - \pi(c_i), r(R(c_i, B_j))\right], 1 \le i \le n, 1 \le j \le m \tag{4}$$

$$X_4(B_j) = \arg\min_{c_i \in X} \max\left[\pi(c_i), r(R(c_i, 1 - B_j))\right], 1 \le i \le n, 1 \le j \le m \tag{5}$$

In the 2nd step, a decision-maker determines valid regret action points of each solution based on his attitude toward possibility grades and regret level. Then a decision-maker considers those regret action points that are most suitable for the scenario and then selects the alternative that will show the better result from all the valid regret action points. There are four types of ideal alternatives corresponding to valid regret action points.

$$B_1 = \arg\min \max_{X(B_1)\in X(B_1)} r(R(X(B_1), B_1)) \tag{6}$$

$$B_2 = \arg\min \max_{X(B_2)\in X(B_2)} r(R(X(B_2), B_2)) \tag{7}$$

$$B_3 = \arg\min \max_{X(B_3)\in X(B_3)} r(R(X(B_3), B_3)) \tag{8}$$

$$B_4 = \arg\min \max_{X(B_4)\in X(B_4)} r(R(X(B_4), B_4)) \tag{9}$$

4 Application of E-Bidding Auction in Decision Making

The solution technique used in this study is described in this section. First of all, the bidder can send out a request for quotations to suppliers. The buyer's specifications are well defined in the quotations. These specifications include product specification, price, delivery time, claim time, creditability/reputation, payment option, trade policy, etc. The quotation clearly states the necessary and negotiable components of different attributes. The suppliers who concur to attain the required components of the particular criteria are eligible to participate in this auction. The various phases of the developed technique are depicted in Fig. 2.

4.1 Working Rule

In this section, a procedure for e-bidding auction is shown. The following steps show the algorithm of auction problems.

Step 1: Construct the decision matrix.
Step 2: Calculate the regret levels for all bids using Definition 3.
Step 3: Determine the possibility grades of the each attributes.
Step 4: Construct the regret action points using Eqs. 2, 3, 4 and 5.
Step 5: Determine regret level for each regret action points.
Step 6: Calculate optimal bidding price using Eqs. 6, 7, 8 and 9 (based on regret levels of valid regret action points).

The conceptualization of the suggested strategy is shown in Fig. 2.

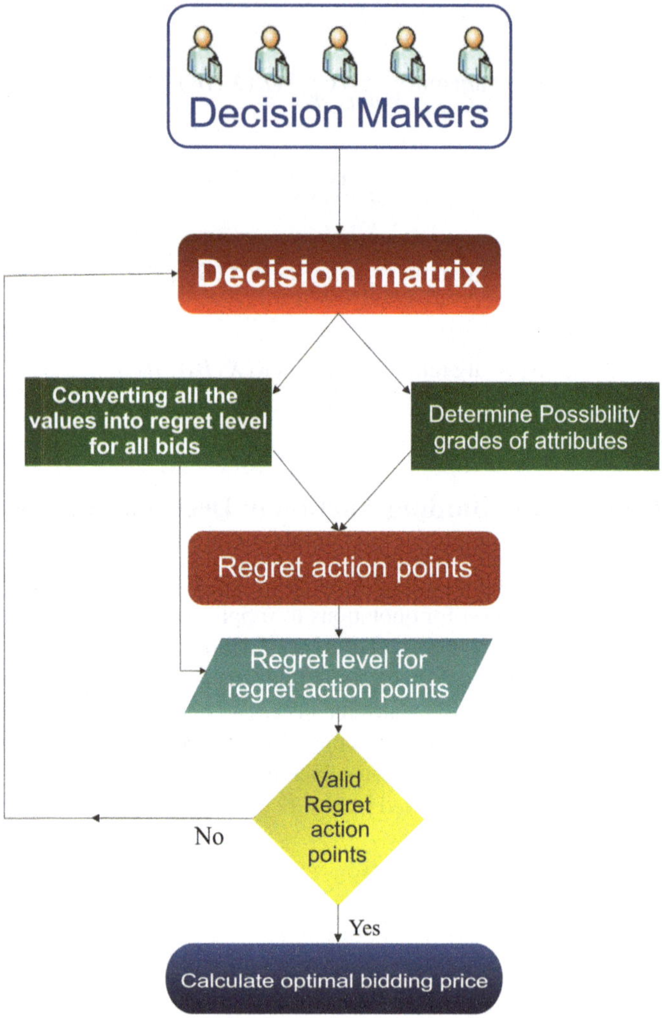

Fig. 2 Algorithm of the proposed strategies of e-bidding auction

4.2 Case Description

In this section, we present an example to show how to apply the regret theory. A sound and safe analysis evaluation based on comparing different providers is required to reduce the online purchase risk. The best use of resources and money is only possible if one can make the right and efficient decisions.

Let's take an example for its elaboration. A decision-maker wants to buy an online Samsung Galaxy smartphone. He will need to observe the price offered and published by various online suppliers, e.g. B_1, B_2, B_3 and B_4. However, not only the price is the

Table 1 Payoffs associated with each state

Suppliers	Price (currencydollar)	Delay (Days)	Claim (Days)	Creditability	Shipping charges
	c_1	c_2	c_3	c_4	c_5
B_1	2000	14	21	5	427
B_2	1900	15	24	4	400
B_3	2020	12	21	3	300
B_4	1980	10	25	4	400

Table 2 Possibility grades for each attributes

	c_1	c_2	c_3	c_4	c_5
Possibility	0.83	0.5	0.33	1	0.67

Table 3 Regret levels for each attributes

	c_1	c_2	c_3	c_4	c_5
B_1	0.167	0.2	1	0	0
B_2	1	0	0.25	0.5	0.213
B_3	0	0.6	1	1	1
B_4	0.33	1	0	0.5	0.213

determinant factor also. The buyer further needs to analyse and evaluate the delivery time, terms of payment, warranty claims, creditability and reputation of the provider. All these factors are given in Table 1. The possibility grades of each attributes as given in Table 2.

Let us calculate the regret level of each attributes, according to the definition of regret level as shown in Eq. 1. The regret level for each attributes are shown in Table 3.

Next, the regret action points are calculated with the help of Eqs. 2, 3, 4, 5 and results are listed in Table 4. The regret level of action points are shown in Table 5. Let us analyze the detail for obtaining the regret action points as follows: Using Eq. 2, $X_1(B_1)$ calculated as: $\max\Big\{\min[0.83, 0.167], \min[0.5, 0.2], \min[0.3, 1], \min[1, 0], \min[0.67, 0]\Big\} = \max\Big\{[0.167, 0.167], [0.2, 0.2], [0.3, 0.3], [0, 0], [0, 0]\Big\} = [0.3, 0.3]$, because $[0.3, 0.3]$ corresponds to c_3, therefore we select c_3 for $X_1(B_1)$, so $X_1(B_1) = c_3$.

Using Eq. 4, for $X_3(B_2)$: $\min\Big\{\max[0.17, 1], \max[0.5, 0], \max[0.67, 0.25], \max[0, 0.5], \max[0.33, 0.213]\Big\} = \min\Big\{[1, 1], [0.5, 0.5], [0.67, 0.67], [0.5, 0.5], [0.33, 0.33]\Big\} = [0.33, 0.33]$, because $[0.33, 0.33]$ corresponds to c_5, therefore we choose c_5 for $X_3(B_2)$, so $X_3(B_2) = c_5$.

Using Eq. 5, for $X_4(B_4)$: $\min\Big\{\max[0.83, 0.67], \max[0.5, 0], \max[0.33, 1], \max[1, 0.5], \max[0.67, 0.79]\Big\} = \min\Big\{[0.83, 0.83], [0.5, 0.5], [1, 1], [1, 1], [0.79,

Table 4 Regret action points

	B_1	B_2	B_3	B_4
1st type	c_3	c_1	c_4	c_2, c_4
2nd type	c_2	c_3	c_2	c_3
3rd type	c_4	c_5	c_1	c_1, c_5
4th type	c_3	c_3	c_3	c_2

Table 5 Regret level of regret action points

	B_1	B_2	B_3	B_4
1st type	1	1	1	1,0.5
2nd type	0.2	0.25	0.6	0
3rd type	0	0.213	0	0.33, 0.213
4th type	1	0.25	1	1

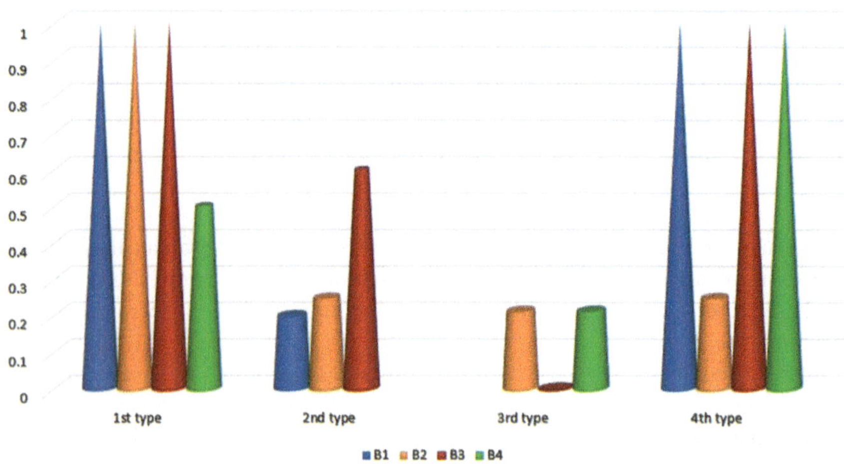

Fig. 3 Valid regret level of action points

$0.79]\} = [0.5, 0.5]$, because $[0.5, 0.5]$ corresponds to c_2, therefore we choose c_2 for $X_4(B_4)$, so $X_4(B_4) = c_2$.

According to Table 5, the optimal decisions are selected based on regret levels of valid regret action points, Fig. 3 shows valid regret action points. Therefore, in 1st type, the optimal action is B_4. and for 2nd type, optimal action is B_4. For 3rd type the optimal action is B_1 and B_3 and for 4th type the optimal action is B_2.

4.3 Comparison Analysis

In this section, we compare our proposed theory with existing auction theories. Previous approaches to address the reverse auction problems have mostly concentrated on the supplier's winner evaluation process. As a result, the researchers design and develops a trading scheme that gives a suggested price to suppliers [2, 17, 38]. But the price proposal process was only one basic goal to reduce the buyer's cost. As a result, this research has established that decision-making techniques not represent only the particular concern of the parties. This research discusses all the other attributes addressed in the proposed example.

Furthermore, there were many limitations in the previous work; for example, the utility function of gain and loss is either convex or concave [20]. According to some recent prospect theories, the utility function is S-shaped, meaning convex in the loss domain and concave in the benefits domain [17]. And in the case of no profit and no loss, then the utility function is linear. But the regret function has no connection with the above situation; it simply reflects the relative position of regret level. In this suggested auction theory, we consider a sequence of regret action points representing the decision-makers attitude; even there is an ability to discuss the behavior of zero possibility degree.

In subjective expected utility and other decision theories, evaluating each alternative with the weighted average method and then selecting the optimal alternative with maximum average [33]. Mahboob et al. [26] proposed preference based approach in group decision-making. We investigate the effect of repeated values. Every decision in this procedure has only one effect on a single auction. We have observed that many repetitions values in higher average bids give the wrong decision of bid. In contrast, the proposed theory is scenario-based for selecting an optimal alternative. A minimal regret level will be used in the suggested approach among all the selected regret levels and show an accurate picture of bids.

The nature of our experiment is different from the traditional auction studies, and evaluating the impact of these differences provides our insight into e-auction bidding behavior. As a result, increased supply-side competition improves efficiency and reduces the gap between auction structures. Furthermore, the use of our research work improves the practice benefits. We present a new vision on the proficiency of auction compared to other articles [30, 32, 35, 38].

5 Conclusion

In this article, we propose a concept of regret theory to look at multiple issues of e-bidding auctions. In this auction, a buyer forwards a request of quotation to many suppliers. The suppliers reply by sending a bid that includes many attributes. All the recommendations are visible to the buyer only and he is supposed to select one bid. In the proposed model, the regret function is used to solve the decision-making problems of suppliers and buyers. The main objective of this theory is to find out the bid which is equally suitable for the supplier and the buyers. We reframe this problem by assuming that a buyer is interested only in the lowest price offered by suppliers when determining a bidding price. Our model is scenario-oriented, which distinguishes it from the existing models which are based on expected utility. The results display that our proposed model is the best because we find the real-world results and further analyzed the effect of overbidding on the optimal bidding strategy which shows the actual picture of e-bidding auction. Due to the multi-attribute decision-making nature of e-auction problems, most e-auction structures are dependent on pricing and criteria evaluated in terms of price, resulting in relatively poor bid selection. A variety of quantitative attributes are considered in the suggested approach in formulating the e-bidding problems.

In the field of management like a manager of a textile mill, the proposed theory provides a comprehensive, structured framework for modelling of rational thoughts. While it is well-known that behavioral variables are very significant in the study of research, however, it is complicated to integrate the characteristics qualities of the bidders into mathematical models due to the absence of proper theories. The proposed approach gives a conceptual framework for the development of behavioral models.

References

1. Allais, M.: Le comportement de l'homme rationnel devant le risque: critique des postulats et axiomes de l'école américaine. Econometrica **21**, 503–546 (1953)
2. Bell, D.E.: Regret in decision making under uncertainty. Oper. Res. **30**(5), 961–981 (1982)
3. Bell, D.E.: Risk premiums for decision regret. Manage. Sci. **29**(10), 1156–1166 (1983)
4. Bichler, M., Lee, J., Kim, C.H., Lee, H.S.: Design and implementation of an intelligent decision analysis system for e-sourcing. Int. Conf. Arti, Intell (2001)
5. Bichler, M., Kalagnanam, J.: Configurable offers and winner determination in multi-attribute auctions. Eur. J. Oper. Res. **160**(2), 380–394 (2005)
6. Chandrashekar, T.S., Narahari, Y., Rosa, C.H., Kulkarni, D.M., Tew, J.D., Dayama, P.: Computer simulation to aid robot selection. Robotic. Tech. J., Peter Perigrinus Ltd, London **4**(3), 297–321 (2007)
7. Chaturvedi, A.: Procurement auctions with capacity constrained suppliers. Eur. J. Oper. Res. **247**(3), 987–995 (2005)
8. Chetan, T.G., Jenamani, M., Sarmah, S.: Two-stage multi-attribute auction mechanism for price discovery and winner determination. IEEE Trans. Engin. Manage. **66**(1), 112–126 (2018)
9. Chen, W., Goh, M., Zou, Y.: Logistics provider selection for omni-channel environment with fuzzy axiomatic design and extended regret theory. App. Soft Comput. **71**, 353–363 (2018)
10. Che, Y.K.: Design competition through multidimensional auctions. RAND J. Econ. **24**(4), 668–680 (1993)
11. Croom, S.R.: The impact of web-based procurement on the management of operating resources supply. J. Supp. Chain Manage. **36**(4), 4–13 (2000)
12. Ellsberg, D.: Risk, ambiguity, and the Savage axioms. Quart. J. Econ., 643–669 (1961)
13. Emiliani, M.L.: Business-to-business online auctions: key issues for purchasing process improvement. Int. J. Supp. Chain Manage. **5**(4), 176–186 (2000)
14. Guo, P.: Focus theory of choice and its application to resolving the St. Petersburg, Allais, and Ellsberg paradoxes and other anomalies. Eur. J. Oper. Res. **276**(3), 1034–1043 (2019)

15. Hwang, C.L., Masud, A.S.M.: Multiple objective decision making—methods and applications: a state-of-the-art survey. Sprin. Sci. Busin. Med., **164**, (2012)
16. Jahan, A., Mustapha, F., Ismail, M.Y., Sapuan, S.M., Bahraminasab, M.: A comprehensive VIKOR method for material selection. Materi. Desi. **32**(3), 1215–1221 (2011)
17. Kahneman, D., Tversky, A.: Prospect theory: an analysis of decision under risk. Econometrica **47**(2), 363–391 (1979)
18. Kameshwaran, S., Benyoucef, L., Xie, X.: Design of progressive auctions for procurement based on Lagrangian relaxation. In: Seventh IEEE International Conference on E-commerce Technology (CEC'05), IEEE, pp 9–16 (2005)
19. Katehakis, M., Puranam, K.: On optimal bidding in sequential procurement auctions. Oper. Res. Lett. **40**(4), 244–249 (2012)
20. Laciana, C.E., Weber, E.U.: Correcting expected utility for comparisons between alternative outcomes: a unified parameterization of regret and disappointment. J. Risk. Uncer. **36**(1), 1–17 (2008)
21. Liang, D., Wang, M., Xu, Z., Chen, X.: Risk interval-valued three-way decisions model with regret theory and its application to project resource allocation. J. Oper. Res. Soci. **72**(1), 180–199 (2021)
22. Loomes, G., Sugden, R.: Regret theory: an alternative theory of rational choice under uncertainty. Eco. J. **92**(368), 805–824 (1982)
23. Lorentziadis, P.: Pricing in a supply chain for auction bidding under information asymmetry. Eur. J. Oper. Res. **237**(3), 871–886 (2014)
24. Mahboob, A., Rashid, T., Salabun, W.: An intuitionistic approach for one-shot decision making. Open. J. Math. Sci. **2**(1), 164–178 (2018)
25. Mahboob, A., Rashid, T.: Multistage decision approach for short life cycle products using Pythagorean fuzzy set. J. Inte. Fuzzy Sys. **40**(1), 1–14 (2020)
26. Mahboob, A., Rashid, T., Sindhu, M.S.: An optimization preference based approach with hesitant intuitionistic linguistic distribution in group decision making. Expert Sys. App. **187**, 115965 (2022)
27. von Neumann, J., Morgenstern, O.: Theory of games and economic behavior. Princeton University Press, Princeton (1994)
28. Peng, H.G., Shen, K.W., He, S.S., Zhang, H.Y., Wang, J.Q.: Investment risk evaluation for new energy resources: an integrated decision support model based on regret theory and ELECTRE III. Ener. Conv. Manage. **183**, 332–348 (2019)
29. Peng, X., Dai, J.: Approaches to Pythagorean fuzzy stochastic multi-criteria decision making based on prospect theory and regret theory with new distance measure and score function. Int. J. Inte. Sys. **32**(11), 1187–1214 (2017)
30. Pham, L., Teich, J., Wallenius, H., Wallenius, J.: Multi-attribute online reverse auctions: recent research trends. Eur. J. Oper. Res. **242**(1), 1–9 (2015)
31. Puranam, K., Katehakis, M.: On optimal bidding and inventory control in sequential procurement auctions: the multi period case. Annals Oper. Res. **217**(1), 447–462 (2014)
32. Qian, X., Fang, S.C., Huang, M., Wang, X.: Winner determination of loss-averse buyers with incomplete information in multiattribute reverse auctions for clean energy device procurement. Energy **177**, 276–292 (2019)
33. Savage, L.J.: The foundations of statistics. Wiley, New York, vol. 253, pp. 254–316 (1954)
34. Shimanoff, S.B.: Commonly named emotions in everyday conversations. Perce. Moto, Skil (1984)
35. Smeltzer, L.R., Carr, A.S.: Electronic reverse auctions: promises, risks and conditions for success. Indus. Mark. Manage. **32**, 481–488 (2003)
36. Strecker, S., Seifert, S.: Electronic sourcing with multi-attribute auctions. In: Proceedings of the 37th Hawaii International Conference on System Science (2004)
37. Sugden, R.: Regret, recrimination and rationality. Theor. Decis. **19**(1), 77–99 (1985)
38. Teich, J.E., Wallenius, H., Wallenius, J., Zaitsev, A.: A multi-attribute e-auction mechanism for procurement: theoretical foundations. Eur. J. Oper. Res. **175**(1), 90–100 (2006)

39. Wang, C., Guo, P.: Behavioral models for first-price sealed-bid auctions with the one-shot decision theory. Eur. J. Oper. Res. **261**(3), 994–1000 (2017)
40. Wang, S., Qu, S., Goh, M., Wahab, M.I.M., Zhou, H.: Integrated multi-stage decision-making for winner determination problem in online multi-attribute reverse auctions under uncertainty. Int. J. Fuzzy Sys. **21**(8), 2354–2372 (2019)
41. Zhang, D., Wu, C., Liu, J.: Ranking products with online reviews: a novel method based on hesitant fuzzy set and sentiment word framework. J. Oper. Res. Soci. **71**(3), 528–542 (2020)
42. Zhang, S., Zhu, J., Liu, X., Chen, Y.: Regret theory-based group decision-making with multi-dimensional preference and incomplete weight information. Inf Fusion **31**, 1–13 (2016)
43. Zhang, X., Liang, H.M., Fang, F., Chen, X.: A method for risky multiple attribute decision making considering regret and rejoicing of the decision maker. Comput. Indus. Engin. **124**, 422–434 (2018)

Evaluation of Sustainable Green Building Indicators by Fuzzy Multi-Criteria Decision Making

Gülay Demir

Abstract The relevance of the notion of sustainability today is revealed by global warming, pollution, and the depletion of water resources in a world that has become a village as a result of fast globalization. The number of buildings in which many natural resources are consumed during the construction phase and severe damage to the environment caused by the wastes generated is constantly increasing. The construction of green buildings can reduce these negativities caused by the construction. Environmental sustainability in Turkey is still in its infancy. The goal of this research is to define and rank sustainability indicators to evaluate green building production in Turkey as a tool for assessing sustainability. To evaluate green buildings, 12 criteria in four dimensions have been presented. The weights of the dimensions and criteria addressed in the assessment framework were determined using the novel fuzzy level-based weight assessment (LBWA), which can represent human reasoning. The results show that ecology and energy-saving are the most critical dimensions in evaluating green building production in Turkey while reducing waste and health are the least essential dimensions. Research output can be a valuable reference for green building builders in the country. It can serve as a guide to selecting and promoting optimal applications in green building products. In addition, the proposed framework can be adapted for countries that are similar to Turkey's climate and environmental conditions.

Keywords Building manufacturing · Fuzzy logic · LBWA

1 Introduction

Green buildings are currently defined as structures that are environmentally friendly and effectively utilize natural resources throughout their life cycle, including "planning, design, construction, operation, maintenance, renovation, and destruction". The

G. Demir (✉)
Sivas Cumhuriyet University, Sivas 58040, Turkey
e-mail: gulaydemir@cumhriyet.edu.tr

L. Sahoo et al. (eds.), *Real Life Applications of Multiple Criteria Decision Making Techniques in Fuzzy Domain*, Studies in Fuzziness and Soft Computing 420,
https://doi.org/10.1007/978-981-19-4929-6_16

purpose of green building is to manage the relationships between nature, architecture, and people [1]. The green building strategy aims to reduce negative effects on the environment and resources by expanding the life cycle of a project [2]. A building must satisfy specific standards to be termed "green." Governments in major economies use a variety of techniques to promote the use and spread of green technology in the construction sector [3]. Countries in the world implement the concept of green buildings and develop their standards and building evaluation certificates or adapt existing certificates to them. The first ones that come to mind from these countries are the USA, England, Germany, Australia, and Japan.

"LEED (Leadership in Energy and Environmental Design)", "DGNB (Deutsche Gesellschaft Für Nachhaltiges)", "CASBEE (Comprehensive Assessment System for Built Environment Efficiency)", "BREEAM (Building Research Establishment's Environmental Assessment Method)", "IISBE (International Initiative for a Sustainable Built Environment)", and "Greenstar" are all still widely used certification systems. According to the ÇEDBİK (Environmentally Friendly Green Buildings) statistics, 495 projects in Turkey have finished the process and earned LEED certification, with 425 receiving LEED certification and 70 receiving BREEAM certification [4].

Traditional constructions require 70% of energy and materials, 17% of pure water, and 25% of forest resources. This results in a 33% increase in carbon dioxide emissions. In these classic structures, recycled materials account for 5% of the total. Both energy and water are saved in green buildings. Since the internal air quality in green buildings increases, operating and maintenance costs are also reduced. Green structures use less energy and water than typical structures. Furthermore, utilizing fewer water facilities in the landscape of green buildings may save 50% of water. CO2 emissions are decreased by 33–39%, energy use is reduced by 24–50%, water consumption is reduced by 30–50%, conservation expenses are lowered by 13%, and solid waste is reduced by 70% because of research on green buildings [5].

There is no legal regulation regarding green buildings in Turkey yet. For this reason, regulations related to energy efficiency are guiding the green building process. Energy efficiency: It is a concept that guarantees supply, focuses on national resources, is low cost, is based on the fight against greenhouse gas emissions, etc., and defines national strategic targets with environmentally friendly policies. Energy efficiency has also gained value with the increasing importance of sustainable development. The number of green buildings is increasing in Turkey. Since there is no national certification system in the country, LEED and BREEAM, which are the most accepted certification systems globally, are applied. However, since these systems are evaluated with the same criteria throughout the world and do not reflect the characteristics of other countries, they have not been successful at the desired rate. Since the requirements of the countries are different, the scores given to the criteria should be arranged by taking these requirements into account. For example, while energy conservation is prominent in some countries, ensuring the sustainability of water resources may be considered important in others. Evaluating Turkey's position in the world, its climatic characteristics, and its different regions, it is inevitable to develop a national certificate of social, economic, legal, and ecological quality that can meet its needs. Within the Ministry of Environment, Urbanization, and Climate Change, studies are carried out for the domestic National Green Certificate System (YeS-TR), which will be developed to disseminate energy-efficient, environmentally friendly building, and settlement practices at the national and local levels. It is planned to issue certificates to buildings and settlements by authorized institutions.

Structure of this study: Chap. 2 proposes a literature review on green building rating systems and their associated criteria, studies of MCDM applications in green building assessment, and a framework for rating green buildings. The studies conducted with the LBWA method are also included in this section. In Chap. 3, the research method is explained, and data analysis and results are given. Chapter 4 includes the findings, analysis results, and study's limitations.

2 Literature Review

2.1 Green Building Rating Systems

Green building practices are a complete technique proposed by the construction industry, private consulting groups, and international rating bodies to assess and verify the green and sustainability standards of buildings. The internationally accepted LEED, BREEAM, CASBEE, SBTool, and GBI certification tools were introduced in this part of the study.

The US Green Building Council developed LEED. Professionals in the US building business have constantly been modifying and personalizing LEED. This system is divided into six categories. The highest possible level is 69. Each release has a set of performance criteria that are divided into five categories. Depending on the style of construction, the quantity of credits and points available varies. The buildings are awarded platinum, gold, and silver certificates [6].

GBI was created by GBI Innovation Sdn Bhd in 2009. GBI is Malaysia's first comprehensive building evaluation system, with 14 unique editions based on various building kinds. GBI has four primary criteria, based on which buildings are rated as "Platinum, Gold, Silver, or Certified" [6].

BREEAM was developed in the United Kingdom to assess the greenness of buildings throughout the construction and usage stages. BREEAM is divided into ten levels, with each level receiving points depending on its performance. After that, the building is given a rating of "Pass, Good, Very Good, or Excellent, and a certificate" is given to the owner [7].

CASBEE originated in Japan. The CASBEE system, which is suitable for the life cycle of buildings, consists of four evaluation categories. Unlike other green building rating systems, CASBEE does not allocate credit to every factor. It is made up of groups in which weights are calculated and displayed on a radar chart, with each credit score being assessed on a scale of 1 to 5 and each credit score being rated on a scale of 1 to 5. In Japan, CASBEE has shown to be a good GBRS in terms of increasing green building [7].

SBTool was introduced by Green Building Challenge (GBC) in 1996. Later, when Natural Resources Canada took over this process, full responsibility was passed to International Sustainable Built Environment Initiative (IISBE). SBTool certificate structures are evaluated out of 100 points in seven main categories.

2.2 Green Building and Its Applications in Multi-Criteria Decision-Making Methods

MCDM methods are also used to evaluate and select various green building rating systems, showing evaluation performance for complex systems. Some recent studies relevant to the assessment of green building indicators are given [8], the criteria based on building information modeling in Malaysia were weighted with ANP. Yadegaridehkordi et al. [9] used the Green Building Index (GBI) and DEMATEL to define and rank sustainability factors to assess green building output in Malaysia [10], Kazakhstan's Building Sustainability Assessment indicators were evaluated with SWARA. Liang [11] applied the IF-EDAS method to evaluate green building energy-efficient design projects [12], with the newly created green building index system, DEMATEL, BWM, and ANP were used to evaluate green buildings in Taiwan.

2.3 Studies with LBWA Method and Fuzzy LBWA Method

LBWA, one of the subjective criterion weighting techniques based on the pairwise comparison, is summarized in certain research [13], used it for transportation mode for Istanbul Airport [14], used it in school selection [15] applied it choosing the most suitable site for offshore wind farms [16], applied weighting criteria that determine the effect of possible residues on products in Raspberry and Blackberry production, in addition, for campsite selection [17], evaluation of health sectors in Eastern Europe [18], and assessment of renewable energy sources [19].

2.4 Proposed Green Building Rating Framework

Experts were asked how to create a green building rating system that considers Turkey's geographical location. Two architects with at least 10 years of expertise in this field, a building structural designer with 20 years of experience, and four specialists with ten years of experience in green building research created a decision-making committee. In addition, the opinions of experts from the Ministry of Environment,

Table 1 Recommended main criteria and sub-criteria for green buildings in Turkey

Main criteria	Sub-criteria
Ecology (MC_1)	"Biodiversity" (C_{11})
	"Land area and degree of greenery" (C_{12})
	"The water content of the site" (C_{13})
	"Outdoor environment" (C_{14})
Energy-saving (MC_2)	"Thermal insulation of the building" (C_{21})
	"Intelligent control of air conditioning and lighting" (C_{22})
	"Water resources recycling" (C_{23})
Waste reduction (MC_3)	"Reducing CO_2 emissions from building materials" (C_{31})
	"Waste reduction during building construction and demolition" (C_{32})
Health (MC_4)	"Indoor environment" (C_{41})
	"Water storage conditions" (C_{42})
	"Sewage and waste disposal" (C_{43})

Urbanization, and Climate Change's YeS-TR team were solicited. Liu and colleagues were agreed to use the framework suggested in the research (2020). Table 1 shows the suggested technique, the green building rating framework, which includes twelve evaluation sub-criteria and four primary criteria.

3 Method

In this research, the fuzzy LBWA technique from MCDM was used to evaluate the relative significance of the four primary criteria and 12 sub-criteria for green building manufacturing in Turkey.

3.1 Fuzzy Theory

Lotfi A. Zadeh's essay "Fuzzy Sets," published in the journal "Information and Control" in 1965, was the first to mention fuzzy set theory in the literature [20]. Fuzzy set theory is a mechanism for expressing uncertainty and evaluating decisions using linguistic variables. Fuzzy sets can be a generalized form of the classical set concept. Membership function values that take $\{0, 1\}$ value in classical sets can take infinite values in the range of $[0, 1]$ in fuzzy sets. According to the characteristics of classical and fuzzy sets, the main difference between the two sets is that their membership functions are different. Theoretically, fuzzy sets have only one membership function, although the membership function of fuzzy sets is infinite. Fuzzy numbers form special subsets of fuzzy sets. Fuzzy numbers can take place

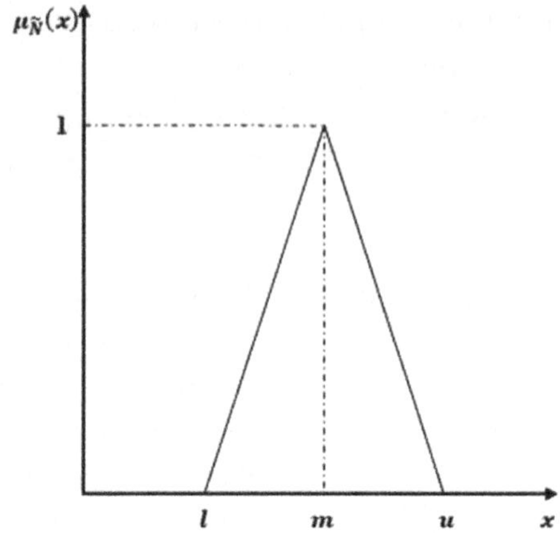

Fig. 1 Triangle fuzzy number $\tilde{N}$

in theory and practice in various ways. These are forms of expression developed to express indefinite quantities. However, the most common fuzzy numbers are fuzzy triangular numbers. In a fuzzy set $\tilde{N}$, fuzzy triangular numbers are expressed as l, m, u. Considering the values expressed.

l : Minimum possible value.
m : Most expected value.
u : It is defined as the highest possible value.

Figure 1 shows a triangular fuzzy number.
Some basic definitions of fuzzy logic are given below.

Definition 1 In a fuzzy set $\tilde{N}$, the triangular fuzzy number l, m, u is expressed by three values, and the membership function is expressed by Eq. (1).

$$\mu_{\tilde{N}}(x) = \begin{cases} 0 & x \leq l \\ \frac{x-l}{m-l} & l \leq x \leq m \\ \frac{u-x}{u-m} & m \leq x \leq u \\ 0 & x > u \end{cases} \tag{1}$$

where l, m, and u are real numbers and must be $l \leq x \leq m$.

Definition 2 Let two fuzzy sets be X and Y. Some operations of these fuzzy sets are defined as follows.

1. $(a_1, a_2, a_3) \oplus (b_1, b_2, b_3) = (a_1 + b_1, a_2 + b_2, a_3 + b_3)$ (2)

2. $(a_1, a_2, a_3) \otimes (b_1, b_2, b_3) = (a_1.b_1, a_2.b_2, a_3.b_3)$ (3)

3. $(a_1, a_2, a_3) \ominus (b_1, b_2, b_3) = (a_1 - b_3, a_2 - b_2, a_3 - b_1)$ (4)

4. $k \otimes (a_1, a_2, a_3) = (k.a_1, k.a_2, k.a_3), k > 0$ (5)

5. $(a_1, a_2, a_3)/(b_1, b_2, b_3) \cong (a_1/b_3, a_2/b_2, a_3/b_1)$ (6)

Since the values obtained are fuzzy numbers, the process called defuzzification, converting the fuzzy value to normal value or converting to crisp value vertex method Eq. (7) and the (center of area) method called Chen or COA, with the help of Eq. (8) can be realized [21, 22].

$$Y_1 = \frac{l + 4m + u}{6} \tag{7}$$

$$Y_2 = \frac{(u - l) + (m - l)}{3} + l = \frac{l + m + u}{3} \tag{8}$$

3.2 Fuzzy LBWA (Level-Based Weight Assessment) Method

Žižović and Pamučar were the first to present the LBWA approach to the literature (2019). The mathematical algorithm for the procedure is quite straightforward. It may be utilized to make decisions both individually and in groups. Criterion weights are determined by following the method's procedural stages [23–26].

Step 1. Identifying the Most Important Criteria

In this step, the decision-maker(s) from the criteria set $S = \{C_1, C_2, \ldots, C_n\}$ determine the most important criterion. Let C_1 be the criterion in the criteria $S = \{C_1, C_2, \ldots, C_n\}$ that is the most important criterion for the decision-making process.

Step 2. Grouping of Criteria

The decision-maker groups the levels/subsets of criteria according to their importance as follows:

Level 1: The most important criterion, C_1, is either equally important or at most 2 times more important than the criteria in this group (except exactly 2 times).
Level 2: The most important criterion, C_1, has at least 2 times (including 2) and at most 3 times more importance (except exactly 3 times) than the criteria in this group.
Level 3: The most important criterion, has at least 3 times (including 3) and at most 4 times more importance (except exactly 4 times) than the criteria in this group.
…
Level k: The most important criterion, C_1, is at least k times (including k) and at most $k + 1$ times more important than the criteria in this group.

The decision-makers decide the approximate categorization of the observed criteria by following the guidelines described above. That is, they group the criteria according to their level of significance. If the importance of the C_j criterion is denoted by $S(C_j)$, then $j \in \{1, 2, \ldots, n\}$ and $S = S_1 \cup S_2 \cup \ldots \cup S_k$ can be written and with the help of $i \in \{1, 2, \ldots, k\}$ and Eq. (9) below can be found.

$$S_i = \{C_{i_1}, C_{i_2}, \ldots, C_{i_s}\} = \{C_j \in S : i \leq s(C_j) < i + 1\} \tag{9}$$

In addition, $p, q \in \{1, 2, \ldots, k\}$ for each of them, if $p \neq q$ is $S_p \cap S_q = \emptyset$. This defines the S criterion sets.

Step 3. Assigning Values to Criteria

The criteria at the levels in the previous step are compared according to their importance. The criteria at each level are assigned an integer $(\gamma_{i_p} \in \{0, 1, \ldots, \lambda\})$ among themselves. Integer assignments are made as "0" for the most important criterion at that level and as "1" … for the criterion considered as secondary importance. The most important criterion for C_i is the integer assignment $I_i = 0$. In this case, p. criterion q. integer assignments are $\gamma_p < \gamma_q$ if they are more important than the criteria, or integer assignments $\gamma_p = \gamma_q$ if these criteria are of equal importance. Equation (10) is obtained, where λ is the maximum integer assignment.

$$\lambda = \max\{|S_1|, |S_2|, \ldots, |S_k|\} \tag{10}$$

Values are computed for each decision-maker based on the comparison criteria at each observed level. Exceptions are predicted in criterion comparison, expressed as fuzzy numbers, since each decision-maker analyzes criteria within subsets (levels) individually. Equation (6) is used to translate accurate outcomes of expert criteria comparisons inside subsets (levels) into fuzzy values. Equation (11) is also used to gather the decision preferences.

$$\tilde{\gamma}_{i_p} = (a, b, c) \Rightarrow a = \min_m \left\{\gamma_{i_p}^m\right\} \quad b = \frac{1}{m}\sum_{e=1}^{m} \gamma_{i_p}^e \quad c = \max_m \left\{\gamma_{i_p}^m\right\} \tag{11}$$

m shows the total number of decision-makers who took part in the study.

Step 4. Determination of Coefficient of Elasticity

The λ value obtained in the previous step is the basis for determining the elasticity coefficient. It is determined as $\varphi > \lambda$, with $\varphi(\varphi \in R)$ being the elasticity coefficient.

Step 5. Calculation of Impact Functions

Impact function $f : S \to R$

i: level order.

φ : coefficient of elasticity.

$\tilde{\gamma}_{i_p}$: fuzzy number value assigned to criteria.

C_{i_p}: observed level.

Equation (12) calculates the impact functions of the criterion. In each $C_{i_p} \in S_i$;

$$\tilde{f}(C_{i_p}) = \frac{\varphi}{i.\varphi + \tilde{\gamma}_{i_p}} \tag{12}$$

Step 6. Determine the best fuzzy values for the criteria's weight coefficients

Finally, the fuzzy weight coefficient of the most significant criterion is calculated using Eq. (13).

$$w_1 = \frac{1}{1 + f(C_2) + \ldots + f(C_i) + \ldots + f(C_n)} \tag{13}$$

Equation (14) is used to compute the weighting coefficients of the remaining criteria once the weight of the most significant criterion has been established.

$$w_j = f(C_j).w_1 \tag{14}$$

where $j = 2, 3, \ldots, n$, and n represent the total number of criteria.

3.3 Application of the F-LBWA Method to the Proposed Framework

The evaluation of green buildings is divided into two hierarchical stages, as indicated in Table 1. Dimensions make up the first hierarchical level, while criteria make up the second level. While determining the weight coefficients, the fuzzy LBWA model's method was used at the first hierarchical level to generate criterion weights.

Step 1. From the criteria list, choose the most significant criterion. As the most important/effective dimension, the decision-makers identified the MC_2 dimension. Then, for the criteria groups, the decision-makers defined the most effective criteria.

The most effective criterion for MC_1 is C_{11}

The most effective criterion for MC_2 is C_{22}

The most effective criterion for MC_3 is C_{31}

The most effective criterion for MC_4 is C_{43}

Step 2. The criteria are grouped by importance for the decision-makers. The dimensions and criteria are divided into subsets/levels based on the decision-maker's choices, as shown in Table 2. Table 2's arrangement of dimensions may alternatively be interpreted as follows.

$$S_1 = \{\mathrm{MC}_1, \mathrm{MC}_2\}$$

$$S_2 = \emptyset$$

$$S_3 = \{\mathrm{MC}_3\}$$

$$S_4 = \emptyset$$

$$S_5 = \{\mathrm{MC}_4\}$$

Table 2 shows how the sub-criteria are organized.

Step 3. To compare the criterion with Eq. (10), the highest value of the scale is established.

Table 2 Grouping of dimensions and criteria by significance

Main criteria/sub-criteria	Level
MC_1	Level 1
MC_2	Level 1
MC_3	Level 3
MC_4	Level 5
Ecology	
C_{11}	Level 1
C_{12}	Level 1
C_{13}	Level 2
C_{14}	Level 2
Energy-saving	
C_{21}	Level 1
C_{22}	Level 3
C_{23}	Level 2
Waste reduction	
C_{31}	Level 1
C_{32}	Level 2
Health	
C_{41}	Level 1
C_{42}	Level 2
C_{43}	Level 2

$$\left.\begin{array}{l} S_1 = \{MC_1, MC_2\} \\ S_2 = \emptyset \\ S_3 = \{MC_3\} \\ S_4 = \emptyset \\ S_5 = \{MC_4\} \end{array}\right\} \Rightarrow \lambda = \max\{|S_1|, |S_3|, |S_5|\} = 2$$

Based on the comparison scale's highest value, it can be concluded that the scale for comparing criteria falls within the $I_{i_p} \in \{0, 1, 2\}$ range.

A main criteria/sub-criteria comparison is made using values at each level of criteria significance. Expert comparisons of $\{0, 1, 2, 3\}$ Table 3 lists the main and sub-criteria at the significance level of the criterion.

Step 4. Table 3 according to the λ value, and under conditions $\varphi > \lambda$ and $\lambda = \max\{|S_1|, |S_2|, \ldots, |S_k|\}$, this study shows that φ =4.

The fuzzy preferences calculated using Eq. (11) are given in Table 4.

Step 5. Equation (12) was used to determine the fuzzy effect functions of the criterion, which are listed in Table 5.

Table 3 Importance ratings of criteria according to expert comparisons

Main criteria/sub-criteria	Level	DM1	DM2	DM3	DM4	λ
MC_1	Level 1	0	1.5	1	0	2
MC_2	Level 1	1.5	2	1.8	1.2	
MC_3	Level 3	2	2	1.8	1.5	
MC_4	Level 5	1	1.8	1	2	
Ecology						
C_{11}	Level 1	0	0.5	0.8	0	2
C_{12}	Level 1	1.5	2	2	2	
C_{13}	Level 2	2	1.5	2	2.5	
C_{14}	Level 2	1	1.5	1.5	1	
Energy-saving						
C_{21}	Level 1	0	0.5	0.5	1	3
C_{22}	Level 3	2.5	2	3	3	
C_{23}	Level 2	1	1.5	1.5	1.5	
Waste reduction						
C_{31}	Level 1	0	0.2	0	0.3	1
C_{32}	Level 2	1	1	1	0.5	
Health						
C_{41}	Level 1	0	0.5	0	0.5	3
C_{42}	Level 2	2.5	3	3	1.4	
C_{43}	Level 2	1.5	3	2	1.5	

Table 4 Fuzzy preferences

Main criteria/sub-criteria	Level	I_{i_p}	Triangle fuzzy numbers
MC_1	Level 1	I_1	(0, 0.63, 1.5)
MC_2	Level 1	I_2	(1.2, 1.63, 2)
MC_3	Level 3	I_3	(1.5, 1.83, 2)
MC_4	Level 5	I_4	(1, 1.45, 2)
Ecology			
C_{11}	Level 1	I_{11}	(0, 0.33, 0.8)
C_{12}	Level 1	I_{12}	(1.5, 1.88, 2)
C_{13}	Level 2	I_{13}	(1.5, 2, 2.5)
C_{14}	Level 2	I_{14}	(1, 1.25, 1.5)
Energy-saving			
C_{21}	Level 1	I_{21}	(0, 0.5, 1)
C_{22}	Level 3	I_{22}	(2, 2.63, 3)
C_{23}	Level 2	I_{23}	(1, 1.38, 1.5)
Waste reduction			
C_{31}	Level 1	I_{31}	(0, 0.13, 0.3)
C_{32}	Level 2	I_{32}	(0.5, 0.88, 1)
Health			
C_{41}	Level 1	I_{41}	(0, 0.25, 0.5)
C_{42}	Level 2	I_{42}	(1.4, 2.48, 3)
C_{43}	Level 2	I_{43}	(1.5, 2, 3)

Step 6. Equation (13) is used to calculate the fuzzy value of the weighting coefficient of the most important criterion relevant to the primary criteria.

$$\tilde{w}_1 = \left(w_1^l, w_1^m, w_1^u\right) = \begin{cases} w_1^l = \frac{1}{1+0.77+0.30+0.19} = 0.443 \\ w_1^m = \frac{1}{1+0.71+0.29+0.19} = 0.457 \\ w_1^u = \frac{1}{1+0.67+0.29+0.18} = 0.468 \end{cases}$$

In the sub-criteria group, the most important criteria are determined in the same way.

$$\tilde{w}_{11} = \left(w_{11}^l, w_{11}^m, w_{11}^u\right) = \begin{cases} w_{11}^l = \frac{1}{1+0.73+0.42+0.44} = 0.386 \\ w_{11}^m = \frac{1}{1+0.68+0.40+0.43} = 0.398 \\ w_{11}^u = \frac{1}{1+0.67+0.38+0.42} = 0.405 \end{cases}$$

$$\tilde{w}_{22} = \left(w_{22}^l, w_{22}^m, w_{22}^u\right) = \begin{cases} w_{22}^l = \frac{1}{1+1+0.44} = 0.409 \\ w_{22}^m = \frac{1}{1+0.89+0.43} = 0.432 \\ w_{22}^u = \frac{1}{1+0.8+0.42} = 0.450 \end{cases}$$

Table 5 Fuzzy effect functions of dimensions/criteria

Main criteria/sub-criteria	$\tilde{f}(C_j)$	Triangle fuzzy numbers
MC_1	$\tilde{f}(C_1)$	(0.73, 0.86, 1)
MC_2	$\tilde{f}(C_2)$	(0.67, 0.71, 0.77)
MC_3	$\tilde{f}(C_3)$	(0.29, 0.29, 0.30)
MC_4	$\tilde{f}(C_4)$	(0.18, 0.19, 0.19)
Ecology		
C_{11}	$\tilde{f}(C_{11})$	(0.83, 0.92, 1)
C_{12}	$\tilde{f}(C_{12})$	(0.67, 0.68, 0.73)
C_{13}	$\tilde{f}(C_{13})$	(0.38, 0.40, 0.42)
C_{14}	$\tilde{f}(C_{14})$	(0.42, 0.43, 0.44)
Energy-saving		
C_{21}	$\tilde{f}(C_{21})$	(0.8, 0.89, 1)
C_{22}	$\tilde{f}(C_{22})$	(0.27, 0.27, 0.29)
C_{23}	$\tilde{f}(C_{23})$	(0.42, 0.43, 0.44)
Waste reduction		
C_{31}	$\tilde{f}(C_{31})$	(0.93, 0.97, 1)
C_{32}	$\tilde{f}(C_{32})$	(0.44, 0.45, 0.47)
Health		
C_{41}	$\tilde{f}(C_{41})$	(0.89, 0.94, 1)
C_{42}	$\tilde{f}(C_{42})$	(0.36, 0.38, 0.43)
C_{43}	$\tilde{f}(C_{43})$	(0.36, 0.40, 0.42)

$$\tilde{w}_{31} = \left(w_{31}^{l}, w_{31}^{m}, w_{31}^{u}\right) = \begin{cases} w_{31}^{l} = \frac{1}{1+0.47} = 0.680 \\ w_{31}^{m} = \frac{1}{1+0.45} = 0.689 \\ w_{31}^{u} = \frac{1}{1+0.44} = 0.692 \end{cases}$$

$$\tilde{w}_{43} = \left(w_{43}^{l}, w_{43}^{m}, w_{43}^{u}\right) = \begin{cases} w_{43}^{l} = \frac{1}{1+1+0.43} = 0.412 \\ w_{43}^{m} = \frac{1}{1+0.94+0.38} = 0.430 \\ w_{43}^{u} = \frac{1}{1+0.89+0.36} = 0.579 \end{cases}$$

Weight coefficients of Eq. (14) and other sub-criteria were calculated and given in Table 6.

$$\tilde{w}_2 = \tilde{f}(C_2).\tilde{w}_1 = (0.67, 0.71, 0.77).(0.442, 0.457, 0.467) = (0.296, 0.325, 0.360)$$

$$\tilde{w}_3 = \tilde{f}(C_3).\tilde{w}_1 = (0.29, 0.29, 0.30).(0.442, 0.457, 0.467) = (0.127, 0.132, 0.139)$$

$$\tilde{w}_4 = \tilde{f}(C_4).\tilde{w}_1 = (0.18, 0.19, 0.19).(0.442, 0.457, 0.467) = (0.081, 0.085, 0.089)$$

The weighting coefficients of the criterion are multiplied by the weighting coefficients of the sub-criteria to generate the global values used to evaluate the alternatives using fuzzy ranking methods.

If non-fuzzy methods rank the alternatives, the weights must be clarified with Eq. (7).

$$\omega_1 = \frac{w_1^l + 4w_1^m + w_1^u}{6} = \frac{0.443 + 4(0.457) + 0.468}{6} = 0.457$$

Table 6 Criterion weights with calculated fuzzy values

Main criteria/sub-criteria	Local fuzzy weight	Global fuzzy weight
w_1	(0.443, 0.457, 0.468)	–
w_{11}	(0.386, 0.398, 0.405)	(0.171, 0.182, 0.190)
w_{12}	(0.296, 0.311, 0.341)	(0.131, 0.142, 0.160)
w_{13}	(0.169, 0.183, 0.197)	(0.075, 0.084, 0.092)
w_{14}	(0.187, 0.198, 0.208)	(0.083, 0.090, 0.098)
w_2	(0.296, 0.325, 0.360)	–
w_{21}	(0.355, 0.406, 0.469)	(0.105, 0.132, 0.169)
w_{22}	(0.409, 0.432, 0.450)	(0.121, 0.140, 0.162)
w_{23}	(0.187, 0.195, 0.208)	(0.055, 0.063, 0.075)
w_3	(0.127, 0.132, 0.139)	–
w_{31}	(0,680, 0.689 0.692)	(0.086, 0.091, 0.096)
w_{32}	(0.197, 0.206, 0.220)	(0.025, 0.027, 0.031)
w_4	(0.081, 0.085, 0.089)	–
w_{41}	(0.394, 0.430, 0.469)	(0.032, 0.037, 0.042)
w_{42}	(0.162, 0.175, 0.199)	(0.013, 0.015, 0.018)
w_{43}	(0.412, 0.430, 0.579)	(0.033, 0.037, 0.052)

$$\omega_2 = \frac{w_2^l + 4w_2^m + w_2^u}{6} = \frac{0.296 + 4(0.325) + 0.360}{6} = 0.326$$

$$\omega_3 = \frac{w_3^l + 4w_3^m + w_3^u}{6} = \frac{0.127 + 4(0.132) + 0.139}{6} = 0.132$$

$$\omega_4 = \frac{w_4^l + 4w_4^m + w_4^u}{6} = \frac{0.081 + 4(0.085) + 0.089}{6} = 0.085$$

Following this method of determining the weights of the primary criterion, the weights of the sub-criteria were calculated and are presented in Table 7.

Ecology is the most critical dimension in evaluating green buildings, with a weight value of 0.457. Of the criteria of this dimension, biodiversity, the degree of the greenness of the land area, the external environment, and the site's water content are important, respectively.

The energy-saving dimension was the second most crucial dimension, with a weight value of 0.326. Intelligent use of air conditioning and lighting, thermal insulation of the building, and water resources recycling are essential criteria, respectively.

The waste reduction dimension is the third most crucial dimension, with a weight value of 0.132. Reducing CO_2 emissions and reducing waste during the construction and demolition of buildings are essential criteria.

The health dimension is the last dimension with a weight value of 0.085. The critical criteria are the indoor environment, waste disposal, and water storage conditions.

Table 7 Weight values of criteria

Sub-criteria	Defuzzification of local fuzzy weights	Ranking
C_{11}	0.181	1
C_{12}	0.143	2
C_{13}	0.084	7
C_{14}	0.090	6
C_{21}	0.134	4
C_{22}	0.141	3
C_{23}	0.064	8
C_{31}	0.091	5
C_{32}	0.027	11
C_{41}	0.037	10
C_{42}	0.015	12
C_{43}	0.039	9

4 Discussion

This research proposes a fuzzy version of the LBWA model that is implemented using a simple mathematical approach. The F-LBWA model demonstrated how the values of weight coefficients are acquired in the group decision-making process under uncertain situations in terms of technique. F-LBWA uses basic mathematical methods to create fuzzy weighting coefficient values, which eliminates discrepancies in expert preferences in specific subjective models.

The F-LBWA paradigm offers a few benefits. (1) With a limited number of criterion comparisons, the F-LBWA model is utilized to derive their weight (importance) coefficients. (2) As the number of criteria increases, F-LBWA model's calculation does not grow more difficult. Because of this model characteristic, it may be employed in more complicated MCDM models with additional assessment criteria. (3) F-LBWA model allows decision-makers to use a logical method to order their preferences while analyzing the criteria. (4) Using the elasticity coefficient in the model, decision-makers change the values of the weight coefficients to suit their preferences. This elasticity coefficient may be used to investigate the MCDM model's sensitivity to changes in the weights of the criteria and their impact on the final choice.

Green building manufacturing in Turkey has been evaluated. According to the results, ecology with a weight of 0.457 and energy-saving with a weight of 0.326 are the two most important main criteria in evaluating green building production. With a weight of 0.132 for waste reduction and 0.085 for health, it ranks third and fourth in terms of importance in evaluating green building production. In this study, studies can be carried out based on similar sectors in different sectors and other countries in the future. This study's findings can be compared to those of future investigations.

Since there may be different evaluation, criteria, or MCDM methods (BWM, IDOCRIW, MEREC, etc.) and their various extensions (spherical fuzzy sets, Pythagorean fuzzy sets, fuzzy heuristic sets, etc.) may be the subject of research. In any country, different criteria can be observed in solving the evaluation of green building manufacturing. The proposed scientific methodology can be easily applied to different countries or cities.

References

1. Li, Q., Long, R., Chen, H., Chen, F., Wang, J.: Visualized analysis of global green buildings: development, barriers and future directions. J. Clean. Prod. **245**, 118775 (2020). https://doi.org/10.1016/j.jclepro.2019.118775
2. Zhang, L., Wu, J., Liu, H.: Turning green into gold: a review on the economics of green buildings. J. Clean. Prod. **172**, 2234–2245 (2018). https://doi.org/10.1016/j.jclepro.2017.11.188
3. Hall, J., Matos, S., Bachor, V.: From green technology development to green innovation: inducing regulatory adoption of pathogen detection technology for sustainable forestry. Small Bus. Econ. **52**(4), 877–889 (2019). https://doi.org/10.1007/s11187-017-9940-0
4. Çevre Dostu Yeşil Binalar Derneği. (2021). https://cedbik.org/. (21 December 2021)

5. Erten D.: Yeşil Binalar. Sürdürülebilir Üretim ve Tüketim Yayınları V. Çevre ve Şehircilik Bakanlığı, Ankara (2011)
6. Shan, M., Hwang, B.-G.: Green building rating systems: Global reviews of practices and research efforts. Sustain. Cities Soc. **39**, 172–180 (2018). https://doi.org/10.1016/j.scs.2018.02.034
7. Zarghami, E., Azemati, H., Fatourehchi, D., Karamloo, M.: Customizing well-known sustainability assessment tools for Iranian residential buildings using Fuzzy Analytic Hierarchy Process. Build. Environ, **128**, (2018). https://doi.org/10.1016/j.buildenv.2017.11.032
8. Haruna, A., Shagiq, N., Montasir, O.A.: Building information modeling application for developing sustainable building (Multi-criteria decision-making approach). Ain Shams Eng. J. **12**(1), 293–302 (2021). https://doi.org/10.1016/j.asej.2020.06.006
9. Yadegaridehkordi, E., Hourmand, M., Nilashi, M., Alsolami, E., Samad, S., Mahmoud, M., Alarood, Ala A., Zainol, A., Majeed, H. D., Shuib, L.: Assessment of sustainability indicators for green building manufacturing using fuzzy multi-criteria decision-making approach. J. Cleaner Prod. **277**, 122905 (2020). https://doi.org/10.1016/j.jclepro.2020.122905
10. Akhanovaa, G., Abid, N., Kim, J.R., Azhar, S.: A multi-criteria decision-making framework for building sustainability assessment in Kazakhstan. Sustain. Cities Soc. **52**, 101842 (2020). https://doi.org/10.1016/j.scs.2019.101842
11. Liang, Y.: An EDAS method for multiple attribute group decision-making under intuitionistic fuzzy environment and its application for evaluating green building energy-saving design projects. Symmetry **12**(3), 484 (2020). https://doi.org/10.3390/sym12030484
12. Liu, P.C.Y., Lo, H.-W., Liou, J.J.H.: A Combination of DEMATEL and BWM-based ANP methods for exploring the green building rating system in Taiwan. Sustainability **12**, 3216 (2020). https://doi.org/10.3390/su12083216
13. Pamučar, D., Deveci, M., Canıtez, F., Lukovac, V.: Selecting an airport ground access mode using novel Fuzzy LBWA-WASPAS-H decision making model. Eng. Appl. Artif. Intell. **93**, 103703 (2020). https://doi.org/10.1016/j.engappai.2020.103703
14. Biswas, S., Pamučar, D.: Facility location selection for B-schools in Indian context: a multi-criteria group decision based analysis. Axioms **9**(77), 2–18 (2020). https://doi.org/10.3390/axioms9030077
15. Deveci, M., Ozcan, E., John, R., Covrig, C., Pamučar, D.: A study on offshore wind farm siting criteria using a novel interval-valued fuzzy-rough based delphi method. J. Environ. Manage. **270**, 110916 (2020). https://doi.org/10.1016/j.jenvman.2020.110916
16. Ćurĉić, S., Titović, M., Koprivica, S.M., Vesković, M., Turanjanin, D.: Multicriteria analysis of production alternatives of raspberry and blackberry production residues. Quaestus Multidiscip. Res. J. **3**(3), 155–158 (2020)
17. Božanić, D., Jurišić, D. & Erkić, D.: LBWA—Z-MAIRCA model supporting decision making in the army. Oper. Res. Eng. Sci.: Theor. Appl. **3(2)**, 87–110 (2020). https://doi.org/10.31181/oresta2003087b
18. Torkayesh, A.E., Pamučar, D., Ecer, F., Chatterjee, P.: An integrated BWM-LBWA-CoCoSo framework for evaluation of healthcare sectors in Eastern Europe. Socio-Econ. Plan. Sci. 101052 (2021). https://doi.org/10.1016/j.seps.2021.101052
19. Ecer, F., Pamučar, D., Mardani, A., Alrasheedi, M.: Assessment of renewable energy resources using new interval rough number extension of the level based weight assessment and combinative distance-based assessment. Renew. Energy **170**, 1156–1177 (2021). https://doi.org/10.1016/j.renene.2021.02.004
20. Zadeh, L.A.: Fuzzy Sets. Inf. Control **8**, 338–353 (1965). https://doi.org/10.1016/S0019-9958(65)90241-X
21. Chen, T.C.: Extensions of the TOPSIS for group decision-making under fuzzy environment. Fuzzy Sets Syst. **114**(1), 1–9 (2020). https://doi.org/10.1016/S0165-0114(97)00377-1
22. Rezaei, J., Ortt, R.: Supplier segmentation using fuzzy logic. Ind. Mark. Manag. **42**(4), 507–517 (2013). https://doi.org/10.1016/j.indmarman.2013.03.003Get rights and content
23. Žižović, M., Pamučar, D.: New model for determining criteria weights: Level Based Weight Assessment (LBWA) model. Decis. Mak. Appl. Manag. Eng. **2**(2), 126–137 (2019). https://doi.org/10.31181/dmame1902102z

24. Demir, G.: LBWA Yöntemi. In: Bircan, H. (Ed.). Çok Kriterli Karar Verme Problemlerinde Kriter Ağırlıklandırma Yöntmleri, pp.137–158. Nobel, Ankara (2020)
25. Pamučar, D., Žižović, M., Marinković, D., Doljanica, D., Jovanović, S.V., Brzaković, P.: Development of a multi-criteria model for sustainable reorganization of a healthcare system in an emergency situation caused by the COVID-19 pandemic. Sustainability **12**(18), 7504 (2020). https://doi.org/10.3390/su12187504
26. Žižović, M., Pamučar, D.: The fuzzy model for determining criteria weights based on level construction. In: Dinesh, C.S.B., Mangey, R. (eds.) Computational Intelligence, pp. 81–84. De Gruyter, Berlin (2020)

Solving Multi-objective Linear Fractional Programming Problem Utilizing (α, β)-Cut in Triangular Intuitionistic Fuzzy Setup

Rakesh Kumar Bajaj, Saurabh Srivastava, and Abhishek Guleria

Abstract In the present communication, a new model of multi-objective linear fractional programming problem (MOLFPP) with triangular intuitionistic fuzzy parameters has been proposed for finding the permissible deviations in the objective values in view of the constraints under consideration. Here, (α, β)-cuts of the triangular intuitionistic fuzzy numbers have been utilized in the proposed model to find out the level of satisfaction and to convert the fuzzy parameters into closed intervals. Here, such fuzzification has been incorporated to encounter the uncertainty and inexactness that arises in the available information. Using the variable transformation technique/Taylor's series, the interval-valued fractional objectives have been mathematically estimated by the intervals of linear functions. The objective functions have also been assigned appropriate weights using the analytic hierarchy process. In order to consolidate the multiple objectives into a single objective, the weighting sum method has been applied. It may be observed that the MOLFPP in the interval-valued form has been analogously reduced to a pair of linear problems which provide the acceptable deviations in the objective values. For the sake of illustration and implementation of the proposed work, a numerical example related to a manufacturing company has been solved in detail.

R. K. Bajaj · S. Srivastava
Department of Mathematics, Jaypee University of Information Technology, Solan, Himachal Pradesh, India
e-mail: rakesh.bajaj@juit.ac.in

S. Srivastava
e-mail: saurabh.srivastava@juit.ac.in

A. Guleria (✉)
Department of Physical Sciences and Languages, CSKHPKV, Palampur, Himachal Pradesh, India
e-mail: abhishekguleriahappy@gmail.com; abhishekguleria@hillagric.ac.in

L. Sahoo et al. (eds.), *Real Life Applications of Multiple Criteria Decision Making Techniques in Fuzzy Domain*, Studies in Fuzziness and Soft Computing 420,
https://doi.org/10.1007/978-981-19-4929-6_17

1 Introduction

In view of the gradually increasing complexity of any modular system of a real world problem, various researchers have studied different approaches to handle the uncertainty and fuzziness inherent in the imprecise and incomplete data of the problem. The mathematical formulation of different optimization and decision making problems under the fuzzy environment consists of rationally synchronizing multiple disagreements and inter-related objectives bounded to a given numbers of physical/economical constraints. The notion of fuzzy multi-objective linear fractional programming problem (MOLFPP) in the area of fractional optimization is an important problem which closely deals with the fractional objectives such as profit/cost, inventory/sale, risk/capital, debit/equity. A range of application problems in the field of engineering economics, finance risk analysis, marketing management, transportation problems, supply chain management, etc. have been studied by different researchers.

In literature, the theory of fuzzy set [1] and its various extensions have been widely utilized and applied in numerous real world applications for handling the uncertainty and fuzziness with the help of a systematic framework. The decision makers/experts are supposed to optimize two or more than two objectives simultaneously in a multi-objective inventory problems in the uncertain environment. Different methods, applications and concepts on fractional programming problems have been studied by Stancu-Minasian [2–5]. Charnes and Cooper [6] devised a methodology for solving the linear fractional objective problem along with some extra constraints by utilizing the variable transformation technique. Various researchers studied the problem of linear fractional programming problem with fuzzy parameters and coefficients in different capacities. Stancu and Pop [7] presented a modified fuzzy approach over the methodology of Dutta et al. [8] for getting more efficient solution to the problem of multi-objective linear fraction programming. The MOLFPP has also been solved using various other approaches, viz. the approach of Taylor's series [9], linear form fuzzy approach [10], weighting sum approach to obtain non-dominated solutions [11], etc. Roy and Maiti [12] were the first one who initiated a novel multi-objective inventory model with fuzzy approach and solved it with the help of geometric programming. Using fuzzy goal programming approach, Dutta and Kumar [13] studied a multi-objective linear fractional inventory model in detail. Also, in view of various lead-time distributions, Garg [14] developed the fuzzy inventory model for the deteriorating items.

Different types of possible extensions/generalizations of fuzzy set have been studied in literature. An important generalization of fuzzy set has been introduced by Atanassov [15] with the incorporation of the idea of hesitancy/indeterminacy, called intuitionistic fuzzy set (IFS), which is based on two characterizing functions, i.e., membership and non-membership function such that their sum must not exceed 1. It may be noted that all the fuzzy sets are IFSs but all IFSs are not fuzzy sets.

In the real world, it is difficult for the decision makers to achieve the finest alternative/attribute/object from the set of feasible ones, due to the increasing complications in the existing system. However, it is hard to summarize but not incredible to achieve the best single objective out of the available ones. There are large number of MOLF-PPs where the various parameters are found to be uncertain, ambiguous, imprecise, and vague. Therefore, to handle this uncertainty and impreciseness in the information, the crisp set seems to be ineffective and hence can be suitably handled by using the intuitionistic fuzzy information concept.

Payan and Noora [16] presented a linearization technique for solving MOLFPP with fuzzy parameters along with interval parameters using α-cut method and reduced to a linear programming problem with interval parameters to obtain the optimal interval values for the objective functions.

In the present work, for the sake of dealing the uncertainty and ambiguity in a more efficient way, we use the notion of a popular and widely used sets termed as 'triangular intuitionistic fuzzy numbers' with (α, β)-cuts which effectively enlarges the span of the preciseness of the information in the system than fuzzy sets or crisp sets. The triangular intuitionistic fuzzy numbers are capable to handle the uncertainty/ambiguity sufficiently close to human nature/opinion, where all the three parameters of uncertainty, that is, degree of membership, indeterminacy (hesitation) and non-membership have been properly taken into account. This approach allows the experts/decision makers to express themselves in a more flexible and realistic way for solving the multi-objective linear fractional programming problems.

The work in the present manuscript has been structured as follows. The basic definitions and fundamental concepts related to the proposed work have been provided in Sect. 2. In Sect. 3, we have briefly presented the analytic hierarchy process and variable transformation methodology for studying the decision making phenomenon in a systematic and defined way. The problem formulation for the multi-objective linear fractional programming has been done in a joint sense by utilizing the (α, β)-cuts in Sect. 4. Further, the proposed problem has been split into two linear programming problems using the variable transformation technique and Taylor's series expansion. Additionally, a numerical example has been presented for the sake of illustration and implementation of the proposed work in Sect. 5. Finally, the paper has been concluded in Sect. 6 with a scope of possible future work.

2 Preliminaries

Some basic preliminaries are briefly presented in this section which are easily available in literature.

Let $U = \{u_1, u_2, \ldots, u_m\}$ be the universe of discourse and $\mu : U \rightarrow [0, 1], \eta : U \rightarrow [0, 1]$ and $\nu : U \rightarrow [0, 1]$ denote the characterizing function for the membership degree, the neutral membership (abstain) degree and the non-membership degree, respectively.

Definition 1 [17] "A Triangular Intuitionistic Fuzzy Number $\tilde{a} = < (a_1, a_2, a_3); w_{\tilde{a}}, u_{\tilde{a}} >$ is a special intuitionistic fuzzy set on a real number set R, whose membership function and non-membership function are defined as follows:

$$\mu_{\tilde{a}}(x) = \begin{cases} \frac{(x-a_1)w_{\tilde{a}}}{(a_2-a_1)}, & \text{if } a_1 \le x \le a_2 \\ w_{\tilde{a}}, & \text{if } x = a, \\ \frac{(a_3-x)w_{\tilde{a}}}{(a_2-a_1)}, & \text{if } a_2 < x \le a_3 \\ 0, & \text{if } x < a_1 \, or \, x > a_3 \end{cases} \quad (1)$$

and

$$\nu_{\tilde{a}}(x) = \begin{cases} \frac{[a_2-x+\mu_{\tilde{a}}(x-a_1)]}{(a_2-a_1)}, & \text{if } a_1 \le x \le a_2 \\ \mu_{\tilde{a}}, & \text{if } x = a \\ \frac{[x-a_2+\mu_{\tilde{a}}(a_3-x)]}{(a_3-a_2)}, & \text{if } a_2 < x \le a_3 \\ 1, & \text{if } x < a_1 \, or \, x > a_3 \end{cases} \quad (2)$$

respectively."

The values $w_{\tilde{a}}$ and $u_{\tilde{a}}$ denote the max membership and the min non-membership, respectively, such that

$$0 \le w_{\tilde{a}} \le 1 , 0 \le u_{\tilde{a}} \le 1 \text{ and } 0 \le w_{\tilde{a}} + u_{\tilde{a}} \le 1.$$

The degree of indeterminancy/hesitancy of the element x in $\tilde{a}$ is given by

$$\pi_{\tilde{a}}(x) = 1 - \mu_{\tilde{a}}(x) - \nu_{\tilde{a}}(x).$$

Definition 2 [17] The α**-cut set** of the TIFN $\tilde{a} = < (a_1, a_2, a_3); w_{\tilde{a}}, u_{\tilde{a}} >$ is a crisp subset of $\mathbb{R}$, defined as

$$\tilde{a}_\alpha = \{x \,|\, \mu_{\tilde{a}}(x) \ge \alpha\}; \text{ where } 0 \le \alpha \le w_{\tilde{a}}.$$

Using the Eq. (1) and Definition 2, the α-cut set may be viewed as a closed interval which is represented as:

$$[\tilde{a}_\alpha^L, \tilde{a}_\alpha^R] = \left[a_1 + \frac{\alpha(a_2 - a_1)}{w_{\tilde{a}}}, a_3 - \frac{\alpha(a_3 - a_2)}{w_{\tilde{a}}}\right]. \quad (3)$$

Definition 3 [17] "The β**-cut set** of a TIFN $\tilde{a} = < (a_1, a_2, a_3); w_{\tilde{a}}, u_{\tilde{a}} >$ is a crisp subset of $\mathbb{R}$, defined as

$$\tilde{a}_\beta = \{x \,|\, \upsilon_{\tilde{a}}(x) \le \beta\} \text{ where } \mu_{\tilde{a}} \le \beta \le 1." \quad (4)$$

Using the Eq. (2) and Definition 3, the β-cut set may similarly be viewed as a closed interval as follows:

$$[\tilde{a}_\beta^L, \tilde{a}_\beta^R] = \left[\frac{(1-\beta)a_2 + (\beta - \mu_{\tilde{a}})a_1}{(1-\mu_{\tilde{a}})}, \frac{(1-\beta)a_2 + (\beta - \mu_{\tilde{a}})a_3}{(1-\mu_{\tilde{a}})}\right]. \quad (5)$$

Definition 4 [17] "The (α, β)**-cut set** of the TIFN $\tilde{a} = < (a_1, a_2, a_3); w_{\tilde{a}}, u_{\tilde{a}} >$ is a crisp subset of $\mathbb{R}$, defined as

$$\tilde{a}_{\alpha,\beta} = \{x | \mu_{\tilde{a}}(x) \geq \alpha, \nu_{\tilde{a}}(x) \leq \beta\};$$

where $0 \leq \alpha \leq w_{\tilde{a}}, u_{\tilde{a}} \leq \beta \leq 1, 0 \leq \alpha + \beta \leq 1$."

Next, we present some fundamental concepts and definitions related to the interval arithmetic and optimization. As we have equivalently expressed the TIFNs as closed intervals using α-cuts and β-cuts, therefore, the arithmetic operations on the TIFNs can be represented using the interval analysis as follows:

- $\tilde{a}_\alpha + \tilde{b}_\alpha = [\tilde{a}^L{}_\alpha + \tilde{b}^L{}_\alpha, \tilde{a}^R{}_\alpha + \tilde{b}^R{}_\alpha]$ and $\tilde{a}_\beta + \tilde{b}_\beta = [\tilde{a}^L{}_\beta + \tilde{b}^L{}_\beta, \tilde{a}^R{}_\beta + \tilde{b}^R{}_\beta]$.
- $p.\tilde{a}_\alpha = \begin{Bmatrix} \left[p\tilde{a}_\alpha^L, p\tilde{a}_\alpha^R\right], \text{ if } p > 0 \\ \left[p\tilde{a}_\alpha^R, p\tilde{a}_\alpha^L\right], \text{ if } p < 0 \end{Bmatrix}$ for $p \in \mathbb{R}$.
- $q.\tilde{a}_\beta = \begin{Bmatrix} \left[q\tilde{a}_\beta^L q\tilde{a}_\beta^R\right], \text{ if } q > 0 \\ \left[q\tilde{a}_\beta^R, q\tilde{a}_\beta^L\right], \text{ if } q < 0 \end{Bmatrix}$ for $q \in \mathbb{R}$.
- $\tilde{a}_\alpha \tilde{b}_\beta = [\min S_1, \max S_1]$, where $S_1 = \left\{a_\alpha^L b_\alpha^L, a_\alpha^L b_\alpha^R, a_\alpha^R b_\alpha^L, a_\alpha^R b_\alpha^R\right\}$.
- $\tilde{a}_\beta \tilde{b}_\beta = [\min S_2, \max S_2]$, where $S_2 = \left\{a_\beta^L b_\beta^L, a_\beta^L b_\beta^R, a_\beta^R b_\beta^L, a_\beta^R b_\beta^R\right\}$.
- $\frac{\tilde{a}_\alpha}{\tilde{b}_\alpha} = [\min S_3, \max S_3]$, where $S_3 = \left\{\frac{a_\alpha^L}{b_\alpha^L}, \frac{a_\alpha^L}{b_{\alpha,}^R}, \frac{a_\alpha^R}{b_{\alpha,}^L}, \frac{a_\alpha^R}{b_{\alpha,}^R}\right\}$.
- $\frac{\tilde{a}_\beta}{\tilde{b}_\beta} = [\min S_4, \max S_4]$, where $S_4 = \left\{\frac{a_\beta^L}{b_\beta^L}, \frac{a_\beta^L}{b_{\beta,}^R}, \frac{a_\beta^R}{b_{\beta,}^L}, \frac{a_\beta^R}{b_{\beta,}^R}\right\}$.
- $\tilde{a}_\alpha \leq \tilde{b}_\alpha$ if $a_\alpha^L \leq b_\alpha^L$ and $a_\alpha^R \leq b_\alpha^R$.
- $\tilde{a}_\beta \leq \tilde{b}_\beta$ If $a_\beta^L \leq b_\beta^L$ and $a_\beta^R \leq b_\beta^R$.

3 Methodology for MOLFPP with AHP and VTM

In this section, we present some basic decision making approaches in developing the proposed problem formulation which are popularly known as Analytic Hierarchy Process and Variable Transformation Method.

Analytic Hierarchy Process (AHP):

The analytic hierarchy process is a well designed methodology proposed by Thomas L. Saaty in the 1970s s for managing the real world multi-criteria decision making (MCDM) problem with a feasible complexity. The prime focus of AHP is to quantify the most appropriate weights for the objectives of the MCDM problems [11, 18]. On the basis of the experiences of the decision makers, the importance of the factors (decision affecting factors) are estimated in the form of pairwise comparison matrix. In the hierarchy process, we fix up the priorities among the elements with the help of sequence of judgments on the basis of pairwise comparisons of the elements, e.g., while comparing during purchases of some real estate property, the investors may have their choice that they prefer the location over the budget and the budget over the timing. Thus, the weights are determined by the experts by keeping the relative importance between the two items for a n-objective optimization problem by implementing the following procedure:

- **Step 1**. First of all we find the pairwise comparison matrix $C = [c_{ij}]_{n\times n}$ where i, j are integers and $1 \leq i, j \leq n$.
- **Step 2**. Next, we construct the normalized pairwise comparison matrix
$$N = [\eta_{ij}]_{n\times n} \text{ where } \eta_{ij} = \frac{c_{ij}}{\sum_{i=1}^{n} c_{ij}};$$
where i, j are integers and $1 \leq i, j \leq n$.
- **Step 3**. The desired weights are evaluated by
$$w_i = \frac{\sum_{j=1}^{n} \eta_{ij}}{n};$$
where i, j are integers and $1 \leq i, j \leq n$.
- **Step 4**. Next, we examine the consistency of the approach [18] by finding ratio called 'consistency ratio (CR)' defined by $CR = \frac{CI}{RI}$; where CI represents the consistency index and RI represents the random index.
- **Step 5**. If the value of $CR \leq 0.1$, then the ongoing process is consistent for achieving the final decision, otherwise we restructure the matrix C.

Variable Transformation Method (VTM):
The motivation behind developing the variable transformation method [19] is to determine the optimal solution of single objective linear fractional programming problem by taking an additional variable. In this method, we first take LFPP-$P1$ and then LPP-$P2$ into the account.

Consider the following LFPP:

$$\begin{array}{lrcl} P1 : \max & f(x) & = & \frac{cx+q}{dx+r} \\ \text{subject to} & S_{P1} & = & \{Ax \leq b, x \geq 0\} \end{array} \tag{6}$$

Next, by using the transformation $z = \frac{1}{dx+r}$, and $y = xz$ in the Eq. (6), $P1$ can be transformed to the LPP-$P2$ as

$$\max\ g(y, z) = cy + qz\ ; \tag{7}$$

$$S_{P2} = \{dy + rz = 1, Ay - bz \leq 0, y \geq 0, z > 0\}.$$

The following are the important remarks based on the fundamental knowledge of fractional programming [2]:

- **Remark 1**: In case of LPP-$P2$, for the feasible solution $(y, z) \in S_{P2}$, the variable $z > 0$.
- **Remark 2**: Also, for (LPP-$P2$), if (y^*, z^*) is the desired optimal solution then $x^* = \frac{y^*}{z^*}$ is the desired optimal solution of LFPP-$P1$.

4 Problem Formulation and Methodology for MOLFPP

In reference to the real world decision making problems, the experts are not always precise in providing the necessary data and the system parameters because of the incompleteness and vagueness inherited in the information. It may also be noted that the probabilistic oriented approach is also not fully effective to handle such uncertainty. In this section, we formulate a new model of multi-objective linear fractional programming problem by utilizing the notion of triangular intuitionistic fuzzy numbers for the system parameters. The mathematical formulation can be given as follows:

$$\begin{aligned} \max \;\; f(x) &= (f_1(x), f_2(x), \ldots, f_k(x)) \\ \text{subject to} & \\ S &= \{x \in \mathbb{R}^n | Ax \leq b, x \geq 0\}. \end{aligned} \tag{8}$$

where

$$f_i(x) = \frac{\tilde{c}_i x + \tilde{q}_i}{\tilde{d}_i x + \tilde{r}_i}; \; i = 1, 2, \ldots, k;$$

$$A = (\tilde{a}_{tj})_{m \times n} \text{ and } b = (\tilde{b}_t)_{m \times 1}, \text{ where } t \text{ is an integer } 1 \leq t \leq m.$$

Suppose that $\tilde{c}_i, \tilde{d}_i, \tilde{q}_i, \tilde{r}_i, \tilde{a}_{tj}, \tilde{b}_t$ are positive triangular intuitionistic fuzzy numbers defined on $\mathbb{R}$ as

$$\tilde{c}_i = (c_i^1, c_i^2, c_i^3); \quad \tilde{q}_i = (q_i^1, q_i^2, q_i^3); \quad \tilde{a}_{tj} = (a_{tj}^1, a_{tj}^2, a_{tj}^3);$$

$$\tilde{d}_i = (d_i^1, d_i^2, d_i^3); \quad \tilde{r}_i = (r_i^1, r_i^2, r_i^3); \quad \tilde{b}_t = (b_t^1, b_t^2, b_t^3).$$

The fuzzified parameters of the objective functions and constraints of MOLFPP can be converted into closed intervals using α-cut and β-cut of TIFN respectively. The outline of the proposed methodology has been presented with the help of the following flow chart (Fig. 1).

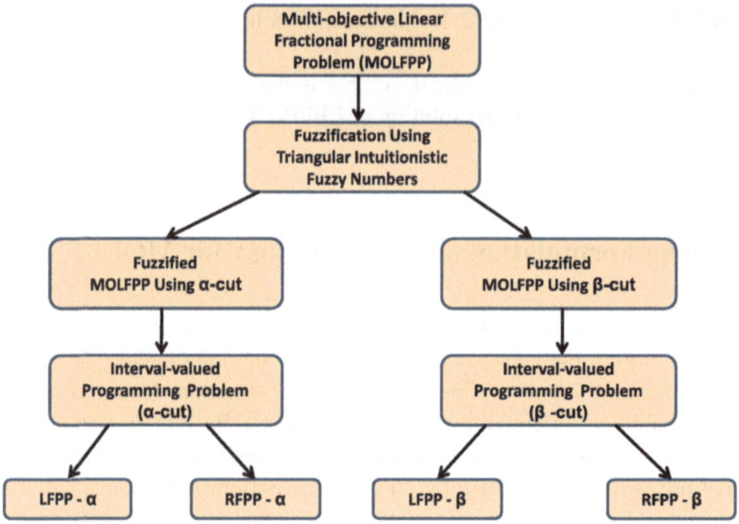

Fig. 1 Flow chart of methodology framework

Objective Function: In terms of α-cut

$$f_{i\,,\,\alpha}(x) = \frac{\left[c_i^1 + \frac{\alpha(c_i^2 - c_i^1)}{w_{\tilde{a}}}\,,\ c_i^3 - \frac{\alpha(c_i^3 - c_i^2)}{w_{\tilde{a}}}\right]x + \left[q_i^1 + \frac{\alpha(q_i^2 - q_i^1)}{w_{\tilde{a}}}\,,\ q_i^3 - \frac{\alpha(q_i^3 - q_i^2)}{w_{\tilde{a}}}\right]}{\left[d_i^1 + \frac{\alpha(d_i^2 - d_i^1)}{w_{\tilde{a}}}\,,\ d_i^3 - \frac{\alpha(d_i^3 - d_i^2)}{w_{\tilde{a}}}\right]x + \left[r_i^1 + \frac{\alpha(r_i^2 - r_i^1)}{w_{\tilde{a}}}\,,\ r_i^3 - \frac{\alpha(r_i^3 - r_i^2)}{w_{\tilde{a}}}\right]}. \tag{9}$$

Using the basic arithmetic operations of interval analysis, we can rewrite the above expression (9) as follows:

$$f_{i,\alpha}(x) = \frac{\left[\left(c_i^1 + \frac{\alpha(c_i^2 - c_i^1)}{w_{\tilde{a}}}\right)x + q_i^1 + \frac{\alpha(q_i^2 - q_i^1)}{w_{\tilde{a}}},\ \left(c_i^3 - \frac{\alpha(c_i^3 - c_i^2)}{w_{\tilde{a}}}\right)x + q_i^3 - \frac{\alpha(q_i^3 - q_i^2)}{w_{\tilde{a}}}\right]}{\left[\left(d_i^1 + \frac{\alpha(d_i^2 - d_i^1)}{w_{\tilde{a}}}\right)x + r_i^1 + \frac{\alpha(r_i^2 - r_i^1)}{w_{\tilde{a}}},\ \left(d_i^3 - \frac{\alpha(d_i^3 - d_i^2)}{w_{\tilde{a}}}\right)x + r_i^3 - \frac{\alpha(r_i^3 - r_i^2)}{w_{\tilde{a}}}\right]};$$

$$\Rightarrow f_{i,\alpha}(x) = \left[\frac{\left(c_i^1 + \frac{\alpha(c_i^2 - c_i^1)}{w_{\tilde{a}}}\right)x + q_i^1 + \frac{\alpha(q_i^2 - q_i^1)}{w_{\tilde{a}}}}{\left(d_i^3 - \frac{\alpha(d_i^3 - d_i^2)}{w_{\tilde{a}}}\right)x + r_i^3 - \frac{\alpha(r_i^3 - r_i^2)}{w_{\tilde{a}}}},\ \frac{\left(c_i^3 - \frac{\alpha(c_i^3 - c_i^2)}{w_{\tilde{a}}}\right)x + q_i^3 - \frac{\alpha(q_i^3 - q_i^2)}{w_{\tilde{a}}}}{\left(d_i^1 + \frac{\alpha(d_i^2 - d_i^1)}{w_{\tilde{a}}}\right)x + r_i^1 + \frac{\alpha(r_i^2 - r_i^1)}{w_{\tilde{a}}}}\right];$$

$$\Rightarrow f_{i,\alpha}(x) = \left[{f^L}_{i,\alpha}(x), {f^R}_{i,\alpha}(x)\right], i = 1, 2, \ldots, k.$$

In terms of β-cut, we have,

$$f_{i,\beta}(x) = \frac{\left[\frac{(1-\beta)c_i^2+(\beta-u_{\tilde{a}})c_i^1}{(1-u_{\tilde{a}})}, \frac{(1-\beta)c_i^2+(\beta-u_{\tilde{a}})c_i^3}{(1-u_{\tilde{a}})}\right]x + \left[\frac{(1-\beta)q_i^2+(\beta-u_{\tilde{a}})q_i^1}{(1-u_{\tilde{a}})}, \frac{(1-\beta)q_i^2+(\beta-u_{\tilde{a}})q_i^3}{(1-u_{\tilde{a}})}\right]}{\left[\frac{(1-\beta)d_i^2+(\beta-u_{\tilde{a}})d_i^1}{(1-u_{\tilde{a}})}, \frac{(1-\beta)d_i^2+(\beta-u_{\tilde{a}})d_i^3}{(1-u_{\tilde{a}})}\right]x + \left[\frac{(1-\beta)r_i^2+(\beta-u_{\tilde{a}})r_i^1}{(1-u_{\tilde{a}})}, \frac{(1-\beta)r_i^2+(\beta-u_{\tilde{a}})r_i^3}{(1-u_{\tilde{a}})}\right]};$$

$$\Rightarrow f_{i,\beta}(x) = \frac{\left[\left(\frac{(1-\beta)c_i^2+(\beta-u_{\tilde{a}})c_i^1}{(1-u_{\tilde{a}})}\right)x + \frac{(1-\beta)q_i^2+(\beta-u_{\tilde{a}})q_i^1}{(1-u_{\tilde{a}})}, \left(\frac{(1-\beta)c_i^2+(\beta-u_{\tilde{a}})c_i^3}{(1-u_{\tilde{a}})}\right)x + \frac{(1-\beta)q_i^2+(\beta-u_{\tilde{a}})q_i^3}{(1-u_{\tilde{a}})}\right]}{\left[\left(\frac{(1-\beta)d_i^2+(\beta-u_{\tilde{a}})d_i^1}{(1-u_{\tilde{a}})}\right)x + \frac{(1-\beta)r_i^2+(\beta-u_{\tilde{a}})r_i^1}{(1-u_{\tilde{a}})}, \left(\frac{(1-\beta)d_i^2+(\beta-u_{\tilde{a}})d_i^3}{(1-u_{\tilde{a}})}\right)x + \frac{(1-\beta)r_i^2+(\beta-u_{\tilde{a}})r_i^3}{(1-u_{\tilde{a}})}\right]}; \tag{10}$$

$$\Rightarrow f_{i,\beta}(x) = \frac{\left[\left(\frac{(1-\beta)c_i^2+(\beta-u_{\tilde{a}})c_i^1}{(1-u_{\tilde{a}})}\right)x + \frac{(1-\beta)q_i^2+(\beta-u_{\tilde{a}})q_i^1}{(1-u_{\tilde{a}})}, \left(\frac{(1-\beta)c_i^2+(\beta-u_{\tilde{a}})c_i^3}{(1-u_{\tilde{a}})}\right)x + \frac{(1-\beta)q_i^2+(\beta-u_{\tilde{a}})q_i^3}{(1-u_{\tilde{a}})}\right]}{\left[\left(\frac{(1-\beta)d_i^2+(\beta-u_{\tilde{a}})d_i^3}{(1-u_{\tilde{a}})}\right)x + \frac{(1-\beta)r_i^2+(\beta-u_{\tilde{a}})r_i^3}{(1-u_{\tilde{a}})}, \left(\frac{(1-\beta)d_i^2+(\beta-u_{\tilde{a}})d_i^1}{(1-u_{\tilde{a}})}\right)x + \frac{(1-\beta)r_i^2+(\beta-u_{\tilde{a}})r_i^1}{(1-u_{\tilde{a}})}\right]}; \tag{11}$$

$$f_{i,\beta}(x) = \left[f^{L}{}_{i,\beta}(x), f^{R}{}_{i,\beta}(x)\right], i = 1, 2, \ldots, k.$$

Calculations for the Constraints:

Using α-cut, the constraints would be transformed as,

$$\sum_{j=1}^{n} \tilde{a}_{tj}x_j \leq \tilde{b}_{t,\alpha}, \text{ where } t \text{ is an integer } 1 \leq t \leq m;$$

$$\sum_{j=1}^{n}\left[a^1{}_{t,j} + \frac{\alpha\left(a_{tj}^2 - a_{tj}^1\right)}{w_{\tilde{a}}}, a_{tj}^3 - \frac{\alpha(a_{tj}^3 - a_{tj}^2)}{w_{\tilde{a}}}\right]x_j$$

$$\leq \left[b^1{}_t + \frac{\alpha\left(b_t^2 - b_t^1\right)}{w_{\tilde{a}}}, b_t^3 - \frac{\alpha(b_t^3 - b_t^2)}{w_{\tilde{a}}}\right];$$

$$\sum_{j=1}^{n}\left[a_{tj,\alpha}^L, a_{tj,\alpha}^R\right]x_j \leq \left[b_{t,\alpha}^L, b_{t,\alpha}^R\right], \text{ where } t \text{ is an integer } 1 \leq t \leq m.$$

Similarly, using β-cut, we have,

$$\sum_{j=1}^{n} \tilde{a}_{tj,\beta} \leq \tilde{b}_{t,\beta}, \text{ where } t \text{ is an integer } 1 \leq t \leq m;$$

$$\sum_{j=1}^{n} \left[\frac{(1-\beta)\, a_{tj}^2 + (\beta - u_{\tilde{a}})\, a_{tj}^1}{(1-u_{\tilde{a}})}, \frac{(1-\beta)\, a_{tj}^2 + (\beta - u_{\tilde{a}})\, a_{tj}^3}{(1-u_{\tilde{a}})} \right] x_j$$

$$\leq \left[\frac{(1-\beta)\, b^2{}_t + (\beta - u_{\tilde{a}})\, b^1{}_t}{(1-u_{\tilde{a}})} \right];$$

$$\sum_{j=1}^{n} \left[a_{tj,\beta}^L, a_{tj,\beta}^R \right] x_j \leq \left[b_{t,\beta}^L, b_{t,\beta}^R \right] \left[\frac{(1-\beta)\, b_t^2 + (\beta - u_{\tilde{a}})\, b_t^3}{(1-u_{\tilde{a}})} \right],$$

where t is an integer $1 \leq t \leq m$.

The MOLFPP can be expressed in terms of α-cut and β-cut as the following scheme:

$$\max f_{i,\alpha}(x) = [f_{i,\alpha}^L(x), f_{i,\alpha}^R(x)],\ i = 1, 2, \ldots, k$$

$$\text{subject to} \tag{12}$$

$$\sum_{j=1}^{n} a_{ij,\alpha}^L x_j \leq b_{t,\alpha}^L, \text{ where } t \text{ is an integer } 1 \leq t \leq m \tag{13}$$

$$\sum_{j=1}^{n} a_{ij,\alpha}^R x_j \leq b_{t,\alpha}^R, \text{ where } t \text{ is an integer } 1 \leq t \leq m$$

$$x \geq 0$$

and

$$\max f_{i,\beta}(x) = [f_{i,\beta}^L(x), f_{i,\beta}^R(x)],\ i = 1, 2, \ldots, k$$

$$\text{subject to} \tag{14}$$

$$\sum_{j=1}^{n} a_{ij,\beta}^L x_j \leq b_{t,\beta}^L, \text{ where } t \text{ is an integer } 1 \leq t \leq m \tag{15}$$

$$\sum_{j=1}^{n} a_{ij,\beta}^R x_j \leq b_{t,\beta}^R, \text{ where } t \text{ is an integer } 1 \leq t \leq m.$$

The fractional objective functions of Eqs. (12) and (14) can be linearly approximated utilizing the Taylor's series expansion of the objective functions about their individual desired solution. The mathematical expression for the lower and upper bands of the ith objective in terms of α and β-cut are as follows:

$$\max \left(c_i^1 + \frac{\alpha \left(c_i^2 - c_i^1 \right)}{w_{\tilde{a}}} \right) y + \left(q_i^1 + \frac{\alpha \left(q_i^2 - q_i^1 \right)}{w_{\tilde{a}}} \right) z$$

subject to (16)

$$\left(d_i^3 - \frac{\alpha \left(d_i^3 - d_i^2 \right)}{w_{\tilde{a}}} \right) y + \left(r_i^3 - \frac{\alpha \left(r_i^3 - r_i^2 \right)}{w_{\tilde{a}}} \right) z = 1,$$

$$\sum_{j=1}^{n} \left(a_{tj}^1 + \frac{\alpha \left(a_{tj}^2 - a_{tj}^1 \right)}{w_{\tilde{a}}} \right) y_i$$

$$\leq \left(b_t^1 + \frac{\alpha \left(b_t^2 - b_t^1 \right)}{w_{\tilde{a}}} \right) z, \text{ where } t \text{ is an integer } 1 \leq t \leq m \quad (17)$$

$$\sum_{j=1}^{n} \left(a_{tj}^1 + \frac{\alpha \left(a_{tj}^2 - a_{tj}^1 \right)}{w_{\tilde{a}}} \right) y_i$$

$$\leq \left(b_t^3 + \frac{\alpha \left(b_t^3 - b_t^2 \right)}{w_{\tilde{a}}} \right) z, \text{ where } t \text{ is an integer } 1 \leq t \leq m$$

$y \geq 0,\ z \geq 0;$

where $y_j = x_j z,\ j = 1, 2, \ldots, n$. This is for lower bounds by using α-cut.

Similarly for lower band using β-cut, we have;

$$\max \left(\frac{(1-\beta)c^2 + (\beta - u_{\tilde{a}})c_i^1}{(1 - u_{\tilde{a}})} \right) y + \left(\frac{(1-\beta)q_i^2 + (\beta - u_{\tilde{a}})q_i^1}{(1 - u_{\tilde{a}})} \right) z$$

subject to (18)

$$\left(\frac{(1-\beta)d_i^2 + (\beta - u_{\tilde{a}})d_i^3}{(1 - u_{\tilde{a}})} \right) y + \left(\frac{(1-\beta)r_i^2 + (\beta - u_{\tilde{a}})r_i^3}{(1 - u_{\tilde{a}})} \right) z = 1$$

$$\sum_{j=1}^{n} \left(\frac{(1-\beta)\, a_{tj}^2 + (\beta - u_{\tilde{a}})\, a_{tj}^1}{(1 - u_{\tilde{a}})} \right) y_j$$

$$\leq \left(\frac{(1-\beta)\, b^2{}_t + (\beta - u_{\tilde{a}})\, b^1{}_t}{(1 - u_{\tilde{a}})} \right) z, t =, 2, \ldots, m \quad (19)$$

$$\sum_{j=1}^{n} \left(\frac{(1-\beta)\, a_{tj}^2 + (\beta - u_{\tilde{a}})\, a_{tj}^3}{(1 - u_{\tilde{a}})} \right) y_j$$

$$\leq \left(\frac{(1-\beta)\, r^2{}_t + (\beta - u_{\tilde{a}})\, r^3{}_t}{(1 - u_{\tilde{a}})} \right) z,\ \ t =, 2, \ldots, m y_j \geq 0,\ \ z > 0$$

where $y_j = x_j z$, $j = 1, 2, \ldots, n$. Solving Eqs. (16) and (18), the individual optimal solutions (IOS) of $f^L{}_{i,\beta}(x)$ can be obtained as

$$x = \left(x_j \frac{y_j}{z}, j = 1, 2, \ldots, n\right).$$

Let us consider

$$\begin{aligned}
X_{i,\alpha}^{L*} &= \left(X_{i1,\alpha}^{L*}, X_{i2,\alpha}^{L*}, \ldots, X_{in,\alpha}^{L*}\right)\\
X_{i,\alpha}^{R*} &= \left(X_{i1,\alpha}^{R*}, X_{i2,\alpha}^{R*}, \ldots, X_{in,\alpha}^{R*}\right)\\
X_{i,\beta}^{L*} &= \left(X_{i1,\beta}^{L*}, X_{i2,\beta}^{L*}, \ldots, X_{in,\beta}^{L*}\right)\\
X_{i,\beta}^{R*} &= \left(X_{i1,\beta}^{R*}, X_{i2,\beta}^{R*}, \ldots, X_{in,\beta}^{R*}\right), i = 1, 2, \ldots, k
\end{aligned}$$

as the individual optimal solutions (IOS) of ${f^L}_{i,\alpha}(x), {f^R}_{i,\alpha}(x)$ *and* ${f^L}_{i,\beta}(x)$, ${f^R}_{i,\beta}(x)$ respectively.

Using first order Taylor's expansion, we have

$$\begin{aligned}
{f^L}_{i,\alpha}(x) &\cong {\bar{f}^L}_{i,\alpha}(x) = {f^L}_{i,\alpha}(X_{i,\alpha}^{L*}) + \textstyle\sum_{k=1}^{n}\left(x_k - X_{iK,\alpha}^{L*}\right)\frac{\partial {f^L}_{i,\alpha}(X_{i,\alpha}^{L*})}{\partial x_k}, i = 1, 2, \ldots, n\\
{f^R}_{i,\alpha}(x) &\cong {\bar{f}^R}_{i,\alpha}(x) = {f^R}_{i,\alpha}(X_{i,\alpha}^{R*}) + \textstyle\sum_{k=1}^{n}\left(x_k - X_{iK,\alpha}^{R*}\right)\frac{\partial {f^R}_{i,\alpha}(X_{i,\alpha}^{R*})}{\partial x_k}, i = 1, 2, \ldots, n\\
{f^L}_{i,\beta}(x) &\cong {\bar{f}^L}_{i,\beta}(x) = {f^L}_{i,\beta}(X_{i,\beta}^{L*}) + \textstyle\sum_{k=1}^{n}\left(x_k - X_{iK,\beta}^{L*}\right)\frac{\partial {f^L}_{i,\beta}(X_{i,\beta}^{L*})}{\partial x_k}, i = 1, 2, \ldots, n\\
{f^R}_{i,\beta}(x) &\cong {\bar{f}^R}_{i,\beta}(x) = {f^R}_{i,\beta}(X_{i,\beta}^{R*}) + \textstyle\sum_{k=1}^{n}\left(x_k - X_{iK,\beta}^{R*}\right)\frac{\partial {f^R}_{i,\beta}(X_{i,\beta}^{R*})}{\partial x_k}, i = 1, 2, \ldots, n
\end{aligned}$$

Using weighted sum method, the MOLFPP can be expressed as

$$\begin{aligned}
&\max f(x) = \left[\sum_{i=1}^{n} w_i f_{i,\alpha}^{L}(x), \sum_{i=1}^{n} w_i f_{i,\alpha}^{R}(x)\right]\\
&\text{subject to} \qquad (20)\\
&\sum_{j=1}^{n} a_{tj,\alpha}^{L} x_j \le b_{t,\alpha}^{L}, \text{ where } t \text{ is an integer } 1 \le t \le m\\
&\sum_{j=1}^{n} a_{tj,\alpha}^{R} x_j \le b_{t,\alpha}^{R}, \text{ where } t \text{ is an integer } 1 \le t \le m \qquad (21)\\
&\qquad x \ge 0
\end{aligned}$$

and

$$\begin{aligned}
&\max f(x) = \left[\sum_{i=1}^{n} w_i f_{i,\beta}^{L}(x), \sum_{i=1}^{n} w_i f_{i,\beta}^{R}(x)\right]\\
&\text{s.t} \sum_{j=1}^{n} a_{tj,\beta}^{L} x_j \le b_{t,\beta}^{L}, \text{ where } t \text{ is an integer } 1 \le t \le m \qquad (22)
\end{aligned}$$

$$\sum_{j=1}^{n} a^R_{tj,\beta} x_j \le b^R_{t,\beta}, \text{ where } t \text{ is an integer } 1 \le t \le m$$

$$x \ge 0 \tag{23}$$

in terms of α-cut and β-cut, respectively, where w_i are weight assigned to the objectives $f_i(x)$ as per the decision maker's preference.

Now these combined MOLFPP can be stated in terms of left and right fractional programming problems as follows:

(LFPP-α)

$$\begin{aligned}
&\max f^L_\alpha(x) = w_i \bar{f}^L_{i,\alpha}(x) \\
&\text{s.t} \sum_{j=1}^{n} a^L_{tj,\alpha} x_j \le b^L_{t,\alpha}, \text{ where } t \text{ is an integer } 1 \le t \le m \\
&\sum_{j=1}^{n} a^R_{tj,\alpha} x_j \le b^R_{t,\alpha}, \text{ where } t \text{ is an integer } 1 \le t \le m \\
&x \ge 0
\end{aligned} \tag{24}$$

(RFPP-α)

$$\begin{aligned}
&\max f^R_\alpha(x) = w_i \bar{f}^R_{i,\alpha}(x) \\
&\text{s.t} \sum_{j=1}^{n} a^L_{tj,\alpha} x_j \le b^L_{t,\alpha}, \text{ where } t \text{ is an integer } 1 \le t \le m \\
&\sum_{j=1}^{n} a^R_{tj,\alpha} x_j \le b^R_{t,\alpha}, \text{ where } t \text{ is an integer } 1 \le t \le m \\
&x \ge 0
\end{aligned} \tag{25}$$

(LFPP-β)

$$\begin{aligned}
&\max f^L_\beta(x) = \sum_{i=1}^{n} w_i \bar{f}^L_{i,\beta}(x) \\
&\text{s.t} \sum_{j=1}^{n} a^L_{tj,\beta} x_j \le b^L_{t,\beta}, \text{ where } t \text{ is an integer } 1 \le t \le m \\
&\sum_{j=1}^{n} a^R_{tj,\beta} x_j \le b^R_{t,\beta}, \text{ where } t \text{ is an integer } 1 \le t \le m \\
&x \ge 0
\end{aligned} \tag{26}$$

(RFPP-β)

$$
\begin{aligned}
&\max f_{\beta}^{R}(x) = \sum_{i=1}^{n} w_i \bar{f}_{i,\beta}^{R}(x) \\
&\text{s.t} \sum_{j=1}^{n} a_{tj,\beta}^{L} x_j \le b_{t,\beta}^{L}, \text{ where } t \text{ is an integer } 1 \le t \le m \\
&\sum_{j=1}^{n} a_{tj,\beta}^{R} x_j \le b_{t,\beta}^{R}, \text{ where } t \text{ is an integer } 1 \le t \le m \\
&x \ge 0
\end{aligned}
\tag{27}
$$

Suppose $x_{\alpha}^{L*}, x_{\alpha}^{R*}, x_{\beta}^{L*}$, *and* x_{β}^{R*} be the solution of Eqs. (24), (25), (26) and (27) respectively. Also, x_{α}^{R*} and x_{β}^{R*} generates the acceptable ranges of objective function values as it provides large interval compared to x_{α}^{L*} and x_{β}^{L*}.

5 Application in a Manufacturing Company

In this section, we present an application example of a manufacturing company for the sake of proper illustration of the proposed methodology. In order to have the multi-objectivity into account, we assume that a manufacturing company manufactures three types of products *P*, *Q* and *R* in two different versions. Suppose that the first version of three products brings the profits of USD 9, USD 13, and USD 7 per unit, respectively, and the second version of product brings the profits of USD 15, USD 12, and USD 8, respectively. However, the costs of production for each unit of product's first version are USD 3, USD 9, and USD 5, respectively and the costs of production for each unit of product's second version are USD 12, USD 9, and USD 7, respectively. Let us suppose that a specified cost of USD 2 and USD 5 are augmented to the cost function because of the projected duration in the process of production. This gives rise to the following objective function:

$$
\max \left\{ \frac{9x_1 + 13x_2 + 7x_3}{3x_1 + 9x_2 + 5x_3 + 2}, \frac{15x_1 + 12x_2 + 8x_3}{12x_1 + 9x_2 + 7x_3 + 5} \right\};
$$

subject to

$$
\begin{aligned}
2x_1 + 3x_2 + x_3 &\le 27, \\
2x_1 + x_2 + x_3 &\le 28, \\
x_1, x_2, x_3 &\ge 0.
\end{aligned}
$$

In addition to this, assume that the raw material required to produce both the versions of the products P, Q and R costs 2 units, 3 units and 1 unit per pound respectively. Also, it is being assumed that the availability of the raw material is limited to 27 pounds. The number of man-hours needed to produce each unit of product P are 2 h, 1 h for product Q and 1 h per unit of product R, provided that the total given man-hours per day are 28 h. We are supposed to find the quantity of product P, Q and R to be produced for the sake of maximizing the total profit. The parameters such as the profit, cost, raw materials (due to wastage) and limited man-hours are kind of inexact numbers having triangular possibility distributions in the inexact information. As a hypothetical prospects, suppose that the profits from product P are (7, 9, 11) dollars, (12, 13, 15) from Q and (6, 7, 9) from R and the cost for each unit of products P, Q and R are (1, 3, 5), (8, 9, 10), and (3, 5, 7), respectively with fixed cost (1, 2, 3). On the similar lines, we assume that rest of the parameters and variables are triangular fuzzy numbers with α-cut and β-cut.

Hence, the problem under consideration as stated above can be expressed as the following MOLFPP:
max

$$f_1(x) = \frac{(7,9,11;0.5,0.3)x_1 + (12,13,15;0.6,0.2)x_2 + (6,7,9;0.3,0.4)x_3}{(1,3,5;0.25,0.6)x_1 + (8,9,10;0.8,0.1)x_2 + (3,5,7;0.4,0.4)x_3 + (1,2,3;0.75,0.25)}$$

$$f_2(x) = \frac{(14,15,16;0.55,0.3)x_1 + (10,12,14;0.4,0.55)x_2 + (7,8,9;0.6,0.3)x_3}{(11,12,13;0.6,0.2)x_1 + (7,9,11;0.4,0.4)x_2 + (6,7,8;0.45,0.4)x_3 + (3,5,7;0.6,0.3)}$$

s.t.

$$(1,2,3;0.4,0.5)x_1 + (2,3,5;0.25,0.3)x_2 + (1,1,4;0.5,0.4)x_3 \le (25,28,30;0.6,0.3)$$
$$x = (x_1, x_2, x_3) \ge 0$$

By mentioning the degree of satisfaction as $\alpha = \beta = 0.5$, the statement of the problem can be formulated as
max

$$f_{1,\alpha}(x) = \left[\frac{9x_1 + 12.83x_2 + 7.67x_3}{5x_1 + 8.63x_2 + 5.5x_3 + 1.67}, \frac{9x_1 + 13.33x_2 + 5.67x_3}{x_1 + 9.38x_2 + 4.5x_3 + 2.33}\right]$$

$$f_{2,\alpha}(x) = \left[\frac{14.91x_1 + 12.50x_2 + 7.83x_3}{11.83x_1 + 9.50x_2 + 7.11x_3 + 4.67}, \frac{15.09x_1 + 11.50x_2 + 8.17x_3}{12.17x_1 + 8.50x_2 + 6.89x_3 + 5.33}\right]$$

s.t.

$$2.55x_1 + 4x_2 + x_3 \le 27.50$$
$$1.75x_1 + x_2 + x_3 \le 28.33$$
$$x = (x_1, x_2, x_3) \ge 0$$

and
max

$$f_{1,\beta}(x) = \left[\frac{8.43x_1 + 14.50x_2 + 7.25x_3}{2.67x_1 + 13x_2 + 5.5x_3 + 3}, \frac{9.57x_1 + 10x_2 + 6.5x_3}{3.33x_1 + 5x_2 + 4.5x_3 + 1}\right]$$

$$f_{2,\beta}(x) = \left[\frac{15.67x_1 + 11.82x_2 + 8.67x_3}{13.50x_1 + 9.50x_2 + 7.25x_3 + 6.33}, \frac{14.33x_1 + 12.18x_2 + 7.33x_3}{10.50x_1 + 8.50x_2 + 6.75x_3 + 3.67}\right]$$

s.t.

$$2x_1 + 3.67x_2 + x_3 \leq 30$$
$$2x_1 + 1.67x_2 + 0.25x_3 \leq 26.67$$
$$x = (x_1, x_2, x_3) \geq 0$$

Next, with the help of VTM and Taylor series, the lower/upper bounds of the interval-valued fractional objectives can be estimated as:

$$f_{1,\alpha}^L(x) = -0.00726x_1 - 0.05503x_2 - 0.04303x_3 + 1.8427120525$$
$$f_{1,\alpha}^R(x) = 0.059181x_1 + 0.14856x_2 + 0.000914x_3 + 1.248757676$$
$$f_{2,\alpha}^L(x) = 0.002572x_1 + 0.004879x_2 - 0.0045x_3 + 1.18923016875$$
$$f_{2,\alpha}^R(x) = 0.003054x_1 + 0.00703x_2 - 0.0037x_3 + 1.1583577495$$
$$f_{1,\beta}^L(x) = 0.513248x_1 + 4.264326x_2 + 1.72996x_3 + 7.2543743$$
$$f_{1,\beta}^R(x) = 0.014019x_1 - 0.06794x_2 - 0.11207x_3 + 2.612847164$$
$$f_{2,\beta}^L(x) = 0.000996x_1 - 0.00304x_2 - 0.00067x_3 + 1.21164369$$
$$f_{2,\beta}^R(x) = 0.00207x_1 + 0.005721x_2 - 0.01181x_3 + 1.302280455$$

The pairwise comparison matrix for $f_1(x)$ and $f_2(x)$ can further be narrated as per the opinion of the decision makers given below:

$$C = \begin{pmatrix} 1 & 3 \\ 1/3 & 1 \end{pmatrix}$$

With the help of AHP, the weights are being obtained as $w_1 = 0.75$ and $w_1 = 0.25$.

Further, utilizing the weighting method, the MOLFPP is converted accordingly by using the α-cut and β-cut as follows:

LFPP-α $\quad \max\ f_\alpha^L(x) = 0.75f_{1,\alpha}^L(x) + 0.25f_{2,\alpha}^L(x)$

RFPP-α $\quad \max\ f_\alpha^R(x) = 0.75f_{1,\alpha}^R(x) + 0.25f_{2,\alpha}^R(x)$

s.t.

$$2.55x_1 + 4x_2 + x_3 \leq 27.50$$
$$1.75x_1 + x_2 + x_3 \leq 28.33$$
$$x = (x_1, x_2, x_3) \geq 0$$

LFPP-β $\quad \max\ f_\beta^L(x) = 0.75 f_{1,\beta}^L(x) + 0.25 f_{2,\beta}^L(x)$

RFPP-β $\quad \max\ f_\beta^R(x) = 0.75 f_{1,\beta}^R(x) + 0.25 f_{2,\beta}^R(x)$

s.t.

$$2x_1 + 3.67x_2 + x_3 \leq 30$$
$$2x_1 + 1.67x_2 + 0.25x_3 \leq 26.67$$
$$x = (x_1, x_2, x_3) \geq 0$$

On solving these objective functions with the help of mathematical software MATLAB, the range of optimal objective values for α-cut is [1.753749, 1.77978] and for β-cut is [44.662752, 2.432317].

Remarks:

- For α-cut both LFPP and RFPP leads to the optimal solution, i.e., $x_1 = 12.2222$, $x_2 = 0$ and $x_3 = 0$ which yields optimal objective values 1.753749 and 1.77978 respectively.
- For β-cut, LFPP leads to the optimal solution, i.e., $x_1 = 0$, $x_2 = 0$ and $x_3 = 0$ which yields optimal objective value 44.662752 and RFPP leads to the optimal solution, i.e., $x_1 = 13.3350$, $x_2 = 0$ and $x_3 = 0$ which yields optimal objective value 2.432317.
- For β-cut both LFPP and RFPP have different optimal values and the change in optimal value, i.e.,44.662753 - 2.432317 = 44.230435.

6 Conclusions and Scope for Future Work

A new model of multi-objective linear fractional programming problem (MOLFPP) with triangular intuitionistic fuzzy parameters has been successfully proposed. The model helps in finding the permissible/ ranges of the objective values under available constraints. The incorporation of the (α, β)-cuts used in the objectives and constraints gives an additional flexibility and freedom for a wider span of the uncertain information effectively. The use of the variable transformation technique and Taylor series expansion helped in handling the inexactness arising due to incompleteness of the information.

In future, the proposed technique under triangular intuitionistic fuzzy with (α, β)-cuts can be extended for generalized multi-objective non-linear optimization problem. Additionally, the proposed work can be translated using next level geometrical extension to trapezoidal intuitionistic fuzzy numbers.

References

1. Zadeh, L.A.: Fuzzy sets. Inf. Control. **8**(3), 338–353 (1965)
2. Stancu-Minasian, I.M.: Fractional Programming: Theory. Methods and Applications. Kluwer Academic Publishers, Dordrecht (1997)
3. Mehra, A., Chandra, S., Bector, C.R.: Acceptable optimality in linear fractional programming with fuzzy coefficients. Fuzzy Optim. Decis. Making **6**, 5–16 (2007)
4. Chinnadurai, V., Muthukumar, S.: Solving the linear fractional programming problem in a fuzzy environment: numerical approach. Appl. Math. Model. **40**, 6148–6164 (2016)
5. Stanojevic, B., Stanojevic, M.: Solving method for linear fractional optimization problem with fuzzy coefficients in the objective function. Int. J. Comp. Commun. Control. **8**, 146–152 (2012)
6. Charnes, A., Cooper, W.W.: An explicit general solution in linear fractional programming. Naval Res. Logist. Q. **20**(3), 449–467 (1973)
7. Stancu-Minasian, I.M., Pop, B.: On a fuzzy set approach to solving multiple objective linear fractional programming problem. Fuzzy Sets Syst. **134**, 397–405 (2003)
8. Dutta, D., Tiwari, R.N., Rao, J.R.: Multiple objective linear fractional programming a fuzzy set theoretic approach. Fuzzy Sets Syst. **52**, 39–45 (1992)
9. Toksari, M.D.: Taylor series approach to fuzzy multi objective linear fractional programming. Inf. Sci. **178**, 1189–1204 (2008)
10. Chakraborty, M., Gupta, S.: Fuzzy mathematical programming for multi objective linear fractional programming problem. Fuzzy Sets Syst. **125**, 335–342 (2002)
11. Mishra, S.: Weighting method for bi-level linear fractional programming problems. Eur. J. Oper. Res. **183**, 296–302 (2007)
12. Roy, K.T., Maiti, M.: Multi-objective inventory models of deteriorating items with some constraints in a fuzzy environment. Comput. Oper. Res. 1085–1095 (1998)
13. Dutta, D., Kumar, P.: Application of fuzzy goal programming approach to multi-objective linear fractional inventory model **46**, 2269–2278 (2015)
14. Garg, H.: Fuzzy inventory models for deteriorating items under different types of lead-time distributions. In: Kahraman, C., Onar, S.C. (eds.) Intelligent Techniques in Engineering Management, pp. 247–274. Springer, Cham (2015)
15. Atanassov, K.T.: Intuitionistic fuzzy sets. Fuzzy Sets Syst. **20**, 87–96 (1986)
16. Payan, A., Noora, A.A.: A linear modelling to solve multi-objective linear fractional programming problem with fuzzy parameters. Int. J. Math. Modell. Numer. Optimisation **5**(3), 210 (2014)
17. Dubois, D., Prade, H.: Fuzzy Sets and Systems: Theory and Applications. Academic Press, New York (1980)
18. Golden, B.L., Wasil, E.A., Harker, P.T.: The Analytic Hierarchy Process. Springier, New York (1989)
19. Charnes, A., Cooper, W.W.: Management models and industrial applications of linear programming. Mang. Sci. **4**, 38–91 (1957)

q-Rung Orthopair Fuzzy 2-Tuple Linguistic Hamy Mean Operators for MAGDM with Modified EDAS Method

Sumera Naz, Muhammad Akram, Areej Fatima, and Ahmad Nadeem

Abstract In this chapter, we investigate a multiple attribute group decision-making (MAGDM) approach based on q-rung orthopair fuzzy 2-tuple linguistic set (q-ROFTLS) to increase the capability of 2-tuple linguistic terms to describe ambiguous information. The Hamy mean (HM) operator can capture the significant connection between multiple integration parameters as an effective aggregation tool. Then, the q-ROFTL Hamy mean (q-ROFTLHM) operator, the q-ROFTL weighted Hamy mean (q-ROFTLWHM) operator, the q-ROFTL dual Hamy mean (q-ROFTLDHM) operator, and the q-ROFTL weighted dual Hamy mean (q-ROFTLWDHM) operator are presented in q-ROFTL environment. In the MAGDM environment, the framework of proposed aggregation operators can efficiently combine attribute values. Aspects of these operators, properties are also investigated. Moreover, in this research study, we use the evaluation based on the distance from the average solution (EDAS) approach to MAGDM with q-ROFTLNs. The developed q-ROFTL-EDAS approach is one of the most successful ways for solving MAGDM issues with the choice of the best option. In a summary, the framework key steps with the graphical representation for the ranking of alternatives are presented. To be precise, the proposed technique can adequately respond to the unpredictability of MAGDM and increase the rationality of group decision-making since it completely incorporates the psychological aspects of decision experts. Finally, to demonstrate the suggested decision procedure, we present an innovative case for evaluating the ecological value of a forest bio-diversity vacation area.

S. Naz · A. Fatima
Department of Mathematics, Division of Science and Technology, University of Education, Lahore, Pakistan
e-mail: sumera.naz@ue.edu.pk

M. Akram (✉)
Department of Mathematics, University of the Punjab, New Campus, Lahore 54590, Pakistan
e-mail: m.akram@pucit.edu.pk

A. Nadeem
School of Mathematics, Minhaj University Lahore, Lahore, Pakistan

L. Sahoo et al. (eds.), *Real Life Applications of Multiple Criteria Decision Making Techniques in Fuzzy Domain*, Studies in Fuzziness and Soft Computing 420,
https://doi.org/10.1007/978-981-19-4929-6_18

Keywords q-rung orthopair fuzzy 2-tuple linguistic set · MAGDM · HM operator · Dual HM operator · q-ROFTL-EDAS method

1 Introduction

Group decision-making [1–3] is an effective method for identifying the appropriate solution from a collection of elective alternatives. Because of the uncertainty and fuzziness of mankind's imagination and the factual matter, fuzzy MAGDM (also known as multiple criteria group decision-making) has become such a warm research area in recent years as an essential constituent part of group decision-making. Zadeh [4] developed the definition of a fuzzy set (FS), which is very useful to present fuzzy data. As FS just has the membership degree (MD), describing certain complex situations may be difficult. Intuitionistic fuzzy set (IFS) suggested through Atanassov [5], as an extension of the FS, facilitated decision experts (DEs) with pair form of MD and non-membership degree (NMD) to demonstrate their fuzziness. IFS assumes that the containing of each ordered pair of the MD and NMD is restricted to one ($0 \leq \mu + \nu \leq 1$). Yager [6] introduced the Pythagorean fuzzy set (PyFS) to give more space to DEs for dealing with uncertainties. The sum of squared MD and NMD is restricted to one ($0 \leq \mu^2 + \nu^2 \leq 1$) in the PyFS, which is the generalization of the IFS. Yager [7] proposed the q-ROFS, which provides the sum of the qth power of both the MD and NMD is restricted to one ($0 \leq \mu^q + \nu^q \leq 1$). The q-ROFS is more efficient and useful than IFS and PyFS because it has a wide membership and non-membership representation space. Seikh and Mandal [8] extended Frank t-norm and t-conorm ideas to the q-ROF context and developed some aggregation operators. Akram et al. [9] presented aggregation operators such as q-ROF Einstein weighted geometric, q-ROF Einstein ordered weighted geometric, generalized q-ROF Einstein weighted geometric, and generalized q-ROF Einstein ordered weighted geometric operators. Zeng et al. [10] provided a new method for dealing with MADM problems in a q-ROFS environment where the attribute weights are entirely unknown, based on the developed q-ROFS distance measurements. Peng et al. [11] later elaborated the exponential operational laws as well as a novel q-ROFS score function. Further scholars expanded the q-ROFS in different decision-making problems [12–16].

MADM has great success with FS theory. However, the MAGDM achievements only focus on DEs' quantitative assessment ideas and disregard their qualitative characteristics evaluation data. The linguistic variable (LV) of Zadeh [17] is an excellent method for expressing qualitative assessment knowledge from DEs. Wang and Li [18] have suggested the intuitionistic linguistic set (ILS), which combines the LV and the IFS, provides MD and NMD to the LV. Extensive methodologies for ILS are available, including inter-valued intuitionistic fuzzy linguistic set, linguistic Pythagorean fuzzy set (LPyFS), and linguistic q-ROFS. In the LPyFS, the LPyF number is used to characterize MD and NMD using a pair of linguistic words from the linguistic term set (LTS). The squares of the LTS's cardinality are smaller than the sum of squares of both MD and NMD for each LPyFN. Li et al. [19] suggested the q-rung picture linguistic set by combining the q-ROFS, the picture fuzzy set, and the linguistic terms.

Aggregation is a significant field of information process systems, including information fusion, pattern recognition, and decision support systems. It is frequently used to decrease the dimensions of assessment data during data aggregation procedures, such as triangular norms [20] and arithmetic means [21]. It is critical in the process of obtaining related assessment information from various assessors when dealing with MADM difficulties. The primary goal of the aggregation operator is to combine the assessment information of alternatives based on their many attributes in order to obtain the entire sustainability assessment information of all available alternatives. Conventional aggregation operators such as the weighted average, the weighted geometric, and the ordered weighted average have been widely employed in a variety of FSs. Unfortunately, these operators make the assumption that the characteristics are independent, which might result in data loss and inconsistent behavior in actual concerns. As the evaluation data and decision-making parameters get more complicated, the interrelation of all input factors should be considered in practical applications. In response to this circumstance, Hara et al. [22] developed the HM operator, which can capture the correlation between many reasons and has an adjustable parameter that makes the decision method more versatile. Feng et al. [23] used the HM operator to combine hesitant fuzzy linguistic (HFL) data with several novel HFL aggregation operators, with the HFLHM operator, the HFLWHM operator, the HFLDHM operator, and the HFLWDHM operator. Garg et al. [24] explored the complex q-ROF-HM and complex q-ROF weighted HM operators, which was based on the flexible complex q-ROF setup, as well as some of its desired qualities. To eliminate aggregation restrictions, Wu et al. [25] presented some HM operators. As a result, how to aggregate interval-valued intuitionistic fuzzy numbers using typical HM operators is a fascinating topic. Based on interaction operations, Xing et al. [26] introduced a family of q-ROF-HM operators and described interaction operational rules for q-ROF numbers. The q-ROF interaction HM and q-ROF interaction weighted HM operators were suggested based on the new operational rules.

In addition, for aggregating data, different MADM methods are used, and a more advanced and applicable method are the EDAS method. Keshavarz Ghorabaee et al. [27] initially described the EDAS technique as a MADM method for ranking alternatives. The EDAS technique evaluates alternatives using an average solution (AS) instead of the TODIM, VIKOR, or TOPSIS methods. In other terms, positive distance from AS (PDAS) and negative distance from AS (NDAS) are used to determine the best choice. The EDAS method has the advantage of determining the attribute weights based on the AS, which makes the method more robust when rating alternatives. Huang et al. [28] proposed an EDAS method based on prospect theory that has improved qualities, in reality, applied the traditional EDAS method and the enhancement EDAS method to the same case and analyzed their performances using sensitivity analysis and comparative analysis. Wei et al. [29] utilized the EDAS approach to MAGDM with probabilistic LTSs. Jana and Pal [30] used bipolar fuzzy numbers to apply the EDAS approach to MAGDM.

Scholars have certainly paid attention to EDAS, but no work has been done to expand it to the q-ROFTLS environment. According to the analysis above, the publication about the q-ROFTL-EDAS method based on the HM operator does not exist. We propose the HM aggregation operators for the q-ROFTLS. The main aim of this research study is to extend the EDAS methodology in the q-ROFTLS context, to solve the MAGDM problems. The manuscript's organization can be summarized as follows: (1) the HM aggregation operator is extended by q-ROFTLS; (2) the q-ROFTL-EDAS technique is proposed to deal with decision-making problems that have q-ROFTLS characteristics; (3) a calculation example is performed to test the designed approach; and (4) some comparison studies with the q-ROFTL-CODAS and q-ROFTL-TOPSIS techniques are provided to check the rational thinking of DEs about the q-ROFTL-EDAS.

The remaining chapter is organized in the following manner to accomplish the goal of this research study. In Sect. 2, we recap the idea of LTSs, q-ROFTLS with its score, accuracy functions, and operational laws. In Sect. 3, we combine the q-ROFTLS with HM operators and develop the q-ROFTLHM operator, the q-ROFTLWHM operator, the q-ROFTLDHM operator, the q-ROFTLWDHM operator. In Sect. 4, we present a novel approach for MAGDM known as the q-ROFTL-EDAS method. In Sect. 5, in order to explain the techniques suggested in this research, a numerical example to select the best bio-diversity vacation places in Pakistan is described. Section 6 concludes the research article by giving future directions.

2 Preliminaries

The 2TL representation model, q-ROFTLS, and the HM operator and its dual form are briefly discussed in this section.

2.1 2-Tuple Linguistic Representation Model

Tai and Chen [31] proposed the concept of 2TL representation model as follows:

Definition 1 [31] Let $S = \{s_0, s_1, s_2, \ldots, s_\tau\}$ be a linguistic term set (LTS) and $\xi \in [0, 1]$ be a number value that represents the aggregation result of linguistic symbolic. The function Δ is then defined to obtain the 2TL information equivalent to ξ as follows:

$$\Delta : [0, 1] \longrightarrow S \times \left[-\frac{1}{2\tau}, \frac{1}{2\tau}\right),$$

$$\Delta(\xi) = (s_\jmath, \gamma) \text{ with } \begin{cases} s_\jmath, & \jmath = \text{round}(\xi \times \tau), \\ \gamma = \xi - \frac{\jmath}{\tau}, & \gamma \in [-\frac{1}{2\tau}, \frac{1}{2\tau}). \end{cases}$$

where γ is the value of the symbolic translation and s_J has the closest index label to ξ.

Definition 2 [31] Let $S = \{s_0, s_1, s_2, \ldots, s_\tau\}$ be an LTS with the 2TL information (s_J, γ). Then there exists an inverse function Δ^{-1}, that transforms 2TL information into its equivalent numerical value $\xi \in [0, 1]$. The inverse function Δ^{-1} is defined as follows:

$$\Delta^{-1} : S \times \left[-\frac{1}{2\tau}, \frac{1}{2\tau}\right) \longrightarrow [0, 1],$$

$$\Delta^{-1}(s_J, \gamma) = \frac{J}{\tau} + \gamma = \xi.$$

Definition 3 [32] Let R be a non-empty universe of discourse and S be an LTS with odd cardinality. Then the definition of q-ROFTLSs is as follows:

$$D = \{\langle (s_{\psi(r)}, \gamma), (\zeta_D(r), \delta_D(r))\rangle | r \in R\},$$

where $s_{\psi(r)} \in S$, $\gamma \in [-\frac{1}{2\tau}, \frac{1}{2\tau})$, the numbers $\zeta_D(r)$, $\delta_D(r)$ represent the MD and NMD of r to the 2TL variable $(s_{\psi(r)}, \gamma)$, respectively, which satisfies the following conditions: $\zeta_D(r) \in [0, 1]$, $\delta_D(r) \in [0, 1]$, $0 \leq (\zeta_D(r))^q + (\delta_D(r))^q \leq 1$, for all $r \in R$ and $q \geq 1$.

For $r \in R$, $\pi_D(r) = (1 - (\zeta_D(r))^q - (\delta_D(r))^q)^{\frac{1}{q}}$ is degree of indeterminacy of the element r to 2TL variable $(s_{\psi(r)}, \gamma)$. For simplicity, $\eta = ((s_{\psi(\eta)}, \gamma), (\zeta_\eta, \delta_\eta))$ will be called a q-ROFTLN.

Definition 4 [32] Let $\eta_1 = ((s_1, \gamma_1), (\zeta_1, \delta_1))$ and $\eta_2 = ((s_2, \gamma_2), (\zeta_2, \delta_2))$ be two q-ROFTLNs. Then the operational laws of q-ROFTLNs are described as follows:

1. $\eta_1 \oplus \eta_2 = \left(\Delta(\Delta^{-1}(s_1, \gamma_1) + \Delta^{-1}(s_2, \gamma_2)), (((\zeta_1)^q + (\zeta_2)^q - (\zeta_1)^q \times (\zeta_2)^q)^{\frac{1}{q}}, \delta_1 \times \delta_2)\right)$;
2. $\eta_1 \otimes \eta_2 = \left(\Delta(\Delta^{-1}(s_1, \gamma_1) \times \Delta^{-1}(s_2, \gamma_2)), (\zeta_1 \times \zeta_2, ((\delta_1)^q + (\delta_2)^q - (\delta_1)^q \times (\delta_2)^q)^{\frac{1}{q}})\right)$;
3. $\rho\eta_1 = \left(\Delta(\rho \times \Delta^{-1}(s_1, \gamma_1)), ((1 - (1 - (\zeta_1)^q)^\rho)^{\frac{1}{q}}, (\delta_1)^\rho)\right)$, $\rho > 0$;
4. $\eta_1^\rho = \left(\Delta((\Delta^{-1}(s_1, \gamma_1))^\rho), ((\zeta_1)^\rho, (1 - (1 - (\delta_1)^q)^\rho)^{\frac{1}{q}})\right)$, $\rho > 0$.

Definition 5 [33] Let $\eta = ((s_{\psi(\eta)}, \gamma), (\zeta_\eta, \delta_\eta))$ be a q-ROFTLN. Then the score value $\mathcal{S}(\eta)$ and the accuracy value $\mathcal{S}^*(\eta)$ of q-ROFTLN η are defined as:

$$\mathcal{S}(\eta) = \frac{\Delta^{-1}(s_{\psi(\eta)}, \gamma) \times (1 + \zeta_\eta^q - \delta_\eta^q)}{2}, \tag{1}$$

$$\mathcal{S}^*(\eta) = \Delta^{-1}(s_{\psi(\eta)}, \gamma) \times (\zeta_\eta^q + \delta_\eta^q). \tag{2}$$

Definition 6 [33] Let $\eta_1 = ((s_1, \gamma_1), (\zeta_1, \delta_1))$ and $\eta_2 = ((s_2, \gamma_2), (\zeta_2, \delta_2))$ be two q-ROFTLNs, then we can compare them using the rules below:

(1) If $\mathcal{S}(\eta_1) > \mathcal{S}(\eta_2)$, then $\eta_1 \succ \eta_2$;
(2) If $\mathcal{S}(\eta_1) = \mathcal{S}(\eta_2)$, then

- If $\mathcal{S}^*(\eta_1) > \mathcal{S}^*(\eta_2)$, then $\eta_1 \succ \eta_2$;
- If $\mathcal{S}^*(\eta_1) = \mathcal{S}^*(\eta_2)$, then $\eta_1 \sim \eta_2$.

2.2 HM Operators and Its Weighted Form

Definition 7 Let $\mathfrak{a}_\jmath(\jmath = 1, 2, \ldots, \mathfrak{n})$ be a set of non-negative real numbers. Some HM aggregation operators are defined as follows:

1. Hamy mean [22]:

$$\mathrm{HM}^{(\mathfrak{z})}(\mathfrak{a}_1, \mathfrak{a}_2, \ldots, \mathfrak{a}_\mathfrak{n}) = \frac{\sum_{1 \le \imath_1 < \cdots < \imath_\mathfrak{z} \le \mathfrak{n}} \left(\prod_{\jmath=1}^{\mathfrak{z}} \mathfrak{a}_{\imath_\jmath}\right)^{\frac{1}{\mathfrak{z}}}}{C_\mathfrak{n}^\mathfrak{z}};$$

2. Weighted Hamy mean [22]:

$$\mathrm{WHM}_{\varpi}^{(\mathfrak{z})}(\mathfrak{a}_1, \mathfrak{a}_2, \ldots, \mathfrak{a}_\mathfrak{n}) = \frac{\sum_{1 \le \imath_1 < \cdots < \imath_\mathfrak{z} \le \mathfrak{n}} \left(\prod_{\jmath=1}^{\mathfrak{z}} (\mathfrak{a}_{\imath_\jmath})^{\varpi_{\imath_\jmath}}\right)^{\frac{1}{\mathfrak{z}}}}{C_\mathfrak{n}^\mathfrak{z}};$$

3. Dual Hamy mean [34]:

$$\mathrm{DHM}^{(\mathfrak{z})}(\mathfrak{a}_1, \mathfrak{a}_2, \ldots, \mathfrak{a}_\mathfrak{n}) = \left(\prod_{1 \le \imath_1 < \cdots < \imath_\mathfrak{z} \le \mathfrak{n}} \left(\frac{\sum_{\jmath=1}^{\mathfrak{z}} \mathfrak{a}_{\imath_\jmath}}{\mathfrak{z}}\right)\right)^{\frac{1}{C_\mathfrak{n}^\mathfrak{z}}};$$

4. Weighted dual Hamy mean [34]:

$$\mathrm{WDHM}_{\varpi}^{(\mathfrak{z})}(\mathfrak{a}_1, \mathfrak{a}_2, \ldots, \mathfrak{a}_\mathfrak{n}) = \left(\prod_{1 \le \imath_1 < \cdots < \imath_\mathfrak{z} \le \mathfrak{n}} \left(\frac{\sum_{\jmath=1}^{\mathfrak{z}} \varpi_{\imath_\jmath} \mathfrak{a}_{\imath_\jmath}}{\mathfrak{z}}\right)\right)^{\frac{1}{C_\mathfrak{n}^\mathfrak{z}}}$$

where $\mathfrak{z}$ is a parameter and $\mathfrak{z} = 1, 2, \ldots, \mathfrak{n}$, $\imath_1, \imath_2, \ldots, \imath_\mathfrak{z}$ are $\mathfrak{z}$ integer values taken from the set $\{1, 2, \ldots, \mathfrak{n}\}$ of $\jmath$ integer values, $C_\mathfrak{n}^\mathfrak{z}$ denotes the binomial coefficient, and $C_\mathfrak{n}^\mathfrak{z} = n!/\mathfrak{z}!(n - \mathfrak{z})!$.

The nomenclature of the proposed research work is given in Table 1.

Table 1 Nomenclature of proposed work

Notation	Description
q-ROFS	q-rung orthopair fuzzy set
q-ROFTLS	q-rung orthopair fuzzy 2-tuple linguistic set
q-ROFTLHM	q-ROFTL Hamy mean
q-ROFTLWHM	q-ROFTL weighted Hamy mean
q-ROFTLDHM	q-ROFTL dual Hamy mean
q-ROFTLWDHM	q-ROFTL weighted dual Hamy mean
$\eta = \left((s_J, \gamma_J), (\zeta_J, \delta_J)\right)$	q-ROFTL number
(s_J, γ_J)	2-Tuple linguistic term
ζ_J	Membership degree of q-ROFTLN
δ_J	Non-membership degree of q-ROFTLN
$\Im$	Alternative
$\hbar$	Attribute
$\mathcal{S}(\eta)$	Score function of q-ROFTLN
$\mathcal{S}^*(\eta)$	Accuracy function of q-ROFTLN
$\pi_D(r)$	Uncertainty of q-ROFTLN

3 Some q-ROFTLHM Aggregation Operators

Since the q-ROFTLS is a powerful tool for expressing values in the decision-making process, we augment the HM and DHM operators to q-ROFTLSs and establish the q-ROFTLHM and q-ROFTLDHM operators for aggregating q-ROFTLNs, as well as study their desirable characteristics.

3.1 q-ROFTLHM Aggregation Operator

This subsection demonstrates the HM operator with q-ROFTLSs to develop and study the q-ROFTLHM operator for analyzing q-ROFTLNs.

Definition 8 Let $\eta_J = ((s_J, \gamma_J), (\zeta_J, \delta_J))(J = 1, 2, \ldots, \mathfrak{n})$ be a collection of q-ROFTLNs. The q-ROFTLHM operator is a mapping $T^{\mathfrak{n}} \rightarrow T$ such that

$$q\text{-ROFTLHM}^{(3)}(\eta_1, \eta_2, \ldots, \eta_{\mathfrak{n}}) = \frac{\bigoplus_{1 \le \iota_1 < \cdots < \iota_3 \le \mathfrak{n}} \left(\bigotimes_{J=1}^{3} \eta_{\iota_J}\right)^{\frac{1}{3}}}{C_{\mathfrak{n}}^{3}}. \tag{3}$$

Theorem 1 *The aggregated value by utilizing q-ROFTLHM operator is also a q-ROFTLN, where*

$$q\text{-}ROFTLHM^{(3)}(\eta_1, \eta_2, \ldots, \eta_n) = \left(\begin{array}{l} \Delta\left(\frac{\sum_{1\le\iota_1<\cdots<\iota_3\le n}\left(\prod_{j=1}^{3}\Delta^{-1}(s_{\iota_j},\gamma_{\iota_j})\right)^{\frac{1}{3}}}{C_n^3} \right), \\ \left(\sqrt[q]{1-\prod_{1\le\iota_1<\ldots<\iota_3\le n}\left(1-\left(\prod_{j=1}^{3}\zeta_j\right)^{\frac{q}{3}}\right)^{\frac{1}{C_n^3}}}, \right. \\ \left. \left(\prod_{1\le\iota_1<\ldots<\iota_3\le n}\sqrt[q]{1-\prod_{j=1}^{3}(1-(\delta_j)^q)^{\frac{1}{3}}}\right)^{\frac{1}{C_n^3}} \right) \end{array} \right). \tag{4}$$

Proof Utilizing definition 4, we get

$$\otimes_{j=1}^{3}\eta_{\iota_j} = \left(\Delta\left(\prod_{j=1}^{3}\Delta^{-1}\left(s_{\iota_j},\gamma_{\iota_j}\right)\right), \left(\prod_{j=1}^{3}\zeta_j, \sqrt[q]{1-\prod_{j=1}^{3}\left(1-(\delta_j)^q\right)}\right)\right).$$

Thus,

$$\left(\otimes_{j=1}^{3}\eta_{\iota_j}\right)^{\frac{1}{3}} = \left(\begin{array}{l} \Delta\left(\prod_{j=1}^{3}\Delta^{-1}\left(s_{\iota_j},\gamma_{\iota_j}\right)\right)^{\frac{1}{3}}, \\ \left(\left(\prod_{j=1}^{3}\zeta_j\right)^{\frac{1}{3}}, \sqrt[q]{1-\prod_{j=1}^{3}(1-(\delta_j)^q)^{\frac{1}{3}}}\right) \end{array} \right).$$

Therefore,

$$\oplus_{1\le\iota_1<\cdots<\iota_3\le n}\left(\otimes_{j=1}^{3}\eta_{\iota_j}\right)^{\frac{1}{3}} = \left(\begin{array}{l} \Delta\left(\sum_{1\le\iota_1<\cdots<\iota_3\le n}\left(\prod_{j=1}^{3}\Delta^{-1}\left(s_{\iota_j},\gamma_{\iota_j}\right)\right)^{\frac{1}{3}}\right), \\ \left(\sqrt[q]{1-\prod_{1\le\iota_1<\cdots<\iota_3\le n}\left(1-\left(\prod_{j=1}^{3}\zeta_j\right)^{\frac{q}{3}}\right)}, \right. \\ \left. \prod_{1\le\iota_1<\cdots<\iota_3\le n}\sqrt[q]{1-\prod_{j=1}^{3}\left(1-(\delta_j)^q\right)^{\frac{1}{3}}} \right) \end{array} \right).$$

Furthermore,

$$q\text{-ROFTLHM}^{(3)}(\eta_1, \eta_2, \ldots, \eta_n) = \left(\begin{array}{l} \Delta\left(\frac{\sum_{1 \le \iota_1 < \cdots < \iota_3 \le n} \left(\prod_{j=1}^{3} \Delta^{-1}(s_{\iota_j}, \gamma_{\iota_j}) \right)^{\frac{1}{3}}}{C_n^3} \right), \\ \left(\begin{array}{l} \sqrt[q]{1 - \prod_{1 \le \iota_1 < \cdots < \iota_3 \le n} \left(1 - \left(\prod_{j=1}^{3} \zeta_j \right)^{\frac{q}{3}} \right)^{\frac{1}{C_n^3}}}, \\ \left(\prod_{1 \le \iota_1 < \cdots < \iota_3 \le n} \sqrt[q]{1 - \prod_{j=1}^{3} (1 - (\delta_j)^q)^{\frac{1}{3}}} \right)^{\frac{1}{C_n^3}} \end{array} \right) \end{array} \right).$$

Further, some desirable properties of the q-ROFTLHM operator are investigated as follows:

Proerty 1 (Idempotency) *If all* $\eta_j = ((s_j, \gamma_j), (\zeta_j, \delta_j))(j = 1, 2, \ldots, n)$ *are equal, for all* j, *then*

$$q\text{-}ROFTLHM^{(3)}(\eta_1, \eta_2, \ldots, \eta_n) = \eta.$$

Proof

$$q\text{-ROFTLHM}^{(3)}(\eta_1, \eta_2, \ldots, \eta_n) = \frac{\oplus_{1 \le \iota_1 < \cdots < \iota_3 \le n} \left(\otimes_{j=1}^{3} \eta \right)^{\frac{1}{3}}}{C_n^3}$$

$$= \left(\begin{array}{l} \Delta\left(\frac{\sum_{1 \le \iota_1 < \cdots < \iota_3 \le n} \left(\prod_{j=1}^{3} \Delta^{-1}(s_\star, \gamma_\star) \right)^{\frac{1}{3}}}{C_n^3} \right), \\ \left(\begin{array}{l} \sqrt[q]{1 - \prod_{1 \le \iota_1 < \cdots < \iota_3 \le n} \left(1 - \left(\prod_{j=1}^{3} \zeta_\star \right)^{\frac{q}{3}} \right)^{\frac{1}{C_n^3}}}, \\ \left(\prod_{1 \le \iota_1 < \cdots < \iota_3 \le n} \sqrt[q]{1 - \prod_{j=1}^{3} (1 - \delta_\star^q)^{\frac{1}{3}}} \right)^{\frac{1}{C_n^3}} \end{array} \right) \end{array} \right)$$

$$= \left(\begin{array}{l} \Delta\left(\left((\Delta^{-1}(s_\star, \gamma_\star))^3 \right)^{\frac{1}{3}} \right), \\ \left(\sqrt[q]{1 - \left(\left(1 - (\zeta_\star^3)^{\frac{q}{3}} \right)^{C_n^3} \right)^{\frac{1}{C_n^3}}}, \left(\left(\sqrt[q]{1 - ((1 - \delta_\star^q)^3)^{\frac{1}{3}}} \right)^{C_n^3} \right)^{\frac{1}{C_n^3}} \right) \end{array} \right)$$

$$= ((s_\star, \gamma_\star), (\zeta_\star, \delta_\star)) = \eta.$$

Proerty 2 (Monotonicity) *Let* $\eta_j = ((s_j, \gamma_j), (\zeta_j, \delta_j))$, $\eta'_j = ((s'_j, \gamma'_j), (\zeta'_j, \delta'_j))(j = 1, 2, \ldots, n)$ *be two sets of q-ROFTLNs, if* $\eta_j \le \eta'_j$, *for all* j, *then*

$$q\text{-}ROFTLHM^{(3)}(\eta_1, \eta_2, \ldots, \eta_n) \le q\text{-}ROFTLHM^{(3)}(\eta'_1, \eta'_2, \ldots, \eta'_n).$$

Proof Let $\eta_j = ((s_j, \gamma_j), (\zeta_j, \delta_j)), \eta'_j = ((s'_j, \gamma'_j), (\zeta'_j, \delta'_j)), j = 1, 2, \ldots, \mathrm{n}$ be two sets of q-ROFTLNs, let

$$(s_j, \gamma_j) = \Delta\left(\frac{\sum_{1\le \iota_1<\cdots<\iota_3\le \mathrm{n}}\left(\prod_{j=1}^{3}\Delta^{-1}\left(s_{\iota_j}, \gamma_{\iota_j}\right)\right)^{\frac{1}{3}}}{C_{\mathrm{n}}^{3}}\right),$$

$$\zeta = \left(\sqrt[q]{1-\prod_{1\le \iota_1<\cdots<\iota_3\le \mathrm{n}}\left(1-\left(\prod_{j=1}^{3}\zeta_j\right)^{\frac{q}{3}}\right)^{\frac{1}{C_{\mathrm{n}}^{3}}}}\right),$$

$$\delta = \left(\left(\prod_{1\le \iota_1<\cdots<\iota_3\le \mathrm{n}}\sqrt[q]{1-\prod_{j=1}^{3}(1-(\delta_j)^q)^{\frac{1}{3}}}\right)^{\frac{1}{C_{\mathrm{n}}^{3}}}\right).$$

Thus, we have

$$\Delta^{-1}(s_{\iota_j}, \gamma_{\iota_j}) \le \Delta^{-1}(s'_{\iota_j}, \gamma'_{\iota_j}),$$

$$\prod_{j=1}^{3}\Delta^{-1}(s_{\iota_j}, \gamma_{\iota_j}) \le \prod_{j=1}^{3}\Delta^{-1}(s'_{\iota_j}, \gamma'_{\iota_j}),$$

$$\Delta\left(\frac{\sum_{1\le \iota_1<\cdots<\iota_3\le \mathrm{n}}\left(\prod_{j=1}^{3}\Delta^{-1}\left(s_{\iota_j}, \gamma_{\iota_j}\right)\right)^{\frac{1}{3}}}{C_{\mathrm{n}}^{3}}\right)$$
$$\le \Delta\left(\frac{\sum_{1\le \iota_1<\cdots<\iota_3\le \mathrm{n}}\left(\prod_{j=1}^{3}\Delta^{-1}\left(s'_{\iota_j}, \gamma'_{\iota_j}\right)\right)^{\frac{1}{3}}}{C_{\mathrm{n}}^{3}}\right).$$

given that $\zeta_j \le \zeta'_j$ then

$$\left(\prod_{j=1}^{3}\zeta_j\right)^{\frac{q}{3}} \le \left(\prod_{j=1}^{3}\zeta'_j\right)^{\frac{q}{3}}.$$

Moreover,

$$\prod_{1\le \iota_1<\cdots<\iota_3\le n}\left(1-\left(\prod_{j=1}^{3}\zeta_j\right)^{\frac{q}{3}}\right)^{\frac{1}{C_n^3}} \ge \prod_{1\le \iota_1<\cdots<\iota_3\le n}\left(1-\left(\prod_{j=1}^{3}\zeta_j'\right)^{\frac{q}{3}}\right)^{\frac{1}{C_n^3}}.$$

Furthermore,

$$\sqrt[q]{1-\prod_{1\le \iota_1<\cdots<\iota_3\le n}\left(1-\left(\prod_{j=1}^{3}\zeta_j\right)^{\frac{q}{3}}\right)^{\frac{1}{C_n^3}}} \le \sqrt[q]{1-\prod_{1\le \iota_1<\cdots<\iota_3\le n}\left(1-\left(\prod_{j=1}^{3}\zeta_j'\right)^{\frac{q}{3}}\right)^{\frac{1}{C_n^3}}}.$$

Therefore, $\zeta \le \zeta'$. Similarly, we can show that $\delta \ge \delta'$.
Hence, q-ROFTLHM$^{(3)}(\eta_1, \eta_2, \ldots, \eta_n) \le q$-ROFTLHM$^{(3)}(\eta_1', \eta_2', \ldots, \eta_n')$.

Proerty 3 (Boundedness) *Let* $\eta_j = ((s_j, \gamma_j), (\zeta_j, \delta_j))(j = 1, 2, \ldots, n)$ *be a collection of q-ROFTLNs, and let* $\eta^- = \min_j((s_j, \gamma_j), (\zeta_j, \delta_j))$ *and* $\eta^+ = \max_j((s_j, \gamma_j), (\zeta_j, \delta_j))$*, then*

$$\eta^- \le q\text{-}ROFTLHM^{(3)}(\eta_1, \eta_2, \ldots, \eta_n) \le \eta^+.$$

From Property 1,

$$q\text{-}ROFTLHM^{(3)}(\eta_1^-, \eta_2^-, \ldots, \eta_n^-) = \eta^-,$$
$$q\text{-}ROFTLHM^{(3)}(\eta_1^+, \eta_2^+, \ldots, \eta_n^+) = \eta^+.$$

From Property 2,

$$\eta^- \le q\text{-}ROFTLHM^{(3)}(\eta_1, \eta_2, \ldots, \eta_n) \le \eta^+.$$

3.2 q-ROFTLWHM Aggregation Operator

The significance of the analyzed suppositions is ignored by the q-ROFTLHM operator. In particular case, the attribute weights play a vital role in the analyzing process in a wide range of real situations, especially in MADM problems. To overcome limitations of q-ROFTLHM operator, we put forward the q-ROFTLWHM operator as follows:

Definition 9 Let $\eta_j = ((s_j, \gamma_j), (\zeta_j, \delta_j))(j = 1, 2, \ldots, \mathrm{n})$ be a collection of q-ROFTLNs. The q-ROFTLWHM operator is a mapping $T^{\mathrm{n}} \longrightarrow T$ such that

$$q\text{-ROFTLWHM}_{\varpi}^{(3)}(\eta_1, \eta_2, \ldots, \eta_{\mathrm{n}}) = \frac{\bigoplus_{1\le\iota_1<\cdots<\iota_3\le\mathrm{n}}\left(\bigotimes_{j=1}^{3}(\eta_{\iota_j})^{\varpi_{\iota_j}}\right)^{\frac{1}{3}}}{C_{\mathrm{n}}^{3}}. \quad (5)$$

Theorem 2 *The aggregated value by utilizing q-ROFTLWHM operator is also a q-ROFTLN, where*

$$q\text{-}ROFTLWHM_{\varpi}^{(3)}(\eta_1, \eta_2, \ldots, \eta_{\mathrm{n}})$$
$$= \left(\begin{array}{c} \Delta\left(\dfrac{\sum_{1\le\iota_1<\cdots<\iota_3\le\mathrm{n}}\left(\prod_{j=1}^{3}\Delta^{-1}(s_{\iota_j}, \gamma_{\iota_j})^{\varpi_{\iota_j}}\right)^{\frac{1}{3}}}{C_{\mathrm{n}}^{3}}\right), \\ \left(\left(\sqrt[q]{1 - \prod_{1\le\iota_1<\cdots<\iota_3\le\mathrm{n}}\left(1-\left(\prod_{j=1}^{3}(\zeta_j)^{\varpi_{\iota_j}}\right)^{\frac{q}{3}}\right)}\right)^{\frac{1}{C_{\mathrm{n}}^{3}}}, \\ \left(\prod_{1\le\iota_1<\cdots<\iota_3\le\mathrm{n}}\sqrt[q]{1-\prod_{j=1}^{3}((1-(\delta_j)^q)^{\varpi_{\iota_j}})^{\frac{1}{3}}}\right)^{\frac{1}{C_{\mathrm{n}}^{3}}} \right) \end{array} \right). \quad (6)$$

Proof Utilizing Definition 4, we get

$$(\eta_{\iota_j})^{\varpi_{\iota_j}} = \left(\Delta\left(\Delta^{-1}\left(s_{\iota_j}, \gamma_{\iota_j}\right)^{\varpi_{\iota_j}}\right), \left((\zeta_j)^{\varpi_{\iota_j}}, \sqrt[q]{1-(1-(\delta_j)^q)^{\varpi_{\iota_j}}}\right)\right).$$

Then,

$$\otimes_{j=1}^{3}(\eta_{\iota_j})^{\varpi_{\iota_j}} = \left(\begin{array}{c} \Delta\left(\prod_{j=1}^{3}\Delta^{-1}\left(s_{\iota_j}, \gamma_{\iota_j}\right)^{\varpi_{\iota_j}}\right), \\ \left(\prod_{j=1}^{3}(\zeta_j)^{\varpi_{\iota_j}}, \sqrt[q]{1-\prod_{j=1}^{3}(1-(\delta_j)^q)^{\varpi_{\iota_j}}}\right) \end{array} \right).$$

Thus,

$$\left(\otimes_{j=1}^{3}(\eta_{\iota_j})^{\varpi_{\iota_j}}\right)^{\frac{1}{3}} = \left(\begin{array}{c} \Delta\left(\prod_{j=1}^{3}\Delta^{-1}\left(s_{\iota_j}, \gamma_{\iota_j}\right)^{\varpi_{\iota_j}}\right)^{\frac{1}{3}}, \\ \left(\left(\prod_{j=1}^{3}(\zeta_j)^{\varpi_{\iota_j}}\right)^{\frac{1}{3}}, \sqrt[q]{1-\left(\prod_{j=1}^{3}(1-(\delta_j)^q)^{\varpi_{\iota_j}}\right)^{\frac{1}{3}}}\right) \end{array} \right).$$

Therefore,

$$\oplus_{1\le\iota_1<\cdots<\iota_3\le n}\left(\otimes_{j=1}^{3}\left(\eta_{\iota_j}\right)^{\varpi_{\iota_j}}\right)^{\frac{1}{3}}$$

$$=\left(\begin{array}{c}\Delta\left(\sum_{1\le\iota_1<\cdots<\iota_3\le n}\left(\prod_{j=1}^{3}\Delta^{-1}\left(s_{\iota_j},\gamma_{\iota_j}\right)^{\varpi_{\iota_j}}\right)^{\frac{1}{3}}\right),\\ \left(\sqrt[q]{1-\prod_{1\le\iota_1<\cdots<\iota_3\le n}\left(1-\left(\prod_{j=1}^{3}(\zeta_j)^{\varpi_{\iota_j}}\right)^{\frac{q}{3}}\right)},\right.\\ \left.\prod_{1\le\iota_1<\cdots<\iota_3\le n}\sqrt[q]{1-\left(\prod_{j=1}^{3}(1-(\delta_j)^q)^{\varpi_{\iota_j}}\right)^{\frac{1}{3}}}\right)\end{array}\right).$$

Furthermore,

$$q\text{-ROFTLWHM}_{\varpi}^{(3)}(\eta_1,\eta_2,\ldots,\eta_n)$$

$$=\left(\begin{array}{c}\Delta\left(\frac{\sum_{1\le\iota_1<\cdots<\iota_3\le n}\left(\prod_{j=1}^{3}\Delta^{-1}\left(s_{\iota_j},\gamma_{\iota_j}\right)^{\varpi_{\iota_j}}\right)^{\frac{1}{3}}}{C_n^3}\right),\\ \left(\sqrt[q]{1-\prod_{1\le\iota_1<\cdots<\iota_3\le n}\left(1-\left(\prod_{j=1}^{3}(\zeta_j)^{\varpi_{\iota_j}}\right)^{\frac{q}{3}}\right)^{\frac{1}{C_n^3}}},\right.\\ \left.\left(\prod_{1\le\iota_1<\cdots<\iota_3\le n}\sqrt[q]{1-\prod_{j=1}^{3}((1-(\delta_j)^q)^{\varpi_{\iota_j}})^{\frac{1}{3}}}\right)^{\frac{1}{C_n^3}}\right)\end{array}\right).$$

Proerty 4 (Monotonicity) *Let* $\eta_j=((s_j,\gamma_j),(\zeta_j,\delta_j))$, $\eta'_j=((s'_j,\gamma'_j),(\zeta'_j,\delta'_j))$ $(j=1,2,\ldots,n)$ *be two sets of q-ROFTLNs, if* $\eta_j\le\eta'_j$, *for all* j, *then*

$$q\text{-}ROFTLWHM_{\varpi}^{(3)}(\eta_1,\eta_2,\ldots,\eta_n)\le q\text{-}ROFTLWHM_{\varpi}^{(3)}(\eta'_1,\eta'_2,\ldots,\eta'_n).$$

Proof Let $\eta_j=((s_j,\gamma_j),(\zeta_j,\delta_j))$, $\eta'_j=((s'_j,\gamma'_j),(\zeta'_j,\delta'_j))$, $j=1,2,\ldots,n$ be two sets of q-ROFTLNs; let

$$(s_\jmath, \gamma_\jmath) = \Delta\left(\frac{\sum_{1\le\iota_1<\cdots<\iota_3\le \mathrm{n}}\left(\prod_{\jmath=1}^{3}\Delta^{-1}\left(s_{\iota_\jmath}, \gamma_{\iota_\jmath}\right)^{\varpi_{\iota_\jmath}}\right)^{\frac{1}{3}}}{C_{\mathrm{n}}^{3}}\right),$$

$$\zeta = \left(\sqrt[q]{1-\prod_{1\le\iota_1<\cdots<\iota_3\le \mathrm{n}}\left(1-\left(\prod_{\jmath=1}^{3}(\zeta_\jmath)^{\varpi_{\iota_\jmath}}\right)^{\frac{q}{3}}\right)^{\frac{1}{C_{\mathrm{n}}^{3}}}}\right),$$

$$\delta = \left(\left(\prod_{1\le\iota_1<\cdots<\iota_3\le \mathrm{n}}\sqrt[q]{1-\prod_{\jmath=1}^{3}((1-(\delta_\jmath)^q)^{\varpi_{\iota_\jmath}})^{\frac{1}{3}}}\right)^{\frac{1}{C_{\mathrm{n}}^{3}}}\right).$$

Thus we have

$$\Delta^{-1}(s_{\iota_\jmath}, \gamma_{\iota_\jmath})^{\varpi_{\iota_\jmath}} \le \Delta^{-1}(s'_{\iota_\jmath}, \gamma'_{\iota_\jmath})^{\varpi_{\iota_\jmath}}$$

$$\left(\prod_{\jmath=1}^{3}\Delta^{-1}\left(s_{\iota_\jmath}, \gamma_{\iota_\jmath}\right)^{\varpi_{\iota_\jmath}}\right)^{\frac{1}{3}} \le \left(\prod_{\jmath=1}^{3}\Delta^{-1}\left(s'_{\iota_\jmath}, \gamma'_{\iota_\jmath}\right)^{\varpi_{\iota_\jmath}}\right)^{\frac{1}{3}},$$

$$\Delta\left(\frac{\sum_{1\le\iota_1<\cdots<\iota_3\le \mathrm{n}}\left(\prod_{\jmath=1}^{3}\Delta^{-1}\left(s_{\iota_\jmath}, \gamma_{\iota_\jmath}\right)^{\varpi_{\iota_\jmath}}\right)^{\frac{1}{3}}}{C_{\mathrm{n}}^{3}}\right)$$
$$\le \Delta\left(\frac{\sum_{1\le\iota_1<\cdots<\iota_3\le \mathrm{n}}\left(\prod_{\jmath=1}^{3}\Delta^{-1}\left(s'_{\iota_\jmath}, \gamma'_{\iota_\jmath}\right)^{\varpi_{\iota_\jmath}}\right)^{\frac{1}{3}}}{C_{\mathrm{n}}^{3}}\right).$$

Since $\zeta_\jmath \le \zeta'_\jmath$

$$\left(\prod_{\jmath=1}^{3}(\zeta_\jmath)^{\varpi_{\iota_\jmath}}\right)^{\frac{q}{3}} \le \left(\prod_{\jmath=1}^{3}(\zeta'_\jmath)^{\varpi_{\iota_\jmath}}\right)^{\frac{q}{3}}.$$

Moreover,

$$\prod_{1\le\iota_1<\cdots<\iota_3\le n}\left(1-\left(\prod_{j=1}^{3}(\zeta_j)^{\varpi_{\iota_j}}\right)^{\frac{q}{3}}\right)^{\frac{1}{C_n^3}} \ge \prod_{1\le\iota_1<\cdots<\iota_3\le n}\left(1-\left(\prod_{j=1}^{3}(\zeta_j')^{\varpi_{\iota_j}}\right)^{\frac{q}{3}}\right)^{\frac{1}{C_n^3}}.$$

Furthermore,

$$\sqrt[q]{1-\prod_{1\le\iota_1<\cdots<\iota_3\le n}\left(1-\left(\prod_{j=1}^{3}(\zeta_j)^{\varpi_{\iota_j}}\right)^{\frac{q}{3}}\right)^{\frac{1}{C_n^3}}} \le \sqrt[q]{1-\prod_{1\le\iota_1<\cdots<\iota_3\le n}\left(1-\left(\prod_{j=1}^{3}(\zeta_j')^{\varpi_{\iota_j}}\right)^{\frac{q}{3}}\right)^{\frac{1}{C_n^3}}}.$$

Therefore, $\zeta \le \zeta'$. Similarly, we can show that $\delta \ge \delta'$.

Hence $q\text{-ROFTLWHM}_{\varpi}^{(3)}(\eta_1, \eta_2, \ldots, \eta_n) \le q\text{-ROFTLWHM}_{\varpi}^{(3)}(\eta_1', \eta_2', \ldots, \eta_n')$.

Proerty 5 (Boundedness) *Let* $\eta_j = ((s_j, \gamma_j), (\zeta_j, \delta_j))(j = 1, 2, \ldots, n)$ *be a collection of q-ROFTLNs, and let* $\eta^- = \min_j((s_j, \gamma_j), (\zeta_j, \delta_j))$, $\eta^+ = \max_j((s_j, \gamma_j), (\zeta_j, \delta_j))$, *then*

$$\eta^- \le q\text{-}ROFTLWHM_{\varpi}^{(3)}(\eta_1, \eta_2, \ldots, \eta_n) \le \eta^+.$$

Proof From the Theorem 2, we get

$$q\text{-ROFTLWHM}_{\varpi}^{(3)}(\eta_1^-, \eta_2^-, \ldots, \eta_n^-) = \frac{\oplus_{1\le\iota_1<\cdots<\iota_3\le n}\left(\otimes_{j=1}^{3}\left(\min\eta_{\iota_j}\right)^{\varpi_{\iota_j}}\right)^{\frac{1}{3}}}{C_n^3}$$

$$= \left(\begin{array}{l}\Delta\left(\frac{\sum_{1\le\iota_1<\cdots<\iota_3\le n}\left(\prod_{j=1}^{3}\min(\Delta^{-1}(s_{\iota_j},\gamma_{\iota_j}))^{\varpi_{\iota_j}}\right)^{\frac{1}{3}}}{C_n^3}\right),\\ \left(\begin{array}{l}\sqrt[q]{1-\prod_{1\le\iota_1<\cdots<\iota_3\le n}\left(1-\left(\prod_{j=1}^{3}(\min\zeta_j)^{\varpi_{\iota_j}}\right)^{\frac{q}{3}}\right)^{\frac{1}{C_n^3}}},\\ \left(\prod_{1\le\iota_1<\cdots<\iota_3\le n}\sqrt[q]{1-\prod_{j=1}^{3}((1-(\max\delta_j)^q)^{\varpi_{\iota_j}})^{\frac{1}{3}}}\right)^{\frac{1}{C_n^3}}\end{array}\right)\end{array}\right).$$

$$q\text{-ROFTLWHM}_{\varpi}^{(3)}(\eta_1^+, \eta_2^+, \ldots, \eta_n^+) = \frac{\oplus_{1\leq\iota_1<\cdots<\iota_3\leq n}\left(\otimes_{j=1}^{3}(\max \eta_{\iota_j})^{\varpi_{\iota_j}}\right)^{\frac{1}{3}}}{C_n^3}$$

$$= \left(\begin{array}{l} \Delta\left(\frac{\sum_{1\leq\iota_1<\cdots<\iota_3\leq n}\left(\prod_{j=1}^{3}\max(\Delta^{-1}(s_{\iota_j},\gamma_{\iota_j}))^{\varpi_{\iota_j}}\right)^{\frac{1}{3}}}{C_n^3}\right), \\ \left(\left(\sqrt[q]{1-\prod_{1\leq\iota_1<\cdots<\iota_3\leq n}\left(1-\left(\prod_{j=1}^{3}(\max \zeta_j)^{\varpi_{\iota_j}}\right)^{\frac{q}{3}}\right)}\right)^{\frac{1}{C_n^3}}, \\ \left(\prod_{1\leq\iota_1<\cdots<\iota_3\leq n}\sqrt[q]{1-\prod_{j=1}^{3}((1-(\min\delta_j)^q)^{\varpi_{\iota_j}})^{\frac{1}{3}}}\right)^{\frac{1}{C_n^3}} \right) \end{array} \right).$$

From Property 4,

$$\eta^- \leq q\text{-ROFTLWHM}_{\varpi}^{(3)}(\eta_1, \eta_2, \ldots, \eta_n) \leq \eta^+.$$

The q-ROFTLWHM operator loses the idempotency property.

3.3 q-ROFTLDHM Aggregation Operator

This subsection analyzes the DHM operator with q-ROFTLSs to establish the q-ROFTLDHM operator for analyzing q-ROFTLNs and investigates its desirable characteristics.

Definition 10 Let $\eta_j = ((s_j, \gamma_j), (\zeta_j, \delta_j))(j = 1, 2, \ldots, n)$ be a collection of q-ROFTLNs. The q-ROFTLDHM operator is a mapping $T^n \longrightarrow T$ such that

$$q\text{-ROFTLDHM}^{(3)}(\eta_1, \eta_2, \ldots, \eta_n) = \left(\otimes_{1\leq\iota_1<\cdots<\iota_3\leq n}\left(\frac{\oplus_{j=1}^{3}\eta_{\iota_j}}{3}\right)\right)^{\frac{1}{C_n^3}}. \quad (7)$$

Theorem 3 *The aggregated value by utilizing the q-ROFTLDHM operator is also a q-ROFTLN, where*

$$q\text{-}ROFTLDHM^{(3)}(\eta_1, \eta_2, \ldots, \eta_n) = \left(\begin{array}{c} \Delta\left(\prod_{1 \le \iota_1 < \cdots < \iota_3 \le n} \left(\frac{\sum_{j=1}^{3} \Delta^{-1}(s_{\iota_j}, \gamma_{\iota_j})}{3} \right) \right)^{\frac{1}{C_n^3}}, \\ \left(\left(\prod_{1 \le \iota_1 < \cdots < \iota_3 \le n} \sqrt[q]{1 - \prod_{j=1}^{3} (1 - (\zeta_j)^q)^{\frac{1}{3}}} \right)^{\frac{1}{C_n^3}}, \sqrt[q]{1 - \prod_{1 \le \iota_1 < \cdots < \iota_3 < n} \left(1 - \left(\prod_{j=1}^{3} \delta_j \right)^{\frac{q}{3}} \right)^{\frac{1}{C_n^3}}} \right) \end{array} \right). \tag{8}$$

Proof Utilizing Definition 4, we get

$$\oplus_{j=1}^{3} \eta_{\iota_j} = \left(\Delta\left(\sum_{j=1}^{3} \Delta^{-1}\left(s_{\iota_j}, \gamma_{\iota_j}\right) \right), \left(\sqrt[q]{1 - \prod_{j=1}^{3} (1 - (\zeta_j)^q)}, \prod_{j=1}^{3} \delta_j \right) \right).$$

Thus,

$$\left(\frac{\oplus_{j=1}^{3} \eta_{\iota_j}}{3} \right) = \left(\Delta\left(\frac{\sum_{j=1}^{3} \Delta^{-1}(s_{\iota_j}, \gamma_{\iota_j})}{3} \right), \left(\sqrt[q]{1 - \prod_{j=1}^{3} (1 - (\zeta_j)^q)^{\frac{1}{3}}}, \left(\prod_{j=1}^{3} \delta_j \right)^{\frac{1}{3}} \right) \right).$$

Therefore,

$$\otimes_{1 \le \iota_1 < \cdots < \iota_3 \le n} \left(\frac{\oplus_{j=1}^{3} \eta_{\iota_j}}{3} \right) = \left(\begin{array}{c} \Delta\left(\prod_{1 \le \iota_1 < \cdots < \iota_3 \le n} \left(\frac{\sum_{j=1}^{3} \Delta^{-1}(s_{\iota_j}, \gamma_{\iota_j})}{3} \right) \right), \\ \left(\prod_{1 \le \iota_1 < \cdots < \iota_3 \le n} \sqrt[q]{1 - \prod_{j=1}^{3} (1 - (\zeta_j)^q)^{\frac{1}{3}}}, \sqrt[q]{1 - \prod_{1 \le \iota_1 < \cdots < \iota_3 \le n} \left(1 - \left(\prod_{j=1}^{3} \delta_j \right)^{\frac{q}{3}} \right)} \right) \end{array} \right).$$

Furthermore,

$$q\text{-ROFTLDHM}^{(3)}(\eta_1, \eta_2, \ldots, \eta_n) = \left(\begin{array}{l} \Delta\left(\prod_{1 \le \iota_1 < \cdots < \iota_3 \le n} \left(\frac{\sum_{j=1}^{3} \Delta^{-1}(s_{\iota_j}, \gamma_{\iota_j})}{3} \right) \right)^{\frac{1}{C_n^3}}, \\ \left(\left(\prod_{1 \le \iota_1 < \cdots < \iota_3 \le n} \sqrt[q]{1 - \prod_{j=1}^{3} (1 - (\zeta_j)^q)^{\frac{1}{3}}} \right)^{\frac{1}{C_n^3}}, \sqrt[q]{1 - \prod_{1 \le \iota_1 < \cdots < \iota_3 \le n} \left(1 - \left(\prod_{j=1}^{3} \delta_j \right)^{\frac{q}{3}} \right)^{\frac{1}{C_n^3}}} \right) \end{array} \right).$$

Proerty 6 (Idempotency) *If all* $\eta_j = ((s_j, \gamma_j), (\zeta_j, \delta_j))(j = 1, 2, \ldots, n)$ *are equal, for all* j, *then*

$$q\text{-}ROFTLDHM^{(3)}(\eta_1, \eta_2, \ldots, \eta_n) = \eta.$$

Proerty 7 (Monotonicity) *Let* $\eta_j = ((s_j, \gamma_j), (\zeta_j, \delta_j)), \eta'_j = ((s'_j, \gamma'_j), (\zeta'_j, \delta'_j))(j = 1, 2, \ldots, n)$ *be two sets of q-ROFTLNs, and if* $\eta_j \le \eta'_j$, *for all* j, *then*

$$q\text{-}ROFTLDHM^{(3)}(\eta_1, \eta_2, \ldots, \eta_n) \le q\text{-}ROFTLDHM^{(3)}(\eta'_1, \eta'_2, \ldots, \eta'_n).$$

Proerty 8 (Boundedness) *Let* $\eta_j = ((s_j, \gamma_j), (\zeta_j, \delta_j))(j = 1, 2, \ldots, n)$ *be a set of q-ROFTLNs, and let* $\eta^- = \min_j((s_j, \gamma_j), (\zeta_j, \delta_j))$ *and* $\eta^+ = \max_j((s_j, \gamma_j), (\zeta_j, \delta_j))$, *then*

$$\eta^- \le q\text{-}ROFTLDHM^{(3)}(\eta_1, \eta_2, \ldots, \eta_n) \le \eta^+.$$

3.4 *q*-ROFTLWDHM Aggregation Operator

The significance of the analyzed attributes is ignored by the q-ROFTLDHM operator. In any case, attribute weights play an important role in the aggregation process in a variety of real-world situations, especially in MADM problems. To overcome limitations of q-ROFTLDHM operator, we propose the q-ROFTLWDHM operator as follows:

Definition 11 Let $\eta_j = ((s_j, \gamma_j), (\zeta_j, \delta_j))(j = 1, 2, \ldots, n)$ be a collection of q-ROFTLNs. The q-ROFTLWDHM operator is a mapping $T^n \longrightarrow T$ such that

$$q\text{-ROFTLWDHM}_{\varpi}^{(3)}(\eta_1, \eta_2, \ldots, \eta_n) = \left(\otimes_{1 \le \iota_1 < \cdots < \iota_3 \le n}\left(\frac{\oplus_{j=1}^{3} \varpi_{\iota_j}\eta_{\iota_j}}{3}\right)\right)^{\frac{1}{C_n^3}}. \tag{9}$$

Theorem 4 *The aggregated value by utilizing the q-ROFTLWDHM operator is also a q-ROFTLN, where*

$$q\text{-}ROFTLWDHM_{\varpi}^{(3)}(\eta_1, \eta_2, \ldots, \eta_n) = \left(\begin{array}{c} \Delta\left(\prod_{1 \le \iota_1 < \cdots < \iota_3 \le n}\left(\frac{\sum_{j=1}^{3} \varpi_{\iota_j}(\Delta^{-1}(s_{\iota_j}, \gamma_{\iota_j}))}{3}\right)\right)^{\frac{1}{C_n^3}}, \\ \left(\begin{array}{c} \left(\prod_{1 \le \iota_1 < \cdots < \iota_3 \le n} \sqrt[q]{1 - \prod_{j=1}^{3}((1 - (\zeta_j)^q)^{\varpi_{\iota_j}})^{\frac{1}{3}}}\right)^{\frac{1}{C_n^3}}, \\ \sqrt[q]{1 - \prod_{1 \le \iota_1 < \cdots < \iota_3 \le n}\left(1 - \left(\prod_{j=1}^{3}(\delta_j)^{\varpi_{\iota_j}}\right)^{\frac{q}{3}}\right)^{\frac{1}{C_n^3}}} \end{array}\right) \end{array}\right). \tag{10}$$

Proof Utilizing Definition 4, we get

$$\varpi_{\iota_j}\eta_{\iota_j} = \left(\begin{array}{c} \Delta\left(\varpi_{\iota_j}\left(\Delta^{-1}\left(s_{\iota_j}, \gamma_{\iota_j}\right)\right)\right), \\ \left(\sqrt[q]{1 - (1 - (\zeta_j)^q)^{\varpi_{\iota_j}}}, (\delta_j)^{\varpi_{\iota_j}}\right) \end{array}\right).$$

Then,

$$\oplus_{j=1}^{3} \varpi_{\iota_j}\eta_{\iota_j} = \left(\Delta\left(\sum_{j=1}^{3} \varpi_{\iota_j}\left(\Delta^{-1}\left(s_{\iota_j}, \gamma_{\iota_j}\right)\right)\right), \left(\sqrt[q]{1 - \prod_{j=1}^{3}(1 - (\zeta_j)^q)^{\varpi_{\iota_j}}}, \prod_{j=1}^{3}(\delta_j)^{\varpi_{\iota_j}}\right)\right).$$

Thus,

$$\left(\frac{\oplus_{j=1}^{3} \varpi_{\iota_j}\eta_{\iota_j}}{3}\right) = \left(\begin{array}{c} \Delta\left(\frac{\sum_{j=1}^{3} \varpi_{\iota_j}(\Delta^{-1}(s_{\iota_j}, \gamma_{\iota_j}))}{3}\right), \\ \left(\sqrt[q]{1 - \left(\prod_{j=1}^{3}(1 - (\zeta_j)^q)^{\varpi_{\iota_j}}\right)^{\frac{1}{3}}}, \left(\prod_{j=1}^{3}(\delta_j)^{\varpi_{\iota_j}}\right)^{\frac{1}{3}}\right) \end{array}\right).$$

Therefore,

$$\otimes_{1\le\imath_1<\cdots<\imath_3\le n}\left(\frac{\oplus_{\jmath=1}^{3}\varpi_{\imath_\jmath}\eta_{\imath_\jmath}}{3}\right)$$

$$=\left(\begin{array}{l}\Delta\left(\prod_{1\le\imath_1<\cdots<\imath_3\le n}\left(\frac{\sum_{\jmath=1}^{3}\varpi_{\imath_\jmath}(\Delta^{-1}(s_{\imath_\jmath},\gamma_{\imath_\jmath}))}{3}\right)\right),\\ \left(\prod_{1\le\imath_1<\cdots<\imath_3\le n}\sqrt[q]{1-\left(\prod_{\jmath=1}^{3}(1-(\zeta_\jmath)^q)^{\varpi_{\imath_\jmath}}\right)^{\frac{1}{3}}},\ \sqrt[q]{1-\prod_{1\le\imath_1<\cdots<\imath_3\le n}\left(1-\left(\prod_{\jmath=1}^{3}(\delta_\jmath)^{\varpi_{\imath_\jmath}}\right)^{\frac{q}{3}}\right)}\right)\end{array}\right).$$

Furthermore,

$$q\text{-ROFTLWDHM}_{\varpi}^{(3)}(\eta_1,\eta_2,\ldots,\eta_n)$$

$$=\left(\begin{array}{l}\Delta\left(\prod_{1\le\imath_1<\cdots<\imath_3\le n}\left(\frac{\sum_{\jmath=1}^{3}\varpi_{\imath_\jmath}(\Delta^{-1}(s_{\imath_\jmath},\gamma_{\imath_\jmath}))}{3}\right)\right)^{\frac{1}{C_n^3}},\\ \left(\left(\prod_{1\le\imath_1<\cdots<\imath_3\le n}\sqrt[q]{1-\prod_{\jmath=1}^{3}((1-(\zeta_\jmath)^q)^{\varpi_{\imath_\jmath}})^{\frac{1}{3}}}\right)^{\frac{1}{C_n^3}},\ \sqrt[q]{1-\prod_{1\le\imath_1<\cdots<\imath_3\le n}\left(1-\left(\prod_{\jmath=1}^{3}(\delta_\jmath)^{\varpi_{\imath_\jmath}}\right)^{\frac{q}{3}}\right)^{\frac{1}{C_n^3}}}\right)\end{array}\right).$$

Proerty 9 (Monotonicity) *Let* $\eta_\jmath=((s_\jmath,\gamma_\jmath),(\zeta_\jmath,\delta_\jmath)),\eta'_\jmath=((s'_\jmath,\gamma'_\jmath),(\zeta'_\jmath,\delta'_\jmath))(\jmath=1,2,\ldots,n)$ *be two sets of q-ROFTLNs, and if* $\eta_\jmath\le\eta'_\jmath$, *for all* $\jmath$, *then*

$$q\text{-}ROFTLWDHM_{\varpi}^{(3)}(\eta_1,\eta_2,\ldots,\eta_n)\le q\text{-}ROFTLWDHM_{\varpi}^{(3)}(\eta'_1,\eta'_2,\ldots,\eta'_n).$$

Proerty 10 (Boundedness) *Let* $\eta_\jmath=((s_\jmath,\gamma_\jmath),(\zeta_\jmath,\delta_\jmath))(\jmath=1,2,\ldots,n)$ *be a set of q-ROFTLNs, and let* $\eta^-=\min_\jmath((s_\jmath,\gamma_\jmath),(\zeta_\jmath,\delta_\jmath))$ *and* $\eta^+=\max_\jmath((s_\jmath,\gamma_\jmath),(\zeta_\jmath,\delta_\jmath))$, *then*

$$\eta^-\le q\text{-}ROFTLWDHM_{\varpi}^{(3)}(\eta_1,\eta_2,\ldots,\eta_n)\le\eta^+.$$

The q-ROFTLWDHM operator loses the idempotency property.

4 EDAS Model for MAGDM with q-ROFTLNs

In many MAGDM scenarios, the classical EDAS method can take into account the competing qualities. By measuring the AS depending on two distance measurements, PDAS and NDAS, this framework can explain the variance among all options. The alternative with a greater PDAS and a lower NDAS is the best.

Under the q-ROFTL environment, we develop an interconnected framework that combines the EDAS with q-ROFTLNs. Therefore, based on the q-ROFTLWHM (q-ROFTLWDHM) operator, we propose the q-ROFTL-EDAS approach in order to deal with MAGDM problems. The goal of the MAGDM process is to choose the best alternative from a set of $\mathfrak{m}$ alternatives $\Im = \{\Im_1, \Im_2, \ldots, \Im_\mathfrak{m}\}$ under the attribute set $\hbar = \{\hbar_1, \hbar_2, \ldots, \hbar_\mathfrak{n}\}$. Assume a group of DEs appointed to serve on a panel $\mathfrak{D} = \{\mathfrak{D}_1, \mathfrak{D}_2, \ldots, \mathfrak{D}_\mathfrak{l}\}$, which was formulated in order to find the optimal alternative(s). Let $\varpi = (\varpi_1, \varpi_2, \ldots, \varpi_\mathfrak{n})^\mathrm{T}$ and $\varpi* = (\varpi_1^*, \varpi_2^*, \ldots, \varpi_\mathfrak{l}^*)^\mathrm{T}$ be the weighting vector of the attributes, and the DEs satisfying $\varpi_\jmath \in [0, 1]$, $\varpi_\lambda^* \in [0, 1]$, $\sum_{\jmath=1}^{\mathfrak{n}} \varpi_\jmath = 1$ and $\sum_{\lambda=1}^{\mathfrak{l}} \varpi_\lambda^* = 1$. The technique of implementing the q-ROFTL-EDAS approach is described in the following steps:

Phase 1. Construct the evaluation matrix by utilizing the q-ROFTLNs.

$$\mathfrak{L}^\lambda = [\eta_{\imath\jmath}^\lambda]_{\mathfrak{m}\times\mathfrak{n}} = \begin{bmatrix} \eta_{11}^\lambda & \eta_{12}^\lambda & \cdots & \eta_{1\mathfrak{n}}^\lambda \\ \eta_{21}^\lambda & \eta_{22}^\lambda & \cdots & \eta_{2\mathfrak{n}}^\lambda \\ \vdots & \vdots & \vdots & \vdots \\ \eta_{\mathfrak{m}1}^\lambda & \eta_{\mathfrak{m}2}^\lambda & \cdots & \eta_{\mathfrak{m}\mathfrak{n}}^\lambda \end{bmatrix}_{\mathfrak{m}\times\mathfrak{n}}, \tag{11}$$
$$\imath = 1, 2, \ldots, \mathfrak{m},\ \jmath = 1, 2, \ldots, \mathfrak{n},\ \lambda = 1, 2, \ldots, \mathfrak{l}.$$

where $\eta_{\imath\jmath}^\lambda$ represents the q-ROFTL information of alternatives $\Im_\imath$ on attributes $\hbar_\jmath$ by DEs $\mathfrak{D}_\lambda$.

Phase 2. Normalize the evaluation matrix $\mathfrak{L}^\lambda = [\eta_{\imath\jmath}^\lambda]_{\mathfrak{m}\times\mathfrak{n}}$ by using a formula determined by the kind of each attribute:

For benefit attributes:

$$\eta'_{\imath\jmath} = \eta_{\imath\jmath},\ \imath = 1, 2, \ldots, \mathfrak{m},\ \jmath = 1, 2, \ldots, \mathfrak{n}. \tag{12}$$

For cost attributes:

$$\eta'_{\imath\jmath} = \eta_{\imath\jmath}^c,\ \imath = 1, 2, \ldots, \mathfrak{m},\ \jmath = 1, 2, \ldots, \mathfrak{n}. \tag{13}$$

Phase 3. By utilizing the aggregation operators of Eqs. (6) and (10), the fused matrix can be obtained as follows:

$$\mathfrak{L} = [\eta'_{\imath\jmath}]_{\mathrm{m}\times\mathrm{n}} = \begin{bmatrix} \eta'_{11} & \eta'_{12} & \cdots & \eta'_{1\mathrm{n}} \\ \eta'_{21} & \eta'_{22} & \cdots & \eta'_{2\mathrm{n}} \\ \vdots & \vdots & \vdots & \vdots \\ \eta'_{\mathrm{m}1} & \eta'_{\mathrm{m}2} & \cdots & \eta'_{\mathrm{mn}} \end{bmatrix}_{\mathrm{m}\times\mathrm{n}}, \imath = 1, 2, \ldots, \mathrm{m}, \ \jmath = 1, 2, \ldots, \mathrm{n}. \tag{14}$$

where $\eta'_{\imath\jmath}$ indicates the aggregated q-ROFTL information of alternatives $\Im_\imath(\imath = 1, 2, \ldots, \mathrm{m})$ on attributes $\hbar_\jmath(\jmath = 1, 2, \ldots, \mathrm{n})$.

Phase 4. Compute the AS of all the alternatives under each attribute.

$$\mathrm{AS} = (\mathrm{AS}_1, \mathrm{AS}_2, \ldots, \mathrm{AS}_\mathrm{n}),$$

where $\mathrm{AS}_\jmath = \eta'_\jmath = ((s'_\jmath, \gamma'_\jmath), (\zeta'_\jmath, \delta'_\jmath)) = \frac{1}{\mathrm{m}}\eta'_1 \oplus \frac{1}{\mathrm{m}}\eta'_2 \oplus \cdots \oplus \frac{1}{\mathrm{m}}\eta'_\mathrm{m}$

$$= \left(\Delta\left(\frac{1}{\mathrm{m}}\sum_{\imath=1}^{\mathrm{m}}\Delta^{-1}\left(s'_\jmath, \gamma'_\jmath\right)\right), \left(\sqrt[q]{1 - \prod_{\imath=1}^{\mathrm{m}}(1-(\zeta'_{\imath\jmath})^q)^{\frac{1}{\mathrm{m}}}}, \prod_{\imath=1}^{\mathrm{m}}\left(\delta'_{\imath\jmath}\right)^{\frac{1}{\mathrm{m}}}\right)\right). \tag{15}$$

Phase 5. According to different types of attributes, calculate the PDAS matrix and the NDAS matrix.

$$\mathrm{PDAS} = (\mathrm{PDAS}_{\imath\jmath})_{\mathrm{m}\times\mathrm{n}}, \ \mathrm{NDAS} = (\mathrm{NDAS}_{\imath\jmath})_{\mathrm{m}\times\mathrm{n}},$$

$$\begin{aligned} \mathrm{PDAS}_{\imath\jmath} &= \frac{\max\left(0, \left(S\left(\eta'_{\imath\jmath}\right) - S\left(\mathrm{AS}_\jmath\right)\right)\right)}{S\left(\mathrm{AS}_\jmath\right)}, \\ \mathrm{NDAS}_{\imath\jmath} &= \frac{\max\left(0, \left(S\left(\mathrm{AS}_\jmath\right) - S\left(\eta'_{\imath\jmath}\right)\right)\right)}{S\left(\mathrm{AS}_\jmath\right)}, \end{aligned} \tag{16}$$

where $S(\eta'_{\imath\jmath})$ and $S(\mathrm{AS}_\jmath)$ are the score functions.

Phase 6. Compute the positive weighted distance $\mathrm{SP}_\imath(\imath = 1, 2, \ldots, \mathrm{m})$ and the negative weighted distance $\mathrm{SN}_\imath(\imath = 1, 2, \ldots, \mathrm{m})$:

$$\mathrm{SP}_\imath = \sum_{\jmath=1}^{\mathrm{n}} \varpi_\jmath \mathrm{PDAS}_{\imath\jmath}, \ \mathrm{SN}_\imath = \sum_{\jmath=1}^{\mathrm{n}} \varpi_\jmath \mathrm{NDAS}_{\imath\jmath}, \tag{17}$$

where $\varpi_\jmath \in [0, 1]$, $\sum_{\jmath=1}^{\mathrm{n}} \varpi_\jmath = 1$.

Phase 7. Normalize the $\mathrm{SP}_\imath(\imath = 1, 2, \ldots, \mathrm{m})$ and $\mathrm{SN}_\imath(\imath = 1, 2, \ldots, \mathrm{m})$ by following equation:

$$\mathrm{NSP}_{\imath} = \frac{\mathrm{SP}_{\imath}}{\max\left(\mathrm{SP}_1, \mathrm{SP}_2, \ldots, \mathrm{SP}_{\mathfrak{m}}\right)}, \mathrm{NSN}_{\imath} = 1 - \frac{\mathrm{SN}_{\imath}}{\max\left(\mathrm{SN}_1, \mathrm{SN}_2, \ldots, \mathrm{SN}_{\mathfrak{m}}\right)}, \tag{18}$$

where $\max(\mathrm{SP}_1, \mathrm{SP}_2, \ldots, \mathrm{SP}_{\mathfrak{m}})$ and $\max(\mathrm{SN}_1, \mathrm{SN}_2, \ldots, \mathrm{SN}_{\mathfrak{m}})$ are the maximum distances.

Phase 8. Derive the integrative appraisal score ($\mathrm{IAS}_{\imath}$) ($\imath = 1, 2, \ldots, \mathfrak{m}$) by following equation:

$$\mathrm{IAS}_{\imath} = \frac{1}{2}\left(\mathrm{NSP}_{\imath} + \mathrm{NSN}_{\imath}\right), \tag{19}$$

where $\mathrm{IAS}_{\imath} \in [0, 1]$

Phase 9. Derive the ordering in accordance with the results of $\mathrm{IAS}_{\imath}$ ($\imath = 1, 2, \ldots, \mathfrak{m}$). The larger the $\mathrm{IAS}_{\imath}$ ($\imath = 1, 2, \ldots, \mathfrak{m}$), the better the alternative is.

The description of the steps of proposed model is given in the following flowchart (see Fig. 1).

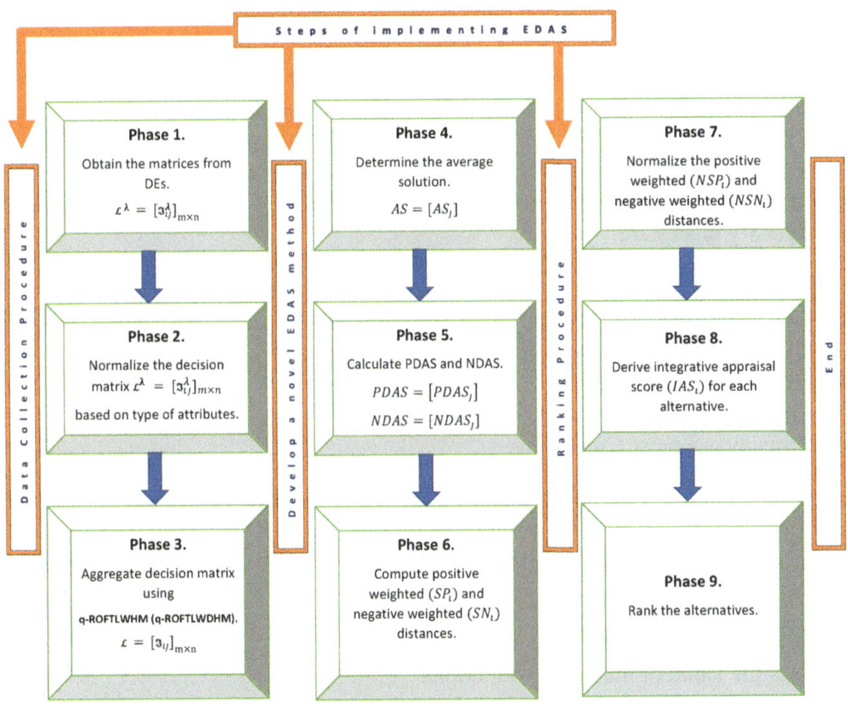

Fig. 1 Flowchart of developed MAGDM approach

5 Numerical Illustration

5.1 The Problem Description

In this part, we present a numerical example to evaluate how well our strategy works. The forest bio-diversity vacation area has many applications, which is an essential aspect of the national bio-diversity vacation area. It serves as an essential place for forest eco-tourism. A connected study of forest bio-diversity vacation area of eco-systems is vital for the progress of human happiness and social sustainable development. That is to say, determining the bio-logical worth of the forests' ecological tourist area is a hot spot and dynamic concern in domestic and worldwide bio-diversity academic circles and society. The current urbanization, increasing industrialization, and environmental pollution situations are dire, and many people are living in less realistic conditions as a result of these three major pressures. This emphasizes the need for greater investigation into the forest bio-diversity vacation area of ecological importance. Our proposed q-ROFTL-EDAS model includes determining the environmental attainability of forest bio-diversity vacation areas.

By above analysis, let $\{\Im_1, \Im_2, \Im_3, \Im_4, \Im_5, \Im_6\}$ be a set of six forest bio-diversity vacation places and let $\{\hbar_1, \hbar_2, \hbar_3, \hbar_4, \hbar_5\}$ be a set of five attributes with weighting vector $\varpi = (0.16, 0.31, 0.27, 0.11, 0.15)^T$. Suppose, six forest bio-diversity vacation places are evaluated by four experts (Ecologists) $\mathfrak{D} = \{\mathfrak{D}_1, \mathfrak{D}_2, \mathfrak{D}_3, \mathfrak{D}_4\}$, with weighting vector $\varpi* = (0.19, 0.31, 0.17, 0.33)^T$ for choosing the best forest bio-diversity place for vacation. In order to quantify each linguistic term set, four experts $\mathfrak{D}_\mathfrak{l}(\mathfrak{l} = 1, 2, 3, 4)$ provide their opinions. Based on their experience, each DE has opinion for the selection of best forest bio-diversity place for vacation. These forest bio-diversity places are:

(1) Mukshpuri Top Forest ($\Im_1$);
(2) Pain Forest ($\Im_2$);
(3) Rama Meadows Forest ($\Im_3$);
(4) Tropical Dry Forest ($\Im_4$);
(5) Changa Manga Forest ($\Im_5$);
(6) Himalayan Forest ($\Im_6$).

Further details about the six forest bio-diversity vacation places of Pakistan can be seen in Fig. 2.

The six above described forest bio-diversity places are evaluated according to five attributes, including:

(1) $\hbar_1$ is the value of tourism and leisure;
(2) $\hbar_2$ is the value of material production;
(3) $\hbar_3$ is the value of scientific research;
(4) $\hbar_4$ is the value of climatic regulation;
(5) $\hbar_5$ is the value of culture.

Fig. 2 Forest bio-diversity vacation places of Pakistan

Experts should evaluate the effectiveness of forest bio-diversity vacation places concerning all attributes in accordance with their interaction and identify the most suitable one according to the views of ecologists. Each DE uses the q-ROFTLNs to assess each forest bio-diversity vacation place's ability to control the effect of environmental pollution for each attribute.

The evaluation matrix is given in Table 2 by utilizing the q-ROFTLNs given by DEs.

Table 2 The assessing matrix with q-ROFTLNs by four DEs

DEs	Alternatives	Attributes				
		$\hbar_1$	$\hbar_2$	$\hbar_3$	$\hbar_4$	$\hbar_5$
$\mathfrak{D}_1$	$\mathfrak{I}_1$	$((s_5,0),(0.6,0.3))$	$((s_4,0),(0.7,0.5))$	$((s_3,0),(0.8,0.1))$	$((s_2,0),(0.4,0.5))$	$((s_1,0),(0.2,0.3))$
	$\mathfrak{I}_2$	$((s_1,0),(0.7,0.7))$	$((s_2,0),(0.6,0.3))$	$((s_3,0),(0.4,0.6))$	$((s_4,0),(0.4,0.4))$	$((s_5,0),(0.5,0.7))$
	$\mathfrak{I}_3$	$((s_6,0),(0.4,0.5))$	$((s_7,0),(0.6,0.6))$	$((s_8,0),(0.1,0.7))$	$((s_5,0),(0.8,0.3))$	$((s_4,0),(0.8,0.1))$
	$\mathfrak{I}_4$	$((s_8,0),(0.8,0.6))$	$((s_7,0),(0.5,0.3))$	$((s_6,0),(0.8,0.2))$	$((s_4,0),(0.1,0.5))$	$((s_5,0),(0.7,0.4))$
	$\mathfrak{I}_5$	$((s_3,0),(0.4,0.6))$	$((s_2,0),(0.8,0.5))$	$((s_1,0),(0.6,0.6))$	$((s_4,0),(0.5,0.6))$	$((s_5,0),(0.8,0.5))$
	$\mathfrak{I}_6$	$((s_7,0),(0.7,0.3))$	$((s_4,0),(0.5,0.6))$	$((s_2,0),(0.8,0.3))$	$((s_6,0),(0.1,0.8))$	$((s_3,0),(0.6,0.2))$
$\mathfrak{D}_2$	$\mathfrak{I}_1$	$((s_1,0),(0.5,0.2))$	$((s_2,0),(0.4,0.4))$	$((s_3,0),(0.3,0.9))$	$((s_4,0),(0.4,0.5))$	$((s_5,0),(0.5,0.7))$
	$\mathfrak{I}_2$	$((s_1,0),(0.6,0.5))$	$((s_2,0),(0.4,0.5))$	$((s_3,0),(0.2,0.6))$	$((s_4,0),(0.3,0.6))$	$((s_5,0),(0.4,0.5))$
	$\mathfrak{I}_3$	$((s_4,0),(0.4,0.3))$	$((s_5,0),(0.5,0.3))$	$((s_6,0),(0.8,0.6))$	$((s_7,0),(0.3,0.3))$	$((s_8,0),(0.8,0.2))$
	$\mathfrak{I}_4$	$((s_4,0),(0.1,0.9))$	$((s_5,0),(0.2,0.4))$	$((s_8,0),(0.3,0.4))$	$((s_7,0),(0.8,0.6))$	$((s_6,0),(0.2,0.5))$
	$\mathfrak{I}_5$	$((s_2,0),(0.2,0.6))$	$((s_4,0),(0.9,0.1))$	$((s_6,0),(0.7,0.7))$	$((s_8,0),(0.4,0.5))$	$((s_1,0),(0.7,0.1))$
	$\mathfrak{I}_6$	$((s_8,0),(0.6,0.2))$	$((s_6,0),(0.8,0.2))$	$((s_4,0),(0.5,0.4))$	$((s_2,0),(0.9,0.4))$	$((s_2,0),(0.6,0.3))$
$\mathfrak{D}_3$	$\mathfrak{I}_1$	$((s_8,0),(0.2,0.5))$	$((s_7,0),(0.8,0.2))$	$((s_6,0),(0.5,0.6))$	$((s_5,0),(0.8,0.4))$	$((s_4,0),(0.1,0.4))$
	$\mathfrak{I}_2$	$((s_4,0),(0.5,0.7))$	$((s_5,0),(0.3,0.9))$	$((s_8,0),(0.6,0.4))$	$((s_7,0),(0.7,0.5))$	$((s_6,0),(0.5,0.6))$
	$\mathfrak{I}_3$	$((s_2,0),(0.4,0.6))$	$((s_3,0),(0.7,0.7))$	$((s_5,0),(0.7,0.2))$	$((s_7,0),(0.5,0.6))$	$((s_1,0),(0.8,0.4))$
	$\mathfrak{I}_4$	$((s_1,0),(0.5,0.8))$	$((s_3,0),(0.6,0.3))$	$((s_5,0),(0.5,0.5))$	$((s_7,0),(0.1,0.6))$	$((s_9,0),(0.3,0.7))$
	$\mathfrak{I}_5$	$((s_2,0),(0.7,0.2))$	$((s_4,0),(0.7,0.3))$	$((s_6,0),(0.4,0.6))$	$((s_8,0),(0.4,0.7))$	$((s_7,0),(0.6,0.4))$
	$\mathfrak{I}_6$	$((s_8,0),(0.2,0.6))$	$((s_6,0),(0.3,0.8))$	$((s_4,0),(0.9,0.4))$	$((s_2,0),(0.6,0.6))$	$((s_3,0),(0.9,0.1))$
$\mathfrak{D}_4$	$\mathfrak{I}_1$	$((s_2,0),(0.3,0.6))$	$((s_4,0),(0.5,0.7))$	$((s_6,0),(0.7,0.3))$	$((s_8,0),(0.4,0.8))$	$((s_4,0),(0.7,0.4))$
	$\mathfrak{I}_2$	$((s_8,0),(0.9,0.3))$	$((s_6,0),(0.6,0.6))$	$((s_4,0),(0.9,0.2))$	$((s_2,0),(0.5,0.5))$	$((s_2,0),(0.5,0.2))$
	$\mathfrak{I}_3$	$((s_1,0),(0.4,0.6))$	$((s_3,0),(0.8,0.4))$	$((s_5,0),(0.5,0.6))$	$((s_7,0),(0.9,0.1))$	$((s_5,0),(0.8,0.3))$
	$\mathfrak{I}_4$	$((s_5,0),(0.7,0.2))$	$((s_7,0),(0.5,0.5))$	$((s_5,0),(0.6,0.5))$	$((s_3,0),(0.7,0.4))$	$((s_1,0),(0.4,0.5))$
	$\mathfrak{I}_5$	$((s_5,0),(0.3,0.8))$	$((s_4,0),(0.7,0.2))$	$((s_7,0),(0.8,0.6))$	$((s_8,0),(0.9,0.2))$	$((s_6,0),(0.5,0.6))$
	$\mathfrak{I}_6$	$((s_4,0),(0.7,0.2))$	$((s_5,0),(0.6,0.5))$	$((s_2,0),(0.4,0.9))$	$((s_3,0),(0.3,0.7))$	$((s_1,0),(0.3,0.9))$

Table 3 Collective q-ROFTL assessing matrix by the q-ROFTLWHM operator

	$\hbar_1$	$\hbar_2$	$\hbar_3$
$\mathfrak{I}_1$	$((s_6, 0.0011), (0.7922, 0.3083))$	$((s_7, -0.0559), (0.8730, 0.3634))$	$((s_7, -0.0203), (0.8693, 0.4175))$
$\mathfrak{I}_2$	$((s_6, 0.0066), (0.9143, 0.4066))$	$((s_6, 0.0577), (0.8330, 0.4581))$	$((s_7, -0.0378), (0.8555, 0.3460))$
$\mathfrak{I}_3$	$((s_6, -0.0008), (0.7991, 0.3686))$	$((s_7, -0.0258), (0.9008, 0.3656))$	$((s_7, 0.0476), (0.8463, 0.4252))$
$\mathfrak{I}_4$	$((s_7, -0.0382), (0.8436, 0.5206))$	$((s_7, 0.0345), (0.8125, 0.2815))$	$((s_7, 0.0556), (0.8568, 0.3085))$
$\mathfrak{I}_5$	$((s_6, 0.0289), (0.7790, 0.4596))$	$((s_7, -0.0609), (0.9418, 0.2093))$	$((s_7, -0.0063), (0.8980, 0.4525))$
$\mathfrak{I}_6$	$((s_8, -0.0601), (0.8644, 0.2369))$	$((s_7, 0.0245), (0.8697, 0.4081))$	$((s_6, 0.0204), (0.8908, 0.4053))$
	$\hbar_4$	$\hbar_5$	
$\mathfrak{I}_1$	$((s_7, -0.0036), (0.8308, 0.4238))$	$((s_6, 0.0526), (0.7816, 0.3452))$	
$\mathfrak{I}_2$	$((s_7, -0.0583), (0.8239, 0.3658))$	$((s_7, -0.0406), (0.8316, 0.3815))$	
$\mathfrak{I}_3$	$((s_8, -0.0481), (0.8959, 0.2510))$	$((s_7, -0.0207), (0.9466, 0.1962))$	
$\mathfrak{I}_4$	$((s_7, 0.0053), (0.8010, 0.3767))$	$((s_7, -0.0549), (0.7847, 0.3818))$	
$\mathfrak{I}_5$	$((s_8, -0.0319), (0.8698, 0.3807))$	$((s_7, -0.0583), (0.8973, 0.3242))$	
$\mathfrak{I}_6$	$((s_6, 0.0228), (0.8241, 0.4678))$	$((s_6, -0.0522), (0.8712, 0.3189))$	

Table 4 Scores of fused q-ROFTLNs matrix

Alternatives	Scores				
	$\hbar_1$	$\hbar_2$	$\hbar_3$	$\hbar_4$	$\hbar_5$
$\mathfrak{I}_1$	0.5553	0.6643	0.6656	0.6331	0.5466
$\mathfrak{I}_2$	0.6703	0.6107	0.6572	0.6326	0.5860
$\mathfrak{I}_3$	0.5572	0.6971	0.6395	0.7191	0.7244
$\mathfrak{I}_4$	0.5747	0.6074	0.6609	0.6101	0.5461
$\mathfrak{I}_5$	0.5308	0.7584	0.6949	0.6802	0.6585
$\mathfrak{I}_6$	0.6237	0.6562	0.6924	0.6197	0.6297

5.2 *The Outcomes of a Case Study*

5.2.1 Decision-Making Process by Using q-ROFTLWHM Operator

Phase 1. According to the q-ROFTLWHM aggregation operator, we utilize overall $\eta'^{\lambda}_{\iota J}$ to $\eta'_{\iota J}$, the fused q-ROFTLNs matrix $\mathfrak{L} = [\eta'_{\iota J}]_{m \times n}$ given in Table 3. (Suppose $q = 4$, $\tau = 8$, $\varpi* = (0.19, 0.31, 0.17, 0.33)^{T}$).

Phase 2. Determining the scores of fused q-ROFTLNs matrix (see Table 4).

Phase 3. Compute the AS matrix $\text{AS} = [\text{AS}_J]_{5\times 3}$ by utilizing the attributes weights. The results of AS matrix $\text{AS} = [\text{AS}_J]_{5\times 3}$ are given in Table 5.

Phase 4. According to the scores of fused q-ROFTLNs matrix and AS matrix, calculate the PDAS and NDAS. The results of $\text{PDAS} = [\text{PDAS}_\iota]_{6\times 5}$ and the results of $\text{NDAS} = [\text{NDAS}_\iota]_{6\times 5}$ are given in Table 6.

Phase 5. According to the outcomes of PDAS and NDAS, calculate the assessing values of positive weighted distance $\text{SP}_\iota (\iota = 1, 2, \ldots, 6)$ and the negative

Table 5 Assessing AS matrix with q-ROFTLNs

AS's	The q-ROFTLNs for AS's	Scores
AS_1	$((s_6, 0.0521), (0.8426, 0.3713))$	0.5955
AS_2	$((s_7, -0.0252), (0.8818, 0.3370))$	0.6763
AS_3	$((s_7, -0.0110), (0.8712, 0.3892))$	0.6710
AS_4	$((s_7, 0.0019), (0.8456, 0.3711))$	0.6543
AS_5	$((s_7, -0.0036), (0.8705, 0.3173))$	0.6290

Table 6 Assessing values of PDAS and NDAS

Alternatives	PDAS					NDAS				
	$\hbar_1$	$\hbar_2$	$\hbar_3$	$\hbar_4$	$\hbar_5$	$\hbar_1$	$\hbar_2$	$\hbar_3$	$\hbar_4$	$\hbar_5$
$\Im_1$	0.0000	0.0000	0.0000	0.0000	0.0000	0.0675	0.0177	0.0081	0.0324	0.1310
$\Im_2$	0.1255	0.0000	0.0000	0.0000	0.0000	0.0000	0.0970	0.0206	0.0332	0.0684
$\Im_3$	0.0000	0.0307	0.0000	0.0990	0.1517	0.0644	0.0000	0.0469	0.0000	0.0000
$\Im_4$	0.0000	0.0000	0.0000	0.0000	0.0000	0.0350	0.1019	0.0151	0.0676	0.1317
$\Im_5$	0.0000	0.1213	0.0355	0.0395	0.0469	0.1086	0.0000	0.0000	0.0000	0.0000
$\Im_6$	0.0472	0.0000	0.0318	0.0000	0.0012	0.0000	0.0297	0.0000	0.0529	0.0000

Table 7 Assessing values of SP_ι, SN_ι, NSP_ι, and NSN_ι $(\iota = 1, 2, \ldots, 6)$

Alternatives	The outcomes of SP	The outcomes of SN	The outcomes of NSP	The outcomes of NSN
$\Im_1$	0.0000	0.0417	0.0000	0.3907
$\Im_2$	0.0201	0.0495	0.3431	0.2769
$\Im_3$	0.0432	0.0230	0.7375	0.6641
$\Im_4$	0.0000	0.0685	0.0000	0.0000
$\Im_5$	0.0586	0.0174	1.0000	0.7460
$\Im_6$	0.0163	0.0150	0.2786	0.7802

weighted distance SN_ι $(\iota = 1, 2, \ldots, 6)$. The results of SP and SN are given in Table 7.

Phase 6. According to the outcomes of SP and SN, calculate the normalized positive weighted distance NSP_ι $(\iota = 1, 2, \ldots, 6)$ and the normalized negative weighted distance NSN_ι $(\iota = 1, 2, \ldots, 6)$. The results of NSP and NSN are given in Table 7.

Phase 7. Derive the integrative appraisal score (IAS_ι) $(\iota = 1, 2, \ldots, 6)$ by following equation:

$$IAS_\iota = \frac{1}{2}\left(NSP_\iota + NSN_\iota\right), \tag{20}$$

where $\text{IAS}_\iota \in [0, 1]$. The computed results are shown below:

$$\text{IAS}_1 = 0.1955, \text{IAS}_2 = 0.3096, \text{IAS}_3 = 0.7006,$$
$$\text{IAS}_4 = 0, \text{IAS}_5 = 0.8731, \text{IAS}_6 = 0.5294$$

Phase 8. Derive the ordering in accordance with the results of $(\text{IAS}_\iota)(\iota = 1, 2, \ldots, 6)$. The larger the $\text{IAS}_\iota (\iota = 1, 2, \ldots, 6)$, the better the alternative is. The ranking of bio-diversity vacation places is as follows: $\Im_5 > \Im_3 > \Im_6 > \Im_2 > \Im_1 > \Im_4$. So, $\Im_5$ is the best alternative.

5.2.2 Decision-Making Process by Using q-ROFTLWDHM Operator

Phase 1. According to the q-ROFTLWDHM aggregation operator, we utilize overall $\eta'^{\lambda}_{\iota j}$ to $\eta'_{\iota j}$, the fused q-ROFTLNs matrix $\mathfrak{L} = [\eta'_{\iota j}]_{m \times n}$ given in Table 8. (Suppose $q = 4$, $\tau = 8$, $\varpi* = (0.19, 0.31, 0.17, 0.33)^T$).

Phase 2. Determine the scores of fused q-ROFTLNs matrix (see Table 9).

Phase 3. Compute the AS matrix $\text{AS} = [\text{AS}_j]_{5\times3}$ by utilizing the attributes weights. The results of AS matrix $\text{AS} = [\text{AS}_j]_{5\times3}$ are given in Table 10.

Phase 4. According to the scores of fused q-ROFTLNs matrix and AS matrix, calculate the PDAS and NDAS. The results of $\text{PDAS} = [\text{PDAS}_\iota]_{6\times5}$ and the results of $\text{NDAS} = [\text{NDAS}_\iota]_{6\times5}$ are given in Table 11.

Phase 5. According to the outcomes of PDAS and NDAS, calculate the assessing values of positive weighted distance $\text{SP}_\iota (\iota = 1, 2, \ldots, 6)$ and the negative weighted distance $\text{SN}_\iota (\iota = 1, 2, \ldots, 6)$. The results of SP and SN are given in Table 12.

Phase 6. According to the outcomes of SP and SN, calculate the normalized positive weighted distance $\text{NSP}_\iota (\iota = 1, 2, \ldots, 6)$ and the normalized negative weighted distance $\text{NSN}_\iota (\iota = 1, 2, \ldots, 6)$. The results of NSP and NSN are given in Table 12.

Phase 7. Derive the integrative appraisal score $(\text{IAS}_\iota)(\iota = 1, 2, \ldots, 6)$ by following equation:

$$\text{IAS}_\iota = \frac{1}{2}\left(\text{NSP}_\iota + \text{NSN}_\iota\right), \tag{21}$$

where $\text{IAS}_\iota \in [0, 1]$. The computed results are shown below:

$$\text{IAS}_1 = 0.4370, \text{IAS}_2 = 0.1031, \text{IAS}_3 = 0.7333,$$
$$\text{IAS}_4 = 0.2857, \text{IAS}_5 = 0.5354, \text{IAS}_6 = 0.4951.$$

Phase 8. Derive the ordering in accordance with the results of $(\text{IAS}_\iota)(\iota = 1, 2, \ldots, 6)$. The larger the $\text{IAS}_\iota (\iota = 1, 2, \ldots, 6)$, the better the alternative is. The ranking

Table 8 Collective q-ROFTL assessing matrix by the q-ROFTLWDHM operator

	$\hbar_1$	$\hbar_2$	$\hbar_3$
$\Im_1$	$((s_1, -0.0265), (0.3122, 0.7940))$	$((s_1, -0.0053), (0.4398, 0.8235))$	$((s_1, 0.0104), (0.4454, 0.8241))$
$\Im_2$	$((s_1, -0.0305), (0.5269, 0.8537))$	$((s_1, -0.0140), (0.3612, 0.8630))$	$((s_1, 0.0035), (0.4348, 0.8162))$
$\Im_3$	$((s_1, -0.0355), (0.2825, 0.8397))$	$((s_1, 0.0089), (0.4820, 0.8317))$	$((s_1, 0.0567), (0.4532, 0.8524))$
$\Im_4$	$((s_1, 0.0117), (0.4408, 0.8929))$	$((s_1, 0.0469), (0.3393, 0.7901))$	$((s_1, 0.0608), (0.4189, 0.7959))$
$\Im_5$	$((s_1, -0.0326), (0.2978, 0.8681))$	$((s_1, -0.0156), (0.5862, 0.7090))$	$((s_1, 0.0317), (0.4895, 0.8923))$
$\Im_6$	$((s_2, -0.0504), (0.4513, 0.7388))$	$((s_1, 0.0370), (0.4414, 0.8391))$	$((s_1, -0.0353), (0.4963, 0.8489))$
	$\hbar_4$	$\hbar_5$	
$\Im_1$	$((s_1, 0.0221), (0.3679, 0.8693))$	$((s_1, -0.0153), (0.3234, 0.8247))$	
$\Im_2$	$((s_1, -0.0061), (0.3505, 0.8457))$	$((s_1, 0.0035), (0.3360, 0.8320))$	
$\Im_3$	$((s_2, -0.0478), (0.5107, 0.7361))$	$((s_1, 0.0182), (0.5907, 0.6992))$	
$\Im_4$	$((s_1, 0.0299), (0.3622, 0.8514))$	$((s_1, 0.0144), (0.3054, 0.8491))$	
$\Im_5$	$((s_2, -0.0313), (0.4369, 0.8311))$	$((s_1, 0.0063), (0.4768, 0.7893))$	
$\Im_6$	$((s_1, -0.0312), (0.4175, 0.8882))$	$((s_1, -0.0621), (0.4646, 0.7742))$	

Table 9 Scores of fused q-ROFTLNs matrix

Alternatives	Scores				
	$\hbar_1$	$\hbar_2$	$\hbar_3$	$\hbar_4$	$\hbar_5$
$\Im_1$	0.0346	0.0383	0.0411	0.0335	0.0316
$\Im_2$	0.0309	0.0306	0.0421	0.0377	0.0307
$\Im_3$	0.0288	0.0381	0.0365	0.0579	0.0508
$\Im_4$	0.0228	0.0413	0.0447	0.0368	0.0281
$\Im_5$	0.0249	0.0573	0.0301	0.0419	0.0382
$\Im_6$	0.0421	0.0359	0.0385	0.0305	0.0396

Table 10 Assessing AS matrix with q-ROFTLNs

AS's	The q-ROFTLNs for AS's	Scores
AS_1	$((s_1, -0.0118), (0.3744, 0.8407))$	0.0294
AS_2	$((s_1, 0.0075), (0.4344, 0.8173))$	0.0391
AS_3	$((s_1, 0.0172), (0.4555, 0.8426))$	0.0383
AS_4	$((s_1, 0.0247), (0.4040, 0.8450))$	0.0387
AS_5	$((s_1, 0.0221), (0.4039, 0.8026))$	0.0352

Table 11 Assessing values of PDAS and NDAS

Alternatives	PDAS					NDAS				
	$\hbar_1$	$\hbar_2$	$\hbar_3$	$\hbar_4$	$\hbar_5$	$\hbar_1$	$\hbar_2$	$\hbar_3$	$\hbar_4$	$\hbar_5$
$\Im_1$	0.1769	0.0000	0.0724	0.0000	0.0000	0.0000	0.0203	0.0000	0.1348	0.1035
$\Im_2$	0.0498	0.0000	0.0980	0.0000	0.0000	0.0000	0.2157	0.0000	0.0256	0.1276
$\Im_3$	0.0000	0.0000	0.0000	0.4984	0.4432	0.0208	0.0000	0.0462	0.0000	0.0000
$\Im_4$	0.0000	0.0580	0.1677	0.0000	0.0000	0.2268	0.0000	0.0000	0.0486	0.2007
$\Im_5$	0.0000	0.4682	0.0000	0.0824	0.0848	0.1540	0.0000	0.2145	0.0000	0.0000
$\Im_6$	0.4297	0.0000	0.0041	0.0000	0.1236	0.0000	0.0800	0.0000	0.2103	0.0000

Table 12 Assessing values of SP_ι, SN_ι, NSP_ι, and NSN_ι $(\iota = 1, 2, \ldots, 6)$

Alternatives	The outcomes of SP	The outcomes of SN	The outcomes of NSP	The outcomes of NSN
$\Im_1$	0.0478	0.0366	0.2866	0.5874
$\Im_2$	0.0344	0.0888	0.2062	0.0000
$\Im_3$	0.1213	0.0231	0.7267	0.7399
$\Im_4$	0.0633	0.0717	0.3789	0.1924
$\Im_5$	0.1669	0.0825	1.0000	0.0708
$\Im_6$	0.0884	0.0479	0.5295	0.4606

of bio-diversity vacation places is as follows: $\Im_3 > \Im_5 > \Im_6 > \Im_1 > \Im_4 > \Im_2$. So, $\Im_3$ is the best alternative.

5.3 Influence of Parameters

In this research study, the parameter $\mathfrak{z}$ is utilized to describe the interrelationship between distinct quantifiable attributes, and the various numerical values of parameter $\mathfrak{z}$ depict the multiple decision-making scenarios and circumstances. The parameter $\mathfrak{z}$ is providing alternate numerical values using the numerical illustration data in Table 2 under the assumption of maintaining q constant ($q = 4$). Distinct appraisal scoring outcomes are obtained by varying the numerical value of the parameter $\mathfrak{z}$ and the results are depicted in Tables 13, 14 and Figs. 3, 4. Tables 13, 14 give the scoring values of each alternative vary depending on the numerical value of parameter $\mathfrak{z}$, and the obtaining results are approximately the same. The best bio-diversity vacation places are $\Im_3$ and $\Im_5$ when we assign distinct numerical values to parameter $\mathfrak{z}$. When $\mathfrak{z} = 1, 2, 3, 4$, the obtaining results order varies from $\Im_5 > \Im_6 > \Im_3 > \Im_2 > \Im_1 > \Im_4$ to $\Im_5 > \Im_3 > \Im_6 > \Im_2 > \Im_1 > \Im_4$ by utilizing q-ROFTLWHM operator, and the other ordering is $\Im_3 > \Im_5 > \Im_4 > \Im_1 > \Im_6 > \Im_2$, $\Im_3 > \Im_5 > \Im_6 > \Im_1 > \Im_4 > \Im_2$, and $\Im_3 > \Im_6 > \Im_5 > \Im_1 > \Im_4 > \Im_2$ by utilizing q-ROFTLWDHM operator. As a result, it can be seen that changing the numerical value of $\mathfrak{z}$ affects the resulting outcomes. The DEs can alternate the numerical value of parameter $\mathfrak{z}$ following the actual demands to accommodate the complicated scenario and acquire the assessment results during the real decision-making procedure. We can also conclude from the appraisal scoring results as given in Tables 15, 16 and Figs. 5, 6, 7, 8, 9, 10, 11, 12, 13, and 14 that the q-ROFTLWHM and q-ROFTLWDHM aggregation operators proposed in this research study are the most preferable approaches to describe aggregated decision information by the q-ROFTLNs as these two operators enable the information aggregation procedure more reliable with parameter q. When $q \in (1, 20)$ the obtaining results varies from $\Im_5 > \Im_3 > \Im_6 > \Im_1 > \Im_2 > \Im_4$ to $\Im_5 > \Im_3 > \Im_6 > \Im_2 > \Im_1 > \Im_4$ by utilizing q-ROFTLWHM operator and the other order values varies in different ways by utilizing q-ROFTLWDHM operator. The best bio-diversity vacation places are $\Im_3$ and $\Im_5$ when we assign distinct numerical values to parameter q. By assigning different numerical values to parameter q, the range of describing the decision information is stronger, and hence decreases the loss of essential aggregated information. As a result, the methodologies describes in this research study are more suited to real-world decision-making problems.

5.4 Comparative Analysis with Existing MAGDM Methods

We compare the new method EDAS based on q-ROFTLNs and q-ROFTLWHM (q-ROFTLWDHM) operator with different methodologies to accurately reflect the

Table 13 Integrative appraisal scores and ranking results of bio-diversity vacation places by q-ROFTLWHM operator ($\mathfrak{z} = 1, 2, 3, 4$ and $q = 4$)

Parameter	IAS($\Im_1$)	IAS($\Im_2$)	IAS($\Im_3$)	IAS($\Im_4$)	IAS($\Im_5$)	IAS($\Im_6$)	Ranking
$\mathfrak{z} = 1$	0.2087	0.2667	0.6937	0.0568	0.8813	0.7165	$\Im_5 > \Im_6 > \Im_3 > \Im_2 > \Im_1 > \Im_4$
$\mathfrak{z} = 2$	0.1955	0.3096	0.7006	0.0000	0.8731	0.5294	$\Im_5 > \Im_3 > \Im_6 > \Im_2 > \Im_1 > \Im_4$
$\mathfrak{z} = 3$	0.1999	0.3377	0.6886	0.0000	0.8779	0.4582	$\Im_5 > \Im_3 > \Im_6 > \Im_2 > \Im_1 > \Im_4$
$\mathfrak{z} = 4$	0.1954	0.3558	0.6877	0.0000	0.8857	0.4009	$\Im_5 > \Im_3 > \Im_6 > \Im_2 > \Im_1 > \Im_4$

Table 14 Integrative appraisal scores and ranking results of bio-diversity vacation places by q-ROFTLWDHM operator ($\mathfrak{z} = 1, 2, 3, 4$ and $q = 4$)

Parameter	IAS($\Im_1$)	IAS($\Im_2$)	IAS($\Im_3$)	IAS($\Im_4$)	IAS($\Im_5$)	IAS($\Im_6$)	Ranking
$\mathfrak{z} = 1$	0.5492	0.1184	0.8241	0.5730	0.7973	0.2714	$\Im_3 > \Im_5 > \Im_4 > \Im_1 > \Im_6 > \Im_2$
$\mathfrak{z} = 2$	0.4370	0.1031	0.7333	0.2857	0.5354	0.4951	$\Im_3 > \Im_5 > \Im_6 > \Im_1 > \Im_4 > \Im_2$
$\mathfrak{z} = 3$	0.3941	0.1482	0.6756	0.3123	0.5000	0.5729	$\Im_3 > \Im_6 > \Im_5 > \Im_1 > \Im_4 > \Im_2$
$\mathfrak{z} = 4$	0.3624	0.2041	0.6416	0.3319	0.5000	0.6071	$\Im_3 > \Im_6 > \Im_5 > \Im_1 > \Im_4 > \Im_2$

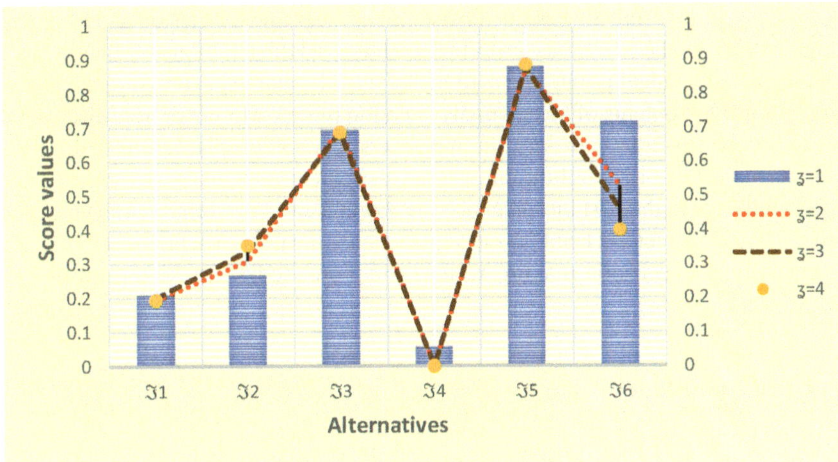

Fig. 3 Scores of bio-diversity vacation places $\Im_\iota (\iota = 1, 2, \ldots, 6)$ when $q = 4$ and $\mathfrak{z} \in (1, 4)$ based on the q-ROFTLWHM operator

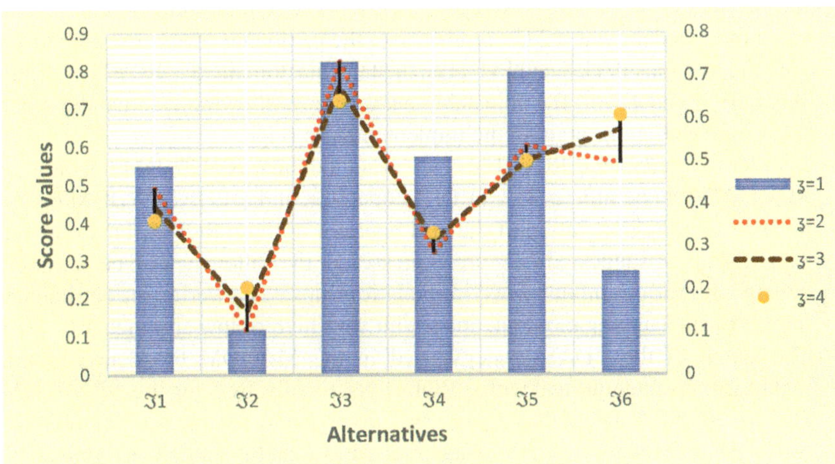

Fig. 4 Scores of bio-diversity vacation places $\Im_\iota (\iota = 1, 2, \ldots, 6)$ when $q = 4$ and $\mathfrak{z} \in (1, 4)$ based on the q-ROFTLWDHM operator

reasonability and effectiveness of the new method introduced in this research paper. However, there are some minor differences in the ordering of judgments, as given by Table 17, 18 and Fig. 15, 16, it is useful to note that the best decisions are consistently different. In rationality, each method has both advantages and disadvantages. To rank the bio-diversity vacation places, the q-ROFTL-EDAS method computes the distances of bio-diversity vacation places from the average ratings of attributes. The q-ROFTL-CODAS method is used to determine the worthiness of bio-diversity vacation places by using the Euclidean distance as the primary criterion and the Hamming distance as the secondary criterion, both are calculated by using the distance from the negative ideal point. The q-ROFTL-TOPSIS method is based on an analysis of all of the problems of bio-diversity vacation places. Moreover, it is discovered that the score function values of the bio-diversity vacation places under different assessments differ only slightly. In general, the given methods can efficiently and successfully select the best bio-diversity vacation places. Furthermore, the proposed q-ROFTL-EDAS method needs to take into account the influence factors of DEs and the uncertainty in decision-making, resulting in more credible ranking results.

6 Conclusions

The following is a summary of the main contributions to the given manuscript:

(1) The goal of this research is to provide a unique MAGDM technique for ranking given q-ROFTLNs in a diverse range of fuzzy environment. To deal with the inconsistencies of data, a q-ROFTLS is used, wherein the information is analyzed independently as a 2TL number with the parameter $q \geq 1$. In comparison with the previous IFTLS and PyFTLS, the feasible domain of q-ROFTLS is substantially broader with the adjustable parameter q, allowing it to handle much more information throughout the implementation stage.

(2) In order to aggregate q-ROFTLNs, we presented the q-ROFTLHM operator, q-ROFTLWHM operator, q-ROFTLDHM operator, and q-ROFTLWDHM operator based on the integration of the HM operator in the q-ROFTL environment. The benefits of suggested AOs are expressed in the following perspectives: (1) coping with various input values' interrelationships; (2) lowering the detrimental impact of final aggregation outcomes with very high or low arguments.

(3) The scope of the EDAS strategy used in the MAGDM broadened in this manuscript by integrating it with q-ROFTLNs. Finally, by using the q-ROFTLW-HM (q-ROFTLWDHM) operator, we proposed a novel decision-making q-ROFTL-EDAS approach. Our technique has benefits and superiorities over various current decision-making approaches, according to the quantitative formulation and comparison assessment.

(4) A numerical example 5.1 is used to demonstrate the advantages of the new MAGDM algorithm over the existing ones. As a result, we conclude that existing techniques are insufficient for distinguishing objects; nevertheless, the suggested MAGDM strategy based on the q-ROFTLS demonstrated to be beneficial. As a result of these contradictory cases, it is demonstrated that the suggested MAGDM approach significantly outperformed the existing methods.

We plan to continue our study from the following aspects. We want to pursue our research in the areas listed as follows: First, we will implement our MAGDM strategy with more realistic decision-making issues, assessment of solid-state disc production [37], also including public analysis of shared bicycles [38], selection of

Table 15 Integrative appraisal scores and ranking results of bio-diversity vacation places by q-ROFTLWHM operator ($\mathfrak{z} = 2$ and $q \in (1, 20)$)

Parameter	IAS($\Im_1$)	IAS($\Im_2$)	IAS($\Im_3$)	IAS($\Im_4$)	IAS($\Im_5$)	IAS($\Im_6$)	Ranking
$q = 1$	0.2064	0.1952	0.7009	0.0834	0.7775	0.4600	$\Im_5 > \Im_3 > \Im_6 > \Im_1 > \Im_2 > \Im_4$
$q = 2$	0.1941	0.2732	0.7207	0.0164	0.8467	0.5164	$\Im_5 > \Im_3 > \Im_6 > \Im_2 > \Im_1 > \Im_4$
$q = 3$	0.1946	0.3002	0.7074	0.0000	0.8618	0.5274	$\Im_5 > \Im_3 > \Im_6 > \Im_2 > \Im_1 > \Im_4$
$q = 4$	0.1955	0.3096	0.7006	0.0000	0.8731	0.5294	$\Im_5 > \Im_3 > \Im_6 > \Im_2 > \Im_1 > \Im_4$
$q = 5$	0.1927	0.3078	0.6958	0.0000	0.8807	0.5267	$\Im_5 > \Im_3 > \Im_6 > \Im_2 > \Im_1 > \Im_4$
$q = 6$	0.1881	0.3007	0.6924	0.0000	0.8865	0.5219	$\Im_5 > \Im_3 > \Im_6 > \Im_2 > \Im_1 > \Im_4$
$q = 7$	0.1827	0.2910	0.6900	0.0000	0.8910	0.5160	$\Im_5 > \Im_3 > \Im_6 > \Im_2 > \Im_1 > \Im_4$
$q = 8$	0.1770	0.2799	0.6883	0.0000	0.8949	0.5094	$\Im_5 > \Im_3 > \Im_6 > \Im_2 > \Im_1 > \Im_4$
$q = 9$	0.1740	0.2715	0.6887	0.0037	0.8990	0.5034	$\Im_5 > \Im_3 > \Im_6 > \Im_2 > \Im_1 > \Im_4$
$q = 10$	0.1722	0.2642	0.6901	0.0084	0.9031	0.4977	$\Im_5 > \Im_3 > \Im_6 > \Im_2 > \Im_1 > \Im_4$
$q = 11$	0.1704	0.2567	0.6918	0.0122	0.9067	0.4920	$\Im_5 > \Im_3 > \Im_6 > \Im_2 > \Im_1 > \Im_4$
$q = 12$	0.1688	0.2494	0.6938	0.0152	0.9099	0.4864	$\Im_5 > \Im_3 > \Im_6 > \Im_2 > \Im_1 > \Im_4$
$q = 13$	0.1673	0.2423	0.6961	0.0177	0.9129	0.4810	$\Im_5 > \Im_3 > \Im_6 > \Im_2 > \Im_1 > \Im_4$
$q = 14$	0.1660	0.2355	0.6986	0.0197	0.9157	0.4760	$\Im_5 > \Im_3 > \Im_6 > \Im_2 > \Im_1 > \Im_4$
$q = 15$	0.1650	0.2291	0.7013	0.0212	0.9183	0.4713	$\Im_5 > \Im_3 > \Im_6 > \Im_2 > \Im_1 > \Im_4$
$q = 16$	0.1642	0.2232	0.7042	0.0225	0.9207	0.4670	$\Im_5 > \Im_3 > \Im_6 > \Im_2 > \Im_1 > \Im_4$
$q = 17$	0.1637	0.2177	0.7072	0.0234	0.9229	0.4633	$\Im_5 > \Im_3 > \Im_6 > \Im_2 > \Im_1 > \Im_4$
$q = 18$	0.1633	0.2127	0.7104	0.0241	0.9250	0.4599	$\Im_5 > \Im_3 > \Im_6 > \Im_2 > \Im_1 > \Im_4$
$q = 19$	0.1632	0.2080	0.7136	0.0246	0.9271	0.4571	$\Im_5 > \Im_3 > \Im_6 > \Im_2 > \Im_1 > \Im_4$
$q = 20$	0.1632	0.2038	0.7169	0.0249	0.9290	0.4548	$\Im_5 > \Im_3 > \Im_6 > \Im_2 > \Im_1 > \Im_4$

Table 16 Integrative appraisal scores and ranking results of bio-diversity vacation places by q-ROFTLWDHM operator ($\mathfrak{z} = 2$ and $q \in (1, 20)$)

Parameter	IAS($\Im_1$)	IAS($\Im_2$)	IAS($\Im_3$)	IAS($\Im_4$)	IAS($\Im_5$)	IAS($\Im_6$)	Ranking
$q = 1$	0.2858	0.1220	0.7006	0.0559	0.7597	0.5963	$\Im_5 > \Im_3 > \Im_6 > \Im_1 > \Im_2 > \Im_4$
$q = 2$	0.3496	0.1086	0.7242	0.1975	0.6592	0.5521	$\Im_3 > \Im_5 > \Im_6 > \Im_1 > \Im_4 > \Im_2$
$q = 3$	0.3964	0.1040	0.7303	0.2566	0.5889	0.5194	$\Im_3 > \Im_5 > \Im_6 > \Im_1 > \Im_4 > \Im_2$
$q = 4$	0.4370	0.1031	0.7333	0.2857	0.5354	0.4951	$\Im_3 > \Im_5 > \Im_6 > \Im_1 > \Im_4 > \Im_2$
$q = 5$	0.4772	0.1107	0.7424	0.3152	0.5000	0.4805	$\Im_3 > \Im_5 > \Im_6 > \Im_1 > \Im_4 > \Im_2$
$q = 6$	0.5234	0.1431	0.7576	0.3656	0.5000	0.4857	$\Im_3 > \Im_1 > \Im_5 > \Im_6 > \Im_4 > \Im_2$
$q = 7$	0.5633	0.1656	0.7736	0.4099	0.5000	0.4919	$\Im_3 > \Im_1 > \Im_5 > \Im_6 > \Im_4 > \Im_2$
$q = 8$	0.5976	0.1838	0.7911	0.4500	0.5000	0.4988	$\Im_3 > \Im_1 > \Im_5 > \Im_6 > \Im_4 > \Im_2$
$q = 9$	0.6286	0.2008	0.8097	0.4864	0.5000	0.5058	$\Im_3 > \Im_1 > \Im_6 > \Im_5 > \Im_4 > \Im_2$
$q = 10$	0.6581	0.2168	0.8290	0.5196	0.5000	0.5127	$\Im_3 > \Im_1 > \Im_4 > \Im_6 > \Im_5 > \Im_2$
$q = 11$	0.6729	0.2236	0.8268	0.5237	0.4756	0.5009	$\Im_3 > \Im_1 > \Im_4 > \Im_6 > \Im_5 > \Im_2$
$q = 12$	0.6757	0.2226	0.8090	0.5024	0.4361	0.4745	$\Im_3 > \Im_1 > \Im_4 > \Im_6 > \Im_5 > \Im_2$
$q = 13$	0.6816	0.2399	0.7997	0.5000	0.4241	0.4660	$\Im_3 > \Im_1 > \Im_4 > \Im_6 > \Im_5 > \Im_2$
$q = 14$	0.6872	0.2595	0.7940	0.5000	0.4201	0.4623	$\Im_3 > \Im_1 > \Im_4 > \Im_6 > \Im_5 > \Im_2$
$q = 15$	0.6919	0.2794	0.7906	0.5000	0.4205	0.4609	$\Im_3 > \Im_1 > \Im_4 > \Im_6 > \Im_5 > \Im_2$
$q = 16$	0.6963	0.2995	0.7891	0.5000	0.4244	0.4618	$\Im_3 > \Im_1 > \Im_4 > \Im_6 > \Im_5 > \Im_2$
$q = 17$	0.7007	0.3193	0.7892	0.5000	0.4310	0.4643	$\Im_3 > \Im_1 > \Im_4 > \Im_6 > \Im_5 > \Im_2$
$q = 18$	0.7053	0.3391	0.7908	0.5000	0.4397	0.4683	$\Im_3 > \Im_1 > \Im_4 > \Im_6 > \Im_5 > \Im_2$
$q = 19$	0.7102	0.3587	0.7938	0.5000	0.4499	0.4737	$\Im_3 > \Im_1 > \Im_4 > \Im_6 > \Im_5 > \Im_2$
$q = 20$	0.7156	0.3781	0.7980	0.5000	0.4614	0.4803	$\Im_3 > \Im_1 > \Im_4 > \Im_6 > \Im_5 > \Im_2$

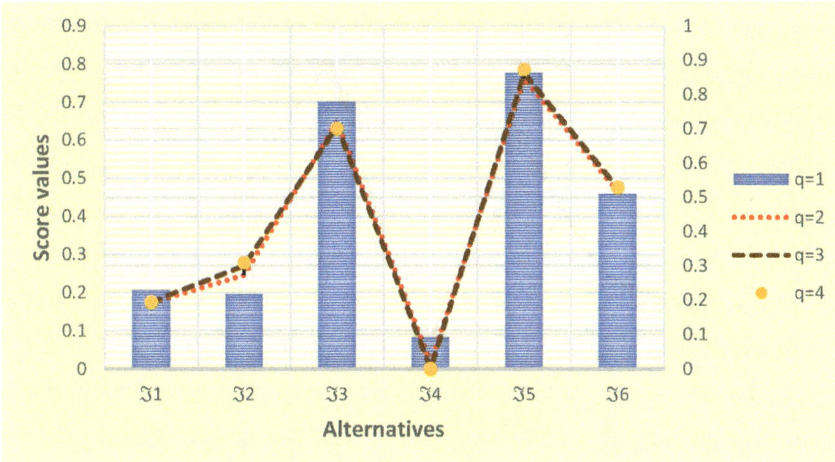

Fig. 5 Scores of bio-diversity vacation places $\Im_\iota$ $(\iota = 1, 2, \ldots, 6)$ when $\mathfrak{z} = 2$ and $q \in (1, 4)$ based on the q-ROFTLWHM operator

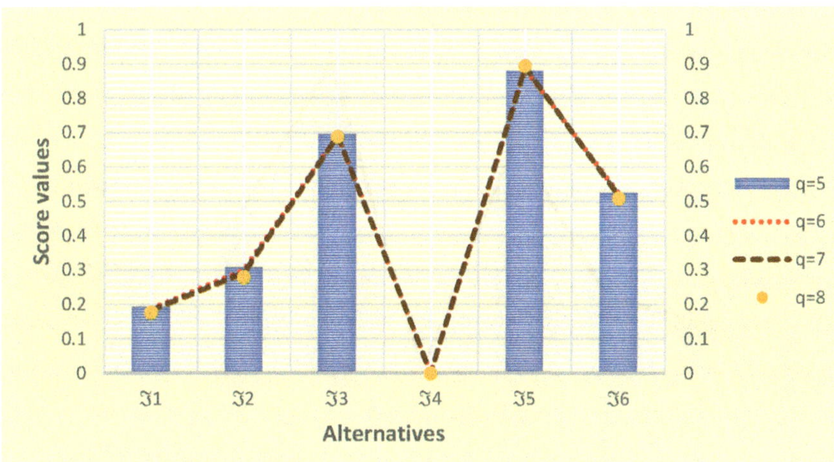

Fig. 6 Scores of bio-diversity vacation places $\Im_\iota$ $(\iota = 1, 2, \ldots, 6)$ when $\mathfrak{z} = 2$ and $q \in (5, 8)$ based on the q-ROFTLWHM operator

influence factors of sustainable supply chain advancement [39], stock investment assessment [40], and so on. Second, we discovered that some scholars have recognized the significance and requirement of evaluating the interconnection between the membership function and the non-membership function when coordinating attribute values [41, 42]. As a result, in the future, we will investigate novel AOs for interval-valued q-rung dual hesitant fuzzy elements that incorporate the interaction between the membership function and the non-membership function. Third, the interdepen-

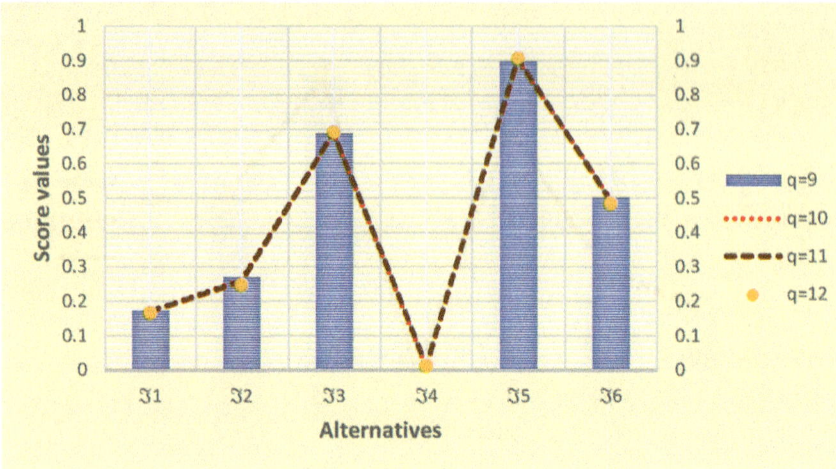

Fig. 7 Scores of bio-diversity vacation places $\Im_\iota (\iota = 1, 2, \ldots, 6)$ when $\mathfrak{z} = 2$ and $q \in (9, 12)$ based on the q-ROFTLWHM operator

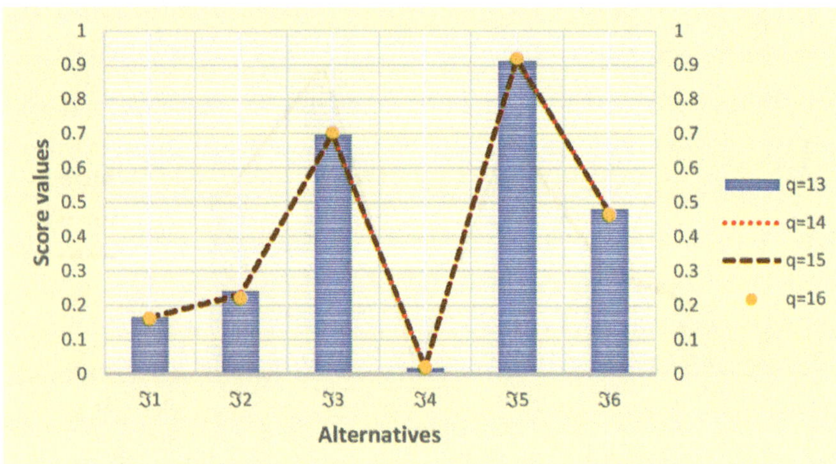

Fig. 8 Scores of bio-diversity vacation places $\Im_\iota (\iota = 1, 2, \ldots, 6)$ when $\mathfrak{z} = 2$ and $q \in (13, 16)$ based on the q-ROFTLWHM operator

dencies between attributes can be heterogeneous at times [38, 43]. As a result, we will investigate some novel MAGDM methods in the context of interval-valued q-rung dual hesitant fuzzy elements with heterogeneous interdependencies.

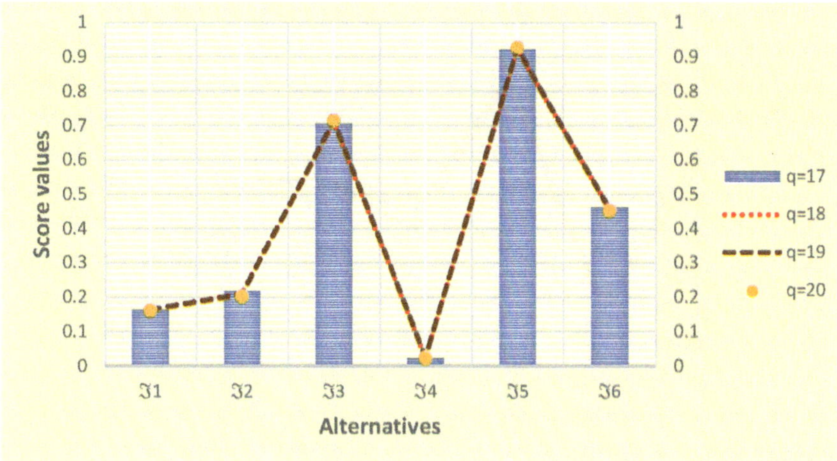

Fig. 9 Scores of bio-diversity vacation places $\Im_\iota (\iota = 1, 2, \ldots, 6)$ when $\mathfrak{z} = 2$ and $q \in (17, 20)$ based on the q-ROFTLWHM operator

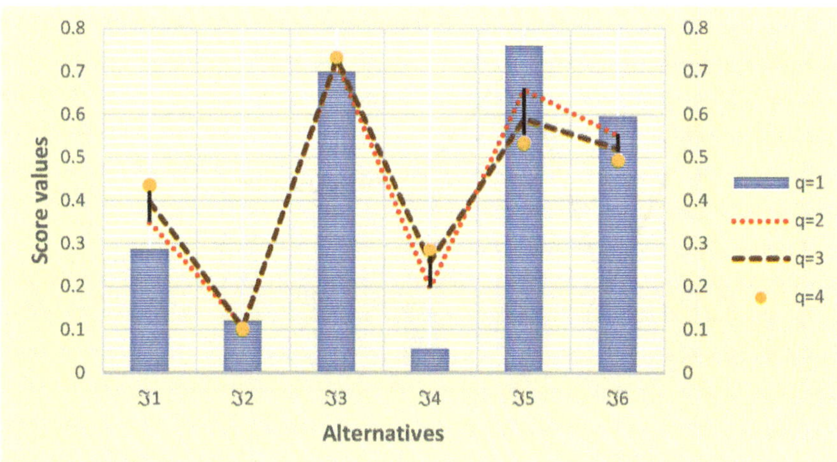

Fig. 10 Scores of bio-diversity vacation places $\Im_\iota (\iota = 1, 2, \ldots, 6)$ when $\mathfrak{z} = 2$ and $q \in (1, 4)$ based on the q-ROFTLWDHM operator

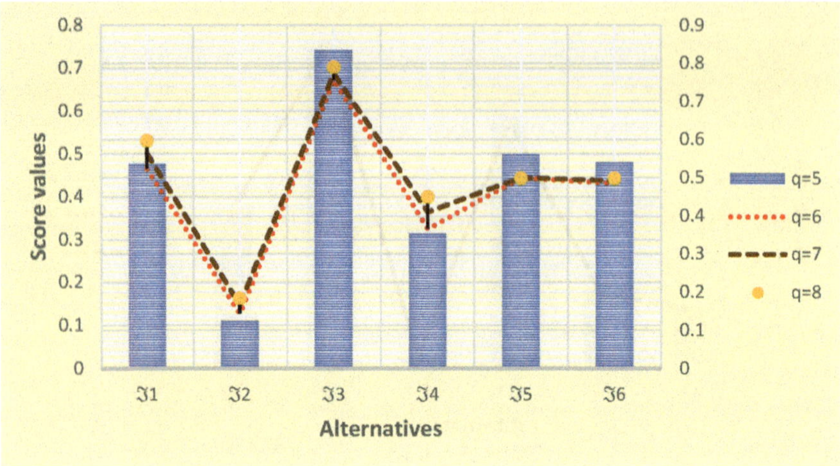

Fig. 11 Scores of bio-diversity vacation places $\Im_\iota (\iota = 1, 2, \ldots, 6)$ when $\mathfrak{z} = 2$ and $q \in (5, 8)$ based on the q-ROFTLWDHM operator

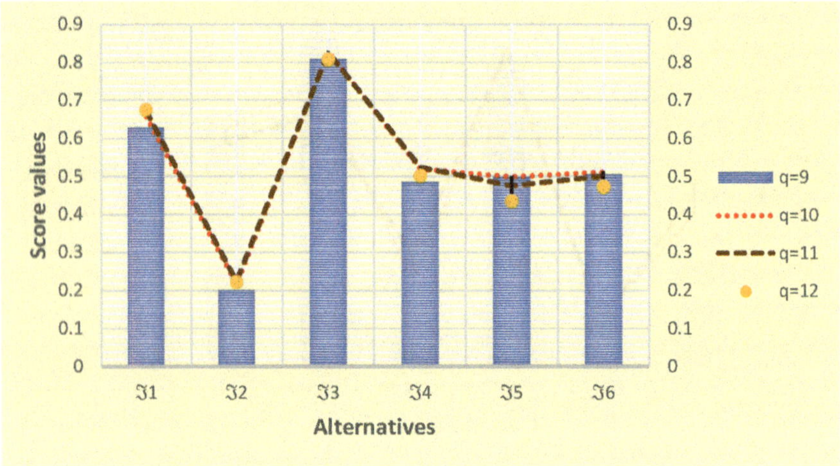

Fig. 12 Scores of bio-diversity vacation places $\Im_\iota (\iota = 1, 2, \ldots, 6)$ when $\mathfrak{z} = 2$ and $q \in (9, 12)$ based on the q-ROFTLWDHM operator

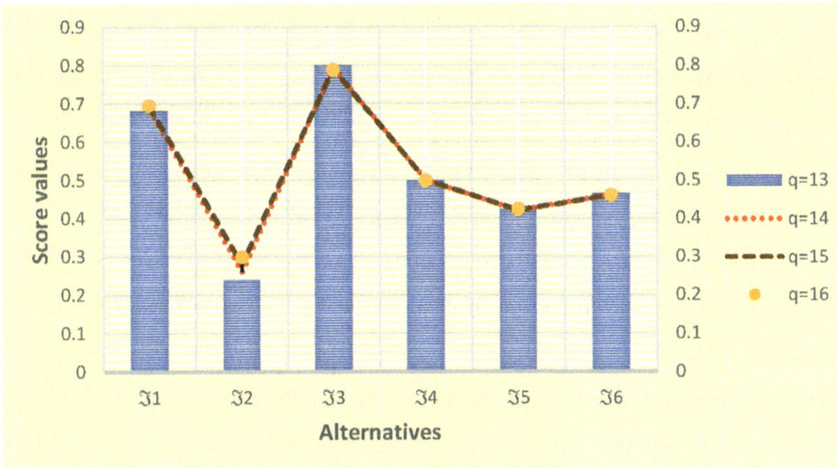

Fig. 13 Scores of bio-diversity vacation places $\Im_\iota(\iota = 1, 2, \ldots, 6)$ when $\mathfrak{z} = 2$ and $q \in (13, 16)$ based on the q-ROFTLWDHM operator

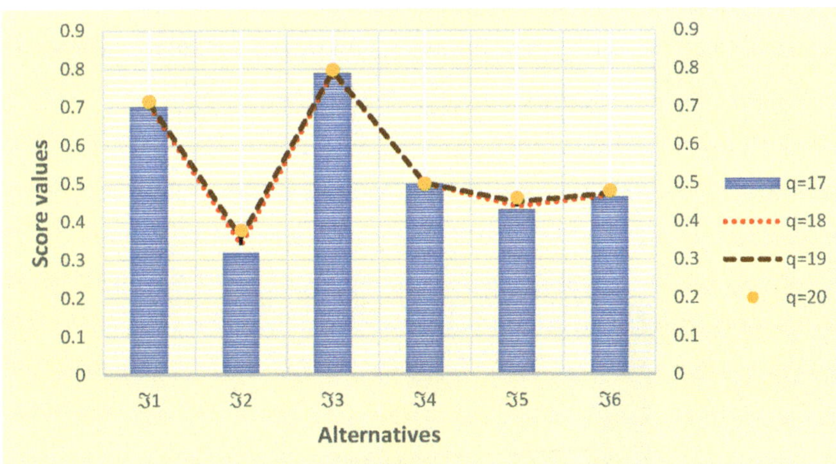

Fig. 14 Scores of bio-diversity vacation places $\Im_\iota(\iota = 1, 2, \ldots, 6)$ when $\mathfrak{z} = 2$ and $q \in (17, 20)$ based on the q-ROFTLWDHM operator

Table 17 Evaluation outcomes by utilizing different methodologies based on q-ROFTLWHM operator

Alternatives	EDAS [27]	Ranking	TOPSIS [35]	Ranking	CODAS [36]	Ranking
$\Im_1$	0.1955	V	0.1249	III	0.0008	I
$\Im_2$	0.3096	III	0.1262	IV	0.0007	II
$\Im_3$	0.7006	VI	0.9530	V	−0.0002	V
$\Im_4$	0.0000	II	0.5688	VI	−0.0021	VI
$\Im_5$	0.8731	I	0.4064	II	0.0006	III
$\Im_6$	0.5294	IV	0.1860	I	0.0001	IV

Table 18 Evaluation outcomes by utilizing different methodologies based on q-ROFTLWDHM operator

Alternatives	EDAS [27]	Ranking	TOPSIS [35]	Ranking	CODAS [36]	Ranking
$\Im_1$	0.4370	III	0.1214	IV	0.0003	V
$\Im_2$	0.1031	V	0.1587	V	0.0004	II
$\Im_3$	0.7333	VI	0.5956	III	−0.0000	I
$\Im_4$	0.2857	I	0.8906	VI	−0.0014	VI
$\Im_5$	0.5354	IV	0.6892	II	0.0006	III
$\Im_6$	0.4951	II	0.3676	I	0.0001	IV

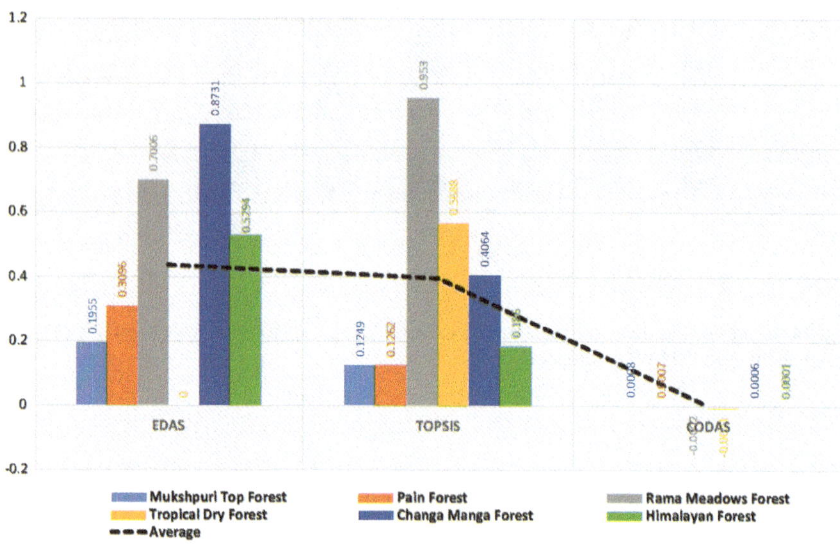

Fig. 15 Comparative outcomes with q-ROFTLWHM operator

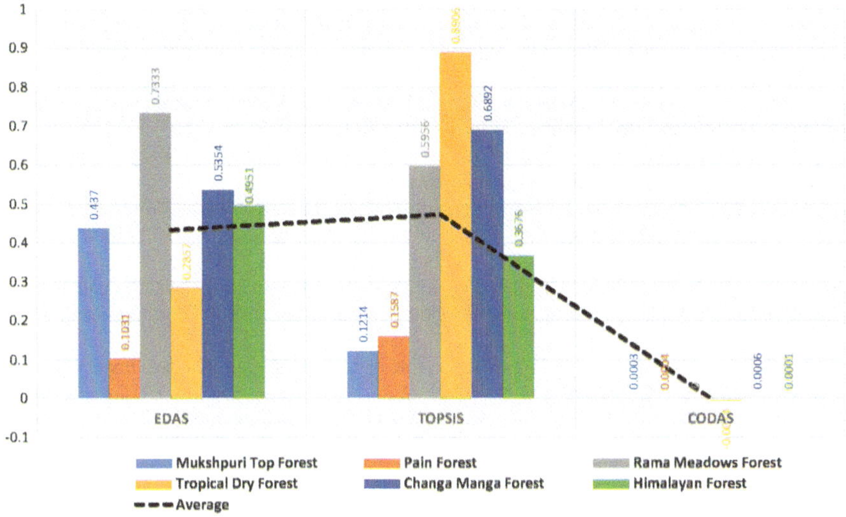

Fig. 16 Comparative outcomes with q-ROFTLWDHM operator

References

1. Liu, P., Naz, S., Akram, M., Muzammal, M.: Group decision-making analysis based on linguistic q-rung orthopair fuzzy generalized point weighted aggregation operators. Int. J. Mach. Learn. Cybern. 1–24 (2021). https://doi.org/10.1007/s13042-021-01425-2
2. Akram, M., Naz, S., Edalatpanah, S.A., Mehreen, R.: Group decision-making framework under linguistic q-rung orthopair fuzzy Einstein models. Soft Comput. **25**, 10309–10334 (2021)
3. Garg, H., Naz, S., Ziaa, F., Shoukat, Z.: A ranking method based on Muirhead mean operator for group decision making with complex interval-valued q-rung orthopair fuzzy numbers. Soft Comput. **25**(22), 14001–14027 (2021)
4. Zadeh, L.A.: Fuzzy sets. Inf. Control **8**(3), 338–353 (1965)
5. Atanassov, K.T.: Intuitionistic fuzzy sets. In: Intuitionistic Fuzzy Sets, pp. 1–137. Physica, Heidelberg
6. Yager, R.R.: Pythagorean membership grades in multicriteria decision making. IEEE Trans. Fuzzy Syst. **22**(4), 958–965 (2013)
7. Yager, R.R.: Generalized orthopair fuzzy sets. IEEE Trans. Fuzzy Syst. **25**(5), 1222–1230 (2016)
8. Seikh, M.R., Mandal, U.: Q-rung orthopair fuzzy Frank aggregation operators and its application in multiple attribute decision-making with unknown attribute weights. Granular Comput. 1–22
9. Akram, M., Shahzadi, G., Peng, X.: Extension of Einstein geometric operators to multi-attribute decision making under q-rung orthopair fuzzy information. Granular Comput. **6**(4), 779–795 (2021)
10. Zeng, S., Hu, Y., Xie, X.: Q-rung orthopair fuzzy weighted induced logarithmic distance measures and their application in multiple attribute decision making. Eng. Appl. Artif. Intell. **100**, 104167 (2021)
11. Peng, X., Dai, J., Garg, H.: Exponential operation and aggregation operator for q-rung orthopair fuzzy set and their decision-making method with a new score function. Int. J. Intell. Syst. **33**(11), 2255–2282 (2018)

12. Naz, S., Akram, M., Alsulami, S., Ziaa, F.: Decision-making analysis under interval-valued q-rung orthopair dual hesitant fuzzy environment. Int. J. Comput. Intell. Syst. **14**(1), 332–357 (2020)
13. Akram, M., Shahzadi, G., Alcantud, J.C.R.: Multi-attribute decision-making with q-rung picture fuzzy information. Granular Comput. 1–19 (2021)
14. Naz, S., Akram, M.: Novel decision-making approach based on hesitant fuzzy sets and graph theory. Comput. Appl. Math. **38**(1), 7 (2019)
15. Akram, M., Naz, S., Ziaa, F.: Novel decision making framework based on complex q-rung orthopair fuzzy information. Sci. Iranica 1–34 (2021). https://doi.org/10.24200/SCI.2021.55413.4209
16. Akram, M., Naz, S., Smarandache, F.: Generalization of maximizing deviation and TOPSIS method for MADM in simplified neutrosophic hesitant fuzzy environment. Symmetry **11**(8), 1058 (2019)
17. Zadeh, L.A.: The concept of a linguistic variable and its application to approximate reasoning I. Inf. Sci. **8**(3), 199–249 (1975)
18. Wang, J.Q., Li, J.J.: The multi-criteria group decision making method based on multi-granularity intuitionistic two semantics. Sci. Technol. Inf. **33**(1), 8–9 (2009)
19. Li, L., Zhang, R., Wang, J., Shang, X., Bai, K.: A novel approach to multi-attribute group decision-making with q-rung picture linguistic information. Symmetry **10**(5), 172 (2018)
20. Klement, E.P., Mesiar, R., Pap, E.: Problems on triangular norms and related operators. Fuzzy Sets Syst. **145**(3), 471–479 (2004)
21. Beliakov, G., Bustince, H., Goswami, D.P., Mukherjee, U.K., Pal, N.R.: On averaging operators for Atanassov's intuitionistic fuzzy sets. Inf. Sci. **181**(6), 1116–1124 (2011)
22. Hara, T., Uchiyama, M., Takahasi, S.E.: A refinement of various mean inequalities. J. Inequalities Appl. **1998**(4), 932025 (1998)
23. Feng, X., Shang, X., Wang, J., Xu, Y.: A multiple attribute decision-making method based on interval-valued q-rung dual hesitant fuzzy power Hamy mean and novel score function. Comput. Appl. Math. **40**(1), 1–32 (2021)
24. Garg, H., Sirbiladze, G., Ali, Z., Mahmood, T.: Hamy mean operators based on complex q-rung orthopair fuzzy setting and their application in multi-attribute decision making. Mathematics **9**(18), 2312 (2021)
25. Wu, L., Wang, J., Gao, H.: Models for competiveness evaluation of tourist destination with some interval-valued intuitionistic fuzzy Hamy mean operators. J. Intell. Fuzzy Syst. **36**(6), 5693–5709 (2019)
26. Xing, Y., Zhang, R., Wang, J., Bai, K., Xue, J.: A new multi-criteria group decision-making approach based on q-rung orthopair fuzzy interaction Hamy mean operators. Neural Comput. Appl. **32**(11), 7465–7488 (2020)
27. Keshavarz Ghorabaee, M., Zavadskas, E.K., Olfat, L., Turskis, Z.: Multi-criteria inventory classification using a new method of evaluation based on distance from average solution (EDAS). Informatica **26**(3), 435–451 (2015)
28. Huang, Y., Lin, R., Chen, X.: An enhancement EDAS method based on prospect theory. Technol. Econ. Dev. Econ. **27**(5), 1019–1038 (2021)
29. Wei, G., Wei, C., Guo, Y.: EDAS method for probabilistic linguistic multiple attribute group decision making and their application to green supplier selection. Soft Comput. **25**(14), 9045–9053 (2021)
30. Jana, C., Pal, M.: Extended bipolar fuzzy EDAS approach for multi-criteria group decision-making process. Comput. Appl. Math. **40**(1), 1–15 (2021)
31. Tai, W.S., Chen, C.T.: A new evaluation model for intellectual capital based on computing with linguistic variable. Expert Syst. Appl. **36**(2), 3483–3488 (2009)
32. Wei, G., Gao, H., Wei, Y.: Some q-rung orthopair fuzzy Heronian mean operators in multiple attribute decision making. Int. J. Intell. Syst. **33**(7), 1426–1458 (2018)
33. Ju, Y., Wang, A., Ma, J., Gao, H., Santibanez Gonzalez, E.D.: Some q-rung orthopair fuzzy 2-tuple linguistic Muirhead mean aggregation operators and their applications to multiple-attribute group decision making. Int. J. Intell. Syst. **35**(1), 184–213 (2020)

34. Wu, S., Wang, J., Wei, G., Wei, Y.: Research on construction engineering project risk assessment with some 2-tuple linguistic neutrosophic Hamy mean operators. Sustainability **10**(5), 1536 (2018)
35. Jahanshahloo, G.R., Lotfi, F.H., Izadikhah, M.: Extension of the TOPSIS method for decision-making problems with fuzzy data. Appl. Math. Comput. **181**(2), 1544–1551 (2006)
36. Keshavarz Ghorabaee, M., Zavadskas, E. K., Turskis, Z., Antucheviciene, J.: A new combinative distance-based assessment (CODAS) method for multi-criteria decision-making. Econ. Comput. Econ. Cybern. Stud. Res. **50**(3) (2016)
37. Huang, C., Lin, M., Xu, Z.: Pythagorean fuzzy MULTIMOORA method based on distance measure and score function: its application in multi-criteria decision making process. Knowl. Inf. Syst. **62**(11), 4373–4406 (2020)
38. Liu, P., Zhu, B., Wang, P., Shen, M.: An approach based on linguistic spherical fuzzy sets for public evaluation of shared bicycles in China. Eng. Appl. Artif. Intell. **87**, 103295 (2020)
39. Shete, P.C., Ansari, Z.N., Kant, R.: A Pythagorean fuzzy AHP approach and its application to evaluate the enablers of sustainable supply chain innovation. Sustain. Prod. Consumption **23**, 77–93 (2020)
40. Tang, G., Chiclana, F., Liu, P.: A decision-theoretic rough set model with q-rung orthopair fuzzy information and its application in stock investment evaluation. Appl. Soft Comput. **91**, 106212 (2020)
41. Wang, J., Wei, G., Wei, C., Wei, Y.: MABAC method for multiple attribute group decision making under q-rung orthopair fuzzy environment. Defence Technol. **16**(1), 208–216 (2020)
42. Wang, L., Garg, H., Li, N.: Pythagorean fuzzy interactive Hamacher power aggregation operators for assessment of express service quality with entropy weight. Soft Comput. **25**(2), 973–993 (2021)
43. Liu, P., Chen, S.M., Wang, Y.: Multiattribute group decision making based on intuitionistic fuzzy partitioned Maclaurin symmetric mean operators. Inf. Sci. **512**, 830–854 (2020)

Determining the Importance Level of Effective Criteria in the Health Information System Selection via Spherical Fuzzy DEMATEL Method

Ahmet Aytekin, Gözde Koca, Basil Oluch Okoth, and Çağlar Karamaşa

Abstract The dynamic structure of the health sector entails its integration with new technologies that transform every aspect of life. According to the developments in computer and communication technologies, health information systems have quickly been used in health institutions. On the other hand, with the spread of health information systems, problems were encountered in ensuring inter-institutional standards and quality. At this point, the selection of health information systems that are compatible with national and international health systems and that will meet the current and future needs of health institutions have gained importance. Purpose of this study is to determine the effective criteria in the selection of health information systems and the importance levels of these criteria. The criteria determined by scanning the literature were reduced to eight (security, cost, system quality and technical support, design, statistical information delivery, decision support system inclusion, accessibility, integration with external systems and subsystem inclusivity) by using the Delphi method with expert opinions. The spherical fuzzy DEMATEL method was applied to determine the importance levels of the criteria. The study will provide

A. Aytekin
Department of Business Administration, Faculty of Hopa Economics and Administrative Sciences, Artvin Çoruh University, 08100 Hopa, Artvin, Turkey
e-mail: ahmetaytekin@artvin.edu.tr

G. Koca
Department of Business Administration, Faculty of Economics and Administrative Sciences, Bilecik Şeyh Edebali University, Bilecik, Turkey
e-mail: gozde.koca@bilecik.edu.tr

B. O. Okoth
Department of Business Administration, Faculty of Economics and Administrative Sciences, Anadolu University, Eskişehir, Turkey
e-mail: bookoth@anadolu.edu.tr

Ç. Karamaşa (✉)
Faculty of Business, Department of Business Administration, Anadolu University, Eskişehir, Turkey
e-mail: ckaramasa@anadolu.edu.tr

L. Sahoo et al. (eds.), *Real Life Applications of Multiple Criteria Decision Making Techniques in Fuzzy Domain*, Studies in Fuzziness and Soft Computing 420,
https://doi.org/10.1007/978-981-19-4929-6_19

important novelties in terms of using the Delphi method at the point of determining the criteria and considering fuzziness and the interaction between the criteria.

Keywords Health information systems · Delphi · Spherical fuzzy numbers · Spherical fuzzy DEMATEL

1 Introduction

The dynamic structure of the health sector necessitated its integration with new technologies that transform every aspect of life. At this point, it has been seen that information systems have started to be used rapidly in health institutions. The developments in the information systems in Turkey are transferred to the health sector almost to the same extent. So much so that shortly after the electronic computer system came to the General Directorate of Highways in 1960, the Integrated Information System Project was implemented at Hacettepe University [19].

Operating systems used for the application of information technologies in the fields of health service delivery are called Health Information Systems. Information technologies and health information systems have benefits in the conscious use of data in the decision-making processes of health institutions, expanding the service area, increasing efficiency, ensuring effective management of resources, reducing costs and improving customer services. These expectations cause significant resources to be allocated to health information systems [2, 17, 19].

On the other hand, it is observed that standardization and quality problems increase with the spread of health information systems. Health Information Management Systems Society (HIMSS); Health Information Management Systems Society, which has carried out important studies in terms of health information systems, has proposed the Electronic Medical Record Compliance Model (EMRAM) system to evaluate the efficiency of the services provided in health institutions. With EMRAM, health institutions are graded from Level 0 to Level 7. In order for a hospital to obtain a valid digital hospital accreditation document in international standards, it must be evaluated at least at Level 6 [15]. At this point, it seems difficult to grasp at first glance which health system will carry health institutions to better points such as having international digital certificates. Increasing the success of the health system can be achieved by choosing the system effectively. System selection to be carried out with multi-criteria decision methods can directly increase the system success of health institutions. In this study, it is purposed to determine the effective criteria for health information system selection and the importance levels of these criteria. Thus, it is aimed to create a basic decision model that health institutions will benefit from in system selection and efficiency analysis.

The process of determining the effective criteria in the selection of health information system in the context of literature and expert opinion will be carried out. In addition, the spherical fuzzy DEMATEL method will be used for determining the importance levels of the mentioned criteria. The study differs from the literature in terms of evaluating the effective criteria for health information system selection according to new and existing conditions and eliminating the uncertainties in these evaluations with the spherical fuzzy DEMATEL method. In the following section, prominent results from the literature on health information systems will be presented.

2 Literature

Hospital information systems are generally the focus of studies on health information systems. In this context, the perceptions and attitudes of the health personnel toward the use of the hospital information system were measured. Table 1 shows the studies carried out on health information systems in Turkey.

It is understood from the literature that many advantages such as standardization, efficiency and ease of use have been achieved simultaneously with the replacement of traditional systems by health information systems. On the other hand, it is observed that the main problem is personnel training and cost. In addition, it is seen that most of the studies focus on the interaction of the personnel with the system. The study discussed at this point differs from other studies by taking into account the uncertainties in the literature and the evaluations of experts.

3 Research Method

In this part of the study, spherical fuzzy sets and the spherical fuzzy DEMATEL method will be discussed.

3.1 *Spherical Fuzzy Sets*

The spherical fuzzy set $\tilde{B}_s$ in the X discourse universe is shown as follows [10–13]:

$$\tilde{B}_s = \left\{x, \mu_{\tilde{B}_s}(x), v_{\tilde{B}_s}(x), \pi_{\tilde{B}_s}(x)|x \in X\right\} \quad (1)$$

For the above notation $\mu_{\tilde{B}_s}(x) : X \to [0, 1], v_{\tilde{B}_s}(x) : X \to [0, 1], \pi_{\tilde{B}_s}(x) : X \to [0, 1]$ with the situation in Eq. 2 is valid:

$$0 \leq \mu^2_{\tilde{B}_s} + v^2_{\tilde{B}_s} + \pi^2_{\tilde{B}_s} \leq 1, \forall x \in X \quad (2)$$

For each x, $\mu_{\tilde{B}_s}(x)$, $v_{\tilde{B}_s}(x)$ and $\pi_{\tilde{B}_s}(x)$ show the degrees of membership, non-membership and hesitation of x to $\tilde{B}_s$. Some numerical operations have been developed by making use of the relationship between spherical and pythagorean fuzzy

Table 1 Major studies on health information systems

Author(s)	Evaluation and conclusion	Method
Ömürbek and Altın [17]	The level of use of health information systems and information technologies by hospitals was examined. According to the results, information technologies usage was aimed to increase the quality, to keep records in electronic environment, to access information quickly and cheaply and to increase productivity. In addition, the lack of adequate training of the employees and the rapid out-of-dateness of the current technology are seen as important problems by the hospitals	t-test
Işık and Akbolat [5]	Opinions of healthcare professionals on the use and benefits of Hospital Information System (HIS) were taken with a questionnaire. Employees stated that HISs provide easy access to information, increase the quality of medical services, save time, facilitate communication, and facilitate patient appointments and assignments	Interpretation of the answers received with the questionnaire form
Bal et al. [2]	In order to analyze the success of the health information systems used by family physicians, the model called Information System Success Model was tested. It has been stated that there is a significant relationship in the same direction between information quality and system usage, system quality and user satisfaction, system usage and usage/user satisfaction, user satisfaction and net benefit	Structural equation model
Ömürbek et al. [18]	It is aimed to determine the clinical information system developments, the situation of meeting the costs of information technology, the status of meeting the purpose of use of clinical information systems and the problems encountered. The most important problem for the use of clinical information systems was expressed as the lack of education of the employees	Wilcoxon sign test, t-test
Uslu et al. [21]	Evaluation of Hospital Information Management System (HIMS) was carried out with the software usability scale. It was stated that the factors with the highest average score were satisfaction and ease of use	Interpretation of the answers received with the questionnaire form
Göktaş et al. [7]	It is aimed to evaluate the opinions of employees in health information systems. Related to health information systems, it has been stated that standardization, information security in the public sector, and Social Security Institution (SSI) and payments in the private sector are at the forefront	Interview and questionnaire applied

(continued)

Table 1 (continued)

Author(s)	Evaluation and conclusion	Method
İleri and Uludağ [6]	It is aimed to evaluate the e-Pulse application in the context of management information systems, patient privacy, personal rights and to identify the deficiencies and present suggestions. It has been suggested to pay attention to the protection of personal health data while developing the technical aspects of the systems	Qualitative study
Küzeci [14]	It has been stated that health information technologies cause the confidential information between the patient and the doctor to be recorded indefinitely. In this context, it has been emphasized that a process management that includes the control of individuals and the transparency of the time is necessary	Qualitative study
Peker et al. [19]	It is purposed to evaluate the digitalization status of hospitals in Turkey	Compilation
Limon [15]	The transition process from the traditional medical record system to electronic systems has been examined	Qualitative study
Yıldırım [26]	It is aimed to introduce personal health information systems. It has been stated that obtaining the expected benefit from the system depends on the factors of data privacy and security, integration, accessibility, usability, health literacy	Qualitative study
Alacadağlı [1]	The effects of information management systems on the health sector were examined	Interpretation of the answers received with the questionnaire form
Kaçmaz Bali et al. [9]	It has been stated that the transmission of consultation requests via HIMS enables the consultation physician to access patient data, that the request can be read easily, and that it is recorded without the need for an additional registration process	Interpretation of semi-structured interview answers
Eriş and Ilıman [4]	Evaluation of hospital personnel's opinions about HIS was carried out according to the answers to the questionnaires prepared within the scope of the Unified Technology Acceptance and Use Theory scale. The importance of ensuring the merit and training in personnel selection has been mentioned	ANOVA
Yeşilyurt et al. [25]	Literature and expert opinions were used for the selection of HIMS. In the study, in which seven alternatives were evaluated, the results obtained with different methods were compared	AHP, TOPSIS and PROMETHEE

(continued)

Table 1 (continued)

Author(s)	Evaluation and conclusion	Method
Ülke and Atilla [22]	A questionnaire form was directed to the patients for measuring the usage level of internet-based health information resources. It has been suggested to organize training programs in neighborhoods in order to improve individuals' Internet usage skills and to reach accurate and reliable health information	Mann Whitney-U test, Kruskal Wallis test

sets. X_1 and X_2 are two universes of discourse and $\tilde{B}_s = \left(\mu_{\tilde{B}_s}, v_{\tilde{B}_s}, \pi_{\tilde{B}_s}\right)$ ile $\tilde{C}_s = \left(\mu_{\tilde{C}_s}, v_{\tilde{C}_s}, \pi_{\tilde{C}_s}\right)$ be two spherical fuzzy sets in this universe. Basic operations for these sets are expressed as:

Addition:

$$\tilde{B}_s \oplus \tilde{C}_s = \left\{\left(\mu^2_{\tilde{B}_s} + \mu^2_{\tilde{C}_s} - \mu^2_{\tilde{B}_s}\mu^2_{\tilde{C}_s}\right)^{1/2}, v_{\tilde{B}_s}\, v_{\tilde{C}_s}, \left(\left(1 - \mu^2_{\tilde{C}_s}\right)\pi^2_{\tilde{B}_s} + \left(1 - \mu^2_{\tilde{B}_s}\right)\pi^2_{\tilde{C}_s} - \pi^2_{\tilde{B}_s}\pi^2_{\tilde{C}_s}\right)^{1/2}\right\} \tag{3}$$

Multiplication:

$$\tilde{B}_s \otimes \tilde{C}_s = \left\{\mu_{\tilde{B}_s}\, \mu_{\tilde{C}_s}, \left(v^2_{\tilde{B}_s} + v^2_{\tilde{C}_s} - v^2_{\tilde{B}_s}v^2_{\tilde{C}_s}\right)^{1/2}, \left(\left(1 - v^2_{\tilde{C}_s}\right)\pi^2_{\tilde{B}_s} + \left(1 - v^2_{\tilde{B}_s}\right)\pi^2_{\tilde{C}_s} - \pi^2_{\tilde{B}_s}\pi^2_{\tilde{C}_s}\right)^{1/2}\right\} \tag{4}$$

Multiplication by scalar ($\lambda > 0$):

$$\lambda * \tilde{B}_s = \left\{\left(1 - \left(1 - \mu^2_{\tilde{B}_s}\right)^{\lambda}\right)^{1/2}, v^{\lambda}_{\tilde{B}_s}, \left(\left(1 - \mu^2_{\tilde{B}_s}\right)^{\lambda} - \left(1 - \mu^2_{\tilde{B}_s} - \pi^2_{\tilde{B}_s}\right)^{\lambda}\right)^{1/2}\right\} \tag{5}$$

Power of $\tilde{B}_s (\lambda > 0)$:

$$\tilde{B}^{\lambda}_s = \left\{\mu^{\lambda}_{\tilde{B}_s}, \left(1 - \left(1 - v^2_{\tilde{B}_s}\right)^{\lambda}\right)^{1/2}, \left(\left(1 - v^2_{\tilde{B}_s}\right)^{\lambda} - \left(1 - v^2_{\tilde{B}_s} - \pi^2_{\tilde{B}_s}\right)^{\lambda}\right)^{1/2}\right\} \tag{6}$$

To provide aggregation, the Spherical Weighted Arithmetic Mean (SWAM) was formed and where the weight vector is defined as $w = (w_1, w_2, \ldots, w_n)$ where $w_i \in [0, 1]$ is $w_i \in [0, 1]$ ile $\sum_{i=1}^{n} w_i = 1$.

$$\text{SWAM}_w\left(\tilde{B}_{s1}, \tilde{B}_{s2}, \ldots, \tilde{B}_{sn}\right) = w_1\tilde{B}_{s1} + w_2\tilde{B}_{s2} + \ldots + w_n\tilde{B}_{sn}$$

$$= \left\{ \left[1 - \prod_{i=1}^{n} \left(1 - \mu_{\tilde{B}_{si}}^{2}\right)^{w_i} \right]^{1/2}, \prod_{i=1}^{n} \upsilon_{\tilde{B}_{si}}^{w_i}, \left[\prod_{i=1}^{n} \left(1 - \mu_{\tilde{B}_{si}}^{2}\right)^{w_i} - \prod_{i=1}^{n} \left(1 - \mu_{\tilde{B}_{si}}^{2} - \pi_{\tilde{B}_{si}}^{2}\right)^{w_i} \right]^{1/2} \right\} \quad (7)$$

3.2 DEMATEL Method

The traditional DEMATEL method consists of six steps and is summarized below [3, 8, 24]:

(a) At the first stage, the attributes to be evaluated are determined. The matrix-based DEMATEL method is effective in solving the decision problem, which is considered to show the effect between the attributes.

(b) Decision makers determine the degree of influence of each attribute on the other, and the effect of the i attribute on j is shown by e_{ij}. The direct impact assessment matrix is related to the pairwise comparison of the components in terms of the degree of impact. The comparison scale includes four levels and each shows the strength of the effect. Here, 3 indicates strong effect, 2 indicates moderate effect, 1 indicates weak effect and 0 indicates ineffective. The direct impact mean matrix is obtained by calculating the mean values for the same attribute pairs in different direct impact assessment matrices and is shown in the equation below:

$$E = \begin{bmatrix} e_{11} & \cdots & e_{1n} \\ \vdots & \ddots & \vdots \\ e_{n1} & \cdots & e_{nn} \end{bmatrix} \quad (8)$$

(c) With the normalization of the E matrix, the initial direct effects matrix X is constructed. Following equation is used to obtain the matrix X:

$$X = pE,\ p = \min\left[\frac{1}{\max_i \sum_{j=1}^{n} |e_{ij}|}, \frac{1}{\max_j \sum_{i=1}^{n} |e_{ij}|} \right] \quad (9)$$

According to the above equation, all the features on the diagonal equal zero and p is called the normalization index.

(iv) The total effect is obtained by adding up the direct and indirect effects. The indirect effect is computed via summing the powers of the X matrix. Because of the indirect effects related to the features decrease with the powers of X, an approximate solution is found by inversion of the matrix and is shown as:

$$X' = X^2 + X^3 + X^4 + \ldots + X^\infty = X^2(I - X)^{-1} \quad (10)$$

The total impact matrix is obtained as in Eq. (11):

$$T = X + X' = X(I - X)^{-1} = \begin{bmatrix} t_{11} & \cdots & t_{1n} \\ \vdots & \ddots & \vdots \\ t_{n1} & \cdots & t_{nn} \end{bmatrix} \quad (11)$$

According to the above equation, the t_{ij} elements in the T matrix show the total effect between each attribute pair, and in addition, the row (f_i) and column totals (g_j) are calculated as follows:

$$f_i = \left[\sum_{j=1}^{n} t_{ij}\right] \quad (12)$$

$$g_j = \left[\sum_{i=1}^{n} t_{ij}\right] \quad (13)$$

The row sum (f_i) indicates the overall effect of the ith attribute on others and indicates the overall strength of the attribute. The column sum (g_j) indicates the jth attribute affected by others and indicates the general weakness related to the attribute.

(e) Under the condition $i = j$, $f_i + g_i$ value is the index showing the strength of the effect of the i attribute and is called the importance value. This value is also considered to determine the degree of the basic role of the ith attribute for the decision problem. The relation value dividing the attribute set into cause-effect groups is calculated by $f_i - g_i$. If the correlation value is positive, $(f_i - g_i > 0)$ belongs to the ith attribute cause group and has an effect on the others. If the relation value is negative $(f_i - g_i < 0)$, the ith attribute belongs to the result group and is affected by other attributes. The creation of cause-effect groups is important in terms of analyzing the problem, determining the effects and weighting the features. In other words, any development or progress in the cause group has a direct impact on the result group.

(f) In the last stage of the DEMATEL method, a network relational map that shows the causal dependencies or effects in the feature set and takes into account the importance and relation values, is constructed. In the relational network map where all the relations can be seen, the importance values are on the x-axis, while the y-axis shows the relation values. The components of the cause group are clustered at the top of the graph, while the elements related to result group are located at the bottom of the graph. The complexity increases as the number of specified relationships increases, and to overcome this, the threshold value β for the total effect values (t_{ij}) can be established.

With DEMATEL, the weight values of the criteria can also be calculated. These calculated weight values can be handled for solving the decision problems such as selection, ranking and classification. In order to compute the weight values of the criteria, firstly, non-normalized weight values (q_j) are obtained by using Eq. (14). Then, the weight values normalized by Eq. (15) are calculated.

$$q_j = \sqrt{\left(d_g + d_j\right)^2 + \left(d_g - d_j\right)^2} \tag{14}$$

$$w_j = \frac{q_j}{\sum_{j=1}^{n} q_j} \tag{15}$$

The criteria weight values calculated by Eq. (15) will be in the range of 0–1. Accordingly, the most important criterion is the one whose weight value is closest to one.

3.3 Spherical Fuzzy DEMATEL Method

The spherical fuzzy DEMATEL method was proposed by Gül [8], and it provides more independence to decision makers by considering the degree of hesitation. For spherical fuzzy sets, each degree of membership (membership, non-membership, hesitancy) takes a value between 0 and 1 independently, like neutrophic clusters, but differently the sum of its squares is between 0 and 1. A wider choice area is offered to decision makers via the geometric spherical surface. The procedure of the scherical fuzzy DEMATEL method can be summarized as below [8]:

(a) In the first stage, the attributes are defined, and the decision makers are selected.
(b) In the second stage, the preferences or opinions of the decision makers regarding the possible effects between the attributes are obtained. In the evaluation, a linguistic expression set developed by Kutlu Gündoğdu and Kahraman [10] and taking into account the score index (SI) is used. SI is calculated according to Eq. (16) and the linguistic expression set is shown in Table 2.

$$\mathrm{SI} = \sqrt{\left|100 * \left[(\mu - \pi)^2 - (\upsilon - \pi)^2\right]\right|} \tag{16}$$

The hth decision maker's assessment of the effect of attribute i on attribute j is shown as $e_{ij}^h = \left(\mu_{ij}^h, \upsilon_{ij}^h, \pi_{ij}^h\right)$ $(i, j = 1, 2, \ldots, n; h = 1, 2, \ldots, \text{for } m)$. In addition, the direct impact evaluation matrix related to the hth decision maker is expressed as the spherical fuzzy matrix $E^h = \left\langle e_{ij}^h \right\rangle_{nxn} = \left\langle \mu_{ij}^h, \upsilon_{ij}^h, \pi_{ij}^h \right\rangle_{nxn}$.

(iii) To aggregate the direct impact assessment matrices of different decision makers, the SWAO is used, and as a result, the aggregated direct impact matrix

Table 2 Linguistic expressions and spherical fuzzy numbers

Definition	μ	v	π	SI
Strong (S)	0.85	0.15	0.45	3
Moderate (M)	0.6	0.2	0.35	2
Weak (W)	0.35	0.25	0.25	1
No influence (NI)	0	0.3	0.15	0

$\left(E^{\text{top}}\right)$ is obtained. The calculations for E^{top} and the resulting matrix are shown in the equations below.

$$\begin{aligned} E^{\text{top}} &= \text{KAAO}_w\left(E^1, E^2, \ldots, E^m\right) = w_1 E^1 + w_2 E^2 + \cdots + w_m E^m \\ &= \left\langle \mu_{ij}^{\text{top}}, v_{ij}^{\text{top}}, \pi_{ij}^{\text{top}} \right\rangle = \left\{ \left[1 - \prod_{h=1}^{m} \left(1 - \left(\mu_{ij}^h\right)^2\right)^{w_h} \right]^{1/2}, \prod_{h=1}^{m} \left(v_{ij}^h\right)^{w_h}, \right. \\ & \left. \left[\prod_{h=1}^{m} \left(1 - \left(\mu_{ij}^h\right)^2\right)^{w_h} - \prod_{h=1}^{m} \left(1 - \left(\mu_{ij}^h\right)^2 - \left(\pi_{ij}^h\right)^2\right)^{w_h} \right]^{1/2} \right\} \end{aligned} \tag{17}$$

$$E^{\text{top}} = \begin{bmatrix} 0 & \cdots & \left\langle \mu_{1n}^{\text{top}}, v_{1n}^{\text{top}}, \pi_{1n}^{\text{top}} \right\rangle \\ \vdots & \ddots & \vdots \\ \left\langle \mu_{n1}^{\text{top}}, v_{n1}^{\text{top}}, \pi_{n1}^{\text{top}} \right\rangle & \cdots & 0 \end{bmatrix} \tag{18}$$

(iv) Each comparison pair includes three elements. Therefore, E^{top} is divided into three submatrices. Necessary matrix operations are performed in these matrices, which are allocated for membership, non-membership and hesitation degrees. By normalizing these matrices according to Eq. (9) and then combining them on the same matrix, the initial direct effects matrix is acquired.

(v) The initial direct impact matrix X forms the basis related to the total impact matrix T. Besides, submatrices of X are transformed into submatrices of T using Eqs. (12 and 13). The resulting total impact matrix T is as follows:

$$T = \begin{bmatrix} \left\langle \mu_{11}^T, v_{11}^T, \pi_{11}^T \right\rangle & \left\langle \mu_{12}^T, v_{12}^T, \pi_{12}^T \right\rangle & \cdots & \left\langle \mu_{1n}^T, v_{1n}^T, \pi_{1n}^T \right\rangle \\ \left\langle \mu_{21}^T, v_{21}^T, \pi_{21}^T \right\rangle & \left\langle \mu_{22}^T, v_{22}^T, \pi_{22}^T \right\rangle & \cdots & \left\langle \mu_{2n}^T, v_{2n}^T, \pi_{2n}^T \right\rangle \\ \vdots & \vdots & \ddots & \vdots \\ \left\langle \mu_{n1}^T, v_{n1}^T, \pi_{n1}^T \right\rangle & \mu_{n2}^T, v_{n2}^T, \pi_{n2}^T & \cdots & \left\langle \mu_{nn}^T, v_{nn}^T, \pi_{nn}^T \right\rangle \end{bmatrix} \tag{19}$$

If T matrix values are not in the range of 0–1, it is recommended to create values in the range of 0–1 with vector normalization [8].

(f) The sum operator as seen in Eq. (3) is used for finding spherical fuzzy row (f_i) and column totals $\left(g_j\right)$.

$$f_i = \sum_{i=1}^{n} \langle \mu_{ij}^T, v_{ij}^T, \pi_{ij}^T \rangle \tag{20}$$

$$g_j = \sum_{j=1}^{n} \langle \mu_{ij}^T, v_{ij}^T, \pi_{ij}^T \rangle \tag{21}$$

The general strength (f_i) and general weakness (g_j) values of each feature were obtained as spherical fuzzy numbers. In the next step, spherical fuzzy numbers are defuzzified in terms of the sum of each row and column, using the score function given in the equation below, to find the significance and correlation values.

$$\text{Score} = (2\mu - \pi)^2 - (v - \pi)^2 \tag{22}$$

(g) By calculating the importance $(f_i + g_i)$ and relation values $(f_i - g_i)$, the attributes are allocated to the cause or effect group according to the sign of the relation value.
(h) A network relationship map is created showing the causal dependence or effects between the attributes. The horizontal axis of the graph presents the importance values of the attributes, while the vertical axis shows the relationship values. By filtering the total impact matrix according to the threshold value, the significant effects between the features can be revealed and these can be shown in the network relationship map.
(i) Equations (14 and 15) are used to establish the weight values of the criteria.

4 Application and Findings

Within the scope of the study, a wide list of criteria was created by making use of the literature. The list of criteria in question is presented below:

- Ease of use [2, 9, 25]
- Maintenance opportunities and technical support [4, 25]
- Statistical information recording [20, 25]
- Appointment systems [25]
- Security (backup, antivirus support, etc.) [7, 25]
- Cost [25]
- Software technologies and system quality [2, 25]
- Patient privacy and personality rights [6]
- Design [21]
- Individualization [21]
- Helping and offering suggestions [21; 26]
- Include a decision support system [16]
- Submission of consultation requests via HIMS [9]

- Educational support [4, 7, 18]
- Accessibility [26]
- Integration with systems other than the health information system (MEDULA, MHRS, other systems that do not contain information technologies, etc.) or covering all sub-systems in the health institution (laboratory, health record, archive, etc.) [4, 16, 26]
- Ensuring standardization [26]
- Improvement of patient/customer interaction [17]
- Constant presence of the authorized representative of the company in the health institution [7].

Then, the list in question was presented to the experts, and the opinions of the experts on the criteria were taken. In line with the opinions received, changes (removal, addition and correction) were made in the criteria set. Accordingly, the following criteria were taken into account in the study:

C1: Security (patient privacy etc.)
C2: Cost
C3: System quality and technical support
C4: Design, individualization and ease of use
C5: Presenting statistical information
C6: Include a decision support system
C7: Providing accessibility
C8: Integration with systems other than the health information system (MEDULA, MHRS, other systems that do not contain information technologies, etc.) or covering all sub-systems in the health institution (laboratory, health record, archive, etc.)

The Spherical Fuzzy DEMATEL method was applied for determining the effects and importance levels related to the criteria that should be considered in the health information system. In this context, the opinions of four experts were consulted and the qualitative evaluations were obtained.

These evaluations were used to create the integrated decision matrix by using the Spherical Fuzzy Linguistic Scale in Table 2 and Eqs. (17 and 18). The integrated Spherical fuzzy decision matrix is presented in Table 3.

To obtain the total effect matrix, the integrated spherical fuzzy decision matrix is decomposed according to μ, v and π values.

The integrated decision matrix is decomposed according to membership, non-membership and hesitation degrees, and following to that the initial (X) and direct (T) impact matrices are formed within the scope of Eqs. (8 and 11). Then, the spherical fuzzy row (f_i) and column sums $\left(g_j\right)$ are acquired via spherical fuzzy operator in Eq. (3). The values obtained in this context are seen in Table 4.

In terms of the sign of the relation value for the criteria, the attributes are allocated to the cause or effect group. In addition, the weight values related to the criteria can be determined via using the Eqs. (14 and 15). In this context, the information obtained

Table 3 Integrated spherical fuzzy decision matrix

	C1	C1	C1	*C2*	*C2*	*C2*	C3	C3	C3	*C4*	*C4*	*C4*	C5	C5	C5	*C6*	*C6*	*C6*	C7	C7	C7	*C8*	*C8*	*C8*
	μ	ν	π	μ	ν	π	μ	ν	π	μ	ν	π	μ	ν	π	μ	ν	π	μ	ν	π	μ	ν	π
C1	0.00	0.00	0.00	0.69	0.19	0.45	0.69	0.19	0.45	0.76	0.17	0.47	0.69	0.19	0.45	0.76	0.17	0.47	0.76	0.17	0.47	0.69	0.19	0.45
C2	0.81	0.16	0.47	0.00	0.00	0.00	0.69	0.19	0.45	0.69	0.19	0.45	0.81	0.16	0.47	0.76	0.17	0.47	0.76	0.17	0.47	0.76	0.17	0.47
C3	0.76	0.17	0.47	0.66	0.20	0.45	0.00	0.00	0.00	0.50	0.22	0.32	0.55	0.21	0.34	0.60	0.20	0.35	0.55	0.21	0.34	0.63	0.21	0.46
C4	0.40	0.25	0.28	0.31	0.26	0.23	0.43	0.24	0.29	0.00	0.00	0.00	0.43	0.24	0.29	0.35	0.25	0.25	0.40	0.25	0.28	0.35	0.25	0.25
C5	0.55	0.21	0.34	0.60	0.20	0.35	0.69	0.19	0.45	0.50	0.22	0.32	0.00	0.00	0.00	0.69	0.19	0.45	0.55	0.21	0.34	0.53	0.22	0.33
C6	0.50	0.22	0.32	0.60	0.20	0.35	0.74	0.18	0.49	0.60	0.20	0.35	0.69	0.19	0.45	0.00	0.00	0.00	0.55	0.21	0.34	0.55	0.21	0.34
C7	0.50	0.22	0.32	0.48	0.23	0.31	0.43	0.24	0.29	0.48	0.23	0.31	0.35	0.25	0.25	0.31	0.26	0.23	0.00	0.00	0.00	0.43	0.24	0.29
C8	0.43	0.24	0.29	0.63	0.21	0.46	0.66	0.20	0.45	0.66	0.20	0.45	0.66	0.20	0.45	0.63	0.21	0.46	0.76	0.17	0.47	0.00	0.00	0.00

Table 4 Sum of rows and columns of total influence matrix

	f values				g values			
	μ	ν	π	Score	μ	ν	π	Score
C1	0.9627	0.0100	0.2705	2.6706	0.9009	0.0293	0.4278	1.7292
C2	0.9730	0.0077	0.2307	2.8926	0.8988	0.0304	0.4339	1.6969
C3	0.9228	0.0249	0.3828	2.0118	0.9314	0.0214	0.3633	2.1318
C4	0.7205	0.0940	0.6134	0.4153	0.9222	0.0250	0.3849	2.0007
C5	0.9104	0.0264	0.4079	1.8510	0.9214	0.0231	0.3869	1.9872
C6	0.9170	0.0229	0.3944	1.9343	0.9141	0.0260	0.4036	1.8865
C7	0.7729	0.0762	0.5804	0.6779	0.9338	0.0208	0.3566	2.1705
C8	0.9270	0.0221	0.3741	2.0663	0.9010	0.0309	0.4299	1.7236

Table 5 Influence groups and weights of criteria

	Prominence	Relation	Influence	$(f+g)^2$	$(f-g)^2$	q_j	w_j
	$f+g$	$f-g$	group				
C1	4.3999	0.9414	Cause	19.3588	0.8863	4.4994	0.1453
C2	4.5895	1.1956	Cause	21.0633	1.4296	4.7427	0.1532
C3	4.1436	−0.1201	Effect	17.1692	0.0144	4.1453	0.1339
C4	2.4160	−1.5854	Effect	5.8369	2.5136	2.8897	0.0933
C5	3.8382	−0.1361	Effect	14.7317	0.0185	3.8406	0.1241
C6	3.8208	0.0477	Cause	14.5983	0.0023	3.8211	0.1234
C7	2.8484	−1.4926	Effect	8.1131	2.2280	3.2158	0.1039
C8	3.7899	0.3428	Cause	14.3632	0.1175	3.8053	0.1229

regarding the criteria by using the f and g score values in Table 4 is presented in Table 5.

From Table 5, it is seen that C1, C2, C6 and C8 are from the cause criteria group, and C3, C4, C5 and C7 are from the effect criteria group. However, $f+g$ values also show the importance of the relevant criterion in the problem or system. Similar to $f+g$ values, the weight values calculated for the criteria show that the order of importance of the criteria is C2 $\succ$ C1 $\succ$ C3 $\succ$ C5 $\succ$ C6 $\succ$ C8 $\succ$ C7 $\succ$ C4. On the other hand, $f-g$ values provide the information necessary for the classification of criteria into groups in the form of affecting-affected. Accordingly, the criteria with a negative $f-g$ value are the affected criteria, and the positive ones are the affecting/cause criteria. DEMATEL allows visualization of relationships and positions between criteria. In this context, firstly, the T matrix is defuzzified and the threshold value is determined for the drawing of the effect diagram showing the dependencies and relational situations between the criteria. The defuzzified T matrix obtained by using Eq. (22) is shown as Table 6.

Table 6 Deffuzzified T matrix

	C1	C2	C3	C4	C5	C6	C7	C8
C1	0.0908	0.1488	0.1537	0.1791	0.1632	0.1597	0.1930	0.1487
C2	0.2188	0.0942	0.1776	0.1945	0.2114	0.1841	0.2210	0.1816
C3	0.1464	0.1137	0.0686	0.1177	0.1259	0.1229	0.1440	0.1050
C4	−0.1046	−0.1241	−0.0507	−0.0698	−0.0661	−0.0949	−0.0603	−0.1141
C5	0.1102	0.1105	0.1433	0.1134	0.0654	0.1295	0.1417	0.0948
C6	0.1068	0.1168	0.1504	0.1402	0.1459	0.0611	0.1476	0.1065
C7	−0.0392	−0.0479	−0.0140	−0.0219	−0.0450	−0.0666	−0.0233	−0.0558
C8	0.0841	0.0905	0.1072	0.1150	0.1112	0.0940	0.1446	0.0432

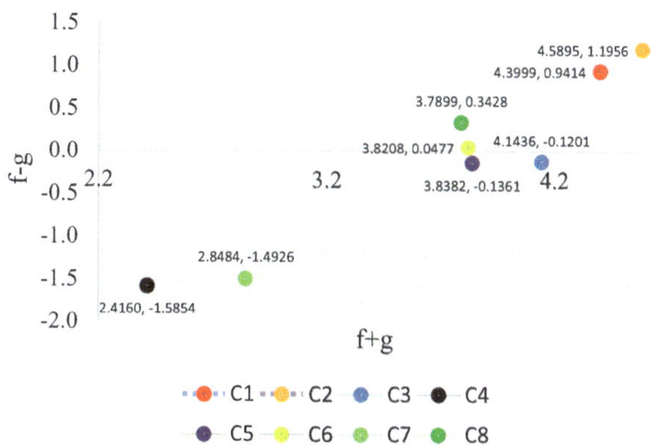

Fig. 1 Network relations map

The mean of the absolute values of the elements in Table 6 was used as the threshold value. In this context, the threshold value $\alpha = 0.1146$ has been determined. The effect network diagram is presented in Fig. 1. In Fig. 1, arrows showing the effects between criteria are not given because the image does not create complexity, only the positions of the criteria in this diagram are shown.

From Fig. 1, it is seen that the criteria C1, C2, C6 and C8, which are above the 0 position according to the $f - g$ axis, are the ones that affect it. In other words, the criteria C3, C4, C5 and C7 located in the lower region of the $f - g$ axis are the affected criteria.

5 Conclusion and Recommendations

Main aim of this study was to determine the effective criteria for the selection of health information system and the importance levels of these criteria, and for this purpose, Delphi-based spherical DEMATEL method was used. The DEMATEL method, which allows interaction between criteria, has been handled in a spherical fuzzy environment, and as a result of taking into account the degree of hesitation, it has offered more independence and wider choice to decision makers. The most important criterion in the selection of health information system is obtained as cost, while the least important criterion is found as design, individualization and ease of use. The order of importance of other criteria is security (patient privacy, etc.); system quality and technical support; provide statistical information; include a decision support system; integration with systems other than the health information system (MEDULA, MHRS, other systems that do not contain information technologies, etc.) or covering all sub-systems in the health institution (laboratory, health record, archive, etc.) and providing accessibility. In addition, the criteria are divided among themselves as affecting and affected. In this context, the criteria that affect or cause the selection of the health information system are security (patient privacy, etc.); cost; decision support system and integration with systems other than the health information system (MEDULA, MHRS, other systems that do not contain information technologies, etc.) or covering all sub-systems in the health institution (laboratory, health record, archive, etc.). On the other hand, if the health information system is affected by the selection or the outcome criteria are system quality and technical support; design, individualization and ease of use; providing statistical information and providing accessibility. In addition, the relations and locations between the criteria were visualized with the effect network diagram, thus facilitating the decision makers. The advantage of the study is that it allows the criteria that are effective in the selection of health information system to be evaluated in an environment that handles uncertainties according to new and existing conditions. The disadvantage or limitation of the study is the low number of experts reached, which can create a problem in the generalization of the results. For future studies, by increasing the number of criteria that are effective in the selection of health information system, comparative analyzes can be performed in different decision-making environments (hesitant, intuitionistic, pythagorean, fermatean, etc.) with different MCDM techniques (PIPRECIA, MARCOS, MAIRCA, MEREC, etc.).

References

1. Alacadağlı, E.: Bilgi yönetimi, dijitalleşme ve Türk sağlik sistemi. Electron. Turkish Stud. **14**(2), 67–86 (2019). https://doi.org/10.7827/TurkishStudies.14918
2. Bal, C.G., Serkan, A.D.A., Çelik, A.: Bilişim sistemleri başarı modeli ve aile hekimliği bilişim sistemleri. Yönetim ve Ekonomi: Celal Bayar Üniversitesi İktisadi ve İdari Bilimler Fakültesi Dergisi **19**(1), 35–46 (2012)
3. Dalalah, D., Hayajneh, M., Batieha, F.: A fuzzy multi-criteria decision making model for supplier selection. Exp. Syst. Appl. **38**(7), 8384–8391 (2011). https://doi.org/10.1016/j.eswa.2011.01.031
4. Eriş, H., Ilıman, E.: Sağlık Çalışanlarının Hastane Bilgi Sistemi Hakkındaki Görüşleri. Avrupa Bilim ve Teknoloji Dergisi (16), 301–309 (2019). https://doi.org/10.31590/ejosat.566750

5. Işık, O., Akbolat, M.: Bilgi teknolojileri ve hastane bilgi sistemleri kullanımı: Sağlık çalışanları üzerine bir araştırma. Bilgi Dünyası **11**(2), 365–389 (2010). https://doi.org/10.15612/BD.2010.242
6. İleri, Y.Y., Uludağ, A.: E-Nabiz uygulamasinin yönetim bilişim sistemleri ve hasta mahremiyeti açisindan değerlendirilmesi. Uluslararası Sağlık Yönetimi ve Stratejileri Araştırma Dergisi **3**(3), 318–325 (2017)
7. Göktaş, B., Önder, Ö.R., Duran, M., Şakar, S., Yilmaz, M., Güler, S., Çınar, İ, Çamlıdağ, T., Şenkal, Y., Özdemir, G.: Türkiye'de sağlık bilgi sistemleri üzerine bir araştırma. Ankara Sağlık Bilimleri Dergisi **6**(1), 125–138 (2017). https://doi.org/10.1501/Asbd_0000000066
8. Gül, S.: Spherical fuzzy extension of DEMATEL (SF-DEMATEL). Int. J. Intell. Syst. **35**, 1329–1353 (2020). https://doi.org/10.1002/int.22255
9. Kaçmaz Bali, K., Kidak, L., Güngör, D.Ö.: Hastane bilgi yönetim sistemlerinin konsültasyon süreci üzerindeki etkileri. Hacettepe Sağlık İdaresi Dergisi **22**(4), 765–780 (2019)
10. Kutlu Gündoğdu, F., Kahraman, C.: A novel spherical fuzzy analytic hierarchy process and its renewable energy application. Soft. Comput. **24**, 4607–4621 (2019). https://doi.org/10.1007/s00500-019-04222-w
11. Kutlu Gündoğdu, F., Kahraman, C.: Spherical fuzzy sets and spherical fuzzy TOPSIS method. J. Intell. Fuzzy Syst. **36**(1), 337–352 (2019)
12. Kutlu Gündoğdu, F., Kahraman, C.: A novel fuzzy TOPSIS method using emerging interval-valued spherical fuzzy sets. Eng. Appl. Artif. Intel. **85**, 307–323 (2019). https://doi.org/10.1016/j.engappai.2019.06.003
13. Kutlu Gündoğdu, F., Kahraman, C.: Extension of WASPAS with spherical fuzzy sets. Informatica **30**(2), 269–292 (2019d). https://doi.org/10.15388/Informatica.2019.206
14. Küzeci, E.: Sağlik bilişim teknolojileri ve yeni hukuksal soru(n)lar. İnönü Üniversitesi Hukuk Fakültesi Dergisi **9**(1), 477–506 (2018). https://doi.org/10.21492/inuhfd.410571
15. Limon, S.: Hastanelerdeki tibbi dokümanlarin geleneksel ortamdan elektronik ortama dönüşümü. Uygulamalı Sosyal Bilimler ve Güzel Sanatlar Dergisi **1**(1), 30–39 (2019)
16. Ministry of Health, Hastane bilgi yönetim sistemi alım kılavuzu (2010). Retrieved from 13 July 2021 https://sbsgm.saglik.gov.tr/TR-1724/hastane-bilgi-yonetim-sistemi-alim-kilavuzu.html
17. Ömürbek, N., Altın, F.G.: Sağlık bilişim sistemlerinin uygulanmasına ilişkin bir araştırma: İzmir örneği. Süleyman Demirel Üniversitesi Fen-Edebiyat Fakültesi Sosyal Bilimler Dergisi **2009**(19), 211–232 (2009)
18. Ömürbek, N., Demirgubuz, M.Ö., Tunca, M.Z.: Hastanelerdeki bilişim sistemlerinden klinik bilgi sistemlerinin kullanimina yönelik bir araştırma: Denizli ve Isparta örneği. Sosyal Ekonomik Araştırmalar Dergisi **13**(25), 301–328 (2013)
19. Peker, S.V., Van Giersbergen, M.Y., Biçersoy, G.: Sağlik bilişimi ve Türkiye'de hastanelerin dijitalleşmesi. Sağlık Akademisi Kastamonu **3**(3), 228–267 (2018). https://doi.org/10.25279/sak.398078
20. Rađenović, Ž, Veselinović, I.: Integrated AHP-TOPSIS method for the assessment of health management information systems efficiency. Econ. Themes **55**(1), 121–142 (2017). https://doi.org/10.1515/ethemes-2017-0008
21. Uslu, D., Toygar, ŞA., Mansur, F.: Hastane bilgi yönetim sisteminin kullanilabilirliğini belirlemeye yönelik bir araştırma. Uluslararası Sağlık Yönetimi ve Stratejileri Araştırma Dergisi **2**(3), 45–57 (2016)
22. Ülke, R., Atilla, E.A.: Sağlık hizmetlerinde bilişim sistemleri ve e-sağlık: Ankara ili örneği. Gazi İktisat ve İşletme Dergisi **6**(1), 86–100 (2020). https://doi.org/10.30855/gjeb.2020.6.1.006
23. Varol, Ş, Orhan, F., Tuncer, S., Akyüz, S.: Sağlık kurumlarında bilgi güvenliği bağlamında biyometrik sistemler. Sağlık Akademisyenleri Dergisi **3**(4), 155–162 (2016)
24. Wang, W.C., Lin, Y.H., Lin, C.L., Chung, C.H., Lee, M.T.: DEMATEL-based model to improve the performance in a matrix organization. Exp. Syst. Appl. **39**, 4978–4986 (2012). https://doi.org/10.1016/j.eswa.2011.10.016
25. Yeşilyurt, B., Karakuş, K., Gür, Ş, Eren, T.: Çok ölçütlü karar verme yöntemleri ile hastane bilgi yönetim sistemleri için paket programı seçimi. Başkent Üniversitesi Ticari Bilimler Fakültesi Dergisi **3**(1), 1–21 (2019)

26. Yıldırım, B.F.: Sağlığın kişiselleşmesi ve kişisel sağlık bilgi sistemleri. Bilgi Yönetimi **2**(2), 127–135 (2019). https://doi.org/10.33721/by.642698

Ahmet Aytekin Dr. Ahmet Aytekin is a Researcher at Business Administration Department, Faculty of Economics and Business Administrations, Artvin Çoruh University, Turkey. Dr. Ahmet Aytekin received his MSc and PhD in Quantitative Methods from Anadolu University in Eskişehir, Turkey. His research interests include decision analysis, multi-criteria decision making, fuzzy set theory, multivariate statistical analysis, and data analysis. Dr. Aytekin has also served on the review boards of several international journals. He has many research papers that have been published in indexed journals or books.

Gözde Koca Dr. Gözde Koca is a Researcher at Business Administration Department, Faculty of Economics and Business Administrations, Bilecik Şeyh Edebali University, Turkey. Dr. Gözde Koca received her MSc in Industrial Engineering and her PhD in Econometrics from Dumlupınar University in Kütahya, Turkey. Her research interests include decision multi-criteria decision making, fuzzy set theory, supply chain management, and data mining. She has many research papers that have been published in indexed journals or books.

Basil Oluoch Okoth Dr. Basil OKOTH recently completed his Ph.D. in the field of Management and Organization at Anadolu University in Turkey. He has a Master's degree in Finance from the same university and has a bachelor's in Finance from Maseno University in Kenya. His research interest includes leadership, organizational performance, organizational culture, organizational ecology, CSR and Corporate governance as well as knowledge management in organizations. His work in the field of finance is mainly concerned with financial inclusion, especially through mobile banking and mobile money. Dr. Okoth is a young researcher who has shown diversity in his ability to conduct research in a variety of fields.

Çağlar Karamaşa Dr. Çağlar Karamaşa is a member of the Faculty of Business at the Anadolu University. He received his Master's degree and PhD in the field of Quantitative Methods from Anadolu University in Turkey. His research interests include fuzzy set theory, multi criteria decision making, data mining and multivariate statistical analysis. Dr. Karamaşa has served on the editorial boards of several international journals. He has also numerous scientific studies that have been published in indexed journals or books.

Score Function-Based Effective Ranking of Interval-Valued Fermatean Fuzzy Sets and Its Applications to Multi-criteria Decision Making Problem

Laxminarayan Sahoo, Akul Rana, Tapan Senapati, and Ronald R. Yager

Abstract Fermatean fuzzy sets (FFSs), an orthopair fuzzy set proposed by Senapati and Yager (Journal of Ambient Intelligence and Humanized Computing 11:663–674, 2020, [24]), can handle the situation with ambiguous and incomplete information in a more effective manner than the Pythagorean fuzzy sets presented by Yager (Pythagorean fuzzy subsets, 2013 Joint IFSA world congress and NAFIPS annual meeting (IFSA/NAFIPS), pp 57–61, 2013, [2]) and the intuitionistic fuzzy sets presented by Atanassov (Fuzzy Sets and Systems 20:87–96, 1986, [3]). Sergi et al. (Journal of Intelligent & Fuzzy Systems 42:365–376, 2022, [40]) initiated interval-valued Fermatean fuzzy sets (IVFFSs) and established IVFFS ordering, as well as some mathematical operations. In addition, Jeevraj (Expert Systems with Applications 185:1–20, 2021, [37]) introduced the concept of a score and accuracy function for IVFFSs. The main objective of this chapter is to suggest some score functions for acceptable ranking of IVFFSs, as well as the interval-valued Fermatean fuzzy TOPSIS method for solving multi-criteria decision making problems. Here, we have proposed six new variants of score functions for effective ranking of interval-valued Fermatean fuzzy sets. Depending on various types of score functions, we have used a TOPSIS-based multi-criteria decision making (MCDM) problem in which decision makers' (DMs') preference knowledge is summarized in the pattern of interval-valued Fermatean fuzzy sets. To illustrate the usefulness of the proposed method, a

L. Sahoo (✉)
Department of Computer and Information Science, Raiganj University, Raiganj 733134, India
e-mail: lxsahoo@gmail.com

A. Rana
Department of Mathematics, Narajole Raj College, Paschim Medinipur, West Bengal 721211, India
e-mail: arnrc79@gmail.com

T. Senapati
School of Mathematics and Statistics, Southwest University, Beibei, Chongqing 400715, China
e-mail: math.tapan@gmail.com

R. R. Yager
Machine Intelligence Institute, Iona College, New Rochelle, NY 10801, USA
e-mail: yager@panix.com

L. Sahoo et al. (eds.), *Real Life Applications of Multiple Criteria Decision Making Techniques in Fuzzy Domain*, Studies in Fuzziness and Soft Computing 420,
https://doi.org/10.1007/978-981-19-4929-6_20

computational paradigm has been considered. Finally, concluding remarks and future scope of research have been mentioned.

Keywords Interval-valued Fermatean fuzzy set · Fermatean fuzzy set · TOPSIS · Multi-criteria decision making · Score function

1 Introduction

Zadeh [1] introduced the theoretical concept of fuzzy set in 1965 to deal with imprecision and vagueness in real-world situations. In the year 1970, Bellman and Zadeh [2] introduced the paradigm of decision making problems under uncertainty. Intuitionistic fuzzy sets (IFSs) were first suggested by Atanassov [3] in 1986 to describe realism using three characteristics: support, opposition and neutrality, and they have now been widely considered and used by different researchers [4, 5]. In addition, several researchers have focused on interval-valued fuzzy sets (IVFSs) [6, 7], and interval-valued intuitionistic fuzzy sets (IVIFSs) [8], which are all suppositions of the fuzzy set suggested by Zadeh [1], and have used them for a variety of decision making problems. On the other hand, the fuzzy set makes off with only a membership value and the degree of non-membership value, which is exactly a compliment to the degree of membership value.

Also, in many instances, the sum of the membership and non-membership values has been found to be more than one. As a result, orthopair fuzzy sets have been suggested, in which an element's membership scores are pairs of values in the unit interval, $\langle \alpha(x), \beta(x) \rangle$, where the first one indicates support for membership in the fuzzy set and the other supports opposing the membership. For instance, Atanassov's classical intuitionistic fuzzy sets [3, 9, 10] and his second kind of intuitionistic fuzzy sets [11, 12]. Yager [13, 14] has proposed another orthopair fuzzy set, known as the Pythagorean fuzzy set (PFS), in which the square sum of the support for membership and support against membership value is equal to or less than one. In addition to that a number of researchers have provided advance notice of interval-valued Pythagorean fuzzy sets (IVPFSs) [15, 16], which are the extension of Yager's PFSs, and have employed these fuzzy sets in various DM problems. In a relatively short amount of time, PFSs and IVPFSs have attracted the attention of many researchers. Extension of TOPSIS [17], similarity measure [18], alternative queuing approach [19], expanded TODIM methods [20], Bonferroni mean [21], enhanced scoring function [16, 20], as well as many other approaches have been suggested in the domain of PFS to address realistic multi-criteria decision making problems. However, if orthopair fuzzy set as <0.9, 0.6> , where 0.9 is the support for membership of a specific parameter's criterion and 0.6 is the support opposing membership, it does not follow the IFS and PFS requirements. However, the cubic aggregate of support for membership and opposition to membership is less than one. So, as a consequence, Senapati and Yager [22] very recently introduced the Fermatean fuzzy set (FFS) into the literature of fuzzy set theory. They also studied and established that FFSs have more uncertainty than IFSs and PFSs, and that they are able to withstand high levels of uncertainty [23] and tackle the MCDM problem easily. Furthermore, many researchers have proposed different solution methodologies and solved real-life application problems under Pythagorean fuzzy as well as Fermatean fuzzy environments. For more details, one may mention the works of Li et al. [24], Zhou et al. [25], Bolturk et al. [26], Qin [27], Wan et al. [28], Lin et al. [29], Chen [30], Sahoo [31–33], Senapati and Yager [34], Senapati and Chen [35], Jana et al. [36]. Jeevaraj [37] established some mathematical operations and proposed interval-valued Fermatean fuzzy sets (IVFFSs) in 2021. In this study, we have proposed several variants of novel scoring functions for ranking of IVFFSs that are acceptable, as well as the interval-valued Fermatean fuzzy TOPSIS technique for handling multi-criteria decision making problems. For interval-valued Fermatean fuzzy sets, we have gone through some basic set operations and arithmetic operations. Different variants of scoring functions have been defined for the ranking of interval-valued Fermatean fuzzy sets. Also, we have used the TOPSIS technique for solving the multi-criteria decision making (MCDM) problem, in which decision makers' preferences are represented in a pattern of interval-valued Fermatean fuzzy sets. A quantitative problem has been considered and solved to explain the usefulness of the suggested method. Finally, concluding remarks and suggestions for future research have been addressed.

Purpose of the chapter

The main purposes of this chapter are as follows:

1. We defined different variants of score functions for effective ranking of interval-valued Fermatean fuzzy sets.
2. We defined the accuracy function in terms of expected value and supremum value for effective ranking of interval-valued Fermatean fuzzy sets.
3. We used the TOPSIS technique for solving the multi-criteria decision making (MCDM) problem, in which decision makers' preferences are represented in a pattern of interval-valued Fermatean fuzzy sets.
4. Concluding remarks have been presented, along with the future direction of the proposed research work.

The definitions and crucial concepts of interval-valued Fermatean fuzzy sets are discussed in Sect. 2. Section 3 presents a ranking of interval-valued Fermatean fuzzy sets. Section 4 describes the TOPSIS method for addressing interval-valued Fermatean fuzzy multi-criteria decision making problems using proposed score functions. For the purpose of illustration, a mathematical model/problem is discussed in Sect. 5. The conclusion, as well as the research's implications, can be found in Sect. 6.

2 Preliminaries

In this section, some basic concepts related to Fermatean fuzzy sets used in this paper are described briefly.

Definition 2.1 Assume that $D[0, 1]$ is a set of all closed subintervals of the interval $[0, 1]$ and $X(\neq \Phi)$ is a pre specified nonempty set. An IVFFS F in X is given by $F = \{\langle \alpha_F(x), \beta_F(x)\rangle : x \in X\}$ where the functions $\alpha_F(x) : X \to D[0, 1]$ and $\beta_F(x) : X \to D[0, 1]$, such that $0 \le (\sup(\alpha_F(x)))^3 + (\sup(\beta_F(x)))^3 \le 1$ for every $x \in X$. Then the two intervals $\alpha_F(x)$ and $\beta_F(x)$ define, respectively, the support for membership and support against membership of the element $x \in X$ to the set F which is a subset of X and every $x \in X$.

Thus, for each $x \in X$, $\alpha_F(x)$ and $\beta_F(x)$ are closed intervals and their lower and upper end points are, respectively, denoted by $\alpha_{FL}(x)$, $\alpha_{FU}(x)$, $\beta_{FL}(x)$ and $\beta_{FU}(x)$.

Now, we denote IVFFS F as $F = \{\langle[\alpha_{FL}(x), \alpha_{FU}(x)], [\beta_{FL}(x), \beta_{FU}(x)]\rangle : x \in X\}$ where $0 \le (\alpha_{FU}(x))^3 + (\beta_{FU}(x))^3 \le 1$ and $\alpha_{FL}(x) \ge 0,\ \beta_{FL}(x) \ge 0$.

Again, each IVFFS F in X, the function $\pi_F(x) = \sqrt[3]{1 - ([\alpha_{FL}(x), \alpha_{FU}(x)])^3 - ([\beta_{FL}(x), \beta_{FU}(x)])^3}$ is called the degree of hesitation or indeterminacy. It is obvious that $\pi_F(x)$ is an interval-valued function and we denote $\pi_F(x) = [\pi_{FL}(x), \pi_{FU}(x)]$.

where $\pi_{FL}(x) = \sqrt[3]{1 - (\alpha_{FU}(x))^3 - (\beta_{FU}(x))^3}$ and $\pi_{FU}(x) = \sqrt[3]{1 - (\alpha_{FL}(x))^3 - (\beta_{FL}(x))^3}$ for all $x \in X$.

It is to be noted that for simplicity, we shall denote the object $F = \langle[\alpha_{FL}, \alpha_{FU}], [\beta_{FL}, \beta_{FU}]\rangle$ instead of $F = \{\langle[\alpha_{FL}(x), \alpha_{FU}(x)], [\beta_{FL}(x), \beta_{FU}(x)]\rangle : x \in X\}$.

Definition 2.2 Let $F_1 = \langle[\alpha_{F_1L}, \alpha_{F_1U}], [\beta_{F_1L}, \beta_{F_1U}]\rangle$ and $F_2 = \langle[\alpha_{F_2L}, \alpha_{F_2U}], [\beta_{F_2L}, \beta_{F_2U}]\rangle$ be two IVFFSs, then their operations are defined as follows:

(i) *Union:* $F_1 \cup F_2 = \left\langle \begin{matrix}[\max(\alpha_{F_{1L}}, \alpha_{F_{2L}}), \max(\alpha_{F_{1U}}, \alpha_{F_{2U}})], \\ [\min(\beta_{F_{1L}}, \beta_{F_{2L}}), \min(\beta_{F_{1U}}, \beta_{F_{2U}})]\end{matrix} \right\rangle$

(ii) *Intersection:* $F_1 \cap F_2 = \left\langle \begin{matrix}[\min(\alpha_{F_{1L}}, \alpha_{F_{2L}}), \min(\alpha_{F_{1U}}, \alpha_{F_{2U}})], \\ [\max(\beta_{F_{1L}}, \beta_{F_{2L}}), \max(\beta_{F_{1U}}, \beta_{F_{2U}})]\end{matrix} \right\rangle$

(iii) *Compliment:* $(F_1)' = \langle[\beta_{F_1L}, \beta_{F_1U}], [\alpha_{F_1L}, \alpha_{F_1U}]\rangle$

Definition 2.3 Let $F_1 = \langle[\alpha_{F_1L}, \alpha_{F_1U}], [\beta_{F_1L}, \beta_{F_1U}]\rangle$ and $F_2 = \langle[\alpha_{F_2L}, \alpha_{F_2U}], [\beta_{F_2L}, \beta_{F_2U}]\rangle$ be two IVFFSs. Then the basic arithmetical operations of two IVFFSs F_1 and F_2 are defined as follows:

(i) *Addition:* $$F_1 \oplus F_2 = \left\langle \begin{bmatrix}\sqrt[3]{(\alpha_{F_{1L}})^3 + (\alpha_{F_{2L}})^3 - (\alpha_{F_{1L}})^3(\alpha_{F_{2L}})^3}, \\ \sqrt[3]{(\alpha_{FU_1})^3 + (\alpha_{F_{2U}})^3 - (\alpha_{F_{1U}})^3(\alpha_{F_{2U}})^3}\end{bmatrix}, \right\rangle$$
$$[\beta_{F_{1L}}\beta_{F_{2L}}, \beta_{F_{1U}}\beta_{F_{2U}}]$$

(ii) *Multiplication:* $$F_1 \otimes F_2 = \left\langle \begin{matrix}[\alpha_{F_{1L}}\alpha_{F_{2L}}, \alpha_{F_{1U}}\alpha_{F_{2U}}], \\ \begin{bmatrix}\sqrt[3]{(\beta_{F_{1L}})^3 + (\beta_{F_{2L}})^3 - (\beta_{F_{1L}})^3(\beta_{F_{2L}})^3}, \\ \sqrt[3]{(\beta_{FU_1})^3 + (\beta_{F_{2U}})^3 - (\beta_{F_{1U}})^3(\beta_{F_{2U}})^3}\end{bmatrix}\end{matrix} \right\rangle$$

(iii) *Scalar Multiplication:* $\lambda \odot F_1 = \left\langle \begin{bmatrix} \sqrt[3]{1-(1-(\alpha_{F_{1L}})^3)^\lambda}, \\ \sqrt[3]{1-(1-(\alpha_{F_{1U}})^3)^\lambda} \end{bmatrix}, \left[(\beta_{F_{1L}})^\lambda, (\beta_{F_{1U}})^\lambda\right] \right\rangle$ provided $\lambda > 0$

(iv) *Exponent* $F_1^\lambda = \left\langle \left[(\alpha_{F_{1L}})^\lambda, (\alpha_{F_{1U}})^\lambda\right], \begin{bmatrix} \sqrt[3]{1-(1-(\beta_{F_{1L}})^3)^\lambda}, \\ \sqrt[3]{1-(1-(\beta_{F_{1U}})^3)^\lambda} \end{bmatrix} \right\rangle$

Theorem 1 *If* $F_1 = \langle[\alpha_{F_1L}, \alpha_{F_1U}], [\beta_{F_1L}, \beta_{F_1U}]\rangle$, $F_2 = \langle[\alpha_{F_2L}, \alpha_{F_2U}], [\beta_{F_2L}, \beta_{F_2U}]\rangle$ *and* $F_3 = \langle[\alpha_{F_3L}, \alpha_{F_3U}], [\beta_{F_3L}, \beta_{F_3U}]\rangle$ *are three IVFFSs, then the following ones are valid:*

(i) $F_1 \oplus F_2 = F_2 \oplus F_1$
(ii) $F_1 \otimes F_2 = F_2 \otimes F_1$
(iii) $\lambda \odot (F_1 \oplus F_2) = \lambda \odot F_1 \oplus \lambda \odot F_2,\ \lambda > 0$
(iv) $(\lambda_1 + \lambda_2) \odot F_1 = \lambda_1 \odot F_1 \oplus \lambda_2 \odot F_1,\ \lambda_1, \lambda_2 > 0$
(v) $(F_1 \otimes F_2)^\lambda = F_1^\lambda \otimes F_2^\lambda,\ \lambda > 0$
(vi) $F_1^{\lambda_1} \otimes F_1^{\lambda_2} = F_1^{\lambda_1+\lambda_2},\ \lambda_1, \lambda_2 > 0$

Proof For two IVFFSs F_1, F_2 and $\lambda_1, \lambda_2 > 0$ according to definition 2.3, we can obtain.

(i)

$$\begin{aligned} F_1 \oplus F_2 &= \left\langle \begin{bmatrix} \sqrt[3]{(\alpha_{F_{1L}})^3 + (\alpha_{F_{2L}})^3 - (\alpha_{F_{1L}})^3(\alpha_{F_{2L}})^3}, \\ \sqrt[3]{(\alpha_{FU_1})^3 + (\alpha_{F_{2U}})^3 - (\alpha_{F_{1U}})^3(\alpha_{F_{2U}})^3} \end{bmatrix}, \left[\beta_{F_{1L}}\beta_{F_{2L}}, \beta_{F_{1U}}\beta_{F_{2U}}\right] \right\rangle \\ &= \left\langle \begin{bmatrix} \sqrt[3]{(\alpha_{F_{2L}})^3 + (\alpha_{F_{1L}})^3 - (\alpha_{F_{2L}})^3(\alpha_{F_{1L}})^3}, \\ \sqrt[3]{(\alpha_{FU_2})^3 + (\alpha_{F_{1U}})^3 - (\alpha_{F_{2U}})^3(\alpha_{F_{1U}})^3} \end{bmatrix}, \left[\beta_{F_{2L}}\beta_{F_{1L}}, \beta_{F_{2U}}\beta_{F_{1U}}\right] \right\rangle \\ &= F_2 \oplus F_1 \end{aligned}$$

(ii)

$$\begin{aligned} F_1 \otimes F_2 &= \left\langle \left[\alpha_{F_{1L}}\alpha_{F_{2L}}, \alpha_{F_{1U}}\alpha_{F_{2U}}\right], \begin{bmatrix} \sqrt[3]{(\beta_{F_{1L}})^3 + (\beta_{F_{2L}})^3 - (\beta_{F_{1L}})^3(\beta_{F_{2L}})^3}, \\ \sqrt[3]{(\beta_{FU_1})^3 + (\beta_{F_{2U}})^3 - (\beta_{F_{1U}})^3(\beta_{F_{2U}})^3} \end{bmatrix} \right\rangle \\ &= \left\langle \left[\alpha_{F_{2L}}\alpha_{F_{1L}}, \alpha_{F_{2U}}\alpha_{F_{1U}}\right], \begin{bmatrix} \sqrt[3]{(\beta_{F_{2L}})^3 + (\beta_{F_{1L}})^3 - (\beta_{F_{2L}})^3(\beta_{F_{1L}})^3}, \\ \sqrt[3]{(\beta_{FU_2})^3 + (\beta_{F_{1U}})^3 - (\beta_{F_{2U}})^3(\beta_{F_{1U}})^3} \end{bmatrix} \right\rangle \\ &= F_2 \otimes F_1 \end{aligned}$$

(iii)

$$\lambda \odot (F_1 \oplus F_2) = \lambda \odot \left\langle \begin{bmatrix} \sqrt[3]{(\alpha_{F_{1L}})^3 + (\alpha_{F_{2L}})^3 - (\alpha_{F_{1L}})^3(\alpha_{F_{2L}})^3}, \\ \sqrt[3]{(\alpha_{FU_1})^3 + (\alpha_{F_{2U}})^3 - (\alpha_{F_{1U}})^3(\alpha_{F_{2U}})^3} \end{bmatrix}, \left[\beta_{F_{1L}}\beta_{F_{2L}}, \beta_{F_{1U}}\beta_{F_{2U}}\right] \right\rangle$$

$$= \left\langle \begin{bmatrix} \sqrt[3]{1 - (1 - (\alpha_{F_{1L}})^3 - (\alpha_{F_{2L}})^3 + (\alpha_{F_{1L}})^3(\alpha_{F_{2L}})^3)^\lambda}, \\ \sqrt[3]{1 - (1 - (\alpha_{FU_1})^3 - (\alpha_{F_{2U}})^3 + (\alpha_{F_{1U}})^3(\alpha_{F_{2U}})^3)^\lambda} \end{bmatrix}, \left[(\beta_{F_{1L}}\beta_{F_{2L}})^\lambda, (\beta_{F_{1U}}\beta_{F_{2U}})^\lambda\right] \right\rangle$$

Now,

$$\lambda \odot F_1 = \left\langle \left[\sqrt[3]{1 - (1 - (\alpha_{F_{1L}})^3)^\lambda}, \sqrt[3]{1 - (1 - (\alpha_{F_{1U}})^3)^\lambda}\right], \left[(\beta_{F_{1L}})^\lambda, (\beta_{F_{1U}})^\lambda\right] \right\rangle$$

$$\lambda \odot F_2 = \left\langle \left[\sqrt[3]{1 - (1 - (\alpha_{F_{2L}})^3)^\lambda}, \sqrt[3]{1 - (1 - (\alpha_{F_{2U}})^3)^\lambda}\right], \left[(\beta_{F_{2L}})^\lambda, (\beta_{F_{2U}})^\lambda\right] \right\rangle$$

$$\lambda \odot F_1 \oplus \lambda \odot F_2 = \left\langle \begin{bmatrix} \sqrt[3]{1 - (1 - (\alpha_{F_{1L}})^3 - (\alpha_{F_{2L}})^3 + (\alpha_{F_{1L}})^3(\alpha_{F_{2L}})^3)^\lambda}, \\ \sqrt[3]{1 - (1 - (\alpha_{FU_1})^3 - (\alpha_{F_{2U}})^3 + (\alpha_{F_{1U}})^3(\alpha_{F_{2U}})^3)^\lambda} \end{bmatrix}, \left[(\beta_{F_{1L}}\beta_{F_{2L}})^\lambda, (\beta_{F_{1U}}\beta_{F_{2U}})^\lambda\right] \right\rangle$$

$$= \lambda \odot (F_1 \oplus F_2)$$

(iv)

$$(\lambda_1 + \lambda_2) \odot F_1 = \left\langle \begin{bmatrix} \sqrt[3]{1 - (1 - (\alpha_{F_{1L}})^3)^{(\lambda_1+\lambda_2)}}, \\ \sqrt[3]{1 - (1 - (\alpha_{F_{1U}})^3)^{(\lambda_1+\lambda_2)}} \end{bmatrix}, \left[(\beta_{F_{1L}})^{(\lambda_1+\lambda_2)}, (\beta_{F_{1U}})^{(\lambda_1+\lambda_2)}\right] \right\rangle$$

$$= \left\langle \begin{bmatrix} \sqrt[3]{1 - (1 - (\alpha_{F_{1L}})^3)^{(\lambda_1)}(1 - (\alpha_{F_{1L}})^3)^{(\lambda_2)}}, \\ \sqrt[3]{1 - (1 - (\alpha_{F_{1U}})^3)^{(\lambda_1)}(1 - (\alpha_{F_{1U}})^3)^{(\lambda_2)}} \end{bmatrix}, \left[(\beta_{F_{1L}})^{(\lambda_1)}(\beta_{F_{1L}})^{(\lambda_2)}, (\beta_{F_{1U}})^{(\lambda_1)}(\beta_{F_{1U}})^{(\lambda_2)}\right] \right\rangle$$

$$= \lambda_1 \odot F_1 \oplus \lambda_2 \odot F_1$$

(v)

$$(F_1 \otimes F_2)^\lambda = \left\langle \begin{matrix} \left[(\alpha_{F_{1L}}\alpha_{F_{2L}})^\lambda, (\alpha_{F_{1U}}\alpha_{F_{2U}})^\lambda\right], \\ \left[\begin{matrix} \sqrt[3]{1-(1-(\beta_{F_{1L}})^3-(\beta_{F_{2L}})^3+(\beta_{F_{1L}})^3(\beta_{F_{2L}})^3)^\lambda}, \\ \sqrt[3]{1-(1-(\beta_{FU_1})^3-(\beta_{F_{2U}})^3+(\beta_{F_{1U}})^3(\beta_{F_{2U}})^3)^\lambda} \end{matrix}\right] \end{matrix} \right\rangle$$

$$= \left\langle \left[(\alpha_{F_{1L}})^\lambda, (\alpha_{F_{1U}})^\lambda\right], \left[\sqrt[3]{1-(1-(\beta_{F_{1L}})^3)^\lambda}, \sqrt[3]{1-(1-(\beta_{F_{1U}})^3)^\lambda}\right] \right\rangle$$

$$\otimes \left\langle \left[(\alpha_{F_{2L}})^\lambda, (\alpha_{F_{2U}})^\lambda\right], \left[\sqrt[3]{1-(1-(\beta_{F_{2L}})^3)^\lambda}, \sqrt[3]{1-(1-(\beta_{F_{2U}})^3)^\lambda}\right] \right\rangle$$

$$= F_1^\lambda \otimes F_2^\lambda, \lambda > 0$$

(vi)

$$F_1^{\lambda_1} \otimes F_1^{\lambda_2} = \left\langle \left[(\alpha_{F_{1L}})^{\lambda_1}, (\alpha_{F_{1U}})^{\lambda_1}\right], \left[\sqrt[3]{1-(1-(\beta_{F_{1L}})^3)^{\lambda_1}}, \sqrt[3]{1-(1-(\beta_{F_{1U}})^3)^{\lambda_1}}\right] \right\rangle$$

$$\otimes \left\langle \left[(\alpha_{F_{1L}})^{\lambda_2}, (\alpha_{F_{1U}})^{\lambda_2}\right], \left[\sqrt[3]{1-(1-(\beta_{F_{1L}})^3)^{\lambda_2}}, \sqrt[3]{1-(1-(\beta_{F_{1U}})^3)^{\lambda_2}}\right] \right\rangle$$

$$= \left\langle \begin{matrix} \left[(\alpha_{F_{1L}})^{\lambda_1}(\alpha_{F_{1L}})^{\lambda_2}, (\alpha_{F_{1U}})^{\lambda_1}(\alpha_{F_{1U}})^{\lambda_2}\right], \\ \left[\sqrt[3]{1-(1-(\beta_{F_{1L}})^3)^{\lambda_1+\lambda_2}}, \sqrt[3]{1-(1-(\beta_{F_{1U}})^3)^{\lambda_1+\lambda_2}}\right] \end{matrix} \right\rangle$$

$$= F_1^{\lambda_1+\lambda_2}, \lambda_1, \lambda_2 > 0$$

Theorem 2 *If* $F_1 = \langle[\alpha_{F_1L}, \alpha_{F_1U}], [\beta_{F_1L}, \beta_{F_1U}]\rangle$, $F_2 = \langle[\alpha_{F_2L}, \alpha_{F_2U}], [\beta_{F_2L}, \beta_{F_2U}]\rangle$ *and* $F_3 = \langle[\alpha_{F_3L}, \alpha_{F_3U}], [\beta_{F_3L}, \beta_{F_3U}]\rangle$ *are three IVFFSs, then the following ones are valid:*

(i) $F_1 \cup F_2 = F_2 \cup F_1$
(ii) $F_1 \cap F_2 = F_2 \cap F_1$
(iii) $F_1 \cup (F_2 \cup F_3) = (F_1 \cup F_2) \cup F_3$
(iv) $F_1 \cap (F_2 \cap F_3) = (F_1 \cap F_2) \cap F_3$
(v) $\lambda \odot (F_1 \cup F_2) = \lambda \odot F_1 \cup \lambda \odot F_2$
(vi) $(F_1 \cup F_2)^\lambda = F_1^\lambda \cup F_2^\lambda$

Proof

(i)

$$F_1 \cup F_2 = \left\langle \begin{matrix} [\max(\alpha_{F_{1L}}, \alpha_{F_{2L}}), \max(\alpha_{F_{1U}}, \alpha_{F_{2U}})], \\ [\min(\beta_{F_{1L}}, \beta_{F_{2L}}), \min(\beta_{F_{1U}}, \beta_{F_{2U}})] \end{matrix} \right\rangle$$

$$= \left\langle \begin{matrix} [\max(\alpha_{F_{2L}}, \alpha_{F_{1L}}), \max(\alpha_{F_{2U}}, \alpha_{F_{1U}})], \\ [\min(\beta_{F_{2L}}, \beta_{F_{1L}}), \min(\beta_{F_{2U}}, \beta_{F_{1U}})] \end{matrix} \right\rangle$$

$$= F_2 \cup F_1$$

(ii)

$$F_1 \cap F_2 = \left\langle \begin{matrix} [\min(\alpha_{F_{1L}}, \alpha_{F_{2L}}), \min(\alpha_{F_{1U}}, \alpha_{F_{2U}})], \\ [\max(\beta_{F_{1L}}, \beta_{F_{2L}}), \max(\beta_{F_{1U}}, \beta_{F_{2U}})] \end{matrix} \right\rangle$$

$$= \left\langle \begin{matrix} [\min(\alpha_{F_{2L}}, \alpha_{F_{1L}}), \min(\alpha_{F_{2U}}, \alpha_{F_{1U}})], \\ [\max(\beta_{F_{2L}}, \beta_{F_{1L}}), \max(\beta_{F_{2U}}, \beta_{F_{1U}})] \end{matrix} \right\rangle$$

$$= F_2 \cap F_1$$

(iii)

$$
\begin{aligned}
F_1 \cup (F_2 \cup F_3) &= F_1 \cup \left\langle \begin{matrix} [\max(\alpha_{F_{2L}}, \alpha_{F_{3L}}), \max(\alpha_{F_{2U}}, \alpha_{F_{3U}})], \\ [\min(\beta_{F_{2L}}, \beta_{F_{3L}}), \min(\beta_{F_{2U}}, \beta_{F_{3U}})] \end{matrix} \right\rangle \\
&= \left\langle \begin{matrix} [\max(\alpha_{F_{1L}}, \max(\alpha_{F_{2L}}, \alpha_{F_{3L}})), \max(\alpha_{F_{1U}}, \max(\alpha_{F_{2U}}, \alpha_{F_{3U}}))], \\ [\min(\beta_{F_{1L}}, \min(\beta_{F_{2L}}, \beta_{F_{3L}})), \min(\beta_{F_{1U}}, \min(\beta_{F_{2U}}, \beta_{F_{3U}}))] \end{matrix} \right\rangle \\
&= \left\langle \begin{matrix} [\max(\max(\alpha_{F_{1L}}, \alpha_{F_{2L}}), \alpha_{F_{3L}}), \max(\max(\alpha_{F_{1U}}, \alpha_{F_{2U}}), \alpha_{F_{3U}})], \\ [\min(\min(\beta_{F_{1L}}, \beta_{F_{2L}}), \beta_{F_{3L}}), \min(\min(\beta_{F_{1U}}, \beta_{F_{2U}}), \beta_{F_{3U}})] \end{matrix} \right\rangle \\
&= (F_1 \cup F_2) \cup F_3
\end{aligned}
$$

(iv)

$$
\begin{aligned}
F_1 \cap (F_2 \cap F_3) &= F_1 \cap \left\langle \begin{matrix} [\min(\alpha_{F_{2L}}, \alpha_{F_{3L}}), \min(\alpha_{F_{2U}}, \alpha_{F_{3U}})], \\ [\max(\beta_{F_{2L}}, \beta_{F_{3L}}), \max(\beta_{F_{2U}}, \beta_{F_{3U}})] \end{matrix} \right\rangle \\
&= \left\langle \begin{matrix} [\min(\alpha_{F_{1L}}, \min(\alpha_{F_{2L}}, \alpha_{F_{3L}})), \min(\alpha_{F_{1U}}, \min(\alpha_{F_{2U}}, \alpha_{F_{3U}}))], \\ [\max(\beta_{F_{1L}}, \max(\beta_{F_{2L}}, \beta_{F_{3L}})), \max(\beta_{F_{1U}}, \max(\beta_{F_{2U}}, \beta_{F_{3U}}))] \end{matrix} \right\rangle \\
&= \left\langle \begin{matrix} [\min(\min(\alpha_{F_{1L}}, \alpha_{F_{2L}}), \alpha_{F_{3L}}), \min(\min(\alpha_{F_{1U}}, \alpha_{F_{2U}}), \alpha_{F_{3U}})], \\ [\max(\max(\beta_{F_{1L}}, \beta_{F_{2L}}), \beta_{F_{3L}}), \max(\max(\beta_{F_{1U}}, \beta_{F_{2U}}), \beta_{F_{3U}})] \end{matrix} \right\rangle \\
&= (F_1 \cap F_2) \cap F_3
\end{aligned}
$$

(v)

$$
\begin{aligned}
\lambda \odot (F_1 \cup F_2) &= \lambda \odot \left\langle \begin{matrix} [\max(\alpha_{F_{1L}}, \alpha_{F_{2L}}), \max(\alpha_{F_{1U}}, \alpha_{F_{2U}})], \\ [\min(\beta_{F_{1L}}, \beta_{F_{2L}}), \min(\beta_{F_{1U}}, \beta_{F_{2U}})] \end{matrix} \right\rangle \\
&= \left\langle \begin{matrix} \left[\begin{matrix} \sqrt[3]{1 - (1 - (\max(\alpha_{F_{1L}}, \alpha_{F_{2L}}))^3)^\lambda}, \\ \sqrt[3]{1 - (1 - (\max(\alpha_{F_{1U}}, \alpha_{F_{2U}}))^3)^\lambda} \end{matrix} \right], \\ \left[(\min(\beta_{F_{1L}}, \beta_{F_{2L}}))^\lambda, (\min(\beta_{F_{1U}}, \beta_{F_{2U}})^\lambda\right] \end{matrix} \right\rangle \\
&= \left\langle \begin{matrix} [\max(\sqrt[3]{1 - (1 - (\alpha_{F_{1L}})^3)^\lambda}, \sqrt[3]{1 - (1 - (\alpha_{F_{2L}})^3)^\lambda}), \\ \max(\sqrt[3]{1 - (1 - (\alpha_{F_{1U}})^3)^\lambda}, \sqrt[3]{1 - (1 - (\alpha_{F_{2U}})^3)^\lambda})], \\ [\min((\beta_{F_{1L}})^\lambda, (\beta_{F_{2L}})^\lambda), \min((\beta_{F_{1U}})^\lambda, (\beta_{F_{2U}})^\lambda)] \end{matrix} \right\rangle \\
&= \left\langle \begin{matrix} \left[\sqrt[3]{1 - (1 - (\alpha_{F_{1L}})^3)^\lambda}, \sqrt[3]{1 - (1 - (\alpha_{F_{1U}})^3)^\lambda}\right], \\ \left[(\beta_{F_{1L}})^\lambda, (\beta_{F_{1U}})^\lambda\right] \end{matrix} \right\rangle \\
&\quad \cup \left\langle \begin{matrix} \left[\sqrt[3]{1 - (1 - (\alpha_{F_{2L}})^3)^\lambda}, \sqrt[3]{1 - (1 - (\alpha_{F_{2U}})^3)^\lambda}\right], \\ \left[(\beta_{F_{2L}})^\lambda, (\beta_{F_{2U}})^\lambda\right] \end{matrix} \right\rangle \\
&= \lambda \odot F_1 \cup \lambda \odot F_2
\end{aligned}
$$

$$(F_1 \cup F_2)^{\lambda} = \left(\left\langle \begin{matrix} [\max(\alpha_{F_{1L}}, \alpha_{F_{2L}}), \max(\alpha_{F_{1U}}, \alpha_{F_{2U}})], \\ [\min(\beta_{F_{1L}}, \beta_{F_{2L}}), \min(\beta_{F_{1U}}, \beta_{F_{2U}})] \end{matrix} \right\rangle\right)^{\lambda}$$

$$= \left\langle \begin{matrix} [(\max(\alpha_{F_{1L}}, \alpha_{F_{2L}}))^{\lambda}, (\max(\alpha_{F_{1U}}, \alpha_{F_{2U}}))^{\lambda}], \\ \begin{bmatrix} \sqrt[3]{1-(1-(\min(\beta_{F_{1L}}, \beta_{F_{2L}}))^3)^{\lambda}}, \\ \sqrt[3]{1-(1-(\min(\beta_{F_{1U}}, \beta_{F_{2U}}))^3)^{\lambda}} \end{bmatrix} \end{matrix} \right\rangle$$

(vi)

$$= \left\langle \begin{matrix} [\max((\alpha_{F_{1L}})^{\lambda}, (\alpha_{F_{2L}})^{\lambda}), \max((\alpha_{F_{1U}})^{\lambda}, (\alpha_{F_{2U}})^{\lambda})], \\ [\min(\sqrt[3]{1-(1-(\beta_{F_{1L}})^3)^{\lambda}}, \sqrt[3]{1-(1-(\beta_{F_{2L}})^3)^{\lambda}}), \\ \min(\sqrt[3]{1-(1-(\beta_{F_{1U}})^3)^{\lambda}}, \sqrt[3]{1-(1-(\beta_{F_{2U}})^3)^{\lambda}})] \end{matrix} \right\rangle$$

$$= \left\langle [(\alpha_{F_{1L}})^{\lambda}, (\alpha_{F_{1U}})^{\lambda}], \left[\sqrt[3]{1-(1-(\beta_{F_{1L}})^3)^{\lambda}}, \sqrt[3]{1-(1-(\beta_{F_{1U}})^3)^{\lambda}}\right] \right\rangle$$

$$\cup F_2^{\lambda} = \left\langle [(\alpha_{F_{2L}})^{\lambda}, (\alpha_{F_{2U}})^{\lambda}], \begin{bmatrix} \sqrt[3]{1-(1-(\beta_{F_{2L}})^3)^{\lambda}}, \\ \sqrt[3]{1-(1-(\beta_{F_{2U}})^3)^{\lambda}} \end{bmatrix} \right\rangle$$

$$= F_1^{\lambda} \cup F_2^{\lambda}$$

Definition 2.4 Let $F = \langle [\alpha_{FL}, \alpha_{FU}], [\beta_{FL}, \beta_{FU}] \rangle$ be an IVFFS, then expected value of $F = \langle [\alpha_{FL}, \alpha_{FU}], [\beta_{FL}, \beta_{FU}] \rangle$ is a FFS denoted by $E(F)$ and is defined as follows: $E(F) = \left\langle \frac{\alpha_{FL}+\alpha_{FU}}{2}, \frac{\beta_{FL}+\beta_{FU}}{2} \right\rangle$.

Definition 2.5 Let $F = \langle [\alpha_{FL}, \alpha_{FU}], [\beta_{FL}, \beta_{FU}] \rangle$ be an IVFFS, then supremum value of $F = \langle [\alpha_{FL}, \alpha_{FU}], [\beta_{FL}, \beta_{FU}] \rangle$ is a FFS denoted by $\sup(F)$ and is defined as follows: $\sup(F) = \langle \max(\alpha_{FL}, \alpha_{FU}), \max(\beta_{FL}, \beta_{FU}) \rangle$.

3 Ranking of IVFFSs

For the purpose of ranking, we define score function of the IVFFS.

Definition 3.1 Suppose that $F = \langle [\alpha_{FL}, \alpha_{FU}], [\beta_{FL}, \beta_{FU}] \rangle$ be an IVFFS, then the score function of F is represented by $S(F)$ and we have introduced different variants of score functions which are as follows:

(i) (V-1): $S(F) = \frac{1}{2}(1 + \left(\frac{\alpha_{FL}+\alpha_{FU}}{2}\right)^3 - \left(\frac{\beta_{FL}+\beta_{FU}}{2}\right)^3$

(ii) (V-2): $S(F) = \frac{1}{2}(1 + (\max(\alpha_{FL}, \alpha_{FU}))^3 - (\max(\beta_{FL}, \beta_{FU}))^3)$

(iii) (V-3): $S(F) = \frac{1}{3}(1 + 2\left(\frac{\alpha_{FL}+\alpha_{FU}}{2}\right)^3 - \left(\frac{\beta_{FL}+\beta_{FU}}{2}\right)$

(iv) (V-4):$S(F) = \frac{1}{3}(1 + 2(\max(\alpha_{FL}, \alpha_{FU}))^3 - (\max(\beta_{FL}, \beta_{FU}))^3)$

(v) (V-5): $S(F) = \dfrac{1-\left(\frac{\beta_{FL}+\beta_{FU}}{2}\right)^2}{2-\left(\frac{\alpha_{FL}+\alpha_{FU}}{2}\right)^2-\left(\frac{\beta_{FL}+\beta_{FU}}{2}\right)^2}$

(vi) (V-6): $S(F) = \dfrac{1-(\max(\beta_{FL},\beta_{FU}))^2}{2-\left((\max(\alpha_{FL},\alpha_{FU}))^2\right)^2-(\max(\beta_{FL},\beta_{FU}))^2}$

Table 1 Score function values of F_1 and F_2

Score function variant	$S(F_1)$	$S(F_2)$	Remark
V-1	$S(F_1) = 0.5490$	$S(F_2) = 0.6295$	$F_1 < F_2$
V-2	$S(F_1) = 0.5760$	$S(F_2) = 0.6580$	$F_1 < F_2$
V-3	$S(F_1) = 0.4077$	$S(F_2) = 0.5112$	$F_1 < F_2$
V-4	$S(F_1) = 0.4560$	$S(F_2) = 0.5530$	$F_1 < F_2$
V-5	$S(F_1) = 0.5265$	$S(F_2) = 0.5757$	$F_1 < F_2$
V-6	$S(F_1) = 0.5442$	$S(F_2) = 0.5969$	$F_1 < F_2$

Here $V - i$ stands for i - th variant.

Definition 3.2 Let $F = \langle[\alpha_{FL}, \alpha_{FU}], [\beta_{FL}, \beta_{FU}]\rangle$ be an IVFFS, then the accuracy function of F is denoted by $H(F)$ and are defined as follows:

(i) [Accuracy function in terms of expected value] $H(F) = \left(\frac{\alpha_{FL}+\alpha_{FU}}{2}\right)^3 + \left(\frac{\beta_{FL}+\beta_{FU}}{2}\right)^3$

(ii) [Accuracy function in terms of supremum value] $H(F) = (\max(\alpha_{FL}, \alpha_{FU}))^3 + (\max(\beta_{FL}, \beta_{FU}))^3$

Definition 3.3 Let $F_1 = \langle[\alpha_{F_1L}, \alpha_{F_1U}], [\beta_{F_1L}, \beta_{F_1U}]\rangle$ and $F_2 = \langle[\alpha_{F_2L}, \alpha_{F_2U}], [\beta_{F_2L}, \beta_{F_2U}]\rangle$ be two IVFFSs, $S(F_1)$ and $S(F_2)$ be the score function of F_1 and F_2, respectively, $H(F_1)$ and $H(F_2)$ be the accuracy function of F_1 and F_2, respectively, then

(i) $S(F_1) < S(F_2) \Rightarrow F_1 < F_2$
(ii) $S(F_1) > S(F_2) \Rightarrow F_1 > F_2$
(iii) $(S(F_1) = S(F_2)) \wedge (H(F_1) < H(F_2)) \Rightarrow F_1 < F_2$
(iv) $(S(F_1) = S(F_2)) \wedge (H(F_1) > H(F_2)) \Rightarrow F_1 > F_2$
(v) $(S(F_1) = S(F_2)) \wedge (H(F_1) = H(F_2)) \Rightarrow F_1 = F_2$

Example 3.1 Let $F_1 = \langle[0.4, 0.6], [0.2, 0.4]\rangle$ and $F_2 = \langle[0.6, 0.7], [0.2, 0.3]\rangle$ be two IVFFSs. Then according to Definition 3.1 we have calculated score function values which are given in Table 1.

4 TOPSIS Method in View of Interval-Valued Fermatean Fuzzy Sets

Here, we shall discuss how to employ the TOPSIS method to solve multi-criteria decision making (MCDM) problems where predilection knowledge furnished by the decision makers (DMs) is articulated in the pattern of interval-valued Fermatean fuzzy sets and the particulars of the criterion weights are known. Here, we use our suggested score functions to compute the separation measures of each alternative from positive and negative ideal solutions to find out the coefficient of relative closeness.

In the MCDM problem concerning interval-valued Fermatean fuzzy scenario, suppose that there are m alternatives $A_i(i = 1, 2, \ldots, m)$ and n criterion $C_j(j = 1, 2, \ldots, n)$ with the criteria weight vector $w = (w_1, w_2, \ldots, w_n)$, where $0 \leq w_j \leq 1$ and $\sum w_j = 1$. The characteristics of an alternative A_i with respect to the criteria C_j can be represented by an IVFFS value $u_{ij} = \langle [a_{ij}, b_{ij}], [c_{ij}, d_{ij}] \rangle (i = 1, 2, \ldots, m; j = 1, 2, \ldots, n)$ and $D = \left(C_j(A_i)\right)_{m \times n}$ is an Interval-valued Fermatean fuzzy matrix. Thus, the MCDM problem can be represented in terms of a decision matrix as follows:

$$D = \left(C_j(A_i)\right)_{m \times n} = \begin{pmatrix} \langle [a_{11}, b_{11}], [c_{11}, d_{11}] \rangle & \langle [a_{12}, b_{12}], [c_{12}, d_{12}] \rangle & \ldots & \langle [a_{1n}, b_{1n}], [c_{1n}, d_{1n}] \rangle \\ \langle [a_{21}, b_{21}], [c_{21}, d_{21}] \rangle & \langle [a_{22}, b_{22}], [c_{22}, d_{22}] \rangle & \ldots & \langle [a_{2n}, b_{2n}], [c_{2n}, d_{2n}] \rangle \\ \vdots & \vdots & \vdots & \vdots \\ \langle [a_{m1}, b_{m1}], [c_{m1}, d_{m1}] \rangle & \langle [a_{m2}, b_{m2}], [c_{m2}, d_{m2}] \rangle & \ldots & \langle [a_{mn}, b_{mn}], [c_{mn}, d_{mn}] \rangle \end{pmatrix} \tag{1}$$

Now, based on score function, we convert the decision matrix $D = \left(C_j(A_i)\right)_{m \times n}$ into the following score matrix $S(D) = \left(S\left(C_j(A_i)\right)\right)_{m \times n}$.

$$S(D) = \left(S\left(C_j(A_i)\right)\right)_{m \times n} = \begin{pmatrix} S(u_{11}) & S(u_{12}) & \ldots & S(u_{1n}) \\ S(u_{21}) & S(u_{22}) & \ldots & S(u_{2n}) \\ \vdots & \vdots & \vdots & \vdots \\ S(u_{m1}) & S(u_{m2}) & \ldots & S(u_{mn}) \end{pmatrix} \tag{2}$$

Let A^+ and A^- denote, respectively, the positive and negative ideal solution for the alternatives $A_i(i = 1, 2, \ldots, m)$, where $A^+ = \left\{\langle C_j, [1, 1], [0, 0] \rangle : C_j \in C\right\}$ and $A^- = \left\{\langle C_j, [0, 0], [1, 1] \rangle : C_j \in C\right\}$, $j = 1, 2, \ldots, n$. Now, the score function-based separation measures $d_i^+ = d_i^+(A^+, A_i)$ and $d_i^- = d_i^-(A^-, A_i)$ of each alternative from positive ideal and negative ideal solutions, are, respectively, derived by using the following formulas:

$$d_i^+ = \sqrt{\sum_{j=1}^{n} (w_j(1 - S(u_{ij})))^2} \tag{3}$$

$$\text{and } d_i^- = \sqrt{\sum_{j=1}^{n} (w_j S(u_{ij}))^2} \tag{4}$$

The relative closeness of an alternative $A_i(i = 1, 2, \ldots, m)$ with respect to the positive ideal solution A^+ is denoted by $C_i(A_i)$ and is calculated by the formula which is as follows:

$$C_i(A_i) = \frac{d_i^-}{d_i^- + d_i^+} \tag{5}$$

where $C_i(A_i), i = 1, 2, \ldots, m$ is the relative closeness coefficient of A_i with respect to A^+. Therefore, the substitute/alternative might be ranked in accordance with the descending order of $C_i(A_i)$ and the alternative with highest rank will be the suitable way out.

5 Numerical Example and Discussion

In this chapter, we have taken an example to discuss the demonstrated approach.

Example 5.1 For future profit, suppose that an animation specialist wants to select a best computer animation task. Let us consider an animation contest with four feasible alternatives $A_i(i = 1, 2, 3, 4)$. The specialist has to adopt a decision in keeping with the following three criteria: (i) C_1 is the inventive attraction (ii) C_2 is the seeing out-turn and (iii) C_3 is the original screenplay. The criterion weight is specified as $w = (w_1 = 0.35, , w_2 = 0.25, w_3 = 0.40)$. The alternative $A_i(i = 1, 2, 3, 4)$ is to be computed using the IVFFS value by decision maker in accordance with the above criteria, as given by the decision matrix $D_{4\times3}$.

$$D_{4\times3} = \begin{pmatrix} \langle[0.4, 0.5], [0.3, 0.4]\rangle & \langle[0.4, 0.6], [0.2, 0.4]\rangle & \langle[0.1, 0.3], [0.5, 0.6]\rangle \\ \langle[0.6, 0.7], [0.2, 0.3]\rangle & \langle[0.6, 0.7], [0.2, 0.3]\rangle & \langle[0.4, 0.7], [0.1, 0.2]\rangle \\ \langle[0.3, 0.6], [0.3, 0.4]\rangle & \langle[0.5, 0.6], [0.3, 0.4]\rangle & \langle[0.5, 0.6], [0.1, 0.3]\rangle \\ \langle[0.7, 0.8], [0.1, 0.2]\rangle & \langle[0.6, 0.7], [0.1, 0.3]\rangle & \langle[0.3, 0.4], [0.1, 0.2]\rangle \end{pmatrix}$$

Here, it is to be noted that the mentioned example 5.1 looks like an example by Zhi-yong Bai [38]. The reason is that in the example of Zhi-yong Bai [38], they considered interval-valued intuitionistic fuzzy sets, but we considered the same as interval-valued Fermatean fuzzy sets as all the elements satisfy the proposed definition of interval-valued Fermatean fuzzy sets defined in definition 2.1. For finding the score matrix, we have used our proposed score function variants, which are defined in definition 3.1. Here, we have proposed six variants of score functions and, based on these variants, we have obtained six score matrices, which are as follows:

Score Matrix 5.1

When the score function is of V-1 variant, i.e., $S(F) = \frac{1}{2}\left(1 + \left(\frac{\alpha_{FL}+\alpha_{FU}}{2}\right)^3 - \left(\frac{\beta_{FL}+\beta_{FU}}{2}\right)^3\right)$.

Using this V-1 variant, the IVFF decision matrix $D_{4\times3}$ converted into the score matrix $S(D_{4\times3})$ which is as follows:

$$S(D_{4\times3}) = \begin{pmatrix} 0.5241 & 0.5490 & 0.4208 \\ 0.6295 & 0.6295 & 0.5815 \\ 0.5241 & 0.5618 & 0.5792 \\ 0.7093 & 0.6333 & 0.5198 \end{pmatrix}$$

Score Matrix 5.2

When the score function is of V-2 variant, i.e., $S(F) = \frac{1}{2}(1 + (\max(\alpha_{FL}, \alpha_{FU}))^3 - (\max(\beta_{FL}, \beta_{FU}))^3)$.

Using this V-2 variant, the IVFF decision matrix $D_{4\times3}$ converted into the score matrix $S(D_{4\times3})$ which is as follows:

$$S(D_{4\times3}) = \begin{pmatrix} 0.5305 & 0.5760 & 0.4055 \\ 0.6580 & 0.6580 & 0.6675 \\ 0.5760 & 0.5760 & 0.5945 \\ 0.7520 & 0.6580 & 0.5280 \end{pmatrix}$$

Score Matrix 5.3

When the score function is of V-3 variant, i.e., $S(F) = \frac{1}{3}(1 + 2(\frac{\alpha_{FL}+\alpha_{FU}}{2})^3 - (\frac{\beta_{FL}+\beta_{FU}}{2})^3)$.

Using this V-3 variant, the IVFF decision matrix $D_{4\times3}$ converted into the score matrix $S(D_{4\times3})$ which is as follows:

$$S(D_{4\times3}) = \begin{pmatrix} 0.3798 & 0.4077 & 0.2832 \\ 0.5112 & 0.5112 & 0.4431 \\ 0.3798 & 0.4300 & 0.4416 \\ 0.6135 & 0.5138 & 0.3608 \end{pmatrix}$$

Score Matrix 5.4

When the score function is of V-4 variant, i.e., $S(F) = \frac{1}{3}(1+2(\max(\alpha_{FL}, \alpha_{FU}))^3 - (\max(\beta_{FL}, \beta_{FU}))^3)$.

Using this V-4 variant, the IVFF decision matrix $D_{4\times3}$ converted into the score matrix $S(D_{4\times3})$ which is as follows:

$$S(D_{4\times3}) = \begin{pmatrix} 0.3953 & 0.4560 & 0.2793 \\ 0.5530 & 0.5530 & 0.5593 \\ 0.4560 & 0.4560 & 0.4683 \\ 0.6720 & 0.5530 & 0.3733 \end{pmatrix}$$

Score Matrix 5.5

When the score function is of V-5 variant, i.e., $S(F) = \frac{1-\left(\frac{\beta_{FL}+\beta_{FU}}{2}\right)^2}{2-\left(\frac{\alpha_{FL}+\alpha_{FU}}{2}\right)^2-\left(\frac{\beta_{FL}+\beta_{FU}}{2}\right)^2}$.

Using this V-5 variant, the IVFF decision matrix $D_{4\times3}$ converted into the score matrix $S(D_{4\times3})$ which is as follows:

$$S(D_{4\times3}) = \begin{pmatrix} 0.5129 & 0.5265 & 0.4566 \\ 0.5757 & 0.5757 & 0.5445 \\ 0.5129 & 0.5776 & 0.5434 \\ 0.6329 & 0.5530 & 0.5101 \end{pmatrix}$$

Score Matrix 5.6

When the score function is of V-6 variant, i.e., $S(F) = \frac{1-(\max(\beta_{FL},\beta_{FU}))^2}{2-\left((\max(\alpha_{FL},\alpha_{FU}))^2\right)^2-(\max(\beta_{FL},\beta_{FU}))^2}$.

Using this V-6 variant, the IVFF decision matrix $D_{4\times3}$ converted into the score matrix $S(D_{4\times3})$ which is as follows:

$$S(D_{4\times3}) = \begin{pmatrix} 0.5168 & 0.5442 & 0.4462 \\ 0.5969 & 0.5969 & 0.6016 \\ 0.5442 & 0.5442 & 0.5538 \\ 0.6703 & 0.5969 & 0.5145 \end{pmatrix}$$

Now, we have computed $d_i^+(i = 1, 2, 3, 4)$ and $d_i^-(i = 1, 2, 3, 4)$ using score functions (V-1 to V-6) which are exhibit in Tables 2 and 3.

Table 2 Values of $d_i^+(i = 1, 2, 3, 4)$ for different variants of score functions

	V-1	V-2	V-3	V-4	V-5	V-6
d_1^+	0.3068	0.3079	0.3889	0.3826	0.3005	0.3011
d_2^+	0.2311	0.1983	0.3063	0.2608	0.2579	0.2355
d_3^+	0.2609	0.2441	0.3425	0.3162	0.2256	0.2651
d_4^+	0.2359	0.2247	0.3138	0.2975	0.2570	0.2474

Table 3 Values of $d_i^-(i = 1, 2, 3, 4)$ for different variants of score functions

	V-1	V-2	V-3	V-4	V-5	V-6
d_1^-	0.2843	0.2855	0.2022	0.2112	0.2880	0.2883
d_2^-	0.3569	0.3891	0.2824	0.3265	0.3298	0.3519
d_3^-	0.3272	0.3434	0.2485	0.2712	0.3120	0.3223
d_4^-	0.3604	0.3754	0.2888	0.3110	0.3340	0.3459

Table 4 Values of $C_i(A_i), i = 1, 2, 3, 4$ for different variants of score functions

	V-1	V-2	V-3	V-4	V-5	V-6
$C_1(A_1)$	0.4810	0.4812	0.3421	0.3557	0.4893	0.4891
$C_2(A_2)$	0.6070	0.6624	0.4797	0.5559	0.5612	0.5991
$C_3(A_3)$	0.5563	0.5846	0.4178	0.4617	0.5309	0.5486
$C_4(A_4)$	0.6044	0.6256	0.4793	0.5111	0.5651	0.5831

Table 5 Order of alternatives with highest rank

Score functions	Alternatives	Descending order of $A_i, i = 1, 2, 3, 4$	Alternative with highest rank
V-1	A_1, A_2, A_3 and A_4	A_2, A_4, A_3, A_1	A_2
V-2	A_1, A_2, A_3 and A_4	A_2, A_4, A_3, A_1	A_2
V-3	A_1, A_2, A_3 and A_4	A_2, A_4, A_3, A_1	A_2
V-4	A_1, A_2, A_3 and A_4	A_2, A_4, A_3, A_1	A_2
V-5	A_1, A_2, A_3 and A_4	A_4, A_2, A_3, A_1	A_4
V-6	A_1, A_2, A_3 and A_4	A_2, A_4, A_3, A_1	A_2

Using (5), we have computed $C_i(A_i), i = 1, 2, 3, 4$ which are given in Table 4.

Now, we have ranked the alternative according to the descending order of $C_i(A_i), i = 1, 2, 3, 4$ and the alternative with highest rank will be the best choice. These are shown in Table 5.

From Table 5 it is seen that A_2 is the best alternative for case 1, case 2, case 3, case 4 and case 6 whereas A_4 is foremost alternative for case 5. These outcomes obtained from our suggested method are compliance with the works mentioned in [31, 39].

6 Conclusion

In this chapter, we have initiated different variants of score functions for interval-valued Fermatean fuzzy sets (IVFFSs). Here, we have discussed some elementary set operations and arithmetic operations of IVFFSs. Based on different variants of score functions, we have employed the TOPSIS method to solve the MCDM problem, where a predilection knowledge furnished by the decision makers (DMs) is presented in the pattern of IVFFSs. To demonstrate the suggested method, a practical numerical example has been considered and solved. The results acquired from our proposed method considering IVFFS coincide with those of other methods that are available in the existing literature. Therefore, we have deduced from the aforesaid that the suggested variant of score functions for IVFFSs can be used to solve MCDM problems considering IVFFSs. In future research, one may apply the proposed fuzzy sets/operations in the fields of science, engineering and management science, etc.

Finally, it may be claimed that the proposed variants of score functions for IVFFSs introduced in this study can be applied to handle more uncertainty and solve realistic decision making problems involving interval-valued Fermatean parameters.

References

1. Zadeh, L.A.: Fuzzy Sets. Inf. Control **8**, 338–353 (1965)
2. Bellman, R.E., Zadeh, L.A.: Decision making in a fuzzy environment. Manage. Sci. **17**, 141–164 (1970)
3. Atanassov, K.T.: Intuitionistic fuzzy sets. Fuzzy Sets Syst. **20**, 87–96 (1986)
4. Fei, L., Wang, H., Chen, L., Deng, Y.: A new vector valued similarity measure for intuitionistic fuzzy sets based on OWA operators. Iran. J. Fuzzy Syst. **15**(5), 31–49 (2018)
5. Zhang, R., Ashuri, B., Deng, Y.: A novel method for forecasting time series based on fuzzy logic and visibility graph. Adv. Data Anal. Classif. **11**(4), 759–783 (2018)
6. Gorzalczany, M.B.: A method of inference in approximate reasoning based on interval-valued fuzzy sets. Fuzzy Sets Syst. **21**(1), 1–17 (1987)
7. Turksen, I.B.: Interval valued fuzzy sets based on normal forms. Fuzzy Sets Syst. **20**(2), 191–210 (1986)
8. Atanassov, K.T., Gargov, G.: Interval valued intuitionistic fuzzy sets. Fuzzy Sets Syst. **31**, 1–17 (1987)
9. Atanassov, K.T.: On Intuitionistic Fuzzy Sets Theory. Springer, Berlin (2012)
10. Atanassov, K.T., Vassilev, P.M., Tsvetkov, R.T.: Intuitionistic Fuzzy Sets, Measures and Integrals. Professor Marin Drinov Academic Publishing House, Sofia (2013)
11. Atanassov, K.T.: A Second type of Intuitionistic fuzzy sets. BUSE-FAL **56**, 66–70 (1983)
12. Atanassov, K.T.: Geometric interpretation of the elements of the intuitionistic fuzzy objects. Int. J. Bioautom. **20**(S1), S27–S42 (2016)
13. Yager, R.R.: Pythagorean fuzzy subsets. In: 2013 Joint IFSA World Congress and NAFIPS Annual Meeting (IFSA/NAFIPS), pp. 57–61 (2013)
14. Yager, R.R.: Pythagorean membership grades in multi-criteria decision making. IEEE Trans. Fuzzy Syst. **22**, 958–965 (2014)
15. Garg, H.: A linear programming method based on an improved score function for interval valued Pythagorean fuzzy numbers and its application to decision-making. Int. J. Uncertainty Fuzziness Knowl. Based Syst. **26**, 67–80 (2018)
16. Garg, H.: A new improved score function of an interval valued Pythagorean fuzzy set based TOPSIS method. Int. J. Uncertainty Quant. **7**, 463–474 (2017)
17. Zhang, X., Xu, Z.: Extention of TOPSIS to multiple criteria decision making with Pythagorean fuzzy sets. Int. J. Intell. Syst. **29**, 1061–1078 (2014)
18. Zhang, X.: A novel approach based on similarity measure for Pythagorean fuzzy multiple criteria group decision making. Int. J. Intell. Syst. **31**, 593–611 (2016)
19. Gou, X., Xu, Z., Liao, H.: Alternative queuing method for multiple criteria decision making with hybrid fuzzy and ranking information. Inf. Sci. **357**, 144–160 (2016)
20. Geng, Y., Liu, P., Teng, F., Liu, Z.: Pythagorean fuzzy uncertain linguistic TODIM method and their application to multiple criteria group decision making. Int. J. Intell. Syst. **33**, 3383–3395 (2017)
21. Jing, N., Xian, S., Xiao, Y.: Pythagorean triangular fuzzy linguistic bonferroni mean operators and their application for multi-attribute decision making, In: 2nd IEEE International Conference on Computational Intelligence and Applications (ICCIA), pp. 435–439 (2017)
22. Senapati, T., Yager, R.R.: Fermatean fuzzy sets. J. Ambient Intell. Humaniz. Comput. **11**, 663–674 (2020)

23. Senapati, T., Yager, R.R.: Fermatean fuzzy weighted averaging/geometric operators and its application in multi-criteria decision-making methods. Eng. Appl. Artif. Intell. **85**, 112–121 (2019)
24. Li, Z., Wei, G., Lu, M.: Pythagorean fuzzy hamy mean operators in multiple attribute group decision making and their application to supplier selection. Symmetry **10**, 505 (2018)
25. Zhou, J., Su, W., Balezentis, T., Streimikiene, D.: Multiple criteria group decision making considering symmetry with regards to the positive and negative ideal solution via the Pythagorean normal cloud model for application to economic decision. Symmetry **10**(5), 140 (2018)
26. Boltruk, E.: Pythagorean fuzzy CODAS and its application to supplier selection in a manufacturing firm. J. Enterp. Inf. Manag. **31**, 550–564 (2018)
27. Qin, J.: Generalized Pythagorean fuzzy Maclaurin symmetric means and its application to multiple attribute SIR group decision model. Int. J. Fuzzy Syst. **20**, 943–957 (2018)
28. Wan, S.-P., Li, S.-Q., Dong, J.-Y.: A three phase method for Pythagorean fuzzy multi attribute group decision making and application to haze management. Comput. Ind. Eng. **123**, 348–363 (2018)
29. Lin, Y.-L., Ho, L.-H., Yeh, S.-L., Chen, T.-Y.: A Pythagorean fuzzy topsis method based on novel correlation measures and its application to multiple criteria decision analysis of inpatient stoke rehabilitiation. Int. J. Comput. Intell. Syst. **12**(1), 410–425 (2018)
30. Chen, T.-Y.: An outranking approach using a risk attitudinal assignment model involving Pythagorean fuzzy information and its application to financial decision making. Appl. Soft. Comput. **71**, 460–487 (2018)
31. Sahoo, L.: Some score functions on Fermatean fuzzy sets and its application to bride selection based on TOPSIS method. Int. J. Fuzzy Syst. Appl. **10**(3), 18–29 (2021)
32. Sahoo, L.: A new score function based Fermatean fuzzy transportation problem. Results Control Opt. **4**, 100040 (2021)
33. Sahoo, L.: Similarity measures for Fermatean fuzzy sets and its applications in group decision-making. Decis. Sci. Lett. **11**, 167–180 (2022)
34. Senapati, T., Yager, R.R.: Some new operations over Fermatean fuzzy numbers and application of Fermatean fuzzy WPM in multiple criteria decision making. Informatica **30**(2), 391–412 (2019)
35. Senapati, T., Chen, G.: Some novel interval-valued Pythagorean fuzzy aggregation operator based on Hamacher triangular norms and their application in MADM issues. Comput. Appl. Math. **40**(4), 1–27 (2021). https://doi.org/10.1007/s40314-021-01502-w
36. Jana, C., Senapati, T., Pal, M.: Pythagorean fuzzy Dombi aggregation operators and its applications in multiple attribute decision-making. Int. J. Intell. Syst. **34**(9), 2019–2038 (2019)
37. Jeevaraj, S.: Ordering of interval-valued Fermatean fuzzy sets and its applications. Expert Syst. Appl. **185**, 1–20 (2021)
38. Bai, Z.-Y.: An interval valued intuitionistic fuzzy TOPSIS method based on an improved score function. Sci. World J. **2013**, 1–6 (2013)
39. Ye, J.: Multicriteria fuzzy-decision making method based on a novel accuracy function under interval valued intuitionistic fuzzy environment. Expert Syst. Appl. **36**(3), 6899–6902 (2009)
40. Sergi, D., Sari, I.U., Senapati, T.: Extension of capital budgeting techniques using interval-valued Fermatean fuzzy sets. J. Intell. Fuzzy Syst. **42**(1), 365–376 (2022)

A Fuzzy EOQ Model for Deteriorating Items Under Trade Credit Policy with Unfaithfulness Nature of Customers

Rituparna Mondal, Ranjan Kumar Jana, Prasenjit Pramanik, and Manas Kumar Maiti

Abstract In this study, a fuzzy EOQ model has been developed for deteriorating product with time-dependent deterioration and fixed expiration date under partial trade credit policy, considering unfaithfulness nature of the base customers. This model is a mixed-integer profit maximization problem designed in a crisp and imprecise (fuzzy) environment for the uncertainty of different costs and banking interest involving the inventory control system. The decision variables and input parameters affect in the model are fuzzified. This research investigates about optimal replenishing time and selling price to maximize the profit. Also, a particle swarm optimization (PSO) algorithm has been utilized to obtain the best suitable marketing plan for the decision-maker. Numerous numerical examples are also cited.

1 Introduction

In the current competitive, volatile marketing situation, it is crucial to draw the optimal decisions for the decision-maker/store manager/retailer. To enhance the demand for a product, the inventory system/supply chain players adopt various promotional activities. Among those, trade credit policy is the best productive activity. But for this credit policy, there is a possibility of cheating the credit amount, i.e., default credit risk. In a supply chain, this type of risk occurs at the retailer-customers level because there is no such certified bonding between retailer and customers; credit is

R. Mondal · R. K. Jana (✉)
Department of Mathematics and Humanities, Sardar Vallabhbhai National Institute of Technology, Surat, Gujarat, India
e-mail: rkj@amhd.svnit.ac.in

P. Pramanik
Department of Applied Mathematics with Oceanology and Computer Programming, Vidyasagar University, Midnapore, Paschim Medinipur, West Bengal, India

M. K. Maiti
Department of Mathematics, Mahishadal Raj College, Mahishadal, Purba Medinipur, West Bengal, India

L. Sahoo et al. (eds.), *Real Life Applications of Multiple Criteria Decision Making Techniques in Fuzzy Domain*, Studies in Fuzziness and Soft Computing 420,
https://doi.org/10.1007/978-981-19-4929-6_21

given based on complete faith. In a local market, it is observed that there is a certain percentage of customers who are not come back to repay the credit amount. To draw an optimal decision, it is mandatory to consider this phenomenon into account.

With the introduction of trade credit policy by Goyal [1], there are so many research works available in literature under different levels of trade credit policy: such as [2–20]. The above studies show that this policy is an essential factor in any inventory control system. But it occurs a default risk. To decrease the default risk, practitioners and the investigator allow partial credit opportunity [3, 21–26] to the customers. Also, to decrease the default risk, Teng [17] considered different credit policies for different customers, i.e., full credit policy for honest (faithful) customers and partial credit policy for dishonest (unfaithful) customers. Although in reality, this is really difficult to find out dishonest customers who will not obey business ethics. Also, different credit opportunities for different customers are considered as a malpractice for any business transaction and affect negative impression to some customers. Though it is tough to mark out the good and bad customers in a regional market. So it is more practical to consider a partial trade credit policy for all customers to decline the credit risk, assuming a rate of total customers is unfaithful customers who do not come back to repay the credit amount. Though it is a natural phenomenon, it is overlooked by the researcher of this field.

On the other hand, proper maintenance of an inventory system of the deteriorated items is another crucial problem for the decision-maker in any business sector. Ghare and Schrader [27] first originate a model for deteriorating products. Afterward, a large number of research articles found with this consideration [2, 21, 28–30]. In the above studies, they have considered that the deterioration rate of an item is constant. But there are so many products, such as fruits, vegetables, pharmaceuticals, etc., that degrades with time, and each product has an expiration date. There are very few research works [22, 23] considered time-dependent deterioration with a fixed expiration date. In all the above studies, it is observed that they have considered a deteriorating item to develop their models under trade credit policy. Still, unfortunately, none has been regarded as the payable interest portion for the deteriorated units in the mathematical development of their model. In other words, a wholesaler allows a credit period L_1 to its retailer with retailer's replenishment time $T_r(> L_1)$ and some units have deteriorated during $(0, L_1)$. As a result, retailer should pay the prices of those deteriorated units at the time L_1 to its wholesaler by a bank loan. Thus the retailer must pay interest for those deteriorated units for a time duration $(T_r - L_1)$. But none has considered this phenomenon in the development of their models; they have only considered the pricing value of the deteriorated units in their model development section. In this study, an attempt has been made to overcome this mentioned flaw.

Base demand or market demand is the key to any inventory control system/supply chain management. Here, since market demand is determined by an item's selling price, the model must include the selling price as a decision variable. Then the optimal business plan should occur with optimal pricing for the retailer. Several studies have been found with this consideration [3, 23, 31–35].

For a business transaction, it is not possible to understand dishonesty from outside unless and until they have fled without paying the credit amount. So this is much suitable to take a specific rate of the customer to be considered as dishonest. But this rate changes over time. So to overcome this difficulty, it will be rational to consider this parameter as a fuzzy parameter. Again, due to the unstable market situation, uncertainty arises in different parameters in the cost function. So the cost function as well as the profit function becomes fuzzy. With the introduction of the fuzzy set [36], it has been well designed and implemented broadly in several areas of science and technology, including inventory control difficulties. In these circumstances, further research work can done in inventory control, taking into consideration the work of [13, 29, 34, 37] and many others.

A fuzzy EOQ model has been designed for the deteriorating product having time-dependent deterioration with a fixed expiration date under a partial trade credit policy, considering the unfaithfulness nature of the base customers. Here it is supposed that a wholesaler grants a full credit period to its retailer, but the retailer gives a partial credit period to their customers. Also, it is considered that a fraction of total customers are unfaithful customers, who will not return to repay the retailer's credit amount. The goal of this investigation can be stated as below:

- In every business transaction with a credit policy, it is observed that few customers do not obey the business ethics; in other words, few customers escaped without repaying the credit amount to the retailer. This occurrence is considered in this study.
- In all the EOQ models for deteriorating products under trade credit policy, deterioration of an item is considered. Still, unfortunately, the amount of charged interest paid for the same is overlooked. For instance, it is considered that supplier provides full credit period L_1 to its retailer with retailer's replenishment time $T_r(> L_1)$. Now some units deteriorate during $(0, L_1)$. Thus, retailer should pay the purchase cost of the units at time L_1 to its supplier by a bank loan whose account is settled at the end of the return of all credit amount, i.e., at $(T_r + L_2)$. Since $T_r > L_1$, retailer must pay interest for those deteriorated units for the duration of $(T_r - L_1)$. Additionally, some units will have deteriorated during $[L_1, T_r]$, and the retailer should pay interest for those units. Also payment of those units are made from a bank loan and account of the retailer will be settled with the bank at time $(T_r + L_2)$. This occurrence is taken into consideration in this research.
- Also, majority of the existing literature on trade credit policy considered that retailer earned interest from the bank on the selling price of the sold units, which is unrealistic because some expenditure is incurred to keep track of the inventory. It comes from a portion of the profit part of the sell units. Usually, the retailer receives bank interest from the purchase cost portion of the selling units before settling his account to the supplier. So retailer earned interest from bank on the purchase cost for the sold units. This phenomenon is incorporated in this study.
- Very few works have been found in a fuzzy environment under trade credit policy. Despite this, no universal approach for solving such problems in a crisp background and imprecise surroundings has been offered.

2 Mathematical Prerequisite

2.1 Credibility Measure of an Event

Let $\tilde{X} = (p_1, p_2, p_3)$ and $\tilde{Y} = (r_1, r_2, r_3)$ be two triangular fuzzy numbers (TFNs), where $0 < p_1 < p_2 < p_3$ and $0 < r_1 < r_2 < r_3$, then credibility of an event $(\tilde{X} < \tilde{Y})$ is given by

$$Cr(\tilde{X} < \tilde{Y}) = \frac{1}{2}[\text{Pos}(\tilde{X} < \tilde{Y}) + Nes(\tilde{X} < \tilde{Y})]$$

$$\text{Pos}(\tilde{X} < \tilde{Y}) = \begin{cases} 0 & ; \quad p_1 \geq r_3 \\ \frac{p_1 - r_3}{r_2 - r_3 + p_1 - p_2} & ; \quad p_1 \leq r_1 \quad \& \quad p_2 \geq r_2 \\ 1 & ; \quad p_2 \leq r_2 \end{cases}$$

$$Nes(\tilde{X} < \tilde{Y}) = \begin{cases} 0 & ; \quad p_2 \geq r_2 \\ \frac{p_3 - r_1}{r_2 - r_1 - p_2 + p_3} & ; \quad p_2 \leq r_2 \quad \& \quad p_3 \geq r_1 \\ 1 & ; \quad p_3 \leq r_1 \end{cases}$$

where $Pos(\tilde{X} < \tilde{Y})$, $Nes(\tilde{X} < \tilde{Y})$ and $Cr(\tilde{X} < \tilde{Y})$ respectively denote possibility, necessity and credibility for the event $(\tilde{X} < \tilde{Y})$.

3 Assumptions and Notations for the Proposed Models

To design this model mathematically, following assumptions and notations are used all over the article.

3.1 Assumptions

1. Time horizon is infinite.
2. Replenishment is instantaneous.
3. Shortages are not acceptable.
4. Market demand of a product is depending upon the retailer's selling price and is defined as $D(s) = as^{-b}$, which is a decreasing function of the retail price s; where $a(> 0)$ is a scaling factor and $b(> 1)$ is a price elasticity coefficient. For notational simplicity, $D(s)$ and D will be used interchangeably in this article.
5. Here it is considered that customers must pay the δ-fraction of the total purchased amount as a collateral deposit at the receiving time of the units of the product and the retailer offers a credit period L_2 to the customers on $(1 - \delta)$-fraction of the total purchased amount.

6. It is taken that α-fraction of the total customers are disloyal or unfaithful customers and the remaining are faithful customers, follow the business rules.
7. All units of the item have own expiration dates. The physical signification of the deterioration rate is the rate to be closed to 1 when time is approaching to the maximum lifetime e_p. The items deteriorates at a rate $\theta_1(t)$ which depends on time as follow:

$$\theta_1(t) = \frac{1}{1 + e_p - t}, 0 \leq t \leq T_r \leq e_p \tag{1}$$

   Now, it is clearly seen from Eq. (1) that the replenishment cycle time T_r must be less than or equal to e_p, and the deterioration rate is not applicable for non-deteriorating items, in which $\theta_1(t) \to 0$ as $e_p \to \infty$.
8. Retailer can collect interest at a rate i_e on the sold units during the credit period provided by the supplier. At the end of the permissible credit period, the retailer pays off the purchase cost to the supplier by a bank loan and pays off the interest to the bank at a rate i_p for the stocked units already sold but have not been paid for yet and for the deteriorated units.

3.2 Notations

1. $R(t)$, retailer's stock level at any instant t.
2. D_m, marketing demand of an item.
3. Q_r, retailer's order amount per cycle.
4. T_r, retailer's replenishment time, a decision variable.
5. p, unit purchasing cost in \$.
6. s, unit selling price in \$ with $(s > p)$.
7. A_c, ordering cost per order in \$.
8. h_c, inventory holding cost in \$ per year excluding interest charges.
9. i_e, earned interest rate per \$ per year.
10. i_p, payable interest rate per \$ per year.
11. L_1, retailer's credit period provided by the supplier per cycle.
12. L_2, customer's credit period provided by the retailer per cycle.
13. $\theta_1(t)$, deterioration rate of the product at time t, $0 \leq \theta_1(t) < 1$.
14. e_p, expiration time of the item.
15. δ, positive real number, $0 \leq \delta \leq 1$.
16. α, positive real number, $0 \leq \alpha \leq 1$.
17. $TP(T_r, s)$ retailer's total profit in \$ per year.

4 Model Development and Analysis

To develop this problem mathematically, it is taken that the inventory starts with Q_r units and then the stock level continuously depletes to zero because of the combined result of demand and deterioration. Thus stock level of the retailer is determined by:

$$\frac{\mathrm{d}R(t)}{\mathrm{d}t} + \theta_1(t)R(t) = -D_m; \quad 0 \le t \le T_r. \tag{2}$$

with boundary condition $R(0) = Q_r$, $R(T_r) = 0$. Solving the above differential Eq. (3) with the boundary condition $R(T_r) = 0$, we get the stock level of the retailer at instant t is

$$R(t) = D_m(1 + e_p - t)\ln\left(\frac{1 + e_p - t}{1 + e_p - T_r}\right) \text{ for } 0 \le t \le T_r \tag{3}$$

As a consequence, the order quantity of the retailer is

$$Q_r = R(0) = D_m(1 + e_p)\ln\left(\frac{1 + e_p}{1 + e_p - T_r}\right) \tag{4}$$

The relevant costs of the inventory system are as follows:

1. **Annual sales revenue (SR)**:

$$\mathrm{SR} = sD_m\{\delta + (1 - \delta)(1 - \alpha)\} \tag{5}$$

2. **Annual ordering cost (OC)**:

$$\mathrm{OC} = \frac{A_c}{T_r} \tag{6}$$

3. **Annual purchase cost (PC)**:

$$\mathrm{PC} = \frac{p}{T_r}R(0) = \frac{pD_m(1 + e_p)}{T_r}\ln\left(\frac{1 + e_p}{1 + e_p - T_r}\right) \tag{7}$$

4. **Annual stock holding cost (AHC), excluding interest charges**:

$$\mathrm{AHC} = \frac{h_c}{T_r}\int_0^{T_r} R(t)\mathrm{d}t = \frac{h_c D_m}{T_r}\left[\frac{(1+e_p)^2}{2}\ln\left(\frac{1+e_p}{1+e_p-T_r}\right) + \frac{T_r^2}{4} - \frac{(1+e_p)T_r}{2}\right] \tag{8}$$

5. **Annual interest charge (payable/earning) for the retailer**:

Depending upon the values of L_1 and L_2, two cases may arise:
Case 1: $L_2 < L_1$ and **Case 2**: $L_2 \geq L_1$.
The above mentioned two cases has been studied under different situations and those are explained in details as below.
Case 1: $L_2 < L_1$
Based on the values of L_1, T_r and $T_r + L_2$, three situations may occur and those are:

- **Situation-1.1** $0 < T_r + L_2 < L_1$
- **Situation-1.2** $T_r < L_1 \leq T_r + L_2$ and
- **Situation-1.3** $L_1 \leq T_r$

For each situation of this case, earned and payable interest of the retailer are presented below:
Situation 1.1: $0 < T_r + L_2 < L_1$ (i.e., $0 < T_r < L_1 - L_2$)
Here, retailer receives all the returns from customers before the delay period L_1. Therefore, retailer not to pay any interest whereas he can earn interest from the customers' payment.
Thus the amount of interest that the retailer must pay each year is given by

$$IP = 0 \tag{9}$$

However, the annual interest generated by the retailer is

$$IE = \frac{1}{T_r}[IE_1 + IE_2], \tag{10}$$

where,
IE_1 = interest earned for the instant payment of the sold units during $(0, T_r]$

$$= \int_0^{T_r} \delta s D_m (L_1 - t)\mathrm{d}t = \delta s D_m \left[L_1 T_r - \frac{T_r^2}{2} \right] \tag{11}$$

IE_2 = interest earned for the credit payment of the sold units during $(0, T_r]$

$$= \int_0^{T_r} (1-\delta)(1-\alpha) s D_m (L_1 - L_2 - t)\mathrm{d}t$$

$$= (1-\delta)(1-\alpha) s D_m T_r (L_1 - L_2 - \frac{T_r}{2}) \tag{12}$$

Situation 1.2: $T_r < L_1 \leq T_r + L_2$ (i.e., $L_1 - L_2 \leq T_r < L_1$)
Here, the retailer should pay interest for the customers' credit amount for the sold units during $[L_1 - L_2, T_r]$. Again there are some units, those are deteriorated during $(0, T_r]$, but the retailer must pay full purchased amount to its supplier at the time L_1 by a bank loan and his account will be settled with the bank at time $t = (T_r + L_2)$.

Thus the retailer must pay interest for those deteriorated units for a time period $(T_r + L_2 - L_1)$. But this phenomenon is overlooked by the researchers.
So, the amo unt of interest that the retailer must pay each year is

$$IP = (IP_1 + IP_2)/T_r, \tag{13}$$

where,
IP_1 interest charged due to credit amount of sold units during $[L_1 - L_2, T_r]$

$$= \int_{L_1-L_2}^{T_r} (1-\delta)pD_m(t - L_1 + L_2)dt = \frac{(1-\delta)pD_m}{2}(T_r + L_2 - L_1)^2 \tag{14}$$

IP_2 = interest charged for the deteriorated units during $(0, T_r]$

$$= p(T_r + L_2 - L_1)\int_0^{T_r} \theta_1(t)R(t)dt$$

$$= pD_m(T_r + L_2 - L_1)\left[(1+e_p)\ln\left(\frac{1+e_p}{1+e_p-T}\right) - T\right] \tag{15}$$

However, the retailer can earn interest for the collateral deposit of the customers for the sold units during $(0, T_r]$ and for the repayment of the customers' credit amount for the sold units during $(0, L_1 - L_2]$.
Hence, the annual interest generated by the retailer is

$$IE = (IE_1 + IE_2)/T_r], \tag{16}$$

where,
IE_1 = interest earned due to instant payment of sold units in time $(0, T_r]$

$$= \int_0^{T_r} \delta s D_m(L_1 - t)dt = \delta s D_m\left[L_1 T_r - \frac{T_r^2}{2}\right] \tag{17}$$

IE_2 = interest earned due to credit payment of sold units in time $(0, L_1 - L_2]$

$$= \int_0^{L_1-L_2} (1-\delta)(1-\alpha)sD_m(L_1 - L_2 - t)dt$$

$$= (1-\delta)(1-\alpha)sD_m\frac{(L_1-L_2)^2}{2} \tag{18}$$

Situation 1.3: $L_1 \leq T_r$
Here, the retailer should pay interest due to

1. the stocked units during $[L_1, T_r]$
2. customers' credit amount for the sold units during $(L_1 - L_2, L_1]$
3. customers' credit amount for the sold units during $[L_1, T_r]$
4. the deteriorated units during $(0, L_1]$
5. the deteriorated units during $(L_1, T_r]$.

Hence, the amount of interest that the retailer must pay each year is

$$IP = (IP_1 + IP_2 + IP_3 + IP_4 + IP_5)/T_r, \tag{19}$$

where,

IP_1 = interest charged due to stock units during $[L_1, T_r] = p\int\limits_{L_1}^{T_r} R(t)\mathrm{d}t$

$$= pD_m\left[\frac{(1+e_p-L_1)^2}{4}\left\{2\ln\left(\frac{1+e_p-L_1}{1+e_p-T_r}\right)-1\right\}+\frac{(1+e_p-T_r)^2}{4}\right] \tag{20}$$

IP_2 = interest charged due to customers' credit of sold units during

$$[L_1 - L_2, L_1] = \int\limits_{L_1-L_2}^{L_1} (1-\delta)pD_m(t - L_1 + L_2)\mathrm{d}t = (1-\delta)pD_m\frac{L_2^2}{2} \tag{21}$$

IP_3 = interest charged due to customers' credit amount of sold units

$$\text{during } [L_1, T_r] = \int\limits_{L_1}^{T_r} (1-\delta)pDL_2\mathrm{d}t = (1-\delta)pDL_2(T_r - L_1) \tag{22}$$

IP_4 = interest charged for the deteriorated units during $(0, L_1]$

$$\begin{aligned} &= p(T_r + L_2 - L_1)\int\limits_0^{L_1} \theta_1(t)R(t)\mathrm{d}t \\ &= pD_m(T_r + L_2 - L_1)\left[(1+e_p)\ln\left(\frac{1+e_p}{1+e_p-L_1}\right)\right. \\ &\left. -L_1\left\{1-\ln\left(\frac{1+e_p-L_1}{1+e_p-T_r}\right)\right\}\right] \end{aligned} \tag{23}$$

IP_5 = interest charged for the deteriorated units during $(L_1, T_r]$

$$= p\int_{L_1}^{T_r} \theta_1(t)R(t)(T_r + L_2 - t)\mathrm{d}t$$

$$= pD_m[\left\{\frac{L_1^2}{2} - \frac{(1+e_p)^2}{2} + (T_r + L_2)(1 + e_p - L_1)\right\} \ln\left(\frac{1+e_p-L_1}{1+e_p-T_r}\right)$$
$$+ \frac{T_r^2 - L_1^2}{4} - (T_r + L_2)(T_r - L_1) + \frac{(1+e_p)(T_r - L_1)}{2}] \quad (24)$$

On the other hand, retailer can earn interest on the initial payment of the customers for the sold items during $(0, L_1]$ and on the credit payment of the customers for the sold items during $(0, L_1 - L_2]$.

Hence, the annual interest generated by the retailer is

$$IE = (IE_1 + IE_2)/T_r, \quad (25)$$

where, IE_1 = interest earned for the instant payment of sold units during $(0, L_1]$

$$= \int_0^{L_1} \delta s D_m(L_1 - t)\mathrm{d}t = \delta s D_m \frac{L_1^2}{2} \quad (26)$$

IE_2 = interest earned for the credit payment of sold units during $(0, L_1 - L_2]$

$$= \int_0^{L_1 - L_2} (1-\delta)(1-\alpha)sD_m(L_1 - L_2 - t)\mathrm{d}t = \frac{1}{2}(1-\delta)(1-\alpha)sD_m(L_1 - L_2)^2 \quad (27)$$

Therefore, the total annual profit of the retailer when $L_2 < L_1$ is given by

$$TP_1(T_r, s) = \begin{cases} TP_{11}(T_r, s), \text{ if } 0 < T_r < L_1 - L_2 \\ TP_{12}(T_r, s), \text{ if } L_1 - L_2 \le T_r < L_1 \\ TP_{13}(T, s), \text{ if } L_1 \le T_r \end{cases} \quad (28)$$

where,

$$TP_{11}(T_r, s) = sD_m[\delta + (1-\delta)(1-\alpha)] - \frac{cD_m(1+e_p)}{T_r} \ln\left(\frac{1+e_p}{1+e_p-T_r}\right) - \frac{A_c}{T_r}$$
$$-\frac{h_c D_m}{T_r}\left[\frac{(1+e_p)^2}{2} \ln\left(\frac{1+e_p}{1+e_p-T_r}\right) + \frac{T_r^2}{4} - \frac{(1+e_p)T_r}{2}\right]$$
$$+\frac{sD_m i_e}{2}[T_r + 2\delta(L_1 - T_r) - (1-\delta)\alpha T_r + 2(1-\delta)(1-\alpha)(L_1 - L_2 - T_r)] \quad (29)$$

$$TP_{12}(T_r, s) = sD_m[\delta + (1-\delta)(1-\alpha)] - \frac{pD_m(1+e_p)}{T_r} \ln\left(\frac{1+e_p}{1+e_p-T_r}\right) - \frac{A_c}{T_r}$$

$$-\frac{h_c D_m}{T_r}\left[\frac{(1+e_p)^2}{2}\ln\left(\frac{1+e_p}{1+e_p-T_r}\right)+\frac{T_r^2}{4}-\frac{(1+e_p)T_r}{2}\right]$$
$$+\frac{sD_m i_e}{2T_r}\left[\delta T^2+2\delta T_r(L_1-T_r)+(1-\delta)(1-\alpha)(L_1-L_2)^2\right]-\frac{pD_m i_p}{T_r}(T_r+L_2-L_1)$$
$$\left[\frac{(1-\delta)(T_r+L_2-L_1)}{2}+(1+e_p)\ln\left(\frac{1+e_p}{1+e_p-T_r}\right)-T_r\right] \tag{30}$$

$$TP_{13}(T_r,s)=sD_m\left[\delta+(1-\delta)(1-\alpha)\right]-\frac{pD_m(1+e_p)}{T_r}\ln\left(\frac{1+e_p}{1+e_p-T_r}\right)-\frac{A_c}{T_r}$$
$$-\frac{h_c D_m}{T_r}\left[\frac{(1+e_p)^2}{2}\ln\left(\frac{1+e_p}{1+e_p-T_r}\right)+\frac{T_r^2}{4}-\frac{(1+e_p)T_r}{2}\right]$$
$$+\frac{sD_m i_e}{2T_r}\left[\delta L_1^2+(1-\delta)(1-\alpha)(L_1-L_2)^2\right]-\frac{pD_m i_p}{T_r}\Bigg[(1+e_p)(T_r+L_2-L_1)$$
$$\ln\left(\frac{1+e_p}{1+e_p-T_r}\right)+(1-\delta)L_2\left(T_r-L_1+\frac{L_2}{2}\right)+\frac{T_r^2+L_1^2}{2}-T_r(T_r+L_2)\Bigg] \tag{31}$$

Case 2: $L_2 \geq L_1$

Here, depending on the values of L_1 and T_r, two situations will occur:

- **Situation-2.1** $T_r < L_1$ and
- **Situation-2.2** $T_r \geq L_1$

Situation 2.1: $T_r < L_1$

Here, the amount of interest that the retailer must pay each year is

$$IP=(IP_1+IP_2)/T_r, \tag{32}$$

where,

IP_1 = interest charged due to credit amount of sold units during $(0, T_r]$

$$=\int_0^{T_r}(1-\delta)pD_m(t-L_1+L_2)\mathrm{d}t=\frac{1}{2}(1-\delta)pD_m T_r\left\{T_r+2(L_2-L_1)\right\} \tag{33}$$

IP_2 = interest charged for the deteriorated units during $(0, T_r]$

$$=p(T_r+L_2-L_1)\int_0^{T_r}\theta_1(t)R(t)\mathrm{d}t$$
$$=pD_m(T_r+L_2-L_1)\left\{(1+e_p)\ln\left(\frac{1+e_p}{1+e_p-T_r}\right)-T\right\} \tag{34}$$

Again, retailer can earn interest for the initial payment of the customers during $(0, T_r]$. Thus the annual interest generated by the retailer is

$$IE = IE_1/T_r, \tag{35}$$

where,

IE_1 = interest earned due to instant payment of sold units during $(0, T_r]$

$$= \int_0^{T_r} \delta s D_m(L_1 - t)\mathrm{d}t = \delta s D_m\left[L_1 T_r - \frac{T_r^2}{2}\right] \tag{36}$$

Situation 2.2: $L_1 \leq T_r$

Here, retailer should pay interest due to

1. the stocked units during $[L_1, T_r]$
2. customers' credit amount for the sold units during $(0, L_1]$
3. customers' credit amount for the sold units during $[L_1, T_r]$
4. the deteriorated units during $(0, L_1]$
5. the deteriorated units during $[L_1, T_r]$.

Thus the amount of interest that the retailer must pay each year is

$$IP = (IP_1 + IP_2 + IP_3 + IP_4 + IP_5)/T_r, \tag{37}$$

where, IP_1 = interest charged for the stocked units during $[L_1, T_r] = p\int_{L_1}^{T_r} R(t)\mathrm{d}t$

$$= \frac{1}{4}pD_m\left[(1+e_p-L_1)^2\left\{2\ln\left(\frac{1+e_p-L_1}{1+e_p-T_r}\right)-1\right\}+(1+e_p-T_r)^2\right] \tag{38}$$

IP_2 = interest charged due to customers' credit amount of sold units

$$\text{during}(0, L_1] = \int_0^{L_1}(1-\delta)pD_m(t-L_1+L_2)\mathrm{d}t = \frac{1}{2}(1-\delta)pD_m L_1(2L_2 - L_1) \tag{39}$$

IP_3 = interest charged due to customers' credit amount of sold units

$$\text{during } [L_1, T_r] = \int_{L_1}^{T_r}(1-\delta)pD_m L_2\mathrm{d}t = (1-\delta)pD_m L_2(T_r - L_1) \tag{40}$$

IP_4 = interest charged for the deteriorated units during $(0, L_1]$

$$= p(T_r + L_2 - L_1)\int_0^{L_1}\theta_1(t)R(t)\mathrm{d}t$$

$$= pD_m(T_r + L_2 - L_1)\left[(1+e_p)\ln\left(\frac{1+e_p}{1+e_p-L_1}\right) - L_1\left\{1 - \ln\left(\frac{1+e_p-L_1}{1+e_p-T_r}\right)\right\}\right] \tag{41}$$

IP_5 = interest charged for the deteriorated units during $[L_1, T_r]$

$$= p\int_{L_1}^{T_r} \theta_1(t)R(t)(T_r + L_2 - t)\mathrm{d}t$$

$$= pD_m\left[\left\{\frac{L_1^2}{2} - \frac{(1+e_p)^2}{2} + (T_r + L_2)(1+e_p-L_1)\right\}\ln\left(\frac{1+e_p-L_1}{1+e_p-T_r}\right) + \frac{T_r^2 - L_1^2}{4} - (T_r+L_2)(T_r-L_1) + \frac{(1+e_p)(T_r-L_1)}{2}\right] \tag{42}$$

Otherwise, the annual interest generated by the retailer is

$$IE = IE_1/T_r, \tag{43}$$

where,

IE_1 = interest earned for the customers' instant payment of sold units

$$\text{during } (0, L_1] = \int_0^{L_1} \delta s D_m(L_1 - t)\mathrm{d}t = \delta s D_m \frac{L_1^2}{2} \tag{44}$$

So, the annual total profit of the retailer for $L_2 \geq L_1$ is given by

$$TP_2(T_r, s) = \begin{cases} TP_{21}(T_r, s), \text{ if } T_r < L_1 \\ TP_{22}(T_r, s), \text{ if } T_r \geq L_1 \end{cases} \tag{45}$$

where,

$$TP_{21}(T_r, s) = sD_m[\delta + (1-\delta)(1-\alpha)] - \frac{pD_m(1+e_p)}{T_r}\ln\left(\frac{1+e_p}{1+e_p-T_r}\right) - \frac{A_c}{T_r} - \frac{h_cD_m}{T_r}\left[\frac{(1+e_p)^2}{2}\ln\left(\frac{1+e_p}{1+e_p-T_r}\right) + \frac{T_r^2}{4} - \frac{(1+e_p)T_r}{2}\right] + \frac{\delta s D_m i_e}{2}(2L_1 - T_r) - \frac{pD_m i_p}{T_r}\left[(1-\delta)T_r\left(\frac{T_r}{2} + L_2 - L_1\right) + (T_r + L_2 - L_1)\left\{(1+e_p)\ln\left(\frac{1+e_p}{1+e_p-T_r}\right) - T_r\right\}\right] \tag{46}$$

$$TP_{22}(T_r, s) = sD_m[\delta + (1-\delta)(1-\alpha)] - \frac{pD_m(1+e_p)}{T_r}\ln\left(\frac{1+e_p}{1+e_p-T_r}\right) - \frac{A_c}{T_r}$$

$$-\frac{h_cD_m}{T_r}\left[\frac{(1+e_p)^2}{2}\ln\left(\frac{1+e_p}{1+e_p-T_r}\right)+\frac{T_r^2}{4}-\frac{(1+e_p)T_r}{2}\right]+\frac{sD_mi_e}{2T_r}\delta L_1^2$$
$$-\frac{pD_mi_p}{T-r}\left[(T_r+L_2-L_1)(1+e_p)\ln\left(\frac{1+e_p}{1+e_p-T_r}\right)+(1-\delta)\left(L_2T_r-\frac{L_1^2}{2}\right)\right.$$
$$\left.+\frac{T_r^2+L_1^2}{2}-T_r(T_r+L_2)\right] \tag{47}$$

Hence the problem is to

$$\text{maximize } TP(T_r,s)=\begin{cases} TP_1(T_r,s), \text{ if } L_2<L_1 \\ TP_2(T_r,s), \text{ if } L_2\geq L_1 \end{cases} \tag{48}$$

where $TP_i(T_r, s), i = 1, 2$ is mentioned in Eqs. (28) and (45) respectively.

It should be noticed that $TP_{11}(L_1 - L_2, s) = TP_{12}(L_1 - L_2, s)$, $TP_{12}(L_1, s) = TP_{13}(L_1, s)$ and $TP_{21}(L_1, s) = TP_{22}(L_1, s)$. Hence $TP_i(T_r, s)$ is a continuous function $\forall$ s and $T_r > 0$ for $i = 1, 2$.

4.1 Crisp Model (Model-1)

If all the parameters are in crisp environment, the mathematical problem becomes

$$\left.\begin{array}{ll} \text{to find} & T_r, s; \\ \text{which maximize} & TP(T_r,s) \end{array}\right\} \tag{49}$$

Here $TP(T_r, s)$ is defined by

$$TP(T_r,s)=SR-OC-PC-AHC+i_eIE+i_pIP \tag{50}$$

where, SR, OC, PC, AHC, are respectively equation number (5), (6), (7), (8). Also IE are (10), (16), (25), (35) and (43) for situation 1.1, 1.2, 1.3, 2.1 and 2.2 respectively and IP are (9), (13), (19), (32) and (37) respectively.

4.2 Fuzzy Model (Model-2)

Due to fluctuating business market situation different cost involving the inventory such as bank interest rate, ordering cost etc. changes frequently. Also it is very difficult to predict the unfaithful customer. So it is more realistic to assume payable interest rate, earning interest, ordering cost and unfaithful customer as fuzzy number. As a result, it is taken that payable interest rate (i_p), earning interest (i_e), ordering cost (A_c) and unfaithful customer (α) as a triangular fuzzy number (TFNs) and represented as $\tilde{i_p} = (i_{p1}, i_{p2}, i_{p3}), \tilde{i_e} = (i_{e1}, i_{e2}, i_{e3}), \tilde{A_c} = (A_1, A_2, A_3)$ and $\tilde{\alpha} = (\alpha_1, \alpha_2, \alpha_3)$ According to the extension principle of fuzzy numbers, total profit per unit time of retailer for the three scenarios are TFNs , $\widetilde{TP(T_r, s)} = (Z_1, Z_2, Z_3)$ Where

$$Z_1 = sD_m\{\delta + (1-\delta)(1-\alpha_3)\} - PC - \tfrac{A_3}{T_r} - HOC - i_{p3}\widetilde{IP} + i_{e1}\widetilde{IE}$$

$$Z_2 = sD_m\{\delta + (1-\delta)(1-\alpha_2)\} - PC - \tfrac{A_2}{T_r} - HOC - i_{p2}\widetilde{IP} + i_{e2}\widetilde{IE}$$

$$Z_3 = sD_m\{\delta + (1-\delta)(1-\alpha_1)\} - PC - \tfrac{A_1}{T_r} - HOC - i_{p1}\widetilde{IP} + i_{e3}\widetilde{IE}$$

where, PC, and HOC are respectively given by Eq. (7) and (8).Also IE are (10), (16), (25), (35) and (35) for situation 1.1, 1.2, 1.3, 2.1 and 2.2 respectively and IP are (9), (13), (19), (32) and (37) respectively. So, the mathematical problem reduces to

$$\left.\begin{array}{ll} \text{to find} & T_r, s; \\ \text{which maximize} & \widetilde{TP(T_r, s)} \end{array}\right\} \tag{51}$$

5 Theoretical Discussion for Existence of a Solution

Here, a theoretical discussion is made for each of the situations of the proposed model to search retailer's optimal replenishment time (T_r) and optimal selling price (s) so that the annual profit of the retailer is maximum.

5.1 Discussion for an Optimal Solution to the Situation-1.1

For a fixed selling price (s), first order partial derivative of $TP_{11}(T_r, s)$ with respect to T_r is

$$\frac{\partial TP_{11}(T_r, s)}{\partial T_r} = \frac{F_{11}(T_r)}{T_r^2}, \text{ where} \tag{52}$$

$$F_{11}(T_r) = A_c + (1+e_p)D_m\left\{p + \frac{h_c(1+e_p)}{2}\right\}\left\{\ln\left(\frac{1+e_p}{1+e_p-T_r}\right) - \frac{T_r}{1+e_p-T_r}\right\}$$
$$-\frac{D_m T_r^2}{4}\{h_c + 2pi_e\{1-(1-\delta)\alpha\}\} \tag{53}$$

Differentiating $F_{11}(T_r)$ with respect to $T_r \in (0, L_1 - L_2]$ we have

$$F_{11}^{'}(T_r) = \frac{dF_{11}(T_r)}{\mathrm{d}t_r}$$
$$= -D_m T_r\Big[\frac{(1+e_p)}{(1+e_p-T_r)^2}\{p + \frac{h_c(1+e_p)}{2}\}$$
$$+\frac{h_c}{2} + pi_e\{1-(1-\delta)\alpha\}\Big] < 0 \tag{54}$$

Thus $F_{11}(T_r)$ is strictly decreasing function with respect to $T_r \in (0, L_1 - L_2]$. Further, $\lim_{T_r \to \infty} F_{11}(T_r) = -\infty$, $F_{11}(0) = A_c > 0$ and

$$F_{11}(L_1 - L_2) = A_c + (1 + e_p)D_m\left\{p + \frac{h_c(1+e_p)}{2}\right\}\left\{\ln\left(\frac{1+e_p}{1+e_p-L_1+L_2}\right) - \frac{L_1-L_2}{1+e_p-L_1+L_2}\right\}$$
$$-\frac{h_c}{4}D_m(L_1 - L_2)^2 - \frac{D_m pi_e}{2}(L_1 - L_2)^2\{1 - (1 - \delta)\alpha\} = \Delta_1(say) \tag{55}$$

If $\Delta_1 \leq 0$ then by Intermediate Value Theorem there must have a unique value of T_r (say $T_{11} \in (0, L_1 - L_2]$) such that $F_{11}(T_{11}) = 0$.

$$\text{Again, } \left[\frac{\partial^2 TP_{11}(T_r, s)}{\partial T^2}\right]_{T_r=T_{11}} = \left[-\frac{2}{T_r{}^3}F_{11}(T) + \frac{1}{T_r{}^2}\frac{dF_{11}(T_r)}{dt}\right]_{T_r=T_{11}}$$
$$= -\frac{2}{T_{11}^3}F_{11}(T_{11}) - \frac{D_m}{T_{11}}\left[\frac{(1+e_P)}{(1+e_p-T_{11})^2}\left\{p + \frac{h_c(1+e_p)}{2}\right\} + \frac{h_c}{2} + pi_e\left\{1-(1-\delta)\alpha\right\}\right]$$
$$= -\frac{D_m}{T_{11}}\left[\frac{(1+e_p)}{(1+e_p-T_{11})^2}\left\{p + \frac{h_c(1+e_p)}{2}\right\} + \frac{h_c}{2} + pi_e\{1-(1-\delta)\alpha\}\right] < 0 \tag{56}$$

Thus $TP_{11}(T_r, s)$ has a global maxima at the point $T_r = T_{11}$.
Conversely, if $\Delta_1 > 0$ we have $F_{11}(T_r) > 0 \ \forall T_r \in (0, L_1 - L_2]$ which implies $TP_{11}(T_r, s)$ is a strictly increasing function of T_r for a fixed s. Hence $TP_{11}(T_r, s)$ has a maximum value at the boundary point $T_r = (L_1 - L_2)$.
From this discussion, it is seen that for a fixed s, there exist an optimal $T_r = T_{11}^*$ such that TP_{11} is maximized, where T_{11}^* is given by the Eq. (57).

$$T_{11}^* = \begin{cases} T_{11}, \text{ if } \Delta_1 \leq 0 \\ (L_1 - L_2), \text{ if } \Delta_1 > 0 \end{cases} \tag{57}$$

Again for the fixed T_{11}^* defined in Eq. (57), first order partial derivative of $TP_{11}(T_{11}^*, s)$ with respect to s gives

$$\frac{\partial TP_{11}(T_{11}^*, s)}{\partial s} = -\frac{(b-1)}{s^b}\eta_1 + \frac{b}{s^{b+1}}\eta_2 + \frac{b}{s^{b+1}}\eta_3 - \frac{b}{s^{b+1}}\eta_4, \tag{58}$$

where,

$$\eta_1 = a\{\delta + (1 - \delta)(1 - \alpha)\} \tag{59}$$

$$\eta_2 = \frac{ap}{T_r}(1 + e_p)\log\left(\frac{1+e_p}{1+e_p-T_r}\right) \tag{60}$$

$$\eta_3 = \frac{ah_c}{T_r}[\frac{(1+e_p)^2}{T_r}\log\left(\frac{1+e_p}{1+e_p-T_r}\right) + \frac{T_r^2}{4} - \frac{(1+e_p)}{2}T_r] \tag{61}$$

$$\eta_4 = \frac{api_e}{2}[T_r + 2\delta(L_1 - T_r) - (1-\delta)\alpha T_r + 2(1-\delta)(1-\alpha)(L_1 - L_2 - T_r)] \tag{62}$$

Now, $\frac{\partial TP_{11}(T_{11}^*, s)}{\partial s} = 0$ implies that

$$s = \frac{b(\eta_2 + \eta_3 - \eta_4)}{(b-1)\eta_1} = s_{11}^* \text{ (say)} \tag{63}$$

Again at $s = s_{11}^*$,

$$\frac{\partial^2 TP_{11}(T_{11}^*, s)}{\partial s^2} = -\frac{(b-1)}{s_{11}^{*b+1}}\eta_1 < 0 \tag{64}$$

Thus from the above discussion it is evident that for an optimal replenishment time (T_{11}^*) there is an optimal retail selling price (s_{11}^*) for which annual retailer's profit ($TP_{11}(T_r, s)$) is maximum and hence there is an optimal solution of the proposed model for the situation-1.1.

5.2 *Discussion for an Optimal Solution to the Situation-1.2*

For a fixed selling price (s), first order partial derivative of $TP_{12}(T_r, s)$ with respect to T_r is

$$\frac{\partial TP_{12}(T_r, s)}{\partial T_r} = \frac{F_{12}}{T_r^2}, \text{ where} \tag{65}$$

$$F_{12}(T_r) = A_c + (1+e_p)D_m\left\{p + \frac{h_c(1+e_p)}{2}\right\}\left\{\ln\left(\frac{1+e_p}{1+e_p-T_r}\right) - \frac{T_r}{1+e_p-T_r}\right\}$$
$$-\frac{h_cD_mT_r^2}{4} - \frac{pD_mi_e}{2}\{\delta T_r^2 + (1-\delta)(1-\alpha)(L_1-L_2)^2\}$$

$$+pD_mi_pT_r^2 - \frac{pD_mi_p(1-\delta)}{2}\{T_r^2 - (L_1-L_2)^2\}$$
$$-pD_mi_p(1+e_p)\left\{(L_1-L_2)\ln\left(\frac{1+e_p}{1+e_p-T_r}\right) + \frac{T(T+L_2-L_1)}{1+e_p-T_r}\right\} \tag{66}$$

Differentiating $F_{12}(T_r)$ with respect to $T_r \in [L_1 - L_2, L_1]$ we have

$$F'_{12}(T_r) = \frac{dF_{12}(T_r)}{dt_r}$$
$$= -D_m T_r \Big[\frac{(1+e_p)}{(1+e_p-T_r)^2} \Big\{ p + \frac{h_c(1+e_p)}{2} \Big\} + \frac{h_c}{2} + pi_e\delta +$$
$$pi_p \Big\{ (1-\delta) + \frac{2T_r}{1+e_p-T_r} + \frac{(1+e_p)(T_r+L_2-L_1)}{(1+e_p-T_r)^2} \Big\} \Big] < 0 \quad (67)$$

Thus $F_{12}(T_r)$ is strictly decreasing function with respect to $T_r \in [L_1 - L_2, L_1]$. Moreover, $\lim_{T_r \to \infty} F_{12}(T_r) = -\infty$. Now,

$$F_{12}(L_1-L_2) = A_c + (1+e_p)D_m \Big\{ p + \frac{h_c(1+e_p)}{2} \Big\} \Big\{ \ln\Big(\frac{1+e_p}{1+e_p-L_1+L_2} \Big) -$$
$$\frac{L_1-L_2}{1+e_p-L_1+L_2} \Big\} - (L_1-L_2)^2 \Big[\frac{h_c D_m}{4} + \frac{p D_m i_e}{2} \{1-\alpha(1-\delta)\} \Big]$$
$$-pD_m i_p (L_1-L_2) \Big\{ (1+e_p)\ln\Big(\frac{1+e_p}{1+e_p-L_1+L_2} \Big) - (L_1-L_2) \Big\} \text{ and} \quad (68)$$
$$F_{12}(L_1) = A_c + (1+e_p)D_m \Big\{ p + \frac{h_c(1+e_p)}{2} \Big\} \Big\{ \ln\Big(\frac{1+e_p}{1+e_p-L_1} \Big) - \frac{L_1}{1+e_p-L_1} \Big\}$$
$$-\frac{h_c D_m L_1^2}{4} - \frac{pDi_e}{2} \Big\{ \delta L_1^2 + (1-\delta)(1-\alpha)(L_1-L_2)^2 \Big\}$$
$$+\frac{pD_m i_p}{2} \Big\{ 2L_1^2 - (1-\delta)L_2(2L_1-L_2) \Big\}$$
$$-pD_m i_p (1+e-p) \Big\{ (L_1-L_2)\ln\Big(\frac{1+e_p}{1+e_p-L_1} \Big) + \frac{L_1 L_2}{1+e_p-L_1} \Big\} \quad (69)$$

Here, three scenarios may arise:
Scenario-1: $F_{12}(L_1) < F_{12}(L_1 - L_2) < 0$
Then, $F_{12}(T_r) < 0$ for all $T \in [L_1 - L_2, L_1]$ and hence $TP_{12}(T_r, s)$ is a decreasing function of $T_r \in [L_1 - L_2, L_1]$ for a fixed s and hence $TP_{12}(T_r, s)$ attain its optimal value at $T_r = (L_1 - L_2)$.
Scenario-2: $F_{12}(L_1) < F_{12}(L_1 - L_2) > 0$
Then, $F_{12}(T_r) > 0$ for all $T_r \in [L_1 - L_2, L_1]$ and hence $TP_{12}(T_r, s)$ is a increasing function of $T_r \in [L_1 - L_2, L_1]$ for a fixed s and hence $TP_{12}(T_r, s)$ attain its optimal value at $T_r = L_1$.
Scenario-3: $F_{12}(L_1) < 0$ but $F_{12}(L_1 - L_2) > 0$
Then, $\exists$ a $T_r = T_{12}(say) \in [L_1 - L_2, L_1]$ such that $F_{12}(T_{12}) = 0$.

$$\text{Now, } \Big[\frac{\partial^2 TP_{12}(T,s)}{\partial T_r^2} \Big]_{T_r=T_{12}} = \Big[-\frac{2}{T_r{}^3} F_{12}(T_r) + \frac{1}{T_r{}^2} \frac{dF_{12}(T_r)}{dt} \Big]_{T_r=T_{12}}$$
$$= -\frac{2}{T_{12}{}^3} F_2(T_{12}) - \frac{D_m}{T_{12}} \Big[\frac{(1+e_p)}{(1+e_p-T_{12})^2} \Big\{ p + \frac{h_c(1+e_p)}{2} \Big\} + (\frac{h_c}{2} + pi_e\delta) +$$
$$pi_p \Big\{ (1-\delta) + \frac{2T_r}{1+e_p-T_{12}} + \frac{(1+e_p)(T_{12}+L_2-L_1)}{(1+e_p-T_{12})^2} \Big\} \Big] \quad (70)$$
$$= -\frac{D_m}{T_{12}} \Big[\frac{(1+e_p)}{(1+e_p-T_{12})^2} \Big\{ p + \frac{h_c(1+e_p)}{2} \Big\} + \Big(\frac{h_c}{2} + pi_e\delta \Big) +$$
$$pi_p \Big\{ (1-\delta) + \frac{2T_r}{1+e_p-T_{12}} + \frac{(1+e_p)(T_{12}+L_2-L_1)}{(1+e_p-T_{12})^2} \Big\} \Big] < 0. \quad (71)$$

Thus for a fixed s, there exist an optimal $T_r = T_{12}^*$ such that $TP_{12}(T_r, s)$ is maximized, where T_{12}^* is given by the Eq. (72).

$$T_{12}^{*} = \begin{cases} (L_1 - L_2), \text{ if } F_{12}(L_1) < F_{12}(L_1 - L_2) < 0 \\ L_1, \text{ if } F_{12}(L_1) < F_{12}(L_1 - L_2) > 0 \\ T_{12}, \text{ if } F_{12}(L_1) < 0 \text{ but } F_{12}(L_1 - L_2) > 0 \end{cases} \tag{72}$$

Again for the fixed T_{12}^{*} defined in Eq. (72), first order partial derivative of $TP_{12}(T_{12}^{*}, s)$ with respect to s gives

$$\frac{\partial TP_{12}(T_{12}^{*}, s)}{\partial s} = -\frac{(b-1)}{s^b}\eta_1 + \frac{b}{s^{b+1}}\eta_2 + \frac{b}{s^{b+1}}\eta_3 - \frac{b}{s^{b+1}}\eta_4 + \frac{b}{s^{b+1}}\eta_5, \tag{73}$$

where, η_1, η_2, η_3 are same as Eqs. (59)–(61) respectively and

$$\eta_4 = \frac{api_e}{2T_r}\left[\delta T_r^2 + 2\delta T_r(L_1 - T_r) + (1-\delta)(1-\alpha)(L_1 - L_2)^2\right] \tag{74}$$

$$\eta_5 = \frac{api_p}{T_r}(T_r + L_2 - L_1)\left[\frac{(1-\delta)}{2}(T_r + L_2 - L_1) + (1+e_p)\log\left(\frac{1+e_p}{1+e_p-T_r}\right) - T_r\right] \tag{75}$$

Now, $\frac{\partial TP_{12}(T_{12}^{*}, s)}{\partial s} = 0$ implies that

$$s = \frac{b(\eta_2 + \eta_3 - \eta_4 + \eta_5)}{(b-1)\eta_1} = s_{12}^{*} \text{ (say)} \tag{76}$$

Again at $s = s_{12}^{*}$,

$$\frac{\partial^2 TP_{12}(T_{12}^{*}, s)}{\partial s^2} = -\frac{(b-1)}{s_{12}^{*b+1}}\eta_1 < 0 \tag{77}$$

Thus from the above discussion it is observed that there is an optimal solution (T_{12}^{*}, s_{12}^{*}) of the proposed model for the situation-1.2, for which the retailer's annual profit $(TP_{12}(T_r, s))$ should be maximum.

5.3 Discussion for an Optimal Solution to the Situation-1.3

Moreover, for a fixed s, first order partial derivative of $TP_{13}(T_r, s)$ with respect to T_r is given by

$$\frac{\partial TP_{13}(T_r)}{\partial T_r} = \frac{F_{13}}{T_r^2} \tag{78}$$

where $F_{13}(T_r) = A_c + (1+e_p)D_m\left\{p + \frac{h_c(1+e_p)}{2}\right\}\left\{\ln\left(\frac{1+e_p}{1+e_p-T_r}\right) - \frac{T_r}{1+e_p-T_r}\right\}$

$$-\frac{h_c D_m T_r^2}{4} - \frac{pD_m i_e}{2}\left\{\delta L_1^2 + (1-\delta)(1-\alpha)(L_1 - L_2)^2\right\}$$

$$+\frac{pD_m i_p}{2}\Big\{T_r^2 + L_1^2 - (1-\delta)L_2(2L_1 - L_2)\Big\}$$
$$-pD_m i_p(1+e_p)\Big\{(L_1 - L_2)\ln\Big(\frac{1+e_p}{1+e_p-T_r}\Big) + \frac{T_r(T_r+L_2-L_1)}{1+e_p-T_r}\Big\} \tag{79}$$

Differentiating $F_{13}(T_r)$ with respect to $T_r \in [L_1, \infty)$ we have

$$F'_{13}(T_r) = \frac{dF_{13}(T)}{dt_r}$$
$$= -D_m T_r\Big[\frac{(1+e_p)}{(1+e_p-T_r)^2}\{p + \frac{h_c(1+e_p)}{2}\} + \frac{h_c}{2} - pi_p$$
$$+\frac{pi_p(1+e_p)}{(1+e_p-T_r)^2}\{(1+e_p-T_r) + (1+e_p-L_1+L_2)\}\Big] < 0 \tag{80}$$

Thus $F_{13}(T_r)$ is strictly decreasing function with respect to $T_r \in [L_1, \infty)$. Moreover, $\lim_{T_r\to\infty} F_{13}(T_r) = -\infty$. Now,

$$F_{13}(L_1) = A_c + (1+e_p)D_m\Big\{p + \frac{h_c(1+e_p)}{2}\Big\}\Big\{\ln\Big(\frac{1+e_p}{1+e_p-L_1}\Big) - \frac{L_1}{1+e_p-L_1}\Big\}$$
$$-\frac{h_c D_m L_1^2}{4} - \frac{pD_m i_e}{2}\Big\{\delta L_1^2 + (1-\delta)(L_1-L_2)^2\Big\} + \frac{pD_m i_p}{2}\Big\{2L_1^2 - (1-\delta)$$
$$L_2(2L_1 - L_2)\Big\} - pD_m i_p(1+e_p)\Big\{(L_1-L_2)\ln\Big(\frac{1+e_p}{1+e_p-L_1}\Big) + \frac{L_1 L_2}{1+e_p-L_1}\Big\} \tag{81}$$

If $F_{13}(L_1) < 0$ we have $F_{13}(T_r) < 0$ for all $T_r \in [L_1, \infty)$, which implies that TP_{13} (T_r, s) is strictly decreasing function of $T_r \in [L_1, \infty)$. Hence $TP_{13}(T_r, s)$ has a maximum value at the boundary point $T_r = L_1$.
Conversely, if $F_{13}(L_1) \geq 0$, then by Intermediate Value Theorem there must have a unique value of $T_r = T_{13}(say) \in [L_1, \infty)$ such that $F_{13}(T_{13}) = 0$. Again at $T_r = T_{13}$,

$$\Big[\frac{d^2 TP_{13}(T_r)}{dt^2}\Big]_{T_r=T_{13}} = \frac{F'_{13}(T_{13})}{T_{13}^2} - \frac{2F_{13}(T_{13})}{T_{13}^3} < 0, \text{ as } F'_{13}(T_r) < 0 \forall T_r \in [L_1, \infty] \tag{82}$$

Thus for a fixed s, there must have an optimal $T_r = T_{13}^*$ such that $TP_{13}(T_r, s)$ is maximized, where T_{13}^* is given by the Eq. (83).

$$T_{13}^* = \begin{cases} L_1, \text{ if } F_{13}(L_1) < 0 \\ T_{13}, \text{ if } F_{13}(L_1) > 0 \end{cases} \tag{83}$$

Again for the fixed T_{13}^* defined in Eq. (83), first order partial derivative of $TP_{13}(T_{13}^*, s)$ with respect to s gives

$$\frac{\partial TP_{13}(T_{13}^*, s)}{\partial s} = -\frac{(b-1)}{s^b}\eta_1 + \frac{b}{s^{b+1}}\eta_2 + \frac{b}{s^{b+1}}\eta_3 - \frac{b}{s^{b+1}}\eta_4 + \frac{b}{s^{b+1}}\eta_5, \tag{84}$$

where η_1, η_2, η_3 are respectively given by Eqs. (59)–(61) and

$$\eta_4 = \frac{api_e}{2T_r}[\delta L_1^2 + (1-\delta)(1-\alpha)(L_1 - L_2)^2] \tag{85}$$

$$\eta_5 = \frac{api_p}{T_r}\left[(1+e_p)(T_r + L_2 - L_1)\log\left(\frac{1+e_p}{1+e_p-T_r}\right) + (1-\delta)L_2\left(T_r - L_1 + \frac{L_2}{2}\right) + \frac{T_r^2 + L_1^2}{2} - T_r(T_r + L_2)\right] \tag{86}$$

Now, $\frac{\partial TP_{13}(T_{13}^*, s)}{\partial s} = 0$ implies that

$$s = \frac{b(\eta_2 + \eta_3 - \eta_4 + \eta_5)}{(b-1)\eta_1} = s_{13}^* \text{ (say)} \tag{87}$$

Again at $s = s_{13}^*$,

$$\frac{\partial^2 TP_{12}(T_{12}^*, s)}{\partial s^2} = -\frac{(b-1)}{s_{13}^{*b+1}}\eta_1 < 0 \tag{88}$$

Thus for the situation-1.3 also, it is observed that there must have an optimal solution (T_{13}^*, s_{13}^*) for which the retailer's total annual profit $(TP_{13}(T_r, s))$ is maximum.

5.4 Discussion for an Optimal Solution to the Situation-2.1

For a fixed s, first order partial derivative of $TP_{21}(T_r, s)$ with respect to T_r is

$$\frac{\partial TP_{21}(T_r)}{\partial T_r} = \frac{1}{T_r^2}\left[A_c + (1+e_p)D_m\left\{p + \frac{h_c(1+e_p)}{2}\right\}\left\{\ln\left(\frac{1+e_p}{1+e_p-T_r}\right) - \frac{T_r}{1+e_p-T_r}\right\} - \frac{h_c D_m T_r^2}{4} - \frac{pD_m i_e}{2}\delta T_r^2 - pD_m i_p\left\{\frac{1}{2}(1-\delta)T_r^2 + \frac{T_r + L_2 - L_1}{1+e_p-T_r}T_r^2 - (L_2 - L_1)\left\{(1+e_p)\ln\left(\frac{1+e_p}{1+e_p-T_r}\right) - T_r\right\}\right\}\right] \tag{89}$$

Proceeding similarly as in situation-1.1, there should have an unique value $T_r = T_{21}(\text{say}) \in (0, L_1]$ at which $\frac{\partial TP_{21}(T_r, s)}{\partial T_r} = 0$.

Hence for a fixed s, the profit function $TP_{21}(T_r, s)$ is maximum at an optimal value $T_r = T_{21}^*(say)$ given by the Eq. (90).

$$T_{21}^* = \begin{cases} L_1, & \text{if } \Delta_2 < 0 \\ T_{21}, & \text{if } \Delta_2 \geq 0, \text{ where} \end{cases} \tag{90}$$

$$\Delta_2 = A_c + (1+e_p)D_m\Big\{p + \frac{h_c(1+e_p)}{2}\Big\}\Big\{\ln\Big(\frac{1+e_p}{1+e_p-L_1}\Big) - \frac{L_1}{1+e_p-L_1}\Big\} -$$
$$\frac{h_c D_m L_1^2}{4} - \frac{pDi_e}{2}\delta L_1^2 - pD_m i_p\Big\{\frac{1}{2}(1-\delta)L_1^2 + \frac{L_2 L_1^2}{1+e_p-L_1} -$$
$$(L_2 - L_1)\Big\{(1+e_p)\ln\Big(\frac{1+e_p}{1+e_p-L_1}\Big) - L_1\Big\}\Big\} \quad (91)$$

Again for the fixed T_{21}^* defined in Eq. (90), first order partial derivative of $TP_{21}(T_{21}^*, s)$ with respect to s gives

$$\frac{\partial TP_{21}(T_{21}^*, s)}{\partial s} = -\frac{(b-1)}{s^b}\eta_1 + \frac{b}{s^{b+1}}\eta_2 + \frac{b}{s^{b+1}}\eta_3 - \frac{b}{s^{b+1}}\eta_4 + \frac{b}{s^{b+1}}\eta_5, \quad (92)$$

where η_1, η_2, η_3 are respectively given by Eqs. (59)–(61) and

$$\eta_4 = \frac{ap\delta i_e}{2}(2L_1 - T_r) \quad (93)$$
$$\eta_5 = \frac{api_p}{T_r}\Big[(1+e_p)\Big(\frac{T_r}{2} + L_2 - L_1\Big) + (T_r + L_2 - L_1)$$
$$\Big\{(1+e_p)\ln\Big(\frac{1+e_p}{1+e_p-T_r}\Big) - T_r\Big\}\Big] \quad (94)$$

Now, $\frac{\partial TP_{21}(T_{21}^*, s)}{\partial s} = 0$ implies that

$$s = \frac{b(\eta_2 + \eta_3 - \eta_4 + \eta_5)}{(b-1)\eta_1} = s_{21}^* \text{ (say)} \quad (95)$$

Again at $s = s_{21}^*$,

$$\frac{\partial^2 TP_{21}(T_{21}^*, s)}{\partial s^2} = -\frac{(b-1)}{s_{21}^{*b+1}}\eta_1 < 0 \quad (96)$$

Hence from the above discussion it is evident that there is an optimal solution of the mentioned model for the situation-2.1.

5.5 *Discussion for an Optimal Solution to Situation-2.2*

Similarly, for a fixed s, first order partial derivative of $TP_{22}(T_r, s)$ with respect to T_r is

$$\frac{\partial TP_{22}(T_r, s)}{\partial T_r} = \frac{1}{T_r^2}\Big[A_c + (1+e_p)D_m\Big\{p + \frac{h_c(1+e_p)}{2}\Big\}\Big\{\ln\Big(\frac{1+e_p}{1+e_p-T_r}\Big) - \frac{T_r}{1+e_p-T_r}\Big\} - \frac{h_c D_m T_r^2}{4} - \frac{pD_m i_e}{2}\delta L_1^2 - pD_m i_p(1+e_p)\Big\{(L_1 - L_2)\ln\Big(\frac{1+e_p}{1+e_p-T_r}\Big) + \frac{T_r(T_r + L_2 - L_1)}{1+e_p-T_r}\Big\} + \frac{pD_m i_p}{2}\Big\{T_r^2 + L_1^2 - (1-\delta)L_1^2\Big\}\Big] \quad (97)$$

Proceeding similarly as in situation-1.3, there should have an unique value $T_r = T_{22}(\text{say}) \in [L_1, \infty)$ at which $\frac{\partial TP_{22}(T_r,s)}{\partial T_r} = 0$.
Hence for a fixed s, the profit function $TP_{22}(T_r, s)$ is maximum at an optimal value $T_r = T_{22}^*(say)$, given by the Eq. (98).

$$T_{22}^* = \begin{cases} L_1, & \text{if } \Delta_3 < 0 \\ T_{22}, & \text{if } \Delta_3 \geq 0, \text{ where} \end{cases} \quad (98)$$

$$\Delta_3 = A_c + (1+e_p)D_m\Big\{p + \frac{h_c(1+e_p)}{2}\Big\}\Big\{\ln\Big(\frac{1+e_p}{1+e_p-L_1}\Big) - \frac{L_1}{1+e_p-L_1}\Big\} - \frac{h_c D_m L_1^2}{4} - \frac{pD_m i_e}{2}\delta L_1^2 - pD_m i_p \Big[(1+e_p)\Big\{\frac{L_2 L_1}{1+e_p-L_1} - (L_2 - L_1)\ln\Big(\frac{1+e_p}{1+e_p-L_1}\Big)\Big\} - \frac{(1+\delta)}{2}L_1^2\Big] \quad (99)$$

Again for the fixed T_{22}^* defined in Eq. (98), first order partial derivative of $TP_{22}(T_{22}^*, s)$ with respect to s gives

$$\frac{\partial TP_{22}(T_{22}^*, s)}{\partial s} = -\frac{(b-1)}{s^b}\eta_1 + \frac{b}{s^{b+1}}\eta_2 + \frac{b}{s^{b+1}}\eta_3 - \frac{b}{s^{b+1}}\eta_4 + \frac{b}{s^{b+1}}\eta_5, \quad (100)$$

where, η_1, η_2, η_3 are same as Eqs. (59)–(61) respectively and

$$\eta_4 = \frac{api_e}{2}\delta L_1^2 \quad (101)$$

$$\eta_5 = \frac{api_p}{T_r}\Big[(T_r + L_2 - L_1)(1+e_p)\ln\Big(\frac{1+e_p}{1+e_p-T_r}\Big) + (1-\delta)\Big(L_2 T_r - \frac{L_1^2}{2}\Big) + \frac{1}{2}(T_r^2 + L_1^2) - T_r(T_r + L_2)\Big] \quad (102)$$

Now, $\frac{\partial TP_{22}(T_{22}^*,s)}{\partial s} = 0$ implies that

$$s = \frac{b(\eta_2 + \eta_3 - \eta_4 + \eta_5)}{(b-1)\eta_1} = s_{22}^* \text{ (say)} \quad (103)$$

Again at $s = s_{22}^*$,

$$\frac{\partial^2 TP_{22}(T_{22}^*, s)}{\partial s^2} = -\frac{(b-1)}{s_{22}^{*b+1}}\eta_1 < 0 \quad (104)$$

Thus from the above discussion it is observed that there exists an optimal replenishing time (T_{22}^*) and an optimal selling price (s_{22}^*) for which annual retailer's profit ($TP_{22}(T_r, s)$) is maximum and hence there is a feasible solution of this model for the situation-2.2.

6 Numerical Experiments

It is clear from Sect. 5 that there is an optimal solution to the given model. A soft computing technique, Particle Swarm Optimization (PSO)[38] algorithm, is built in C-language to discover the optimal solution of the suggested model. PSO is used in this part to run a number of numerical examples to demonstrate this model and acquire some managerial insights.

Here five numerical examples are taken with different parametric values, results are obtained and presented in Table 1.

Example-1: The following parametric values are considered for Example-1: $a = 5 \times 10^6$, $b = 2.3$, $A_c = 120$, $h_c = 7$, $p = 15$, $i_p = 0.15$, $i_e = 0.12$, $e_p = 1$ year, $L_1 = 60/365$, $L_2 = 10/365$, $\delta = 0.5$ and $\alpha = 0.05$.
Example-2: The following parametric values are considered for Example-2: $a = 5 \times 10^6$, $b = 2.3$, $A_c = 100$, $h_c = 6$, $p = 12$, $i_p = 0.10$, $i_e = 0.08$, $e_p = 1$ year, $L_1 = 45/365$, $L_2 = 30/365$, $\delta = 0.3$ and $\alpha = 0.07$.
Example-3: The assumed parametric values of Example-3 are as follows: $a = 5 \times 10^6$, $b = 2.3$, $A_c = 150$, $h_c = 10$, $p = 18$, $i_p = 0.12$, $i_e = 0.10$, $e_p = 1$ year, $L_1 = 25/365$, $L_2 = 15/365$, $\delta = 0.6$ and $\alpha = 0.06$.
Example-4: The assumed parametric values of Example-4 are as follows: $a = 5 \times 10^6$, $b = 2.3$, $A_c = 130$, $h_c = 9$, $p = 10$, $i_p = 0.14$, $i_e = 0.11$, $e_p = 1$ year, $L_1 = 30/365$, $L_2 = 45/365$, $\delta = 0.5$ and $\alpha = 0.04$.

Table 1 Results of the present model for the Examples 1–5

Example	T_r^*	s^*	$TP^*(T_r^*, s^*)$	Result belongs to the
1	0.0746	27.93	26629.31	Situation-1.1
2	0.0620	22.96	33657.67	Situation-1.2
3	0.0684	29.52	19020.37	Situation-1.3
4	0.0532	19.15	44787.07	Situation-2.1
5	0.0392	14.71	62848.43	Situation-2.2

Example-5: The assumed parametric values of Example-5 are as follows: $a = 5 \times 10^6, b = 2.3, A_c = 80, h_c = 5, p = 8, i_p = 0.10, i_e = 0.08, e_p = 1$ year, $L_1 = 10/365, L_2 = 15/365, \delta = 0.6$ and $\alpha = 0.04$.

6.1 Parametric Study on L_1 and L_2

To obtain some managerial insights for the retailer, a parametric study with respect to retailer's credit period (L_1) and customers' credit period (L_2) is performed here. The results are obtained with the parametric values of the Example-1 and presented in Table 2.

Based on the results of Table 2, some managerial insights are found, which are listed below:

- From Table 2, it is noticed that for a fixed L_1, when customers' credit period increases then retail selling price increases. With the increase of selling price, demand decreases and for that retailer's order quantity decreases as well as retailer's annual profit decreases.
- Also from Table 2, it is noticed that for a fixed L_2, when retailer's credit period increases then retail selling price decreases and with less selling price, demand increases as well as the annual profit of the retailer increases.

6.2 Effect of Customers' Collateral Deposit (δ) on the Proposed Model

Here a study is made for the proposed model with respect to customers' collateral deposit (δ) with the same parametric values as taken in Example-1 except $L_1 = 45/365$ and $L_2 = 30/365$. Results are obtained and presented in Table-3, form which some managerial insights are outlined.

From Table 3, it is seen that if the customers' collateral deposit increases then the optimal selling price decreases. Since the selling price decreases so the base demand increases and for that retailer's annual profit increases, which lead to a realistic phenomenon.

6.3 A Sensitivity Analysis with Respect to Expiration Time (e_p) for an Item

Here, a sensitivity analysis is performed in relation to the item's expiration date (e_p) for the proposed model with the parametric values as taken in Example-1 except

Table 2 Parametric study on L_1 and L_2

L_1	L_2	T_r^*	s^*	Q_r^*	TP^*
10/365	10/365	0.0777	28.37	180.39	25,838.24
	15/365	0.0775	28.40	179.61	25,801.80
	30/365	0.0779	28.49	179.11	25,692.10
	45/365	0.0782	28.59	178.47	25,584.96
	60/365	0.0784	28.67	177.62	25,477.70
15/365	10/365	0.0800	28.35	182.36	25,882.46
	15/365	0.0775	28.35	180.23	25,870.27
	30/365	0.0776	28.43	179.38	25,760.97
	45/365	0.0781	28.53	179.06	25,652.46
	60/365	0.0782	28.62	178.02	25,544.73

Table 3 Effect of δ

δ	T_r^*	s^*	Q_r^*	TP^*
0.00	0.0788	28.99	174.26	24,417.25
0.10	0.0785	28.82	176.02	24,761.91
0.20	0.0781	28.63	177.60	25,109.57
0.30	0.0777	28.45	179.33	25,460.24
0.40	0.0772	28.28	180.55	25,813.93
0.50	0.0770	28.10	182.79	26,170.67
0.60	0.0767	27.94	184.42	26,530.43
0.70	0.0761	27.75	185.94	26,893.29
0.80	0.0758	27.58	187.78	27,259.20
0.90	0.0755	27.41	189.70	27,628.19
1.00	0.0751	27.25	191.31	28,000.28

$L_1 = 45/365$ and $L_2 = 30/365$. Table 3 shows the results that were obtained. From the results of Table 4, it is observed that if the expiration time of a product increases then the retailer's replenishment time (T_r) also increases and for that selling price (s) decreases. Due to the less selling price, base demand (D_m) increases as well as the retailer's profit increases. This observation lead to a natural phenomenon.

6.4 Fuzzy Model

Here five numerical examples are taken with different parametric values, results are attained and reported in Table 5.

Table 4 A sensitivity analysis with respect to e_p

e_p	T_r^*	s^*	Q_r^*	$TP^*(T_r^*, s^*)$
0.6	0.0728	28.16	172.52	25,995.36
0.8	0.0748	28.13	177.51	26,091.61
1.0	0.0770	28.10	182.79	26,170.67
1.2	0.0786	28.07	186.69	26,236.85
1.4	0.0802	28.05	190.75	26,293.04

Table 5 Results of the proposed model for the Examples 6–10

Example	T_r^*	s^*	$\widetilde{TP^*(T_r^*, s^*)}$	Result belongs to the
6	0.0746	27.95	(26,120.38, 26,553.05, 26,986.38)	Situation-1.1
7	0.0663	23.19	(32,260.59, 32,878.91, 33,647.96)	Situation-1.2
8	0.0923	29.84	(18,254.96, 18,943.24, 19,823.47)	Situation-1.3
9	0.0521	18.82	(43,541.20, 44,122.01, 44,702.08)	Situation-2.1
10	0.0410	14.45	(61,519.36, 62,187.33, 62,855.28)	Situation-2.2

Example-6: The following parametric values are considered for Example-6: $a = 5 \times 10^6$, $b = 2.3$, $\tilde{A}_c = (120, 122, 124)$, $h_c = 7$, $p = 15$, $\tilde{i}_p = (0.15, 0.16, 0.17)$, $\tilde{i}_e = (0.12, 0.13, 0.14)$, $e_p = 1$, $L_1 = 60/365$, $L_2 = 10/365$, $\delta = 0.5$, $\tilde{\alpha} = (0.05, 0.06, 0.07)$.
Example-7: The following parametric values are considered for Example-7: $a = 5 \times 10^6$, $b = 2.3$, $\tilde{A}_c = (100, 112, 114)$, $h_c = 6$, $p = 12$, $\tilde{i}_p = (0.10, 0.11, 0.12)$, $\tilde{i}_e = (0.08, 0.09, 0.10)$, $e_p = 1$, $L_1 = 45/365$, $L_2 = 30/365$, $\delta = 0.3$, $\tilde{\alpha} = (0.07, 0.08, 0.09)$.
Example-8: The following parametric values are considered for Example-8: $a = 5 \times 10^6$, $b = 2.3$, $\tilde{A}_c = (150, 152, 154)$, $h_c = 10$, $p = 18$, $\tilde{i}_p = (0.12, 0.13, 0.14)$, $\tilde{i}_e = (0.10, 0.11, 0.12)$, $e_p = 1$, $L_1 = 25/365$, $L_2 = 15/365$, $\delta = 0.6$, $\tilde{\alpha} = (0.06, 0.07, 0.08)$.
Example-9: The following parametric values are considered for Example-9: $a = 5 \times 10^6$, $b = 2.3$, $\tilde{A}_c = (130, 132, 134)$, $h_c = 9$, $p = 10$, $\tilde{i}_p = (0.14, 0.15, 0.16)$, $\tilde{i}_e = (0.11, 0.12, 0.13)$, $e_p = 1$, $L_1 = 30/365$, $L_2 = 45/365$, $\delta = 0.5$, $\tilde{\alpha} = (0.04, 0.05, 0.06)$.
Example-10: The following parametric values are considered for Example-10: $a = 5 \times 10^6$, $b = 2.3$, $\tilde{A}_c = (80, 82, 84)$, $h_c = 5$, $p = 8$, $\tilde{i}_p = (0.10, 0.11, 0.12)$, $\tilde{i}_e = (0.08, 0.09, 0.10)$, $e_p = 1$, $L_1 = 10/365$, $L_2 = 15/365$, $\delta = 0.6$, $\alpha = (0.04, 0.05, 0.06)$.

7 Conclusion

Here, a fuzzy EOQ model has been designed considering deteriorating product under two-level partial trade credit period incorporating the interest paid due to the deteriorated units of the item in the profit function. Moreover, in this study, the unfaithfulness nature of the customers is taken into account to decrease the default credit risk of the inventory system. To investigate the solution of this model a soft computing technique is developed and the results are taken for the numerical illustration of the model. Also from the numerical study some managerial insights are framed, which help a decision maker in drawing an effective business plan.

In this study, we have considered a model where certain percentage of the customer will not repay the credit amount to the retailer. But anyone can consider this as a function of the credit period and credit amount offered by the retailer. Another interesting extension of this model is possible under inflation, retailer's discount policy, with various forms of demand etc.

References

1. Goyal, S.K.: Economic order quantity under conditions of permissible delay in payment. J. Operat. Res. Soc. **36**, 335–338 (1985)
2. Aggarwal, S.P., Jaggi, C.K.: Ordering policies of deteriorating items under permissible delay in payments. J. Operat. Res. Soc. **46**, 658–662 (1995)
3. Ouyang, L.Y., Teng, J.T., Goyal, S.K., Yang, C.T.: An economic order quantity model for deteriorating items with partially permissible delay in payments to order quantity. Eur. J. Operat. Res. **194**, 418–431 (2009)
4. Pramanik, P., Maiti, M.K., Maiti, M.: An appropriate business strategy for a sale item. OPSEARCH. **85**(1), 85–106 (2018)
5. Chung, K.J.: Comments on the EOQ model under retailer partial trade credit policy in supply chain. Int. J. Prod. Econom. **114**, 308–312 (2008)
6. Chung, K.J.: The simplified solution procedures for the optimal replenishment decisions under two level trade credit policy depending on the order quantity in a supply chain system. Expert Syst. Appl. **38**, 13482–13486 (2011)
7. Chung, K.J.: The EPQ model under conditions of two levels of trade credit and limited storage capacity in supply chain management. Int. J. Syst. Sci. **44**, 1675–1691 (2013)
8. Chung, K.J., Cardenas-Barron, L.E., Ting, P.S.: An inventory model with non-instantaneous receipt and exponentially deteriorating items for an integrated three layer supply chain system under two levels of trade credit. Int. J. Prod. Econom. **155**, 310–317 (2014)
9. Guchhait, P., Maiti, M.K., Maiti, M.: Multi-item inventory model of breakable items with stock-dependent demand under stock and time dependent breakability rate. Comput. Ind. Eng. **59**, 911–920 (2010)
10. Huang, Y.F.: Optimal retailer's ordering policies in the EOQ model under trade credit financing. J. Operat. Res. Soc. **54**, 1011–1015 (2003)
11. Huang, Y.F.: An inventory model under two levels of trade credit and limited storage space derived without derivatives. Appl. Math. Modell. **30**, 418–436 (2006)
12. Kundu, A., Guchhait, P., Pramanik, P., Maiti, M.K., Maiti, M.: A production inventory model with price discounted fuzzy demand using an interval compared hybrid algorithm. Swarm Evolut. Comput. **34**, 1–17 (2016)

13. Maiti, M.K.: A fuzzy genetic algorithm with varying population size to solve an inventory model with credit-linked promotional demand in an imprecise planning horizon. Eur. J. Operat. Res. **213**, 96–106 (2011)
14. Pramanik, P., Maiti, M.K.: An inventory model with variable demand incorporating unfaithfulness of customers under two-level trade credit. Eur. J. Ind. Eng. **13**(4), 461–488 (2019)
15. Pramanik, P., Das, S.M., Maiti, M.K.: Note on: Supply chain inventory model for deteriorating items with maximum lifetime and partial trade credit to credit risk customers. J. Ind. Manage. Opt. **15**(3), 1289–1315 (2019)
16. Pramanik, P., Maiti, M.K.: Trade credit policy of an inventory model with imprecise variable demand: an ABC-GA approach. Soft Comput. **24**(11), 1–18 (2020)
17. Teng, J.T.: Optimal ordering policies for a retailer who offers distinct trade credits to its good and bad customers. Int. J. Prod. Econom. **119**, 415–423 (2009)
18. Teng, J.T., Chang, C.T.: Optimal manufacture's replenishment policies in the EPQ model under two level trade credit policy. Eur. J. Operat. Res. **195**, 358–363 (2009)
19. Teng, J.T., Cheng, C.T., Chern, M.S., Chan, Y.L.: Retailer's optimal ordering policies with trade credit financing. Int. J. Syst. Sci. **38**, 269–278 (2007)
20. Teng, J.T., Yang, H.L., Chern, M.S.: An inventory model for increasing demand under two levels of trade credit linked to order quantity. Appl. Math. Modell. **37**, 7624–7632 (2013)
21. Guchhait, P., Maiti, M.K., Maiti, M.: Two storage inventory model of a deteriorating item with variable demand under partial credit period. Appl. Soft Comput. **13**, 428–448 (2013)
22. Mahata, G.C.: Partial trade credit policy of retailer in economic order quantity models for deteriorating items with expiration dates and price sensitive demand. J. Math. Modelling Algor. Operat. Res. **14**(4), 363–392 (2015)
23. Mahata, G.C., De, S.K.: Supply chain inventory model for deteriorating items with maximum lifetime and partial trade credit to credit risk customers. Int. J. Manage. Sci. Eng. Manage. **12**(1), 21–32 (2016)
24. Pramanik, P., Maiti, M.K., Maiti, M.: A supply chain with variable demand under three level trade credit policy. Comput. Ind. Eng. **106**, 205–221 (2017)
25. Pramanik, P., Maiti, M.K., Maiti, M.: Three level partial trade credit with promotional cost sharing. Appl. Soft Comput. **58**, 553–575 (2017)
26. Pramanik, P., Maiti, M.K.: An inventory model for deteriorating items with inflation induced variable demand under two level partial trade credit: a hybrid ABC-GA approach. Eng. Appl. Artif. Intell. **85**, 194–207 (2019)
27. Ghare, P.M., Schrader, G.H.: A model for exponentially decaying inventory system. Int. J. Prod. Res. **21**, 449–460 (1963)
28. Chung, K.J., Huang, T.S.: The optimal retailer's ordering policies for deteriorating items with limited storage capacity under trade credit financing. Int. J. Prod. Econom. **106**, 127–145 (2007)
29. Guchhait, P., Maiti, M.K., Maiti, M.: Inventory model of a deteriorating item with price and credit linked fuzzy demand?: a fuzzy differential equation approach. OPSEARCH. **51**, 321–353 (2014)
30. Min, J., Zhou, Y.W., Zhao, J.: An inventory model for deteriorating items under stock dependent demand and two-level trade credit. App. Math. Modelling. **34**, 3273–3285 (2010)
31. Chao, X., Gong, X., Zheng, S.: Optimal pricing and inventory policies with reliable and random-yield suppliers: characterization and comparison. Ann Operat. Res. **241**(1), 35–51 (2016)
32. Das, R.L., Jana, R.K.: Some studies on EPQ model of substitutable products under imprecise environment, In: Shah N, Mittal M (eds) Optimization and Inventory Management. Asset Analytics (Performance and Safety Management). Springer, Singapore (2020). https://doi.org/10.1007/978-981-13-9698-4_18
33. Dye, C.Y.: Joint pricing and ordering policy for a deteriorating inventory with partial backlogging. Omega. **35**, 184–189 (2007)
34. Maiti, M.K., Maiti, M.: Two-storage inventory model with lot-size dependent fuzzy lead-time under possibility constraints via genetic algorithm. Eur J Operat Res **179**, 352–371 (2007)
35. You, P.S.: Inventory policy for product with price and time-dependent demands. J. Operat. Res. Soc. **56**, 870–873 (2005)

36. Zadeh, L.A.: Fuzzy sets. Inf. Cont. **8**, 338–356 (1965)
37. Das, R.L., Jana, R.K., De, M., Maiti, M.: Prices and order quantities of substitutable products in an EPQ model over different uncertain finite budget constraints. J Intell Fuzzy Syst **33**(1), 229–244 (2017)
38. Engelbrecht, A.P.: Fundamentals of Computational Swarm Intelligence. Wiley (2005)

Comparison of Herd Tracking Systems Using Fuzzy Logic-Based Multi-Criteria Decision Making Methods

Merve Kılınç Yılmaz and Nazan Kemaloğlu Alagöz

Abstract Minimizing the manpower in production, mechanization and automation has made its impact felt in the field of agriculture and animal husbandry in recent years. In the livestock sector, information on the health, production and reproduction of animals is automatically collected and interpreted to provide guiding information to the operators, and the software written specifically in this direction is rapidly increasing. In dairy and beef cattle, this type of software is called "herd tracking system". Although there are many software used in the market, a small number of herd tracking systems, which are advantageous with the services they provide or the hardware they have, and which are preferred more intensely, constitute an oligopoly market. The power of being preferred contributes to R&D and investment budgets, and this increases the preferability of software. The criteria that users consider when choosing the program: It varies according to the business structure (dairy/fattening focused) and subjective requests. The criteria accepted as basic are not subject to an objective measurement system. At this point, since the comparison of the alternatives according to the criteria is made based on the linguistic expressions of the users, the ranking process has a fuzzy multi-criteria decision-making problem structure. In this study, four different herd tracking software were ranked according to seven different criteria. The CRITIC method was used to weigh the criteria, and the fuzzy ARAS and fuzzy TOPSIS methods were used to rank the alternatives. As a result of the findings obtained from the analyses, the four major herd tracking software were ranked according to seven different criteria.

Keywords Herd tracking systems · Fuzzy logic · Multi-criteria decision making

M. K. Yılmaz (✉) · N. K. Alagöz
Digital Technologies in Livestock Sector Joint Application and Research Center, Burdur Mehmet Akif Ersoy University, Istiklal Campus, Burdur 15030, Turkey
e-mail: mervekilinc@mehmetakif.edu.tr

N. K. Alagöz
e-mail: nkemaloglu@mehmetakif.edu.tr

L. Sahoo et al. (eds.), *Real Life Applications of Multiple Criteria Decision Making Techniques in Fuzzy Domain*, Studies in Fuzziness and Soft Computing 420,
https://doi.org/10.1007/978-981-19-4929-6_22

1 Introduction

The fuzzy TOPSIS method has found application in many different fields. Considering the studies in different fields in the literature, Boran et al. [8], Büyüközkan and Çiftçi [9], Govindan et al. [22] in the field of supply chain, Amiri [2] and Şengül et al. [44] in the field of energy, Zyoud et al. [58], He et al. [24] in management and [45] and [53] used the fuzzy TOPSIS method in the field of business. The use of the method in different fields and in different fuzzy forms continues today.

Fuzzy ARAS method has been applied in many different fields such as equivalent multi-criteria decision-making methods. Fu et al. [21], Ulutaş [59] and Rostamzadeh et al. [42] in supply chain management, Karagöz et al. [32] and Dahoole et al. [12] in the field of energy and Ghadikolael et al. [43] and Ecer [16] are some of the examples used in the field of finance [4]. The purpose of their technologies used in the livestock sector is real-time monitoring of animals to improve the "farmer's eyes and ears" [4, 17].

Farmers considering expanding their businesses need to provide best practices in herd management, operations, finance, human resources and strategic management [23]. As dairy farms continue to increase in size, monitoring and managing cows becomes more challenging and complex. [7, 18]. The use of automation and sensor systems, commonly referred to as precision technology, provides growers with the tools to increasingly reduce labor requirements and improve the management of large herds [6, 14, 15].

The purpose of herd management systems is to use technologies to measure physiological, behavioral and production indicators on individual animals to improve management strategies and farm performance. Many herd management technologies are currently used by dairy producers, including daily milk yield recording, milk component monitoring, pedometers, automatic temperature recorders, milk conductivity indicators, automatic heat detection monitors and daily body weight measurements [37, 46].

This type of software is often of vital importance for businesses whose assets are alive, and for this reason, the companies that are the developers of these software make large investments and R&D studies. In the sector where there is great competition, materials that will collect data for the software are often developed together with the software and offered for sale as a package with the software. There are two advantages of selling hardware as a package with software. The first of these is the creation of a new income item, and the second is the acquisition of intermediate devices that will communicate with the software and collect data and store it in the database. The obtained data is used to convert the information provided and calculated by the software into a product. In addition to the user's ability to easily access the desired screen, it should present important and critical information, be functional, and beyond that the cost should be acceptable. Most of the software that is used intensively in the sector closely follows the services and products offered by other software in order not to lose its chance to compete.

Since the sector requires technology, studies in this field generally progress to strengthen the technological infrastructure. When we look at the literature, there is no study in which such programs are evaluated in terms of users. On the other hand, there are studies that evaluate the technical features of the programs, albeit limited. Berger and Hovav made an analysis on the farm data and efficiency as a result of using modules on the AfiFarm herd technology system. Indıra and Suresh in 2020 conducted a study on SCR Swarm technology, describing the use of the main modules of the software. Çakıcı in 2006 conducted a research on the modules included in the software of the herd management systems of AfiFarm, GAE and Delaval companies. When the literature is examined, it has been seen that only studies based on program features have been made in the field of herd management systems and software. In the literature, there is a lack of a study in which comparative analyses of these systems are made with each other.

In this study, a comparative quantitative analysis was made using a total of 7 criteria under 4 main headings of herd management software. Thus, the existing systems have been tried to be evaluated from the perspective of the user, not from the perspective of the manufacturer. In the first part, information about herd management software is given and information is given about the 4 most used herd management software in the market. In the second chapter, the methods used and the dataset are introduced. The analyses made in the third section are explained, and the results obtained in the fourth section are interpreted.

2 Herd Tracking Systems

Herd tracking system is the use of technologies to measure physiological, behavioral and production indicators on individual animals to improve management strategies and farm performance Lokhorst et al. [37]. The main purposes in the dairy cattle industry are the growth of herds, increase in productivity, intensification of competition in the dairy market and increase in labor costs [6]. Another development that concerns the sector is the increase in consumer expectations [48]. Since computers are at the center of the main advanced technology elements used within the scope of sensitive herd management applications, these applications are generally gathered under the concept of "computer assisted herd tracking systems". The main advanced technology elements within the scope of herd tracking system applications are

- Electronic animal identification systems
- Automatic milk measurement systems integrated into the milking system (milk amount, milking time, milk flow rate, electrical conductivity of milk, milk temperature)
- Automatic animal weighing system
- Activity meters
- Automatic dense feed units (monitoring and measurement of dense feed consumption)

- Feeder systems measuring roughage consumption
- Coarse-dense feed mixers and dispensers with electronic scales
- Image analysis systems
- Herd tracking system and Internet connections (breeder's association, milk yield and quality control organization, genetic evaluation center) stand out [13, 20, 33, 52].

Computer-aided herd tracking systems are to maximize individual animal potential and to minimize the use of drugs through early detection of disease and preventive health measures [6]. Computer-aided herd management systems frequently make use of informatics, biostatistics, ethology, economics, animal husbandry, animal nutrition and engineering. It is an interdisciplinary field that includes the concepts of [6, 46]. The main elements of herd tracking system are given in Fig. 1.

In computerized herd tracking systems, all equipment that collects data about animals (such as electronic animal recognition, mobility measurement, milk meter, animal weighing) are connected to an expert software [52]. All values (milk yield, milking time, milk flow rate, milk temperature, electrical conductivity of milk, activity, live weight, feed consumption, etc.) related to the measurements made in the parlor or other automatic detection units are recorded by the software to the

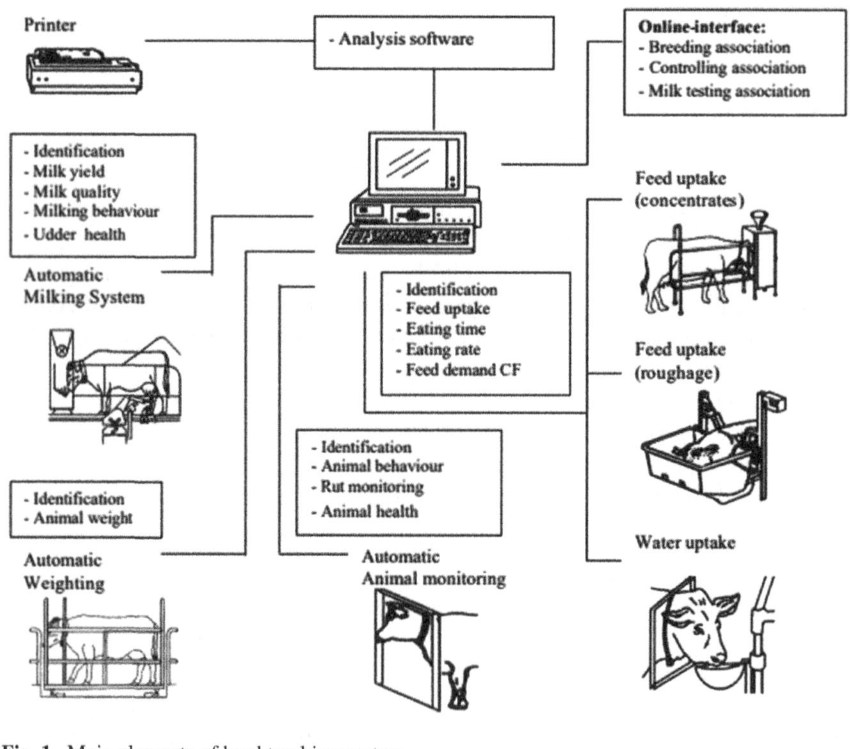

Fig. 1 Main elements of herd tracking system

individual data of the animals concerned [49]. Herd tracking system is particularly important for larger herds where the benefits from data summary reporting are more difficult to monitor individual animals [36]. Today, there are automation and software developed by large companies that produce and develop milking systems and milk meters, as well as various software companies. The main herd tracking systems that are widely used in the world and in Turkey are as follows [39]:

- Alpro (DeLaval company, Sweden),
- AfiFarm (SAE Afikim company, Israel),
- DairyPlan (GEA, WestfaliaSurge company, Germany),
- Data Flow (SCR company, Israel).

Alpro

DeLaval is a dairy and agricultural machinery manufacturer with a head office in Tumba, Sweden, and is part of the Tetra Laval group. The company has 18 factories worldwide, employing more than 4500 people (Delaval.com). In Fig. 2, the standard view of the Alpro software is given.

AfiFarm

Founded in 1977, afimilk Ltd. (S.A.E. Afikim) is one of the world's leading companies in the development, production and marketing of advanced computerized systems for modern dairy farm and herd management. More than 30 years ago, Afimilk launched the world's first electronic milk meter, the first pedometer to monitor cows, the first dairy farm management software and the first online milk

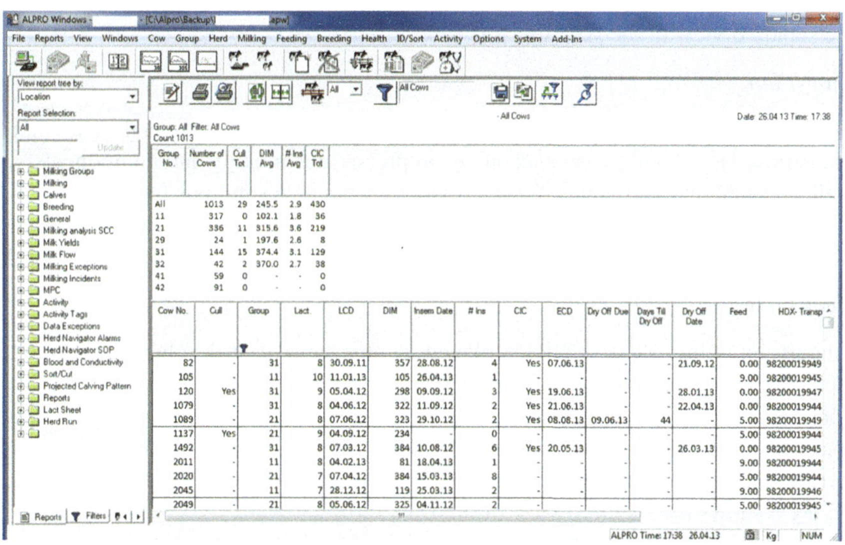

Fig. 2 Standard view of the Alpro software

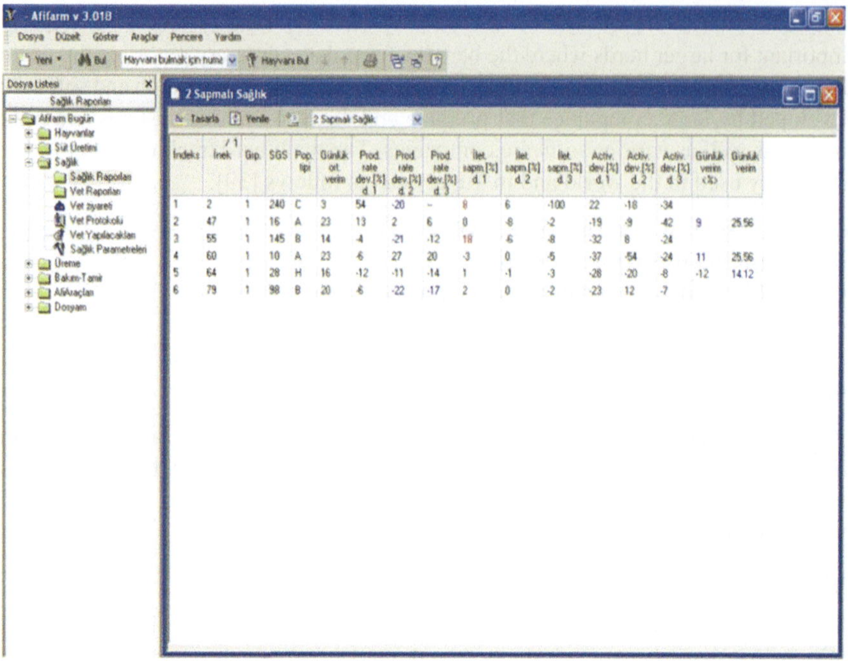

Fig. 3 Standard view of the AfiFarm software

analyzer (AfiLab). The company operates in more than 50 countries, and user interfaces have been translated to over 20 languages (Afimilk.com; [5]). Figure 3 shows the standard view of the AfiFarm software.

DairyPlan

GEA is one of the largest technology suppliers in food processing and many other industries. The global group specializes in process technology and components, as well as machinery and plants. GEA provides sustainable solutions for sophisticated production processes in various end-user markets and offers a comprehensive service portfolio (GEA.com). The standard view of the DairyPlan software is given in Fig. 4.

Data Flow

Since its founding in 1976, SCR has played an important role in developing innovative technological solutions for the global dairy industry. During the first 20 years of operation, the company focused on the development of electromechanical devices for dairy farms. In the last 10 years, SCR has gained a strong reputation among farmers and has become one of the few companies in the sector (ru.scrdairy.com). Figure 5 shows the standard view of the Data Flow software.

When herd tracking system is examined in general, it consist of sections such as personal reports design, graphic formats of data, veterinary works (diseases, drugs, treatment, comments, etc.), warning lists (alarm lists, automatic listing of animals

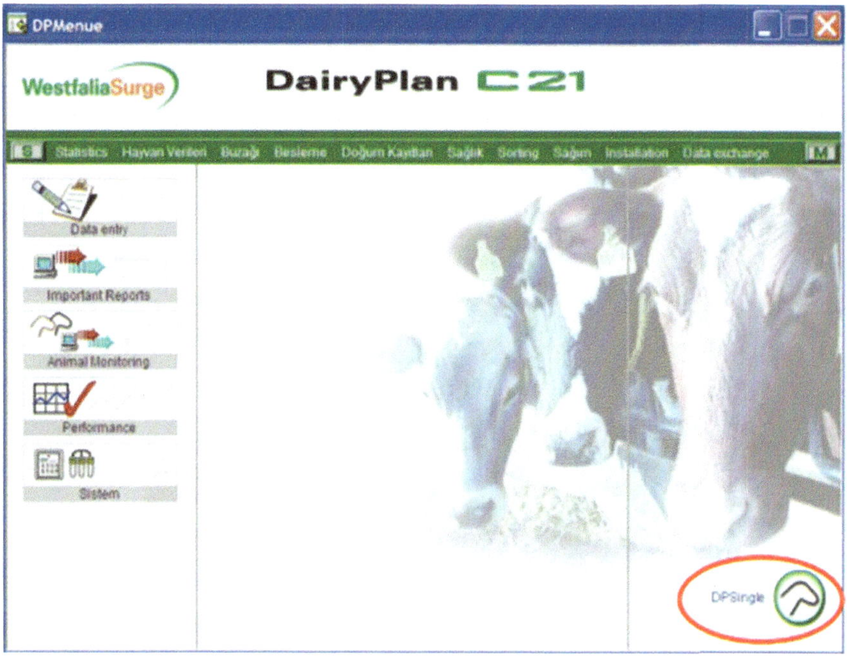

Fig. 4 Standard view of the DairyPlan software

that deviate from the determined parameters), parameter entries and milking. It is seen that important modules in terms of herd tracking such as breeding, activity, health, automatic animal separation, automatic weight measurement and feeding information menus are available in all software.

3 Material and Method

In this study, it is aimed to make a comparison based on subjective judgments. For now, it does not seem possible for users to express their satisfaction level about the software numerically. For example, the evaluation of ease of use with a quantitative scale or the fact that the appreciation of the technical infrastructure is directly related to the expectations prevents this study from being carried out through objective data. When the research purpose, subject and question are examined, many alternatives should be evaluated by more than one decision-maker using many criteria. The study has multi-criteria decision-making characteristics up to this point. In addition, the fact that the subjective judgments of the users will be used as input shows that the fuzzy logic process should be applied due to the variability that may arise from linguistic expressions.

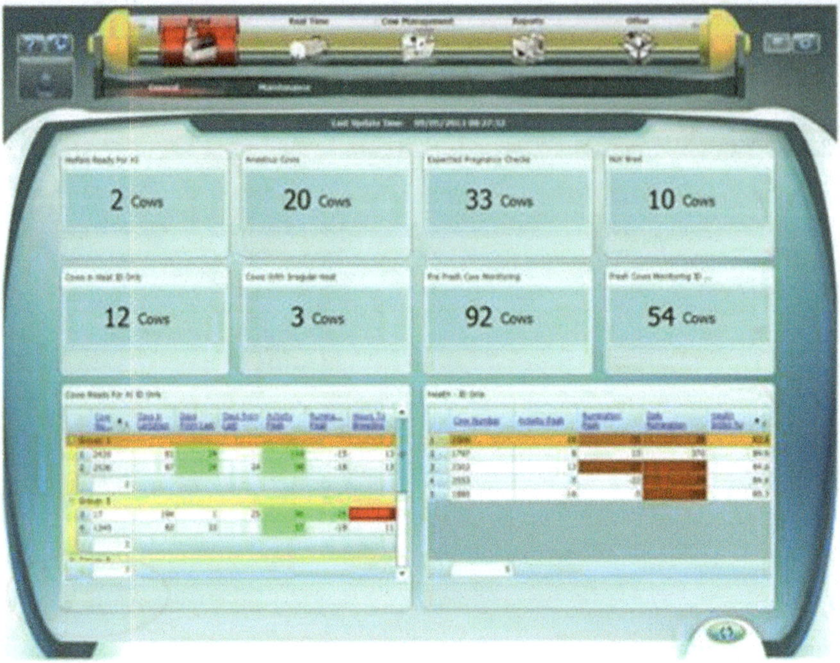

Fig. 5 Standard view of the data flow software

In this study, fuzzy multi-criteria decision making and fuzzy multi-criteria decision-making methods have been applied due to the reasons stated. Fuzzy datasets and the methods used are briefly mentioned below.

3.1 Fuzzy Logic and the Concept of Fuzzy Set

The basic principle of positive sciences is to reach more general inferences by using precise and concrete information. It is much easier to stay within the framework of this principle in the sciences than in the social sciences. Because the data of researchers usually contain objective information. However, the distinction between objectivity and subjectivity in social sciences is not as clear-cut as in science. For this reason, some of the analysis methods arising from positive sciences do not give satisfactory results in research in social sciences. Likes, tastes and preferences violate the unshakeable rules of objectivity.

Expression of personal feelings and thoughts due to linguistic differences may contain differences even if the same scoring is done. The word "many" differs from person to person unless it expresses a concrete size. "Satisfaction", "preference", "ranking scores" also do not always mean the same, as they do not express a concrete

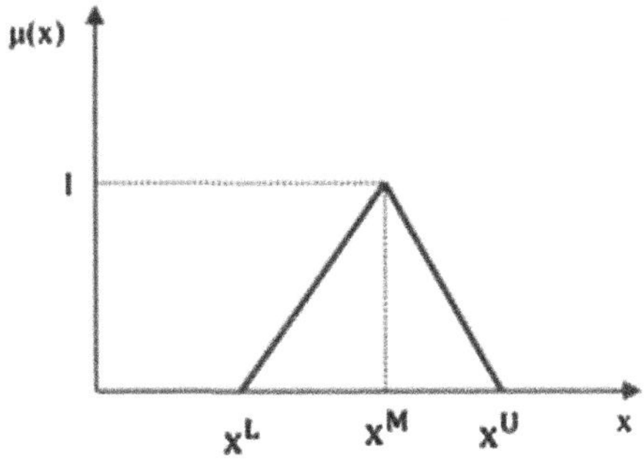

Fig. 6 Triangle membership function

size. In this case, talking about a common multiplicity or satisfaction in studies investigating subjects such as "many", "satisfaction", "preference" and "ranking scores" means presenting biased results. At this point, fuzzy logic and fuzzy sets come into play and give satisfactory results to researchers.

Fuzzy sets were introduced to the literature by Zadeh [57]. Until then, binary or Aristotelian logic was used to solve problems. Aristotelian logic deals with events like true–false, yes–no, black–white, etc. approaches on a dual basis. It is not allowed between these two values as it is not certain. In fuzzy sets, it is possible to talk about different membership degrees varying between 0 and 1 [35, 47]. Therefore, it has been possible to express the values that are not precise in the language of mathematics. If the set A is a fuzzy set, it is denoted as Ã. Fuzzy set theory, which is a kind of multi-valued set theory, is the formulation of uncertainty [3]. It represents the degree of belonging of the element x to the set. The fact that nxu()x approaches 1 indicates that the element belongs to the cluster with a higher membership degree [40, 57]. The membership function can be defined in different ways depending on the characteristics of the problem. The definition of triangular membership functions used in this study is as follows [3] (Fig. 6).

$$\tilde{\mu}(x; x^L, x^M, x^U) = \begin{cases} 0, x \le x^L, x^U, \le x \\ \dfrac{x - x^L}{x^M - x^L}, x^L < x \le x^M \\ \dfrac{x^U - x}{x^U - x^M}, x^M \le x < x^U \end{cases} \tag{1}$$

3.2 Fuzzy TOPSIS

Multi-criteria decision-making methods are frequently used in complex decision-making methods. The TOPSIS method, which is under the title of multi-criteria decision-making methods and is used for the solution of various decision-making problems today, was used for the first time in 1981 [60].

This method tries to determine the positive ideal solution (PIS) and negative ideal solution (NIS) points. In the positive ideal solution, the aim is to minimize the cost and maximize profit, while in the negative solution point, the cost is kept at the maximum and the profit is minimized. However, the TOPSIS method assumes that each criterion has a uniformly increasing or decreasing utility trend. Therefore, it is easy to define ideal and negative ideal solutions [54], (Alpay 2003). Due to the conflict between criteria, it is often not possible to reach the ideal solution. For this reason, "compromise solution" is mentioned in MCDM. In some of the MCDM methods, the "Compromising Model" is used, which aims to reach a solution that is as close to the ideal as possible. One of these methods is "The Technique of Reaching the Order of Preference by Similarity to the Ideal Solution", which is short for TOPSIS (Technique for Order Preference by Similarity to Ideal Solution). The basis of this method is the selection of the alternative with the highest relative closeness to the ideal solution [30, 55].

The fuzzy TOPSIS method is proposed where the weights of criteria and ratings of alternatives are evaluated by linguistic variables represented by fuzzy numbers to deal with the deficiency in the traditional TOPSIS [19]. In such situations, instead of exact numbers, OFNs can be used [31]. The fuzzy TOPSIS method proposed by Chen et al. [10] was used to select the herd tracking system, which is the subject of our study. The steps of the method are as follows. Step 1: In the first step of this method, a jury consisting of expert decision-makers (DM) is formed. It is expressed as a set of N decision-makers. After the jury is formed, the available alternatives (A) and the criteria(C) to be used in evaluating these alternatives are determined. Following this, verbal variables to be used in the evaluation of alternatives and determining the importance weights of the criteria are selected, and the decision-makers evaluate the alternatives and criteria with the help of these verbal variables. The expression of these evaluations in the form of fuzzy numbers is as in Tables 1 and 2 [10]:

Step 2. Criterion values and criterion weights are calculated: Using the formulas below, criterion values are determined using equality set (2) and criterion weights using equality set (3).

Table 1 Linguistic expressions used in determining the criterion weights

Very high (VH)	(0.8, 1, 1)
High (H)	(0.7, 0.8, 0.9)
Little high (LH)	(0.5, 0.65, 0.8)
Medium (M)	(0.4, 0.5, 0.6)
Little low (LL)	(0.2, 0.35, 0.5)
Low (L)	(0.1, 0.2, 0.3)
Very low (VL)	(0, 0, 0.2)

Table 2 Linguistic expressions used in evaluating alternatives

Very good (VG)	(8, 10, 10)
Good (G)	(7, 8, 9)
Some good (SG)	(8, 6.5, 8)
Medium (M)	(4, 5, 6)
Little bad (LB)	(2, 3.5, 5)
Bad (B)	(1, 2, 3)
Very bad (VB)	(0, 0, 2)

$$\begin{aligned} &\tilde{x}_{ij}(a_j, b_j, c_j) \\ &a_{ij} = \min_k\{a_{ijk}\} \\ &b_{ij} = \frac{1}{N}\sum_{k=1}^{N} b_{ijk} \\ &c_{ij} = \max_k\{c_{ijk}\} \end{aligned} \tag{2}$$

$$\begin{aligned} &\tilde{w}_j = (w_{j1}, w_{j2}, w_{j3}) \\ &w_{j1} = \min_k\{w_{jk1}\} \\ &w_{j2} = \frac{1}{N}\sum_{k=1}^{N} w_{jk2} \\ &w_{j3} = \max_k\{w_{jk3}\} \end{aligned} \tag{3}$$

Step 3. Create a fuzzy decision matrix: It shows the fuzzy decision matrix, the fuzzy weight matrix. These matrices are:

$$\tilde{D} = \begin{array}{c|cccc} & K_1 & K_2 & \dots & K_n \\ \hline A_1 & \tilde{x}_{11} & \tilde{x}_{12} & \dots & \tilde{x}_{1n} \\ A_2 & \tilde{x}_{21} & \tilde{x}_{22} & \dots & \tilde{x}_{2n} \\ \dots & \dots & \dots & \dots & \dots \\ A_m & \tilde{x}_{m1} & \tilde{x}_{m2} & \dots & \tilde{x}_{mn} \end{array} \tag{4}$$

$$\tilde{W} = [\tilde{w}_1, \tilde{w}_2, \tilde{w}_3] \tag{5}$$

Step 4. Generating the normalized fuzzy decision matrix: While constructing the normalized fuzzy matrix shown as: Benefit (6) and cost (7) are used.

$$\tilde{r}_j = \left(\frac{a_j}{c_j^*}, \frac{b_j}{c_j^*}, \frac{c_j}{c_j^*}\right), c_j^* = \max_i c_j, \ j \in B \tag{6}$$

$$\tilde{r}_j = \left(\frac{a_j^-}{c_j}, \frac{a_j^-}{b_j}, \frac{a_j^-}{a_j}\right), a_j^- = \min_i a_j, j \in C \tag{7}$$

Step 5. Generating the weighted normalized fuzzy decision matrix: The weighted matrix, shown as, is obtained after the scalar multiplication of the fuzzy weight coefficients found before and the normalized decision matrix.

$$\tilde{V} = \tilde{R} * \tilde{W} \tag{8}$$

Step 6. Finding the positive fuzzy ideal solution (FPIS, A*) and the negative fuzzy ideal solution (FNIS, A-):

$$A^* = (\tilde{v}_1^*, \tilde{v}_2^*, \ldots, \tilde{v}_n^*), \tilde{v}_j^* \max_i(v_j) \tag{9}$$

$$A^- = (\tilde{v}_1^-, \tilde{v}_2^-, \ldots, \tilde{v}_n^-), \tilde{v}_j^- \min_i(v_j) \tag{10}$$

Step 7. Calculating the distances of alternatives from FPIS and FNIS: The distances of each alternative from the positive ideal solution and the negative ideal solution are found as a sum and used as input in the next step to calculate the Euclidean distances.

$$d_i^* = \sum_{j=1}^{n} d_v(\tilde{v}_j, \tilde{v}_j^*), i = 1, 2, \ldots, m \tag{11}$$

$$d_i^- = \sum_{j=1}^{n} d_v(\tilde{v}_j, \tilde{v}_j^-), i = 1, 2, \ldots, m \tag{12}$$

Step 8. Calculation of closeness coefficient: The closeness coefficient is obtained by using negative and positive ideal solutions. Alternatives are ranked according to the coefficient obtained.

$$C_i = \frac{d^-}{d^- + d^*}, i = 1, 2, \ldots, m \tag{13}$$

3.3 *Fuzzy ARAS*

The classical ARAS method is a MCDM method proposed by Zavadskas and Turskis [50]. The basis of the method is based on the argument that complex problems can be solved using simple relative comparisons. The B-ARAS method, on the other hand, is a method proposed by Turskis and Zavadskas [50] based on comparing each alternative with a theoretical ideal value, as in the classical ARAS method [34, 41]. The steps of the fuzzy ARAS method are as follows [50].

Step 1. Creation of fuzzy decision matrix: The first stage is fuzzy decision-making matrix (FDMM) forming. In the FMCDM of the discrete optimization problem any problem which has to be solved is represented by the following DMM of preferences for m reasonable alternatives (rows) rated on n criteria (column).

$$\tilde{X} = \begin{bmatrix} \tilde{x}_{01} & \tilde{x}_{02} & \dots & \tilde{x}_{0n} \\ \tilde{x}_{11} & \tilde{x}_{12} & \dots & \tilde{x}_{1n} \\ \dots & \dots & \dots & \dots \\ \tilde{x}_{m1} & \tilde{x}_{m2} & \dots & \tilde{x}_{mn} \end{bmatrix}, i = 0, j = 1, \ldots, n \quad (14)$$

$\tilde{x}_{0j}$ shown here denotes the optimal value. If the optimal value is not known, the optimal values can be determined as follows:

$$\tilde{x}_{0j} = \max_i \tilde{x}_{ij},\ if \max_i \tilde{x}_{ij} \text{ is preferable} \quad (15)$$

$$\tilde{x}_{0j} = \min_i \tilde{x}_{ij},\ if \max_i \tilde{x}_{ij} \text{ is preferable} \quad (16)$$

Step 2. Performing the normalization process:

$$\bar{\tilde{x}}_{ij} = \frac{\tilde{x}_{ij}}{\sum_{i=1}^{m} \tilde{x}_{ij}} \quad (17)$$

If the optimum value of the criterion is the minimum value rather than the maximum, the following correction is made.

$$\tilde{x}_{ij} = \frac{1}{\tilde{x}_{ij}^{*}},\ \bar{\tilde{x}}_{ij} = \frac{\tilde{x}_{ij}}{\sum_{i=1}^{m} \tilde{x}_{ij}} \quad (18)$$

Step 3. Finding the weighted and normalized fuzzy decision matrix: It is represented by the weighted and normalized matrix. The result is obtained by multiplying the weights with the vectorial matrix with the normalized matrix.

$$\hat{\tilde{x}} = \bar{\tilde{x}}_{ij} * \tilde{w}_j,\ i = 0, \ldots, m,\ j = 1, \ldots, n \quad (19)$$

Step 4. Calculate optimality function values: It is the optimal function value of the alternative.

$$\tilde{S}_i = \sum_{j=1}^{n} \hat{\tilde{x}}_{ij},\ i = 0, \ldots, m \quad (20)$$

Step 5. Calculate the optimal function values of the alternatives: This is the defuzzification phase.

$$S_i = \frac{1}{3}(S_{i\alpha}, S_{i\beta}, S_{i\gamma}) \tag{21}$$

Step 6. Calculating the utility degrees of the alternatives and ranking.

$$K_i = \frac{S_i}{S_0} \tag{22}$$

3.4 Fuzzy CRITIC Method

The CRITIC method was first developed by diakoulaki mavrotas and papayannakis (1995). CRITIC approach concludes that the conflict that characterizes a decision situation can be well resolved. Objective weights are measured using standard statistical measures. These actions represent two fundamental principles of MCDM in mathematical terms. The contrasting intensity of performance of the alternative criteria in each criterion and the conflict between other criteria for evaluation ([63]:112).

Usage frequencies of CRITIC method with respect to the years: After 2017, it has 39 become more popular than ever. The leading countries using CRITIC method is China, USA, Japan, Iran and 43 Netherlands, respectively. The most used subject areas are computers science, engineering and 47 mathematics ([63]:1). Although there are many problems related to the CRITIC method from various fields, the method has begun to be adapted to fuzzy logic in recent years. Kahraman, etc. [61] integrated the method with spherical fuzzy numbers and integrated the method with spherical fuzzy numbers and calculated CRITIC weights for the first time with spherical numbers. Igrawve and Babatubde (2018) used the CRITIC method adapted to type two fuzzy sets to make decisions in macroeconomics. Ali [62] has adapted the spherical fuzzy numbers to the CRITIC method and weighted the criteria to be used in phone selection.

The steps of the fuzzy critic method are as follows ([61]:113–115).

Step 1. Construct of fuzzy decision-making matrix:

Step 2. Construct of average fuzzy decision-making matrix:

$$Y = \left[\tilde{y}_{ij}\right]_{nxm} \tag{23}$$

where

$$\tilde{y}_{ij} = \frac{1}{k}\sum_{p=1}^{k} \tilde{y}_{ij}^{p} \tag{24}$$

Concerning to criterion $c_j(1 \leq j \leq m)$ assigned by the pth decision-maker $(1 \leq p \leq k)$ and the performance value of alternative $D_i(1 \leq i \leq n)$ that is denoted by $\tilde{y}_{ij}^{p}$.

Step 3. Calculate the transformation of performance values, which are used in obtaining the criteria vectors:

$$A_{ijk}^{T} = \left\{ \frac{A_{ijk} - A_{jk}^{-}}{A_{jk}^{*} - A_{jk}^{-}} \, if \, j \in M, \quad A_{ijk}^{T} = \left\{ \frac{A_{jk}^{-} - A_{ijk}}{A_{jk}^{-} - A_{jk}^{*}} \, if \, j \in N, \right. \right.$$

$$A_{jk} = (A_{1jk}^{T} x A_{2jk}^{T} x \ldots x A_{njk}^{T}) \tag{25}$$

A_{jk} denotes the kth vector of jth criterion and A_{jk}^{*} and A_{jk}^{-} are the ideal and anti-ideal values concerning the jth criterion and kth element of $\tilde{A}_{ij}$.

Step 4. Calculate the standard deviation (σ_{jk}) of each vector (A_{jk})

Step 5. Construct four symmetric matrices, with dimension m x m and generic elements $s_{jj'}^{k}$ (j' = 1, 2,…, m and k = 1, 2, 3). The elements of this matrix are the linear correlation coefficients between the vectors A_{jk} and $A_{j'k}$.

Step 6. Calculate the information measures of each criterion as follows:

$$G_{jk} = \sigma_{jk} \sum_{j=1}^{m} (1 - s_{jj'}^{k}) \tag{26}$$

Step 7. Determine the unsorted objective weights

$$w'_{jk} = \frac{G_{jk}}{\sum_{j'=1}^{m} (G_{jj'})} \tag{27}$$

3.5 *Critaeria Used In The Study*

For this study, a total of 7 different evaluation criteria under 3 headings were discussed. These criteria are given in Fig. 7.

Three main headings have been determined for herd management systems and the software included in these systems. These can be listed as the ease of use of the software, sales and after-sales processes and technical details.

Ease of Use

It is expected that the herd management software used in the livestock industry will provide convenience to users in terms of workload and time management, based on the industry's workforce. For this reason, it is expected that the software will be functional in order to contain the data recording modules required for herd management

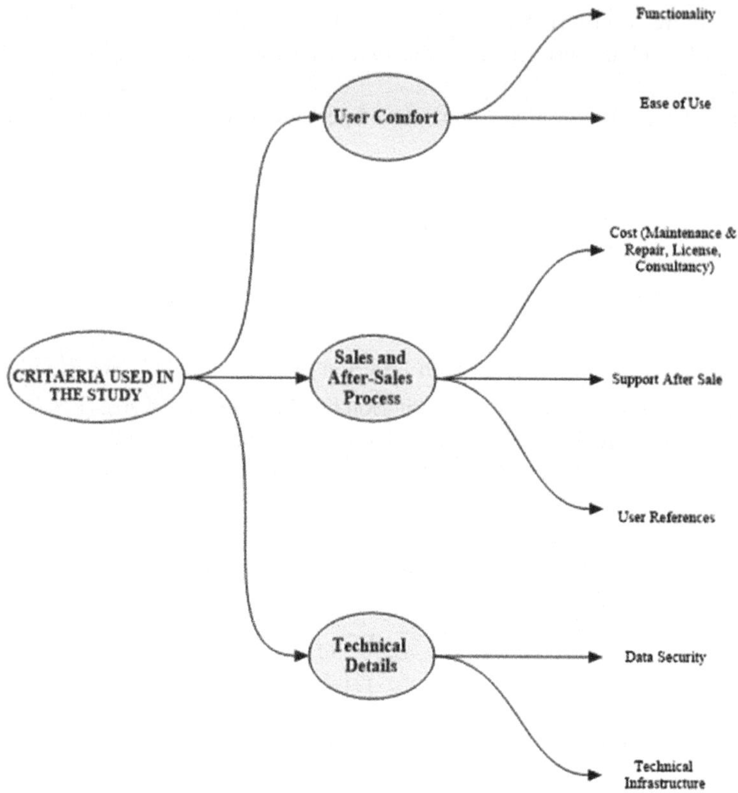

Fig. 7 Criteria used in the study

and to provide quick and easy to understand feedback to the user, and it is expected to consist of screens and modules that are easy to understand and apply, which will not increase the daily workload of the users.

Sales and After-Sales Support

Herd management systems are used in our country by importing them from abroad. Therefore, the operating cost of these systems for the sector, the license term criteria, providing the necessary support to the after-sales users and positive references from other users using the system are among the sought-after features of herd management software.

Technical details

It is expected that the equipment used in herd management systems should be suitable for the business infrastructure and work in accordance with the existing physical conditions of the business. At the same time, in herd management software, the data recorded in the system should not be shared with third parties, should not be opened to the outside without user control and should be recorded correctly.

During the data collection phase, user opinions were collected at the Burdur/TURKEY location. Business owners operating in the specified location and veterinarians serving businesses are the users of these software. Four business owners and 2 veterinarians were asked to evaluate 4 programs according to 7 criteria. The obtained evaluation results were converted into fuzzy numbers (triangular fuzzy numbers), the above-described methods were applied, and the results were obtained. The findings are detailed below.

4 Results

The fuzzy decision matrix with the criteria ($1 \leq j \leq m$) decision-makers ($1 \leq p \leq k$) and alternatives ($1 \leq i \leq n$) was arranged as follows, and the first step of the fuzzy TOPSIS, fuzzy CRITIC and fuzzy ARAS methods was performed (Table 3).

In order to calculate the fuzzy TOPSIS weights, 3 experts were asked to evaluate the criteria and as a result of transforming the obtained data into fuzzy numbers, the fuzzy weight matrix was formed as follows. Table 4 contains the fuzzy TOPSIS weight matrix and the fuzzy TOPSIS weights obtained as a result of the calculation.

For the fuzzy CRITIC weighting method, standard deviations and total correlation coefficients were obtained by using Eqs. (23), (24), (25) and (26) expressed in the explanations above, and the fuzzy weights of the criteria were obtained as follows (Table 5).

As a result of applying fuzzy TOPSIS weights to fuzzy TOPSIS method and fuzzy CRITIC weights to fuzzy ARAS method, alternatives were evaluated and the following findings were reached (Table 6 and Fig. 8).

5 Conclusion

Fuzzy multi-criteria decision-making methods are frequently used in our increasingly complex world. In an environment where tastes, satisfaction-oriented and even niche productions are rapidly becoming popular, manufacturers are trying to expand their customer portfolios by bearing the minimum cost and creating maximum value. Decision support systems, artificial intelligence estimators, customer characterizations and big data analyses work to determine the wishes, expectations and tastes of consumers. Many large companies shape their product lines and depths as a result of these analyses. Herd tracking systems used in the livestock sector also try to make similar analyzes and aim to prevent being away from the competitive environment. However, when it comes to a service that serves intangible expectations rather than a tangible product, it becomes increasingly difficult to conduct a grounded analysis.

In this study, herd tracking systems were evaluated in terms of users. The herd tracking systems, which are popularly used in the Burdur region, were evaluated by the users and the evaluation inputs were analyzed through TOPSIS and ARAS,

Table 3 Fuzzy decision-making matrix

		DM1			DM2			DM3			DM4			DM5			DM6		
CRITERIA 1	**GEA**	2	3,5	5	7	8	9	7	8	9	7	8	9	0	0	2	2	3,5	5
	AFI	5	6,5	8	8	10	10	2	3,5	5	2	3,5	5	2	3,5	5	8	6,5	7
	SCR	8	10	10	8	10	10	0	0	2	0	0	2	1	2	3	1	2	3
	Delaval	7	8	9	5	6,5	8	1	2	3	1	2	3	8	10	10	8	10	10
CRITERIA 2	**GEA**	5	6,5	8	0	0	2	5	6,5	8	8	10	10	0	0	2	4	5	6
	AFI	7	8	9	2	3,5	5	7	8	9	4	5	6	7	8	9	5	6,5	8
	SCR	8	10	10	7	8	9	7	8	9	1	2	3	1	2	3	7	8	9
	Delaval	5	6,5	8	8	10	10	4	5	6	2	3,5	5	8	10	10	1	2	3
CRITERIA 3	**GEA**	4	5	6	1	2	3	8	10	10	7	8	9	0	0	2	4	5	6
	AFI	7	8	9	4	5	6	7	8	9	5	6,5	8	8	10	10	5	6,5	7
	SCR	7	8	9	5	6,5	8	8	10	10	1	2	3	2	3,5	5	8	10	10
	Delaval	2	3,5	5	5	6,5	8	4	5	6	2	3,5	5	7	8	9	1	2	3
CRITERIA 4	**GEA**	7	8	9	7	8	9	7	8	9	8	10	10	4	5	6	2	3,5	5
	AFI	8	10	10	7	8	9	8	10	10	5	6,5	8	5	6,5	8	4	5	6
	SCR	7	8	9	7	8	9	8	10	10	0	0	2	4	5	6	5	6,5	8
	Delaval	2	3,5	5	7	8	9	5	6,5	8	1	2	3	7	8	9	1	2	3
CRITERIA 5	**GEA**	2	3,5	5	8	10	10	2	3,5	5	8	10	10	4	5	6	2	3,5	5
	AFI	7	8	9	7	8	9	8	10	10	4	5	6	7	8	9	4	5	6
	SCR	5	6,5	8	5	6,5	8	5	6,5	8	1	2	3	1	2	3	8	10	10
	Delaval	5	6,5	8	7	8	9	1	2	3	2	3,5	5	8	10	10	0	0	2
CRITERIA 6	**GEA**	2	3,5	5	7	8	9	0	0	2	7	8	9	2	3,5	5	5	6,5	8

(continued)

Table 3 (continued)

		DM1			DM2			DM3			DM4			DM5			DM6		
	AFI	5	6,5	8	5	6,5	8	8	10	10	4	5	6	4	5	6	7	8	9
	SCR	7	8	9	8	10	10	7	8	9	0	0	2	1	2	3	8	10	10
	Delaval	2	3,5	5	4	5	6	0	0	2	2	3,5	5	5	6,5	8	2	3,5	5
CRITERIA 7	**GEA**	1	2	3	4	5	6	5	6,5	8	8	10	10	0	0	2	2	3,5	5
	AFI	7	8	9	2	3,5	5	2	3,5	5	4	5	6	2	3,5	5	4	5	6
	SCR	5	6,5	8	7	8	9	4	5	6	2	3,5	5	1	2	3	7	8	9
	Delaval	1	2	3	5	6,5	8	1	2	3	5	6,5	8	7	8	9	0	0	2

Table 4 Fuzzy weight matrix and fuzzy weights

	DM1			DM2			DM3			Fuzzy weights		
C1	0.8	1	1	0.8	1	1	0.4	0.5	0.6	**0.4**	0.83	1
C2	0.8	1	1	0.8	1	1	0.5	0.65	0.8	**0.5**	0.883	1
C3	0.5	0.65	0.8	0.8	1	1	0.5	0.65	0.8	**0.5**	0.76	1
C4	0.2	0.35	0.5	0.1	0.2	0.3	0.2	0.35	0.5	**0.1**	0.3	0.5
C5	0.7	0.8	0.9	0.5	0.65	0.8	0.4	0.5	0.6	**0.4**	0.65	0.9
C6	0.1	0.2	0.3	0.7	0.8	0.9	0.8	1	1	**0.1**	0.67	1
C7	0.5	0.65	0.8	0.7	0.8	0.9	0.7	0.8	0.9	**0.5**	0.75	0,9

Table 5 Details and weights of the fuzzy CRITIC method

	σ_{jk}			$S_{jj'}^{k}$			Fuzzy weights		
C1	0.4663	0.4410	0.4194	1.8000	2.3333	1.6667	0.1267	0.1693	0.1252
C2	0.4266	0.4778	0.4545	1.8571	1.1429	1.4615	0.1196	0.0898	0.1190
C3	0.4843	0.4881	0.4658	2.4444	2.1613	1.9091	0.1788	0.1736	0.1592
C4	0.4090	0.4338	0.4325	2.0000	1.7500	1.7143	0.1235	0.1249	0.1328
C5	0.4819	0.4139	0.4456	2.4667	2.1935	2.3571	0.1795	0.1494	0.1881
C6	0.4197	0.4524	0.4501	2.0000	1.7632	1.6667	0.1267	0.1313	0.1343
C7	0.4270	0.4250	0.4646	2.2500	2.3125	1.7000	0.1451	0.1617	0.1414

Table 6 Fuzzy TOPSIS and fuzzy ARAS results

	F-TOPSIS (C_i)	Rank	F-ARAS (K_i)	Rank
A1	0.45558135	4	0.67020551	3
A2	0.50566314	1	0.75742997	1
A3	0.48239438	2	0.75324863	2
A4	0.45710334	3	0.64301199	4

which are fuzzy multi-criteria decision-making methods. Finally, it was concluded that the order of alternative 2 and alternative 3 did not change, and the order of alternative 1 and alternative 4 could vary. Alternative 2 and alternative 3 are software that are known for their R&D activities in the sector and have a strong hardware infrastructure. At the same time, there are satisfactory works in terms of technical support and after-sales services. However, in addition to these positive features, they are higher in cost compared to the other 2 alternatives. When this point of view is taken into consideration, it is seen that the services that the users will obtain that is, the benefits they will gain (place utility, time benefit, product/service benefit) are more important than the cost benefit. When evaluated on the benefit/cost axis, it was concluded that the benefit factor is more preferable than the cost factor. It is thought

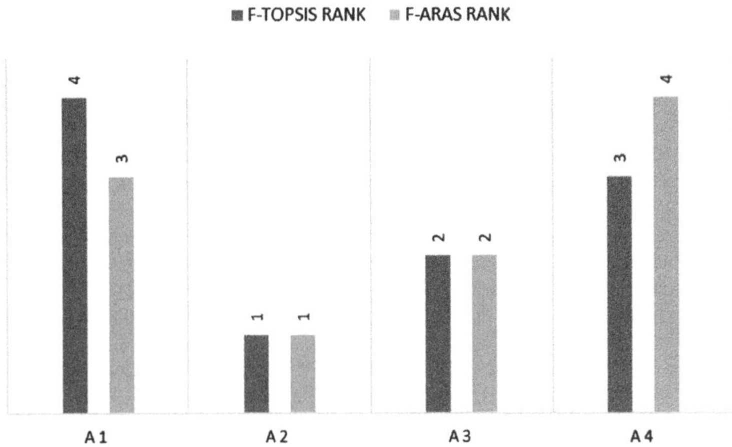

Fig. 8 Graphic display of results

that the software and hardware to be developed in this area and the R&D studies to be carried out will find sufficient demand.

References

1. Alpay, M.: Kredi değerliliğinin ölçülmesinde TOPSIS yöntemi ve bir uygulama. Dokuz Eylül Üniversitesi, Basılmamış Yüksek Lisans Tezi, İzmir (2010)
2. Amiri, M.P.: Project selection for oil-fields development by using the AHP and fuzzy TOPSIS methods. Expert Syst. Appl. **37**(9), 6218–6224 (2010)
3. Baykal, N., Beyan, T.: *Bulanık mantık ilke ve temelleri*. Bıçaklar Kitabevi (2004)
4. Berckmans, D.: 1.2. Smart farming for Europe: value creation through precision livestock farming. In: Precision Livestock Farming Applications. pp. 25–36. Wageningen Academic Publishers (2015)
5. Berger, R., Hovav, A.: Using a dairy management information system to facilitate precision agriculture: the case of the AfiMilk® system. Inf. Syst. Manag. **30**(1), 21–34 (2013)
6. Bewley, J.: Precision dairy farming: advanced analysis solutions for future profitability. In: Proceedings of the First North American Conference on Precision Dairy Management, Toronto, Canada (pp. 2–5) (2010)
7. Bewley, J.M.: Opportunities for monitoring and improving animal welfare using precision dairy monitoring technologies. J. Anim. Sci. **94**, 11–11 (2016)
8. Boran, F.E., Genç, S., Kurt, M., Akay, D.: A multi-criteria intuitionistic fuzzy group decision making for supplier selection with TOPSIS method. Expert Syst. Appl. **36**(8), 11363–11368 (2009)
9. Büyüközkan, G., Çifçi, G.: A novel hybrid MCDM approach based on fuzzy DEMATEL, fuzzy ANP and fuzzy TOPSIS to evaluate green suppliers. Expert Syst. Appl. **39**(3), 3000–3011 (2012)
10. Chen, C.T., Lin, C.T., Huang, S.F.: A fuzzy approach for supplier evaluation and selection in supply chain management. Int. J. Prod. Econ. **102**(2), 289–301 (2006)
11. Çakıcı, İ: The Comparison of Herd Management Softwares Commonly Used at Turkey. Selçuk Universty, Graduate School of Natural and Applied Science (2006)

12. Dahooie, J.H., Zavadskas, E.K., Abolhasani, M., Vanaki, A., Turskis, Z.: A novel approach for evaluation of projects using an interval–valued fuzzy additive ratio assessment (ARAS) method: a case study of oil and gas well drilling projects. Symmetry **10**(2), 45 (2018)
13. Doluschitz, R.: Precision agriculture-applications, economic considerations, experiences and perspectives. In: EFITA 2003 conference, July, pp 5–9. (2003)
14. Eastwood, C.R., Chapman, D.F., Paine, M.S.: Networks of practice for co-construction of agricultural decision support systems: case studies of precision dairy farms in Australia. Agric. Syst. **108**, 10–18 (2012)
15. Eastwood, C.R., Jago, J.G., Edwards, J.P., Burke, J.K.: Getting the most out of advanced farm management technologies: roles of technology suppliers and dairy industry organisations in supporting precision dairy farmers. Animal Prod. Sci. **56**(10), 1752–1760 (2015)
16. Ecer, F.: An integrated Fuzzy AHP and ARAS model to evaluate mobile banking services. Technol. Econ. Dev. Econ. **24**(2), 670–695 (2018)
17. Eckelkamp, E.A.: Invited review: current state of wearable precision dairy technologies in disease detection. Appl. Animal Sci. **35**(2), 209–220 (2019)
18. Edwards, J.P., Rue, B.D., Jago, J.G.: Evaluating rates of technology adoption and milking practices on New Zealand dairy farms. Animal Prod. Sci. **55**(6), 702–709 (2014)
19. Ertuğrul, İ, Karakaşoğlu, N.: Comparison of fuzzy AHP and fuzzy TOPSIS methods for facility location selection. Int. J. Adv. Manuf. Technol. **39**(7–8), 783–795 (2008)
20. Frost, A.R., Schofield, C.P., Beaulah, S.A., Mottram, T.T., Lines, J.A., Wathes, C.M.: A review of livestock monitoring and the need for integrated systems. Comput. Electron. Agric. **17**(2), 139–159 (1997)
21. Fu, Y.K., Wu, C.J., Liao, C.N.: Selection of in-flight duty-free product suppliers using a combination fuzzy AHP, fuzzy ARAS, and MSGP methods. Mathem. Problems Eng. (2021)
22. Govindan, K., Khodaverdi, R., Jafarian, A.: A fuzzy multi criteria approach for measuring sustainability performance of a supplier based on triple bottom line approach. J. Clean. Prod. **47**, 345–354 (2013)
23. Hadley, G.L., Harsh, S.B., Wolf, C.A.: Managerial and financial implications of major dairy farm expansions in Michigan and Wisconsin. J. Dairy Sci. **85**(8), 2053–2064 (2002)
24. He, Y.H., Wang, L.B., He, Z.Z., Xie, M.: A fuzzy TOPSIS and rough set based approach for mechanism analysis of product infant failure. Eng. Appl. Artif. Intell. **47**, 25–37 (2016)
25. https://www.afimilk.com/ , Access Time: (22 Dec 2021)
26. https://www.DeLavalCorporate.com . Access Time: (22 Dec 2021)
27. https://www.gea.com/ . Access Time (22 Dec 2021)
28. http://www.ru.scrdairy.com/ Access Time (22 Dec 2021)
29. Indira, D., Suresh, J.: Accurate heat detection and health monitoring of cattle by SCR sense neck collar. World J. Adv. Res. Rev. **8**(3), 075–079 (2020)
30. Kabak, M., Sağlam, F., Aktas, A.: Farklı uzaklık hesaplama yaklaşımlarının TOPSIS üzerinde kullanılabilirliğinin incelenmesi. J. The Faculty of Eng. Architect. Gazi University **32**(1), 35–43 (2017)
31. Kacprzak, D.: An extended TOPSIS method based on ordered fuzzy numbers for group decision making. Artif. Intell. Rev. **53**(3) (2020)
32. Karagöz, S., Deveci, M., Simic, V., Aydin, N.: Interval type-2 fuzzy ARAS method for recycling facility location problems. Appl. Soft Comput. **102**, 107107 (2021)
33. Kaya, İ., Uzmay, C., Kaya, A.: Precision dairy herd management application. Tarımda Bilgisayar Uygulamaları Sempozyumu, pp 5–7 (1994)
34. Keršulienė, V., Turskis, Z.: Integrated fuzzy multiple criteria decision making model for architect selection. Technol. Econ. Dev. Econ. **17**(4), 645–666 (2011)
35. Küçük, O., Ecer, F.: Assesing suppliers using fuzzy topsis and an application in erzurum. Ekonomik ve Sosyal Araştırmalar Dergisi **3**(1), 45–65 (2007)
36. Lazarus, W.F., Streeter, D., Jofre-Giraudo, E.: Management information systems: impact on dairy farm profitability. Appl. Econ. Perspect. Policy **12**(2), 267–277 (1990)
37. Lokhorst, C., De Mol, R.M., Kamphuis, C.: Invited review: big data in precision dairy farming. Animal **13**(7), 1519–1528 (2019)

38. Mishra, A.R., Rani, P., Krishankumar, R., Ravichandran, K.S., Kar, S.: An extended fuzzy decision-making framework using hesitant fuzzy sets for the drug selection to treat the mild symptoms of Coronavirus disease 2019 (COVID-19). Appl. Soft Comput. **103**, 107155 (2021)
39. Onur, S.: Automation software development for milk monitoring and metering station (Master's thesis, Balıkesir Universty Graduate School of Natural and Applied Science) (2016)
40. Oruç, K.O.: Malmquist productivity index in fuzzy environment and an application in universty hospitals. Uluslararası Yönetim İktisat ve İşletme Dergisi **12**(28), 163–188 (2016)
41. Oruç, K.O., Arıcan, M.: Bulanık Analitik Hiyerarşi Süreci ve Bulanık Aras Yöntemleri ile Polis Merkezi Kuruluş Yeri Seçimi: Isparta Örneği Süleyman Demirel Üniversitesi İktisadi ve İdari Bilimler Fakültesi Dergisi **24**(3), 715–734 (2019)
42. Rostamzadeh, R., Esmaeili, A., Nia, A.S., Saparauskas, J., Ghorabaee, M.K.: A fuzzy aras method for supply chain management performance measurement in SMEs under uncertainty. Transform. Business Econom. **16** (2017)
43. Safaei Ghadikolaei, A., Khalili Esbouei, S., Antucheviciene, J.: Applying fuzzy MCDM for financial performance evaluation of Iranian companies. Technol. Econ. Dev. Econ. **20**(2), 274–291 (2014)
44. Sengül, Ü., Eren, M., Shiraz, S.E., Gezder, V., ¸Sengül, A.B.: Fuzzy TOPSIS method for ranking renewable energy supply systems in Turkey. Renew. Energy **75**, 617–625 (2015)
45. Shen, F., Ma, X., Li, Z., Xu, Z., Cai, D.: An extended intuitionistic fuzzy TOPSIS method based on a new distance measure with an application to credit risk evaluation. Inf. Sci. **428**, 105–119 (2018)
46. Spilke, J., Fahr, R.: Decision support under the conditions of automatic milking systems using mixed linear models as part of a precision dairy farming concept. In: EFITA 2003 Conference, July, pp. 5–9. (2003)
47. Şen, Z.: Bulanık Mantık ve Modelleme İlkeleri, Bilge Kültür Sanat, İstanbul (2001)
48. Tomaszewski, M.A.: Record-keeping systems and control of data flow and information retrieval to manage large high producing herds. J. Dairy Sci. **76**(10), 3188–3194 (1993)
49. Tömek, B.: An evaluation of modern technology applications used by herd management in dairy cattle husbandry. Master's Thesis, Ege Universty. Graduate School of Natural and Applied Science. Bornova-İzmir (2007)
50. Turskis, Z., Zavadskas, E.K.: A new fuzzy additive ratio assessment method (ARAS-F). Case study: the analysis of fuzzy multiple criteria in order to select the logistic centers location. Transport **25**(4), 423–432 (2010)
51. Ulutas, A.: Using of fuzzy SWARA and fuzzy ARAS methods to solve supplier selection problem. In: Theoretical and Applied Mathematics in International Business, pp. 136–148. IGI Global (2020)
52. Uzmay, C., İbrahim, K.A.Y.A., Tömek, B.: Precision dairy herd management application. Hayvansal Üretim **51**(2) (2010)
53. Wu, T., Liu, X., Liu, F.: An interval type-2 fuzzy TOPSIS model for large scale group decision making problems with social network information. Inf. Sci. **432**, 392–410 (2018)
54. Yayar, R., Baykara, H.V.: TOPSIS Yöntemi ile Katilim Bankalarinin Etkinligi ve Verimliligi Üzerine Bir Uygulama/an implementation upon efficiency and productivity of participation banks with TOPSIS method. Business Econom. Res. J. **3**(4), 21 (2012)
55. Yoon, K.: Systems Selection by Multiple Attribute Decision Making. Kansas State University, Doktora Tezi (1980)
56. Yoon, K.P., Hwang, C.L.: In: Multiple Attribute Decision Making: An İntroduction. Sage Publications (1995)
57. Zadeh, L.A.: Fuzzy sets. Inf. Control **8**, 338–353 (1965)
58. Zyoud, S.H., Kaufmann, L.G., Shaheen, H., Samhan, S., Fuchs-Hanusch, D.: A framework for water loss management in developing countries under fuzzy environment: integration of Fuzzy AHP with Fuzzy TOPSIS. Expert Syst. Appl. 61, 86–105 (2016)
59. Ulutaş, A.: Supplier evaluation with BWM and fuzzy CODAS methods. In: Handbook of Research on Recent Perspectives on Management, International Trade, and Logistics, pp. 335–351. IGI Global (2021)

60. Hwang, C.L., Yoon, K.: Basic concepts and foundations. In: Multiple Attribute Decision Making, pp. 16–57. Springer, Berlin, Heidelberg (1981)
61. Kahraman, C., Öztayşi, B., Onar, S.Ç.: Spherical fuzzy CRITIC method: prioritizing supplier selection criteria. In: International Conference on Intelligent and Fuzzy Systems, pp. 705–714. Springer, Cham (2021, August).
62. Ali, J.: A novel score function based CRITIC-MARCOS method with spherical fuzzy information. Comput. Appl. Math. 40(8), 280 (2021).https://doi.org/10.1007/s40314-021-01670-9
63. Haleem, A., Khan, S., Pundir, H., Jain, A., Upadhyay. P., Khan, M.I.: Investigating barriers toward the implementation of circular economy: a fuzzy CRITIC approach. J. Ind. Integr. Manag. 06(01), 107–139 (2021).https://doi.org/10.1142/S2424862220500177

Multi-criteria Decision Making and Its Application to Online Learning Platform Selection During the COVID-19 Pandemic Based on TOPSIS Method

Subhradeep Maitra, Laxminarayan Sahoo, Jayati Lahiri Dey, and Kalishankar Tiwary

Abstract The COVID-19 outbreak has made a huge impact across the nation. It really has caused damage on the education system, as governments across the world partially or completely shutting down educational institutions to combat the transmission of the deadly virus. Classroom teachings have been made obsolete by online classes, leading to a rise of web-based learning, which has sure to meet the students to continue their studies. Students, teachers, administrations, and education professionals have experience gathered of significant challenges as a result of the abrupt shift from classroom teaching to web-based learning. As a result, Learning Management Systems (LMS) and Virtual Learning Environments (VLE) have become widely more popular and attractive, especially as a result of the pandemic's forced educational adjustments. In a web-based learning system, search engines have been found applicable to continue the educational process. Also, the search engines are involved with retrieving necessary documentation (images, text, video, web pages, and so on) from a large database in order to respond to the learner's request and several online platforms have been made with different features and facilities viz. Zoom, Google Meet, Google Classroom, Microsoft Teams, Cisco Webex, GoTo Meet, skype, etc. In this chapter, we have been applied TOPSIS method for selecting suitable online platform from existing online platforms using Multi-Criteria Decision Making (MCDM) technique. Finally, to illustrate the proposed technique, a realistic scenario for selecting the online platform from a web-based learning system has been exhibited.

S. Maitra (✉) · L. Sahoo · J. Lahiri Dey
Department of Computer and Information Science, Raiganj University, Raiganj 733134, India
e-mail: smoitra449@gmail.com

L. Sahoo
e-mail: lxsahoo@gmail.com

J. Lahiri Dey
e-mail: ld.jayatii@gmail.com

K. Tiwary
Department of Mathematics, Raiganj University, Raiganj 733134, India
e-mail: tiwarykalishankar@yahoo.com

L. Sahoo et al. (eds.), *Real Life Applications of Multiple Criteria Decision Making Techniques in Fuzzy Domain*, Studies in Fuzziness and Soft Computing 420,
https://doi.org/10.1007/978-981-19-4929-6_23

Keywords Online learning platform · Linguistic variables · The COVID-19 pandemic · TOPSIS · Multi-criteria decision making · Web-based learning

1 Introduction

In order to combat the transmission of the SARS-COV2 virus [13], all governments across the globe have made a policy which includes the canceling or shutdown of transportation systems viz. flights, trains, bus services etc., shutdown of restaurants and bars, and strict quarantine for symptomatic persons who are suspected to be infected. As a consequence, educational institutions such as schools, colleges and universities, and higher educational institutions have temporally suspended from their regular classroom teaching and more focused on online learning or web-based learning system to continue the education systems [5]. This policy has also been made for protection of health of the students and the education staff community. To prevent the spreading of the deadly virus, educational establishments discontinued conventional/classroom teaching activities, such as laboratory facilities, semesters examinations, and shifted teaching activities to a digital platform. As modules are easily available, shifting educational activity to the online platform allows teachers and learners to increase freedom in teaching and learning. Despite the crisis, courses were posted to the internet at an unprecedented and unparalleled rate [6]. The digital India movement was stared in the year 2015 with the objective of transforming each aspect of our society through the use of Internet age. This present circumstance provides such a possibility to use Internet devices in the face of adversity. In 2020, India had a total of 749 million Internet users. By 2040, this number is expected to rise to nearly 1.5 billion users, indicating a large market for internet services in the south Asian country (https://www.statista.com/statistics/255146/number-of-internet-users-in-india/). In fact, India was listed as the world's second-largest internet market in 2019, trailing to China. The number of Internet users is expected to rise in both urban and rural areas, showing a dynamic increase in Internet access also. It is to be mentioned that like other sectors, the health sector has embedded use of such digital technologies for a lot of services, spanning from quality medical service to ongoing teaching and research. This marked a change in the traditional framework of classroom learning, beyond pen and paper doodling to digital "log in" to the virtual classroom [4, 6, 8]. Traditional classroom teaching is widely acknowledged, but in these challenging times, when meetings and public gatherings were restricted and social distancing was the rule all over the countries, digital teleconference facilities to provided way to obtain in the form of webinars as the accepted norm [13]. In fact, search engines such as Google, YouTube, Amazon, Facebook, Microsoft Bing, Baidu, Yandex, and others are involved in fetching required documents (images, text, video, web pages, etc.) from such a big database in order to respond to the learner's requirement, and many online platforms such as Zoom, Google Meet, Google Classroom, Microsoft Teams, Cisco Webex, GoTo Meet, Skype, and many others have been used with various features and functions. **Zoom**: Zoom is a cloud-based teleconferencing application that provides users to virtually meet with others while conducting live chats, and it also allows users to capture those meetings to stream live. **Google Meet**: Google Meet is a video-communication service developed by Google. It was previously known as Hangouts Meet. It is one of two programs that take the place of Google Hangouts, with the other being Google Chat. **Cisco Webex**: Cisco's Webex is an online teleconference, video conferencing, system integration as a service, and client services as a service application that it produces and sells. It was developed in 1995 at WebEx and purchased by Cisco in 2007. **Skype**: Skype is a commercial telecommunication services tool that is run by Skype Technologies, a Microsoft company, and is well-known for VoIP-based videotelephony, videoconferencing, and voice calls. It also includes services such as instant messaging, file transfer, debit-based calls to landlines and mobile phones through standard telephone networks, and more. Skype is available on a variety of platforms, including desktop, mobile, and video game platforms. **GoTo Meeting**: LogMeIn developed and distributes GoTo Meeting, a web-based application. It is an online conference, screen sharing, and teleconferencing software suite that lets users meet with the other users, customers, employees, or collaborators in real time over the Internet. **Blackboard Collaborate**: Blackboard Collaborate is a complete online learning and collaboration tool created with educators in mind. It already is a part of having a good Schoolcraft Blackboard Courses and Communities. Faculty and students can effortlessly and for freely arrange live sessions to enhance communication and participation for teaching, learning, and other purposes. Some basic features and functions of few popular online platforms like Zoom, Google Meet, Cisco Webex, Skype, GoTo Meeting, and Blackboard Collaborate have been listed in Table 1.

Table 1 shows that online platforms have a variety of features and functions, and that each online platform has some unique and/or common as well as almost similar qualities. As a consequence, finding a suitable digital platform for teaching and learning is a difficult issue for users. So, one can think the selection of online learning platform as a multi-criteria decision making problem (MCDM) [9, 11, 12]. The process of selecting the best option from among all possible options is known as decision making [3]. Almost all such situations have variety of reviewing the alternatives. That is, the decision maker wants to solve a multiple criteria decision making (MCDM) [7] problem from most of these difficulties. Here, we have considered several online platforms as alternatives and different characteristics/features as a criterion. Therefore, our aim is to finding best alternative from all possible characteristics/features. In this chapter, we have applied TOPSIS method for selecting suitable online platform from existing online platforms using Multi-Criteria Decision Making (MCDM) technique. The TOPSIS [1] methods asserts that the preferred alternative should be the nearest to the positive ideal solution and the furthest away from the negative ideal solution. TOPSIS does have the advantage of determining the relevance of attributes and the effectiveness of alternatives with regard to various attributes. The scores and weights of the parameters are known accurately in classical MCMD methods, including classical TOPSIS. On the other hand, precise data

Table 1 Features and functions of few popular online platforms like Zoom, Google Meet, etc.

Zoom	Google meet	Cisco webex	Skype	Goto meeting	Blackboard collaborate
It is accessible from website. App download is not mandatory	It is not necessary to download app	It is necessary to download app	It is not necessary to download app	It is necessary to download app	It is necessary to download app
It has appearance touch up feature	It does not have appearance touch up feature	It does not have appearance touch up feature	It has appearance touch up feature	It has appearance touch up feature	It has appearance touch up feature
It has audio, video and chat functionality	It has availability of phone, video call and chat facility	It has video call, phone calls and chatting functionality	It has video call, audio call and chatting functionality	It has video call, audio call and chatting functionality	It has video call, audio call and chatting functionality
It can be accessed on Desktop, PC, Android and Apple devices	Desktop version or app on Android or iOS devices is available	It can be accessed on PC and mobile	It can be accessed on PC and Mobile	It can be accessed on PC and Mobile	It can be accessed on PC and Mobile
It has a virtual background that is customizable	It does not have background customization feature	It has the facility to blur or replace the chat background	It has no facility to blur or replace the chat background	It has no facility to blur or replace the chat background	It has facility to blur or replace the chat background
It is possible to allow privacy to the room, using the virtual background facility	It does not have privacy in case of work from home or non-office environment	It has to background customization feature to add a layer of privacy	It has background customization feature to add a layer of privacy	It has background customization feature to add a layer of privacy	It has background customization feature to add a layer of privacy
It has a limit of 100 participants	It has a limit of 250 participants	It has a limit of 1 lakh participants	It has limit of 100 participants at a time	It has limit of 150 participants at a time	It has limit of 100 participants at a time
It supports 40 min of video call session	It supports 60 min of video call session	It supports 40 min of video call session	It supports 100 h per month with no more than 10 h per day and a limit of 4 h per individual video call	It supports 40 min of video call session	It supports 720 min of video call session

(continued)

Table 1 (continued)

Zoom	Google meet	Cisco webex	Skype	Goto meeting	Blackboard collaborate
It has paid versions starting from $14.99 monthly	It has paid versions starting from $6 monthly	It has paid versions starting from $13.50 monthly	It has paid versions starting from $5 USD per month	It has paid versions starting from $12 USD per month	It has paid versions $300–$9000
Screen supports 49 people at once	Screen supports 16people at once	Screen supports only 1 person at once. The remaining people can be seen as thumbnails	Screen supports only 1 person at once. The remaining people can be seen as thumbnails	Screen supports only 1 person at once. The remaining people can be seen as thumbnails	Screen supports only 1 person at once. The remaining people can be seen as thumbnails
It is integrated with email and calendar	It is integrated with G Suite, Google Drive, Google Docs, Google notes as well as Google sheets	It can be integrated with Google Drive, Google Tasks and similar applications to enhance its performance and productivity	It can be integrated with up to 5 platforms	It can be integrated with up to 5 platforms	It can be integrated with up to 10 platforms
It came into existence in 2013	It came into existence in 2017	It came into existence in 2007	It came into existence in 2014	It came into existence in 2004	It came into existence in 1997
It focuses at all aspects of society	It aims at people having Google accounts	Its main focus is enterprises	Its main focus in enterprise	Its main focus in enterprise	Its main focus in enterprise
Microsoft documents collaboration is not available	Collaboration in terms of MS Office is not available	Collaboration in terms of MS office is not available	Collaboration in terms if MS Office is available	Collaboration in terms if MS Office is available	Collaboration in terms if MS Office is available
Direct schedule from Gmail inbox is not supported	direct schedule from Gmail inbox is possible	Direct schedule from Gmail inbox is not supported	Direct schedule from Gmail inbox is not supported	GOTO Meeting directly from Chrome Browser	Direct schedule from Gmail inbox is not supported
It is accessible from website. App download is not mandatory	It is not necessary to download app	It is necessary to download app	It is not necessary to download app	It is necessary to download app	It is necessary to download app

are insufficient to represent real-life occurrences in many circumstances since individual evaluations, especially preferences, are often ambiguous and cannot be estimated with a precise numerical value [16]. A far more reasonable approach seems to be to utilize linguistic evaluations rather than quantitative data, presuming that the grades and weights of the problem's criteria are examined using linguistic variables [15]. Very Good, Good, Medium, Fair, and other linguistic terms are recognized as natural expressions of evaluation. These properties highlight the relevance of fuzzy set theory in reflecting the option to enable of decision makers [2]. Fuzzy set theory assists in evaluating the ambiguities of concepts involved with subjective judgment in individuals. In this study, we have utilized triangular fuzzy numbers to represent linguistic parameters [10]. To compute the relative closeness coefficient, we have used Yager ranking index [14] to quantify the separation measure of each alternative from the positive and negative ideal solutions. In this chapter, a comprehensive evaluation model of online platform selection has been carried out by use of the entropy weight and TOPSIS. Finally, to illustrate the proposed technique, a realistic scenario for selecting the online platform from a web-based learning system has been studied.

2 Basics of Fuzzy Set Theory, Linguistic Variables, and Yager Ranking Index

A membership function $\mu_{\tilde{F}}(x)$ that maps to each element x in X to a real number in the interval $0 \leq x \leq 1$, forms a fuzzy set. The function $\mu_{\tilde{F}}(x)$ represents the degree of membership of x in the fuzzy set $\tilde{F}$.

Definition 2.1 The α-cut of a fuzzy set $\tilde{F}$ is a crisp subset of X and is denoted by $F_\alpha = \{x \in X : \mu_{\tilde{F}}(x) \geq \alpha\}$, where $\mu_{\tilde{A}}(x)$ is the membership function of $\tilde{F}$ and $\alpha \in [0, 1]$.

Definition 2.2 A fuzzy set $\tilde{F}$ is called a normal fuzzy set if there exists at least one $x \in X$ such that $\mu_{\tilde{F}}(x) = 1$.

Definition 2.3 A fuzzy set $\tilde{F}$ is called convex if and only if for every pair of $x_1, x_2 \in X$, the membership function of $\tilde{F}$ satisfies $\mu_{\tilde{F}}(\theta x_1 + (1-\theta)x_2) \geq \min\{\mu_{\tilde{F}}(x_1), \mu_{\tilde{F}}(x_2)\}$, where $\theta \in [0, 1]$.

Definition 2.4 A fuzzy number $\tilde{F}$ is a fuzzy set which is both convex and normal.

Definition 2.5 The triangular fuzzy number (TFN) is a normal fuzzy number denoted as $\tilde{F} = (a, b, c)$ where $a \leq b \leq c$ and its membership function $\mu_{\tilde{F}}(x) : X \to [0, 1]$ is defined by.

$$\mu_{\tilde{F}}(x) = \begin{cases} \frac{x-a}{b-a} & \text{if } a \leq x \leq b \\ 1 & \text{if } x = b \\ \frac{c-x}{c-b} & \text{if } b \leq x \leq c \end{cases}$$

2.1 α-Level Set of Triangular Fuzzy Number

Let $\tilde{F} = (a, b, c)$ be a triangular fuzzy number, then α-Level set of the triangular fuzzy number $\tilde{F} = (a, b, c)$ is $F_\alpha = \{x \in X : \mu_{\tilde{F}}(x) \geq \alpha\} = [F_\alpha^l, F_\alpha^r]$ where $F_\alpha^l = a + (b - a)\alpha$ and $F_\alpha^r = c - (c - b)\alpha, \alpha \in [0, 1]$.

2.2 Linguistic Variables

According to Zadeh [15], a linguistic variable is a quintuple $(L, T(L), U, G, M)$ where L is the name of the variable, $T(L)$ is the set of values for the L, U is the universe of discourse, G is a syntactic rule which generates the terms in $T(L)$ and M is a semantic rule which maps each linguistics terms $A \in T(L)$, to its meaning $M(A)$, where $M(A)$ denotes a fuzzy subset of U.

2.3 Yager's Ranking Index

Yager's ranking index [14] for a fuzzy number $\tilde{F}$ is denoted by $Y(\tilde{F})$ and is computed by the formula $Y(\tilde{F}) = 0.5 \int_0^1 \left(F_\alpha^l + A_\alpha^r\right)d\alpha$ where $[F_\alpha^-, F_\alpha^+]$ is the α-level interval of the fuzzy number $\tilde{F}$. If $\tilde{F} = (a, b, c)$ be the triangular fuzzy number then Yager's ranking index $Y(\tilde{F})$ of $\tilde{F} = (a, b, c)$ is $Y(\tilde{F}) = \frac{1}{4}(a + 2b + c)$.

3 TOPSIS Method Based on Linguistic Variables

In this section, we have discussed about how to use the TOPSIS method to solve multi-criteria decision making (MCDM) problems under which the decision makers' (DMs') predilection insight is expressed in the form of linguistic variables and the criterion weights are unspecified, which can be determined using the Shannon Entropy method. To find the coefficient of relative closeness, we have utilized a triangular fuzzy representation of linguistic variables and well-known Yager ranking indexes to quantify the distancing measurements of each alternative from positive and negative ideal solutions.

Suppose that there are m alternatives $A_i (i = 1, 2, ..., m)$ and n criterion $C_j (j = 1, 2, ..., n)$ with the criteria weight vector $w = (w_1, w_2, ..., w_n)$, where $0 \leq w_j \leq 1$ and $\sum w_j = 1$. The characteristics of an alternative A_i with respect to the criteria C_j can be expressed by a linguistic term/value $x_{ij} = (a_{ij}, b_{ij}, c_{ij})$, $(i = 1, 2, ..., m; j = 1, 2, ..., n)$ and $D = \left(C_j(A_i)\right)_{m \times n}$ is a decision matrix. Thus, the MCDM problem can be written in terms of decision matrix as follows:

$$D = \left(C_j(A_i)\right)_{m\times n} = \begin{pmatrix} x_{11} & x_{12} & \cdots & x_{1n} \\ x_{21} & x_{22} & \ldots & x_{2n} \\ \vdots & \vdots & \vdots & \vdots \\ x_{m1} & x_{m2} & \ldots & x_{mn} \end{pmatrix} \tag{1}$$

Now, using Yager Ranking index, we have converted the decision matrix $D = \left(C_j(A_i)\right)_{m\times n}$ into the following matrix $Y(D) = \left(Y\left(C_j(A_i)\right)\right)_{m\times n}$ as follows:

$$Y(D) = \left(Y\left(C_j(A_i)\right)\right)_{m\times n} = \begin{pmatrix} Y(x_{11}) & Y(x_{12}) & \cdots & Y(x_{1n}) \\ Y(x_{21}) & Y(x_{22}) & \ldots & Y(x_{2n}) \\ \vdots & \vdots & \vdots & \vdots \\ Y(x_{m1}) & Y(x_{m2}) & \ldots & Y(x_{mn}) \end{pmatrix} \tag{2}$$

Let A^+ and A^- denote, respectively, the positive and negative ideal solution for the alternatives $A_i(i = 1, 2, ..., m)$, where $A^+ = (1, 1, ..., 1)$ and $A^- = (0, 0, ..., 0)$, $j = 1, 2, ..., n$. Now, the separation measures $d_i^+ = d_i^+(A^+, A_i)$ and $d_i^- = d_i^-(A^-, A_i)$ of each alternative from positive ideal and negative ideal solutions, can be calculated as:

$$d_i^+ = \sqrt{\sum_{j=1}^{n} (w_j(1 - Y(x_{ij})))^2} \tag{3}$$

and

$$d_i^- = \sqrt{\sum_{j=1}^{n} (w_j Y(x_{ij}))^2} \tag{4}$$

The relative closeness of an alternative $A_i(i = 1, 2, ..., m)$ with respect to the positive ideal solution A^+ is denoted by $C_i(A_i)$ and is calculated as follows:

$$C_i(A_i) = \frac{d_i^-}{d_i^- + d_i^+} \tag{5}$$

$C_i(A_i), i = 1, 2, ..., m$ calculated by (5), can determine the ranking order of all alternatives and indicate the best one among a set of given feasible alternatives. As a result, the alternatives might be ordered in descending order of closeness coefficient, with the alternative with the highest rank becoming the best choice.

4 Calculation of Criteria Weight Using Entropy Method

Following are the different steps to calculate the criteria weight w_j, $j = 1, 2, ..., n$, where $0 \leq w_j \leq 1$ and $\sum w_j = 1$.

Step 1: For $j = 1, 2, ..., n$ calculate $p_{ij} = \frac{Y(x_{ij})}{\sum_{i=1}^{m} Y(x_{ij})}$

Step 2: For $j = 1, 2, ..., n$ calculate $e_j = -\frac{1}{\log(m)} \sum_{i=1}^{m} p_{ij} \log(p_{ij})$

Step 3: Evaluate $w_j = \frac{1-e_j}{\sum_{j=1}^{n} 1-e_j}$, for $j = 1, 2, ..., n$

It is to be worth mentioning that $\lim_{p_{ij} \to 0} p_{ij} \log p_{ij} \to 0$.

5 Numerical Example and Discussion

To illustrate our proposed approach, we will discuss a numerical example. Assume that an educational institution "XYZ" desires to select an online platform for teaching their students during COVID-19 Pandemic. Suppose there is six online platforms which are possible to use like Zoom (A_1), Google Meet (A_2), Cisco Webex (A_3), Skype (A_4), GoTo Meeting (A_5), and Blackboard Collaborate (A_6). So possible alternatives are A_i, $i = 1, 2, 3, 4, 5, 6$. Also, each alternative has 15 benefit criteria which are (i) App download is not mandatory (C_1) (ii) Appearance touches up feature (C_2) (iii) Audio, video and chat functionality (C_3) (iv) Accessed on Desktop, PC, Android and Apple devices (C_4) (v) Virtual background that is customizable (C_5) (vi) Privacy to the room, using the virtual background facility (C_6) (vii) Limit of participants (C_7) (viii) Session timing (C_8) (ix) Starting cost for paid version (C_9) (x) Screen supports (C_{10}) (xi) Integrated with email and calendar and other platforms (C_{11}) (xii) Came into existence (C_{12}) (xiii) Focuses at all aspects of society (C_{13}) (xiv) Microsoft documents collaboration (C_{14}), and (xv) Direct schedule from Gmail inbox (C_{15}). It is to be noted that all the alternatives and all the criteria are realistic in nature and we have taken these from respective website using Google search engine. Details have been listed in Table 1.

The proposed method is utilized to solve the problem. The procedure is summarized as follows:

(a) The decision makers use the linguistic variables to represent each criterion based on his/her experience about all the alternatives (see Table 2).
(b) Linguistic variables for the ratings along with triangular fuzzy representation has been made using Table 3 (see Table 4).
(c) Using Yager Ranking Index, $Y(D) = \left(Y\left(C_j(A_i)\right)\right)_{6\times 15}$ matrix has been prepared (see Table 5)

Now, we compute the criteria weight w_j, $j = 1, 2, ..., n$, where $0 \leq w_j \leq 1$ such that $\sum w_j = 1$. Using Sect. 4 we have obtained $w_1 = 0.02$, $w_2 = 0.01$, $w_3 = 0.01$, $w_4 = 0.02$, $w_5 = 0.01$, $w_6 = 0.02$, $w_7 = 0.12$, $w_8 = 0.16$, $w_9 = 0.08$, $w_{10} = 0.15$, $w_{11} = 0.02$, $w_{12} = 0.02$, $w_{13} = 0.13$, $w_{14} = 0.01$, and$w_{15} = 0.15$.

Table 2 Linguistic variables representation of each criterion

Criteria (C_j) and alternative (A_i)	A_1	A_2	A_3	A_4	A_5	A_6
C_1	[VG]	[VG]	[G]	[VG]	[G]	[G]
C_2	[G]	[M]	[M]	[G]	[G]	[G]
C_3	[G]	[VG]	[VG]	[VG]	[VG]	[VG]
C_4	[VG]	[VG]	[G]	[G]	[G]	[G]
C_5	[VG]	[M]	[G]	[F]	[F]	[M]
C_6	[VG]	[M]	[M]	[G]	[G]	[G]
C_7	[F]	[M]	[VG]	[F]	[M]	[F]
C_8	[F]	[VG]	[F]	[VG]	[F]	[G]
C_9	[M]	[VG]	[G]	[VG]	[G]	[F]
C_{10}	[VG]	[G]	[F]	[F]	[F]	[F]
C_{11}	[G]	[VG]	[VG]	[G]	[G]	[VG]
C_{12}	[G]	[G]	[VG]	[G]	[VG]	[VG]
C_{13}	[VG]	[M]	[F]	[F]	[F]	[F]
C_{14}	[M]	[M]	[M]	[G]	[G]	[G]
C_{15}	[F]	[VG]	[F]	[F]	[G]	[F]

Table 3 Linguistic variables for the ratings

Very good	[VG]	(0.9, 1.0, 1.0)
Good	[G]	(0.65, 0.80,0.95)
Medium	[M]	(0.5, 0.7, 0.9)
Fair	[F]	(0.35, 0.5, 0.65)

Using (3), (4), and (5) we have calculatedd_i^+,d_i^-, and $C_i(A_i)$ which have been shown in Table 6.

Now, we have ranked the alternative according to the descending order $C_i(A_i)$, $i = 1, 2, 3, 4, 5, 6$ and the alternative with highest rank will be the suitable choice. These have been shown in Table 7.

From Table 7 it has been sheen that ranking order of six online platforms is$A_2 \succ A_1 \succ A_4 \succ A_5 \succ A_3 \succ A_6$, i.e., Google Meet is the best online platform for online teaching and learning based on 15 criteria's which have been provided in Table 1. Order preference of six online platforms according to their merit/rank has been shown in Table 8.

6 Concluding Remarks

During COVID-19 pandemic, classroom teachings have been made obsolete by online classes, leading to a rise of web-based learning, which is sure to meet the

Table 4 Linguistic variables replaced by triangular fuzzy numbers

Criteria (C_j) and alternative (A_i)	A_1	A_2	A_3	A_4	A_5	A_6
C_1	(0.9, 1.0, 1.0)	(0.9, 1.0, 1.0)	(0.65, 0.80, 0.95)	(0.9, 1.0, 1.0)	(0.65, 0.80, 0.95)	(0.65, 0.80, 0.95)
C_2	(0.65, 0.80, 0.95)	(0.5, 0.7, 0.9)	(0.5, 0.7, 0.9)	(0.65, 0.80, 0.95)	(0.65, 0.80, 0.95)	(0.65, 0.80, 0.95)
C_3	(0.65, 0.80, 0.95)	(0.9, 1.0, 1.0)	(0.9, 1.0, 1.0)	(0.9, 1.0, 1.0)	(0.9, 1.0, 1.0)	(0.9, 1.0, 1.0)
C_4	(0.9, 1.0, 1.0)	(0.9, 1.0, 1.0)	(0.65, 0.80, 0.95)	(0.65, 0.80, 0.95)	(0.65, 0.80, 0.95)	(0.65, 0.80, 0.95)
C_5	(0.9, 1.0, 1.0)	(0.5, 0.7, 0.9)	(0.65, 0.80, 0.95)	(0.35, 0.5, 0.65)	(0.35, 0.5, 0.65)	(0.5, 0.7, 0.9)
C_6	(0.9, 1.0, 1.0)	(0.5, 0.7, 0.9)	(0.5, 0.7, 0.9)	(0.65, 0.80, 0.95)	(0.65, 0.80, 0.95)	(0.65, 0.80, 0.95)
C_7	(0.35, 0.5, 0.65)	(0.5, 0.7, 0.9)	(0.9, 1.0, 1.0)	(0.35, 0.5, 0.65)	(0.5, 0.7, 0.9)	(0.35, 0.5, 0.65)
C_8	(0.35, 0.5, 0.65)	(0.9, 1.0, 1.0)	(0.35, 0.5, 0.65)	(0.9, 1.0, 1.0)	(0.35, 0.5, 0.65)	(0.65, 0.80, 0.95)
C_9	(0.5, 0.7, 0.9)	(0.9, 1.0, 1.0)	(0.65, 0.80, 0.95)	(0.9, 1.0, 1.0)	(0.65, 0.80, 0.95)	(0.35, 0.5, 0.65)
C_{10}	(0.9, 1.0, 1.0)	(0.65, 0.80, 0.95)	(0.35, 0.5, 0.65)	(0.35, 0.5, 0.65)	(0.35, 0.5, 0.65)	(0.35, 0.5, 0.65)
C_{11}	(0.65, 0.80, 0.95)	(0.9, 1.0, 1.0)	(0.9, 1.0, 1.0)	(0.65, 0.80, 0.95)	(0.65, 0.80, 0.95)	(0.9, 1.0, 1.0)
C_{12}	(0.65, 0.80, 0.95)	(0.65, 0.80, 0.95)	(0.9, 1.0, 1.0)	(0.65, 0.80, 0.95)	(0.9, 1.0, 1.0)	(0.9, 1.0, 1.0)
C_{13}	(0.9, 1.0, 1.0)	(0.5, 0.7, 0.9)	(0.35, 0.5, 0.65)	(0.35, 0.5, 0.65)	(0.35, 0.5, 0.65)	(0.35, 0.5, 0.65)
C_{14}	(0.5, 0.7, 0.9)	(0.5, 0.7, 0.9)	(0.5, 0.7, 0.9)	(0.65, 0.80, 0.95)	(0.65, 0.80, 0.95)	(0.65, 0.80, 0.95)
C_{15}	(0.35, 0.5, 0.65)	(0.9, 1.0, 1.0)	(0.35, 0.5, 0.65)	(0.35, 0.5, 0.65)	(0.65, 0.80, 0.95)	(0.35, 0.5, 0.65)

Table 5 $Y(D) = \left(Y(C_j(A_i))\right)_{6\times 15}$ matrix

	C_1	C_2	C_3	C_4	C_5	C_6	C_7	C_8	C_9	C_{10}	C_{11}	C_{12}	C_{13}	C_{14}	C_{15}
A_1	0.98	0.80	0.80	0.98	0.98	0.98	0.50	0.50	0.70	0.98	0.80	0.80	0.98	0.70	0.50
A_2	0.98	0.70	0.98	0.98	0.70	0.70	0.70	0.98	0.98	0.80	0.98	0.80	0.70	0.70	0.98
A_3	0.80	0.70	0.98	0.80	0.80	0.70	0.98	0.50	0.80	0.50	0.98	0.98	0.50	0.70	0.50
A_4	0.98	0.80	0.98	0.80	0.50	0.80	0.50	0.98	0.98	0.50	0.80	0.80	0.50	0.80	0.50
A_5	0.80	0.80	0.98	0.80	0.50	0.80	0.70	0.50	0.80	0.50	0.80	0.98	0.50	0.80	0.80
A_6	0.80	0.80	0.98	0.80	0.70	0.80	0.50	0.80	0.50	0.50	0.98	0.98	0.50	0.80	0.50

Table 6 Calculated values of d_i^+, d_i^- and $C_i(A_i)$

i	1	2	3	4	5	6
d_i^+	0.128	0.062	0.149	0.139	0.137	0.148
d_i^-	0.241	0.289	0.203	0.226	0.208	0.197
$C_i(A_i)$	0.654	0.823	0.577	0.620	0.603	0.572

Table 7 Order of alternative with highest rank

Rank	Rank 2	Rank 1	Rank 5	Rank 3	Rank 4	Rank 6
Alternative	A_1	A_2	A_3	A_4	A_5	A_6

Table 8 Order of preference of alternative (online platform) according to their rank

Rank	Rank 1	Rank 2	Rank 3	Rank 4	Rank 5	Rank 6
Online platform	Google meet	Zoom	Skype	Goto meeting	Cisco Webex	Blackboard collaborate

students to continue their studies. And as a consequence, finding a suitable digital platform for teaching and learning, it is a difficult issue for users. Therefore, in this study, we have examined selection of online learning platforms based on TOPSIS method. Selection of online platform is a multi-criteria decision making (MCDM) problem as there are several alternatives having different characteristics/features. As precise information remains unavailable to simulate real-life scenarios in MCDM, we have implemented TOPSIS technique where all the information provided by the decision makers have been presented in the form of linguistic variables. In this study, we have considered triangular fuzzy number representation of linguistic variables. To compute the relative closeness coefficient, we have also used Yager ranking index to quantify the separation measure of each alternative from the positive and negative ideal solutions. A practical numerical example has been explored and solved to highlight the proposed TOPSIS technique. The results obtained from our study is more realistic as Google Meet is always most popular online platform for all most all the teaching learning community during COVID-19 pandemic. The solution procedure presented here is very simple to implement, and it can be mentioned that the solution technique proposed in this work will perform well enough to overcome several multi-criteria decision making problems where precise information remains unavailable to simulate real-life scenarios.

References

1. Bai, Y.Z.: An interval valued intuitionistic fuzzy TOPSIS method based on an improved scorefunction. Sci. World J. 1–6. (2013)

2. Bellman, R.E., Zadeh, L.A.: Decision making in a fuzzy environment. Manage. Sci. **17**, 141–164 (1970)
3. Bhunia, A.K., Sahoo, L., Shaikh, A.A.: Advanced Optimization and Operations Research Springer Nature, Singapore (2020)
4. Butnaru, G.I., Niță, V., Anichiti, A., Brînză, G.: The effectiveness of online education during Covid 19 pandemic—a comparative analysis between the perceptions of academic students and high school students from romania. Sustainability. **13**(9), 5311 (2021). https://doi.org/10.3390/su13095311
5. Choudhury, S., Pattnaik, S.: Emerging themes in e-learning: a review from the stakeholders' perspective. Comput. Educ. **144**, 1–20 (2020)
6. Dash, S., et al.: Review of online teaching platforms in the current period of COVID-19 pandemic. Indian J. Surg. (2021). https://doi.org/10.1007/s12262-021-02962-4
7. Hwang, C.L., Yoon, K.: Multiple Attributes Decision Making Methods and Applications. Springer, Berlin Heidelberg (1981)
8. Mpungose, C.B.: Emergent transition from face-to-face to online learning in a South African University in the context of the Coronavirus pandemic. Human. Soc. Sci. Commun. **7**, 113 (2020). https://doi.org/10.1057/s41599-020-00603-x
9. Sahoo, L.: Similarity measures for Fermatean fuzzy sets and its applications in group decision-making. Decis. Sci. Lett. **11**, 167–180 (2022)
10. Sahoo, L.: Solving matrix games with linguistic payoffs. Int. J. Syst. Assurance Eng. Manag. **10**, 484–490 (2019)
11. Sahoo, L.: Some score functions on Fermatean Fuzzy sets and its application to bride selection based on TOPSIS method. Int. J. Fuzzy Syst. Appl. **10**(3), 18–29 (2021)
12. Senapati, T., Yager, R.R.: Fermatean fuzzy weighted averaging/geometric operators and its application in multi-criteria decision-making methods. Eng. Appl. Artif. Intell. **85**, 112–121 (2019)
13. WHO, WHO Director-General's opening remarks at the media briefing on COVID-19, 11 March 2020. https://www.who.int/dg/speeches/detail/who-director-general-s-opening-remarks-at-the-media-briefing-on-covid-19--11-march-2020
14. Yager, R.R.: A procedure for ordering fuzzy subsets of the unit interval. Inf. Sci. **24**(2), 143–161 (1981)
15. Zadeh, L.A.: Fuzzy logic = computing with words. IEEE Trans. Fuzzy Syst. **4**(2), 103–111 (1996)
16. Zadeh, L.A.: Fuzzy Sets. Inf. Control **8**, 338–353 (1965)

A Comprehensive Study on Neutrosophic Fuzzy Solid Transportation Model and Its Solution Technique

Amrit Das

Abstract Solid transportation problem mainly deals with the situation of optimizing different objective functions related to many industrial problems considering constraints sets and due to the constraints uncertainty is always present in different parameters of solid transportation problems. This chapter intends to present a comprehensive study on the development of a mathematical model for solid transportation problems considering uncertain parameters and the neutrosophic fuzzy numbers are used to define those uncertain parameters. For the developed model, a solution approach is hereby discussed in this chapter. The feasibility conditions are validated with the case of numerical implementation of the model and its solution technique. A conversion process that converts a neutrosophic fuzzy number to an equivalent crisp number is discussed in this chapter, and using this method, only an equivalent form of the fuzzy model is obtained, which is later on solved with the LINGO software. A discussion is made based on the obtained result to validate the proposed methodology and the models of solid transportation problems.

Keywords Solid transportation problem · Neutrosophic fuzzy number · Neutrosophic ranking approach · LINGO

1 Introduction

Transportation problem (TP) is originated from the linear programming problem (LPP), in which there are mainly two constraint sets have been considered, viz., source constraint and demand constraint to optimize the objective function, i.e., the total transportation cost (TTC) associated with the system. According to Hitchcock [1], a formal definition of TP can be drawn as the distribution of a commodity from different source centers to the respective distribution centers with a linear objective

A. Das (✉)
Department of Mathematics, School of Advanced Sciences, Vellore Institute of Technology, Vellore, India
e-mail: amrit.das@vit.ac.in

L. Sahoo et al. (eds.), *Real Life Applications of Multiple Criteria Decision Making Techniques in Fuzzy Domain*, Studies in Fuzziness and Soft Computing 420,
https://doi.org/10.1007/978-981-19-4929-6_24

function that mainly minimizes the TTC. From the general mathematical perspective, TP is an optimization programming that consists of one linear objective function and a set of linear equalities and inequalities constraints. However, in the process of minimizing the TTC in TP, there are two types of constraints (source and demand) are considered, but in practicality another constraint, i.e., the conveyance used for the transportation also expresses a great impact on the TTC minimizing. To overcome this situation of a TP, later on in the year 1955, Shell [2] added an extra constraint to the TP and named it as conveyance constraint to state the solid transportation problem (STP). So in an STP, one has to deal with the three types of linear equalities and inequalities constraints (source, demand, and conveyance) to minimize the linear objective function, i.e., TTC. It was Haley [3] who developed the complete solution technique for an STP in the year 1962. Since the development of STP, it shows great interest among a larger number of researchers worldwide, and much research work has been developed. Some researchers focus on the development of modeling considering some real-life situation and accordingly build up some mathematical models like single objective [4] and multi-objective [5, 6] STP, fixed charge STP [7, 8], price discounted STP [9], sustainable STP [10], etc.

For a TP or STP demands of a commodity from different destinations, availability of source quantity at different source points and the capacity of conveyance are not always available in exact number or quantity, and the reason behind this is the market fluctuation, geographical impact on the transportation network, environmental impact of production, premature demands for time horizon suspensions, etc. And thus, the unit transportation cost for transporting one unit of a commodity from a particular source point to a particular destination becomes uncertain in nature. Thus, the whole system of transportation network becomes uncertain in nature, and therefore, it is required to define the different parameters of the STP using some uncertain number. The fuzzy is one of the widely used uncertain tools to define any uncertain parameters due to its flexibility over the decision-making process in an uncertain situation. The fuzzy number was first introduced by Zadeh [11] in 1965. A fuzzy number is an extension of the classical number where a member and a corresponding membership value are used to represent it. Based on this membership function, different form of fuzzy number has been developed by many researchers over time to time. Few of these are intuitionistic fuzzy number [12], where two types of membership functions have been considered, type-2 fuzzy number [13] in which the membership functions itself again a fuzzy number of type-1, hesitant fuzzy set [14], which allows the membership grades presuming a set of possible values instead of a single one, neutrosophic fuzzy number [15], which is a generalization of fuzzy as well as the intuitionistic fuzzy numbers, and it consists of three types of membership functions.

For the last two decades, modeling of STP with fuzzy numbers becomes a very interesting research topic for researchers worldwide, and its strong and large literature is the evidence. After the development of neutrosophic fuzzy numbers, it also took an interesting place among the researchers, and thus, much of the research work is undergoing its development. Some researchers [16–18] used this neutrosophic fuzzy number to define some uncertain parameters in the modeling of STP. The process of such modeling of STP using neutrosophic fuzzy numbers can be characterized as a neutrosophic solid transportation problem (NSTP). NSTP is more useful than any other uncertain modeling of STP using different uncertain tools and the reason for it is that the decision-maker in his/her formulation of modeling for an STP is not forced to make a delicate formulation. Therefore, this chapter is dedicated to presenting a comprehensive study by incorporating some variants on the modeling of NSTP. The trapezoidal neutrosophic numbers are used to define NSTP's parameters, and for the conversion, (trapezoidal neutrosophic number to an equivalent classical number) one ranking function is presented. For each of the models in this chapter one, numerical experiment has been done considering neutrosophic parameters. The models for NSTP after conversion to its equivalent classical form need to solve using a soft computing platform due to the 3D structure of the modeling and the conventional method that are available to solve a TP can not be utilized. For that reason, the LINGO software [19] is used to obtain the solution numerically for each of the converted NSTP models.

The chapter is organized as follows: an introduction to STP and neutrosophic fuzzy number following the literature review and hence the NSTP is in Sect. 1. Some basic definitions and properties of a neutrosophic fuzzy number along with a conversion method (neutrosophic fuzzy number to equivalent crisp number) are presented in Sect. 2. A general mathematical model for the NSTP is defined in Sect. 3, also this section interpreted the model in detail. The model solution techniques are discussed in the Sect. 4. Next, Sect. 5 is discussed the numerical experiment of the proposed NSTP. This section presents a particular form of the proposed generalized NSTP model considering some specific configuration for the number of source points, destinations of demand, and the number of conveyance with different capacities. The result and the feasibility discussion of obtained result for the equivalent crisp model are presented and discussed in Sect. 6. The conclusion of this chapter is drawn in Sect. 7. And at the last, a list of references that is used in this chapter is attached.

2 Preliminaries of Neutrosophic Fuzzy Number

This section contains information about the main concepts and definitions relating to fuzzy numbers from a neutrosophic perspective.

Single valued neutrosophic set (SVNS): [20] A SVNS denoted by ζ on U has the form [21] $\zeta=\{\langle x, T_\zeta(u), I_\zeta(u), F_\zeta(u) : u \in U\rangle\}$, where $T_\zeta(u) : U \longrightarrow [0, 1]$, $I_\zeta(u) : U \longrightarrow [0, 1]$, $F_\zeta(u) : U \longrightarrow [0, 1]$ and $0 \leq T_\zeta(u) + I_\zeta(u) + F_\zeta(u) \leq 3 \quad \forall u \in U$. $T_\zeta(u)$ represents the degree of membership for truthiness, $I_\zeta(u)$ represents the degree of membership for indeterminacy, and $F_\zeta(u)$ represents the degree of membership falsity of u to ζ, respectively.

Single valued trapezoidal neutrosophic fuzzy number (SVTNFN): [20] A SVTNFN which is denoted by $\tilde{\zeta}$ and characterized by $\langle(a_1, a_2, a_3, a_4), T_\zeta, I_\zeta, F_\zeta\rangle$ is a SVNS in $\Re$, and the three membership functions (MF), i.e., the truth MF, indeterminacy MF, and the falsity MF are defined as follows:

$$T_{\zeta}(u)=\begin{cases}\alpha_{\tilde{\zeta}}(\frac{u-a_1}{a_2-a_1}); & \text{if } a_1\le u\le a_2\\ \alpha_{\tilde{\zeta}}; & \text{if } a_2\le u\le a_3\\ \alpha_{\tilde{\zeta}}; & \text{if } a_3\le u\le a_4\\ 0; & \text{otherwise}\end{cases}$$

$$I_{\zeta}(u)=\begin{cases}\frac{(a_2-u+\beta_{\tilde{\zeta}}(u-a_1'))}{a_2-a_1'}; & \text{if } a_1'\le u\le a_2\\ \beta_{\tilde{\zeta}}; & \text{if } a_2\le u\le a_3\\ \frac{(u-a_3+\beta_{\tilde{\zeta}}(a_4'-u))}{a_4'-a_3}; & \text{if } a_3\le u\le a_4'\\ 1; & \text{otherwise}\end{cases}$$

$$F_{\zeta}(u)=\begin{cases}\frac{(a_2-u+\gamma_{\tilde{\zeta}}(u-a_1''))}{a_2-a_1''}; & \text{if } a_1''\le u\le a_2\\ \gamma_{\tilde{\zeta}}; & \text{if } a_2\le u\le a_3\\ \frac{(u-a_3+\gamma_{\tilde{\zeta}}(a_4''-u))}{a_4''-a_3}; & \text{if } a_3\le u\le a_4''\\ 1; & \text{otherwise}\end{cases}$$

Note that, in the above definition of MFs $\alpha_{\tilde{\zeta}}$, $\beta_{\tilde{\zeta}}$ and $\gamma_{\tilde{\zeta}}$ represent the highest degree of truth MF, indeterminacy MF, and false MF, respectively.

$\alpha_{\tilde{\zeta}}$, $\beta_{\tilde{\zeta}}$ and $\gamma_{\tilde{\zeta}} \in [0,1]$.

Also, $a_1'' \le a_1 \le a_1' \le a_2 \le a_3 \le a_4' \le a_4 \le a_4''$.

Refer to [20] for further details and the definition of the ranking function as well as the arithmetic operation of the trapezoidal neutrosophic fuzzy number.

Symmetric trapezoidal neutrosophic fuzzy number (STNFN): [22] A neutrosophic fuzzy set denoted by $\tilde{\zeta}$ over $\Re$ will be called a STNFN subject to the existence of real numbers a_1, a_2, with $a_1 \le a_2$ and $d \ge 0$ such that

$$\tilde{a}(u)=\begin{cases}\frac{u}{d}+\frac{d-a_1}{d}; & \text{for } u\in[a_1-d,a_1]\\ 1; & \text{for } u\in[a_1-d,a_1]\\ \frac{-u}{d}+\frac{d-a_1}{d}; & \text{for } u\in[a_2,a_2+d]\\ 0; & \text{otherwise}\end{cases}$$

We denoted it by $\tilde{a}=[a_1,a_2,d,d]$ when h=0; $\tilde{a}=[a_1,a_2]$.

It is worth noting that STNFN has three membership functions and takes the form $\tilde{\zeta}=(\zeta^l,\zeta^u,\alpha,\alpha,T_{\zeta},I_{\zeta},F_{\zeta})$, where ζ^l and ζ^u indicating the two bounds, i.e., lower and uppers bounds, α, α stands for two median (first and second) values associated to the STNFN [22, 23].

2.1 De-neutrosophication Process of SVTNFN

A new approach is developed to seek out equivalent crisp number for a SVTNFN by Basset et al. [20]. In line with their methodology, the crisp conversion formula has been outlined bellow.

Let $\tilde{\zeta}=(\zeta^l,\zeta^{m1},\zeta^{m2},\zeta^u;\alpha_{\tilde{\zeta}},\beta_{\tilde{\zeta}},\gamma_{\tilde{\zeta}})$ be a SVTNFN, where the lower bound is ζ^l, the first and second medians are ζ^{m1}, ζ^{m2}, respectively, and ζ^u is the upper bound for SVTNFN. The degrees of MFs for the SVTNFN are as follows, the truth MF is $\alpha_{\widetilde{\zeta}}$, and the indeterminacy MF is $\beta_{\widetilde{\zeta}}$ and the false MF $\gamma_{\widetilde{\zeta}}$. Now depending on the nature of the optimization problem, there are two cases possible in the process of de-neutrosophication of a SVTNFN, and these are as follows.

C-I: If the optimization problem consists of a maximizing objective function, then $R(\widetilde{\zeta}) = (\frac{\zeta^l + 2(\zeta^{m1} + \zeta^{m2}) + \zeta^u}{2}) + (confirmation)$.
The mathematical definition is as follows.

$$R(\widetilde{\zeta}) = (\frac{\zeta^l + 2(\zeta^{m1} + \zeta^{m2}) + \zeta^u}{2}) + (T_{\widetilde{\zeta}} - I_{\widetilde{\zeta}} - F_{\widetilde{\zeta}}) \tag{1}$$

C-II: If the optimization problem consists of a minimizing objective function, then $R(\widetilde{\zeta}) = (\frac{\zeta^l - 3(\zeta^{m1} + \zeta^{m2}) + \zeta^u}{2}) + (confirmation)$.
The mathematical definition is as follows.

$$R(\widetilde{\zeta}) = (\frac{\zeta^l - 3(\zeta^{m1} + \zeta^{m2}) + \zeta^u}{2}) + (T_{\widetilde{\zeta}} - I_{\widetilde{\zeta}} - F_{\widetilde{\zeta}}) \tag{2}$$

3 Mathematical Models for NSTP

This section defines the mathematical models for the NSTP and its different variants by defining different parameters of models using the neutrosophic fuzzy numbers. To define the mathematical model, some notations are used, and their definitions are as follows.

$\hat{c}_{ijk}$ = the neutrosophic fuzzy unit transportation cost in STP.
$\hat{a}_i$ = the available neutrosophic fuzzy source at source point i.
$\hat{b}_j$ = the neutrosophic fuzzy demand at destination j.
$\hat{e}_k$ = the neutrosophic fuzzy conveyance capacity for the vehicle k.
x_{ijk} = the crisp amount of unknown quantity to be transported from i-th origin to j-th destination by means of k-th conveyance.
where $i = 1, 2, \cdots, m$; $j = 1, 2, \cdots, n$; $k = 1, 2, \cdots, l$, respectively.

Then, the fuzzy mathematical model for NSTP is given by

$$\text{Min } Z = \sum_{i=1}^{m} \sum_{j=1}^{n} \sum_{k=1}^{l} \hat{c}_{ijk} x_{ijk} \tag{3}$$

subject to,

$$\sum_{j=1}^{n} \sum_{k=1}^{l} x_{ijk} \leq \hat{a}_i, \qquad i = 1, 2, \ldots, m \tag{4}$$

$$\sum_{i=1}^{m}\sum_{k=1}^{l} x_{ijk} \geq \hat{b}_j, \qquad j = 1, 2, \ldots, n \tag{5}$$

$$\sum_{i=1}^{m}\sum_{j=1}^{n} x_{ijk} \leq \hat{e}_k, \qquad k = 1, 2, \ldots, l \tag{6}$$

$$x_{ijk} \geq 0, \qquad \forall i, j, k. \tag{7}$$

The above model consists of one objective function (3) which minimizes the TTC for the transportation network. This function is defined by multiplying the neutrosophic fuzzy unit transportation cost by the unknown quantity of transported amounts. The total cost can be obtained by summing the converted fuzzy unit transportation cost to its equivalent crisp number and then by multiplying it to the amount of transported goods. In the minimization process of TTC, there are four fundamental constraints sets are imposed in the modeling to satisfy the feasibility condition of the transportation system. The first constraint (4) guarantee the source availability in all the source points, the second constraint (5) is for the demand for certain goods from the different destinations, the third constraint (6) checking the feasibility of conveyance capacities for smooth transportation using different vehicles, and the last constraint (7) is for the non-negativity restriction in the transportation system which means that the transported amount can not be a negative number. It should be always a positive number, as it is representing an amount of certain goods.

4 Solution Techniques

The mathematical model presented in Sect. 3 cannot be solved using the conventional methods developed to solve a TP because this model has the three-dimensional configuration, and those methods are designed to solve two-dimensional configured TP. Also, the neutrosophic fuzzy parameter present in the mathematical model needs to be converted first in their respective equivalent crisp number, and then by obtaining the equivalent crisp form of the neutrosophic fuzzy model, we can solve it using the LINGO software approach. Therefore, the solution technique is divided into two parts, the first part will be on the conversion process of a neutrosophic fuzzy number to an equivalent crisp number for certain parameters of the model, and the second part is about solving the converted mathematical model using LINGO.

4.1 Solution Part-I

It can be observed that the presented mathematical model consists of some parameters are in the form of neutrosophic fuzzy numbers, and these are $\hat{c}_{ijk}$, $\hat{a}_i$, $\hat{b}_j$, and

$\hat{e}_k$. The first step is to convert all these parameters from their neutrosophic fuzzy version to their respective equivalent crisp numbers, and for this, we have to apply the ranking function presented in Sect. 2.1. Since the model's objective function is of minimization type, we use the ranking function given by the equation (2). Once the conversion process is done, then the equivalent crisp model can be solved using a soft computing approach.

4.2 Solution Part-II

In this chapter, the LINGO software is used to solve the converted equivalent crisp model. LINGO is an iterative software tool designed to solve a linear, nonlinear programming problem, it works based on the generalized reduced gradient method (GRG). This technique is an extension of the reduced gradient technique to accommodate nonlinear inequality constraints. In this technique, a hunt direction is found such for any little move, the present active constraints stay exactly active. Here, it should be noted that the conventional method of solving a crisp TP cannot be utilized to solve an STP due to its complex configuration like non-equality constraints, three-dimensional structure, etc.

5 Numerical Experiment

This section presents a complete framework for the numerical experiment to the proposed model and the conversion process of a neutrosophic fuzzy number to its equivalent crisp number with numerical data in a very comprehensive manner. For the model, we choose 3 source points ($i = 3$), two destinations ($j = 2$), and two types of vehicles ($k = 2$) are available for transportation. The objective is to minimize the total transportation cost by satisfying all the demands and the respective constraints of source and conveyance.

Objective and constraint inputs are shown in the Tables 1 and 2. Uncertain parameters are assumed in the form of symmetric trapezoidal neutrosophic fuzzy numbers, such as [20].

Table 1 Input of the objective function for the model

Transportation cost				
j	$i = 1$	$i = 2$	$i = 3$	k
1	(13,15,2.81,2.81;0.87,0.21,0.21)	(17,23,3.33,3.33;0.85,0.43,0.29)	(23,29,4.95,4.95;0.69,0.13,0.08)	1
2	(20,27,4.51,4.51;0.97,0.32,0.27)	(22,31,5.85,5.85;0.99,0.01,0.08)	(15,23,2.98,2.98;0.67,0.23,0.09)	1
1	(18,26,3.17,3.17,0.82,0.45,0.07)	(11,19,2.03,2.03,0.93,0.21,0.12)	(14,21,2.71,2.71;0.99,0.08,0.12)	2
2	(10,16,1.97,1.97;0.78,0.33,0.41)	(16,25,2.59,2.59;0.45,0.78,0.38)	(22,31,4.94,4.94;0.37,0.56,0.87)	2

Table 2 Input of the constraints for the model

Capacity of each source point		
$i = 1$	$i = 1$	$i = 1$
(123,139,15.91,15.91;0.56,0.48,0.18)	(67,110,10.03,10.03;0.66,0.37,0.21)	(91,142,12.74,12.74;0.71,0.83,0.15)
Demand from each destination		
$j = 1$	$j = 1$	
(126,157,14.69,14.69;0.34,0.58,0.71)	(145,179,19.83,19.83;0.97,0.06,0.12)	
Capacity of each conveyance		
$k = 1$	k=2	
(189,243,27.56,27.56;0.71,0.54,0.31)	(93,134,15.42,15.42;0.57,0.49,0.67)	

5.1 A Particular Form of the General Model

With the neutrosophic fuzzy input presented in Tables 1 and 2, the general NSTP model 3 takes the following particular form.

$$\begin{aligned}
&\text{Min } Z = \hat{c}_{111}x_{111} + \hat{c}_{211}x_{211} + \hat{c}_{311}x_{311} + \hat{c}_{121}x_{121} + \hat{c}_{221}x_{221} + \hat{c}_{321} \\
&x_{321} + \hat{c}_{112}x_{112} + \hat{c}_{212}x_{212} + \hat{c}_{312}x_{312} + \hat{c}_{122}x_{122} + \hat{c}_{222}x_{222} + \hat{c}_{322}x_{322} \\
&\text{subject to,} \\
&x_{111} + x_{121} + x_{112} + x_{122} \le \hat{a}_1, \\
&x_{211} + x_{221} + x_{212} + x_{222} \le \hat{a}_2, \\
&x_{311} + x_{321} + x_{312} + x_{322} \le \hat{a}_3, \\
&x_{111} + x_{211} + x_{311} + x_{112} + x_{212} + x_{312} \ge \hat{b}_1, \\
&x_{121} + x_{221} + x_{321} + x_{122} + x_{222} + x_{322} \ge \hat{b}_2,
\end{aligned}$$

$$\begin{aligned}
&x_{111} + x_{211} + x_{311} + x_{121} + x_{221} + x_{321} \le \hat{e}_1, \\
&x_{112} + x_{212} + x_{312} + x_{122} + x_{222} + x_{322} \le \hat{e}_2, \\
&x_{111} \ge 0, x_{211} \ge 0, x_{311} \ge 0, x_{121} \ge 0, x_{221} \ge 0, x_{321} \ge 0, x_{112} \ge 0, x_{212} \ge 0, \\
&x_{312} \ge 0, x_{122} \ge 0, x_{222} \ge 0, x_{322} \ge 0.
\end{aligned}$$

In the particular form of the model, the parameters are neutrosophic fuzzy numbers, and hence, it is a fuzzy model, and thus to solve this particular model, we need one equivalent crisp model which can be obtained using the ranking function given by the equation (2). After the defuzzification of neutrosophic fuzzy parameters in the above fuzzy model, the following equivalent crisp model is obtained.

$$\text{Min } Z = 6.02 \times x_{111} + 10.14 \times x_{211} + 11.63 \times x_{311} + 10.35 \times x_{121} + 9.85 \times x_{221} + 10.41 \times x_{321} + 12.79 \times x_{112} + 9.51 \times x_{212} + 10.16 \times x_{312} + 7.13 \times x_{122} + 12.02 \times x_{222} + 10.62 \times x_{322}$$

subject to,

$$x_{111} + x_{121} + x_{112} + x_{122} \leq 83.17,$$
$$x_{211} + x_{221} + x_{212} + x_{222} \leq 58.49,$$
$$x_{311} + x_{321} + x_{312} + x_{322} \leq 78.01,$$
$$x_{111} + x_{211} + x_{311} + x_{112} + x_{212} + x_{312} \geq 96.48,$$
$$x_{121} + x_{221} + x_{321} + x_{122} + x_{222} + x_{322} \geq 103.30,$$
$$x_{111} + x_{211} + x_{311} + x_{121} + x_{221} + x_{321} \leq 133.18,$$
$$x_{112} + x_{212} + x_{312} + x_{122} + x_{222} + x_{322} \leq 66.65,$$
$$x_{111} \geq 0, x_{211} \geq 0, x_{311} \geq 0, x_{121} \geq 0, x_{221} \geq 0, x_{321} \geq 0, x_{112} \geq 0, x_{212} \geq 0,$$
$$x_{312} \geq 0, x_{122} \geq 0, x_{222} \geq 0, x_{322} \geq 0.$$

The equivalent crisp model is solved using the LINGO software and obtained the global optimal solution, i.e., the optimal allocation network for the transportation is obtained by satisfying all the constraints and finding the minimum total transportation cost. The following section presented the optimal solutions of the model and discussed the various feasible features of the solutions.

6 Result Discussion

The converted equivalent crisp model is coded in LINGO and solved. LINGO gives the optimal solution which is a global optimal solution for the NSTP. Table 3 represents the solution, i.e., the different allocations for the transportation system with the minimized total transportation cost. Now based on the obtained solution, the following feasibility conditions are hereby discussed.

It can be observed from the Table 3 that the total unit of certain transported goods is equal to 199.78 units, whereas the total demand based on the defuzzified value in equivalent crisp model from the two destinations was 199.78 units. That means each

Table 3 Solution obtained for the equivalent crisp model

Total transportation cost = 1688.505				
The optimal allocations				
j	$i = 1$	$i = 2$	$i = 3$	k
1	83.17	0	0	1
2	0	45.18	4.83	1
1	0	13.31	0	2
2	0	0	53.29	2

demand is fulfilled in this system. Therefore, the demand feasibility condition of an STP is satisfied for the proposed NSTP.

Next from the equivalent crisp model, it is observed that the total amount of certain goods available at the three source points is 219.67 units, and from the obtained solution, we have seen that the fulfilled demand was 199.78 units, therefore, the feasibility condition of an STP that the demand always should be less or equal to the available source at origin is satisfied for the proposed NSTP.

Also from the Table 3, it can be observed that the used vehicle's total capacity is 199.83 units, and it is also greater than the value of total demand (i.e., 199.78 units), which indicates that the feasibility condition for conveyance in an STP is satisfied in the proposed NSTP.

7 Conclusion and Future Direction

This chapter presents an application-based study of neutrosophic fuzzy numbers with the help of STP, which is a generalization of well know TP. The main advantage of an STP over a TP is that an STP contains an extra constraint compared to a TP, and it is the conveyance constraint that has a great impact on the total transportation cost overall. The neutrosophic fuzzy is also a generalization of well known fuzzy number, proposed by Zadeh [11]. The main advantage of using a neutrosophic fuzzy over the normal fuzzy is that the neutrosophic fuzzy consists of three kinds of membership functions (truth, falsity, indeterminacy), whereas in normal fuzzy, we have only one kind of membership function and that is truth only. This chapter discussed all the important information related to a neutrosophic fuzzy number and their conversion process by which one can get an equivalent crisp value of a neutrosophic fuzzy number. In the application segment of neutrosophic fuzzy number, the STP model is derived, and a numerical experiment is also performed. The iterative platform LINGO is utilized to obtain the solution from the equivalent crisp model. The obtained result was discussed and tested for the different feasibility conditions related to STP.

In future, the presented study of this chapter can be extended in several fields of research. For example, this chapter contains a two-ranking function for neutrosophic fuzzy numbers, which can be used in different decision-making problems such as disaster management problems, traffic management problems, data envelopment analysis modeling, and inventory management. Following the proposed NSTP model, one can extend this generalized model to an industrial problem like transportation of cold supply chain, sustainable supply chain, disaster freight transportation, debris removal in disaster management, etc. Overall, this chapter gives a new direction to the study of the inherent uncertainty in our real-life problems.

References

1. Hitchcock, F.L.: The distribution of a product from several sources to numerous localities. J. Math. Phys. **20**(1–4), 224–230 (1941)
2. Shell, E.: Distribution of a product by several properties, directorate of management analysis. In: Proceedings of the Second Symposium in Linear Programming, Vol. 2, pp. 615–642 (1955)
3. Haley, K.: New methods in mathematical programming-the solid transportation problem. Oper. Res. **10**(4), 448–463 (1962)
4. Sengupta, D., Das, A., Bera, U.K.: A gamma type-2 defuzzification method for solving a solid transportation problem considering carbon emission. Appl. Intelli. **48**(11), 3995–4022 (2018)
5. Das, A., Bera, U.K., Maiti, M.: A solid transportation problem in uncertain environment involving type-2 fuzzy variable. Neural Comput. Appl. **31**(9), 4903–4927 (2019)
6. Kundu, P., Kar, S., Maiti, M.: Multi-objective solid transportation problems with budget constraint in uncertain environment. Int. J. Syst. Sci. **45**(8), 1668–1682 (2014)
7. Sengupta, D., Das, A., Dutta, A., Bera, U.K.: A fixed charge solid transportation problem with possibility and expected value approaches in hybrid uncertain environment. In: International Conference on Information Technology and Applied Mathematics. Springer, Heidelberg, pp. 182–193 (2019)
8. Pramanik, S., Jana, D.K., Mondal, S.K., Maiti, M.: A fixed-charge transportation problem in two-stage supply chain network in gaussian type-2 fuzzy environments. Inf. Sci. **325**, 190–214 (2015)
9. Ojha, A., Das, B., Mondal, S., Maiti, M.: A solid transportation problem for an item with fixed charge, vechicle cost and price discounted varying charge using genetic algorithm. Appl. Soft Comput. **10**(1), 100–110 (2010)
10. Mondal, A., Roy, S.K., Midya, S.: Intuitionistic fuzzy sustainable multi-objective multi-item multi-choice step fixed-charge solid transportation problem. J. Ambient Intelli. Humanized Comput. 1–25 (2021)
11. Zadeh, L.: Fuzzy sets. Inf. Control **8**(3), 338–353 (1965)
12. Atanassov, K.: Intuitionistic fuzzy sets. Int. J. Bioautomation **20**, 1 (2016)
13. Mendel, J.M.: Type-2 fuzzy sets and systems: an overview. IEEE Comput. Intelli. Mag. **2**(1), 20–29 (2007)
14. Torra, V.: Hesitant fuzzy sets. Int. J. Intelli. Syst. **25**(6), 529–539 (2010)
15. Smarandache, F.: Neutrosophic set-a generalization of the intuitionistic fuzzy set. Int. J. Pure Appl. Math. **24**(3), 287 (2005)
16. Sarma, D., Das, A., Bera, U.K., Hezam, I.M.: Redistribution for cost minimization in disaster management under uncertainty with trapezoidal neutrosophic number. Comput. Ind. **109**, 226–238 (2019)
17. Ghosh, S., Roy, S.K., Verdegay, J.L.: Fixed-Charge Solid Transportation Problem with Budget Constraints Based on Carbon Emission in Neutrosophic Environment (2021)
18. Singh, A., Das, A., Bera, U.K., Lee, G.M.: Prediction of transportation costs using trapezoidal neutrosophic fuzzy analytic hierarchy process and artificial neural networks. IEEE Access **9**, 103497–103512 (2021)
19. Oepen, S., Flickinger, D., Toutanova, K., Manning, C.D.: Lingo redwoods. Res. Language Comput. **2**(4), 575–596 (2004)
20. Abdel-Basset, M., Gunasekaran, M., Mohamed, M., Smarandache, F.: A novel method for solving the fully neutrosophic linear programming problems. Neural Comput. Appl. 1–11 (2018)
21. Mohamed, M., Abdel-Basset, M., Zaied, A.N.H., Smarandache, F.: Neutrosophic integer programming problem. In: Infinite Study (2017)
22. Ganesan, K., Veeramani, P.: Fuzzy linear programs with trapezoidal fuzzy numbers. Ann. Oper. Res. **143**(1), 305–315 (2006)
23. Ebrahimnejad, A., Tavana, M.: A novel method for solving linear programming problems with symmetric trapezoidal fuzzy numbers. Appl. Math. Modelling **38**(17–18), 4388–4395 (2014)

Cubic Picture Hesitant Fuzzy Linear Spaces and Their Applications in Multi Criteria Decision Making

P. R. Kavyasree and B. Surender Reddy

Abstract Numerous techniques have been developed to confront MCDM problems effectively. This article proposes the notion of cubic picture hesitant fuzzy linear spaces based on the notion of interval-valued picture hesitant fuzzy linear spaces and picture hesitant fuzzy linear spaces. With a few examples, we also present $P(R)-$intersection and $P(R)-$union of internal and external cubic picture hesitant linear spaces, as well as their properties. Secondly, a series of operators like cubic picture hesitant fuzzy weighted averaging aggregation operators, cubic picture hesitant fuzzy order weighted averaging aggregation operators, and cubic picture hesitant fuzzy hybrid order weighted averaging aggregation operators are developed. Third, these aggregation operators are further extended to cubic picture hesitant fuzzy prioritized weighted averaging aggregation operators in order to solve the MCDM problems by assigning different priorities to the given criteria. Eventually, a real-life MCDM problem has been solved with the help of the proposed operator to determine the best alternative.

Keywords Multi-criteria decision-making · Cubic picture hesitant fuzzy linear spaces · Cubic picture hesitant fuzzy weighted averaging aggregation operators · Cubic picture hesitant fuzzy prioritized weighted averaging aggregation operators

Mathematics Subject Classification 46B20 · 46B99 · 46A45 · 46B45 · 46A1

1 Introduction

Despite the introduction of fuzzy sets by Zadeh [1] in 1965 to handle incomplete, vague, and imprecise information, it is helpful only to some extent and cannot solve complex and naturally incomplete data. To further evaluate the information more

P. R. Kavyasree (✉) · B. Surender Reddy
Department of Mathematics, Osmania University, Tarnaka, Hyderabad, Telangana 500007, India
e-mail: kavyasree.anu@gmail.com; kavyasree@alumni.iisertvm.ac.in

B. Surender Reddy
e-mail: bsrmathou@osmania.ac.in

L. Sahoo et al. (eds.), *Real Life Applications of Multiple Criteria Decision Making Techniques in Fuzzy Domain*, Studies in Fuzziness and Soft Computing 420,
https://doi.org/10.1007/978-981-19-4929-6_25

effectively, fuzzy sets were extended to many generalized forms. These generalized forms of fuzzy sets include interval-valued fuzzy sets by Zadeh [2] in 1975, intuitionistic fuzzy sets [3] in 1986, cubic sets [4], picture fuzzy sets [5], hesitant fuzzy sets [6], picture hesitant fuzzy sets [7]. This vast development of fuzzy sets in recent years has helped researchers carry out their studies on complex MCDM problems [8–10]. For instance, if there are six investors interested in investing in setting up a restaurant in this case, choosing a place in order to get good returns is the first priority; now, two investors among six suggest the place is good to start the business; two are totally against the place, and the other two are abstaining from making any decision. Clearly, in such cases, fuzzy sets can only decide the degree of membership after evaluating the information, and thus, the opinion of every investor cannot be taken into consideration using fuzzy sets. But, it is quite important that rejection plays an equally important role. To handle such an issue, Atanassov further advanced the theory of fuzzy sets to intuitionistic fuzzy sets [3] which also includes the degree of non-membership. Subsequently, interval numbers were used instead of crisp numbers. Jun put forward a novel hybrid structure combining interval-valued fuzzy sets and fuzzy sets called cubic sets. Jana et al. [11], Senapati et al. [12–15] applied cubic set theories to different algebraic structures.

Although all these make it easier to handle the uncertainties to a great extent, determining the degree of accurate membership by crisp numbers is sometimes difficult using fuzzy sets. As a result, the hesitant fuzzy sets [6] were introduced, in which the degree of membership was represented by distinct crisp numbers. Dey et al. [16] presented a new technique for hesitant multi-fuzzy-based decision-making issues. Senapati et al. [17] introduced some hesitant fuzzy Aczel-Alsina aggregation operators and applied them in cyclone disaster assessment. These hesitant fuzzy sets are further extended to interval-valued hesitant fuzzy sets and dual-hesitant fuzzy sets [18].

Currently, many scholars have been involved in developing different aggregation operators to address countless complex MCDM problems. For instance, with regard to picture fuzzy sets, Wei [19] proposed the picture fuzzy weighted aggregation operators. Jana et al. [20] proposed picture fuzzy Dombi aggregation operators. Senapati et al. introduced the picture fuzzy WASPAS technique [21] and picture fuzzy Aczel-Alsina average aggregation operators [21] and utilized them in decision-making issues. Abdullah et al. [22] introduced the cubic picture fuzzy weighted average operator. Likewise, hesitant fuzzy aggregation operators [23], generalized picture hesitant fuzzy aggregation operators were developed.

Moreover, MCDM problems are further provided with a criteria set with different priorities. Setting up priorities for the criteria is important because, for example, safety and price are two criteria where safety is given higher priority than price. Hence, Yager [24] put forward the concept of a prioritized averaging operator. Motivated by Yu et al. [25] introduced the intuitionistic fuzzy prioritized averaging operator and interval-valued intuitionistic fuzzy prioritized averaging operator. Apart from this, hesitant fuzzy prioritized aggregation operators [19] are proposed.

In addition to this, Lubczonok and Murali [26] proposed a striking theory of flags and fuzzy subspaces of vector spaces. Recently, researchers have been further extending fuzzy linear spaces and basic operations on these fuzzy linear spaces. These extensions of fuzzy linear spaces include: interval-valued fuzzy linear spaces, cubic linear spaces [27], cubic Γ $n-$normed linear spaces [28], $N-$cubic sets applied to linear spaces [29], cubic picture fuzzy linear spaces [30]. The purpose of this paper is to introduce the notion of cubic picture hesitant fuzzy linear spaces, combining interval-valued picture hesitant fuzzy linear space, and picture hesitant fuzzy linear space, and to discuss prudent properties with examples.

Keeping in mind the abovementioned theory, this paper proposes cubic picture hesitant fuzzy prioritized weighted averaging aggregation operators, cubic picture hesitant fuzzy prioritized order weighted averaging aggregation operators, and cubic picture hesitant fuzzy prioritized hybrid weighted averaging aggregation operators. In addition to this, using the above-referred operators, a real-life MCDM problem has been evaluated.

2 Preliminaries

Definition 1 A hesitant fuzzy set on a non-empty set M is defined as a function $\hat{h}(m)$ when applied to M which returns a finite subset of [0, 1] that is

$$h_f = \{\langle m, \hat{h}(m)\rangle |\ m \in M\}$$

where $\hat{h}(m)$ is a set of some different values in [0, 1] which denotes the possible membership degrees of an element $m \in M$ to the set h_f.

Definition 2 An interval-valued hesitant fuzzy set $\widetilde{h_f}$ on a non-empty set M is defined as

$$\widetilde{h_f} = \{\langle m, \widetilde{\hat{h}}(m)\rangle |\ m \in M\}$$

where $\widetilde{\hat{h}}(m)$ is a set of some different interval values in [0, 1] which denotes the possible membership degrees of an element $m \in M$ to the set $\widetilde{h_f}$.

Definition 3 A cubic hesitant fuzzy set $C_{h_f} = \{\langle m, \widetilde{h_f}(m), h_f(m)\rangle |\ m \in M\}$ where $\widetilde{h_f}$ is an interval-valued hesitant fuzzy set and h_f is an hesitant fuzzy set and is simply denoted by $C_{h_f} = \langle \widetilde{h_f}, h_f\rangle$.

Definition 4 A picture fuzzy set P_f on a non-empty set M is of the form

$$P_f = \{\big(m, \mu(m), \eta(m), \vartheta(m)\big) |\ m \in M\}$$

where $\mu : M \to [0, 1], \eta : M \to [0, 1]$, and $\vartheta : M \to [0, 1]$ are said to be the degree of positive, neutral, and negative membership degrees of $m \in M$ and also satisfy the following condition

$$0 \leq \mu(m) + \eta(m) + \vartheta(m) \leq 1\ \forall\ m \in M.$$

Definition 5 An interval-valued picture fuzzy set on a non-empty set M is of the form

$$\tilde{P_f} = \{(m, \tilde{\mu}(m), \tilde{\eta}(m), \tilde{\vartheta}(m))|\ m \in M\}$$

where $\tilde{\mu}(m), \tilde{\eta}(m), \tilde{\vartheta}(m)$ where each represents interval value of positive, neutral, and negative membership degrees of $m \in M$, respectively. Also, the following condition is satisfied

$$0 \le \sup(\tilde{\mu}(m)) + \sup(\tilde{\eta}(m)) + \sup(\tilde{\vartheta}(m)) \le 1\ \forall\ m \in M.$$

Definition 6 A picture hesitant fuzzy set Ph_f on a non-empty set M is of the form.

$$Ph_f = \{(m, \hat{\mu}(m), \hat{\eta}(m), \hat{\vartheta}(m))|\ m \in M\}$$

where $\hat{\mu}(m), \hat{\eta}(m), \hat{\vartheta}(m)$ represents different values of positive, neutral, and negative membership degrees in [0, 1].

Definition 7 An interval-valued picture hesitant fuzzy set on a non-empty set M is of the form

$$\widetilde{Ph_f} = \{(m, \tilde{\hat{\mu}}(m), \tilde{\hat{\eta}}(m), \tilde{\hat{\vartheta}}(m)|\ m \in M\}$$

where $\tilde{\hat{\mu}}(m), \tilde{\hat{\eta}}(m), \tilde{\hat{\vartheta}}(m)$ gives different interval values of positive, neutral, and negative membership degrees in [0, 1].

Definition 8 A cubic picture fuzzy set on a non-empty set M is of the form

$$C_{P_f} = \{(m, \tilde{P_f}(m), P_f(m)|\ m \in M)\}$$

where $\tilde{P_f}(m)$ is an interval-valued picture fuzzy set in M and $P_f(m)$ is a picture fuzzy set in M. Simply, we denote cubic picture fuzzy set as $C_{P_f} = (\tilde{P_f}, P_f)$.

Definition 9 A cubic picture hesitant fuzzy set on a non-empty set M is of the form

$$C_{Ph_f} = \{m, \widetilde{Ph_f}(m), Ph_f(m)|\ m \in M\}$$

where $\widetilde{Ph_f}(m)$ is an interval-valued picture hesitant fuzzy set in M and $Ph_f(m)$ is a picture hesitant fuzzy set in M.. Simply, we denote cubic picture hesitant fuzzy set as $C_{Ph_f} = (\tilde{Ph_f}, Ph_f)$.

Definition 10 For an interval-valued picture hesitant fuzzy set $\widetilde{Ph_f}$,, we define refined minimum (in short, rmin) of $\tilde{\hat{\mu}}_i = [^{-}\tilde{\hat{\mu}}_i, \tilde{\hat{\mu}}_i^{+}]$ by,

$$[\tilde{\hat{\mu}}_i^{-}, \tilde{\hat{\mu}}_i^{+}]^{-} = r\min\{[\tilde{\hat{\mu}}_i^{-}(m), \tilde{\hat{\mu}}_i^{+}](m)\} = [\min\{\tilde{\hat{\mu}}_i^{-}(m)\}, \min\{\tilde{\hat{\mu}}_i^{+}(m)\}].$$

we define refined maximum (in short, rmax) of $\tilde{\hat{\mu}}_i = [^{-}\tilde{\hat{\mu}}_i, \tilde{\hat{\mu}}_i^{+}]$ by,

$$[\tilde{\hat{\mu}}_i^{-}, \tilde{\hat{\mu}}_i^{+}]^{+} = r\max\{[\tilde{\hat{\mu}}_i^{-}(m), \tilde{\hat{\mu}}_i^{+}(m)]\} = [\max\{\tilde{\hat{\mu}}_i^{-}(m)\}, \max\{\tilde{\hat{\mu}}_i^{+}(m)\}].$$

Definition 11 Let $C_{Ph_{f_1}} = \{\widetilde{Ph_{f_1}}, Ph_{f_1}\}$ and $C_{Ph_{f_2}} = \{\widetilde{Ph_{f_2}}, Ph_{f_2}\}$ be two cubic picture hesitant fuzzy sets then,

(i) **P-Union**

$C_{Ph_{f_1}} \cup_p C_{Ph_{f_2}} = \Big\{\big(l, \gamma \in \widetilde{Ph_{f_1}} \cup \widetilde{Ph_{f_2}}(l), \kappa \in Ph_{f_1} \cup Ph_{f_2}(l) |\ \gamma = [\gamma^{-}, \gamma^{+}] \geq \max\{[\tilde{\hat{\mu}}_i^{-}(l), \tilde{\hat{\mu}}_i^{+}(l)]^{-}, [\tilde{\hat{\eta}}_i^{-}(l), \tilde{\hat{\eta}}_i^{+}(l)]^{-}, [\tilde{\hat{\vartheta}}_i^{-}(l), \tilde{\hat{\vartheta}}_i^{+}(l)]^{-}\}, \kappa \geq \max\{\hat{\mu}^{-}(l), \hat{\eta}^{-}(l), \hat{\vartheta}^{-}(l)\}\big)\Big\}.$

(ii) **P-Intersection**

$C_{Ph_{f_1}} \cap_p C_{Ph_{f_2}} = \Big\{\big(l, \gamma \in \widetilde{Ph_{f_1}} \cup \widetilde{Ph_{f_2}}(l), \kappa \in Ph_{f_1} \cup Ph_{f_2}(l) |\quad \gamma = [\gamma^{-}, \gamma^{+}] \leq \min\{[\tilde{\hat{\mu}}_i^{-}(l), \tilde{\hat{\mu}}_i^{+}(l)]^{+}, [\tilde{\hat{\eta}}_i^{-}(l), \tilde{\hat{\eta}}_i^{+}(l)]^{+}, [\tilde{\hat{\vartheta}}_i^{-}(l), \tilde{\hat{\vartheta}}_i^{+}(l)]^{+}\}, \kappa \leq \min\{\hat{\mu}^{+}(l), \hat{\eta}^{+}(l), \hat{\vartheta}^{+}(l)\}\big)\Big\}.$

(iii) **R-Union**

$C_{Ph_{f_1}} \cup_R C_{Ph_{f_2}} = \Big\{\big(l, \gamma \in \widetilde{Ph_{f_1}} \cup \widetilde{Ph_{f_2}}(l), \kappa \in Ph_{f_1} \cup Ph_{f_2}(l) |\quad \gamma = [\gamma^{-}, \gamma^{+}] \geq \max\{[\tilde{\hat{\mu}}_i^{-}(l), \tilde{\hat{\mu}}_i^{+}(l)]^{-}, [\tilde{\hat{\eta}}_i^{-}(l), \tilde{\hat{\eta}}_i^{+}(l)]^{-}, [\tilde{\hat{\vartheta}}_i^{-}(l), \tilde{\hat{\vartheta}}_i^{+}(l)]^{-}\}, \kappa \leq \min\{\hat{\mu}^{+}(l), \hat{\eta}^{+}(l), \hat{\vartheta}^{+}(l)\}\big)\Big\}.$

(iii) **R-Intersection**

$C_{Ph_{f_1}} \cap_R C_{Ph_{f_2}} = \Big\{\big(l, \gamma \in \widetilde{Ph_{f_1}} \cup \widetilde{Ph_{f_2}}(l), \kappa \in Ph_{f_1} \cup Ph_{f_2}(l) |\quad \gamma = [\gamma^{-}, \gamma^{+}] \leq \min\{[\tilde{\hat{\mu}}_i^{-}(l), \tilde{\hat{\mu}}_i^{+}(l)]^{+}, [\tilde{\hat{\eta}}_i^{-}(l), \tilde{\hat{\eta}}_i^{+}(l)]^{+}, [\tilde{\hat{\vartheta}}_i^{-}(l), \tilde{\hat{\vartheta}}_i^{+}(l)]^{+}\}, \kappa \geq \max\{\hat{\mu}^{-}(l), \hat{\eta}^{-}(l), \hat{\vartheta}^{-}(l)\}\big)\Big\}.$

3 Results

3.1 *Cubic Picture Hesitant Fuzzy Linear Spaces*

Definition 12 For a non-empty linear space £ over a field F, a hesitant fuzzy linear space is defined as a pair $£^{hf}$=(£, $\hat{h}$) and $\hat{h} : £ \to \phi([0, 1])$ where $\phi([0, 1])$ denote set all finite subsets of [0, 1] and satisfy the following condition

$$£^{hf}(\alpha l_1 * \beta l_2) \geq £^{hf}(l_1) \cap £^{hf}(l_2).$$

for $l_1, l_2 \in £$ and $\alpha, \beta \in F$.

Definition 13 For a non-empty linear space £ over a field F, an interval hesitant fuzzy linear space is defined as a pair $\widetilde{£^{hf}}$=(£, $\tilde{\hat{h}}$) and $\tilde{\hat{h}} : £ \to D(\phi([0, 1]))$ where $D(\phi([0, 1]))$ denote collection of all finite sub-intervals of [0, 1] and satisfy following condition

$$\widetilde{£^{hf}}(\alpha l_1 * \beta l_2) \geq \min\{\widetilde{£^{hf}}(l_1), \widetilde{£^{hf}}(l_2)\}$$

for $l_1, l_2 \in Ł$ and $\alpha, \beta \in F$.

Definition 14 For a non-empty linear space Ł over a field F, a cubic hesitant fuzzy set $C_{h_f} = \{\langle m, \widetilde{h_f}(m), h_f(m)\rangle | \ m \in M\}$ is said to be a cubic hesitant fuzzy linear space of £ if the following conditions are satisfied

$$\widetilde{h_f}(\alpha l_1 * \beta l_2) \geq \min\{\widetilde{h_f}(l_1), \widetilde{h_f}(l_1)\},$$
$$h_f(\alpha l_1 * \beta l_2) \leq \max\{h_f(l_1), h_f(l_2)\}.$$

Definition 15 For a non-empty linear space £ over a field F, a picture hesitant fuzzy linear space is defined as a pair $£^{ph_f} = (£, \hat{\mu}, \hat{\eta}, \hat{\vartheta})$ where $\hat{\mu} : £ \rightarrow \phi([0, 1])$ is the degree of positive membership, $\hat{\eta} : £ \rightarrow \phi([0, 1])$ is the degree of neutral membership and $\hat{\vartheta} : £ \rightarrow \phi([0, 1])$ is the degree of negative membership and satisfy the following condition

$$£^{ph_f}(\alpha l_1 * \beta l_2 * \gamma l_3) \geq £^{ph_f}(l_1) \cap £^{ph_f}(l_2) \cap £^{ph_f}(l_3)$$

for any $l_1, l_2, l_3 \in £$ and $\alpha, \beta, \gamma \in F$.

Definition 16 For a non-empty linear space £ over a field F, an interval-valued picture hesitant fuzzy linear space is defined as pair $\widetilde{£^{ph_f}} = (£, \widetilde{\hat{\mu}}, \widetilde{\hat{\eta}}, \widetilde{\hat{\vartheta}})$ where $\widetilde{\hat{\mu}} : £ \rightarrow \phi(D[0, 1]), \widetilde{\hat{\eta}} : £ \rightarrow \phi(D[0, 1])$ and $\widetilde{\hat{\vartheta}} : £ \rightarrow \phi(D[0, 1])$ are the degree of positive, neutral, and negative membership degrees, respectively, and satisfy the following condition.

$$\widetilde{£^{ph_f}}(\alpha l_1 * \beta l_2 * \gamma l_3) \geq min\{\widetilde{£^{ph_f}}(l_1), \widetilde{£^{ph_f}}(l_2), \widetilde{£^{ph_f}}(l_3)\}$$

for any $l_1, l_2, l_3 \in £$ and $\alpha, \beta, \gamma \in F$.

Definition 17 For a non-empty linear space £ over a field F, a cubic picture hesitant fuzzy set $C_{Ph_f} = \{m, \widetilde{Ph_f}(m), Ph_f(m) | \ m \in M\}$ is said to be a cubic picture hesitant

Table 1 Values of interval-valued picture hesitant fuzzy sets and picture hesitant fuzzy sets

Ł	$\widetilde{Ph_f}$	Ph_f
l_1	([0.40, 0.53], [0.25, 0.39], [0.39, 0.41]), ([0.31, 0.50], [0.50, 0.60], [0.35, 0.45])	(0.45, 0.35, 0.40), (0.33, 0.51, 0.36)
l_1	([0.33, 0.43], [0.15, 0.28], [0.27, 0.29]), ([0.21, 0.40], [0.40, 0.50], [0.25, 0.35])	(0.42, 0.27, 0.28), (0.33, 0.49, 0.34)
l_3	([0.43, 0.53], [0.25, 0.40], [0.15, 0.16]), ([0.31, 0.50], [0.50, 0.60], [0.35, 0.45])	(0.52, 0.39, 0.16), (0.43, 0.59, 0.44)

fuzzy linear space of £ if the following conditions are satisfied

$$\widetilde{Ph_f}(\alpha l_1 * \beta l_2 * \gamma l_3) \geq \min\left(\widetilde{Ph_f}(l_1), \widetilde{Ph_f}(l_2), \widetilde{Ph_f}(l_3)\right),$$
$$Ph_f(\alpha l_1 * \beta l_2 * \gamma l_3) \leq \max\left(Ph_f(l_1), Ph_f(l_2), Ph_f(l_3)\right).$$

Example 1 The example provided below explains the above definition (Table 1). From the above table, we notice that $\widetilde{Ph_f}$ satisfies the condition required for it to be an inter-valued picture hesitant fuzzy linear space and likewise Ph_f is a picture hesitant fuzzy linear space of £ over field $GF(2)$ with the binary operation $l_2 * l_3 = l_1$ and $\beta = \gamma = 1, \alpha = 0$. And hence, the above example in fact satisfied the conditions required for the cubic picture hesitant fuzzy set to be a cubic picture hesitant fuzzy linear space.

3.2 Internal and External Cubic Picture Hesitant Fuzzy Linear Spaces

Definition 18 A cubic picture hesitant fuzzy set $C_{Ph_f} = \{m, \widetilde{Ph_f}(m), Ph_f(m) | \ m \in M\}$ is said to be an internal cubic picture hesitant fuzzy linear space (shortly, ICPH-FLS) if

$$(\widetilde{Ph_f})^-(\alpha l_1 * \beta l_2) \leq Ph_f(\alpha l_1 * \beta l_2) \leq (\widetilde{Ph_f})^+(\alpha l_1 * \beta l_2)$$

for all $l_1, l_2 \in$ Ł and $\alpha, \beta \in F$.

Example 2 Let us consider the values of interval-valued picture hesitant fuzzy set and picture hesitant fuzzy set as in Table 1. Now for $\beta = \gamma = 1, \alpha = 0$ and $l_2 * l_3 = l_1$ in Definition 18 we have

$$(\widetilde{Ph_f})^-(l_2 * l_3) \leq Ph_f(l_2 * l_3) \leq (\widetilde{Ph_f})^+(l_2 * l_3)$$
$$(\widetilde{Ph_f})^-(l_1) \leq Ph_f(l_1) \leq (\widetilde{Ph_f})^+(l_1)$$

$\Rightarrow 0.45 \in [0.40, 0.53], 0.35 \in [0.25, 0.39], 0.40 \in [0.39, 0.41], 0.33 \in [0.21, 0.40], 0.51 \in [0.50, 0.60], 0.36 \in [0.35, 0.45]$. Hence $C_{Ph_f} = \{m, \widetilde{Ph_f}(m), Ph_f(m) | \ m \in M\}$ is an internal cubic picture hesitant fuzzy linear space.

Definition 19 A cubic picture hesitant fuzzy set $C_{Ph_f} = \{m, \widetilde{Ph_f}(m), Ph_f(m) | \ m \in M\}$ in a linear space Ł over a field F is said to be an internal cubic picture hesitant fuzzy linear space (shortly, ECPHFLS) if

$$Ph_f(\alpha l_1 * \beta l_2) \notin \left((\widetilde{Ph_f})^-(\alpha l_1 * \beta l_2), (\widetilde{Ph_f})^+(\alpha l_1 * \beta l_2)\right)$$

for all $l_1, l_2 \in$ Ł and $\alpha, \beta \in F$.

Table 2 Values of interval-valued picture hesitant fuzzy sets and picture hesitant fuzzy sets

Ł	$\widetilde{Ph_f}$	Ph_f
l_1	$([0.11, 0.30], [0.30, 0.40], [0.15, 0.25])$	$(0.31, 0.21, 0.14)$
l_2	$([0.11, 0.20], [0.20, 0.30], [0.05, 0.15])$	$(0.10, 0.19, 0.16)$
l_3	$([0.11, 0.30], [0.30, 0.40], [0.15, 0.25])$	$(0.33, 0.29, 0.14)$

Example 3 Let us consider the values of interval-valued picture hesitant fuzzy set and picture hesitant fuzzy set as in table below. Now for $\beta = \gamma = 1, \alpha = 0$ and $l_2 * l_3 = l_1$ in Table 2.

$$Ph_f(l_2 * l_3) \notin \left((\widetilde{Ph_f})^{-}(l_2 * l_3), (\widetilde{Ph_f})^{+}(l_2 * l_3) \right)$$

$$Ph_f(l_1) \notin \left((\widetilde{Ph_f})^{-}(l_1), (\widetilde{Ph_f})^{+}(l_1) \right)$$

$\Rightarrow 0.31 \notin [0.11, 0.30], 0.21 \notin [0.30, 0.40]$ and $0.14 \notin [0.15, 0.25]$. Hence $C_{Ph_f} = \{m, \widetilde{Ph_f}(m), Ph_f(m) | \, m \in M\}$ is an external cubic picture hesitant fuzzy linear space.

Proposition 1 *Let* $C_{Ph_{f_1}} = \{\widetilde{Ph_{f_1}}, Ph_{f_1}\}$ *and* $C_{Ph_{f_2}} = \{\widetilde{Ph_{f_2}}, Ph_{f_2}\}$ *be two ICPHFLS. Then, their P−union and P−intersection is again an ICPHFLS.*

Proof Since $C_{Ph_{f_1}} = \{\widetilde{Ph_{f_1}}, Ph_{f_1}\}$ and $C_{Ph_{f_2}} = \{\widetilde{Ph_{f_2}}, Ph_{f_2}\}$ are ICPHFLS in £, we have

$$(\widetilde{Ph_{f_1}})^{-}(\alpha l_1 * \beta l_2 * \gamma l_3) \leq Ph_{f_1}(\alpha l_1 * \beta l_2 * \gamma l_3) \leq (\widetilde{Ph_{f_1}})^{+}(\alpha l_1 * \beta l_2 * \gamma l_3).$$
$$(\widetilde{Ph_{f_2}})^{-}(\alpha l_1 * \beta l_2 * \gamma l_3) \leq Ph_{f_2}(\alpha l_1 * \beta l_2 * \gamma l_3) \leq (\widetilde{Ph_{f_2}})^{+}(\alpha l_1 * \beta l_2 * \gamma l_3).$$

for all $l_1, l_2 \in$ Ł and $\alpha, \beta \in F$.

Thus, we obtain that

$$\begin{aligned} &(\widetilde{Ph_{f_1}} \cup \widetilde{Ph_{f_2}})^{-}(\alpha l_1 * \beta l_2 * \gamma l_3) \\ &\leq Ph_{f_1} \cup Ph_{f_2}(\alpha l_1 * \beta l_2 * \gamma l_3) \\ &\leq (\widetilde{Ph_{f_1}} \cup \widetilde{Ph_{f_2}})^{+}(\alpha l_1 * \beta l_2 * \gamma l_3). \end{aligned}$$

$$\begin{aligned} &(\widetilde{Ph_{f_1}} \cap \widetilde{Ph_{f_2}})^{-}(\alpha l_1 * \beta l_2 * \gamma l_3) \\ &\leq Ph_{f_1} \cap Ph_{f_2}(\alpha l_1 * \beta l_2 * \gamma l_3) \\ &\leq (\widetilde{Ph_{f_1}} \cap \widetilde{Ph_{f_2}})^{+}(\alpha l_1 * \beta l_2 * \gamma l_3). \end{aligned}$$

Table 3 Values of interval-valued picture hesitant fuzzy sets and picture hesitant fuzzy sets

Ł	$\widetilde{Ph_f}$	Ph_f
l_1	([0.1, 0.3], [0.16, 0.26], [0.22, 0.34])	(0.08, 0.15, 0.21)
l_2	([0.1, 0.4], [0.15, 0.25], [0.2, 0.31])	(0.05, 0.14, 0.15)
l_3	([0.05, 0.2], [0.1, 0.2], [0.2, 0.3])	(0.01, 0.22, 0.32)

Table 4 Values of interval-valued picture hesitant fuzzy sets and picture hesitant fuzzy sets

Ł	$\widetilde{Ph_f}$	Ph_f
l_1	([0.1, 0.2], [0.15, 0.28], [0.2, 0.4])	(0.22, 0.11, 0.19)
l_2	([0.05, 0.15], [0.1, 0.2], [0.2, 0.51])	(0.18, 0.3, 0.52)
l_3	([0.1, 0.5], [0.04, 0.14], [0.2, 0.35])	(0.25, 0.16, 0.43)

Hence, $P-$union and $P-$intersection of $C_{Ph_{f_1}} = \{\widetilde{Ph_{f_1}}, Ph_{f_1}\}$ and $C_{Ph_{f_2}} = \{\widetilde{Ph_{f_2}}, Ph_{f_2}\}$ is again an ICPHFLS.

Remark 1 The example given below tells us that P-union of ECPHFLS need not again be an ECPHFLS.

Example 4 Let us consider the values of interval-valued picture fuzzy linear spaces and picture fuzzy linear space as in Table 6 and also values tabulated below (Tables 3 and 4).

$$\begin{aligned}
\widetilde{Ph_{f_1}} \cup \widetilde{Ph_{f_2}}(l_1) &= ([0.1, 0.3], [0.2, 0.4]),\\
\widetilde{Ph_{f_1}} \cup \widetilde{Ph_{f_2}}(l_2) &= ([0.2, 0.51], [0.1, 0.4]),\\
\widetilde{Ph_{f_1}} \cup \widetilde{Ph_{f_2}}(l_3) &= ([0.05, 0.2], [0.1, 0.5]).
\end{aligned}$$

For $\beta = \gamma = 1$ and $\alpha = 0$ in Definition 14 we have,

$$\widetilde{Ph_{f_1}} \cup \widetilde{Ph_{f_2}}(l_2 * l_3) \geq \min\left\{\widetilde{Ph_{f_1}} \cup \widetilde{Ph_{f_2}}(l_2), \widetilde{Ph_{f_1}} \cup \widetilde{Ph_{f_2}}(l_3)\right\}$$

$$\begin{aligned}
\widetilde{Ph_{f_1}} \cup \widetilde{Ph_{f_2}}(l_1) \geq \min\left\{([0.2, 0.51], [0.1, 0.4]), ([0.05, 0.2], [0.1, 0.5]])\right\}\\
= ([0.05, 0.2], [0.1, 0.4]).
\end{aligned}$$

$\Rightarrow \widetilde{Ph_{f_1}} \cup \widetilde{Ph_{f_2}}(l_1) = ([0.1, 0.3], [0.2, 0.4]) \geq ([0.05, 0.2], [0.1, 0.4])$ which is incorrect.

$$\begin{aligned}
Ph_{f_1} \cup Ph_{f_2}(l_1) &= (0.11, 0.15, 0.19, 0.21, 0.22),\\
Ph_{f_1} \cup Ph_{f_2}(l_2) &= (0.18, 0.22, 0.3, 0.32, 0.52),\\
Ph_{f_1} \cup Ph_{f_2}(l_3) &= (0.16, 0.22, 0.25, 0.32, 0.43).
\end{aligned}$$

Table 5 Values of interval-valued picture hesitant fuzzy sets and picture hesitant fuzzy sets

Ł	$\widetilde{Ph_f}$	Ph_f
l_1	([0.1, 0.3], [0.16, 0.26], [0.22, 0.44])	(0.08, 0.15, 0.21)
l_2	([0.2, 0.4], [0.15, 0.25], [0.2, 0.31])	(0.05, 0.04, 0.15)
l_3	([0.05, 0.08], [0.1, 0.2], [0.2, 0.3])	(0.01, 0.22, 0.32)

Table 6 Values of interval-valued picture hesitant fuzzy sets and picture hesitant fuzzy sets

Ł	$\widetilde{Ph_f}$	Ph_f
l_1	([0.1, 0.2], [0.15, 0.28], [0.2, 0.4])	(0.22, 0.11, 0.19)
l_2	([0.05, 0.15], [0.1, 0.2], [0.2, 0.51])	(0.03, 0.08, 0.52)
l_3	([0.05, 0.15], [0.04, 0.14], [0.2, 0.35])	(0.25, 0.16, 0.43)

For $\beta = \gamma = 1$ and $\alpha = 0$ in Definition 14 we have,

$$Ph_{f_1} \cup Ph_{f_1}(l_2 * l_3) \leq \max\left\{Ph_{f_1} \cup Ph_{f_2}(l_2), Ph_{f_1} \cup Ph_{f_2}(l_3)\right\}$$

$$Ph_{f_1} \cup Ph_{f_2}(l_1) \leq \max\left\{(0.18, 0.22, 0.3, 0.32, 0.52), (0.16, 0.22, 0.25, 0.32, 0.43)\right\}$$
$$= (0.18, 0.22, 0.3, 0.32, 0.52).$$

$\Rightarrow Ph_{f_1} \cup Ph_{f_2}(l_1) = (0.11, 0.15, 0.19, 0.21, 0.22) \leq (0.18, 0.22, 0.3, 0.32, 0.52)$ which is correct.

Therefore, the $P-$union of two ECPHFLS need not be ECPHFLS.

Remark 2 Underneath example shows that $P-$intersection of any two ECPHFLS need not be ECPHFLS.

Example 5 Let us consider the values of interval-valued picture fuzzy linear spaces and picture fuzzy linear space as in Table 6 and also values tabulated below (Tables 5 and 6).

$$\widetilde{Ph_{f_1}} \cap \widetilde{Ph_{f_2}}(l_1) = ([0.1, 0.2], [0.15, 0.28], [0.16, 0.26], [0.1, 0.3], [0.2, 0.4]),$$
$$\widetilde{Ph_{f_1}} \cap \widetilde{Ph_{f_2}}(l_2) = ([0.05, 0.15], [0.1, 0.2], [0.15, 0.25], [0.2, 0.31], [0.2, 0.4]),$$
$$\widetilde{Ph_{f_1}} \cap \widetilde{Ph_{f_2}}(l_3) = ([0.04, 0.14], [0.05, 0.15], [0.05, 0.08], [0.1, 0.2], [0.2, 0.3]).$$

For $\beta = \gamma = 1$ and $\alpha = 0$ in definition 14 we have,

$$\widetilde{Ph_{f_1}} \cap \widetilde{Ph_{f_2}}(l_2 * l_3) \geq \min\left\{\widetilde{Ph_{f_1}} \cap \widetilde{Ph_{f_2}}(l_2), \widetilde{Ph_{f_1}} \cap \widetilde{Ph_{f_2}}(l_3)\right\}$$

Table 7 Values of interval-valued picture hesitant fuzzy sets and picture hesitant fuzzy sets

Ł	$\widetilde{Ph_f}$	Ph_f
l_1	([0.1, 0.3], [0.16, 0.26], [0.22, 0.34])	(0.18, 0.19, 0.24)
l_2	([0.1, 0.4], [0.15, 0.25], [0.2, 0.31])	(0.35, 0.20, 0.21)
l_3	([0.05, 0.2], [0.1, 0.2], [0.2, 0.3])	(0.19, 0.15, 0.25)

$$\widetilde{Ph_{f_1}} \cap \widetilde{Ph_{f_2}}(l_1) \geq \min\Big\{([0.05, 0.15], [0.1, 0.2], [0.15, 0.25], [0.2, 0.31], [0.2, 0.4]),$$
$$([0.04, 0.14], [0.05, 0.15], [0.05, 0.08], [0.1, 0.2], [0.2, 0.3])\Big\}$$
$$= ([0.04, 0.14], [0.05, 0.15], [0.05, 0.08], [0.1, 0.2], [0.2, 0.4]).$$

$\Rightarrow \widetilde{Ph_{f_1}} \cap \widetilde{Ph_{f_2}}(l_1) = ([0.1, 0.2], [0.15, 0.28], [0.16, 0.26], [0.1, 0.3], [0.2, 0.4]) \geq ([0.04, 0.14], [0.05, 0.15], [0.05, 0.08], [0.1, 0.2], [0.2, 0.4])$ which is correct.

$$Ph_{f_1} \cap Ph_{f_2}(l_1) = (0.08, 0.11, 0.15, 0.19, 0.21),$$
$$Ph_{f_1} \cap Ph_{f_2}(l_2) = (0.03, 0.04, 0.05, 0.08, 0.15),$$
$$Ph_{f_1} \cap Ph_{f_2}(l_3) = (0.01, 0.16, 0.15, 0.22, 0.32)$$

For $\beta = \gamma = 1$ and $\alpha = 0$ in Definition 14 we have,

$$Ph_{f_1} \cap Ph_{f_1}(l_2 * l_3) \leq \max\Big\{Ph_{f_1} \cap Ph_{f_2}(l_2), Ph_{f_1} \cap Ph_{f_2}(l_3)\Big\}$$
$$Ph_{f_1} \cap Ph_{f_2}(l_1) \leq \max\Big\{(0.03, 0.04, 0.05, 0.08, 0.15), (0.01, 0.16, 0.15, 0.22, 0.32)\Big\}$$
$$= (0.03, 0.16, 0.15, 0.22, 0.32).$$

$\Rightarrow Ph_{f_1} \cap Ph_{f_2}(l_1) = (0.08, 0.11, 0.15, 0.19, 0.21) \leq (0.03, 0.16, 0.15, 0.22, 0.32)$ which is incorrect.

Therefore, the $P-$intersection of two ECPHFLS need not be ECPHFLS.

Remark 3 The upcoming example shows that R-union of any two ICPHFLS need not be ICPHFLS.

Example 6 Let us consider the values of interval-valued picture fuzzy linear spaces and picture fuzzy linear space as in Table 6 and also values tabulated below (Tables 7 and 8).

$$\widetilde{Ph_{f_1}} \cup \widetilde{Ph_{f_2}}(l_1) = ([0.1, 0.3], [0.2, 0.4]),$$
$$\widetilde{Ph_{f_1}} \cup \widetilde{Ph_{f_2}}(l_2) = ([0.2, 0.51], [0.1, 0.4]),$$
$$\widetilde{Ph_{f_1}} \cup \widetilde{Ph_{f_2}}(l_3) = ([0.05, 0.2], [0.1, 0.5]).$$

Table 8 Values of interval-valued picture hesitant fuzzy sets and picture hesitant fuzzy sets

Ł	$\widetilde{Ph_f}$	Ph_f
l_1	([0.1, 0.2], [0.15, 0.28], [0.2, 0.4])	(0.12, 0.17, 0.25)
l_2	([0.05, 0.15], [0.1, 0.2], [0.2, 0.51])	(0.08, 0.13, 0.32)
l_3	([0.1, 0.5], [0.04, 0.14], [0.2, 0.35])	(0.24, 0.12, 0.33)

For $\beta = \gamma = 1$ and $\alpha = 0$ in definition 14 we have,

$$\widetilde{Ph_{f_1}} \cup \widetilde{Ph_{f_2}}(l_2 * l_3) \geq \min\left\{\widetilde{Ph_{f_1}} \cup \widetilde{Ph_{f_2}}(l_2), \widetilde{Ph_{f_1}} \cup \widetilde{Ph_{f_2}}(l_3)\right\}$$

$$\widetilde{Ph_{f_1}} \cup \widetilde{Ph_{f_2}}(l_1) \geq \min\left\{([0.2, 0.51], [0.1, 0.4]), ([0.05, 0.2], [0.1, 0.5])\right\}$$
$$= ([0.05, 0.2], [0.1, 0.4]).$$

$\Rightarrow \widetilde{Ph_{f_1}} \cup \widetilde{Ph_{f_2}}(l_1) = ([0.1, 0.3], [0.2, 0.4]) \geq ([0.05, 0.2], [0.1, 0.4])$ which is incorrect.

$$Ph_{f_1} \cap Ph_{f_2}(l_1) = (0.12, 0.17, 0.18, 0.19, 0.24),$$
$$Ph_{f_1} \cap Ph_{f_2}(l_2) = (0.08, 0.13, 0.18, 0.21, 0.32),$$
$$Ph_{f_1} \cap Ph_{f_2}(l_3) = (0.12, 0.15, 0.19, 0.24, 0.25).$$

For $\beta = \gamma = 1$ and $\alpha = 0$ in Definition 14 we have,

$$Ph_{f_1} \cap Ph_{f_1}(l_2 * l_3) \leq \max\left\{Ph_{f_1} \cap Ph_{f_2}(l_2), Ph_{f_1} \cap Ph_{f_2}(l_3)\right\}$$

$$Ph_{f_1} \cap Ph_{f_2}(l_1) \leq \max\left\{(0.08, 0.13, 0.18, 0.21, 0.32), (0.12, 0.15, 0.19, 0.24, 0.25)\right\}$$
$$= (0.12, 0.15, 0.19, 0.24, 0.32).$$

$\Rightarrow Ph_{f_1} \cap Ph_{f_2}(l_1) = (0.12, 0.17, 0.18, 0.19, 0.24) \leq (0.12, 0.15, 0.19, 0.24, 0.32)$ which is incorrect.

Therefore, the $R-$union of two ICPHFLS need not be ICPHFLS.

Remark 4 The below example shows that R-union of any two ECPHFLS need not be ECPHFLS.

Example 7 Let us consider the values of interval-valued picture hesitant fuzzy linear space and picture hesitant fuzzy linear space as tabulated below (Tables 9 and 10).

$$\widetilde{Ph_{f_1}} \cup \widetilde{Ph_{f_2}}(l_1) = ([0.1, 0.3], [0.2, 0.4]),$$
$$\widetilde{Ph_{f_1}} \cup \widetilde{Ph_{f_2}}(l_2) = ([0.2, 0.51], [0.1, 0.4]),$$
$$\widetilde{Ph_{f_1}} \cup \widetilde{Ph_{f_2}}(l_3) = ([0.05, 0.2], [0.1, 0.5]).$$

Table 9 Values of interval-valued picture hesitant fuzzy sets and picture hesitant fuzzy sets

Ł	$\widetilde{Ph_f}$	Ph_f
l_1	([0.1, 0.3], [0.16, 0.26], [0.22, 0.34])	(0.08, 0.15, 0.21)
l_2	([0.1, 0.4], [0.15, 0.25], [0.2, 0.31])	(0.05, 0.14, 0.15)
l_3	([0.05, 0.2], [0.1, 0.2], [0.2, 0.3])	(0.01, 0.22, 0.32)

Table 10 Values of interval-valued picture hesitant fuzzy sets and picture hesitant fuzzy sets

Ł	$\widetilde{Ph_f}$	Ph_f
l_1	([0.1, 0.2], [0.15, 0.28], [0.2, 0.4])	(0.22, 0.11, 0.19)
l_2	([0.05, 0.15], [0.1, 0.2], [0.2, 0.51])	(0.18, 0.03, 0.52)
l_3	([0.1, 0.5], [0.04, 0.14], [0.2, 0.35])	(0.25, 0.16, 0.43)

For $\beta = \gamma = 1$ and $\alpha = 0$ in Definition 14 we have,

$$\widetilde{Ph_{f_1}} \cup \widetilde{Ph_{f_2}}(l_2 * l_3) \geq \min\left\{\widetilde{Ph_{f_1}} \cup \widetilde{Ph_{f_2}}(l_2), \widetilde{Ph_{f_1}} \cup \widetilde{Ph_{f_2}}(l_3)\right\}$$
$$\widetilde{Ph_{f_1}} \cup \widetilde{Ph_{f_2}}(l_1) \geq \min\left\{([0.2, 0.51], [0.1, 0.4]), ([0.05, 0.2], [0.1, 0.5]])\right\}$$
$$= ([0.05, 0.2], [0.1, 0.4]).$$

$\Rightarrow \widetilde{Ph_{f_1}} \cup \widetilde{Ph_{f_2}}(l_1) = ([0.1, 0.3], [0.2, 0.4]) \geq ([0.05, 0.2], [0.1, 0.4])$ which is incorrect.

$$Ph_{f_1} \cup Ph_{f_2}(l_1) = (0.08, 0.15, 0.11, 0.19, 0.21),$$
$$Ph_{f_1} \cup Ph_{f_2}(l_2) = (0.03, 0.05, 0.14, 0.15, 0.18),$$
$$Ph_{f_1} \cup Ph_{f_2}(l_3) = (0.01, 0.16, 0.22, 0.25, 0.32).$$

For $\beta = \gamma = 1$ and $\alpha = 0$ in Definition 14 we have,

$$Ph_{f_1} \cup Ph_{f_1}(l_2 * l_3) \leq \max\left\{Ph_{f_1} \cup Ph_{f_2}(l_2), Ph_{f_1} \cup Ph_{f_2}(l_3)\right\}$$
$$Ph_{f_1} \cup Ph_{f_2}(l_1) \leq \max\left\{(0.03, 0.05, 0.14, 0.15, 0.18), (0.01, 0.16, 0.22, 0.25, 0.32)\right\}$$
$$= (0.03, 0.16, 0.22, 0.25, 0.32).$$

$\Rightarrow Ph_{f_1} \cup Ph_{f_2}(l_1) = (0.08, 0.15, 0.11, 0.19, 0.21) \leq (0.03, 0.16, 0.22, 0.25, 0.32)$ which is incorrect. Therefore, the $R-$union of two ECPHFLS need not be ECPHFLS.

Remark 5 The below example shows that R-intersection of any two ICPHFLS need not be ICPHFLS.

Example 8 Let us consider the values of interval-valued picture fuzzy linear spaces and picture fuzzy linear space as in Table 6 and also values tabulated below (Tables 11 and 12).

Table 11 Values of interval-valued picture hesitant fuzzy sets and picture hesitant fuzzy sets

Ł	$\widetilde{Ph_f}$	Ph_f
l_1	([0.1, 0.3], [0.16, 0.26], [0.22, 0.44])	(0.28, 0.18, 0.31)
l_2	([0.2, 0.4], [0.15, 0.25], [0.2, 0.31])	(0.3, 0.18, 0.25)
l_3	([0.05, 0.08], [0.1, 0.2], [0.2, 0.3])	(0.07, 0., 0.22)

Table 12 Values of interval-valued picture hesitant fuzzy sets and picture hesitant fuzzy sets

Ł	$\widetilde{Ph_f}$	Ph_f
l_1	([0.1, 0.2], [0.15, 0.28], [0.2, 0.4])	(0.12, 0.19, 0.39)
l_2	([0.05, 0.15], [0.1, 0.2], [0.2, 0.51])	(0.13, 0.19, 0.42)
l_3	([0.05, 0.15], [0.04, 0.14], [0.2, 0.35])	(0.18, 0.13, 0.32)

$$\widetilde{Ph_{f_1}} \cap \widetilde{Ph_{f_2}}(l_1) = ([0.1, 0.2], [0.15, 0.28], [0.16, 0.26], [0.1, 0.3], [0.2, 0.4]),$$
$$\widetilde{Ph_{f_1}} \cap \widetilde{Ph_{f_2}}(l_2) = ([0.05, 0.15], [0.1, 0.2], [0.15, 0.25], [0.2, 0.31], [0.2, 0.4]),$$
$$\widetilde{Ph_{f_1}} \cap \widetilde{Ph_{f_2}}(l_3) = ([0.04, 0.14], [0.05, 0.15], [0.05, 0.08], [0.1, 0.2], [0.2, 0.3]).$$

For $\beta = \gamma = 1$ and $\alpha = 0$ in Definition 14 we have,

$$\widetilde{Ph_{f_1}} \cap \widetilde{Ph_{f_2}}(l_2 * l_3) \geq \min\left\{\widetilde{Ph_{f_1}} \cap \widetilde{Ph_{f_2}}(l_2), \widetilde{Ph_{f_1}} \cap \widetilde{Ph_{f_2}}(l_3)\right\}$$

$$\begin{aligned}\widetilde{Ph_{f_1}} \cap \widetilde{Ph_{f_2}}(l_1) \geq \min &\left\{([0.05, 0.15], [0.1, 0.2], [0.15, 0.25], [0.2, 0.31], [0.2, 0.4]),\right.\\ &\left.([0.04, 0.14], [0.05, 0.15], [0.05, 0.08], [0.1, 0.2], [0.2, 0.3])\right\}\\ &= ([0.04, 0.14], [0.05, 0.15], [0.05, 0.08], [0.1, 0.2], [0.2, 0.4]).\end{aligned}$$

$\Rightarrow \widetilde{Ph_{f_1}} \cap \widetilde{Ph_{f_2}}(l_1)$=([0.1, 0.2], [0.15, 0.28], [0.16, 0.26], [0.1, 0.3], [0.2, 0.4]) $\geq$ ([0.04, 0.14], [0.05, 0.15], [0.05, 0.08], [0.1, 0.2], [0.2, 0.4]) which is correct.

$$Ph_{f_1} \cap Ph_{f_2}(l_1) = (0.18, 0.19, 0.28, 0.31, 0.39),$$
$$Ph_{f_1} \cap Ph_{f_2}(l_2) = (0.18, 0.19, 0.25, 0.3, 0.42),$$
$$Ph_{f_1} \cap Ph_{f_2}(l_3) = (0.13, 0.18, 0.2, 0.22, 0.32).$$

For $\beta = \gamma = 1$ and $\alpha = 0$ in Definition 14 we have,

$$Ph_{f_1} \cap Ph_{f_1}(l_2 * l_3) \leq \max\left\{Ph_{f_1} \cap Ph_{f_2}(l_2), Ph_{f_1} \cap Ph_{f_2}(l_3)\right\}$$
$$Ph_{f_1} \cap Ph_{f_2}(l_1) \leq \max\left\{(0.18, 0.19, 0.25, 0.3, 0.42), (0.13, 0.18, 0.2, 0.22, 0.32)\right\}$$
$$= (0.18, 0.19, 0.25, 0.3, 0.42).$$

Table 13 Values of interval-valued picture hesitant fuzzy sets and picture hesitant fuzzy sets

Ł	$\widetilde{Ph_f}$	Ph_f
l_1	$([0.1, 0.3], [0.16, 0.26], [0.22, 0.44])$	$(0.08, 0.15, 0.21)$
l_2	$([0.2, 0.4], [0.15, 0.25], [0.2, 0.31])$	$(0.05, 0.04, 0.15)$
l_3	$([0.05, 0.08], [0.1, 0.2], [0.2, 0.3])$	$(0.01, 0.22, 0.32)$

Table 14 Values of interval-valued picture hesitant fuzzy sets and picture hesitant fuzzy sets

Ł	$\widetilde{Ph_f}$	Ph_f
l_1	$([0.1, 0.2], [0.15, 0.28], [0.2, 0.4])$	$(0.22, 0.11, 0.19)$
l_2	$([0.05, 0.15], [0.1, 0.2], [0.2, 0.51])$	$(0.03, 0.08, 0.52)$
l_3	$([0.05, 0.15], [0.04, 0.14], [0.2, 0.35])$	$(0.25, 0.16, 0.43)$

$\Rightarrow Ph_{f_1} \cap Ph_{f_2}(l_1) = (0.18, 0.19, 0.28, 0.31, 0.39) \leq (0.18, 0.19, 0.25, 0.3, 0.42)$ which is incorrect.

Therefore, the $R-$intersection of two ICPHFLS need not be ICPHFLS.

Remark 6 Underneath example shows that $R-$intersection of any two ECPHFLS need not be ECPHFLS.

Example 9 Let us consider the values of interval-valued picture fuzzy linear spaces and picture fuzzy linear space as in Table 6 and also values tabulated below (Tables 13 and 14).

$$\widetilde{Ph_{f_1}} \cap \widetilde{Ph_{f_2}}(l_1) = ([0.1, 0.2], [0.15, 0.28], [0.16, 0.26], [0.1, 0.3], [0.2, 0.4]),$$
$$\widetilde{Ph_{f_1}} \cap \widetilde{Ph_{f_2}}(l_2) = ([0.05, 0.15], [0.1, 0.2], [0.15, 0.25], [0.2, 0.31], [0.2, 0.4]),$$
$$\widetilde{Ph_{f_1}} \cap \widetilde{Ph_{f_2}}(l_3) = ([0.04, 0.14], [0.05, 0.15], [0.05, 0.08], [0.1, 0.2], [0.2, 0.3]).$$

For $\beta = \gamma = 1$ and $\alpha = 0$ in Definition 14 we have,

$$\widetilde{Ph_{f_1}} \cap \widetilde{Ph_{f_2}}(l_2 * l_3) \geq \min\left\{\widetilde{Ph_{f_1}} \cap \widetilde{Ph_{f_2}}(l_2), \widetilde{Ph_{f_1}} \cap \widetilde{Ph_{f_2}}(l_3)\right\}.$$

$$\begin{aligned}\widetilde{Ph_{f_1}} \cap \widetilde{Ph_{f_2}}(l_1) \geq \min\Big\{&([0.05, 0.15], [0.1, 0.2], [0.15, 0.25], [0.2, 0.31], [0.2, 0.4]),\\ &([0.04, 0.14], [0.05, 0.15], [0.05, 0.08], [0.1, 0.2], [0.2, 0.3])\Big\}\\ = &([0.04, 0.14], [0.05, 0.15], [0.05, 0.08], [0.1, 0.2], [0.2, 0.4]).\end{aligned}$$

$\Rightarrow \widetilde{Ph_{f_1}} \cap \widetilde{Ph_{f_2}}(l_1) = ([0.1, 0.2], [0.15, 0.28], [0.16, 0.26], [0.1, 0.3], [0.2, 0.4]) \geq ([0.04, 0.14], [0.05, 0.15], [0.05, 0.08], [0.1, 0.2], [0.2, 0.4])$ which is correct.

$$Ph_{f_1} \cup Ph_{f_2}(l_1) = (0.11, 0.15, 0.19, 0.21, 0.22),$$
$$Ph_{f_1} \cup Ph_{f_2}(l_2) = (0.04, 0.05, 0.08, 0.15, 0.52),$$
$$Ph_{f_1} \cup Ph_{f_2}(l_3) = (0.16, 0.22, 0.25, 0.32, 0.43).$$

For $\beta = \gamma = 1$ and $\alpha = 0$ in Definition 14 we have,

$$Ph_{f_1} \cup Ph_{f_1}(l_2 * l_3) \leq \max\left\{Ph_{f_1} \cup Ph_{f_2}(l_2), Ph_{f_1} \cap Ph_{f_2}(l_3)\right\}$$
$$Ph_{f_1} \cup Ph_{f_2}(l_1) \leq \max\left\{(0.04, 0.05, 0.08, 0.15, 0.52), (0.16, 0.22, 0.25, 0.32, 0.43)\right\}$$
$$= (0.16, 0.22, 0.25, 0.32, 0.52).$$

$\Rightarrow Ph_{f_1} \cup Ph_{f_2}(l_1) = (0.11, 0.15, 0.19, 0.21, 0.22) \leq (0.16, 0.22, 0.25, 0.32, 0.52)$ which is incorrect.

Therefore, the $P-$intersection of two ECPHFLS need not be ECPHFLS.

4 Score Function and Cubic Picture Hesitant Fuzzy Weighted Averaging Aggregation Operators

Let $C_{Ph_f} = \{m, \widetilde{Ph_f}(m), Ph_f(m) | \ m \in M\}$ be a cubic picture hesitant fuzzy linear space of £ over field F. Then,

$$Sc(C_{Ph_f}) = \frac{1}{n(\widetilde{Ph_f})}\left(\hat{\mu}^+ + \hat{\mu}^- - \hat{\eta}^+ - \hat{\eta}^- - \hat{\vartheta}^+ - \hat{\vartheta}^- + \hat{\mu} - \hat{\eta} - \hat{\vartheta} + \frac{n(\widetilde{Ph_f})}{2}\right)$$

Definition 20 Let $C_{Ph_{f_k}} = (\widetilde{Ph_f}_k, Ph_{f_k})(k = 1, 2, \ldots n)$ be a collection of cubic picture hesitant fuzzy elements of linear space £ over field F, then their CPHFWA is described as follows

$$\text{CPHFWA}(C_{Ph_{f_1}}, C_{Ph_{f_2}}, \ldots, C_{Ph_{f_n}}) = \sum_{k=1}^{n} \tau_k A_k$$

where τ_k $(k = 1, 2, \ldots, n)$ be the weight vector of $C_{Ph_{f_k}} = (\widetilde{Ph_f}_k, Ph_{f_k})$ with $\tau_k \geq 0$ and $\sum_{k=1}^{n} \tau_k = 1$.

Definition 21 Let $C_{Ph_{f_k}} = (\widetilde{Ph_f}_k, Ph_{f_k})(k = 1, 2, \ldots n)$ be a collection of cubic picture hesitant fuzzy elements of linear space £ over field F, then by using Definition 20 the aggregation of CPHFWA is described as follows

$$\text{CPHFWA}(C_{Ph_{f_1}}, C_{Ph_{f_2}}, \ldots, C_{Ph_{f_n}}) = \left\{ \begin{array}{l} \left\{ \left[1 - \prod_{k=1}^{n} (1 - \hat{\mu_k}^-)^{\tau_k}, 1 - \prod_{k=1}^{n} (1 - \hat{\mu_k}^+)^{\tau_k} \right], \right. \\ \left[\prod_{k=1}^{n} (\hat{\eta_k}^-)^{\tau_k}, \prod_{k=1}^{n} (\hat{\eta_k}^+)^{\tau_k} \right], \\ \left. \left[\prod_{k=1}^{n} (\hat{\vartheta}_k^-)^{\tau_k}, \prod_{k=1}^{n} (\hat{\vartheta}_k^+)^{\tau_k} \right] \right\}, \\ \left\{ \left[1 - \prod_{k=1}^{n} (1 - \mu_k)^{\tau_k}) \right], \left[\prod_{k=1}^{n} (\eta_k)^{\tau_k} \right], \left[\prod_{k=1}^{n} (\vartheta_k)^{\tau_k} \right] \right\}. \end{array} \right.$$

Definition 22 Let $C_{Ph_{f_k}} = (\widetilde{Ph_f}_k, Ph_{f_k})(k = 1, 2, \ldots n)$ be a collection of cubic picture hesitant fuzzy elements of linear space £ over field F, then by using Definition 20 the aggregation of CPHFOWA is described as follows

$$\text{CPHFOWA}(C_{Ph_{f_1}}, C_{Ph_{f_2}}, \ldots, C_{Ph_{f_n}}) = \left\{ \begin{array}{l} \left\{ \left[1 - \prod_{k=1}^{n} (1 - \hat{\mu}^-_{\omega(k)})^{\tau_k}, 1 - \prod_{k=1}^{n} (1 - \hat{\mu}^+_{\omega(k)})^{\tau_k} \right], \right. \\ \left[\prod_{k=1}^{n} (\hat{\eta}^-_{\omega(k)})^{\tau_k}, \prod_{k=1}^{n} (\hat{\eta}^+_{\omega(k)})^{\tau_k} \right], \\ \left. \left[\prod_{k=1}^{n} (\hat{\vartheta}^-_{\omega(k)})^{\tau_k}, \prod_{k=1}^{n} (\hat{\vartheta}^+_{\omega(k)})^{\tau_k} \right] \right\}, \\ \left\{ \left[1 - \prod_{k=1}^{n} (1 - \mu_{\omega(k)})^{\tau_k}) \right], \right. \\ \left. \left[\prod_{k=1}^{n} (\eta_{\omega(k)})^{\tau_k} \right], \left[\prod_{k=1}^{n} (\vartheta_{\omega(k)})^{\tau_k} \right] \right\}. \end{array} \right.$$

where $C_{Ph_{f_{\omega(k)}}}$ is the kth largest value consequently by total order $C_{Ph_{f_{\omega(1)}}} \geq C_{Ph_{f_{\omega(2)}}} \geq \ldots \geq C_{Ph_{f_{\omega(n)}}}$.

Definition 23 Let $C_{Ph_{f_k}} = (\widetilde{Ph_f}_k, Ph_{f_k})(k = 1, 2, \ldots n)$ be a collection of cubic picture hesitant fuzzy elements of linear space £ over field F, then by using definition 20 the aggregation of CPHFHWA is described as follows

$$\text{CPHFHWA}(C_{Ph_{f_1}}, C_{Ph_{f_2}}, \ldots, C_{Ph_{f_n}}) = \left\{ \begin{array}{l} \left\{ \left[1 - \prod_{k=1}^{n} (1 - \hat{\mu'}^-_{\omega(k)})^{\tau_k}, 1 - \prod_{k=1}^{n} (1 - \hat{\mu'}^+_{\omega(k)})^{\tau_k} \right], \right. \\ \left[\prod_{k=1}^{n} (\hat{\eta'}^-_{\omega(k)})^{\tau_k}, \prod_{k=1}^{n} (\hat{\eta'}^+_{\omega(k)})^{\tau_k} \right], \\ \left. \left[\prod_{k=1}^{n} (\hat{\vartheta'}^-_{\omega(k)})^{\tau_k}, \prod_{k=1}^{n} (\hat{\vartheta'}^+_{\omega(k)})^{\tau_k} \right] \right\}, \\ \left\{ \left[1 - \left(1 - \prod_{k=1}^{n} (1 - (1 - \mu'_{\omega(k)})^{\tau_k}) \right) \right], \right. \\ \left. \left[1 - \prod_{k=1}^{n} (1 - \eta'_{\omega(k)})^{\tau_k} \right], \left[1 - \prod_{k=1}^{n} (1 - \vartheta'_{\omega(k)})^{\tau_k} \right] \right\}. \end{array} \right.$$

where $C'_{Ph_{f\omega(k)}}$ is the kth largest value consequently by total order $C'_{Ph_{f\omega(1)}} \geq C'_{Ph_{f\omega(2)}} \geq \ldots \geq C'_{Ph_{f\omega(n)}}$ and $C'_{Ph_{f\omega(k)}} = n\tau_k C_{Ph_{f\omega(k)}}$ where τ_k are the weight vectors such that $\tau_k \geq 0$.

5 Cubic Picture Hesitant Fuzzy Prioritized Weighted Averaging Aggregation Operators

Definition 24 Let $C_{Ph_{fk}} = (\widetilde{Ph_{f}}_{k}, Ph_{f_k})(k = 1, 2, \ldots n)$ be a collection of cubic picture hesitant fuzzy elements of linear space £ over field F, then their CPHFPWA is described as follows

$$\text{CPHFWA}(C_{Ph_{f1}}, C_{Ph_{f2}}, \ldots, C_{Ph_{fn}}) = \sum_{k=1}^{n} p_k A_k$$

where $p_k = \frac{S_k}{\sum_{j=1}^{n} S_j}$ $(k = 1, 2, \ldots, n)$ and $S_j = \prod_{K=1}^{j-1} Sc(C_{Ph_f})(j = 2, \ldots, n)$, $S_1 = 1$.

Definition 25 Let $C_{Ph_{fk}} = (\widetilde{Ph_{f}}_{k}, Ph_{f_k})(k = 1, 2, \ldots n)$ be a collection of cubic picture hesitant fuzzy elements of linear space £ over field F, then by using the definition 20 the aggregation of CPHFPWA is described as follows

$$\text{CPHFPWA}(C_{Ph_{f1}}, C_{Ph_{f2}}, \ldots, C_{Ph_{fn}})$$
$$= \begin{Bmatrix} \left\{ \left[1 - \prod_{k=1}^{n}(1 - \hat{\mu_k}^{-})^{p_k}, 1 - \prod_{k=1}^{n}(1 - \hat{\mu_k}^{+})^{p_k} \right], \right. \\ \left[\prod_{k=1}^{n}(\hat{\eta_k}^{-})^{p_k}, \prod_{k=1}^{n}(\hat{\eta_k}^{+})^{p_k} \right], \\ \left. \left[\prod_{k=1}^{n}(\hat{\vartheta_k}^{-})^{p_k}, \prod_{k=1}^{n}(\hat{\vartheta_k}^{+})^{p_k} \right] \right\}, \\ \left\{ \left[(1 - \prod_{k=1}^{n}(1 - \mu_k)^{p_k} \right], \right. \\ \left. \left[1 - \prod_{k=1}^{n}(1 - \eta_k)^{p_k} \right], \left[1 - \prod_{k=1}^{n}(1 - \vartheta_k)^{p_k} \right] \right\}. \end{Bmatrix}$$

Definition 26 Let $C_{Ph_{fk}} = (\widetilde{Ph_{f}}_{k}, Ph_{f_k})(k = 1, 2, \ldots n)$ be a collection of cubic picture hesitant fuzzy elements of linear space £ over field F, then by using definition 20 the aggregation of CPHFPOWA is described as follows

$$\text{CPHFPOWA}(C_{Ph_{f_1}}, C_{Ph_{f_2}}, \ldots, C_{Ph_{f_n}}) = \left\{ \begin{array}{l} \left\{ \left[1 - \prod_{k=1}^{n} (1 - \hat{\mu}^{-}_{\omega(k)})^{p_k}, 1 - \prod_{k=1}^{n} (1 - \hat{\mu}^{+}_{\omega(k)})^{p_k} \right], \right. \\ \left[\prod_{k=1}^{n} (\hat{\eta}^{-}_{\omega(k)})^{p_k}, \prod_{k=1}^{n} (\hat{\eta}^{+}_{\omega(k)})^{p_k} \right], \\ \left. \left[\prod_{k=1}^{n} (\hat{\vartheta}^{-}_{\omega(k)})^{p_k}, \prod_{k=1}^{n} (\hat{\vartheta}^{+}_{\omega(k)})^{\tau_k} \right] \right\}, \\ \left\{ \left[1 - \prod_{k=1}^{n} (1 - \mu_{\omega(k)})^{p_k} \right], \right. \\ \left. \left[1 - \prod_{k=1}^{n} (1 - \eta_{\omega(k)})^{p_k} \right], \left[1 - \prod_{k=1}^{n} (1 - \vartheta_{\omega(k)})^{p_k} \right] \right\}. \end{array} \right.$$

where $C_{Ph_{f_{\omega(k)}}}$ is the kth largest value consequently by total order $C_{Ph_{f_{\omega(1)}}} \geq C_{Ph_{f_{\omega(2)}}} \geq \ldots \geq C_{Ph_{f_{\omega(n)}}}$.

Definition 27 Let $C_{Ph_{f_k}} = (\widetilde{Ph_f}_k, Ph_{f_k})(k = 1, 2, \ldots n)$ be a collection of cubic picture hesitant fuzzy elements of linear space £ over field F, then by using definition 20 the aggregation of CPHFPHOWA is described as follows

$$\text{CPHFHPWA}(C_{Ph_{f_1}}, C_{Ph_{f_2}}, \ldots, C_{Ph_{f_n}}) = \left\{ \begin{array}{l} \left\{ \left[1 - \prod_{k=1}^{n} (1 - \hat{\mu'}^{-}_{\omega(k)})^{p_k}, 1 - \prod_{k=1}^{n} (1 - \hat{\mu'}^{+}_{\omega(k)})^{p_k} \right], \right. \\ \left[\prod_{k=1}^{n} (\hat{\eta'}^{-}_{\omega(k)})^{p_k}, \prod_{k=1}^{n} (\hat{\eta'}^{+}_{\omega(k)})^{p_k} \right], \\ \left. \left[\prod_{k=1}^{n} (\hat{\vartheta'}^{-}_{\omega(k)})^{p_k}, \prod_{k=1}^{n} (\hat{\vartheta'}^{+}_{\omega(k)})^{p_k} \right] \right\}, \\ \left\{ \left[1 - \prod_{k=1}^{n} (1 - \mu'_{\omega(k)})^{p_k} \right], \right. \\ \left. \left[1 - \prod_{k=1}^{n} (1 - \eta'_{\omega(k)})^{p_k} \right], \left[1 - \prod_{k=1}^{n} (1 - \vartheta'_{\omega(k)})^{p_k} \right] \right\}. \end{array} \right.$$

where $C'_{Ph_{f_{\omega(k)}}}$ is the kth largest value consequently by total order $C'_{Ph_{f_{\omega(1)}}} \geq C'_{Ph_{f_{\omega(2)}}} \geq \ldots \geq C'_{Ph_{f_{\omega(n)}}}$ and $C'_{Ph_{f_{\omega(k)}}} = n\tau_k C_{Ph_{f_{\omega(k)}}}$ where τ_k are the weight vectors such that $\tau_k \geq 0$.

6 MCDM Method Utilizing Cubic Picture Hesitant Fuzzy Aggregation Operators

In this section, we utilize the abovementioned operators to deal with the different MCDM problems under CPHF environment. Consider a group of investors looking to invest their money in the best financial platforms F_1, F_2 and F_3, respectively. The required criteria are as follows:

1. Low-risk factor(C_1).
2. High returns(C_2).
3. Duration of investment(C_3).
4. Ease of settlement(C_4).

and the average weights of criteria from the investors are $\tau = (0.2, 0.1, 0.3, 0.4)$. Now, we list the ratings of three alternatives F_1, F_2 and F_3 under four criteria C_1, C_2, C_3 and C_4 in the table as below.

	C_1	C_1
F_1	([0.1, 0.3], [0.16, 0.26], [0.22, 0.34])	(0.08, 0.15, 0.21)
F_2	([0.1, 0.4], [0.15, 0.25], [0.2, 0.31])	(0.05, 0.14, 0.15)
F_3	([0.05, 0.2], [0.1, 0.2], [0.2, 0.3])	(0.01, 0.22, 0.32)

	C_2	C_2
F_1	([0.1, 0.2], [0.15, 0.28], [0.2, 0.4])	(0.12, 0.17, 0.25)
F_2	([0.05, 0.15], [0.1, 0.2], [0.2, 0.51])	(0.08, 0.13, 0.32)
F_3	([0.1, 0.5], [0.04, 0.14], [0.2, 0.35])	(0.24, 0.12, 0.33)

	C_3	C_3
F_1	([0.20, 0.30], [0.15, 0.26], [0.29, 0.39])	(0.21, 0.17, 0.35)
F_2	([0.14, 0.15], [0.29, 0.39], [0.18, 0.23])	(0.18, 0.23, 0.12)
F_3	([0.41, 0.43], [0.09, 0.18], [0.27, 0.33])	(0.26, 0.12, 0.43)

	C_4	C_4
F_1	([0.39, 0.41], [0.16, 0.27], [0.22, 0.29])	(0.30, 0.08, 0.31)
F_2	([0.18, 0.19], [0.17, 0.33], [0.12, 0.15])	(0.24, 0.15, 0.12)
F_3	([0.27, 0.33], [0.08, 0.17], [0.25, 0.35])	(0.34, 0.12, 0.36)

Step 1

See Tables 15, 16 and 17.

Table 15 Case 1: Operate the aggregation operator utilizing CPHFWA operator

F_1	([0.26, 0.33], [0.15, 0.25], [0.23, 0.33])	(0.22, 0.12, 0.28)
F_2	([0.14, 0.21], [0.18, 0.31], [0.15, 0.22])	(0.17, 0.16, 0.14)
F_3	([0.26, 0.36], [0.08, 0.17], [0.23, 0.33])	(0.25, 0.13, 0.36)

Table 16 Case 2: Operate the aggregation operator utilizing CPHFOWA operator

F_1	([0.32, 0.42], [0.11, 0.22], [0.22, 0.31])	(0.24, 0.10, 0.37)
F_2	([0.20, 0.30], [0.11, 0.22], [0.24, 0.36])	(0.25, 0.14, 0.38)
F_3	([0.13, 0.17], [0.16, 0.26], [0.15, 0.22])	(0.17, 0.18, 0.16)

Table 17 Case 3: Operate the aggregation operator utilizing CPHFHOWA operator

F_1	([0.87, 0.92], [0.00004, 0.0007], [0.001, 0.003])	(0.7, 0.00002, 0.004)
F_2	([0.7, 0.8], [0.00001, 0.0003], [0.0008, 0.006])	(0.77, 0.00005, 0.003)
lF_3	([0.56, 0.6], [0.0002, 0.003], [0.00006, 0.0004])	(0.6, 0.00004, 0.005)

Table 18 Now, rank all the alternatives

	Score	Ranking
CPHFWA	$Sc(l_3) > Sc(l_1) > Sc(l_2)$	$l_3 > l_1 > l_2$
CPHFOWA	$Sc(l_1) > Sc(l_3) > Sc(l_2)$	$l_1 > l_3 > l_2$
CPHFHOWA	$Sc(l_1) > Sc(l_2) > Sc(l_3)$	$l_1 > l_2 > l_3$

Step 2: Now, we find out the score function as defined above

Case 1	Case 2	Case 3
$Sc(F_1) = 0.408$	$Sc(F_1) = 0.441$	$Sc(F_1) = 0.913$
$Sc(F_2) = 0.393$	$Sc(F_2) = 0.383$	$Sc(F_2) = 0.876$
$Sc(F_3) = 0.428$	$Sc(F_3) = 0.39$	$Sc(F_3) = 0.791$

Step 3: Now, rank all the alternatives

See Table 18.

Now, the abovementioned required criteria are prioritized as $C_1 > C_2 > C_3 > C_4$.

Step 1

See Tables 19 and 20.

Table 19 Case 1: Operate the aggregation operator utilizing CPHFPWA operator

l_1	$([0.43, 0.59], [0.08, 0.03], [0.02, 0.07])$	$(0.4, 0.01, 0.04)$
l_2	$([0.23, 0.43], [0.01, 0.04], [0.01, 0.04])$	$(0.31, 0.01, 0.01)$
l_3	$([0.46, 0.7], [0.001, 0.01], [0.02, 0.06])$	$(0.46, 0.006, 0.07)$

Table 20 Case 2: Operate the aggregation operator utilizing CPHFPOWA operator

l_1	$([0.53, 0.7], [0.006, 0.04], [0.02, 0.05])$	$(0.46, 0.01, 0.04)$
l_2	$([0.37, 0.5], [0.005, 0.03], [0.02, 0.08])$	$(0.41, 0.018, 0.04)$
l_3	$([0.25, 0.38], [0.007, 0.03], [0.02, 0.05])$	$(0.3, 0.012, 0.01)$

Table 21 Now, rank all the alternatives

	Score	Ranking
CPHFPWA	$Sc(l_1) > Sc(l_3) > Sc(l_2)$	$l_1 > l_3 > l_2$
CPHFPOWA	$Sc(l_1) > Sc(l_2) > Sc(l_3)$	$l_1 > l_2 > l_3$

Step 2: Now, we find out the score function as defined above

Case 1	Case 2
$Sc(F_1) = 0.695$	$Sc(F_1) = 0.754$
$Sc(F_2) = 0.641$	$Sc(F_2) = 0.673$
$Sc(F_3) = 0.742$	$Sc(F_3) = 0.63$

Step 3: Now, rank all the alternatives

See Table 21.

7 Conclusion

In our present work, we developed cubic picture hesitant fuzzy linear spaces and defined $P(R)$—intersection, $P(R)$—union of internal and external cubic picture hesitant fuzzy linear spaces and proved different results with few examples. We defined a series of operators like cubic picture hesitant fuzzy weighted averaging aggregation operators, cubic picture hesitant fuzzy order weighted averaging aggregation operators, and cubic picture hesitant fuzzy hybrid order weighted averaging aggregation operators are developed. We presented score functions to find scores of cubic picture hesitant fuzzy linear spaces. We solved multi-criteria decision-making (MCDM) problem on cubic picture hesitant fuzzy linear spaces using series of operators defined above. The criteria given by decision makers are prioritized and further solved the MCDM problem using the above defined prioritized operators. The result in both the cases is compared to choose the best criteria among the given ones. We intend to develop the proposed framework to certain other domains in the future, including Fermatean fuzzy sets [17, 31–33], dual-probabilistic linguistic sets [34], cubic intuitionistic sets [35–39], and so on [40].

Declarations

- Funding: This work was supported by CSIR (Council of Scientific and Industrial Research).
- Conflict of interest: The authors declare that they have no conflict of interest.
- Ethics approval: This article does not contain any studies with human participants or animals performed by any of the authors.
- Competing Interests: The authors declare that declare they have no financial interests
- Availability of data and materials: The datasets generated during and/or analyzed during the current study are available in the Research gate.
- Authors' contributions: The authors contributed to the study equally. The first draft of the manuscript was written by P.R. Kavyasree and all authors commented on previous versions of the manuscript. All authors read and approved the final manuscript.

Acknowledgements The authors would like to extend their gratitude to CSIR(Council of Scientific and Industrial Research) for their support.

References

1. Zadeh, L.A.: Fuzzy sets. Inf. Comput. **8**, 338–353 (1965)
2. Zadeh, L.A.: The concept of a linguistic variable and its application to approximate reasoning-I. Inf. Sci. **8**(3), 199–249 (1975)
3. Atanassov, K.: Intuitionistic fuzzy sets. Fuzzy Sets Syst. **20**(1), 87–96 (1986)
4. Jun, Y.B., Kim, C.S., M. Kang MS,: Cubic subalgebras and ideals of BCK/BCI-algebras. Far East J. Mathe. Sci. **44**(22), 239–250 (2010)
5. Cuong BC, Kreinovich V (2013) Picture fuzzy sets—A new concept for computational intelligence problems. In: Third World Congress on Information and Communication Technologies (WICT 2013), pp. 1–6
6. Farhadinia, B.: Information measures for hesitant fuzzy sets and interval-valued hesitant fuzzy sets. Inf. Sci. **240**, 129–144 (2013)
7. Wang, R., Li, Y.: Picture hesitant fuzzy set and its application to multiple criteria decision-making. Symmetry **10**(7), 295 (2018)
8. Senapati, T., Chen, G., Mesiar, R., Yager, R.R., Saha, A.: Novel Aczel-Alsina operations-based hesitant fuzzy aggregation operators and their applications in cyclone disaster assessment. Int. J. General Syst. (2022). https://doi.org/10.1080/03081079.2022.2036140
9. Senapati, T.: Approaches to multi-attribute decision making based on picture fuzzy Aczel-Alsina average aggregation operators. Comput. Appl. Math. **41**, 40 (2022). https://doi.org/10.1007/s40314-021-01742-w
10. Senapati, T., Yager, R.R., Chen, G.: Cubic intuitionistic WASPAS technique and its application in multi-criteria decision-making. J. Amb. Intell. Human. Comput. **12**, 8823–8833 (2021)

11. Jana, C., Senapati, T., Pal, M., Saeid, A.B.: Different types of cubic ideals in BCI-algebras based on fuzzy points. Afrika Matematika **31**, 367–381 (2020)
12. Senapati, T., Kim, C.S., Bhowmik, M., Pal, M.: Cubic subalgebras and cubic closed ideals of B-algebras. Fuzzy Inf. Eng. **7**(2), 129–149 (2015)
13. Senapati, T., Shum, K.P.: Cubic implicative ideals of BCK-algebras. Missouri J. Math. Sci. **29**(2), 125–138 (2017)
14. Senapati, T., Shum, K.P.: Cubic commutative ideals of BCK-algebras. Missouri J. Math. Sci. **30**(1), 5–19 (2018)
15. Senapati, T., Jun, Y.B., Jun Shum, K.P.: Cubic set structure applied in UP-algebras. Discrete Mathematics, Algor. Appl. 10(4) (2018). https://doi.org/10.1142/S1793830918500490
16. Dey, A., Senapati, T., Pal, M., Chen, G.: A novel approach to hesitant multi-fuzzy based decision making. AIMS Math. **5**(3), 1985–2008 (2020)
17. Sergi, D., Sari, I.U., Senapati, T.: Extension of capital budgeting techniques using interval-valued Fermatean fuzzy sets. J. Intell. Fuzzy Syst. **42**, 365–376 (2022)
18. Zhu, B., Xu, Z., Xia, M.: Dual hesitant fuzzy sets. J. Appl. Math. Article ID 879629, 1–13 (2012)
19. Wei, G.: Picture fuzzy aggregation operators and their application to multiple attribute decision making. J. Intell. Fuzzy Syst. **33**(2), 713–724 (2017)
20. Jana, C., Senapati, T., Pal, M., Yager, R.R.: Picture fuzzy Dombi aggregation operators: application to MADM process. Appl. Soft Comput. **74**, 99–109 (2019)
21. Senapati, T., Chen, G.: Picture fuzzy WASPAS technique and its application in multi-criteria decision-making. Soft Comput. (2022). https://doi.org/10.1007/s00500-022-06835-0
22. Qiyas, M., Abdullah, S., Liu, Y., Naeem, M.: Multi-criteria decision support systems based on linguistic intuitionistic cubic fuzzy aggregation operators. J. Amb. Intell. Humanized Comput. **12**(8), 8285–8303 (2021)
23. Xia, M., Xu, Z., Chen, N.: Some hesitant fuzzy aggregation operators with their application in group decision making. Group Dec. Negot. **22**(2), 259–279 (2013)
24. Yager, R.R.: Prioritized aggregation operators. Int. J. Approx. Reason. **48**(1), 263–274 (2008)
25. Yu, X., Xu, Z.: Prioritized intuitionistic fuzzy aggregation operators. Inf. Fusion **14**(1), 108–116 (2013)
26. Lubczonok, G., Murali, V.: On flags and fuzzy subspaces of vector spaces. Fuzzy Sets Syst. **125**(2), 201–207 (2002)
27. Vijayabalaji, S., Sivaramakrishnan, S.: A cubic set theoretical approach to linear space. Abst. Appl. Anal. 1-8 (2015)
28. Kavyasree, P.R., Surender Reddy, B.: Cubic $Gamma$-n normed linear spaces. Malaya J. Matematik **6**, 643–647 (2018)
29. Kavyasree, P.R., Surender Reddy, B.: N-Cubic sets applied to linear spaces. Kragujevac J. Math. **46**(4), 575–594 (2022)
30. Kavyasree, P.R., Surender Reddy, B.: Cubic picture fuzzy linear spaces. AIP Conf. Proc. **2378**(1), 020012 (2021)
31. Senapati, T., Yager, R.R.: Fermatean fuzzy weighted averaging/geometric operators and its application in multi-criteria decision-making methods. Eng. Appl. Artif. Intell. **85**, 112–121 (2019)
32. Senapati, T., Yager, R.R.: Some new operations over Fermatean fuzzy numbers and application of Fermatean fuzzy WPM in multiple criteria decision making. Informatica **30**(2), 391–412 (2019)
33. Senapati, T., Yager, R.R.: Fermatean fuzzy sets. J. Amb. Intell. Human. Comput. **11**(2), 663–674 (2020)
34. Saha, A., Senapati, T., Yager, R.R.: Hybridizations of generalized Dombi operators and Bonferroni mean operators under dual probabilistic linguistic environment for group decision-making. Int. J. Intell. Syst. **36**(11), 6645–6679 (2021)
35. Senapati, T., Jana, C., Pal, M., Jun, Y.B.: Cubic intuitionistic q-ideals of BCI-algebras. Symmetry **10**(12), 752 (2018). https://doi.org/10.3390/sym10120752

36. Senapati, T., Jun, Y.B., Muhiuddin, G., Shum, K.P.: Cubic intuitionistic structures applied to ideals of BCI-algebras. Analele Stiintifice ale Universitatii Ovidius Constanta, Seria Matematica **27**(2), 213–232 (2019)
37. Senapati, T., Jun, Y.B., Shum, K.P.: Cubic intuitionistic subalgebras and closed cubic intuitionistic ideals of B-algebras. J. Intell. Fuzzy Syst. **36**(2), 1563–1571 (2019)
38. Senapati, T., Jun, Y.B., Shum, K.P.: Cubic intuitionistic structure of KU-algebras. Afrika Matematika **31**, 237–248 (2020)
39. Senapati, T., Jun, Y.B., Shum, K.P.: Cubic intuitionistic implicative ideals of BCK-algebras. Proc. Nat. Acad. Sci. India Sect. A: Phys. Sci. **91**, 273–282 (2021)
40. Senapati, T., Chen, G., Yager, R.R.: Aczel-Alsina aggregation operators and their application to intuitionistic fuzzy multiple attribute decision making. Int. J. Intell. Syst. **37**(2), 1529–1551 (2021)

New Ranking Approach to Solve MCDM Problems with Generalized Intuitionistic Fuzzy Information

Amal Kumar Adak, Nilkamal, and Kaushila Nandan Srivastava

Abstract Ranking of Generalized intuitionistic fuzzy sets (GIFSs) assume an unmistakable part, considering the fact that its applicability and appeal to mannequin with uncertainty in "Multi-criteria Decision Making" (MCDM) problems. Intention of this discussion is to inspect the more than one attribute selection issues with GIF data in which priority vectors are absolutely known. Here we lengthen "technique for order preference similarity to ideal solution (TOPSIS) method" for the generalized intuitionistic fuzzy information. Possibility degree approach are used to positioned the alternatives.

Keywords Possibility degree formula · Intuitionistic fuzzy set · Generalized intuitionistic fuzzy set · TOPSIS method

1 Introduction

"Multi-attribute decision making" (MCDM) problems are used to identify the appropriate alternative among all available alternatives. In real life situation, due to the inherent vagueness of decision maker's preferences in practice, Zadeh [1] invented fuzzy sets (FSs) theory to deal with such vague data. FSs gives an advantageous device to describing impression by membership function in [0,1].

A few times the enrollment capacity of a common fuzzy sets are adequately not to portray the characters of appraisal due to human abstract decisions. To defeat this weakness and broaden standard fuzzy sets, Atanassov [2, 3] developed the idea of intuitionistic fuzzy sets (IFSs). IFSs characterized by communicating membership grade (α) and non-membership grade (β) of element, satisfying condition that sum of membership grade and non-membership grade must lie between 0 and 1. IFSs

A. K. Adak (✉)
Department of Mathematics, Ganesh Dutt College, Begusarai, India
e-mail: amaladak17@gmail.com

Nilkamal · K. N. Srivastava
Lalit Narayan Mithila University, Darbhanga, India

L. Sahoo et al. (eds.), *Real Life Applications of Multiple Criteria Decision Making Techniques in Fuzzy Domain*, Studies in Fuzziness and Soft Computing 420,
https://doi.org/10.1007/978-981-19-4929-6_26

are more remarkable and valuable than other non-standard fuzzy models. Now a days IFSs theory being paid more attention for modeling problem containing uncertainty and applied in decision making, medical diagnosis, pattern recognition, fuzzy optimization, etc. [4–9].

An element in IFSs is significant if the of membership value $\alpha > 0.5$, otherwise insignificant. The non-membership value (β) can not be significant. Considering this situation, Mondal and Samanta [10] represent the idea of generalized intuitionistic fuzzy sets (GIFSs). GIFSs extend the condition of IFSs $0 \leq \alpha + \beta \leq 1$ to $0 \leq \alpha \wedge \beta \leq 0.5$ for all elements in the sets. The space of all intuitionistic fuzzy values is also generalized intuitionistic fuzzy values, but converse is not necessary true [11–15].

More than one multi-criteria decision making (MCDM) [16–19] procedures developed and utilized to various fields. "Technique for order preference similarity to ideal solution" (TOPSIS) is the basic method, was developed with the aid of Hwang and Yoon [20] to solve MCDM problems. The main feature of TOPSIS method is that the most favored choice need to no longer solely have the shortest distance from the fine best answer however additionally have the farthest distance from the terrible perfect solution.

Ranking of choices in TOPSIS is a core trouble in fuzzy choices making problems. A lot of effort is made for rating fuzzy number. The ordering of fuzzy numbers with the aid of defining indices. Utilize rating diploma feature for rating IFSs. In this chapter, we think about the opportunity diploma components and applied to clear up multi-criteria selection making problems.

The chapter is equipped as beneath. In Sect. 2, discussed definitions associated to IFSs and GIFSs. In Sect. 3, proposed new methodology to remedy multi-criteria selection making issues to a chosen initiatives with uncertain data. Section 4, applies the our methodology to a actual trouble to show its feasibility and practicality. At last, at the quit of this chapter conclusion is given.

2 Some Definitions

Here, we present essential thoughts, for example, the IFSs, GIFSs. Also, we include elementary operations on these sets that are necessary for this study.

Definition 1 [1] (*Fuzzy set (FS)*) A fuzzy set F in set E is defined as

$$F = \{\langle y, \alpha_F(y)\rangle | y \in E\},$$

where $\alpha_F : E \to [0, 1]$ is membership grade of fuzzy set F.

Definition 2 [2] (*Intuitionistic fuzzy set (IFS)*) An intuitionistic fuzzy set (IFS) I over E is of the form

$$I = \langle y, \alpha(y), \beta(y)\, | y \in E\rangle;$$

where and are denote the membership grade $\alpha(y) : E \to [0, 1]$ and non-membership grade $\beta(y) : E \to [0, 1]$ satisfies condition $0 \le \alpha(y) + \beta(y) \le 1$.

Definition 3 [10] (*Generalized intuitionistic fuzzy set (GIFS)*) A generalized intuitionistic fuzzy set (GIFS) A of E is of the form

$$A = \{\langle y, \alpha_A(y), \beta_A(y)\rangle | y \in E\},$$

where membership grade and non-membership grade lie in between 0 and 1 satisfy the condition

$$\alpha_A(y) \wedge \beta_A(y) \le 0.5, \text{ for every } y \in E.$$

which is known as "generalized intuitionistic fuzzy condition (GIFC)".

GIF index or degree of hesitation is defined as:

$$\gamma_A(y) = 1.5 - \alpha_A(y) - \beta_A(y).$$

It can be observed, $0 \le \gamma_A(y) \le 1$ for every $y \in E$.

The knowledge about an element will be certain if $\gamma_A(y)$ is small and uncertain if $\gamma_A(y)$ is large.

Definition 4 Let us take two GIFSs R and S, where

$$R = \{y, \langle\alpha_R(y), \beta_R(y)\rangle : y \in E\} \text{ and}$$
$$S = \{y, \langle\alpha_S(y), \beta_S(y)\rangle : y \in E\}.$$

For two GIFSs R and S the Hamming Distances are defined as follows:

$$\text{HD}(R, S) = \frac{1}{2}[|\alpha_R(y) - \alpha_S(y)| + |\beta_R(y) - \beta_S(y)| + |\gamma_R(y) - \gamma_S(y)|]$$

3 Proposed Methodology

Here, we introduce TOPSIS method to solve MCDM problems under generalized intuitionistic fuzzy information and apply possibility formula to rank alternatives. Priority vector of criteria are known.

Let the set of alternatives be $\{P_1, P_2, \ldots, P_m\}$ with criteria $\{v_1, v_2, \ldots, v_n\}$. The following steps are consider to discussed are proposed method:

1. Formation decision matrix:
   With the help of information obtained from attributes and criterion, GIF decision matrix are constructed as below:

$$M = \begin{pmatrix} d_{11} & d_{12} & \cdots & d_{1n} \\ d_{21} & d_{22} & \cdots & d_{2n} \\ \vdots & \vdots & \vdots & \vdots \\ d_{m1} & d_{m2} & \cdots & d_{mn} \end{pmatrix},$$

where $d_{ij} = \langle \alpha_{ij}, \beta_{ij}, \gamma_{ij} \rangle$.

2. Determine GIF positive and GIF negative ideal solution: Consider $\omega = (\omega_1, \omega_2, \ldots, \omega_n)^{\mathrm{T}}$ be priority vector of criteria, where $\omega_j = [\omega^l_j, \omega^u_j]$ are known.
Let us consider set of benefit criteria as J_1 and cost criteria as J_2. P^* and P^- denote GIF positive ideal solution and negative ideal solution. The values of P^* and P^- determined by following formulas:

$$P^* = (r^*_1, r^*_2, \cdots, r^*_n), \quad r^*_j = \langle \alpha^*_j, \beta^*_j, \gamma^*_j \rangle \tag{1}$$

$$P^- = (r^-_1, r^-_2, \cdots, r^-_n), \quad r^-_j = \langle \alpha^-_j, \beta^-_j, \gamma^-_j)\rangle \tag{2}$$

$$\alpha^*_{ij} = \{(\max_i\{\alpha_{ij}\}|j \in J_1), \quad (\min_i\{\alpha_{ij}\}|j \in J_2)\} \tag{3}$$

$$\beta^*_{ij} = \{(\min_i\{\beta_{ij}\}|j \in J_1), \quad (\max_i\{\beta_{ij}\}|j \in J_2)\} \tag{4}$$

$$\alpha^-_{ij} = \{(\min_i\{\alpha_{ij}\}|j \in J_1), \quad (\max_i\{\alpha_{ij}\}|j \in J_2)\} \tag{5}$$

$$\beta^-_{ij} = \{(\max_i\{\beta_{ij}\}|j \in J_1), \quad (\min_i\{\beta_{ij}\}|j \in J_2)\} \tag{6}$$

$$\gamma^*_{ij} = \{(1.5 - \max_i\{\alpha_{ij}\} - \min_i\{\beta_{ij}\}|j \in J_1), \quad (1.5 - \min_i\{\alpha_{ij}\} - \max_i\{\beta_{ij}\}|j \in J_2)\} \tag{7}$$

$$\gamma^-_{ij} = \{(1.5 - \min_i\{\alpha_{ij}\} - \max_i\{\beta_{ij}\}|j \in J_1), \quad (1.5 - \max_i\{\alpha_{ij}\} - \min_i\{\beta_{ij}\}|j \in J_2)\} \tag{8}$$

3. Determine weighted separation measure:
To obtain separation measures we use weighted Hamming Distance formula. Lower separation and upper separation measures $(\mathrm{SM}^*_i)^l$, $(\mathrm{SM}^*_i)^u$ and $(\mathrm{SM}^-_i)^l$, $(\mathrm{SM}^-_i)^u$ from ideal solutions are calculated as :

$$(\mathrm{SM}^*_i)^l = \frac{1}{2}\sum_{j=1}^{n} \omega^l_j[|\alpha_{ij} - \alpha^*_j| + |\beta_{ij} - \beta^*_j| + |\gamma_{ij} - \gamma^*_j|] \tag{9}$$

$$(\mathrm{SM}^*_i)^u = \frac{1}{2}\sum_{j=1}^{n} \omega^u_j[|\alpha_{ij} - \alpha^*_j| + |\beta_{ij} - \beta^*_j| + |\gamma_{ij} - \gamma^*_j|] \tag{10}$$

$$(\mathrm{SM}^-_i)^l = \frac{1}{2}\sum_{j=1}^{n} \omega^l_j[|\alpha_{ij} - \alpha^-_j| + |\beta_{ij} - \beta^-_j| + |\gamma_{ij} - \gamma^-_j|] \tag{11}$$

$$(\mathrm{SM}^-_i)^U = \frac{1}{2}\sum_{j=1}^{n} \omega^u_j[|\alpha_{ij} - \alpha^-_j| + |\beta_{ij} - \beta^-_j| + |\gamma_{ij} - \gamma^-_j|] \tag{12}$$

4. Determine relative closeness coefficient:
   The values relative closeness coefficient are determine by P^* and P^- and formulas are

$$((C_i^*)^l, (C_i^*)^u) = \left\{\left(\frac{(\mathrm{SM}_i^-)^l}{(\mathrm{SM}_i^*)^u + (\mathrm{SM}_i^-)^u}\right), \left(\frac{(\mathrm{SM}_i^-)^u}{(\mathrm{SM}_i^*)^l + (\mathrm{SM}_i^-)^l}\right)\right\} \quad (13)$$

5. Positioned the alternative according to the descending order of the relative closeness coefficients $C_i^* = ((C_i^*)^l, (C_i^*)^u)$. Utilizing possibility formula arranged alternatives.

Definition 5 Let us consider two interval numbers $x = [x^l, x^u]$ and $y = [y^l, y^u]$, where $0 \leq x^l \leq x^u \leq 1$ and $0 \leq y^l \leq y^u \leq 1$, possibility degree formula for $x \geq y$ is represent as:

$$p(x \geq y) = \max\left\{1 - \max\left(\frac{y^u - x^l}{y^u - y^l + x^u - x^l}, 0\right), 0\right\} \quad (14)$$

and it is denoted as $x \succ^{p(x \geq y)} y$.

Using the concept of possibility degree formula, we construct complementary generalized intuitionistic fuzzy matrix $B = b_{ij} = p(b_i \geq b_j)$ as follows:

$$B = \begin{bmatrix} b_{11} & b_{12} & \cdots & b_{1n} \\ b_{21} & b_{22} & \cdots & b_{2n} \\ \vdots & \vdots & \vdots & \vdots \\ b_{n1} & b_{n2} & \cdots & b_{nn} \end{bmatrix}.$$

where $b_{ij} \geq 0$, $b_{ij} + b_{ji} = 1$, $b_{ii} = 0.5$ and $i, j = 1, 2, \ldots, n$.

Summing all components in each row of matrix B, i.e.,

$$b_i = \sum_{j=1}^{n} b_{ij}, \quad i = 1, 2, \ldots, n. \quad (15)$$

Positioned the characteristics as per descending order of b_i.

4 Numerical Illustration

An organization choose to fabricate another production line. He had four alternatives. To locate a factory, consider four criteria such as expansion (v_1), material availability (v_2), distance to market (v_3) and cost of labor v_4. Then we apply our proposed method to solve this multi-criteria decision making problems.

1. Construct GIF decision matrix.
   First we consider GIF decision matrix as

Candidates	Criteria			
	v_1	v_2	v_3	v_4
P_1	⟨0.74, 0.44, 0.32⟩	⟨0.72, 0.38, 0.40⟩	⟨0.64, 0.44, 0.62⟩	⟨0.69, 0.45, 0.36⟩
P_2	⟨0.85, 0.42, 0.23⟩	⟨0.78, 0.45, 0.27⟩	⟨0.71, 0.36, 0.43⟩	⟨0.82, 0.42, 0.26⟩
P_3	⟨0.76, 0.36, 0.38⟩	⟨0.66, 0.38, 0.46⟩	⟨0.48, 0.65, 0.37⟩	⟨0.70, 0.40, 0.40⟩
P_4	⟨0.78, 0.46, 0.26⟩	⟨0.75, 0.42, 0.30⟩	⟨0.68, 0.48, 0.34⟩	⟨0.75, 0.35, 0.40⟩

2. Calculate GIF positive and the GIF negative ideal solutions.
   Consider priority vector of criteria as $\omega_1 = [0.228, 0.659]$, $\omega_2 = [0.215, 0.620]$, $\omega_3 = [0.208, 0.733]$ and $\omega_4 = [0.148, 0.653]$.
   Considering that expansion, material availability are the benefit criteria $J_1 = \{v_1, v_2\}$; market distance and labor cost are the cost criteria $J_2 = \{v_3, v_4\}$. Then the GIF positive and GIF negative ideal calculated by utilizing Eqs. 1–8.

$$P^* = \begin{cases} \langle 0.85, 0.23, 0.42\rangle, \ \langle 0.78, 0.38, 0.34\rangle; \\ \langle 0.48, 0.65, 0.37\rangle, \ \langle 0.69, 0.45, 0.36\rangle. \end{cases}$$

$$\text{and} \quad P^- = \begin{cases} \langle 0.74, 0.46, 0.30\rangle, \ \langle 0.66, 0.45, 0.39\rangle; \\ \langle 0.71, 0.36, 0.43\rangle, \ \langle 0.82, 0.35, 0.33\rangle. \end{cases}$$

3. Calculate separation measures.
   Utilizing Eqs. 9–12, determine GIF negative and positive separation measures

Candidates	$(SM^*)^l$	$(SM^*)^u$	$(SM^-)^l$	$(SM^-)^u$
P_1	0.176	0.598	0.089	0.315
P_2	0.155	0.514	0.090	0.317
P_3	0.191	0.641	0.154	0.449
P_4	0.254	0.812	0.146	0.452

4. Determine relative closeness coefficient:
   Utilizing Eq. 13 calculate the GIF positive and GIF negative ideal solutions as follows:

$$\left((C_1^*)^l, (C_1^*)^u\right) = \left(\left(\frac{0.089}{0.598 + 0.315}\right), \left(\frac{0.315}{0.176 + 0.089}\right)\right) = (0.097, 1.189)$$

$$\left((C_2^*)^l, (C_2^*)^u\right) = \left(\left(\frac{0.090}{0.514 + 0.317}\right), \left(\frac{0.317}{0.155 + 0.090}\right)\right) = (0.108, 1.294)$$

$$\left((C_3^*)^l, (C_3^*)^u\right) = \left(\left(\frac{0.154}{0.641 + 0.449}\right), \left(\frac{0.449}{0.191 + 0.154}\right)\right) = (0.141, 1.301)$$

$$\left((C_4^*)^l, (C_4^*)^u\right) = \left(\left(\frac{0.146}{0.812 + 0.452}\right), \left(\frac{0.452}{0.146 + 0.254}\right)\right) = (0.088, 1.130)$$

5. To rank the alternative, utilize possibility formula and formulate complementary GIF matrix as below:

$$B = \begin{pmatrix} 0.50 \; 0.47 \; 0.47 \; 0.52 \\ 0.53 \; 0.50 \; 0.49 \; 0.54 \\ 0.53 \; 0.51 \; 0.50 \; 0.55 \\ 0.48 \; 0.47 \; 0.45 \; 0.50 \end{pmatrix}$$

Add all values in every row of matrix P, then:

$$b_1 = 1.96, \quad b_2 = 2.06, \quad b_3 = 2.09 \quad \text{and} \quad b_4 = 1.90.$$

The candidates have been ranked as P_3, P_2, P_1 and P_4 according to the descending order $b_i, i = 1, \ldots, 4$. Therefore, P_3 selected as most appropriate location among the available alternatives to set up new factory.

5 Conclusion

By taking proper strategic selections any agency can obtain success. Selection of location is one of the strategic decisions. In this chapter, we mentioned TOPSIS technique to choose the most ideal location. Our proposed method has significant probabilities of accomplishment for multi-models decision making issues because of amazing predominance comprising vulnerability information. This strategy will be considered, specifically in provide chain administration and logistics, enterprise and marketing, environmental, human resources, and water aid management, engineering, manufacturing systems.

References

1. Zadeh, L.A.: Fuzzy sets. Inf. Control **8**, 338–353 (1965)
2. Atanassov, K.: Intuitionistic fuzzy sets. Fuzzy Sets Syst. **20**, 87–96 (1986)
3. Atanassov, K.: New operations defined over the intuitionistic fuzzy sets. Fuzzy Sets Syst. **61**, 137–142 (1994)
4. Adak, A.K., Bhowmik, M.: Interval cut-set of interval-valued intuitionistic fuzzy sets. Afr. J. Math. Comput. Sci. **4**(4), 192–200 (2011)
5. Adak, A.K., Bhowmik, M., Pal, M.: Semiring of interval-valued intuitionistic fuzzy matrices. Glob. J. Comput. Appl. Technol. **1**(3), 340–347 (2011)

6. Adak, A.K., Bhowmik, M., Pal, M.: Intuitionistic fuzzy block matrix and its some properties. Ann. Pure Appl. Math. **1**(1), 13–31 (2012)
7. Manna, D., Adak, A.K.: Interval-valued intuitionistic fuzzy R-subgroup of near-rings. J. Fuzzy Math. **24**(4), 985–994 (2016)
8. Adak, A.K.: Interval-valued intuitionistic fuzzy partition matrices. In: Emerging Research on Applied Fuzzy Sets and Intuitionistic Fuzzy Matrices, pp. 64–81. IGI-Global, USA (2017)
9. Ebrahimnejad, A., Adak, A.K., Jamkhaneh, E.B.: Eigenvalue of intuitionistic fuzzy matrices over distributive lattice. Int. J. Fuzzy Syst. Appl. (IJFSA) **8**(1), 1–18 (2019)
10. Mondal, T.K., Samanta, S.K.: Generalized intuitionistic fuzzy sets. J. Fuzzy Math. **10**(4), 839–862 (2002)
11. Adak, A.K., Bhowmik, M., Pal, M.: Interval cut-set of generalized interval-valued intuitionistic fuzzy sets. Int. J. Fuzzy Syst. Appl. **2**(3), 35–50 (2012)
12. Adak, A.K., Bhowmik, M., Pal, M.: Some properties of generalized intuitionistic fuzzy nilpotent matrices over distributive lattice. Int. J. Fuzzy Inf. Eng. **4**(4), 371–387 (2012)
13. Adak, A.K., Bhowmik, M., Pal, M.: Semiring of generalized interval-valued intuitionistic fuzzy matrices. World Appl. Sci. J. **16**(special issue of Applied Math), 07–16 (2012)
14. Adak, A.K., Salookolaei, D.D.: Some properties of rough pythagorean fuzzy sets. Fuzzy Inf. Eng. **13**(4), 420–435 (2021)
15. Bhowmik, M., Pal, M.: Generalized intuitionistic fuzzy matrices. Far-East J. Math. Sci. **29**(3), 533–554 (2008)
16. Boran, F.E.: An integrated intuitionistic fuzzy multi criteria decision making method for facility location selection. Math. Comput. Appl. **16**(2), 487–496 (2011)
17. Xu, Z.S.: Some similarity measure of intuitionistic fuzzy sets and their appliations to multi attribute decision making. Fuzzy Optim. Decis. Making **6**, 109–121 (2007)
18. Yeh, C.H.: A problem based selection of multi attribute decision making methods. Int. Trans. Oper. Res. **9**, 169–181 (2002)
19. Zhang, X., Yue, G., Teng, Z.: Possilbility degree of interval-valued fuzzy numbers and its application. In: International Symposium on Information Processing, pp. 033–036, 21–23 Aug 2009
20. Hwang, C.L., Yoon, K.: Multiple attribute decision methods and applications: a state of the art survey. Springer Verlag, New York (1981)
21. Shyamal, A.K., Pal, M.: Distances between intuitionistics fuzzy matrices. V.U.J. Phys. Sci. **8**, 81–91 (2002)

A Note on "Parameter Estimation and Optimization of Multi-objective Capacitated Stochastic Transportation Problem for Gamma Distribution"

Tanveen Kaur Bhatia, Amit Kumar, and S. S. Appadoo

Abstract Gupta et al. (Complex and Intelligent Systems (2020) 6:651–667) proposed three mathematical models for three different types of multi-objective capacitated interval type-2 fuzzy stochastic transportation problems as well as proposed an approach to solve the proposed mathematical models. In this note, it is pointed out that out of these three mathematical models, the first two mathematical models are not appropriate. Also, the appropriate mathematical models corresponding to the first two mathematical models are proposed. Furthermore, it is pointed out that some mathematical incorrect assumptions are considered in the above-mentioned approach. Therefore, it is inappropriate to use this approach to solve the proposed mathematical models as well as the third existing mathematical model. To propose an appropriate approach for solving the proposed mathematical models as well as the third existing mathematical model is a challenging open research problem.

Keywords Multi-objective capacitated interval type-2 fuzzy stochastic transportation problem · Trapezoidal interval type-2 fuzzy numbers

1 Introduction

Chen and Lee [1] proposed the following arithmetic operations of trapezoidal interval type-2 fuzzy numbers.

T. K. Bhatia (✉) · A. Kumar
School of Mathematics, Thapar Institute of Engineering and Technology (Deemed to Be University), Patiala, Punjab 147004, India
e-mail: shineybhatia25029@gmail.com

S. S. Appadoo
Department of Supply Chain Management, Asper School of Business, University of Manitoba, Winnipeg, Canada
e-mail: SS.Appadoo@umanitoba.ca

L. Sahoo et al. (eds.), *Real Life Applications of Multiple Criteria Decision Making Techniques in Fuzzy Domain*, Studies in Fuzziness and Soft Computing 420,
https://doi.org/10.1007/978-981-19-4929-6_27

Let

$$\tilde{A}_1 =(((a_{11}^U, a_{12}^U, a_{13}^U, a_{14}^U); H_1(A_1^U), H_2(A_1^U)), ((a_{11}^L, a_{12}^L, a_{13}^L, a_{14}^L); H_1(A_1^L), H_2(A_1^L)))$$

and
$\tilde{A}_2 = (((a_{21}^U, a_{22}^U, a_{23}^U, a_{24}^U); H_1(A_2^U), H_2(A_2^U)), ((a_{21}^L, a_{22}^L, a_{23}^L, a_{24}^L); H_1(A_2^L), H_2(A_2^L)))$ be two trapezoidal interval type-2 fuzzy numbers. Then,

(i)
$$\tilde{A}_1 \oplus \tilde{A}_2 =(((a_{11}^U + a_{21}^U, a_{12}^U + a_{22}^U, a_{13}^U + a_{23}^U, a_{14}^U + a_{24}^U); \min(H_1(A_1^U), H_1(A_2^U)), \min(H_2(A_1^U), H_2(A_2^U)),) ((a_{11}^L + a_{21}^L, a_{12}^L + a_{22}^L, a_{13}^L + a_{23}^L, a_{14}^L + a_{24}^L); \min(H_1(A_1^L), H_1(A_2^L)), \min(H_2(A_1^L), H_2(A_2^L)).$$

(ii)
$$\tilde{A}_1 \ominus \tilde{A}_2 =(((a_{11}^U - a_{24}^U, a_{12}^U - a_{23}^U, a_{13}^U - a_{22}^U, a_{14}^U - a_{21}^U); \min(H_1(A_1^U), H_1(A_2^U)), \min(H_2(A_1^U), H_2(A_2^U)),) ((a_{11}^L + a_{21}^L, a_{12}^L + a_{22}^L, a_{13}^L + a_{23}^L, a_{14}^L + a_{24}^L); \min(H_1(A_1^L), H_1(A_2^L)), \min(H_2(A_1^L), H_2(A_2^L)).$$

(iii)
$$\tilde{A}_1 \otimes \tilde{A}_2 =(((a_{11}^U \times a_{21}^U, a_{12}^U \times a_{22}^U, a_{13}^U \times a_{23}^U, a_{14}^U \times a_{24}^U); \min(H_1(A_1^U), H_1(A_2^U)), \min(H_2(A_1^U), H_2(A_2^U)),) ((a_{11}^L \times a_{21}^L, a_{12}^L \times a_{22}^L, a_{13}^L \times a_{23}^L, a_{14}^L \times a_{24}^L); \min(H_1(A_1^L), H_1(A_2^L)), \min(H_2(A_1^L), H_2(A_2^L))$$
if $a_{11}^U \geq 0$ and$a_{21}^U \geq 0$.

Sinha et al. [2] pointed out that the existing arithmetic operations in [1] are not appropriate. To resolve the inappropriateness of the existing arithmetic operations [1], Sinha et al. [2] proposed the following arithmetic operations of trapezoidal interval type-2 fuzzy numbers.

Let $\tilde{A}_1 = (((a_{11}^U, a_{12}^U, a_{13}^U, a_{14}^U); H_1(A_1^U), H_2(A_1^U)), ((a_{11}^L, a_{12}^L, a_{13}^L, a_{14}^L); H_1(A_1^L), H_2(A_1^L)))$ and.

$\tilde{A}_2 = (((a_{21}^U, a_{22}^U, a_{23}^U, a_{24}^U); H_1(A_2^U), H_2(A_2^U)), ((a_{21}^L, a_{22}^L, a_{23}^L, a_{24}^L); H_1(A_2^L), H_2(A_2^L)))$ be two trapezoidal interval type-2 fuzzy numbers. Then,

(i)
$$\tilde{A}_1 \oplus \tilde{A}_2 = (((a_{11}^U + a_{21}^U, a_{12}^U + a_{22}^U, a_{13}^U + a_{23}^U, a_{14}^U + a_{24}^U) ; H_1(A_1^U) + H_1(A_2^U) - H_1(A_1^U) \times H_1(A_2^U), H_2(A_1^U) + H_2(A_2^U) - H_2(A_1^U) \times H_2(A_2^U)), ((a_{11}^L + a_{21}^L, a_{12}^L + a_{22}^L, a_{13}^L + a_{23}^L, a_{14}^L + a_{24}^L); H_1(A_1^L) + H_1(A_2^L) - H_1(A_1^L) \times H_1(A_2^L), H_2(A_1^L) + H_2(A_2^L) - H_2(A_1^L) \times H_2(A_2^L))).$$

(ii)
$$\begin{aligned}\tilde{A}_1 \ominus \tilde{A}_2 = & (((a_{11}^U - a_{24}^U, a_{12}^U - a_{23}^U, a_{13}^U - a_{22}^U, a_{14}^U - a_{21}^U)\ ; H_1(A_1^U) \\ & + H_1(A_2^U) - H_1(A_1^U) \times H_1(A_2^U), H_2(A_1^U) + H_2(A_2^U) \\ & - H_2(A_1^U) \times H_2(A_2^U)), \\ & ((a_{11}^L - a_{24}^L, a_{12}^L - a_{23}^L, a_{13}^L - a_{22}^L, a_{14}^L - a_{21}^L)\ ; H_1(A_1^L) \\ & + H_1(A_2^L) - H_1(A_1^L) \times H_1(A_2^L), H_2(A_1^L) \\ & + H_2(A_2^L) - H_2(A_1^L) \times H_2(A_2^L))).\end{aligned}$$

(iii)
$$\begin{aligned}\tilde{A}_1 \otimes \tilde{A}_2 = & (((a_{11}^U \times a_{21}^U, a_{12}^U \times a_{22}^U, a_{13}^U \times a_{23}^U, a_{14}^U \times a_{24}^U); \\ & H_1(A_1^U) \times H_1(A_2^U), H_2(A_1^U) \times H_2(A_2^U)),) \\ & ((a_{11}^L \times a_{21}^L, a_{12}^L \times a_{22}^L, a_{13}^L \times a_{23}^L, a_{14}^L \times a_{24}^L); \\ & H_1(A_1^L) \times H_1(A_2^L), H_2(A_1^L) \times H_2(A_2^L)))\end{aligned}$$
if $a_{11}^U \geq 0$ and $a_{21}^U \geq 0$.

Using the existing arithmetic operations [2], Gupta et al. [3] proposed the mathematical model (M_1), the mathematical model (M_2) and the mathematical model (M_3) for three different types of multi-objective capacitated fuzzy stochastic transportation problems.

Model (M_1)

$$\text{Minimize}\left(\sum_{i=1}^{m}\sum_{j=1}^{n}\begin{pmatrix}\left(\left(d_{ij1}^U, d_{ij2}^U, d_{ij3}^U, d_{ij4}^U\right); H_1\left(d_{ij}^U\right), H_2\left(d_{ij}^U\right)\right), \\ \left(\left(d_{ij1}^L, d_{ij2}^L, d_{ij3}^L, d_{ij4}^L\right); H_1\left(d_{ij}^L\right), H_2\left(d_{ij}^L\right)\right)\end{pmatrix} x_{ij}\right),$$

$$\text{Minimize}\left(\sum_{i=1}^{m}\sum_{j=1}^{n}\begin{pmatrix}\left(\left(l_{ij1}^U, l_{ij2}^U, l_{ij3}^U, l_{ij4}^U\right); H_1\left(l_{ij}^U\right), H_2\left(l_{ij}^U\right)\right), \\ \left(\left(l_{ij1}^L, l_{ij2}^L, l_{ij3}^L, l_{ij4}^L\right); H_1\left(l_{ij}^L\right), H_2\left(l_{ij}^L\right)\right)\end{pmatrix} x_{ij}\right),$$

$$\text{Minimize}\left(\frac{\sum_{i=1}^{m}\sum_{j=1}^{n}\begin{pmatrix}\left(\left(t_{ij1}^{aU}, t_{ij2}^{aU}, t_{ij3}^{aU}, t_{ij4}^{aU}\right); H_1\left(t_{ij}^{aU}\right), H_2\left(t_{ij}^{aU}\right)\right), \\ \left(\left(t_{ij1}^{aL}, t_{ij2}^{aL}, t_{ij3}^{aL}, t_{ij4}^{aL}\right); H_1\left(t_{ij}^{aL}\right), H_2\left(t_{ij}^{aL}\right)\right)\end{pmatrix} x_{ij}}{\sum_{i=1}^{m}\sum_{j=1}^{n}\begin{pmatrix}\left(\left(t_{ij1}^{sU}, t_{ij2}^{sU}, t_{ij3}^{sU}, t_{ij4}^{sU}\right); H_1\left(t_{ij}^{sU}\right), H_2\left(t_{ij}^{sU}\right)\right), \\ \left(\left(t_{ij1}^{sL}, t_{ij2}^{sL}, t_{ij3}^{sL}, t_{ij4}^{sL}\right); H_1\left(t_{ij}^{sL}\right), H_2\left(t_{ij}^{sL}\right)\right)\end{pmatrix} x_{ij}}\right),$$

$$\text{Minimize}\left(\frac{\sum_{i=1}^{m}\sum_{j=1}^{n}\begin{pmatrix}\left(\left(c_{ij1}^{aU}, c_{ij2}^{aU}, c_{ij3}^{aU}, c_{ij4}^{aU}\right); H_1\left(c_{ij}^{aU}\right), H_2\left(c_{ij}^{aU}\right)\right), \\ \left(\left(c_{ij1}^{aL}, c_{ij2}^{aL}, c_{ij3}^{aL}, c_{ij4}^{aL}\right); H_1\left(c_{ij}^{aL}\right), H_2\left(c_{ij}^{aL}\right)\right)\end{pmatrix} x_{ij}}{\sum_{i=1}^{m}\sum_{j=1}^{n}\begin{pmatrix}\left(\left(c_{ij1}^{sU}, c_{ij2}^{sU}, c_{ij3}^{sU}, c_{ij4}^{sU}\right); H_1\left(c_{ij}^{sU}\right), H_2\left(c_{ij}^{sU}\right)\right), \\ \left(\left(c_{ij1}^{sL}, c_{ij2}^{sL}, c_{ij3}^{sL}, c_{ij4}^{sL}\right); H_1\left(c_{ij}^{sL}\right), H_2\left(c_{ij}^{sL}\right)\right)\end{pmatrix} x_{ij}}\right),$$

$$\text{Maximize}\left(\frac{\sum_{i=1}^{m}\sum_{j=1}^{n}\begin{pmatrix}\left(\left(d_{ij1}^{sU}, d_{ij2}^{sU}, d_{ij3}^{sU}, d_{ij4}^{sU}\right); H_1\left(d_{ij}^{sU}\right), H_2\left(d_{ij}^{sU}\right)\right),\\ \left(\left(d_{ij1}^{sL}, d_{ij2}^{sL}, d_{ij3}^{sL}, d_{ij4}^{sL}\right); H_1\left(d_{ij}^{sL}\right), H_2\left(d_{ij}^{sL}\right)\right)\end{pmatrix}x_{ij}}{\sum_{i=1}^{m}\sum_{j=1}^{n}\begin{pmatrix}\left(\left(c_{ij1}^{aU}, c_{ij2}^{aU}, c_{ij3}^{aU}, c_{ij4}^{aU}\right); H_1\left(c_{ij}^{aU}\right), H_2\left(c_{ij}^{aU}\right)\right),\\ \left(\left(c_{ij1}^{aL}, c_{ij2}^{aL}, c_{ij3}^{aL}, c_{ij4}^{aL}\right); H_1\left(c_{ij}^{aL}\right), H_2\left(c_{ij}^{aL}\right)\right)\end{pmatrix}x_{ij}}\right)$$

Subject to

$$\Pr\left(\sum_{j=1}^{n} x_{ij}\{\le, =, \ge\}a_i\right) \ge 1-\gamma_i; i = 1, 2, \ldots, m,$$

$$\Pr\left(\sum_{i=1}^{m} x_{ij}\{\le, =, \ge\}b_j\right) \ge 1-\delta_j; j = 1, 2, \ldots, n,$$

$$0 \le x_{ij} \le r_{ij}; i = 1, 2, \ldots, m, j = 1, 2, \ldots, n.$$

where,

(i) m : number of sources.

(ii) n : number of destinations.

(iii) $\begin{pmatrix}\left(\left(d_{ij1}^{U}, d_{ij2}^{U}, d_{ij3}^{U}, d_{ij4}^{U}\right); H_1\left(d_{ij}^{U}\right), H_2\left(d_{ij}^{U}\right)\right),\\ \left(\left(d_{ij1}^{L}, d_{ij2}^{L}, d_{ij3}^{L}, d_{ij4}^{L}\right); H_1\left(d_{ij}^{L}\right), H_2\left(d_{ij}^{L}\right)\right)\end{pmatrix}$ is an interval type-2 fuzzy number.

(iv) $\begin{pmatrix}\left(\left(l_{ij1}^{U}, l_{ij2}^{U}, l_{ij3}^{U}, l_{ij4}^{U}\right); H_1\left(l_{ij}^{U}\right), H_2\left(l_{ij}^{U}\right)\right),\\ \left(\left(l_{ij1}^{L}, l_{ij2}^{L}, l_{ij3}^{L}, l_{ij4}^{L}\right); H_1\left(l_{ij}^{L}\right), H_2\left(l_{ij}^{L}\right)\right)\end{pmatrix}$ is an interval type-2 fuzzy number.

(v) $\begin{pmatrix}\left(\left(t_{ij1}^{aU}, t_{ij2}^{aU}, t_{ij3}^{aU}, t_{ij4}^{aU}\right); H_1\left(t_{ij}^{aU}\right), H_2\left(t_{ij}^{aU}\right)\right),\\ \left(\left(t_{ij1}^{aL}, t_{ij2}^{aL}, t_{ij3}^{aL}, t_{ij4}^{aL}\right); H_1\left(t_{ij}^{aL}\right), H_2\left(t_{ij}^{aL}\right)\right)\end{pmatrix}$ is an interval type-2 fuzzy number.

(vi) $\begin{pmatrix}\left(\left(t_{ij1}^{sU}, t_{ij2}^{sU}, t_{ij3}^{sU}, t_{ij4}^{sU}\right); H_1\left(t_{ij}^{sU}\right), H_2\left(t_{ij}^{sU}\right)\right),\\ \left(\left(t_{ij1}^{sL}, t_{ij2}^{sL}, t_{ij3}^{sL}, t_{ij4}^{sL}\right); H_1\left(t_{ij}^{sL}\right), H_2\left(t_{ij}^{sL}\right)\right)\end{pmatrix}$ is an interval type-2 fuzzy number.

(vii) $\begin{pmatrix}\left(\left(c_{ij1}^{aU}, c_{ij2}^{aU}, c_{ij3}^{aU}, c_{ij4}^{aU}\right); H_1\left(c_{ij}^{aU}\right), H_2\left(c_{ij}^{aU}\right)\right),\\ \left(\left(c_{ij1}^{aL}, c_{ij2}^{aL}, c_{ij3}^{aL}, c_{ij4}^{aL}\right); H_1\left(c_{ij}^{aL}\right), H_2\left(c_{ij}^{aL}\right)\right)\end{pmatrix}$ is an interval type-2 fuzzy number.

(viii) $\begin{pmatrix}\left(\left(c_{ij1}^{sU}, c_{ij2}^{sU}, c_{ij3}^{sU}, c_{ij4}^{sU}\right); H_1\left(c_{ij}^{sU}\right), H_2\left(c_{ij}^{sU}\right)\right),\\ \left(\left(c_{ij1}^{sL}, c_{ij2}^{sL}, c_{ij3}^{sL}, c_{ij4}^{sL}\right); H_1\left(c_{ij}^{sL}\right), H_2\left(c_{ij}^{sL}\right)\right)\end{pmatrix}$ is an interval type-2 fuzzy number.

(ix) $\left(\begin{array}{c}\left(\left(d_{ij1}^{sU}, d_{ij2}^{sU}, d_{ij3}^{sU}, d_{ij4}^{sU}\right); H_1\left(d_{ij}^{sU}\right), H_2\left(d_{ij}^{sU}\right)\right), \\ \left(\left(d_{ij1}^{sL}, d_{ij2}^{sL}, d_{ij3}^{sL}, d_{ij4}^{sL}\right); H_1\left(d_{ij}^{sL}\right), H_2\left(d_{ij}^{sL}\right)\right)\end{array}\right)$ is an interval type-2 fuzzy number.

(x) a_i represents the availability of the product at the ith source.

(xi) b_j represents the demand of the product at the jth destination.

(xii) x_{ij} represents the quantity of the product to be transported from the ith source to the jth destination.

(xiii) r_{ij} represents the maximum restriction on quantity of the product to be transported from the ith source to the jth destination.

Model (M_2)

$$\text{Minimize}\left(\frac{\sum_{i=1}^{m}\sum_{j=1}^{n}\left(\begin{array}{c}\left(\left(t_{ij1}^{aU}, t_{ij2}^{aU}, t_{ij3}^{aU}, t_{ij4}^{aU}\right); H_1\left(t_{ij}^{aU}\right), H_2\left(t_{ij}^{aU}\right)\right), \\ \left(\left(t_{ij1}^{aL}, t_{ij2}^{aL}, t_{ij3}^{aL}, t_{ij4}^{aL}\right); H_1\left(t_{ij}^{aL}\right), H_2\left(t_{ij}^{aL}\right)\right)\end{array}\right)x_{ij}}{\sum_{i=1}^{m}\sum_{j=1}^{n}\left(\begin{array}{c}\left(\left(t_{ij1}^{sU}, t_{ij2}^{sU}, t_{ij3}^{sU}, t_{ij4}^{sU}\right); H_1\left(t_{ij}^{sU}\right), H_2\left(t_{ij}^{sU}\right)\right), \\ \left(\left(t_{ij1}^{sL}, t_{ij2}^{sL}, t_{ij3}^{sL}, t_{ij4}^{sL}\right); H_1\left(t_{ij}^{sL}\right), H_2\left(t_{ij}^{sL}\right)\right)\end{array}\right)x_{ij}}\right),$$

$$\text{Minimize}\left(\frac{\sum_{i=1}^{m}\sum_{j=1}^{n}\left(\begin{array}{c}\left(\left(c_{ij1}^{aU}, c_{ij2}^{aU}, c_{ij3}^{aU}, c_{ij4}^{aU}\right); H_1\left(c_{ij}^{aU}\right), H_2\left(c_{ij}^{aU}\right)\right), \\ \left(\left(c_{ij1}^{aL}, c_{ij2}^{aL}, c_{ij3}^{aL}, c_{ij4}^{aL}\right); H_1\left(c_{ij}^{aL}\right), H_2\left(c_{ij}^{aL}\right)\right)\end{array}\right)x_{ij}}{\sum_{i=1}^{m}\sum_{j=1}^{n}\left(\begin{array}{c}\left(\left(c_{ij1}^{sU}, c_{ij2}^{sU}, c_{ij3}^{sU}, c_{ij4}^{sU}\right); H_1\left(c_{ij}^{sU}\right), H_2\left(c_{ij}^{sU}\right)\right), \\ \left(\left(c_{ij1}^{sL}, c_{ij2}^{sL}, c_{ij3}^{sL}, c_{ij4}^{sL}\right); H_1\left(c_{ij}^{sL}\right), H_2\left(c_{ij}^{sL}\right)\right)\end{array}\right)x_{ij}}\right),$$

$$\text{Maximize}\left(\frac{\sum_{i=1}^{m}\sum_{j=1}^{n}\left(\begin{array}{c}\left(\left(d_{ij1}^{sU}, d_{ij2}^{sU}, d_{ij3}^{sU}, d_{ij4}^{sU}\right); H_1\left(d_{ij}^{sU}\right), H_2\left(d_{ij}^{sU}\right)\right), \\ \left(\left(d_{ij1}^{sL}, d_{ij2}^{sL}, d_{ij3}^{sL}, d_{ij4}^{sL}\right); H_1\left(d_{ij}^{sL}\right), H_2\left(d_{ij}^{sL}\right)\right)\end{array}\right)x_{ij}}{\sum_{i=1}^{m}\sum_{j=1}^{n}\left(\begin{array}{c}\left(\left(c_{ij1}^{aU}, c_{ij2}^{aU}, c_{ij3}^{aU}, c_{ij4}^{aU}\right); H_1\left(c_{ij}^{aU}\right), H_2\left(c_{ij}^{aU}\right)\right), \\ \left(\left(c_{ij1}^{aL}, c_{ij2}^{aL}, c_{ij3}^{aL}, c_{ij4}^{aL}\right); H_1\left(c_{ij}^{aL}\right), H_2\left(c_{ij}^{aL}\right)\right)\end{array}\right)x_{ij}}\right)$$

Subject to.

Constraints of the mathematical model (M_1).

Model (M_3)

$$\text{Minimize}\left(\sum_{i=1}^{m}\sum_{j=1}^{n}\left(\begin{array}{c}\left(\left(d_{ij1}^{U}, d_{ij2}^{U}, d_{ij3}^{U}, d_{ij4}^{U}\right); H_1\left(d_{ij}^{U}\right), H_2\left(d_{ij}^{U}\right)\right), \\ \left(\left(d_{ij1}^{L}, d_{ij2}^{L}, d_{ij3}^{L}, d_{ij4}^{L}\right); H_1\left(d_{ij}^{L}\right), H_2\left(d_{ij}^{L}\right)\right)\end{array}\right)x_{ij}\right),$$

$$\text{Minimize}\left(\sum_{i=1}^{m}\sum_{j=1}^{n}\left(\begin{array}{c}\left(\left(l_{ij1}^{U}, l_{ij2}^{U}, l_{ij3}^{U}, l_{ij4}^{U}\right); H_1\left(l_{ij}^{U}\right), H_2\left(l_{ij}^{U}\right)\right), \\ \left(\left(l_{ij1}^{L}, l_{ij2}^{L}, l_{ij3}^{L}, l_{ij4}^{L}\right); H_1\left(l_{ij}^{L}\right), H_2\left(l_{ij}^{L}\right)\right)\end{array}\right)x_{ij}\right)$$

Subject to.

Constraints of the mathematical model (M_1).

Furthermore, Gupta et al. [3] proposed an approach to solve the mathematical models (M_1), (M_2) and (M_3).

It is obvious that to define the objective functions of the mathematical model (M_1) and the mathematical model (M_2), Gupta et al. [3] have assumed that the existing division of two trapezoidal interval type-2 fuzzy numbers [2] is appropriate. While, in this note, it is pointed out that the existing division of two trapezoidal interval type-2 fuzzy numbers [2] is not appropriate. Hence, the mathematical model (M_1) and the mathematical model (M_2) are not appropriate.

Also, it is pointed out that the mathematical model (M_3) is appropriate as no division operation is used to define the mathematical model (M_3). But, Gupta et al.'s approach [3] to solve the mathematical model (M_3) is not valid.

This paper is organized as follows: In Sect. 2, some existing arithmetic operations are discussed. In Sect. 3, the inappropriateness of the existing arithmetic operations i.e., existing multiplicative inverse of a trapezoidal interval type-2 fuzzy number and existing division of two trapezoidal interval type-2 fuzzy numbers are discussed. In Sect. 4, the appropriate arithmetic operations i.e., appropriate multiplicative inverse of a trapezoidal interval type-2 fuzzy number and appropriate division of two trapezoidal interval type-2 fuzzy numbers are proposed. In Sect. 5, the inappropriateness of the first two mathematical models (M_1) and (M_2) is discussed. In Sect. 6, the appropriate mathematical models corresponding to the existing mathematical models (M_1) and (M_2) are proposed. In Sect. 7, Gupta et al.'s approach [3] to solve the proposed mathematical model (M_1) is discussed. In Sect. 8, some mathematical incorrect assumptions, considered in Gupta et al.'s approach [3] are pointed out. Section 9 concludes the paper.

2 Some Existing Arithmetic Operations

It is pertinent to mention that, Sinha et al. [2] have not proposed the additive inverse of a trapezoidal interval type-2 fuzzy number, the multiplicative inverse of a trapezoidal interval type-2 fuzzy number and the division of two trapezoidal interval type-2 fuzzy numbers. However, using the arithmetic operations, proposed by Sinha et al. [2], these operations can be easily defined. The same is done in this section.

2.1 Additive Inverse of a Trapezoidal Interval Type-2 Fuzzy Number

Assuming.

$$\tilde{A}_1 = \begin{pmatrix} ((a_{11}^U, a_{12}^U, a_{13}^U, a_{14}^U); H_1(A_1^U), H_2(A_1^U)), \\ ((a_{11}^L, a_{12}^L, a_{13}^L, a_{14}^L); H_1(A_1^L), H_2(A_1^L)) \end{pmatrix}$$
$$= (((0, 0, 0, 0); 0, 0), ((0, 0, 0, 0); 0, 0))$$

in the existing arithmetic operation (ii), defined in Sect. 1, the following additive inverse of the trapezoidal interval type-2 fuzzy number

$$\tilde{A}_2 = \begin{pmatrix} ((a_{21}^U, a_{22}^U, a_{23}^U, a_{24}^U); H_1(A_2^U), H_2(A_2^U)), \\ ((a_{21}^L, a_{22}^L, a_{23}^L, a_{24}^L); H_1(A_2^L), H_2(A_2^L)) \end{pmatrix}$$

is obtained.

$$\ominus\tilde{A}_2 = \begin{pmatrix} ((-a_{24}^U, -a_{23}^U, -a_{22}^U, -a_{21}^U); H_1(A_2^U), H_2(A_2^U)), \\ ((-a_{24}^L, -a_{23}^L, -a_{22}^L, -a_{21}^L); H_1(A_2^L), H_2(A_2^L)) \end{pmatrix}.$$

2.2 Multiplicative Inverse of a Trapezoidal Interval Type-2 Fuzzy Number

It is obvious from Sect. 2.1 that the positions of $H_1(A_2^U), H_2(A_2^U), H_1(A_2^L), H_2(A_2^L)$ are same in a trapezoidal interval type-2 fuzzy number and its additive inverse. While, the order of $a_{11}^U, a_{12}^U, a_{13}^U, a_{14}^U$ and the order of $a_{11}^L, a_{12}^L, a_{13}^L, a_{14}^L$ have been reversed to maintain the inequality $-a_{24}^U \leq -a_{23}^U \leq -a_{22}^U \leq -a_{21}^U$ and $-a_{24}^L \leq -a_{23}^L \leq -a_{22}^L \leq -a_{21}^L$.

On the same direction, it can be easily concluded that according to the existing arithmetic operations [2], the multiplicative inverse of a trapezoidal interval type-2 fuzzy number $\tilde{A}_2 = (((a_{21}^U, a_{22}^U, a_{23}^U, a_{24}^U); H_1(A_2^U), H_2(A_2^U)), ((a_{21}^L, a_{22}^L, a_{23}^L, a_{24}^L); H_1(A_2^L), H_2(A_2^L)))$ where $a_{21}^U > 0$, will be

$$\frac{1}{\tilde{A}_2} = \Bigg(\bigg(\Big(\frac{1}{a_{24}^U}, \frac{1}{a_{23}^U}, \frac{1}{a_{22}^U}, \frac{1}{a_{21}^U}\Big); H_1(A_2^U), H_2(A_2^U)\bigg),$$
$$\bigg(\Big(\frac{1}{a_{24}^L}, \frac{1}{a_{23}^L}, \frac{1}{a_{22}^L}, \frac{1}{a_{21}^L}\Big); H_1(A_2^L), H_2(A_2^L)\bigg)\Bigg).$$

2.3 Division of Two Trapezoidal Interval Type-2 Fuzzy Numbers

If $\tilde{A}_1 = (((a_{11}^U, a_{12}^U, a_{13}^U, a_{14}^U); H_1(A_1^U), H_2(A_1^U)), ((a_{11}^L, a_{12}^L, a_{13}^L, a_{14}^L); H_1(A_1^L), H_2(A_1^L)))$ and.

$\tilde{A}_2 = (((a_{21}^U, a_{22}^U, a_{23}^U, a_{24}^U); H_1(A_2^U), H_2(A_2^U)), ((a_{21}^L, a_{22}^L, a_{23}^L, a_{24}^L); H_1(A_2^L), H_2(A_2^L)))$ where $a_{21}^U > 0$, are two trapezoidal interval type-2 fuzzy numbers. Then, using Sect. 2.2,

$$\frac{1}{\tilde{A}_2} = \left(\left(\left(\frac{1}{a_{24}^U}, \frac{1}{a_{23}^U}, \frac{1}{a_{22}^U}, \frac{1}{a_{21}^U}\right); H_1(A_2^U), H_2(A_2^U)\right), \left(\left(\frac{1}{a_{24}^L}, \frac{1}{a_{23}^L}, \frac{1}{a_{22}^L}, \frac{1}{a_{21}^L}\right); H_1(A_2^L), H_2(A_2^L)\right)\right).$$

Using the existing arithmetic operation (iii) defined in Sect. 1, $\tilde{A}_1 \otimes \left(\frac{1}{\tilde{A}_2}\right) = \frac{\tilde{A}_1}{\tilde{A}_2} = \left(\left(\left(a_{11}^U \times \frac{1}{a_{24}^U}, a_{12}^U \times \frac{1}{a_{23}^U}, a_{13}^U \times \frac{1}{a_{22}^U}, a_{14}^U \times \frac{1}{a_{21}^U}\right); H_1(A_1^U) \times H_1(A_2^U), H_2(A_1^U) \times H_2(A_2^U)\right), \left(\left(a_{11}^L \times \frac{1}{a_{24}^L}, a_{12}^L \times \frac{1}{a_{23}^L}, a_{13}^L \times \frac{1}{a_{22}^L}, a_{14}^L \times \frac{1}{a_{21}^L}\right); H_1(A_1^L) \times H_1(A_2^L), H_2(A_1^L) \times H_2(A_2^L)\right)\right)$.

3 Inappropriateness of the Existing Arithmetic Operations of Trapezoidal Interval Type-2 Fuzzy Numbers

In this section, it is shown that it is inappropriate to use the existing arithmetic operations of trapezoidal interval type-2 fuzzy numbers.

3.1 Inappropriateness of the Existing Multiplicative Inverse of a Trapezoidal Interval Type-2 Fuzzy Number

It is obvious from Sect. 2.2 that if $\tilde{A} = \begin{pmatrix} ((a_1^U, a_2^U, a_3^U, a_4^U); H_1(A^U), H_2(A^U)), \\ ((a_1^L, a_2^L, a_3^L, a_4^L); H_1(A^L), H_2(A^L)) \end{pmatrix}$ is a trapezoidal interval type-2 fuzzy number such that $a_1^U > 0$ as shown in Fig. 1. Then, $\frac{1}{\tilde{A}} = \left(\left(\left(\frac{1}{a_4^U}, \frac{1}{a_3^U}, \frac{1}{a_2^U}, \frac{1}{a_1^U}\right); H_1(A^U), H_2(A^U)\right), \left(\left(\frac{1}{a_4^L}, \frac{1}{a_3^L}, \frac{1}{a_2^L}, \frac{1}{a_1^L}\right); H_1(A^L), H_2(A^L)\right)\right)$ as shown in Fig. 2.

It is obvious from Figs. 1 and 2 that.

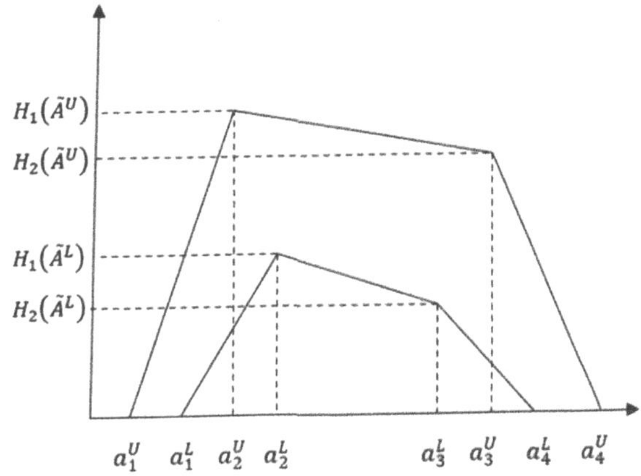

Fig. 1 Graphical representation of $\tilde{A}$

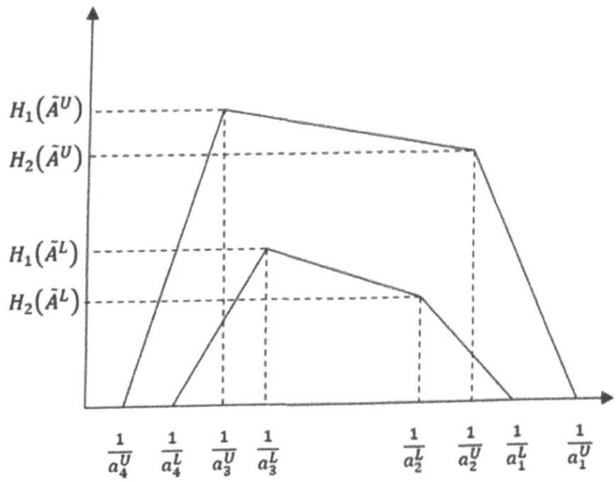

Fig. 2 Graphical representation of $\frac{1}{\tilde{A}}$

(i) The membership value corresponding to a_2^U in $\tilde{A}$ is $H_1(A^U)$. While, the membership value corresponding to $\frac{1}{a_2^U}$ in $\frac{1}{\tilde{A}}$ is $H_2(A^U)$ i.e., the membership value corresponding to a_2^U and $\frac{1}{a_2^U}$ are not equal.

(ii) The membership value corresponding to a_3^U in $\tilde{A}$ is $H_2(A^U)$. While, the membership value corresponding to $\frac{1}{a_3^U}$ in $\frac{1}{\tilde{A}}$ is $H_1(A^U)$ i.e., the membership value corresponding to a_3^U and $\frac{1}{a_3^U}$ are not equal.

(iii) The membership value corresponding to a_2^L in $\tilde{A}$ is $H_1(A^L)$. While, the membership value corresponding to $\frac{1}{a_2^L}$ in $\frac{1}{\tilde{A}}$ is $H_2(A^L)$ i.e., the membership value corresponding to a_2^L and $\frac{1}{a_2^L}$ are not equal.

(iv) The membership value corresponding to a_3^L in $\tilde{A}$ is $H_2(A^L)$. While, the membership value corresponding to $\frac{1}{a_3^L}$ in $\frac{1}{\tilde{A}}$ is $H_1(A^L)$ i.e., the membership value corresponding to a_3^L and $\frac{1}{a_3^L}$ are not equal.

While, as mentioned in the literature [4] that according to Zadeh's extension principle [5], the membership values at $a_2^U, a_2^L, a_3^L, a_3^U$ and $\frac{1}{a_2^U}, \frac{1}{a_2^L}, \frac{1}{a_3^L}, \frac{1}{a_3^U}$ should be equal. This clearly indicates that the existing multiplicative inverse $\frac{1}{\tilde{A}}$ is not appropriate.

3.2 Inappropriateness of the Existing Division of Two Trapezoidal Interval Type-2 Fuzzy Numbers

It is obvious from Sect. 2.3 that the multiplicative inverse $\frac{1}{\tilde{A}_2}$ has been used to define $\frac{\tilde{A}_1}{\tilde{A}_2}$, i.e.,$\tilde{A}_1 \otimes \left(\frac{1}{\tilde{A}_2}\right)$. However, as discussed in Sect. 3.1, that the multiplicative inverse $\frac{1}{\tilde{A}_2}$ is not appropriate. Therefore, the existing division of two trapezoidal interval type-2 fuzzy numbers, discussed in Sect. 2.3, is also not appropriate.

4 Appropriate Arithmetic Operations of Trapezoidal Interval Type-2 Fuzzy Numbers

In this section, appropriate arithmetic operations of trapezoidal interval type-2 fuzzy numbers are proposed.

4.1 Appropriate Multiplicative Inverse of a Trapezoidal Interval Type-2 Fuzzy Number

It is obvious from Sect. 3.1, that the existing multiplicative inverse of a trapezoidal interval type-2 fuzzy number, discussed in Sect. 2.2, is not appropriate.

In this section, the appropriate multiplicative inverse of a trapezoidal interval type-2 fuzzy number is defined.

As discussed in Sect. 3.1, if the membership values corresponding to a_2^U, a_3^U, a_2^L and a_3^L in $\tilde{A} = (((a_1^U, a_2^U, a_3^U, a_4^U); H_1(A^U), H_2(A^U)), ((a_1^L, a_2^L, a_3^L, a_4^L); H_1(A^L), H_2(A^L)))$ are $H_1(A^U), H_2(A^U), H_1(A^L)$ and $H_2(A^L)$ respectively. Then, according to Zadeh's extension principle [5], the membership values corresponding to $\frac{1}{a_2^U}, \frac{1}{a_2^L}, \frac{1}{a_3^L}$

Fig. 3 Appropriate graphical representation of $\frac{1}{\tilde{A}}$

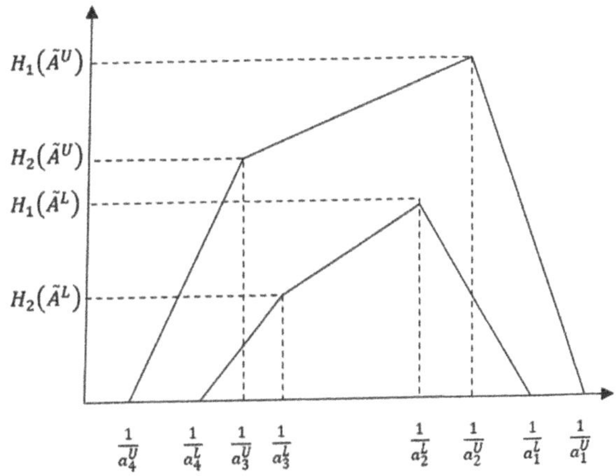

and $\frac{1}{a_3^U}$ in $\frac{1}{\tilde{A}}$ should also be $H_1(A^U)$, $H_2(A^U)$, $H_1(A^L)$ and $H_2(A^L)$ respectively. Hence, the appropriate multiplicative inverse of $\tilde{A}$ is.

$$\frac{1}{\tilde{A}} = \left(\left(\left(\frac{1}{a_4^U}, \frac{1}{a_3^U}, \frac{1}{a_2^U}, \frac{1}{a_1^U}\right); H_2(A^U), H_1(A^U)\right),\right.$$
$$\left.\left(\left(\frac{1}{a_4^L}, \frac{1}{a_3^L}, \frac{1}{a_2^L}, \frac{1}{a_1^L}\right); H_2(A^L), H_1(A^L)\right)\right)$$

as shown in Fig. 3.

4.2 Appropriate Division of Two Trapezoidal Interval Type-2 Fuzzy Numbers

If

$$\tilde{A}_1 = (((a_{11}^U, a_{12}^U, a_{13}^U, a_{14}^U); H_1(A_1^U), H_2(A_1^U)),$$
$$((a_{11}^L, a_{12}^L, a_{13}^L, a_{14}^L); H_1(A_1^L), H_2(A_1^L)))$$

and

$$\tilde{A}_2 = (((a_{21}^U, a_{22}^U, a_{23}^U, a_{24}^U); H_1(A_2^U), H_2(A_2^U)),$$
$$((a_{21}^L, a_{22}^L, a_{23}^L, a_{24}^L); H_1(A_2^L), H_2(A_2^L)))$$

where $a_{21}^U > 0$, are two trapezoidal interval type-2 fuzzy numbers. Then, using Sect. 4.1,

$$\frac{1}{\tilde{A}_2}=\left(\left(\left(\frac{1}{a_{24}^U},\frac{1}{a_{23}^U},\frac{1}{a_{22}^U},\frac{1}{a_{21}^U}\right);H_2(A_2^U),H_1(A_2^U)\right),\right.$$
$$\left.\left(\left(\frac{1}{a_{24}^L},\frac{1}{a_{23}^L},\frac{1}{a_{22}^L},\frac{1}{a_{21}^L}\right);H_2(A_2^L),H_1(A_2^L)\right)\right).$$

Using the existing arithmetic operation (iii) defined in Sect. 1, $\tilde{A}_1\otimes\left(\frac{1}{\tilde{A}_2}\right)=\frac{\tilde{A}_1}{\tilde{A}_2}=\left(\left(\left(a_{11}^U\times\frac{1}{a_{24}^U},a_{12}^U\times\frac{1}{a_{23}^U},a_{13}^U\times\frac{1}{a_{22}^U},a_{14}^U\times\frac{1}{a_{21}^U}\right);\ H_1(A_1^U)\times H_2(A_2^U),H_2(A_1^U)\times H_1(A_2^U)\right),\ \left(\left(a_{11}^L\times\frac{1}{a_{24}^L},a_{12}^L\times\frac{1}{a_{23}^L},a_{13}^L\times\frac{1}{a_{22}^L},a_{14}^L\times\frac{1}{a_{21}^L}\right);H_1(A_1^L)\times H_2(A_2^L),H_2(A_1^L)\times H_1(A_2^L)\right)\right)$.

5 Inappropriateness of the Existing Mathematical Models (M_1) and (M_2)

It is pertinent to mention that the third objective function of mathematical model (M_1) as well as the first objective function of the mathematical model (M_2), has been obtained as follows:

$$\begin{pmatrix}\left(\left(t_{ij1}^{aU},t_{ij2}^{aU},t_{ij3}^{aU},t_{ij4}^{aU}\right);H_1\left(t_{ij}^{aU}\right),H_2\left(t_{ij}^{aU}\right)\right),\\ \left(\left(t_{ij1}^{aL},t_{ij2}^{aL},t_{ij3}^{aL},t_{ij4}^{aL}\right);H_1\left(t_{ij}^{aL}\right),H_2\left(t_{ij}^{aL}\right)\right)\end{pmatrix}$$
$$\otimes\frac{1}{\begin{pmatrix}\left(\left(t_{ij1}^{sU},t_{ij2}^{sU},t_{ij3}^{sU},t_{ij4}^{sU}\right);H_1\left(t_{ij}^{sU}\right),H_2\left(t_{ij}^{sU}\right)\right),\\ \left(\left(t_{ij1}^{sL},t_{ij2}^{sL},t_{ij3}^{sL},t_{ij4}^{sL}\right);H_1\left(t_{ij}^{sL}\right),H_2\left(t_{ij}^{sL}\right)\right)\end{pmatrix}}$$
$$=\begin{pmatrix}\left(\left(t_{ij1}^{aU},t_{ij2}^{aU},t_{ij3}^{aU},t_{ij4}^{aU}\right);H_1\left(t_{ij}^{aU}\right),H_2\left(t_{ij}^{aU}\right)\right),\\ \left(\left(t_{ij1}^{aL},t_{ij2}^{aL},t_{ij3}^{aL},t_{ij4}^{aL}\right);H_1\left(t_{ij}^{aL}\right),H_2\left(t_{ij}^{aL}\right)\right)\end{pmatrix}$$
$$\otimes\begin{pmatrix}\left(\left(\frac{1}{t_{ij4}^{sU}},\frac{1}{t_{ij3}^{sU}},\frac{1}{t_{ij2}^{sU}},\frac{1}{t_{ij1}^{sU}}\right);H_1\left(t_{ij}^{sU}\right),H_2\left(t_{ij}^{sU}\right)\right),\\ \left(\left(\frac{1}{t_{ij4}^{sL}},\frac{1}{t_{ij3}^{sL}},\frac{1}{t_{ij2}^{sL}},\frac{1}{t_{ij1}^{sL}}\right);H_1\left(t_{ij}^{sL}\right),H_2\left(t_{ij}^{sL}\right)\right)\end{pmatrix}$$
$$=\begin{pmatrix}\left(\left(\frac{t_{ij1}^{aU}}{t_{ij4}^{sU}},\frac{t_{ij2}^{aU}}{t_{ij3}^{sU}},\frac{t_{ij3}^{aU}}{t_{ij2}^{sU}},\frac{t_{ij4}^{aU}}{t_{ij1}^{sU}}\right);H_1\left(t_{ij}^{aU}\right)\times H_1\left(t_{ij}^{sU}\right),H_2\left(t_{ij}^{aU}\right)\times H_2\left(t_{ij}^{sU}\right)\right),\\ \left(\left(\frac{t_{ij1}^{aL}}{t_{ij4}^{sL}},\frac{t_{ij2}^{aL}}{t_{ij3}^{sL}},\frac{t_{ij3}^{aL}}{t_{ij2}^{sL}},\frac{t_{ij4}^{aL}}{t_{ij1}^{sL}}\right);H_1\left(t_{ij}^{aL}\right)\times H_1\left(t_{ij}^{sL}\right),H_2\left(t_{ij}^{aL}\right)\times H_2\left(t_{ij}^{sL}\right)\right)\end{pmatrix}$$

$$= \frac{\begin{pmatrix} \left(\left(t^{aU}_{ij1}, t^{aU}_{ij2}, t^{aU}_{ij3}, t^{aU}_{ij4}\right); H_1\left(t^{aU}_{ij}\right), H_2\left(t^{aU}_{ij}\right)\right), \\ \left(\left(t^{aL}_{ij1}, t^{aL}_{ij2}, t^{aL}_{ij3}, t^{aL}_{ij4}\right); H_1\left(t^{aL}_{ij}\right), H_2\left(t^{aL}_{ij}\right)\right) \end{pmatrix}}{\begin{pmatrix} \left(\left(t^{sU}_{ij1}, t^{sU}_{ij2}, t^{sU}_{ij3}, t^{sU}_{ij4}\right); H_1\left(t^{sU}_{ij}\right), H_2\left(t^{sU}_{ij}\right)\right), \\ \left(\left(t^{sL}_{ij1}, t^{sL}_{ij2}, t^{sL}_{ij3}, t^{sL}_{ij4}\right); H_1\left(t^{sL}_{ij}\right), H_2\left(t^{sL}_{ij}\right)\right) \end{pmatrix}}$$

It is obvious that the existing multiplicative inverse, discussed in Sect. 2.2, is used to define the third objective function of mathematical model (M_1) as well as the first objective function of the mathematical model (M_2). While, as discussed in Sect. 3.1, that the existing multiplicative inverse, discussed in Sect. 2.2, is not appropriate. Hence, the existing mathematical models (M_1) and (M_2) are not appropriate.

Remark: The inappropriateness, mentioned in the third objective function of mathematical model (M_1) as well as the first objective function of the mathematical model (M_2), are also occurring in the fourth and fifth objective function of mathematical model (M_1) as well as the second and third objective function of the mathematical model (M_2).

6 Appropriate Mathematical Models (M'_1) and (M'_2) Corresponding to the Existing Mathematical Models (M_1) and (M_2)

It is obvious from Sect. 5, that the appropriate mathematical models (M'_1) and (M'_2) corresponding to the existing mathematical models (M_1) and (M_2) respectively, can be obtained by replacing $H_1\left(t^{sU}_{ij}\right)$, $H_2\left(t^{sU}_{ij}\right)$, $H_1\left(t^{sL}_{ij}\right)$, $H_2\left(t^{sL}_{ij}\right)$, $H_1\left(c^{sU}_{ij}\right)$, $H_2\left(c^{sU}_{ij}\right)$, $H_1\left(c^{sL}_{ij}\right)$, $H_2\left(c^{sL}_{ij}\right)$, $H_1\left(c^{aU}_{ij}\right)$, $H_2\left(c^{aU}_{ij}\right)$, $H_1\left(c^{aL}_{ij}\right)$ and $H_2\left(c^{aL}_{ij}\right)$ in denominator with $H_2\left(t^{sU}_{ij}\right)$, $H_1\left(t^{sU}_{ij}\right)$, $H_2\left(t^{sL}_{ij}\right)$, $H_1\left(t^{sL}_{ij}\right)$, $H_2\left(c^{sU}_{ij}\right)$, $H_1\left(c^{sU}_{ij}\right)$, $H_2\left(c^{sL}_{ij}\right)$, $H_1\left(c^{sL}_{ij}\right)$, $H_2\left(c^{aU}_{ij}\right)$, $H_1\left(c^{aU}_{ij}\right)$, $H_2\left(c^{aL}_{ij}\right)$ and $H_1\left(c^{aL}_{ij}\right)$ respectively.

Model (M'_1)

$$\text{Minimize}\left(\sum_{i=1}^{m}\sum_{j=1}^{n}\begin{pmatrix} \left(\left(d^{U}_{ij1}, d^{U}_{ij2}, d^{U}_{ij3}, d^{U}_{ij4}\right); H_1\left(d^{U}_{ij}\right), H_2\left(d^{U}_{ij}\right)\right), \\ \left(\left(d^{L}_{ij1}, d^{L}_{ij2}, d^{L}_{ij3}, d^{L}_{ij4}\right); H_1\left(d^{L}_{ij}\right), H_2\left(d^{L}_{ij}\right)\right) \end{pmatrix} x_{ij}\right),$$

$$\text{Minimize}\left(\sum_{i=1}^{m}\sum_{j=1}^{n}\begin{pmatrix} \left(\left(l^{U}_{ij1}, l^{U}_{ij2}, l^{U}_{ij3}, l^{U}_{ij4}\right); H_1\left(l^{U}_{ij}\right), H_2\left(l^{U}_{ij}\right)\right), \\ \left(\left(l^{L}_{ij1}, l^{L}_{ij2}, l^{L}_{ij3}, l^{L}_{ij4}\right); H_1\left(l^{L}_{ij}\right), H_2\left(l^{L}_{ij}\right)\right) \end{pmatrix} x_{ij}\right),$$

$$\text{Minimize}\left(\frac{\sum_{i=1}^{m}\sum_{j=1}^{n}\begin{pmatrix}\left(\left(t_{ij1}^{aU},t_{ij2}^{aU},t_{ij3}^{aU},t_{ij4}^{aU}\right);H_1\left(t_{ij}^{aU}\right),H_2\left(t_{ij}^{aU}\right)\right),\\\left(\left(t_{ij1}^{aL},t_{ij2}^{aL},t_{ij3}^{aL},t_{ij4}^{aL}\right);H_1\left(t_{ij}^{aL}\right),H_2\left(t_{ij}^{aL}\right)\right)\end{pmatrix}x_{ij}}{\sum_{i=1}^{m}\sum_{j=1}^{n}\begin{pmatrix}\left(\left(t_{ij1}^{sU},t_{ij2}^{sU},t_{ij3}^{sU},t_{ij4}^{sU}\right);H_2\left(t_{ij}^{sU}\right),H_1\left(t_{ij}^{sU}\right)\right),\\\left(\left(t_{ij1}^{sL},t_{ij2}^{sL},t_{ij3}^{sL},t_{ij4}^{sL}\right);H_2\left(t_{ij}^{sL}\right),H_1\left(t_{ij}^{sL}\right)\right)\end{pmatrix}x_{ij}}\right),$$

$$\text{Minimize}\left(\frac{\sum_{i=1}^{m}\sum_{j=1}^{n}\begin{pmatrix}\left(\left(c_{ij1}^{aU},c_{ij2}^{aU},c_{ij3}^{aU},c_{ij4}^{aU}\right);H_1\left(c_{ij}^{aU}\right),H_2\left(c_{ij}^{aU}\right)\right),\\\left(\left(c_{ij1}^{aL},c_{ij2}^{aL},c_{ij3}^{aL},c_{ij4}^{aL}\right);H_1\left(c_{ij}^{aL}\right),H_2\left(c_{ij}^{aL}\right)\right)\end{pmatrix}x_{ij}}{\sum_{i=1}^{m}\sum_{j=1}^{n}\begin{pmatrix}\left(\left(c_{ij1}^{sU},c_{ij2}^{sU},c_{ij3}^{sU},c_{ij4}^{sU}\right);H_2\left(c_{ij}^{sU}\right),H_1\left(c_{ij}^{sU}\right)\right),\\\left(\left(c_{ij1}^{sL},c_{ij2}^{sL},c_{ij3}^{sL},c_{ij4}^{sL}\right);H_2\left(c_{ij}^{sL}\right),H_1\left(c_{ij}^{sL}\right)\right)\end{pmatrix}x_{ij}}\right),$$

$$\text{Maximize}\left(\frac{\sum_{i=1}^{m}\sum_{j=1}^{n}\begin{pmatrix}\left(\left(d_{ij1}^{sU},d_{ij2}^{sU},d_{ij3}^{sU},d_{ij4}^{sU}\right);H_1\left(d_{ij}^{sU}\right),H_2\left(d_{ij}^{sU}\right)\right),\\\left(\left(d_{ij1}^{sL},d_{ij2}^{sL},d_{ij3}^{sL},d_{ij4}^{sL}\right);H_1\left(d_{ij}^{sL}\right),H_2\left(d_{ij}^{sL}\right)\right)\end{pmatrix}x_{ij}}{\sum_{i=1}^{m}\sum_{j=1}^{n}\begin{pmatrix}\left(\left(c_{ij1}^{aU},c_{ij2}^{aU},c_{ij3}^{aU},c_{ij4}^{aU}\right);H_2\left(c_{ij}^{aU}\right),H_1\left(c_{ij}^{aU}\right)\right),\\\left(\left(c_{ij1}^{aL},c_{ij2}^{aL},c_{ij3}^{aL},c_{ij4}^{aL}\right);H_2\left(c_{ij}^{aL}\right),H_1\left(c_{ij}^{aL}\right)\right)\end{pmatrix}x_{ij}}\right)$$

Subject to.

Constraints of the mathematical model (M_1).

Model $\left(M_2'\right)$

$$\text{Minimize}\left(\frac{\sum_{i=1}^{m}\sum_{j=1}^{n}\begin{pmatrix}\left(\left(t_{ij1}^{aU},t_{ij2}^{aU},t_{ij3}^{aU},t_{ij4}^{aU}\right);H_1\left(t_{ij}^{aU}\right),H_2\left(t_{ij}^{aU}\right)\right),\\\left(\left(t_{ij1}^{aL},t_{ij2}^{aL},t_{ij3}^{aL},t_{ij4}^{aL}\right);H_1\left(t_{ij}^{aL}\right),H_2\left(t_{ij}^{aL}\right)\right)\end{pmatrix}x_{ij}}{\sum_{i=1}^{m}\sum_{j=1}^{n}\begin{pmatrix}\left(\left(t_{ij1}^{sU},t_{ij2}^{sU},t_{ij3}^{sU},t_{ij4}^{sU}\right);H_2\left(t_{ij}^{sU}\right),H_1\left(t_{ij}^{sU}\right)\right),\\\left(\left(t_{ij1}^{sL},t_{ij2}^{sL},t_{ij3}^{sL},t_{ij4}^{sL}\right);H_2\left(t_{ij}^{sL}\right),H_1\left(t_{ij}^{sL}\right)\right)\end{pmatrix}x_{ij}}\right),$$

$$\text{Minimize}\left(\frac{\sum_{i=1}^{m}\sum_{j=1}^{n}\begin{pmatrix}\left(\left(c_{ij1}^{aU},c_{ij2}^{aU},c_{ij3}^{aU},c_{ij4}^{aU}\right);H_1\left(c_{ij}^{aU}\right),H_2\left(c_{ij}^{aU}\right)\right),\\\left(\left(c_{ij1}^{aL},c_{ij2}^{aL},c_{ij3}^{aL},c_{ij4}^{aL}\right);H_1\left(c_{ij}^{aL}\right),H_2\left(c_{ij}^{aL}\right)\right)\end{pmatrix}x_{ij}}{\sum_{i=1}^{m}\sum_{j=1}^{n}\begin{pmatrix}\left(\left(c_{ij1}^{sU},c_{ij2}^{sU},c_{ij3}^{sU},c_{ij4}^{sU}\right);H_2\left(c_{ij}^{sU}\right),H_1\left(c_{ij}^{sU}\right)\right),\\\left(\left(c_{ij1}^{sL},c_{ij2}^{sL},c_{ij3}^{sL},c_{ij4}^{sL}\right);H_2\left(c_{ij}^{sL}\right),H_1\left(c_{ij}^{sL}\right)\right)\end{pmatrix}x_{ij}}\right),$$

$$\text{Maximize}\left(\frac{\sum_{i=1}^{m}\sum_{j=1}^{n}\left(\begin{array}{c}\left(\left(d_{ij1}^{sU}, d_{ij2}^{sU}, d_{ij3}^{sU}, d_{ij4}^{sU}\right); H_1\left(d_{ij}^{sU}\right), H_2\left(d_{ij}^{sU}\right)\right),\\ \left(\left(d_{ij1}^{sL}, d_{ij2}^{sL}, d_{ij3}^{sL}, d_{ij4}^{sL}\right); H_1\left(d_{ij}^{sL}\right), H_2\left(d_{ij}^{sL}\right)\right)\end{array}\right)x_{ij}}{\sum_{i=1}^{m}\sum_{j=1}^{n}\left(\begin{array}{c}\left(\left(c_{ij1}^{aU}, c_{ij2}^{aU}, c_{ij3}^{aU}, c_{ij4}^{aU}\right); H_2\left(c_{ij}^{aU}\right), H_1\left(c_{ij}^{aU}\right)\right),\\ \left(\left(c_{ij1}^{aL}, c_{ij2}^{aL}, c_{ij3}^{aL}, c_{ij4}^{aL}\right); H_2\left(c_{ij}^{aL}\right), H_1\left(c_{ij}^{aL}\right)\right)\end{array}\right)x_{ij}}\right)$$

Subject to.
Constraints of the mathematical model (M_1).

7 Gupta et al.'s Approach to Solve the Mathematical Model (M_1)

It is obvious that the mathematical models (M_2) and (M_3) are special cases of the mathematical model (M_1). Therefore, an approach which can be used to solve the mathematical model (M_1) can also be used to solve the mathematical models (M_2) and (M_3). Hence, in this section, Gupta et al.'s approach [3] to solve the mathematical model (M_1) is discussed.

Step 1: Using the relation (Hu et al. [6]), $E(((a_1^U, a_2^U, a_3^U, a_4^U); H_1(A^U), H_2(A^U)), ((a_1^L, a_2^L, a_3^L, a_4^L); H_1(A^L), H_2(A^L))) = \frac{1}{2}\left(\frac{1}{4}\left(\sum_{r=1}^{4}(a_r^L + a_r^U)\right)\right) \times \frac{1}{4}\left(\sum_{r=1}^{2}(H_r(A^L) + H_r(A^U))\right)$, transform the mathematical model (M_1) into its equivalent mathematical model (M_4).

Model (M_4)

$$\text{Minimize}\left(E\left(\sum_{i=1}^{m}\sum_{j=1}^{n}\left(\begin{array}{c}\left(\left(d_{ij1}^{U}, d_{ij2}^{U}, d_{ij3}^{U}, d_{ij4}^{U}\right); H_1\left(d_{ij}^{U}\right), H_2\left(d_{ij}^{U}\right)\right),\\ \left(\left(d_{ij1}^{L}, d_{ij2}^{L}, d_{ij3}^{L}, d_{ij4}^{L}\right); H_1\left(d_{ij}^{L}\right), H_2\left(d_{ij}^{L}\right)\right)\end{array}\right)x_{ij}\right)\right),$$

$$\text{Minimize}\left(E\left(\sum_{i=1}^{m}\sum_{j=1}^{n}\left(\begin{array}{c}\left(\left(l_{ij1}^{U}, l_{ij2}^{U}, l_{ij3}^{U}, l_{ij4}^{U}\right); H_1\left(l_{ij}^{U}\right), H_2\left(l_{ij}^{U}\right)\right),\\ \left(\left(l_{ij1}^{L}, l_{ij2}^{L}, l_{ij3}^{L}, l_{ij4}^{L}\right); H_1\left(l_{ij}^{L}\right), H_2\left(l_{ij}^{L}\right)\right)\end{array}\right)x_{ij}\right)\right),$$

$$\text{Minimize}\left(E\left(\frac{\sum_{i=1}^{m}\sum_{j=1}^{n}\left(\begin{array}{c}\left(\left(t_{ij1}^{aU}, t_{ij2}^{aU}, t_{ij3}^{aU}, t_{ij4}^{aU}\right); H_1\left(t_{ij}^{aU}\right), H_2\left(t_{ij}^{aU}\right)\right),\\ \left(\left(t_{ij1}^{aL}, t_{ij2}^{aL}, t_{ij3}^{aL}, t_{ij4}^{aL}\right); H_1\left(t_{ij}^{aL}\right), H_2\left(t_{ij}^{aL}\right)\right)\end{array}\right)x_{ij}}{\sum_{i=1}^{m}\sum_{j=1}^{n}\left(\begin{array}{c}\left(\left(t_{ij1}^{sU}, t_{ij2}^{sU}, t_{ij3}^{sU}, t_{ij4}^{sU}\right); H_1\left(t_{ij}^{sU}\right), H_2\left(t_{ij}^{sU}\right)\right),\\ \left(\left(t_{ij1}^{sL}, t_{ij2}^{sL}, t_{ij3}^{sL}, t_{ij4}^{sL}\right); H_1\left(t_{ij}^{sL}\right), H_2\left(t_{ij}^{sL}\right)\right)\end{array}\right)x_{ij}}\right)\right),$$

$$\text{Minimize}\left(E\left(\frac{\sum_{i=1}^{m}\sum_{j=1}^{n}\left(\begin{array}{l}\left(\left(c_{ij1}^{aU},c_{ij2}^{aU},c_{ij3}^{aU},c_{ij4}^{aU}\right);H_1\left(c_{ij}^{aU}\right),H_2\left(c_{ij}^{aU}\right)\right),\\ \left(\left(c_{ij1}^{aL},c_{ij2}^{aL},c_{ij3}^{aL},c_{ij4}^{aL}\right);H_1\left(c_{ij}^{aL}\right),H_2\left(c_{ij}^{aL}\right)\right)\end{array}\right)x_{ij}}{\sum_{i=1}^{m}\sum_{j=1}^{n}\left(\begin{array}{l}\left(\left(c_{ij1}^{sU},c_{ij2}^{sU},c_{ij3}^{sU},c_{ij4}^{sU}\right);H_1\left(c_{ij}^{sU}\right),H_2\left(c_{ij}^{sU}\right)\right),\\ \left(\left(c_{ij1}^{sL},c_{ij2}^{sL},c_{ij3}^{sL},c_{ij4}^{sL}\right);H_1\left(c_{ij}^{sL}\right),H_2\left(c_{ij}^{sL}\right)\right)\end{array}\right)x_{ij}}\right)\right),$$

$$\text{Maximize}\left(E\left(\frac{\sum_{i=1}^{m}\sum_{j=1}^{n}\left(\begin{array}{l}\left(\left(d_{ij1}^{sU},d_{ij2}^{sU},d_{ij3}^{sU},d_{ij4}^{sU}\right);H_1\left(d_{ij}^{sU}\right),H_2\left(d_{ij}^{sU}\right)\right),\\ \left(\left(d_{ij1}^{sL},d_{ij2}^{sL},d_{ij3}^{sL},d_{ij4}^{sL}\right);H_1\left(d_{ij}^{sL}\right),H_2\left(d_{ij}^{sL}\right)\right)\end{array}\right)x_{ij}}{\sum_{i=1}^{m}\sum_{j=1}^{n}\left(\begin{array}{l}\left(\left(c_{ij1}^{aU},c_{ij2}^{aU},c_{ij3}^{aU},c_{ij4}^{aU}\right);H_1\left(c_{ij}^{aU}\right),H_2\left(c_{ij}^{aU}\right)\right),\\ \left(\left(c_{ij1}^{aL},c_{ij2}^{aL},c_{ij3}^{aL},c_{ij4}^{aL}\right);H_1\left(c_{ij}^{aL}\right),H_2\left(c_{ij}^{aL}\right)\right)\end{array}\right)x_{ij}}\right)\right)$$

Subject to.
Constraints of the mathematical model (M_1).

Step 2: Using the relation $E\left(\frac{\tilde{A}_1}{\tilde{A}_2}\right) = \frac{E(\tilde{A}_1)}{E(\tilde{A}_2)}$, transform the mathematical model (M_4) into its equivalent mathematical model (M_5).

Model (M_5)

$$\text{Minimize}\left(E\left(\sum_{i=1}^{m}\sum_{j=1}^{n}\left(\begin{array}{l}\left(\left(d_{ij1}^{U},d_{ij2}^{U},d_{ij3}^{U},d_{ij4}^{U}\right);H_1\left(d_{ij}^{U}\right),H_2\left(d_{ij}^{U}\right)\right),\\ \left(\left(d_{ij1}^{L},d_{ij2}^{L},d_{ij3}^{L},d_{ij4}^{L}\right);H_1\left(d_{ij}^{L}\right),H_2\left(d_{ij}^{L}\right)\right)\end{array}\right)x_{ij}\right)\right),$$

$$\text{Minimize}\left(E\left(\sum_{i=1}^{m}\sum_{j=1}^{n}\left(\begin{array}{l}\left(\left(l_{ij1}^{U},l_{ij2}^{U},l_{ij3}^{U},l_{ij4}^{U}\right);H_1\left(l_{ij}^{U}\right),H_2\left(l_{ij}^{U}\right)\right),\\ \left(\left(l_{ij1}^{L},l_{ij2}^{L},l_{ij3}^{L},l_{ij4}^{L}\right);H_1\left(l_{ij}^{L}\right),H_2\left(l_{ij}^{L}\right)\right)\end{array}\right)x_{ij}\right)\right),$$

$$\text{Minimize}\left(\frac{E\left(\sum_{i=1}^{m}\sum_{j=1}^{n}\left(\begin{array}{l}\left(\left(t_{ij1}^{aU},t_{ij2}^{aU},t_{ij3}^{aU},t_{ij4}^{aU}\right);H_1\left(t_{ij}^{aU}\right),H_2\left(t_{ij}^{aU}\right)\right),\\ \left(\left(t_{ij1}^{aL},t_{ij2}^{aL},t_{ij3}^{aL},t_{ij4}^{aL}\right);H_1\left(t_{ij}^{aL}\right),H_2\left(t_{ij}^{aL}\right)\right)\end{array}\right)x_{ij}\right)}{E\left(\sum_{i=1}^{m}\sum_{j=1}^{n}\left(\begin{array}{l}\left(\left(t_{ij1}^{sU},t_{ij2}^{sU},t_{ij3}^{sU},t_{ij4}^{sU}\right);H_1\left(t_{ij}^{sU}\right),H_2\left(t_{ij}^{sU}\right)\right),\\ \left(\left(t_{ij1}^{sL},t_{ij2}^{sL},t_{ij3}^{sL},t_{ij4}^{sL}\right);H_1\left(t_{ij}^{sL}\right),H_2\left(t_{ij}^{sL}\right)\right)\end{array}\right)x_{ij}\right)}\right),$$

$$\text{Minimize}\left(\frac{E\left(\sum_{i=1}^{m}\sum_{j=1}^{n}\left(\begin{array}{l}\left(\left(c_{ij1}^{aU},c_{ij2}^{aU},c_{ij3}^{aU},c_{ij4}^{aU}\right);H_1\left(c_{ij}^{aU}\right),H_2\left(c_{ij}^{aU}\right)\right),\\ \left(\left(c_{ij1}^{aL},c_{ij2}^{aL},c_{ij3}^{aL},c_{ij4}^{aL}\right);H_1\left(c_{ij}^{aL}\right),H_2\left(c_{ij}^{aL}\right)\right)\end{array}\right)x_{ij}\right)}{E\left(\sum_{i=1}^{m}\sum_{j=1}^{n}\left(\begin{array}{l}\left(\left(c_{ij1}^{sU},c_{ij2}^{sU},c_{ij3}^{sU},c_{ij4}^{sU}\right);H_1\left(c_{ij}^{sU}\right),H_2\left(c_{ij}^{sU}\right)\right),\\ \left(\left(c_{ij1}^{sL},c_{ij2}^{sL},c_{ij3}^{sL},c_{ij4}^{sL}\right);H_1\left(c_{ij}^{sL}\right),H_2\left(c_{ij}^{sL}\right)\right)\end{array}\right)x_{ij}\right)}\right),$$

$$\text{Maximize}\left(\frac{E\left(\sum_{i=1}^{m}\sum_{j=1}^{n}\left(\begin{array}{l}\left(\left(d_{ij1}^{sU},d_{ij2}^{sU},d_{ij3}^{sU},d_{ij4}^{sU}\right);H_1\left(d_{ij}^{sU}\right),H_2\left(d_{ij}^{sU}\right)\right),\\ \left(\left(d_{ij1}^{sL},d_{ij2}^{sL},d_{ij3}^{sL},d_{ij4}^{sL}\right);H_1\left(d_{ij}^{sL}\right),H_2\left(d_{ij}^{sL}\right)\right)\end{array}\right)x_{ij}\right)}{E\left(\sum_{i=1}^{m}\sum_{j=1}^{n}\left(\begin{array}{l}\left(\left(c_{ij1}^{aU},c_{ij2}^{aU},c_{ij3}^{aU},c_{ij4}^{aU}\right);H_1\left(c_{ij}^{aU}\right),H_2\left(c_{ij}^{aU}\right)\right),\\ \left(\left(c_{ij1}^{aL},c_{ij2}^{aL},c_{ij3}^{aL},c_{ij4}^{aL}\right);H_1\left(c_{ij}^{aL}\right),H_2\left(c_{ij}^{aL}\right)\right)\end{array}\right)x_{ij}\right)}\right)$$

Subject to.
Constraints of the mathematical model (M_1).

Step 3: Using the relation $E\left(\tilde{A}_1 \oplus \tilde{A}_2\right) = E\left(\tilde{A}_1\right) + E\left(\tilde{A}_2\right)$, transform the mathematical model (M_5) into its equivalent mathematical model (M_6).

Model (M_6)

$$\text{Minimize}\left(\sum_{i=1}^{m}\sum_{j=1}^{n} E\left(\begin{array}{c}\left(\left(d_{ij1}^{U}, d_{ij2}^{U}, d_{ij3}^{U}, d_{ij4}^{U}\right); H_1\left(d_{ij}^{U}\right), H_2\left(d_{ij}^{U}\right)\right), \\ \left(\left(d_{ij1}^{L}, d_{ij2}^{L}, d_{ij3}^{L}, d_{ij4}^{L}\right); H_1\left(d_{ij}^{L}\right), H_2\left(d_{ij}^{L}\right)\right)\end{array}\right) x_{ij}\right),$$

$$\text{Minimize}\left(\sum_{i=1}^{m}\sum_{j=1}^{n} E\left(\begin{array}{c}\left(\left(l_{ij1}^{U}, l_{ij2}^{U}, l_{ij3}^{U}, l_{ij4}^{U}\right); H_1\left(l_{ij}^{U}\right), H_2\left(l_{ij}^{U}\right)\right), \\ \left(\left(l_{ij1}^{L}, l_{ij2}^{L}, l_{ij3}^{L}, l_{ij4}^{L}\right); H_1\left(l_{ij}^{L}\right), H_2\left(l_{ij}^{L}\right)\right)\end{array}\right) x_{ij}\right),$$

$$\text{Minimize}\left(\frac{\sum_{i=1}^{m}\sum_{j=1}^{n} E\left(\begin{array}{c}\left(\left(t_{ij1}^{aU}, t_{ij2}^{aU}, t_{ij3}^{aU}, t_{ij4}^{aU}\right); H_1\left(t_{ij}^{aU}\right), H_2\left(t_{ij}^{aU}\right)\right), \\ \left(\left(t_{ij1}^{aL}, t_{ij2}^{aL}, t_{ij3}^{aL}, t_{ij4}^{aL}\right); H_1\left(t_{ij}^{aL}\right), H_2\left(t_{ij}^{aL}\right)\right)\end{array}\right) x_{ij}}{\sum_{i=1}^{m}\sum_{j=1}^{n} E\left(\begin{array}{c}\left(\left(t_{ij1}^{sU}, t_{ij2}^{sU}, t_{ij3}^{sU}, t_{ij4}^{sU}\right); H_1\left(t_{ij}^{sU}\right), H_2\left(t_{ij}^{sU}\right)\right), \\ \left(\left(t_{ij1}^{sL}, t_{ij2}^{sL}, t_{ij3}^{sL}, t_{ij4}^{sL}\right); H_1\left(t_{ij}^{sL}\right), H_2\left(t_{ij}^{sL}\right)\right)\end{array}\right) x_{ij}}\right)$$

$$\text{Minimize}\left(\frac{\sum_{i=1}^{m}\sum_{j=1}^{n} E\left(\begin{array}{c}\left(\left(c_{ij1}^{aU}, c_{ij2}^{aU}, c_{ij3}^{aU}, c_{ij4}^{aU}\right); H_1\left(c_{ij}^{aU}\right), H_2\left(c_{ij}^{aU}\right)\right), \\ \left(\left(c_{ij1}^{aL}, c_{ij2}^{aL}, c_{ij3}^{aL}, c_{ij4}^{aL}\right); H_1\left(c_{ij}^{aL}\right), H_2\left(c_{ij}^{aL}\right)\right)\end{array}\right) x_{ij}}{\sum_{i=1}^{m}\sum_{j=1}^{n} E\left(\begin{array}{c}\left(\left(c_{ij1}^{sU}, c_{ij2}^{sU}, c_{ij3}^{sU}, c_{ij4}^{sU}\right); H_1\left(c_{ij}^{sU}\right), H_2\left(c_{ij}^{sU}\right)\right), \\ \left(\left(c_{ij1}^{sL}, c_{ij2}^{sL}, c_{ij3}^{sL}, c_{ij4}^{sL}\right); H_1\left(c_{ij}^{sL}\right), H_2\left(c_{ij}^{sL}\right)\right)\end{array}\right) x_{ij}}\right)$$

$$\text{Maximize}\left(\frac{\sum_{i=1}^{m}\sum_{j=1}^{n} E\left(\begin{array}{c}\left(\left(d_{ij1}^{sU}, d_{ij2}^{sU}, d_{ij3}^{sU}, d_{ij4}^{sU}\right); H_1\left(d_{ij}^{sU}\right), H_2\left(d_{ij}^{sU}\right)\right), \\ \left(\left(d_{ij1}^{sL}, d_{ij2}^{sL}, d_{ij3}^{sL}, d_{ij4}^{sL}\right); H_1\left(d_{ij}^{sL}\right), H_2\left(d_{ij}^{sL}\right)\right)\end{array}\right) x_{ij}}{\sum_{i=1}^{m}\sum_{j=1}^{n} E\left(\begin{array}{c}\left(\left(c_{ij1}^{aU}, c_{ij2}^{aU}, c_{ij3}^{aU}, c_{ij4}^{aU}\right); H_1\left(c_{ij}^{aU}\right), H_2\left(c_{ij}^{aU}\right)\right), \\ \left(\left(c_{ij1}^{aL}, c_{ij2}^{aL}, c_{ij3}^{aL}, c_{ij4}^{aL}\right); H_1\left(c_{ij}^{aL}\right), H_2\left(c_{ij}^{aL}\right)\right)\end{array}\right) x_{ij}}\right)$$

Subject to.
Constraints of the mathematical model (M_1).

Step 4: Using an appropriate method, solve the mathematical model (M_6).

8 Inappropriateness of Gupta Et Al.'s Approach

Gupta et al. [3] proposed an approach to solve the mathematical models(M_1), (M_2) and(M_3). However, as discussed in Sect. 5, that the mathematical models (M_1) and (M_2) are not appropriate. Therefore, one may apply Gupta et al.'s approach [3] to solve the proposed mathematical models $\left(M_1'\right)$ and $\left(M_2'\right)$. Also, one may feel that it is appropriate to use Gupta et al.'s approach [3] for solving the mathematical model (M_3). However, Gupta et al. [3] have used the following mathematical incorrect assumptions in their proposed approach. Therefore, it is neither appropriate to use Gupta et al.'s approach [3] for solving the proposed mathematical models $\left(M_1'\right)$, $\left(M_2'\right)$ nor the existing mathematical model(M_3).

(i) It is obvious from Step 2 of Gupta et al.'s approach [3], discussed in Sect. 7, that Gupta et al. [3] have used the relation $E\left(\frac{\tilde{A}_1}{\tilde{A}_2}\right) = \frac{E(\tilde{A}_1)}{E(\tilde{A}_2)}$, to transform the third, fourth and fifth objective functions of the mathematical model (M_4) into the third, fourth and fifth objective functions of the mathematical model (M_5).

While, the following example clearly indicates that $E\left(\frac{\tilde{A}_1}{\tilde{A}_2}\right) \neq \frac{E(\tilde{A}_1)}{E(\tilde{A}_2)}$. Therefore, the mathematical model (M_4) cannot be transformed into the mathematical model (M_5). Hence, Gupta et al.'s approach [3] is not appropriate.

Let $\tilde{A}_1 = (((1, 5, 8, 12); 0.9, 0.7), ((3, 6, 7, 10); 0.5, 0.2))$ and.

$\tilde{A}_2 = (((2, 5, 10, 14); 0.5, 0.4), ((4, 7, 9, 12); 0.2, 0.1))$ be two trapezoidal interval type-2 fuzzy numbers.

Using the arithmetic operation discussed in Sect. 4.2,

$$\frac{\tilde{A}_1}{\tilde{A}_2} = \left(\left(\left(\frac{1}{14}, \frac{5}{10}, \frac{8}{5}, \frac{12}{2}\right); 0.9 \times 0.4, 0.7 \times 0.5\right),\right.$$
$$\left.\left(\left(\frac{3}{12}, \frac{6}{9}, \frac{7}{7}, \frac{10}{4}\right); 0.5 \times 0.1, 0.2 \times 0.2\right)\right)$$
$$= (((0.07, 0.5, 1.6, 6); 0.36, 0.35), ((0.25, 0.67, 1, 2.5); 0.05, 0.04)).$$

Then, using the relation (Hu et al. [6]),

$$E(((a_1^U, a_2^U, a_3^U, a_4^U); H_1(A^U), H_2(A^U)), ((a_1^L, a_2^L, a_3^L, a_4^L); H_1(A^L), H_2(A^L)))$$
$$= \frac{1}{2}\left(\frac{1}{4}\left(\sum_{r=1}^{4}(a_r^L + a_r^U)\right)\right) \times \frac{1}{4}\left(\sum_{r=1}^{2}(H_r(A^L) + H_r(A^U))\right),$$

$$E\left(\frac{\tilde{A}_1}{\tilde{A}_2}\right) = E(((0.07, 0.5, 1.6, 6); 0.36, 0.35), ((0.25, 0.67, 1, 2.5); 0.05, 0.04))$$
$$= 0.31475 \tag{1}$$

While,

$$E\left(\tilde{A}_1\right) = E(((1, 5, 8, 12); 0.9, 0.7), ((3, 6, 7, 10); 0.5, 0.2))$$
$$= 3.7375$$

$$E\left(\tilde{A}_2\right) = E(((2, 5, 10, 14); 0.5, 0.4), ((4, 7, 9, 12); 0.2, 0.1))$$
$$= 2.3625$$

Hence,

$$\frac{E\left(\tilde{A}_1\right)}{E\left(\tilde{A}_2\right)} = \frac{3.7375}{2.3625} = 1.5820 \quad (2)$$

It is obvious from (1) and (2) that $E\left(\frac{\tilde{A}_1}{\tilde{A}_2}\right) \neq \frac{E(\tilde{A}_1)}{E(\tilde{A}_2)}$.

One may claim that the relation $E\left(\frac{\tilde{A}_1}{\tilde{A}_2}\right) = \frac{E(\tilde{A}_1)}{E(\tilde{A}_2)}$ is not satisfying for the proposed division operation while, it may be satisfied for the existing division operation [3]. Therefore, to avoid any confusion, it is shown that the relation $E\left(\frac{\tilde{A}_1}{\tilde{A}_2}\right) = \frac{E(\tilde{A}_1)}{E(\tilde{A}_2)}$ is also not satisfying for the existing division operation [3].

Using the existing arithmetic operation [3], discussed in Sect. 2.3,

$$\frac{\tilde{A}_1}{\tilde{A}_2} = \left(\left(\left(\frac{1}{14}, \frac{5}{10}, \frac{8}{5}, \frac{12}{2}\right); 0.9 \times 0.5, 0.7 \times 0.4\right),\right.$$
$$\left.\left(\left(\frac{3}{12}, \frac{6}{9}, \frac{7}{7}, \frac{10}{4}\right); 0.5 \times 0.2, 0.2 \times 0.1\right)\right)$$
$$= (((0.07, 0.5, 1.6, 6); 0.45, 0.28), ((0.25, 0.67, 1, 2.5); 0.1, 0.02)).$$

Then, using the relation (Hu et al. [6]),

$$E(((a_1^U, a_2^U, a_3^U, a_4^U); H_1(A^U), H_2(A^U)), ((a_1^L, a_2^L, a_3^L, a_4^L); H_1(A^L), H_2(A^L)))$$
$$= \frac{1}{2}\left(\frac{1}{4}\left(\sum_{r=1}^{4}(a_r^L + a_r^U)\right)\right) \times \frac{1}{4}\left(\sum_{r=1}^{2}(H_r(A^L) + H_r(A^U))\right),$$

$$E\left(\frac{\tilde{A}_1}{\tilde{A}_2}\right) = E(((0.07, 0.5, 1.6, 6); 0.45, 0.28), ((0.25, 0.67, 1, 2.5); 0.1, 0.02))$$
$$= 0.33442 \quad (3)$$

While,

$$E\left(\tilde{A}_1\right) = E(((1, 5, 8, 12); 0.9, 0.7), ((3, 6, 7, 10); 0.5, 0.2))$$
$$= 3.7375$$

$$E\left(\tilde{A}_2\right) = E(((2, 5, 10, 14); 0.5, 0.4), ((4, 7, 9, 12); 0.2, 0.1))$$
$$= 2.3625$$

Hence,

$$\frac{E\left(\tilde{A}_1\right)}{E\left(\tilde{A}_2\right)} = \frac{3.7375}{2.3625} = 1.5820 \quad (4)$$

It is obvious from (3) and (4) that $E\left(\frac{\tilde{A}_1}{\tilde{A}_2}\right) \neq \frac{E(\tilde{A}_1)}{E(\tilde{A}_1)}$..

(ii) It is obvious from Step 3 of Gupta et al.'s approach [3], discussed in Sect. 7, that Gupta et al. [3] have used the relation $E\left(\tilde{A}_1 \oplus \tilde{A}_2\right) = E\left(\tilde{A}_1\right) + E\left(\tilde{A}_2\right)$, to transform all the objective functions of the mathematical model (M_5) into all the objective functions of the mathematical model (M_6).

While, the following example clearly indicates that $E\left(\tilde{A}_1 \oplus \tilde{A}_2\right) \neq E\left(\tilde{A}_1\right) + E\left(\tilde{A}_2\right)$. Therefore, the mathematical model (M_5) cannot be transformed into the mathematical model (M_6). Hence, Gupta et al.'s approach [3] is not appropriate.

Let $\tilde{A}_1 = (((1, 5, 8, 12); 0.9, 0.7), ((3, 6, 7, 10); 0.5, 0.2))$ and.

$\tilde{A}_2 = (((2, 5, 10, 14); 0.5, 0.4), ((4, 7, 9, 12); 0.2, 0.1))$ be two trapezoidal interval type-2 fuzzy numbers.

Then, using the existing arithmetic operation [3], defined in Sect. 1,

$$\begin{aligned}
&\tilde{A}_1 \oplus \tilde{A}_2 \\
&= (((1+2, 5+5, 8+10, 12+14); 0.9+0.5-(0.9)(0.5), 0.7+0.4-(0.7)(0.4)), \\
&\quad ((3+4, 6+7, 7+9, 10+12); 0.5+0.2-(0.5)(0.2), 0.2+0.1-(0.2)(0.1))) \\
&= (((3, 10, 18, 26); 0.95, 0.82), ((7, 13, 16, 22); 0.6, 0.28)).
\end{aligned}$$

Also, using the relation (Hu et al. [6]),

$$\begin{aligned}
&E(((a_1^U, a_2^U, a_3^U, a_4^U); H_1(A^U), H_2(A^U)), ((a_1^L, a_2^L, a_3^L, a_4^L); H_1(A^L), H_2(A^L))) \\
&= \frac{1}{2}\left(\frac{1}{4}\left(\sum_{r=1}^{4}(a_r^L + a_r^U)\right)\right) \times \frac{1}{4}\left(\sum_{r=1}^{2}(H_r(A^L) + H_r(A^U))\right),
\end{aligned}$$

$$E\left(\tilde{A}_1 \oplus \tilde{A}_2\right) = E(((3, 10, 18, 26); 0.95, 0.82), ((7, 13, 16, 22); 0.6, 0.28))$$
$$= 9.5234 \quad (5)$$

While,

$$E\left(\tilde{A}_1\right) = E(((1,5,8,12);0.9,0.7),((3,6,7,10);0.5,0.2))$$
$$= 3.7375$$

$$E\left(\tilde{A}_2\right) = E(((2,5,10,14);0.5,0.4),((4,7,9,12);0.2,0.1))$$
$$= 2.3625$$

Hence,

$$E\left(\tilde{A}_1\right) + E\left(\tilde{A}_2\right) = 3.7375 + 2.3625 = 6.1 \quad (6)$$

It is obvious from (5) and (6) that $E\left(\tilde{A}_1 \oplus \tilde{A}_2\right) \neq E\left(\tilde{A}_1\right) + E\left(\tilde{A}_2\right)$.

9 Conclusions

It is pointed out that the mathematical models (M_1) and (M_2), proposed by Gupta et al. [3], are not appropriate. Also, the appropriate mathematical models $\left(M_1'\right)$ and $\left(M_2'\right)$ corresponding to the existing mathematical models (M_1) and (M_2) are proposed. Furthermore, it is pointed out that as Gupta et al. [3] have used some mathematical incorrect assumptions in their proposed approach. Therefore, it is inappropriate to use Gupta et al.'s approach [3] for solving the proposed mathematical models $\left(M_1'\right)$ and $\left(M_2'\right)$ as well as for solving the existing mathematical model (M_3).

To propose an appropriate approach for solving the proposed mathematical models $\left(M_1'\right)$ and $\left(M_2'\right)$ as well as existing mathematical model (M_3) may be considered as an open challenging research problem. Also, with the help of these models and the existing models [7, 8], new models for multi-objective capacitated fuzzy stochastic fixed-charge solid transportation problems may be proposed.

Acknowledgements The authors would like to thank the anonymous reviewers for their insightful comments and suggestions.

References

1. Chen, S.M., Lee, L.W.: Fuzzy multiple attributes group decision-making based on the ranking values and the arithmetic operations of interval type-2 fuzzy sets. Expert Syst Appl **37**(1), 824–833 (2010)
2. Sinha, B., Das, A., Bera, U.K.: Profit maximization solid transportation problem with trapezoidal interval type-2 fuzzy numbers. Int. J. Appl. Comput. Math. **2**(1), 41–56 (2016)

3. Gupta, S., Garg, H., Chaudhary, S.: Parameter estimation and optimization of multi-objective capacitated stochastic transportation problem for gamma distribution. Complex Intell. Syst. **6**(3), 651–667 (2020)
4. Khalil, A.M., Hassan, N.: A note on “A novel approach to multi attribute group decision making based on trapezoidal interval type-2 fuzzy soft sets.” Appl. Math. Model. **41**, 684–690 (2017)
5. Zimmermann, H.J.: Fuzzy set theory and its applications, 4th edn. Kluwer Academic Publisher, Dordrecht (2001)
6. Hu, J., Zhang, Y., Chen, X., Liu, Y.: Multi-criteria decision making method based on possibility degree of interval type-2 fuzzy number. Knowl.-Based Syst. **43**, 21–29 (2013)
7. Bera, S., Giri, P.K., Jana, D.K., Basu, K., Maiti, M.: Fixed charge 4D-TP for a breakable item under hybrid random type-2 uncertain environments. Inf. Sci. **527**, 128–158 (2020)
8. Jana, S.H., Jana, B.: Application of random triangular and Gaussian type-2 fuzzy variable to solve fixed charge multi-item four dimensional transportation problem. Appl. Soft. Comput. **96**, 106589 (2020)

Assigning Mexican Regions to Categories of Marginality with MR-Sort

Pavel Anselmo Alvarez

Abstract Marginality is a relevant social problem around the world. It shows the inequality between the population in rural and urban areas. In this work, the marginality is studied based on sociodemographic indicators showing the lack of access to basic service and quality of life. The paper analyzes the level of marginality of 32 regions of Mexico by assigning each region to one of five predefined ordered classes. The MR-sort sorting method is applied to assign the regions of Mexico regarding their level of marginality. The outcomes of the current works show three regions in the south of the country with a very high level of marginality. In addition, it is shown the regions with the potential to reduce or increase the risk of the marginality situation.

Keywords Ordered classification · Sorting · MR-sort · Outranking · Regions of Mexico · Marginality

1 Introduction

It is stated that marginality exists around the world, even in developed countries. Nevertheless, marginality seems to be more ingrained in developing countries. A definition stated by Gatzweiler et al. [1] considers marginalization as an involuntary position and condition of an individual or group on the margins of the social, political, economic, ecological, and physical system that prevents them from accessing resources, goods, and services. It limits their freedom of selection, reduces capacities, and eventually causes extreme poverty. It is identifiable that marginality meets a kind of situation of poverty.

The study of marginality is a crucial topic to investigate. The marginalization survey can help to understand the systemic contributors of poverty and exclusion that can overlap with the lack of resources and opportunities needed to achieve the

P. A. Alvarez (✉)
Department of Management and Economic Sciences, Universidad Autónoma de Occidente, 80020 Culiacan, Mexico
e-mail: pavel.alvarez@uadeo.mx

L. Sahoo et al. (eds.), *Real Life Applications of Multiple Criteria Decision Making Techniques in Fuzzy Domain*, Studies in Fuzziness and Soft Computing 420,
https://doi.org/10.1007/978-981-19-4929-6_28

desired conditions in life [2]. In [3] studied the competitiveness of regions of Mexico regarding some factors of inequality. In [4] was developed a study of inequality in Latin America, it is in a context of extreme inequalities and rampant street violence backed by a patrimonial state that tolerates routine judicial discrimination [5]. Some similarities of conjoint rise of inequality and festering of marginality are observed in South America [6].

As a developing country, Mexico presents some inequalities in access to basic services such as education, health, and monetary incomes. Besides that, in Mexico, inclusion policies need to be developed to improve the situation of the marginalized population. The analysis of marginalization in Mexico can be developed with the sociodemographic characteristics of the population. Each rural and urban location presents some elements representing the quality of life based on their access to basic services. In this sense, the marginality can be analyzed in any region of Mexico.

The multiple criteria decision-making (MCDM) is an accepted methodology that helps analysts and experts to analyze the elements (alternatives) to make decisions. The MCDM deals with different problematics to support decision makers [7–10], such as choice, sorting, description, and ranking. In sorting problem (P_β), each alternative is assigned to predefined ordered categories. In this sense, the categories are previously defined by an expert or a decision maker. The marginalization of the regions can be addressed as a sorting problem, assigning each region to a predefined category of marginalization. This kind of analysis allows the characterization by region of a specific grade of marginality and shows the geographical representation of the country.

The work aims to analyze the marginality level of 32 regions of Mexico, regarding their sociodemographic conditions of the population. This study assigns each region to a marginality category as very low, low, medium, high, and very high. The assignment is carried out by the multiple criteria decision-making method, the MR-sort sorting method. The categorization of marginality in the Mexican territory allows the geographical representation of each level of marginality.

The MR-sort method is a sorting method that supports decision makers to assign regions to a predefined ordered class of marginality. As a sorting problematic is dealt in the present paper, the MR-sort method is proper for classifying marginalization and identifying the marginality level of each region in Mexico. A primary characteristic of the method is related to its applicability. MR-sort is not difficult to implement and is a non-compensatory method, which will support the marginalization problem concerning sociodemographic characteristics of the population. The main contribution of the present study is related to the marginalization problem dealt in regions of Mexico as a sorting problematic. To the best of our knowledge, it is the first time the regions of Mexico have been categorized in marginalized categories with the MR-sort method.

The paper is structured as follows. Section 2 is described the marginalization studies and the current proposals of fuzzy sorting methods. Section 3 is described the multiple criteria decision-making MR-sort method. The data about the sociodemographic characteristic of Mexico are described in Sect. 4. Section 5 presents the preference information and the classification of regions of Mexico in marginalization categories. Finally, concluding remarks are provided in Sect. 6.

2 Literature Review

2.1 Studies of Marginalization

The marginality condition is expanded in different regions of the territory. The studies developed by [11–13] show particular situations of marginality. The analysis of marginalization expresses how people are in societal positions, where they are, and what services they have.

Regarding the role that social enterprises play in the integration of socially excluded people into the labor market, it is analyzed within the existing theoretical and political discourses on exclusion [14]. In [15], developed an analysis of Italian and German marginalized areas for solid biomass production planning. It is stated that this kind of process is a very complex process to implement more in marginal areas, where socioeconomic weaknesses are critical obstacles to the sustainable development of these territories. In [16] is stated that the presence of marginalized communities that are at risk due to the strong trend toward the abandonment of rural areas for big cities.

In Mexico, marginalized people are disabled to access and rights to use essential resources. The study realized by [11] is considered the level of happiness and marginalization for born and migrated persons in two cities of Baja California, Mexico. Roldán et al. [13] analyzed Mexico's public health problem across indigenous, rural and urban areas concerning the degree of marginalization and health service coverage. The study found the indigenous is the most affected population sector, followed by the rural sector. The urban sector is the least affected.

In [17], social sustainability indicators related to education level, early school leavers, male employees, and female employees were studied. They applied the MCDM method named PROMETHEE II to compare and rank 11 case study areas based on the social sustainability indicators.

The ELECTRE III method for ranking was applied to study the competitiveness of regions of Mexico. The practical applications considered the dimension of an inclusive, prepared and healthy society to measure the inhabitants' quality of life [3]. On the other hand, the academic performance indicators, medical offer and health services, socioeconomic conditions, poverty and inequality indicators were used.

2.2 Fuzzy Extensions of Ranking Methods

In [18] is developed an extended hesitant fuzzy set to solve digital health interventions adoption problems. It is used the HF-SWARA–WASPAS approach to evaluate the challenges of digital health. Health information systems are ranked in the adoption of digital health interventions during the COVID-19 outbreak.

In [19] is proposed an integrated framework within the context of Hesitanting fuzzy sets (HFSs) to handle MCDM problems. In the study is found the most critical challenges of adapting online education during the COVID-19 outbreak using the SWARA model. And it is ranked higher education institutions during the COVID-19 outbreak using the MULTIMOORA method.

In [20] is identified the challenges to the adoption of Industry 4.0 in Fintech companies and developed the Fermatean fuzzy CRITIC-COPRAS method to rank the identified challenges and evaluate the performance of companies concerning the weighted challenges based on three decision experts' support.

2.3 Fuzzy Sorting Methods

In [21], a survey of sorting methods is developed to identified contribution in the knowledge from novel and applied method developed in the last 30 years. The most recent sorting methods using a fuzzy approach are Fuzzy-Flowsort [22], SMAA Fuzzy-FlowSort [23], ELECTRE TRI-C and trapezoidal fuzzy numbers [24], continuous cooperation fuzzy sorting [25], AHP-FuzzySort [26], and Intuitionistic fuzzy outranking sorting GDM [27].

The Fuzzy-Flowsort [22] uses triangular fuzzy numbers to deal with imprecise data. The SMAA Fuzzy-FlowSort method [23] is an extended version of the Fuzzy-FlowSort method that allows for multiple types of information imperfections (stochastic, ordinal, and interval data, and linguistic variables) to evaluate alternatives and model parameters, along with partial or missing criteria weights.

The ELECTRE TRI-C and trapezoidal fuzzy numbers method [24] allows the definition of imprecision and uncertainty parameters. In this method, the reference alternative information is provided in the form of linguistic terms. Continuous cooperation fuzzy sorting method [25] implements the triangular membership function. The membership function has several classes of cooperation for each criterion used in assessing cooperation.

AHP-FuzzySort method [26] is the fuzzy version that is reached in the pairwise comparison to deal with the uncertainty. The AHP-FuzzySort uses proportional linguistic two-tuples in the sorting process related to soft boundaries between classes. It provides additional information about the alternatives' membership to the corresponding classes. The intuitionistic Fuzzy outranking sorting GDM method [27] combines the outranking relation and the characteristics of the intuitionistic fuzzy sets to develop an intuitionistic fuzzy outranking relation.

3 Methodology to Sorting Regions

The methodology to analyze the marginality situation of regions of Mexico is depicted in Fig. 1. Stage 1 regards the collection of data about the sociodemographic characteristics of the Mexican population. Stage 2 requires the definition of the multiple criteria decision-making problem, where the federal entities (regions) are considered decision alternatives and the sociodemographic characteristic are decision criteria. In Stage 3, the delimitation of marginality categories is needed to represent marginality levels. Moreover, the application of the multiple criteria sorting method is carried out. Finally, in Stage 4 is carried out the analysis of regions assigned in the marginality levels. In the following, the MR-sort method is explained.

The ordered classification corresponds to the decision-making problematic called sorting. The multi-criteria decision-making sorting consists in assign a set of alternatives $A = \{a_1, a_2, \ldots, a_m\}$ in a predefined ordered classes $C_1, C_2, \ldots, C_k$, where the class C_h is preferred to class $C_l (C_h \succ C_l)$. Each alternative a_i is evaluated by criterion in $G = \{g_1, g_2, \ldots, g_n\}$ and compared against a limit profile or central

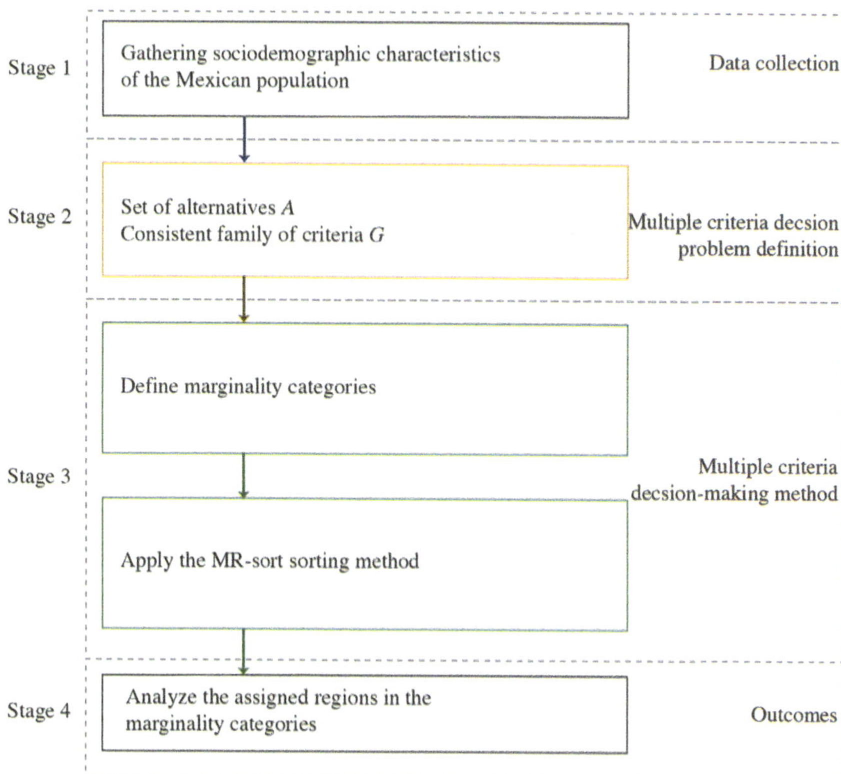

Fig. 1 Methodological procedure for analyzing marginality

profile. When classes are defined by limiting profiles like MR-sort, the profile lp is the vector of attributes values $(b_{h,1}, \ldots, b_{h,j}, \ldots, b_{h,n})$.

The sorting method based on a majority rule (MR-sort) is a non-compensatory sorting method studied by [28, 29]. It is an outranking approach similar to the ELECTRE TRI method. In the MR-sort model, the class C_h is delimited by its lower limit profile b_{h-1} and its upper limit profile b_h. The profile b_h is the vector of attribute values $\left(b_{h,1}, \ldots, b_{h,j}, \ldots, b_{h,n}\right)$. An alternative a_i is assigned to a category C_h when the following situation is fulfilled. The attribute values are at least as good as the lower profile values of b_{h-1} (lower profile of C_h) and having the weighted majority of criteria, and the same condition is not true when the alternative's attribute values are compared to the upper profile values of b_h (upper profile of C_h) [30].

An alternative is a n-tuple of values representing n criteria. It is assumed X is a Cartesian product of attributes ranges, $X = X_1 \times X_2 \times \cdots \times X_n = \prod_{j=1}^{n} X_j$. $g_j(a_i)$ is the evaluation of the criterion g_j associated to the alternative $a_i \in A$.

When an alternative a_i is preferred to a profile, b_h is expressed as $a_i \succsim b_h$. In this case, the following condition is fulfilled:

$$a_i \succsim b_h \Leftrightarrow \sum_{j:g_j(a_i) \geq b_{h,j}} w_j \geq \lambda \tag{1}$$

w_j is the weight of the criterion g_j, and it is considered the weights normalization with $\sum_{j=1}^{n} w_j$. λ is the majority threshold. $\sum_{j:g_j(a_i) \geq b_{h,j}} w_j$ corresponds to the criteria in favor of a_i is preferred to a profile b_h ($a_i \succsim b_h$). The sum of those weights can be called concordance index from the ELECTRE TRI and denoted by $\sigma(a_i, b_h)$. It measures the strength of the coalition of criteria in favor of the assertion a_i is as good as b_h. The definition in (1) is an outranking relation. In MR-sort, the condition of the alternative $a_i \in X$ to be assigned to the category C_h is defined as follows.

$$\sum_{\substack{j \in N: \\ g_j(a_i) \geq b_{h-1,j}}} w_j \geq \lambda \text{ and } \sum_{\substack{j \in N: \\ g_j(a_i) \geq b_{h,j}}} w_j < \lambda \tag{2}$$

where $N = \{1, \ldots, n\}$, and it is assumed that $b_{h-1,j} \leq b_{h,j}$, $h = 1, \ldots, k$; $h < k$. Clearly, the criteria are considered as benefit criteria, where the direction of preference is increasing the value. If cost criteria are considered (decreasing preference), the definition change to $g_j(a_i) \leq b_{h-1,j}$ and $g_j(a_i) \leq b_{h,j}$ to express alternative a_i is preferred to a profile b_h ($a_i \succsim b_h$).

The λ is the majority threshold indicating the coalitions that are strong enough to conclude that a_i is at least as good as b_h. To carry out the assignment rule in the MR-sort method, it is necessary the definition of $kn + 1$ parameters. It is n weights, $(k - 1)$ profiles evaluations, and one majority threshold (λ).

4 Data of Marginalization

The geographical political division of Mexico is constituted in 32 states that are called regions in the present work. In the north, six regions border the United States of America. In the south, one region borders Guatemala, and two regions border Belize. The marginality is measured with indices of sociodemographic characteristics of the population of Mexico. The indices are built by the National Population Council (CONAPO, from the acronym in Spanish) and generated by the National Institute of Statistic and Geography (INEGI, from the acronym in Spanish) [31]. Those indices are expressed in percentages and represented nine ways of excluding the population regarding four dimensions, education, house condition and services, territorial distribution, and monetary income. The criteria decisions are listed as follows.

- Criteria (g_1): illiterate population aged 15 years or over
- Criteria (g_2): population without complete primary education aged 15 years or over
- Criteria (g_3): occupants in private homes without drainage or sanitary service
- Criteria (g_4): occupants in private homes without electricity
- Criteria (g_5): occupants in private homes without piped water
- Criteria (g_6): private homes with some level of overcrowding
- Criteria (g_7): occupants in private dwellings with dirt floors
- Criteria (g_8): population residing in towns with less than 5000 inhabitants
- Criteria (g_9): employed population with incomes of up to two times the minimum wage.

The education factor is measured by the criteria g_1 and g_2. The house condition and services factor is measured by the criteria g_3 to g_7. The evaluation of criterion g_6 (overcrowding) is generated by the Microdata Laboratory in INEGI [32]. A home is considered overcrowded when there are more than 2.5 occupants per room. The territorial distribution is evaluated with criterion g_8. Population with less than 5000 habitants assumes population dispersion and some cases, inaccessibility. The monetary income regards the criterion g_9. Two minimum wages are considered the income that constitutes the lower limit for people to have access to basic articles related to state social spending (such as education and health) and the possibilities of achieving competitive participation in labor markets. Table 1 shows the valuation matrix of the regions of Mexico.

Table 1 Sociodemographic characteristics of the population by region

Label	Region	g_1	g_2	g_3	g_4	g_5	g_6	g_7	g_8	g_9
R_1	Aguascalientes	2.11	23.58	0.35	0.23	0.55	0.77	13.13	21.27	58.50
R_2	Baja California	1.83	24.68	0.20	0.58	2.10	1.91	14.59	8.46	73.55
R_3	Baja California Sur	2.34	23.98	0.42	0.96	5.39	5.06	18.60	10.28	45.49
R_4	Campeche	5.86	29.78	2.52	1.05	3.98	2.69	29.97	29.92	70.01
R_5	Coahuila	1.67	21.49	0.30	0.17	0.94	0.75	13.48	10.04	60.03
R_6	Colima	3.37	27.82	0.27	0.33	0.66	2.62	15.31	13.50	59.73
R_7	Chiapas	13.70	48.12	2.46	1.80	10.68	12.39	36.09	57.64	85.57
R_8	Chihuahua	2.63	27.30	1.42	1.66	1.66	2.18	13.60	14.39	66.70
R_9	Ciudad de México	1.43	17.64	0.05	0.05	1.24	0.63	14.40	1.01	56.13
R_{10}	Durango	2.73	27.49	2.84	2.09	2.31	4.26	16.21	32.50	69.26
R_{11}	Guanajuato	5.29	33.53	1.93	0.38	2.91	2.42	16.90	33.26	67.09
R_{12}	Guerrero	12.47	42.55	9.38	1.41	11.55	15.27	32.86	48.15	80.28
R_{13}	Hidalgo	6.62	29.91	1.90	0.64	3.87	2.85	18.52	53.83	73.95
R_{14}	Jalisco	2.90	29.54	0.57	0.30	0.75	1.66	14.02	16.16	55.86
R_{15}	México	2.90	24.96	1.21	0.26	2.78	2.08	20.70	19.15	66.00
R_{16}	Michoacán	7.05	42.41	1.37	0.49	2.82	5.59	18.83	37.53	70.62
R_{17}	Morelos	4.45	27.53	0.73	0.30	4.22	4.10	19.28	26.13	73.59
R_{18}	Nayarit	4.49	29.05	3.97	2.13	3.02	3.99	18.01	36.47	64.82
R_{19}	Nuevo León	1.47	19.07	0.10	0.11	0.70	0.79	13.20	5.14	46.79
R_{20}	Oaxaca	11.82	45.28	1.94	1.92	10.00	13.99	29.45	59.40	78.85
R_{21}	Puebla	6.97	36.87	1.13	0.61	4.66	5.34	25.53	36.08	77.30
R_{22}	Querétaro	3.48	23.63	1.66	0.47	2.13	2.02	15.13	28.83	57.08
R_{23}	Quintana Roo	3.07	22.53	1.14	0.82	1.66	2.42	26.13	11.47	57.26
R_{24}	San Luis Potosí	5.02	29.32	1.52	1.18	7.64	4.97	16.77	36.03	64.88
R_{25}	Sinaloa	3.56	28.86	1.37	0.40	1.40	2.26	18.87	29.26	61.77
R_{26}	Sonora	1.99	22.41	0.66	0.66	1.30	2.42	16.62	15.16	63.22
R_{27}	Tabasco	5.09	29.17	1.76	0.49	6.18	3.37	26.09	53.27	71.75
R_{28}	Tamaulipas	2.58	25.79	0.25	0.43	1.62	1.42	17.43	11.24	76.27
R_{29}	Tlaxcala	3.35	26.94	0.90	0.41	0.89	1.80	22.24	32.13	78.81
R_{30}	Veracruz	8.50	39.88	1.26	1.07	8.67	5.98	23.09	45.86	78.12
R_{31}	Yucatán	6.00	31.55	5.87	0.69	1.06	1.41	26.17	23.32	69.68
R_{32}	Zacatecas	3.76	32.31	2.32	0.46	1.83	1.25	16.25	44.42	71.92

5 Results

5.1 Preferences Information

The marginalization is categorized into five classes ($k = 5$). The MR-sort method requires $kn + 1$ parameters. It is n weights, $(k - 1)$ profiles evaluations and the majority threshold (λ). For the study of marginalization in Mexico, nine criteria describe the performance of sociodemographic characteristics of the population of the regions. In this case, there are 46 parameters to be defined, nine criteria, four profiles $(k - 1)$ to limit the five classes and one majority threshold (λ).

The weights defined in Table 2 give more importance to the education, territorial distribution factors, and monetary income. Other relevant criteria belonging to house condition and services present relative importance as private homes with some level of overcrowding (g_6) and occupants in private dwellings with dirt floors (g_7). Five classes (categories) are limited by the profiles defined in Table 2 (b_1, b_2, b_3, b_4).

A schematic representation of the profiles and classes defined for the level of marginalization in the regions is shown in Fig. 2. The illustration includes the categories and the profiles where the profiles b_0 and b_5 are fictitious profiles to define the lower limiting profile of Class 1 and the upper limiting profile of Class 5. Class C_1 is the best class because it is considered regions with less level of marginalization. Based on the graphical representation of Fig. 2, profile b_1 is the upper limiting profile of class C_1 and the lower limiting profile of class C_2. C_5 is the worst class and is constituted of the most marginalized regions. Profile b_4 is the lower limiting profile of class C_5 and the upper limiting profile of C_4. In this sense, the order of classes is the following, $C_1 \succ C_2 \succ C_3 \succ C_4 \succ C_5$, $(C_h \succ C_{h+1})$.

For the analysis of marginalization, each region a_i is compared against each class C_h to define is a_i is at least as good as b_h, $(a_i \succsim b_h)$. If the assertion is fulfill, a_i is assigned to class C_h. If not, a_i is assigned to class C_{h+1} only if $(a_i \prec b_{h+1})$.

Table 2 Parameters of the MR-sort

Criteria	Min/Max	Weights	Profile classes			
			b_1	b_2	b_3	b_4
g_1	Min	0.14	2.34	2.70	4.40	11.82
g_2	Min	0.14	24.00	26.94	33.60	42.00
g_3	Min	0.07	0.42	0.90	1.52	1.94
g_4	Min	0.09	0.31	0.60	1.18	1.41
g_5	Min	0.09	1.30	1.60	2.00	6.50
g_6	Min	0.11	1.70	2.00	4.97	12.30
g_7	Min	0.1	14.50	16.00	18.00	29.00
g_8	Min	0.13	11.00	16.00	40.00	45.00
g_9	Min	0.13	57.00	60.00	73.00	79.00

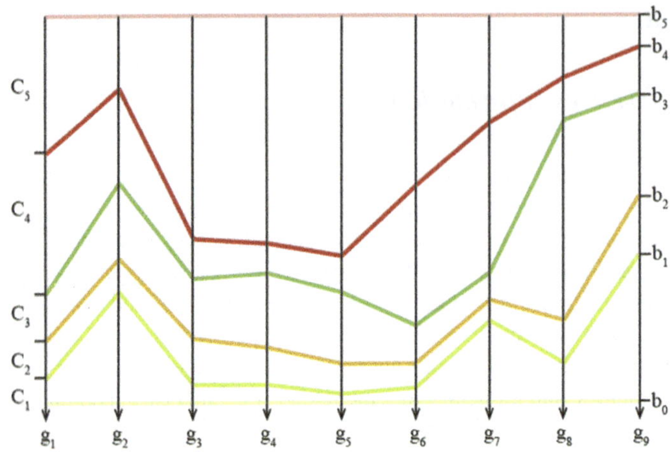

Fig. 2 Profiles and classes of marginality in Mexico

5.2 *Discussion*

The analysis of sociodemographic characteristics performed by nine criteria in the regions shows the level of marginality in Mexico. Based on the profiles and the categories defined, class C_1 is the best class with the lowest level of marginalization. Class C_5 is the worst class with the highest level of marginalization ($C_1 \succ C_2 \succ C_3 \succ C_4 \succ C_5$). The application of MR-sort sorting method in the regions of Mexico regarding the performance of the sociodemographic characteristics of the population results in the assigned regions of marginality level in Table 3.

Six regions are assigned to the class C_1, the less marginality level: Aguascalientes, Baja California Sur, Coahuila, Ciudad de México, Jalisco, and Nuevo León. The regions present close values to the criteria most important related to profile b_1. Criteria g_1 and g_2 are among the most important, illiterate population aged 15 years or over and population without complete primary education aged 15 years or over, respectively. The profile $b_{1,1} = 2.34$ corresponds to criterion g1. The best value is presented by Ciudad de Mexico (1.43) and the worst by Jalisco (2.9). The profile $b_{1,2} = 24$ corresponds to criterion g_2. The best value is presented by Ciudad de Mexico (17.64) and the worst by Jalisco (29.54).

The regions with highest level of marginalization are Chiapas, Guerrero, and Oaxaca; they are assigned to class C_5. The profile $b_{4,1} = 11.82$ corresponds to the criterion g_1. Oaxaca presents 11.82; it is followed by Guerrero (12.49) and Chiapas (13.70). At criterion g_2, Guerrero presents 42.55; it is followed by Oaxaca (45.28) and Chiapas (13.70). The corresponding profile is $b_{4,2} = 42$.

The regions with the highest level of marginalization are Chiapas, Guerrero, and Oaxaca. They are assigned to class C_5. The profile $b_{4,1} = 11.82$ corresponds to criterion g_1. Oaxaca presents 11.82; it is followed by Guerrero (12.49) and Chiapas

Table 3 Ordered classification of regions

Label	Region	Class
R_1	Aguascalientes	C_1
R_3	Baja California Sur	C_1
R_5	Coahuila	C_1
R_9	Ciudad de México	C_1
R_{14}	Jalisco	C_1
R_{19}	Nuevo León	C_1
R_2	Baja California	C_2
R_6	Colima	C_2
R_{26}	Sonora	C_2
R_{28}	Tamaulipas	C_2
R_{29}	Tlaxcala	C_2
R_4	Campeche	C_3
R_8	Chihuahua	C_3
R_{10}	Durango	C_3
R_{11}	Guanajuato	C_3
R_{15}	México	C_3
R_{17}	Morelos	C_3
R_{18}	Nayarit	C_3
R_{22}	Querétaro	C_3
R_{23}	Quintana Roo	C_3
R_{24}	San Luis Potosí	C_3
R_{25}	Sinaloa	C_3
R_{31}	Yucatán	C_3
R_{32}	Zacatecas	C_3
R_{13}	Hidalgo	C_4
R_{16}	Michoacán	C_4
R_{21}	Puebla	C_4
R_{27}	Tabasco	C_4
R_{30}	Veracruz	C_4
R_7	Chiapas	C_5
R_{12}	Guerrero	C_5
R_{20}	Oaxaca	C_5

(13.70). At criterion g_2, Guerrero presents 42.55, followed by Oaxaca (45.28) and Chiapas (13.70). The corresponding profile is $b_{4,2} = 42$.

Figure 3 illustrates the geographical marginality categorization of the regions. The regions with less marginality level are geographical located at the center or north of the country. Nuevo Leon and Coahuila are on the border with the United States. Baja

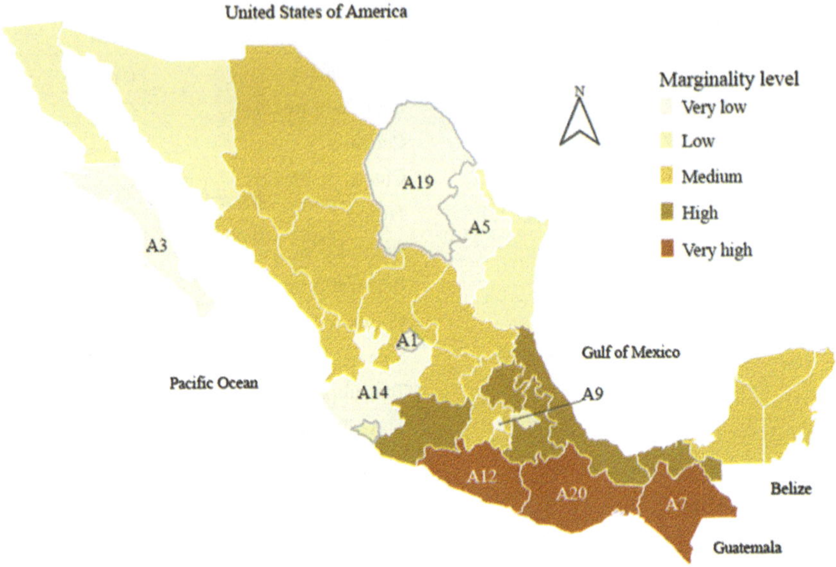

Fig. 3 Geographical categorization of marginality in Mexico

California Sur is an important touristic destination based on sun and beach. Ciudad de Mexico, Jalisco, and Aguascalientes are industrial developed regions.

The regions with the highest level of marginalization are located in the south of Mexico, Oaxaca, Guerrero, and Chiapas. The national competitiveness index in Mexico from 2021 generated by the Mexican Competitiveness Institution (IMCO by its acronym in Spanish) [33] reported some interesting data about the more marginalized regions. The migration in those regions is the highest in the country. Guerrero presents 0.9% of population that migrated, Oaxaca 0.5% and Chiapas 0.3%. The same regions present a high coefficient of wage inequality, Chiapas 4.66, Oaxaca 3.42, and Guerrero 3.33%.

In the intermediate marginality levels, low C_2, medium C_3, and high marginality (C_4) classes are shown. The low class includes five regions; the medium class shows 13 regions, and the high class presents five regions. The most crowded class is the medium marginality level (C_3). Due to the performance of those regions, they have the opportunity to increase their sociodemographic characteristic to be included in a better marginality level (Low, C_2). However, the same regions are at risk of decreasing their performance and being more marginalized (High C_4) if politics and development are not made properly.

6 Conclusions

The applied sorting method shows the regions assigned in the predefined ordered classes of marginality. The analysis evidences the level of marginality of the regions. In this paper, the MR-sort method is applied to assign Mexico's regions to one marginality category. The five categories show three regions with a very high degree of marginality (Class C_5). Those regions show very high levels of illiterate people and a high percentage of the population with very low incomes. The current work shows some of the inequalities evidenced by the different categories of marginalization in some regions of Mexico.

The analysis of marginalization in Mexico may be used for decision-making to help improve policy on marginalized regions regarding their context and particular situation. The proposal by [34] can be replicated when the authors propose a multi-criteria decision-making approach that helps improve policy formulation to identify accessible poor rural areas. A managerial implementation to reduce the marginalization would regard some public politics in regions with higher marginality levels. Public politics can be focused on the outcome of the sorting procedure due it highlights the higher marginality level in Class C_5.

It is stated that MR-sort method is same to the ELECTRE TRI sorting method; however, the first is a simplified version. In this sense, defining some parameters like indifference, preference, and veto thresholds is not required. Instead, only, weights, limiting profiles, and majority threshold were defined. In total, 46 parameters were defined to the categorization of marginality levels of regions of Mexico. The main limitation of the proposed procedure is related with the sorting method. It requires the definition of the limiting profiles of the categories. It means the outcomes depend of that preference information from the decision maker.

For future research related to the marginality analysis, the comparison of methods and data from previous years can be carried out. It will help to understand the increment of reduction of marginalization of the regions and find proper methods to analyze them.

The suggested methods than can be used in future analysis of marginalization with fuzzy sorting are Fuzzy-Flowsort [22], SMAA Fuzzy-FlowSort [23], ELECTRE TRI-C and trapezoidal fuzzy numbers [24], continuous cooperation fuzzy sorting [25], AHP-FuzzySort [26], and intuitionistic Fuzzy outranking sorting GDM [27]. Those methods work in different ways with the fuzzy approaches for the data or for the fuzzyfication of the profiles or the process assignment.

References

1. Gatzweiler, F.W., Baumüller, H., Husmann, C., von Braun, J.: Marginality: addressing the root causes of extreme poverty. SSRN Electron. J. (2011). https://doi.org/10.2139/ssrn.2235654

2. von Braun, J., Gatzweiler, F.W.: Marginality—an overview and implications for policy. In: Marginality: Addressing the Nexus of Poverty, Exclusion and Ecology, pp. 1–23. Springer, The Netherlands (2014)
3. Muñoz-Palma, M., Alvarez-Carrillo, P.A., Miranda-Espinoza, E.L., Avilés-Ochoa, E., León-Castro, E.: Multicriteria analysis model for the evaluation of the competitiveness of the states in Mexico. In: León-Castro, E., Blanco-Mesa, F., Gil-Lafuente, A.M., Merigó, J.M., Kacprzyk, J. (eds.) Intelligent and Complex Systems in Economics and Business. Advances in Intelligent Systems and Computing, vol. 1249, pp. 1–20. Springer, Cham (2021)
4. Wacquant, L.: The militarization of urban marginality: lessons from the Brazilian metropolis. Int. Polit. Sociol. **2**(1), 56–74 (2008). https://doi.org/10.1111/j.1749-5687.2008.00037.x
5. Wacquant, L.: Class, ethnicity and state in the making of urban marginality, pp. 23–50 (2019). https://doi.org/10.1007/978-3-030-16222-1_2
6. Müller, M.-M.: The rise of the penal state in Latin America. Contemp. Justice Rev. **15**(1), 57–76 (2012). https://doi.org/10.1080/10282580.2011.590282
7. Roy, B.: Méthodologie multicritère d'aide à la décision. Economica, Paris (1985)
8. Roy, B.: Multicriteria methodology for decision aiding. In: Nonconvex Optimization and Its Applications, vol. 12. Springer, Boston (1996)
9. Roy, B.: The optimisation problem formulation: criticism and overstepping. J. Oper. Res. Soc. **32**(6), 427–436 (1981). https://doi.org/10.1057/jors.1981.93
10. Doumpos, M., Zopounidis, C.: Multicriteria Decision Aid Classification Methods. Applied Optimization, vol. 73. Kluwer Academic Publishers, New York (2004)
11. Collins, K., Ley García, J.: Happiness and marginalization rates for internal Mexican migrants and the native-born population in Baja California, Mexico. Soc. Sci. J. **51**(4), 598–606 (2019). https://doi.org/10.1016/j.soscij.2014.07.004
12. Peña, S.: Recent developments in urban marginality along Mexico's northern border. Habitat Int. **29**(2), 285–301 (2005). https://doi.org/10.1016/j.habitatint.2003.10.002
13. Roldán, J., Álvarez, M., Carrasco, M., Guarneros, N., Ledesma, J., Cuchillo-Hilario, M., Chávez, A.: Marginalization and health service coverage among Indigenous, rural, and urban populations: a public health problem in Mexico. Rural Remote Health **17**(4) (2017). https://doi.org/10.22605/rrh3948
14. Džunić, M., Stanković, J., Janković-Milić, V.: Multi-criteria approach in evaluating contribution of social entrepreneurship to the employment of socially-excluded groups. Technol. Econ. Dev. Econ. **24**(5), 1885–1908 (2018). https://doi.org/10.3846/20294913.2017.1347906
15. Arbolino, R., De Simone, L., Yigitcanlar, T., Ioppolo, G.: Facilitating solid biomass production planning: insights from a comparative analysis of Italian and German marginalized areas. J. Clean. Prod. **181**, 819–828 (2018). https://doi.org/10.1016/j.jclepro.2018.01.154
16. Ferretti, V., Gandino, E.: Co-designing the solution space for rural regeneration in a new World Heritage site: a choice experiments approach. Eur. J. Oper. Res. **268**(3), 1077–1091 (2018). https://doi.org/10.1016/j.ejor.2017.10.003
17. Chatzinikolaou, P., Bournaris, T., Manos, B.: Multicriteria analysis for grouping and ranking European Union rural areas based on social sustainability indicators. Int. J. Sustain. Dev. **16**(3/4), 335 (2013). https://doi.org/10.1504/ijsd.2013.056559
18. Mardani, A., Saraji, M.K., Mishra, A.R., Rani, P.: A novel extended approach under hesitant fuzzy sets to design a framework for assessing the key challenges of digital health interventions adoption during the COVID-19 outbreak. Appl. Soft Comput. **96**, 106613 (2020). https://doi.org/10.1016/j.asoc.2020.106613
19. Saraji, M.K., Mardani, A., Köppen, M., Mishra, A.R., Rani, P.: An extended hesitant fuzzy set using SWARA-MULTIMOORA approach to adapt online education for the control of the pandemic spread of COVID-19 in higher education institutions. Artif. Intell. Rev. (2021). https://doi.org/10.1007/s10462-021-10029-9
20. Kamali Saraji, M., Streimikiene, D., Kyriakopoulos, G.L.: Fermatean fuzzy CRITIC-COPRAS method for evaluating the challenges to industry 4.0 adoption for a sustainable digital transformation. Sustainability **13**(17), 9577 (2021). https://doi.org/10.3390/su13179577

21. Alvarez, P.A., Ishizaka, A., Martínez, L.: Multiple-criteria decision-making sorting methods: a survey. Expert Syst. Appl. **183**, 115368 (2021). https://doi.org/10.1016/j.eswa.2021.115368
22. Campos, A.C.S.M., Mareschal, B., de Almeida, A.T.: Fuzzy FlowSort: an integration of the FlowSort method and fuzzy set theory for decision making on the basis of inaccurate quantitative data. Inf. Sci. **293**, 115–124 (2015). https://doi.org/10.1016/j.ins.2014.09.024
23. Pelissari, R., Oliveira, M.C., Ben Amor, S., Abackerli, A.J.: A new FlowSort-based method to deal with information imperfections in sorting decision-making problems. Eur. J. Oper. Res. **276**(1), 235–246 (2019). https://doi.org/10.1016/j.ejor.2019.01.006
24. Pereira, J., de Oliveira, E.C.B., Gomes, L.F.A.M., Araújo, R.M.: Sorting retail locations in a large urban city by using ELECTRE TRI-C and trapezoidal fuzzy numbers. Soft. Comput. **23**(12), 4193–4206 (2018). https://doi.org/10.1007/s00500-018-3068-2
25. Silva, M.M., Costa, A., de Gusmao, A.P.H.: Continuous cooperation: a proposal using a fuzzy multicriteria sorting method. Int. J. Prod. Econ. **151**, 67–75 (2014). https://doi.org/10.1016/j.ijpe.2014.01.022
26. Krejčí, J., Ishizaka, A.: FAHPSort: a fuzzy extension of the AHPSort method. Int. J. Inf. Technol. Decis. Mak. **17**(04), 1119–1145 (2018). https://doi.org/10.1142/s0219622018400011
27. Shen, F., Xu, J.P., Xu, Z.S.: An outranking sorting method for multi-criteria group decision making using intuitionistic fuzzy sets. Inf. Sci. **334**, 338–353 (2016). https://doi.org/10.1016/j.ins.2015.12.003
28. Bouyssou, D., Marchant, T.: An axiomatic approach to noncompensatory sorting methods in MCDM, I: The case of two categories. Eur. J. Oper. Res. **178**(1), 217–245 (2007). https://doi.org/10.1016/j.ejor.2006.01.027
29. Bouyssou, D., Marchant, T.: An axiomatic approach to noncompensatory sorting methods in MCDM, II: More than two categories. Eur. J. Oper. Res. **178**(1), 246–276 (2007). https://doi.org/10.1016/j.ejor.2006.01.033
30. Sobrie, O., Mousseau, V., Pirlot, M.: Learning monotone preferences using a majority rule sorting model. Int. T. Oper. Res. (2018). https://doi.org/10.1111/itor.12512
31. INEGI: Censo de Población y Vivienda 2020: Tabulados interactivos. https://www.inegi.org.mx/programas/ccpv/2020/-Tabulados (2021). Accessed Oct 2021
32. INEGI: Censo de Población y Vivienda 2020: Laboratorio de Microdatos (2021). Accessed Oct 2021
33. IMCO: Índice de Competitividad Estatal 2021. https://imco.org.mx/indices/indice-de-competitividad-estatal-2021/ (2021). Accessed Sept 2021
34. Zafri, N.M., Sameen, I., Jahangir, A., Tabassum, N., Hasan, M.M.U.: A multi-criteria decision-making approach for quantification of accessibility to market facilities in rural areas: an application in Bangladesh. GeoJournal (2020). https://doi.org/10.1007/s10708-020-10161-z

Clustering Faculty Members for the Betterment of Research Outcomes: A Fuzzy Multi-criteria Decision-Making Approach in Team Formation

M. Umme Salma, Samayan Narayanamoorthy, and Joseph Varghese Kureethara

Abstract From a talent-pool of people, choosing an efficient team is tough. Faculty members of a higher education institution constitute the talent-pool. Teams have to be formed from them so that research output of each team is maximum. Amongst numerous research skills, thirteen are identified as most desirable skills. The level of these thirteen skills, viz., concept articulation, formatting according to templates/style sheets, identifying the relevant literature, initiative, logical reasoning, patience, problem formulation/problem finding, proof reading skills/identifying mistakes in written communication, searching/browsing skills/quick search techniques, sense of positive criticism, statistical knowledge, the ability to stay calm, and written communication skills, varies from person to person. Historical ranking of these skills and self-evaluation of the level of acquisition of these skills is used along with the years of experience, educational qualification, gender, marital status, etc., to rank individual faculty members. The fuzzy ranking of the faculty members thus obtained is used to cluster them in to teams that are efficient in complementary skills. Each team thus formed is involved in collaborative research leading to research publication. The model is successfully implemented in a university department with 40 faculty members.

1 Introduction

Research is a systematic way of exploring knowledge. Research can be quantitative or qualitative, fundamental or applied, descriptive or empirical. Irrespective of the

---

M. Umme Salma (✉) · J. V. Kureethara
Christ University, Bangalore, India
e-mail: ummesalma.m@christuniversity.in

J. V. Kureethara
e-mail: frjoseph@christuniversity.in

S. Narayanamoorthy
Bharathiar University, Coimbatore, India
e-mail: snmphd@gmail.com

L. Sahoo et al. (eds.), *Real Life Applications of Multiple Criteria Decision Making Techniques in Fuzzy Domain*, Studies in Fuzziness and Soft Computing 420,
https://doi.org/10.1007/978-981-19-4929-6_29

type, the research is executed systematically in an organised manner. The main aim of the research is to come up with information that can benefit the community better. As a researcher, one needs to perform various activities such as data collection, data preparation, data analysis, data interpretation, data modelling, documentation, and proofreading.

A researcher is a person who indulges in research. Regardless of age, gender, or occupation, any person can be a researcher. As a beginner, a researcher may face various challenges like motivational issues, lack of time management, language barrier, limited support, lack of experience in multiple aspects of research, and understanding research misconceptions [2, 8]. All the challenges and misunderstandings can be ruled out by acquiring the required skills. For a beginner, it is essential to know and possess the skills that benefit in research [18]. Specific skills like problem identification, logical reasoning, concept articulation, etc., can improve the quality of research. The skill set requirement may vary depending upon the type of research; however, specific skills are essential to be developed by a researcher. The research community's particular skills widely accepted are given by Mahasneh [13]. Mastering basic research skills contribute to the betterment of the research.

The literature provides details related to various works contributed to education and research using mathematical models. A systematic review to identify the potential insights and research gap in improving the quality of school education was given by Gumus [6]. Carlucci [5] proposed a framework for assessing teaching performance. A fuzzy model was used to personalise mobile game-based learning experience in higher education [19]. Based upon the literature, we aim to propose a new multi-criteria decision-making (MCDM) model to improve the research quality in higher education.

The structure of the article includes a basic introduction to research and challenges faced by researchers in Sect. 1, followed by the objective of the proposed work in Sect. 2, development of instrument and framework for proposed research in Sect. 3, data collection in Sect. 4, fuzzy clustering model in Sect. 5 analysis, and interpretation in Sect. 6, concluded by summary and discussion in Sect. 7.

2 The Objective of the Proposed Work

The main aim of the proposed work is to develop a hybrid model by incorporating the concepts of multi-criteria decision-making and fuzzy logic to improve the quality of research in higher education. To achieve the objective, we emphasise a major key factor called collaboration and its contribution to improving the quality of research. The details of the significance of collaboration are discussed in Sect. 2.1

2.1 Significance of Collaboration in Research

After industry and research and development (R & D) centres, academicians are the people globally contributing to research. The role of academicians in the study is immense and is projected in all fields. Most of the researches in the industry and the R & D centres are done in collaboration. Collaboration in the industry is almost like a natural thing. A business and industry look at an increasing profit in the R & D centres, researchers work collaboratively for quicker and efficient results. Many studies have proved that collaborative research contributed more effectively than individual research [12]. The stronger the collaboration, the better the results are.

Although collaboration is highly recommended in many science disciplines, it is not the case with social sciences, arts, and humanities. It is a fact that collaboration generally leads to greater productivity in research and enhances the mobility and visibility of the researchers. The average number of authors per paper experienced steady growth, from 2.20 in 1980 to 7.00 in 2019 [7].

Having a good team of collaborators is crucial in improving the productivity of publications. There are several reasons for a successful collaboration in research [9]. In this era of greater collaboration in academic research, we propose a fuzzy MCDM model for team selection. Clustering faculty members to better research outcomes is the key to improved collaborative research. This article aims to develop a fuzzy clustering model that helps cluster the faculty members who can contribute effectively and produce better research outcomes. Katz and Martin had tried to distinguish between collaboration in research and co-authorship in research [11].

3 Instrument for the Proposed Research

The instrument used for the proposed research is a questionnaire used to get the participants' preference to rank the thirteen research skills based upon their understanding and preference. Along with the ranking, gender, age, qualification and affiliation details of the participants were also collected. The details related to the selected skills are provided in the next section.

3.1 Identification of Skill Set for Researchers

Many instruments to study the quality of research have been proposed in the literature [14]. The contribution of academicians in research is discussed in Reis-Jorge [16]. Based on the literature available and from the authors' experience, thirteen essential research skills are shortlisted to build the instrument for the proposed research work. The shortlisted skills are concept articulation, formatting according to templates/style sheets, identifying the relevant literature, initiative, logical reasoning, patience, prob-

Table 1 Historical ranking of the research skills

No.	Code	Research skills	Rank
1	A	Concept articulation	5
2	B	Formatting according to templates/style sheets	13
3	C	Identifying the relevant literature	2
4	D	Initiative	3
5	E	Logical reasoning	4
6	F	Patience	7
7	G	Problem formulation/problem finding	1
8	H	Proof reading skills/identifying mistakes in written communication	10
9	I	Searching/browsing skills/quick search techniques	6
10	J	Sense of positive criticism	8
11	K	Statistical knowledge	12
12	L	The ability to stay calm	9
13	M	Written communication skills (writing long and short sentences in correct English)	11

lem formulation/problem finding, proof reading skills/identifying mistakes in written communication, searching/browsing skills/quick search techniques, sense of positive criticism, statistical knowledge, the ability to stay calm, and written communication skills (writing long and short sentences in correct English). Historical ranking based on various literature is given in Table 1.

An individual is endowed with many of these skills. Selecting a team with complementary skills give depth to sports teams, and it is vital for great success [1, 3, 4, 10, 15, 17]. Similar to that of sports, if the faculty members are grouped to form teams with complementary skills, the performance is expected to be enhanced. We present a fuzzy MCDM model to create research teams in the academic department consisting of at least 25 members.

3.2 Framework

The framework for the proposed research encompasses five phases—data collection, data transformation, data modelling, data analysis, and data interpretation. Figure 1 represents the block diagram for the proposed framework.

The importance of each phase is discussed in detail in the later sections.

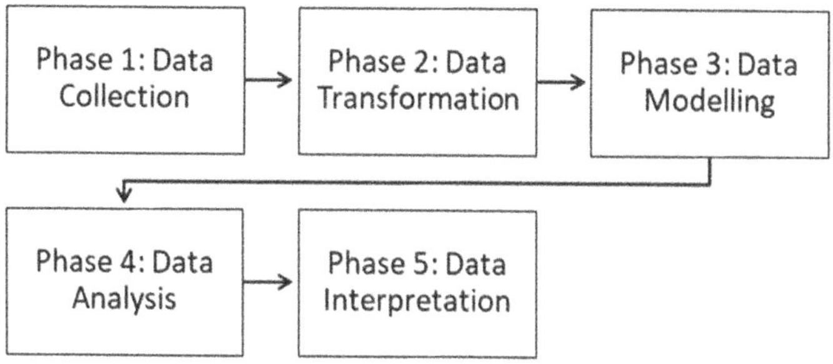

Fig. 1 Block diagram for the proposed framework

4 Data Collection

The instrument is circulated to the academicians of various age groups (25–60) in multiple departments and includes both males and females for self-evaluation. The data was collected from around 240 individuals, out of which 109 research scholars (RS), 50 postgraduates (PG), 59 faculty members(FM), and 32 random persons (MX), from various institutions. Based upon their individual preferences, the shortlisted skills are to be ranked in the range of 1–13.

4.1 Data Description

The collected data consisted of self-evaluation of 13 research characteristics by 240 individuals. The details of the data are provided in Table 2.

Table 2 Data description

Data	Count	Description
Number of samples	240	109 (research scholars) + 50 (PG) + 59 (Faculty)+ 32 (Mixed)
Number of features	13	As given in the instrument
Gender	2	Male (43) and female (106)
Education qualification	3	Masters, M. Phil, Ph.D.
Disciplines covered	19	Biotechnology, Botany, Chemistry, Commerce, Computer Science, Economics, Education, English, International Studies, Law, Management, Mathematics, Media Studies, Physics, Psychology, Social Work, Sociology, Tourism, and Zoology

Table 3 Cumulative ranking of the research skills by various groups

Code	RS	PG	FM	MX	Cumulative
A	6	1	3	3	2
B	13	12	11	13	13
C	3	5	6	2	3
D	2	9	2	5	5
E	5	4	4	4	4
F	4	6	8	7	6
G	1	2	1	1	1
H	12	8	12	10	12
I	9	11	7	6	8
J	11	7	13	9	11
K	7	13	10	8	9
L	10	3	5	12	7
M	8	10	9	11	10

Legend: *RS* research scholar, *PG* postgraduate, *FM* faculty member, *MX* mixed group

The cumulative ranking by each group is given in Table 3. The cumulative rank correlates with the rank given by the mixed group for these 13 research skills.

Hence, as per the cumulative ranking, the top five skills are problem formulation/problem finding, concept articulation, identifying the relevant literature, logical reasoning, and initiative. These are the same five top 5 research skills in the historical ranking.

4.2 Data Transformation

The collected data involving the information related to 13 research skills obtained from 240 participants is converted into a data matrix used for data modelling. A fuzzy model is used to model the data and get the ranking of teachers. The details about the fuzzy model are discussed in Sect. 5.

5 Data Modelling

The proposed work is built on fuzzy data modelling to extract meaningful insight from vague and imprecise data. The flow chart of the data model is given in Fig. 2.

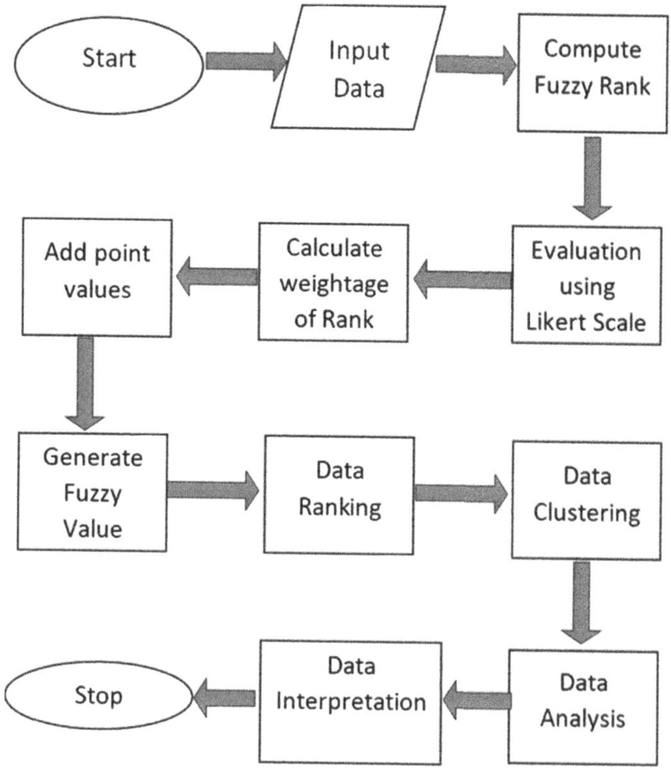

Fig. 2 Flow chart of the proposed data model

5.1 *Fuzzy Clustering Model for Ranking of Teachers*

The proposed model consists of the following steps.

Step 1 Compute the fuzzy rank value for each skill by taking the inverse of the rank. $(1/r)$

Step 2 Conduct a self-evaluation of the faculty member on the Likert scale 5 (high)–1 (low).

Step 3 Compute the rank weight by multiplying the self-evaluation with the fuzzy value of the skill. The maximum of such value is $13 \times 5 = 65$.

Step 4 Add point values based on the status of the faculty members as given in Table 4.

Step 5 Take the total point out of the total sum to get the fuzzy skill value of the faculty.

Table 4 Point values of the faculty members based on their background

Particulars	Weightage
Male	0.6
Female	0.5
Married	0.3
Single	0.4
MPhil	1
PhD	2
Research experience up to 5 years	1
Research experience above 5 years	2

Step 6 Rank the faculty members in the descending order of the rank obtained from Step 5.

Step 7 Cluster the faculty members based on their skill rank such that the aggregate skill value is almost equal for each team.

5.2 *Fuzzy Clustering Algorithm*

Suppose that we need k teams from a set of n faculty members of a department. Assume that n is a multiple of $2k$. Then, as per the rank obtained in Step 6 of the clustering model, label the faculty members as $1, 2, \ldots, (k-1), k, k, (k-1), \ldots, 2, 1$, repeatedly.

If n is not a multiple of $2k$, and if n is a multiple of k, then the leftover k members are labelled $k, (k-1), \ldots, 2, 1$, based on the ascending order of their rank.

If n is not a multiple of $2k$ and k, then the $n \mod k$ faculty members are labelled as $k, (k-1), \ldots, n \mod k$ in the ascending order of their ranks. Then, cluster the faculty members who have the same labels.

6 Analysis and Interpretation

As per the model presented here, a department from an Indian university was identified. There were 40 faculty members in the department. The task is to form 8 teams consisting of 5 members each.

The instrument was distributed to them for self-evaluation. Faculty members conducted the self-evaluation of their skills on a scale 5–1. This data is given in Table 5.

The fuzzy values of the self-evaluation of the skills of the faculty members are given in Table 6. Table 6 also provides the values of other supporting factors such as gender, marital status, educational qualification, and experience.

Table 5 Point values of the faculty members based on their background

Skill No.	1	2	3	4	5	6	7	8	9	10	11	12	13	G	Md	MPhil	PhD	Exp
Faculty 1	4	4	4	3	3	4	3	5	3	2	4	5	4	F	Y	Y	Y	6
Faculty 2	3	3	3	3	3	4	4	4	4	1	4	4	4	F	N	Y	Y	25
Faculty 3	5	5	5	5	5	5	5	5	5	5	4	5	5	M	Y	Y	Y	17
Faculty 4	4	4	3	3	3	4	4	3	4	4	3	3	3	M	Y	Y	Y	17
Faculty 5	5	2	1	3	4	2	1	5	2	4	3	5	5	F	N	N	P	11
Faculty 6	0	3	3	3	4	2	3	4	4	2	2	3	4	F	Y	Y	P	9
Faculty 7	4	4	3	2	5	5	5	4	5	2	4	5	5	F	Y	N	P	2
Faculty 8	4	3	4	3	3	4	4	3	4	4	4	3	3	F	Y	Y	P	13
Faculty 9	4	5	5	5	4	5	4	5	4	4	5	5	5	M	Y	Y	P	7
Faculty 10	3	3	4	4	3	3	3	3	4	3	4	4	3	F	Y	Y	Y	16
Faculty 11	4	3	4	4	4	4	4	4	3	2	4	4	4	M	Y	Y	Y	7
Faculty 12	4	4	4	5	3	4	3	5	5	5	3	4	2	M	N	N	Y	3
Faculty 13	3	4	3	3	3	3	4	3	2	3	3	3	4	M	Y	Y	P	11
Faculty 14	3	3	3	4	4	3	3	4	3	2	2	3	3	F	Y	Y	Y	22
Faculty 15	2	3	4	4	3	3	3	3	2	2	2	3	3	M	N	Y	Y	16
Faculty 16	3	4	3	4	4	4	2	3	5	4	4	3	3	M	N	N	N	4
Faculty 17	3	4	3	3	3	3	2	4	3	3	3	4	4	M	N	N	P	6
Faculty 18	3	4	3	5	4	5	3	4	3	3	5	4	5	F	Y	Y	P	7
Faculty 19	4	3	4	3	4	4	4	3	3	3	4	3	3	F	Y	Y	Y	7
Faculty 20	4	3	3	3	3	4	2	3	4	2	3	4	4	M	N	N	N	4
Faculty 21	4	5	4	3	3	4	3	4	5	2	5	3	4	M	Y	Y	P	6
Faculty 22	3	3	3	4	4	5	3	3	4	3	5	3	5	F	Y	Y	Y	14
Faculty 23	3	5	3	3	4	5	4	5	5	5	5	4	4	F	N	N	P	5
Faculty 24	3	3	3	3	3	4	4	4	3	3	3	3	3	F	Y	Y	P	17
Faculty 25	3	3	3	4	3	4	3	3	5	2	4	3	3	M	N	N	N	2
Faculty 26	2	4	4	4	4	3	4	3	3	2	4	3	3	M	Y	Y	Y	17
Faculty 27	2	2	2	3	3	3	3	1	1	2	3	3	3	F	Y	Y	P	8
Faculty 28	4	5	5	5	5	5	4	4	4	5	5	5	5	M	N	Y	P	11
Faculty 29	3	3	3	5	4	4	3	4	5	3	5	3	5	M	N	N	P	4
Faculty 30	3	3	3	4	3	5	3	4	4	3	5	3	5	F	N	Y	P	20
Faculty 31	4	0	5	5	4	4	4	4	4	4	4	3	5	F	Y	Y	Y	21
Faculty 32	3	3	3	4	3	5	3	4	4	3	5	4	3	F	Y	N	Y	8
Faculty 33	3	4	3	4	4	3	4	3	4	5	4	5	5	M	N	Y	Y	8
Faculty 34	4	4	4	4	4	4	3	3	2	3	3	3	3	F	Y	Y	Y	24
Faculty 35	3	4	4	3	4	3	4	4	3	4	4	3	4	F	Y	Y	Y	17
Faculty 36	0	3	4	1	4	0	4	4	0	1	3	3	3	F	Y	Y	P	5
Faculty 37	3	4	5	3	4	3	4	2	3	2	4	3	3	F	Y	Y	P	12
Faculty 38	4	5	5	5	5	3	5	4	5	3	4	5	4	M	N	N	P	4
Faculty 39	3	4	3	4	4	4	4	4	4	3	4	3	3	M	Y	N	P	4
Faculty 40	3	1	4	4	4	2	3	5	3	1	2	5	4	M	Y	Y	Y	27

Legend: *G* gender, *F* female, *M* male, *Md* married, *Y* yes, *N* no, *P* pursuing, *Exp* experience

Table 6 Scores of the individual faculty members based on the self-evaluation

	0.5	0.08	0.33	0.2	0.25	0.17	1	0.08	0.13	0.09	0.11	0.14	0.1							
	1	2	3	4	5	6	7	8	9	10	11	12	13	G	Md	Ml	PhD	Exp	Tot	FT
F1	0.5	0.15	1	0.8	1.25	1	7	0.67	1.13	0.91	1.22	1.71	1.3	0.5	0.3	1	2	2	24.44	0.34
F2	2.5	0.31	1.33	0.6	0.75	0.67	3	0.42	0.38	0.18	0.44	0.71	0.4	0.5	0.3	1	2	2	17.49	0.25
F3	1.5	0.23	1	0.6	0.75	0.67	4	0.33	0.5	0.09	0.44	0.57	0.4	0.6	0.3	1	2	2	16.99	0.24
F4	2	0.23	1.33	0.8	1	0.67	4	0.33	0.38	0.18	0.44	0.57	0.4	0.6	0.4	0	2	1	16.34	0.23
F5	2	0.31	1.33	1	0.75	0.67	3	0.42	0.63	0.45	0.33	0.57	0.2	0.6	0.3	1	1	2	16.56	0.23
F6	2	0.38	1.33	0.6	0.75	0.67	3	0.33	0.63	0.18	0.56	0.43	0.4	0.5	0.3	1	2	2	17.06	0.24
F7	2	0	1.67	1	1	0.67	4	0.33	0.5	0.36	0.44	0.43	0.5	0.5	0.3	1	1	1	16.7	0.24
F8	0	0.23	1.33	0.2	1	0	4	0.33	0	0.09	0.33	0.43	0.3	0.5	0.3	1	1	2	13.05	0.18
F9	1.5	0.31	1.67	0.6	1	0.5	4	0.17	0.38	0.18	0.44	0.43	0.3	0.5	0.3	0	2	2	16.27	0.23
F10	1.5	0.23	1	0.8	0.75	0.83	3	0.33	0.5	0.27	0.56	0.57	0.3	0.6	0.3	1	2	2	16.55	0.23
F11	2.5	0.38	1.67	1	1.25	0.83	5	0.42	0.63	0.45	0.44	0.71	0.5	0.5	0.3	1	1	2	20.59	0.29
F12	0	0.23	1	0.6	1	0.33	3	0.33	0.5	0.18	0.22	0.43	0.4	0.6	0.3	1	1	2	13.13	0.18
F13	1.5	0.31	1	0.6	0.75	0.5	4	0.25	0.25	0.27	0.33	0.43	0.4	0.6	0.4	0	0	1	12.59	0.18
F14	1.5	0.31	1	0.8	1	0.67	2	0.25	0.63	0.36	0.44	0.43	0.3	0.5	0.3	0	1	1	12.49	0.18
F15	2	0.31	1	0.4	1.25	0.83	5	0.33	0.63	0.18	0.44	0.71	0.5	0.6	0.3	1	2	2	19.49	0.27
F16	1	0.31	1.33	0.8	1	0.5	4	0.25	0.38	0.18	0.44	0.43	0.3	0.5	0.3	1	2	2	16.72	0.24
F17	1.5	0.23	1	0.8	1	0.83	3	0.25	0.5	0.27	0.56	0.43	0.5	0.6	0.4	0	1	1	13.87	0.2
F18	2	0.38	1.67	1	1.25	0.5	5	0.33	0.63	0.27	0.44	0.71	0.4	0.5	0.4	0	1	1	17.49	0.25
F19	1.5	0.38	1	0.6	1	0.83	4	0.42	0.63	0.45	0.56	0.57	0.4	0.6	0.4	1	2	2	18.34	0.26
F20	1.5	0.31	1	0.8	1	0.5	4	0.25	0.5	0.45	0.44	0.71	0.5	0.5	0.3	1	1	2	16.77	0.24

(continued)

Table 6 (continued)

	0.5	0.08	0.33	0.2	0.25	0.17	1	0.08	0.13	0.09	0.11	0.14	0.1							
	1	2	3	4	5	6	7	8	9	10	11	12	13	G	Md	Ml	PhD	Exp	Tot	FT
F21	1	0.15	0.67	0.6	0.75	0.5	3	0.08	0.13	0.18	0.33	0.43	0.3	0.5	0.3	1	2	2	13.92	0.2
F22	2	0.31	1.33	0.8	1	0.67	3	0.25	0.25	0.27	0.33	0.43	0.3	0.6	0.4	1	1	2	15.94	0.22
F23	2	0.38	1.67	1	1.25	0.83	4	0.33	0.5	0.45	0.56	0.71	0.5	0.6	0.4	0	1	2	18.19	0.26
F24	1.5	0.31	1	0.6	0.75	0.5	2	0.33	0.38	0.27	0.33	0.57	0.4	0.5	0.3	1	2	2	14.74	0.21
F25	1.5	0.23	1	0.8	1	0.5	3	0.33	0.38	0.18	0.22	0.43	0.3	0.5	0.3	1	1	2	14.67	0.21
F26	2	0.23	1.33	0.6	0.75	0.67	4	0.25	0.5	0.36	0.44	0.43	0.3	0.6	0.3	0	1	1	14.77	0.21
F27	1.5	0.31	1	0.8	1	0.67	4	0.33	0.5	0.27	0.44	0.43	0.3	0.5	0.3	1	1	2	16.35	0.23
F28	1.5	0.31	1	1	1	0.83	3	0.33	0.38	0.27	0.56	0.57	0.5	0.6	0.3	1	1	2	16.15	0.23
F29	2	0.38	1.67	1	1	0.83	4	0.42	0.5	0.36	0.56	0.71	0.5	0.6	0.4	1	2	2	19.93	0.28
F30	1	0.23	1.33	0.8	0.75	0.5	3	0.25	0.25	0.18	0.22	0.43	0.3	0.6	0.3	1	2	2	15.15	0.21
F31	2	0.31	1	0.6	0.75	0.67	4	0.25	0.5	0.36	0.33	0.43	0.3	0.5	0.3	1	2	2	17.3	0.24
F32	1.5	0.23	1.33	0.8	0.75	0.5	3	0.25	0.5	0.27	0.44	0.57	0.3	0.5	0.3	1	1	2	15.25	0.21
F33	1.5	0.23	1	0.6	0.75	0.67	4	0.33	0.38	0.27	0.33	0.43	0.3	0.5	0.4	0	1	2	14.69	0.21
F34	2.5	0.15	0.33	0.6	1	0.33	1	0.42	0.25	0.36	0.33	0.71	0.5	0.6	0.4	0	1	1	11.5	0.16
F35	1.5	0.23	1	1	1	0.67	3	0.33	0.63	0.27	0.56	0.43	0.5	0.6	0.3	1	2	2	17.01	0.24
F36	1.5	0.08	1.33	0.8	1	0.33	3	0.42	0.38	0.09	0.22	0.71	0.4	0.5	0.3	1	2	2	16.06	0.23
F37	1.5	0.31	1.33	0.6	1	0.5	4	0.33	0.38	0.36	0.44	0.43	0.4	0.5	0.3	1	2	2	17.39	0.24
F38	2	0.23	1.33	0.6	1	0.67	4	0.25	0.38	0.27	0.44	0.43	0.3	0.5	0.4	1	1	2	16.8	0.24
F39	1.5	0.23	1	0.8	0.75	0.83	3	0.33	0.5	0.27	0.56	0.43	0.5	0.6	0.4	0	0	1	12.7	0.18
F40	2	0.23	1	0.6	0.75	0.67	2	0.25	0.5	0.18	0.33	0.57	0.4	0.6	0.4	0	0	1	11.48	0.16

Legend: *F* faculty, *G* gender, *Md* married, *Ml* MPhil, *Exp* experience, *Tot* total, *FT* fuzzy total

Table 7 Fuzzy ranking and grouping

Faculty	Total score	Fuzzy total	Fuzzy rank	Group
F1	24.44111167	0.34	1	1
F11	20.58955766	0.29	2	2
F29	19.93475968	0.28	3	3
F15	19.48990731	0.27	4	4
F19	18.34114497	0.26	5	5
F23	18.19233544	0.26	6	6
F18	17.49107282	0.25	7	7
F2	17.48990731	0.25	8	8
F37	17.38601121	0.24	9	8
F31	17.2999001	0.24	10	7
F6	17.05889388	0.24	11	6
F35	17.01262349	0.24	12	5
F3	16.98755134	0.24	13	4
F38	16.80151238	0.24	14	3
F20	16.77096792	0.24	15	2
F16	16.7208597	0.24	16	1
F7	16.7033189	0.24	17	8
F5	16.55866633	0.23	18	7
F10	16.5471473	0.23	19	6
F27	16.35343545	0.23	20	5
F4	16.33679376	0.23	21	4
F9	16.2708597	0.23	22	3
F28	16.14907037	0.23	23	2
F36	16.06267344	0.23	24	1
F22	15.94232434	0.22	25	1
F32	15.25270285	0.21	26	2
F30	15.1467144	0.21	27	3
F26	14.76742147	0.21	28	4
F24	14.74351482	0.21	29	5
F33	14.69040127	0.21	30	6
F25	14.6717144	0.21	31	7
F21	13.9225691	0.2	32	8
F17	13.87095682	0.2	33	8
F12	13.13004773	0.18	34	7
F8	13.05024975	0.18	35	6
F39	12.70429015	0.18	36	5
F13	12.59232434	0.18	37	4
F14	12.48601121	0.18	38	3
F34	11.4984349	0.16	39	2
F40	11.48401598	0.16	40	1

Table 8 Cumulative skill value of the teams

Team	Cumulative skill value
1	1.192267
2	1.130433
3	1.135773
4	1.129211
5	1.114859
6	1.120268
7	1.114808
8	1.117926

The cumulative fuzzy value of each faculty member is available in the last column of Table 7. The faculty members are ranked according to the cumulative fuzzy values. Rank is computed based on the descending order of the cumulative fuzzy values of the faculty members.

Eight teams are required. Hence, we label the faculty members based on their rank as $1, 2, \ldots, 7, 8, 8, 7, \ldots, 2, 1$. Since there are forty faculty members, 16 of the faculty members with the top ranks and 16 of the faculty members with the bottom ranks are labelled as $1, 2, \ldots, 7, 8, 8, 7, \ldots, 2, 1$. The middle-ranked 8 faculty members are labelled $8, 7, \ldots, 2, 1$. We group the faculty members with the same label. This data is available in Table 8.

The cumulative value for each group is between 1.11 and 1.19.

The conclusion of the results is discussed in the next section.

7 Conclusion

The teams formed using this fuzzy MCDM clustering model are with varied skills sets. As the cumulative value of each team is almost equal, the teams are with relatively equal strength. Hence, there would be healthy competition amongst the teams. Research and publication could be animated to the level of group games. This would improve productivity. The result of the experimentation described in this paper proves it. As each team has faculty members with various skills required for research and publication, every team is competent and actively involved in the research. Research and publication are stalled in many situations due to individuals' insufficient skills. Further, even in a group, the effect is significantly less if the skills are not distributed complementarily. This model helps to group the researchers based upon multiple research skills identified based on self-evaluation and the historical or contextualised ranking.

References

1. Ahmed, F., Deb, K., Jindal, A.: Multi-objective optimisation and decision making approaches to cricket team selection. Appl. Soft Comput. **13**(1), 402–414 (2013)
2. Al-Maamari, F.S.: Re-conceptualizing research misconceptions: top ten myths demystified. Inquiry Educ **8**(2), 6 (2016)
3. Bhattacharjee, D., Saikia, H.: An objective approach of balanced cricket team selection using binary integer programming method. Opsearch **53**(2), 225–247 (2016)
4. Budak, G., Kara, İ., İc, Y.T., Kasımbeyli, R.: New mathematical models for team formation of sports clubs before the match. Central Eur. J. Oper. Rese. **27**(1), 93–109 (2019)
5. Carlucci, D., Renna, P., Izzo, C., Schiuma, G.: Assessing teaching performance in higher education: a framework for continuous improvement. Manage. Decis. **57**(2), 461–479 (2019)
6. Gumus, S., Bellibas, M.S., Esen, M., Gumus, E.: A systematic review of studies on leadership models in educational research from 1980 to 2014. Educ. Manage. Admin. Leadersh. **46**(1), 25–48 (2018)
7. Hu, Z., Tian, W., Guo, J., Wang, X.: Mapping research collaborations in different countries and regions: 1980–2019. Scientometrics **124**(1), 729–745 (2020)
8. Hwang, K.: Sine qua non-conditions for research. J. Craniofacial Surg. **28**(3), 594–595 (2017)
9. Jung, J., Horta, H., Zhang, L.F., Postiglione, G.A.: Factors fostering and hindering research collaboration with doctoral students among academics in Hong Kong. Higher Education, 1–22 (2021)
10. Kalayathankal, S.J., Kureethara, J.V., Narayanamoorthy, S.: A modified fuzzy approach to project team selection. Soft Comput. Lett. **3**, 100012 (2021)
11. Katz, J.S., Martin, B.R.: What is research collaboration? Res. Policy **26**(1), 1–18 (1997)
12. Lee, S., Bozeman, B.: The impact of research collaboration on scientific productivity. Soc. Stud. Sci. **35**(5), 673–702 (2005)
13. Mahasneh, O.M.: The effectiveness of flipped learning strategy in the development of scientific research skills in procedural research course among higher education diploma students. Res. Learn. Technol. **28** (2327) (2020)
14. Omari, A.: Pacififa: a new measure for evaluating authors productivity and activity over time. Int. J. Adv. Inf. Technol. **3**(3), 11 (2013)
15. Ozceylan, E.: A mathematical model using AHP priorities for soccer player selection: a case study. South Afr. J. Ind. Eng. **27**(2), 190–205 (2016)
16. Reis-Jorge, J.M.: Developing teachers' knowledge and skills as researchers: a conceptual framework. Asia-Pacific J. Teacher Educ. **33**(3), 303–319 (2005)
17. Saikia, H., Bhattacharjee, D., Radhakrishnan, U.K.: A new model for player selection in cricket. Int. J. Performance Anal. Sport **16**(1), 373–388 (2016)
18. Toledo-Pereyra, L.H.: Ten qualities of a good researcher. J. Investigative Surg. **25**(4), 201–202 (2012)
19. Troussas, C., Krouska, A., Sgouropoulou, C.: Collaboration and fuzzy-modeled personalisation for mobile game-based learning in higher education. Comput. Educ. **144**, 103698 (2020)

Freight Village Efficiency Criteria Evaluation via Fuzzy Multi-criteria Decision-Making Methods

Mehmet Pekkaya, Nuh Keleş, and Fatma Sönmez Çakır

Abstract The logistics industry is significantly important in international trade; European logistics market is worth over 1.1 trillion euros. There are 240 freight villages established in Europe, and the logistics sector ranked third. The value of freight village is increasing in the economic system and the logistics industry. Presenting information by measuring the efficiency of freight village is important to help companies investing in this area. It is deemed necessary to determine the level of efficient operation of the freight village, to increase their performance, and to analyze the efficient use of their resources. This study aims to determine the interactions between criteria and the criteria priorities for the freight village efficiency evaluation. After selecting the significant criteria from related literature, pairwise comparisons as data are obtained from experts' views. The interactions between criteria via fuzzy DEMATEL method and priorities via fuzzy AHP method are determined. According to findings, "annual amount of freight transport" is determined as the general affecting and overwhelmingly the most important criterion; then, it is recommended to decision makers/researchers to consider that criterion with a high weight in the freight village efficiency evaluation. On the other hand, "number of employees" is observed as both the highly impressed and least important criterion; then, it may be ignored in freight village assessments or represented with a significantly too small weight, with its controversial position in the output or input. One of the originalities of this study is that considering the DEMATEL-AHP results together at the evaluation stage of each criterion; it provides an opportunity to present unique perspectives on the process of analyzing the criteria.

M. Pekkaya (✉)
Business Administration, Zonguldak Bülent Ecevit University, Zonguldak, Turkey
e-mail: mehmet.pekkaya@beun.edu.tr

N. Keleş
Kayseri Customs Directorate, Trade Ministry TR, Kayseri, Turkey

F. S. Çakır
Management Information Systems, Bartın University, Agdacı, Bartın, Turkey
e-mail: fsonmez@bartin.edu.tr

L. Sahoo et al. (eds.), *Real Life Applications of Multiple Criteria Decision Making Techniques in Fuzzy Domain*, Studies in Fuzziness and Soft Computing 420,
https://doi.org/10.1007/978-981-19-4929-6_30

Keywords Freight village efficiency criteria · Fuzzy MCDM · F-DEMATEL · F-AHP

JEL Classification C02 · L91 · M21 · P25

1 Introduction

The logistics industry is one of the drivers of international trade worth over 5.5 trillion euros, and the volume of the European logistics market is worth over 1.1 trillion euros in 2018. The logistics industry, whose development has come a long way in the last 50 years, has shown the importance of logistics centers/freight villages, with a more holistic approach in the twenty-first century. Freight village first appeared in America. Germany, France, Spain and Italy represent 47.9% of the total shipping village and 62.1% of their land area in Europe. Almost, 2400 transport companies operate in about 240 freight villages established in Europe, and the logistics sector ranked third [36: 2]. The value of freight village is increasing in the economic system and the logistics industry. Freight village acts as an economic driver and business generator that helps to improving supply chain efficiency and economic growth in the region. Freight village provides various benefits to reduce unemployment, increase employment, regional investment, per capita income, develop trade, transportation and storage sectors, and related sectors.

Growing markets emphasize the significance of logistics, the effectiveness, and efficiency of which rely on an extensive/complex array of interdependent supply, transportation, storage, inventory, material handling, and packaging, etc., topics [3: 192]. Recently, logistics has been increasingly explored, as the prominence of logistics for businesses/organizations, and national economies are increasingly recognized [36: 1]. During the last 5 decades, a more variety of freight and logistics processes have been expanded to the concept of European freight villages thanks to the borderless market of Europe and the growth of international trade, especially from Asia [41: 1, 35].

International supply and distribution of goods cause changes in logistic developments that produce major problems for logistics firms. If a company wants to overcome complexity and ensure that logistics services are adjusted flexibly/efficiently according to changing demands, it must have competitive activities and a logistics infrastructure that can provide operational efficiency within a network [43: 271]. One of the prominent components of the logistics sector is freight village. The freight village importance where logistics activities are carried out as a whole is increasing day by day. Freight village is used to promote regional economic activities improve local distribution and develop land use, which contributes to combined transportation by combining a number of auxiliary distribution activities that as [37: 87–93] and [23: 9–17] stated. Freight villages are logistic interconnections within the network that mission as an interface between local and long-distance goods transportation, freight villages may cause rationalization of storage, transportation, and logistics services, provide cost savings along the whole logistic chain [43: 272]. Freight village is a center where companies participate in logistics and transportation-related activities in a broad sense [31: 106]. Freight village can also be expressed as a center in the described area where activities related to transportation and goods distribution in (inter)national crossings are conducted commercially via different operators [13: 3, 24: 50]. We can refer to freight village as specialty-operated places where logistics activities are conducted in a unique area for the national or international transport or shipment of goods. For all that combining logistics activities in freight village has been an important way of sustainable development, growth, and strategic competitive advantage. Freight villages are used to promote regional economic activities by combining a number of auxiliary distribution activities, improving local distribution, and developing land use that contributes to combined transport. In fact, firms that want to be ahead of their competitors today shift their logistics activities to freight village where logistics services are performed together, where logistics activities are conducted more holistically and in a way that will benefit the parties.

Freight village authorities are increasingly under pressure to increase freight village productivity by ensuring that logistics services are provided on the basis of international competition, to support trade-oriented economic development [40: 107]. Evaluating the efficiency of freight village has significant results for logistics companies as well as governments [44: 91]. Firms must organize their activities efficiently in order to be able to operate in the market and be profitable. Actually, efficiency is a very prominent indicator of logistics firms' operations analysis, and it is frequently and basic used performances [2: 3926]. Unfortunately, not many academic researchers have been focused on logistics efficiency/performance analysis. Most of the focused ones analyzed their efficiency via data envelopment analysis (DEA). However, concentrating on analyzing the criteria priority and criteria interactions that are suitable for use in the efficiency evaluation of freight villages have not been considered in related academic literature. This research anticipate to contribute to this prominent assignment by performing a supportive approach to logistics efficiency criteria. With the research gaps in these subjects, being the motivation of this study, and it is thought that determining the priorities and interactions of the criteria to be used in efficiency measurement of freight village will benefit the companies in practice and those working in the academy. Moreover, these can be also considered as highlights of the study.

Accordingly, this study aims to determine the criteria interactions and criteria priorities are taken into account in the freight village efficiency evaluation and to introduce the findings to the researchers and decision makers. Significant criteria are selected for the study from the literature on logistics performance and efficiency evaluations, and expert views are obtained by making one-way and two-way pairwise comparisons. The pairwise comparisons are analyzed by using multi-criteria decision-making (MCDM) methods/analysis, namely DEMATEL, AHP, since there are many criteria that exist in freight village efficiency evaluation. By using one-way pairwise comparisons, the interaction between criteria is determined, and the interaction diagram is sketched via the fuzzy DEMATEL (F-DEMATEL) method. By using two-way pairwise comparisons, the criteria priorities are determined via the fuzzy AHP (F-AHP) method. Moreover, criteria taken into account in the freight village efficiency calculations are evaluated by taking into account criteria interactions, the interaction diagram, and their priorities.

The organization of the study can be summarized as follows: Sect. 2 introduces the literature with logistics and efficiency used in the present study. Section 3 describes the data, F-AHP, and F-DEMATEL methods. Section 4 presents the results of the empirical analysis, with the inclusion of efficiency criteria from the F-AHP and F-DEMATEL methods. Section 5 provides relevant conclusions from the study, along with recommendations for future research.

2 Literature Review About Freight Village Efficiency

Some potential performance indicators such as efficiency and effectiveness can be found in academic studies. Efficiency and effectiveness can be measured in a variety of ways. [45: 1006]. Efficiency and effectiveness are important terms used in evaluating and measuring the performance of organizations/firms. Efficiency can be defined as "a weighted sum of outputs divided by a weighted sum of its inputs" [12: 567]. Efficiency is accepted as the measure of how effectively the resources/inputs are used in the activities of the enterprises for the production of goods/services.

Efficiency, from a different perspective, is stated as a necessary condition the firm's managing boundary reflect and represent the term effectiveness as the firm's capability to create sustainable business growth in earnings in its surroundings. Effectiveness can be expressed using the discounted current value of all future earnings. Efficiency is a measure of operational merit or productivity [32: 1124–1126]. It is not only important but also a critical factor for firms to improve the efficiency of the freight village.

Available studies in literature have not been observed to provide a general solution to the problem of determining criteria priority and interactions for logistics efficiency. Most of the studies conducted in the past focus on efficiency analysis for freight village, distribution centers, ports, etc. Tongzon [40] analyzed the efficiency measurement of selected Australian and other international ports via DEA. Kapros et al. [25] worked on the evaluation of intermodal freight village via multi-criteria analysis component. Cullinane et al. [11] compared the container ports' technical efficiency via DEA and stochastic frontier analysis. Liu [28] evaluated the operational efficiency of major ports in the Asia–Pacific region via DEA. Yu and erwin [45] measure efficiency/effectiveness in railway performance with the DEA model. Carvalho et al. [4] examined logistics platforms existing in Europe and USA based on performance analyzes via DEA. Cruijssen et al. [10] describe an application of freight transportation efficiency through DEA. Chakraborty et al. [5] evaluated distribution centers using the AHP and DEA. Markovits-Somogyi et al. [30] evaluated the efficiency of Hungarian logistics centers using DEA. Andrejic et al. [2] analyzed Serbia distribution centers with principal component analysis and DEA. Vrochidis [41] determined regional and economic indicators for freight villages. Yang et al. [44] analyzed the efficiency analysis of European logistics villages with DEA. Bajec et al.

[3] examined Slovenia as a logistics platform using Delphi-AHP and DEA. Indeed, the use of DEA in efficiency analysis allows performing useful analysis. However, it is a gap in the literature that the priorities and interactions of the criteria are not determined when performing an efficiency analysis. In the studies conducted for the criteria to be used in the efficiency analysis, no study using the MCDM methods F-AHP and F-DEMATEL has been observed in the related literature. In fact, there is no study observed concentrating on determining the criteria and their priorities in efficiency of freight village and likewise topics.

In this study, we examine the criteria used for logistics/freight village efficiency measurement. By collecting and comparing various data on how freight village can operate better, efficiency analysis can be done to provide ideas to companies, stakeholders, government units, researchers, and decision makers operating in the logistics sector. For this, the criteria used in the literature can be examined. The criteria selected using the relevant literature are presented in Table 1.

If the criteria used in the literature are examined in Table 1, it can be said that many different criteria are used in efficiency evaluation studies such as freight villages/distribution centers/ports, etc., but there is no common consensus in the criteria/studies. In fact, it can be stated that there are not many studies observed on this subject. For this reason, efficiency studies of similar places such as distribution centers and ports were also examined. In general, it can be said that the existence and characteristics of the area, income, employees, and companies come to the fore in the studies. Considering that it is more inclusive, rational, and beneficial in practice compared to other studies, we found it appropriate to use the criteria defined (but not all of) in the study of [44]. These criteria are arranged and presented: total area, amount of investment/capital, storage and stock area, annual amount of freight transport, annual revenue, number of employees, and number of companies settled.

Evaluation of the priorities and interactions of the criteria to be used in the efficiency analysis of freight village is based on MCDM. A wide variety of MCDM methods has found application in many areas in the literature. In this study, F-AHP and F-DEMATEL methods are used to determine the priorities and interactions of the criteria in the evaluation of freight village.

3 Material and Method

The analysis aims to determine the criteria interaction matrix via F-DEMATEL and construct an interaction diagram for criteria and criteria priorities via F-AHP that are taken into account in the freight village efficiency evaluation by using Turkish experts' views. Evaluation of the priorities and interactions of the criteria to be used in the efficiency analysis of freight village is based on multi-criteria decision methods/analysis (MCDM). Many different MCDM methods have been found application in many areas in the literature. In this study, F-AHP and F-DEMATEL methods, which were not used to determine the priorities and interactions of the criteria determined in the evaluation of freight village activity, were used.

Table 1 Criteria used in efficiency analysis

Study	Criteria take into consideration
Tongzon [40]	TEUs handled, shipcalls, shiprate, crane prod, no. of cranes, no. of container berths, no. of tugs, terminal area, delay time, labor
Kapros et al. [25]	Attractiveness for private financing, contribution to national economy, contribution to local economy, environmental quality, contribution to land use reorganization, complementarity with other policy plans
Cullinane et al. [11]	Container throughput, terminal length, terminal area, quayside gantry, yard gantry, straddle carrier
Liu [28]	Labor, funding, other production, actual volume of output, customer response
Carvalho et al. [4]	Platform area, invested capital, enterprise number, annual load handling
[10]	Total assets, hours worked, added value, profit/loss for the year before taxes
[5]	Fill rate, sales, service time, number of employees, general expenses, space, inventory
Somogyi et al. [30]	Surface size of offices, number of employees, surface of available storage space, volume of total sales revenue, the tons of freight
Andrejic et al. [2]	Equipment and capacity indicators, energy, operational, utilization, quality, with 33 sub-criterion)
Vrochidis [41]	Number of employees, on-site transport and logistics firms, available transportation modes, unemployment rate, GDP per capita, business investment rate, number of patents, annual road, rail, air and maritime transport of freight, number of persons employed, number of local units, number of trucks, wages, and salaries
Yang et al. [44]	Total area, intermodal area, warehouse area, number of jobs, annual load handling, number of companies settled
Bajec et al. [3]	Railway infrastructure, logistics distribution centers, modern IT, logistics staff, modern suprastructure, administrative support, annual goods transshipment, total gross domestic product (GDP), annual revenue from the logistics industry

The usage of the fuzzy approach for MCDM in this study can be summarized as follows. Decisions made by experts or decision makers may contain some uncertainty and ambiguity due to their subjective nature. It can be said that the reason for this is the absence of precise numerical data in some cases and the presence of intermediate values in the decision [1]. When it comes to decisions that are expressed in linguistic terms, that is, they do not have numerical equivalents, it is necessary to develop a different perspective for them. Fuzzy logic, which was introduced by [46], started to be used. Linguistic terms are converted to fuzzy numbers. Since our main data imply experts' views and their pair comparisons can be thought of as linguistic expressions, we decided to use the fuzzy approach in the calculations.

Accordingly, after significant criteria selection from the related literature on logistics performance and efficiency evaluations, and expert views are obtained by making one-way and two-way pairwise comparisons. The one-way pairwise comparisons let to determine the interaction between criteria via the F-DEMATEL method and to sketch interaction diagrams. The two-way pairwise comparisons let to determine the criteria priorities via the F-AHP method. In the following subparts, firstly, experts' characteristics and criteria/variables used in the analyses mentioned. Then, the F-DEMATEL and F-AHP methods' calculation steps and their preference reasons are explained.

3.1 Data

The names and their definitions of selected seven criteria used in the analyses are listed at Table 2. These criteria are selected according to their usage in related literature.

Expert views on pair comparisons of these criteria are the main data of the analyses. One-way pair comparisons consist of 5-point scaled 42 items from these seven criteria are prepared for F-DEMATEL calculations, and two-way pair comparisons consist of 9-point scaled 21 items are prepared for F-AHP calculations, along with six items for characteristics of experts, namely gender, experience, age etc.

Table 2 Criteria used in efficiency analysis of academic studies

Criteria; Abbreviation	Definition	Researcher (s)
Total area; **C1-Area**	Necessary and sufficient area for the freight village to efficient operate	[4, 5, 11, 30, 40, 44]
Amount of investment/capital; **C2-Investment**	The total amount of investments to be made before and after the freight village starts their operations	[4, 10, 41, 44]
Storage and stock area; **C3-Storage**	Stocking and storage area large enough to meet the needs of companies that receive services in the freight village or from outside	[2, 30, 44]
Annual amount of freight transport; **C4-Transport**	The annual amount of freight transport from the freight village in all modes	[2–4, 30, 41, 44]
Annual revenue; **C5-Revenue**	The amount of revenue obtained from the activities and fixed earnings in the freight village	[3, 25, 30, 41]
Number of employees; **C6-Employees**	Number of employees for the activities of the freight village	[2, 3, 5, 28, 30, 40, 41, 44]
Number of companies settled; **C7-Companies**	Number of companies operating in the freight village	[4, 41, 44]

The number of experts whose views were taken in the analysis is six. The experience of experts is at least 10 years to at most 45 years, 25.8 years at the mean. All the experts are male, and they worked in the logistics industry of Turkey in their working life. The experts' views on one/two-way pairwise criteria comparisons which are obtained in November/December 2021 in Turkey are evaluated as data for the analysis of the study.

3.2 *Fuzzy DEMATEL Method*

The DEMATEL method was initially used by The Battelle Memorial Institute through its Geneva Research Center in 1973 [15]. DEMATEL method is a MCDM methods used to solve complex and unsolvable problems. With this method, the cause-effect relationships of all the criteria that are the subject of the analysis can be determined and analyzed more easily. Presenting the results graphically provides convenience and comfort in terms of seeing the interactions. The DEMATEL method is based on graphs that separate the relevant factors as cause group and effect group. Directional graphs, known as digraphs, are more useful than undirected graphs. This is because digraphs show directed relationships [8]. With the DEMATEL technique, a researcher may determine the significant criteria affecting other criteria.

F-DEMATEL method has been used by so many analyses in a variety of academic studies. Vehicle technologies [33], investment assessment [21], energy [17], supplier selection [19], service innovation [14], supply chain management [27], etc., can be observed as application.

The F-DEMATEL method applications are performed in the following steps [8, 33].

Step 1. Paired comparison matrices are created.

The effect of the element in each row applied to the element in each column of this matrix may be represented as a fuzzy number. In case of more than one expert opinion is used, all experts views should be taken for constructing matrixes. The arithmetic mean of all experts' view scores is used to construct the direct relationship matrix 'z'. The linguistic terms [8, 33] are given in Table 3.

Step 2. Normalization of the fuzzy direct-relation matrix

Table 3 Triangular fuzzy numbers scale for DEMATEL calculations

Value	Definition	Linguistic terms scale
1	No influence	(0, 0, 0.25)
2	Very low influence	(0, 0.25, 0.50)
3	Low influence	(0.25, 0.50, 0.75)
4	High influence	(0.50, 0.75, 1)
5	Very high influence	(0.75, 1, 1)

$$\tilde{x}_{ij} = \frac{\tilde{z}_{ij}}{r} = \left(\frac{l_{ij}}{r}, \frac{m_{ij}}{r}, \frac{u_{ij}}{r}\right) \quad (1)$$

where

$$r = \max_{i,j}\left\{\max_{i}\sum_{j=1}^{n} u_{ij}, \max_{j}\sum_{i=1}^{n} u_{ij}\right\} \quad i, j \in \{1, 2, 3, \ldots, n\} \quad (2)$$

Step 3. Calculation of the fuzzy total-relation matrix.

Equation (3) lets to calculate the fuzzy total-relation matrix.

$$\tilde{T} = \lim_{k\to\infty}(\tilde{x}^1 \oplus \tilde{x}^2 \oplus \tilde{x}^3 \oplus \ldots \oplus \tilde{x}^k) \quad (3)$$

When each cell of the fuzzy total-relation matrix is expressed as $\tilde{t}_{ij} = (l''_{ij}, m''_{ij}, u''_{ij})$, it can be calculated as follows Eq. (4):

$$\begin{aligned} [l''_{ij}] &= x_l \times (I - x_l)^{-1} \\ [m''_{ij}] &= x_m \times (I - x_m)^{-1} \\ [u''_{ij}] &= x_u \times (I - x_u)^{-1} \end{aligned} \quad (4)$$

Step 4. After obtaining the total-relationship matrix, $Di + Ri$ and $Di - Ri$ values should be calculated. Di values show row totals, while Ri values show column totals. It is calculated with the following equations.

$$D_i = \sum_{1\le j\le n}^{n} t_{ij} \quad (5)$$

$$R_i = \sum_{1\le i\le n}^{n} t_{ij} \quad (6)$$

Step 5. Defuzzified process is performed. Equation (7) is used for this calculation.

$$(D_i + R_i)_{\text{def}} = \frac{l + 2m + u}{4} \quad (7)$$

$$(D_i - R_i)_{\text{def}} = \frac{l + 2m + u}{4} \quad (8)$$

Step 6. Drawing the cause-effect relationship diagram.

The values obtained from the defuzzified are analyzed by drawing a cause-effect relationship diagram. $Di + Ri$ values are on the horizontal axis, and $Di - Ri$ values are on the vertical axis of the cause-effect relationship diagram.

Step 7. DEMATEL method can sometimes be preferred for criteria priority weight calculation. However, in the DEMATEL criteria weight calculation, the total effects approach or commonly used vector sum approach of the total-net effects can be used. The vectorial sum approach calculation for criteria weights (*wi*) of i = j can be summarized as follows [34: 298].

$$y_i = \sqrt{(D_i + R_i)^2 + (D_i - R_i)^2} \quad w_i = \frac{y_i}{\sum y_i} \tag{9}$$

Due to its advantages/capabilities, DEMATEL is highly preferred in the literature for solving different complex system problems in various fields. The DEMATEL technique produces interaction tables so that causal relationships, effects, net effects, and total interactions can be seen [39]. Moreover, DEMATEL technique was used to determine the structural model for the main criteria for the freight village efficiency criteria evaluation problem. When the related literature was examined, no other study was observed in evaluating the freight village efficiency criteria with the F-DEMATEL method. In this way, since there is no variety of methods in MCDM literature for evaluating the criteria interactions, DEMATEL method is decided to use, which is the only widely known method in the literature. In addition, since linguistic expressions are used for criteria, the use of F-DEMATEL was deemed more appropriate as a method.

3.3 Fuzzy Analytic Hierarchy Process Method

The significant differentiation of fuzzy logic from other logic systems is that verbal variables can be used. Verbal variables let the notion that may not be expressed fairly to be described. Chang [7] presented a new approach for the pairwise comparison scale of F-AHP in which he suggested using triangular fuzzy numbers and making pairwise comparisons with these numbers and introduced F-AHP. Actually, F-AHP method is a modified version of the AHP. Although the purpose of the AHP is to determine the criteria priority weights by using the knowledge of the experts, the traditional AHP may fail to reflect the human thinking style. Therefore, F-AHP, which is a fuzzy extension of AHP, was developed. F-AHP can produces more accurate results for the decision-making [18].

F-AHP method: risk assessment [16], supplier selection [6], financial performance [38] evaluation personnel selection [22], risk analysis [29], resources selection [42], etc., has found application in many fields. F-AHP uses fuzzy numbers or linguistic expressions (equally important, absolute important, etc.) for calculations [39].

The F-AHP method applications are performed in the following steps [39].

Step 1. Pair comparison matrices are created.

Table 4 Fuzzy triangle scale

Value	Definition	Fuzzy triangle scale
1	Equal importance	(1, 1, 1)
3	Moderate importance	(2, 3, 4)
5	Strong importance	(4, 5, 6)
7	Very strong importance	(6, 7, 8)
9	Extreme importance	(9, 9, 9)
2;4;6;8	Intermediate values	(1, 2, 3); (3, 4, 5); (5, 6, 7); (8, 8, 9)

$$\tilde{A}^k = \begin{bmatrix} \tilde{d}_{11}^k & \tilde{d}_{12}^k & \cdots & \tilde{d}_{1n}^k \\ \tilde{d}_{21}^k & \tilde{d}_{22}^k & \cdots & \tilde{d}_{2n}^k \\ \vdots & \vdots & \ddots & \vdots \\ \tilde{d}_{n1}^k & \tilde{d}_{n2}^k & \cdots & \tilde{d}_{nn}^k \end{bmatrix} \tag{10}$$

A pairwise comparison matrix is created by making pairwise comparisons of the criteria. While creating this matrix, fuzzy equivalents of linguistic expressions are used. The membership function can be expressed as follows "with the fuzzy number $\breve{N}$ and the triangular fuzzy number TFN" [9]:

$$\mu_{\tilde{N}}(x) = \begin{cases} (x-1)/(m-1), & l \le x \le m \\ (u-x)/(u-m, & m \le x \le u \\ 0, & \text{others} \end{cases} \tag{11}$$

In the function, 'l' indicates the lower, 'm' mean, and 'u' the upper value. Mathematical operations (multiplication, division, addition etc.) may be conducted between two fuzzy numbers. The triangular fuzzy number equivalents of the comparison expressions between 1 and 9 used in AHP are given in Table 4.

Step 2. The geometric mean approach is preferred to obtain the fuzzy geometric mean and each criterion priorities. Equation (1) is used for this calculation.

$$\tilde{r}_i = \left(\prod_{j=1}^{n} \tilde{d}_{ij}\right)^{1/n} \quad i = 1, 2, \ldots, n \tag{12}$$

Step 3. The fuzzy criteria weights are defined. Equation (10) is used for this calculation.

$$\tilde{w}_i = \tilde{r}_i \otimes (\tilde{r}_i \oplus \cdots \oplus \tilde{r}_n)^{-1} \tag{13}$$

Step 4. The criteria average (M_i) and normalized (N_i) weight can be calculated with using Eqs. (11) and (12).

$$M_i = (\tilde{w}_1 \oplus \cdots \oplus \tilde{w}_n)/n \tag{14}$$

$$N_i = M_i/(M_1 \oplus \cdots \oplus M_n) \quad (15)$$

After these processes are done for each expert, the ranking is determined by taking the average of the ranking scores obtained.

AHP has advantages over other MCDM methods by calculating cross-consistency of pairwise comparisons. A significant advantage of AHP is the pairwise comparisons made to determine the hierarchical structures of the criteria and their relative importance to each other. The AHP method is also considered as converting pairwise comparisons judgments to priorities that is ratio-scaled weights for each criteria [35: 8137]. Moreover, the AHP method is the most preferred MCDM method in determining criteria priority weights. Accordingly, these features make the AHP method for us to prefer as a basic technique for priority calculations. The fact that it can be used in determining both the criteria priorities and the ordering of the criteria enabled the comparison of DEMATEL results and AHP results in our study.

4 Determining the Criteria Interaction, and Priorities for the Freight Village Efficiency

4.1 Fuzzy DEMATEL Results

The first step of the DEMATEL calculation is to reveal the interaction matrix values or criteria interactions. For this, the F-DEMATEL method was applied. Again, operations were performed according to the given steps, and the results are given in tables and interpreted. Pairwise comparison matrices were created and normalized according to the scores of each expert on the DEMATEL scale. The normalized expert tables are combined with the arithmetic mean, and the remaining operations were performed with the obtained matrix. The pairwise comparison matrix of the first expert is given in Table 5.

Table 5 Pairwise comparison matrix of the first expert

CRI	C1			C2			C3			…	C7		
	l	u	m	l	u	m	l	u	m	…	l	u	m
C1-Area	0.00	0.00	0.00	0.25	0.50	0.75	0.75	1.00	1.00	…	0.25	0.50	0.75
C2-Investment	0.50	0.75	1.00	0.00	0.00	0.00	0.75	1.00	1.00		0.00	0.25	0.50
C3-Storage	0.75	1.00	1.00	0.00	0.25	0.50	0.00	0.00	0.00		0.25	0.50	0.75
C4-Transport	0.00	0.00	0.25	0.00	0.25	0.50	0.50	0.75	1.00		0.75	1.00	1.00
C5-Revenue	0.00	0.00	0.25	0.25	0.50	0.75	0.00	0.25	0.50		0.25	0.50	0.75
C6-Employees	0.00	0.00	0.25	0.00	0.00	0.25	0.00	0.00	0.25		0.25	0.50	0.75
C7-Companies	0.00	0.00	0.25	0.00	0.25	0.50	0.25	0.50	0.75		0.00	0.00	0.00

Table 6 Normalized pairwise comparison matrix of the first expert

CR	C1			C2			C3			…	C7		
	l	m	U	l	m	u	l	m	u	…	l	m	u
C1-Area	0.00	0.00	0.00	0.05	0.09	0.14	0.14	0.18	0.18	…	0.05	0.09	0.14
C2-Investment	0.09	0.14	0.18	0.00	0.00	0.00	0.14	0.18	0.18		0.00	0.05	0.09
C3-Storage	0.14	0.18	0.18	0.00	0.05	0.09	0.00	0.00	0.00		0.05	0.09	0.14
C4-Transport	0.00	0.00	0.05	0.00	0.05	0.09	0.09	0.14	0.18		0.14	0.18	0.18
C5-Revenue	0.00	0.00	0.05	0.05	0.09	0.14	0.00	0.05	0.09		0.05	0.09	0.14
C6-Employees	0.00	0.00	0.05	0.00	0.00	0.05	0.00	0.00	0.05		0.05	0.09	0.14
C7-Companies	0.00	0.00	0.05	0.00	0.05	0.09	0.05	0.09	0.14		0.00	0.00	0.00

Table 7 Normalized pairwise comparison matrix of the all expert

CR	C1			C2			C3			…	C7		
	l	m	u	l	m	u	l	m	u	…	l	m	u
C1-Area	0.00	0.00	0.00	0.10	0.14	0.17	0.12	0.16	0.18	…	0.04	0.08	0.12
C2-Investment	0.07	0.11	0.14	0.00	0.00	0.00	0.10	0.15	0.18		0.01	0.04	0.08
C3-Storage	0.09	0.14	0.16	0.06	0.10	0.14	0.00	0.00	0.00		0.04	0.09	0.13
C4-Transport	0.07	0.10	0.13	0.07	0.10	0.14	0.06	0.10	0.15		0.10	0.14	0.16
C5-Revenue	0.03	0.06	0.10	0.04	0.08	0.12	0.03	0.07	0.12		0.02	0.05	0.10
C6-Employees	0.01	0.03	0.07	0.00	0.02	0.07	0.00	0.02	0.07		0.02	0.05	0.09
C7-Companies	0.04	0.07	0.11	0.04	0.08	0.12	0.05	0.10	0.14		0.00	0.00	0.00

Accordingly, the normalized pairwise comparison matrix of the first expert created is presented in Table 6.

Pairwise comparison matrices obtained from all expert views were combined by calculating their mean and after normalization; it is reported in Table 7.

At this stage, the fuzzy total-relation matrix was calculated. Fuzzy total-relation matrix elements were divided into three different groups for '*l*', '*m*', and '*u*' values, subtraction from the unit matrix, inverting the matrix, and multiplying with the normalized decision matrix were done separately, and finally, the values obtained were combined. The combined matrix is given in Table 8.

After calculating the total-relation matrix, total-relation scores of $Di + Ri$ and net relation scores of $Di - Ri$ values for '*l*', '*m*', and '*u*' values were calculated and reported in Table 9. In order to draw the DEMATEL interaction graph, the data were defuzzified and reported in that table.

The defuzzified interaction matrix, total/net affecting/being affected values, and calculated priority weights via F-DEMATEL for each criterion are reported in Table 10. In the calculation of priority weights, the vector sum approach is used. The obtained criteria priorities via F-DEMATEL will be compared and evaluated with obtained criteria priorities via F-AHP. In this study, obtained criteria priorities via

Table 8 Combined matrix

CR	C1			C2			C3			…	C7		
	l	m	u	l	m	u	l	m	u	…	l	m	u
C1-Area	0.03	0.09	0.33	0.11	0.22	0.49	0.14	0.25	0.53	…	0.05	0.14	0.42
C2-Investment	0.09	0.20	0.47	0.02	0.10	0.36	0.12	0.24	0.54		0.02	0.12	0.41
C3-Storage	0.11	0.22	0.51	0.08	0.20	0.51	0.03	0.12	0.42		0.06	0.17	0.47
C4-Transport	0.10	0.21	0.52	0.09	0.21	0.54	0.09	0.23	0.58		0.11	0.23	0.53
C5-Revenue	0.04	0.13	0.40	0.05	0.14	0.43	0.04	0.15	0.45		0.03	0.11	0.38
C6-Employees	0.01	0.07	0.31	0.01	0.07	0.32	0.01	0.07	0.34		0.02	0.08	0.32
C7-Companies	0.07	0.17	0.47	0.06	0.18	0.49	0.08	0.20	0.54		0.02	0.09	0.35

Table 9 Total and net relation values

	l		*m*		*u*			Defuzzified values	
	Di + Ri	Di − Ri	Di + Ri	Di − Ri	Di + Ri	Di − Ri		Di + Ri	Di − Ri
C1-Area	0.853	−0.040	2.164	−0.018	6.038	0.009		2.805	−0.017
C2-Investment	0.858	0.013	2.256	0.033	6.317	0.042		2.922	0.030
C3-Storage	0.988	−0.041	2.502	−0.035	6.795	−0.012		3.197	−0.031
C4-Transport	1.016	0.295	2.527	0.419	6.857	0.612		3.232	0.437
C5-Revenue	0.719	−0.174	2.060	−0.267	6.086	−0.369		2.731	−0.269
C6-Employees	0.449	−0.230	1.596	−0.452	5.344	−0.808		2.246	−0.486
C7-Companies	0.814	0.178	2.197	0.320	6.279	0.526		2.872	0.336

F-AHP are evaluated more respectfully; criteria priorities via F-DEMATEL are only for comparisons since F-DEMATEL criteria weights are quite close to each other and represent criteria interactions intensity instead of criteria priority. This study result supports the study findings of [34, 39].

In determining the threshold value of the interaction matrix, the maximum diagonal value approach (0.172) and the mean value approach (0.204) of interaction matrix cells are taken into consideration. While the maximum diagonal value approach lets the interactions more visible in the interaction matrix and transfers them to the interaction diagram, in most cases, the diagram drawings of line/arrow may be evaluated as more complicated than mean value approach drawings. The interaction diagram with significant affecting/being affected values by using Table 10 is reported in Fig. 1. The affecting/being affected values below 0.172, which is a threshold value of the interaction matrix with the maximum diagonal value approach, are not represented in the diagram. Affecting values below 0.204 score which is the mean value approach threshold value are drawn in pale color without declaring magnitude on its curve. The other curves have affecting/being affected values more than that threshold value, at least 0.214, have bold lines; darker drawings are made as to the magnitude of the effect grew. Those ranked in the top six in affecting are the darkest plotted

Table 10 Total/net affecting/being affected values and interaction matrix

	C1	C2	C3	C4	C5	C6	C7	D	D + R	D-R	weight
C1-Area	0.136	**0.260**	**0.293**	*0.179*	0.172	0.165	*0.189*	1.394	2.805	-0.017	0.1394
C2-Investment	**0.239**	0.144	**0.288**	**0.208**	**0.237**	*0.194*	0.166	1.476	2.922	0.030	0.1452
C3-Storage	**0.265**	**0.247**	0.172	**0.247**	**0.233**	*0.201*	**0.217**	1.583	3.197	-0.031	0.1589
C4-Transport	**0.260**	**0.265**	**0.283**	0.169	**0.312**	**0.272**	**0.273**	1.834	3.232	0.437	0.1621
C5-Revenue	*0.176*	*0.191*	*0.200*	*0.186*	0.125	*0.194*	0.160	1.231	2.731	-0.269	0.1364
C6-Employees	0.118	0.115	0.124	0.149	0.162	0.087	0.126	0.880	2.246	-0.486	0.1142
C7-Companies	**0.217**	**0.224**	**0.254**	**0.260**	**0.260**	**0.252**	0.137	1.604	2.872	0.336	0.1437
R	1.411	1.446	1.614	1.398	1.500	1.366	1.268				

Note(s) Underlined values: threshold value of 0.172 for maximum diagonal value approach. Bold values: threshold value of 0.204 for mean value approach of interaction matrix cells

and have at least 0.265, at most 0.312 influencing/being affected values. The reason why these dark/light drawings are preferred is the interaction diagram drawings can be traced and evaluated more clearly.

Table 10 and Fig. 1 can be interpreted together since the diagram is a kind of visual map that summarizes the interactions and total/net affecting/being affected values in the table. The highest one-directional affecting value is 0.312, from C4-Transport to

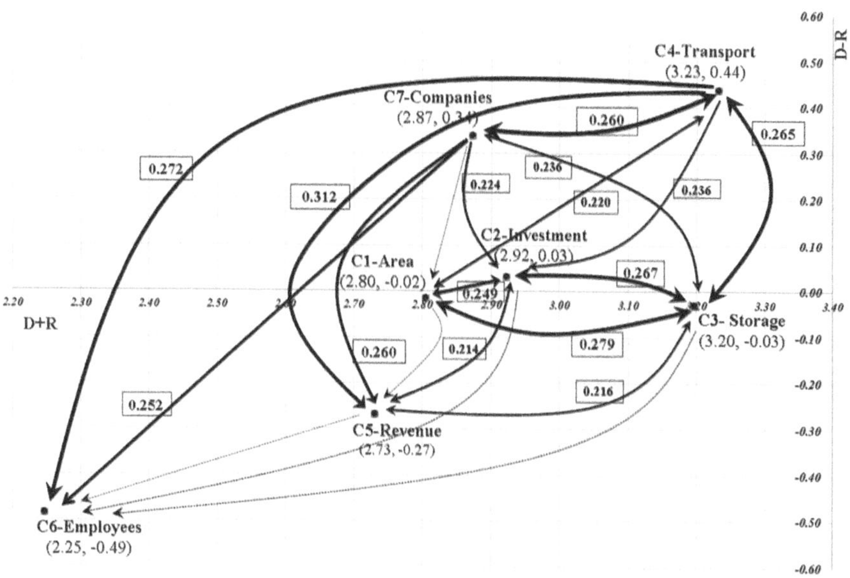

Fig. 1 Interaction diagram with significant affecting/being affected values. *Note* Affecting values below 0,204 are drawn in pale color without declaring magnitude on its curve. Bidirectional effect values written in boxes are the mean of both effects

C5-Income, clearly ahead of other influences. The other highest 5 affecting values are from C1 to C3 (0.293), from C2 to C3 (0.289), from C4 to C3 (0.284), from C4 to C7 (0.273), from C4 to C6 (0.272), respectively.

Then, from C3 to C1 (0.265), from C4 to C2 (0.265), from C1 to C2 (0.260), from C4 to C1 (0.260), from C7 to C5 (0.260), from C7 to C4 (0.260), from C7 to C3 (0.254), and from C7 to C6 (0.252) can be observed as moderately high criteria effect directions. These effect values also show us the intensities of interaction between criteria, as mapped and summarized in the diagram.

General affecting criteria are the annual amount of freight transport from the freight village in all modes, namely, C4-Transport, and then a number of companies operating in the freight village (C7-Companies). While these two criteria with having highest total/net affecting/being affected values, both affect all other criteria, and it can be observed that they are not significantly being affected by the C6-Employees, C5-Income and C2-Investment criteria. According to experts, this feature makes the C4-Transport and C7-Companies criteria the general influencers or general determinants in the formation of the other criteria. C1-Area, C2-Investment, C3-Storage, and C5-Income criteria can be considered as intermediate/transition criteria that generally have strong bidirectional interactions between each other. These criteria are strongly affected by C4-Transport and C7-Companies criteria. The number of employees for the activities of the freight village, represented by C6-Employees criterion with having the lowest net affecting value, is determined as general being affected by other criteria.

4.2 *Fuzzy AHP Results*

First, F-AHP results were obtained in the study. The steps described in the method section for F-AHP calculation were applied for determining criteria priorities. Pairwise comparisons matrices were created for each of them according to the answers given by six subject matter experts whose views were sought in the study. The decision matrix of the first expert is given in Table 11.

The geometric mean approach is used to get the fuzzy geometric mean and each criterion weights. Fuzzy geometric means and fuzzy weight calculated for some experts are presented in Table 12.

The ranking obtained from the experts shows that the most important criterion is C4-Transport with having 32.4% priority weight in the freight village efficiency evaluation. C4-Transport criterion was determined as having the highest priority weight according to the six subject matter experts' views and the F-AHP calculations. This criterion is followed by C2-Investment (17.6%) and C5-Revenue (15.2%) criteria, respectively, in terms of priority weights.

Then, criteria of C7-Companies (10.9%), C1-Area (10.4%), C3-Storage (10.1%), and C6-Employees (3.3%), respectively, can be observed in the priority ranking. Criterion C6-Employees has the lowest priority in the freight village efficiency evaluation.

Table 11 Fuzzy-AHP desicion matrix of the first expert

CRI	C1			C2			C3			…	C6			C7		
	l	*m*	*u*	*l*	*m*	*u*	*l*	*m*	*u*	…	*l*	*m*	*u*	*l*	*m*	*u*
C1-Area	1.00	1.00	1.00	0.17	0.20	0.25	4.00	5.00	6.00	…	4.00	5.00	6.00	2.00	3.00	4.00
C2-Investment	4.00	5.00	6.00	1.00	1.00	1.00	4.00	5.00	6.00		4.00	5.00	6.00	6.00	7.00	8.00
C3-Storage	0.17	0.20	0.25	0.13	0.14	0.17	1.00	1.00	1.00		2.00	3.00	4.00	2.00	3.00	4.00
C4-Transport	6.00	7.00	8.00	4.00	5.00	6.00	6.00	7.00	8.00		6.00	7.00	8.00	6.00	7.00	8.00
C5-Revenue	6.00	7.00	8.00	4.00	5.00	6.00	4.00	5.00	6.00		6.00	7.00	8.00	4.00	5.00	6.00
C6-Employees	0.17	0.20	0.25	0.13	0.14	0.17	0.25	0.33	0.50		1.00	1.00	1.00	2.00	3.00	4.00
C7-Companies	0.25	0.33	0.50	0.13	0.14	0.17	0.25	0.33	0.50		0.25	0.33	0.50	1.00	1.00	1.00

Table 12 Average and normalized criteria priority weights

CRI	Fuzzy geometric means			Fuzzy weights			Weights		Rank
	l	*m*	*u*	*l*	*m*	*u*	*Mi*	*Ni*	
C1-Area	0.754	0.877	1.017	0.078	0.104	0.138	0.107	0.104	5
C2-Investment	1.305	1.486	1.677	0.136	0.177	0.228	0.180	0.176	2
C3-Storage	0.714	0.849	1.006	0.074	0.101	0.137	0.104	0.101	6
C4-Transport	2.407	2.749	3.092	0.250	0.327	0.420	0.333	0.324	1
C5-Revenue	1.107	1.281	1.480	0.115	0.152	0.201	0.156	0.152	3
C6-Employees	0.234	0.270	0.320	0.024	0.032	0.044	0.033	0.033	7
C7-Companies	0.818	0.920	1.041	0.085	0.110	0.142	0.112	0.109	4

Mi: average priority weight, Ni: normalized priority weight

Consistency, as one of the major features/advantages of the AHP method, expresses whether the judgments of the decision maker(s) are consistent. In many studies where the F-AHP method was applied, it was not possible to see the consistency calculation of the criteria. Although there are not many studies in the literature for the calculation of consistency in the F-AHP method, three different results of consistency ratios (CR) are given in this study. First, according to the calculation steps expressed by [20: 137–141], $\lambda_{max} = 7.467$, CI = 0.078, $CR^m = 0.0605$ ($n = 7$, RI = 1.2874), and $\lambda_{max} = 7.4516$, CI = 0.0753, $CR^g = 0.1840$ ($n = 7$, RI = 0.4090). Second, according to the calculation steps expressed by [26: 622–625], $\lambda_{max} = 7.4937$, CI = 0.0822, CR = 0.0623 ($n = 7$, RI = 1.32). Third, according to classical AHP calculations $\lambda_{max} = 7.559$, CI = 0.093, CR = 0.0694 ($n = 7$, RI = 1.3417). All CR values are observed at acceptable level.

In order for the findings to be seen/analyzed better, the criteria priority weights calculated via F-DEMATEL and F-AHP are presented graphically in Fig. 2.

The criteria priority weights calculated via F-DEMATEL and F-AHP are reported in Fig. 2 allows comparing both results for each criterion. As mentioned above, the criteria priority weights calculated via the DEMATEL method are quite close to each other, and they represent criteria interactions intensity instead of criteria priority; these DEMATEL weights are taken into account only for comparisons. The margin of obtained DEMATEL weights is quite small; criteria weights are changing from 11.4% (C6-Employees) to 16.2% (C4-Transport). Then, it is difficult to separate the criteria priorities sharply from each other, and the AHP priority results are determiners and more respectful. However, the margin of obtained AHP weights is quite big and distinguishable from each other; criteria weights are changing from 3.3% (C6-Employees) to 32.4% (C4-Transport). These two method reveals the most and the least important criteria as same. Accordingly, C4-Transport criteria which is also overwhelmingly the highest affecting factor via the DEMATEL method are determined as the most important criteria with 32.4% priority weight via the AHP method in the freight village efficiency evaluation. The criteria of C2-Investment (17.6%) and C5-Revenue (15.2%) have quite high priority. These three criteria have a total

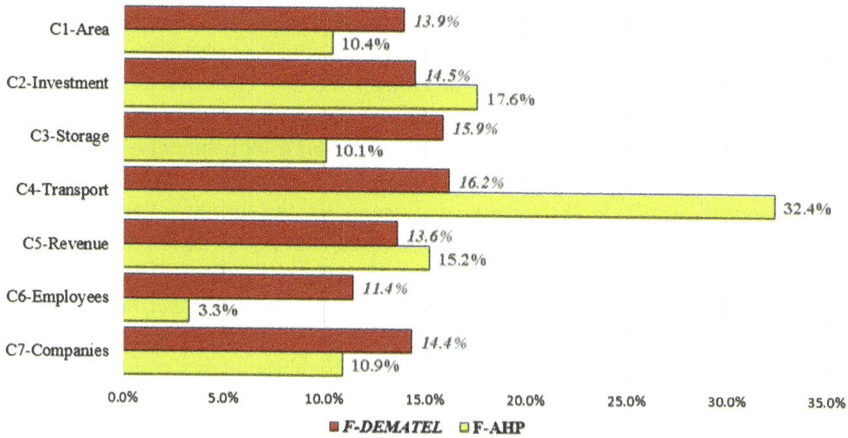

Fig. 2 Comparison of criteria priority weights calculated via F-DEMATEL and F-AHP

priority of 65.5%, which is almost two-thirds of the importance in the freight village efficiency evaluation. The other four criteria have a total priority of 34.5%, which is almost one-third of the importance. C1-Area (10.4%), C3-Storage (10.1%), and C7-Companies (10.9%) have almost similar priority weights. The C6-Employees criterion, which is also observed as the highly impressed factor via the DEMATEL method, is determined as the least important criterion with 3.3% priority weight via the AHP method.

5 Conclusion

The globalization of the world, the incentives for the removal of borders between countries, and the reduction of restrictions enable logistics activities to be carried out on a more general (holistic) scale. Logistics activities carried out throughout the country have important results when companies come together and carry out their activities in logistics centers/freight villages. Choosing a freight village, distribution center, etc., location is not the same as determining the effectiveness of these locations. The determination of the freight village where the companies will do their operations will be shaped according to their future plans and will benefit the company in many ways. For this, evaluating the activities of the freight village operating and presenting their data to the public can be effective in the decisions of the companies. In addition to the absence of an academic study to determine the criteria, criteria priorities, and criteria interactions to be used in the evaluation of the areas specified in the academic literature, where efficiency analysis of freight village, distribution centers, land ports, and similar areas is mostly made if evaluations should be made according to the criteria findings specified in this study in future academic studies, it is thought that more effective and useful results can be obtained.

Freight village subjected in this study is considered as one of the important components of the logistics industry, the importance of which is understood more and more with the Covid-19 period. Freight villages are also considered to be of high importance in terms of the global procurement and distribution of goods. Evaluating freight village sustainable performance as well as their establishment is valuable for its investors and supply chain. Relatedly, the study's aim is to determine the interaction between criteria, and the criteria priorities for the freight village efficiency evaluation, and to present the findings to the researcher and decision makers. After seven main criteria selection from the related literature, F-DEMATEL and F-AHP methods with obtained experts' pair-compared views are used to determine the interactions and priorities of the criteria in the evaluation of freight village. In this study, the input/output distinction was not made for the efficiency analysis, and the criteria were considered as a whole.

There is no study in the literature about choosing the criteria used in the freight village efficiency, defining their priority weights, and determining their interactions, this is also the originality of this study. For this reason, unfortunately, the findings obtained in this study could not be compared with the literature.

According to F-DEMATEL findings, the general affecting criteria are C4-Annual amount of freight transport, and then, C7-Number of companies settled. While these two factors/criteria with having the highest total/net affecting values, they can be named as general influencers or general determinants in the formation of the other criteria. C1-Area, C2-Investment, C3-Storage, and C5-Income criteria are accepted as transition criteria, and C6-Number of employee criterion is determined as general being affected by other criteria. According to F-AHP findings, C4-Transport criterion which is overwhelmingly the most important criterion with C2-Investment and C5-Revenue criteria has a total of almost two-thirds of the importance in the freight village efficiency evaluation. On the other hand, C6-Employees criterion is observed as both the highly impressed and least important criterion.

Since C4-Transport criterion has the significantly highest priority and is called general affecting criterion, it is the undoubtedly dominant factor in the freight village efficiency evaluation. In all efficiency evaluations of freight village, C4-Transport criterion should be considered and weighted quite higher than the other criteria by decision makers and researchers. Since the C6-Employees criterion is strongly both the most being affected factor and the least important criterion, it may be ignored in freight village assessments or represented with a significantly too small weight. In case of not considering C6-Employees criterion in the freight village efficiency evaluation/researches, the other listed criteria priority values may give quite a big impact on the C6-Employees criterion and may indirectly represent it. Decision makers and researchers should take into consider this property of the C6-Employees criterion.

Some other interpretations can be stated from these findings. As C7-Companies is categorized as one of the major affecting criterion, it has quite low priority in the freight village efficiency evaluation. However, as C2-Investment is categorized as a transition criterion, it has second major priority weight in the freight village efficiency evaluation. As C1-Area, C3-Storage, and C5-Income are categorized as transition criteria, C5-Income has one of the major priority weight; C1-Area and C3-Storage have quite low priority in the freight village efficiency evaluation. These findings can be discussed in depth, and with the findings of this study, what should be done for the criteria is cleared.

When the results expressed here are examined, another significant conclusion may be drawn from our study. Considering the DEMATEL-AHP results together during the evaluation phase of the criteria can provide the opportunity to present original perspectives on the process of analyzing the criteria. In this study, the criteria priority levels obtained from the AHP and the interaction structures obtained from the DEMATEL results were examined together from multiple perspectives for each criterion. In addition, by using the fuzzy logic approach, the measurement success level was tried to be clarified/improved further.

In one-on-one interviews with the experts about evaluating the freight village efficiency, it was asked whether the criteria determined in the study should be input or output. Along with significantly too small priority weight, the experts' views for "the number of employees" criterion should be discussed in detail, there was no common decision in the position of the number of employees (it could be input/output). However, they agree on the criteria of amount of annual freight transport and the number of companies settled as outputs and the other criteria as inputs.

The following studies can use the determined criteria weights in the freight village efficiency evaluation. Calculating/evaluating the freight village efficiency is important for its investors and sustainability and performance of the supply chain for enterprises and people.

References

1. Altınırmak, S., Ergün, M., Karamaşa, Ç.: Implementation of the fuzzy DEMATEL method in higher education course selection: the case of Eskişehir vocational school. Atatürk Üniversitesi Sosyal Bilimler Enstitüsü Dergisi **21**(4), 1597–1614 (2017)
2. Andrejic, M., Bojović, N., Kilibarda, M.: Benchmarking distribution centres using principal component analysis and data envelopment analysis: a case study of Serbia. Expert Syst. Appl. **40**(10), 3926–3933 (2013)
3. Bajec, P., Kontelj, M., Groznik, A.: Assessment of logistics platform efficiency using an integrated delphi analytic hierarchy process–data envelopment analysis approach: a novel methodological approach including a case study in Slovenia. Inf. Manage. **23**(3), 191–207 (2020)
4. Carvalho, C.C. de, Lima Jr, O.F.: Efficient logistic platform design: the case of Campinas Platform. In: XVI International Conference on Industrial Engineering and Operations Management, São Carlos, SP, Brasil (2010)
5. Chakraborty, P.S., Majumder, G., Sarkar, B.: Performance measurement of distribution centre combining data envelopment analysis and analytic hierarchy process. Adv. Prod. Eng. Manage. **6**(2), 117–128 (2011)
6. Chan, F.T., Kumar, N., Tiwari, M.K., Lau, H.C., Choy, K.: Global supplier selection: a fuzzy-AHP approach. Int. J. Prod. Res. **46**(14), 3825–3857 (2008)
7. Chang, D.Y.: Applications of the extent analysis method on fuzzy AHP. Eur. J. Oper. Res. **95**(3), 649–655 (1996)

8. Chang, B., Chang, C.W., Wu, C.H.: Fuzzy DEMATEL method for developing supplier selection criteria. Expert Syst. Appl. **38**(3), 1850–1858 (2011)
9. Chou, Y.C., Yen, H.Y., Dang, V.T., Sun, C.C.: Assessing the human resource in science and technology for Asian countries: application of fuzzy AHP and fuzzy TOPSIS. Symmetry **11**(2), 251–267 (2019)
10. Cruijssen, F., Dullaert, W., Joro, T.: Freight transportation efficiency through horizontal cooperation in Flanders. Int. J. Log. Res. Appl. **13**(3), 161–178 (2010)
11. Cullinane, K., Wang, T.-F., Song, D.-W., Ji, P.: The technical efficiency of container ports: comparing data envelopment analysis and stochastic frontier analysis. Transp. Res. Part A: Policy Pract. **40**(4), 354–374 (2006)
12. Doyle, J., Green, R.: Efficiency and cross-efficiency in DEA: derivations, meanings and uses. J. Oper. Res. Soc. **45**(5), 567–578 (1994)
13. Europlatforms: Logistics Centres Directions for Use, EUROPLATFORMS EEIG (2004). https://www.unece.org/fileadmin/DAM/trans/main/eatl/docs/EN-REV-What_is_a_Freight_VillageFinalcorretto.pdf
14. Feng, C., Ma, R.: Identification of the factors that influence service innovation in manufacturing enterprises by using the fuzzy DEMATEL method. J. Clean. Prod. **253**(120002), 1–12 (2020)
15. Fontela, E., Gabus, A.: DEMATEL, Innovative methods. Report No.2 Structural Analysis of the World Problematique, Battelle Geneva Research Institute, pp. 67–69 (1974)
16. Ganguly, K.K., Guin, K.K.: A fuzzy AHP approach for inbound supply risk assessment. Benchmarking: Int. J. **20**(1), 129–146 (2013)
17. Gao, X., Fang, J., Wang, S.: Key risk factors identification for PPP waste-to-energy incineration projects. In: ICCREM 2021, pp. 163–170 (2021). https://doi.org/10.1061/9780784483848.018
18. Gnanavelbabu, A., Arunagiri, P.: Ranking of MUDA using AHP and Fuzzy AHP algorithm. Mater. Today: Proc. **5**(5), 13406–13412 (2018)
19. Gharakhani, D.: The evaluation of supplier selection criteria by fuzzy DEMATEL method. J. Basic Appl. Sci. Res. **2**(4), 3215–3224 (2012)
20. Gogus, O., Boucher, T.O.: Strong transitivity, rationality and weak monotonicity in fuzzy pairwise comparisons. Fuzzy Sets Syst. **94**(1), 133–144 (1998)
21. Guo, F., Gao, J., Liu, H., He, P.: A hybrid fuzzy investment assessment framework for offshore wind-photovoltaic-hydrogen storage project. J. Energy Storage **45**(103757), 1–18 (2022)
22. Güngör, Z., Serhadlıoğlu, G., Kesen, S.E.: A fuzzy AHP approach to personnel selection problem. Appl. Soft Comput. **9**(2), 641–646 (2009)
23. Higgins, C.D., Ferguson, M., Kanaroglou, P.S.: Varieties of logistics centers: developing standardized typology and hierarchy. Transp. Res. Rec. **2288**, 9–18 (2012)
24. Jaržemskis, A.: Research on public logistics centre as tool for cooperation. Transport **22**(1), 50–54 (2007)
25. Kapros, S., Panou, K., Tsamboulas, D.A.: Multicriteria approach to the evaluation of intermodal freight villages. Transp. Res. Rec. **1906**(1), 56–63 (2005)
26. Kwong, C.K., Bai, H.: Determining the importance weights for the customer requirements in QFD using a fuzzy AHP with an extent analysis approach. IIE Trans. **35**(7), 619–626 (2003)
27. Lin, K.P., Tseng, M.L., Pai, P.F.: Sustainable supply chain management using approximate fuzzy DEMATEL method. Resour. Conserv. Recycl. **128**, 134–142 (2018)
28. Liu, C.-C.: Evaluating the operational efficiency of major ports in the Asia-Pacific region using data envelopment analysis. Appl. Econ. **40**(13), 1737–1743 (2008)
29. Mangla, S.K., Kumar, P., Barua, M.K.: Risk analysis in green supply chain using fuzzy AHP approach: a case study. Resour. Conserv. Recycl. **104**, 375–390 (2015)
30. Markovits-Somogyi, R., Gecse, G., Bokor, Z.: Basic efficiency measurement of Hungarian logistics centres using data envelopment analysis. Period. Polytech. Soc. Manag. Sci. **19**(2), 97–101 (2011)
31. Meidute, I.: Comparative analysis of the definitions of logistics centres. Transport **20**(3), 106–110 (2005)
32. Mouzas, S.: Efficiency versus effectiveness in business networks. J. Bus. Res. **59**(10–11), 1124–1132 (2006)

33. Nasrollahi, M., Ghasemi, R., Ghadikolaei, A.S., Sheykhizadeh, M., Abdi, M.: Identification and prioritization of connected vehicle technologies for sustainable development in Iran. Technol. Soc. **68**(101829), 1–16 (2021)
34. Pekkaya, M., Aslan, B.: OSByer seciminde dikkate alınan kriterler arası etkiles¸imin ve kriter önem derecelerinin belirlenmesi. Int. J. Econ. Admin. Stud. **18**(EYI Special Issue), 293–308 (2018)
35. Pekkaya, M., Erol, F.: Generating priority series via AHP for conducting statistical tests on CAMELS dimension priorities in evaluating bank failure risk. J. Intelli. Fuzzy Syst. **37**(6), 8131–8146 (2019)
36. Pekkaya, M., Keleş, N.: Determining criteria interaction and criteria priorities in freight village location selection process: the experts' perspective in Turkey. Asia Pac. J. Mark. Logist. (2021). https://doi.org/10.1108/APJML-05-2021-0338
37. Rimiene, K., Grundey, D.: Logistics centre concept through evolution and definition. Eng. Econ. **4**(54), 87–95 (2007)
38. Shaverdi, M., Heshmati, M.R., Ramezani, I.: Application of fuzzy AHP approach for financial performance evaluation of Iranian petrochemical sector. Proc. Comput. Sci. **31**, 995–1004 (2014)
39. Sönmez Çakır, F., Pekkaya, M.: Determination of interaction between criteria and the criteria priorities in laptop selection problem. Int. J. Fuzzy Syst. **22**(4), 1177–1190 (2020)
40. Tongzon, J.: Efficiency measurement of selected Australian and other international ports using data envelopment analysis. Transp. Res. Part A: Policy Pract. **35**(2), 107–122 (2001)
41. Vrochidis, B.: Logistics Centres as Economic Drivers of Their Regions. Department of Erasmus School of Economics: Erasmus University of Rotterdam, Rotterdam (2013). http://thesis.eur.nl/pub/14748
42. Wang, Y., Xu, L., Solangi, Y.A.: Strategic renewable energy resources selection for Pakistan: based on SWOT-fuzzy AHP approach. Sustain. Cities Soc. **52**(101861), 1–12 (2020)
43. Winkler, H., Seebacher, G.: Management of freight villages: findings from an exploratory study in Germany. Int. J. Log. Res. Appl. **14**(4), 271–283 (2011)
44. Yang, C., Dong, A.T.G.: Efficiency analysis of European freight villages: three peers for benchmarking. Central Eur. J. Oper. Res. **25**(1), 91–122 (2017)
45. Yu, M.-M., Lin, E.T.J.: Efficiency and effectiveness in railway performance using a multi-activity network DEA model. Omega **36**(6), 1005–1017 (2008)
46. Zadeh, L.A.: Fuzzy sets. Inf. Control **8**(3), 338–353 (1965)

The manufacturer's authorised representative in the EU is Springer Nature Customer Service Centre GmbH, Europaplatz 3, 69115 Heidelberg, Germany. If you have any concerns regarding our products, please contact ProductSafety@springernature.com

Printed and bound by CPI Group (UK) Ltd, Croydon, CR0 4YY
30/09/2026
02225244-0020